Competitive Physics

Thermodynamics, Electromagnetism and Relativity

by
Physics Olympiad medalists and trainers

Competitive Physics

Thermodynamics, Electromagnetism and Relativity

by
Physics Olympiad medalists and trainers

Jinhui Wang

Bernard Ricardo

 **World Scientific**

NEW JERSEY · LONDON · SINGAPORE · BEIJING · SHANGHAI · HONG KONG · TAIPEI · CHENNAI · TOKYO

Published by

World Scientific Publishing Co. Pte. Ltd.

5 Toh Tuck Link, Singapore 596224

USA office: 27 Warren Street, Suite 401-402, Hackensack, NJ 07601

UK office: 57 Shelton Street, Covent Garden, London WC2H 9HE

Library of Congress Cataloging-in-Publication Data

Names: Wang, Jinhui (Former Physics Olympiad student), author. | Ricardo, Bernard, 1985–   author.

Title: Competitive physics : thermodynamics, electromagnetism and relativity /
    Wang Jinhui, Bernard Ricardo (Hwa Chong Junior College, Singapore).

Description: Singapore ; Hackensack, NJ : World Scientific, [2018] |
    Includes bibliographical references and index.

Identifiers: LCCN 2018002463| ISBN 9789813239418 (hardcover ; alk. paper) |
    ISBN 9813239417 (hardcover ; alk. paper) | ISBN 9789813238534 (pbk. ; alk. paper) |
    ISBN 9813238534 (pbk. ; alk. paper)

Subjects: LCSH: Physics--Problems, exercises, etc. | Physics--Competitions. |
    Thermodynamics--Problems, exercises, etc. | Electromagnetism--Problems, exercises, etc. |
    Special relativity (Physics)--Problems, exercises, etc.

Classification: LCC QC32 .W217 2018 | DDC 530/.076--dc23

LC record available at https://lccn.loc.gov/2018002463

British Library Cataloguing-in-Publication Data

A catalogue record for this book is available from the British Library.

For any available supplementary material, please visit
https://www.worldscientific.com/worldscibooks/10.1142/10946#t=suppl

Typeset by Stallion Press
Email: enquiries@stallionpress.com

Dedication

The Physics Olympiads were some of my most enjoyable experiences in high school and I hope to share the elegance of problem-solving, prevalent in such competitions, with more people. Writing this book has really been a fulfilling journey that has evoked many moments of nostalgia, ranging from the thrill of solving a problem to my past struggles when learning about physics. I am extremely grateful to the following people who have supported me in the course of writing — during my hardest times in national service — in one way or another.

My co-author, Bernard Ricardo. I can safely say that he is one of the best physics educators in Singapore. His enthusiasm and illuminating pedagogy during his lessons for the Physics Olympiad national training team inspired me to pursue and teach physics. It has been a joy working with him and his many insights have added a novel perspective to this book.

My teachers, Tan Jing Long, Mr Kwek Wei Hong and Mrs Ng Siew Hoon. I still remember the quirky jokes that the class makes during your lessons. Thank you for introducing me to physics and infecting me with your love for the subject.

My students, Cui Zizai, Mao Ziming, Chang Hexiang, Guo Yulong, Andrew Ke Yanzhe, Ryan Wong Jun Hui and many more. It was a pleasure to teach you and your inquisitiveness in our lessons together was a great source of motivation during my conscription. I am also grateful to you for proofreading chapters of this book and providing constructive feedback.

My friends, Jiang Yue and Ma Weijia. Thank you for taking time out from your busy university schedules to draw numerous diagrams for this book as I was preoccupied with National Service. This milestone would literally not have been possible without you.

My parents. Thank you for everything that you have done for me, from cooking meals at home to discussing life decisions with me. I would not be who I am today without you.

Finally, it is with the greatest joy that I present *Competitive Physics* and I hope that you enjoy the book!

— Wang Jinhui

It is a pleasure for me to present our book, *Competitive Physics*. The process of this book's production, that has been thrilling and satisfying, could not be completed without these special people in my life whom I would like to extend my deepest, most heartfelt thanks.

My co-author, Wang Jinhui. This book was birthed out of his vision. He is one of the brightest young men that I know and it was extremely enjoyable to put our thoughts together. I hope this would be the start of many more collaborations and substantial discussions.

My mentor, Professor Yohanes Surya Ph.D. I was first introduced to the world of Physics Olympiads by him. Full of passion, he dedicated himself to impart his knowledge and love for physics to so many people. His slogan, *Physics is Fun*, has encouraged and inspired me to impact the world of Physics Education.

My wife, Yoanna Ricardo, daughter, Evangeline Ricardo, and son, Reinard Ricardo. Thank you so much for always supporting and praying for the completion of this book. The times at home and on the road with you were the times when I was most fruitful in writing and I am truly grateful for those memories.

For all our readers, I hope you will find this book enjoyable and have fun reading it as we have had fun writing it!

— Bernard Ricardo

Preface

Competitive Physics grew out of a Physics Olympiad course taught by Wang Jinhui at Hwa Chong Institution — intended to prepare students for the annual Physics Olympiads and to imbue deeper knowledge in physics beyond the typical high school syllabus. It quickly became a collaboration with his former trainer in the Singapore Physics Olympiad national training team, Bernard Ricardo.

Competitive Physics is meant to be a theory-cum-problem book. The first half of each chapter explores physical theories with illustrations of how they can be creatively applied to problems. The latter half of each chapter revolves around puzzles that we hope will intrigue readers, as we believe that problem-solving is a crucial process in grasping the subtleties of the contents. Therefore, we have included a multitude of problems which are ranked by increasing difficulty from one to four stars. Some problems are original; some are taken from the various Physics Olympiads while the others are instructive classics that have withstood the test of time.

This book is the second part of a two-volume series which will discuss thermodynamics, electromagnetism and special relativity, building on the fundamentals that we have developed in the first volume. A brief overview of geometrical optics is also included.

We envision problem-solving to be a fun process — from the initial excitement of approaching an unfamiliar problem, to the joy of pitting all of one's knowledge against it and finally, the satisfaction earned from solving it after numerous failed attempts. In light of this, our goal is to spread the passion of problem-solving — an infectious hobby. It is difficult to quantity the factors that make a problem interesting or elegant but the following have been our guiding principles in writing *Competitive Physics*:

1. Physical Significance. Quintessentially, physics is about modeling the world around us. Therefore, it is gratifying to be able to analyze everyday

phenomena and to leverage on this knowledge to improve such processes. For example, a problem in Chapter 4 deals with a model of global warming. Meanwhile, we learn how to construct an AC generator in Chapter 8 and a primitive digital-to-analog converter in Chapter 9.

2. Intuition. There are many overarching themes in physics — symmetry, the equivalence of different observational frames of reference, construction of mirror images and many more. Not only are these useful as sleights-of-hand in problem-solving, they reveal crucial aspects of the common structure of physical theories. Developing a strong hunch for them — a gut feel that constantly bugs you to search for ways to exploit them — may prove to be beneficial in one's future physics journey.

3. Insight. Sometimes, a seemingly complex problem can be vastly simplified by making an astute observation — whether mathematical or physical. Perhaps, it is to express the solution in terms of vectors or perhaps it is to observe that two different scenarios "feel" the same to a certain entity and thus conclude that the entity will respond in the same manner in both cases. Maybe it is to draw enlightening analogies between two problems that appear to be completely disparate on the surface. Ultimately, such problems which require perceptive thought do not have cookie-cutter approaches and require the reader to invent an appropriate technique on the spot. They hence implore the reader to really think and are very rewarding to solve.

4. Fundamentals. The objectives above would not be possible without first mastering the fundamentals of a theory — the situations that it can be validly applied to, its assumptions and its ramifications. As such, we have also included many classic problems to reinforce understanding of the basics. To this end, we are extremely grateful to Dr. David J. Morin for allowing us to use some problems from his exemplary textbook: *Introduction to Classical Mechanics*.

In summary, our guiding principles are "PIIF", as in the onomatopoeia "pffft" when, having read this book, you scoff at a future problem after swiftly spotting its trick. Jokes aside, it is paramount for the reader to first attempt the problems before peeking at the solutions. Even when perusing the solution to a problem, the reader should inspect it line by line until he or she reaches an inspiration that sets him or her back on track in attempting the problem again. Only by experiencing the process of problem-solving yourself can you internalize the clues in a problem that hint at a certain approach, understand why certain approaches are incorrect or desirable and ultimately, improve. There is no short-cut to developing an intuition for problem-solving besides, trudging through an arduous but fulfilling journey of enigmas.

Despite out best efforts the probability of this book being error-free is, unfortunately, akin to the odds of observing a car plate that reads "PHY51C". Therefore, if the reader does spot any mistakes or dubious points in our discussions, we would appreciate if they are highlighted to us via the email competitivephysicsguide@gmail.com.

Contents

Chapter 1

Geometrical Optics

Geometrical optics can be applied to situations where the scale of observation is much larger than the wavelength of light. For example, the wavelength of visible light ranges from 400–700 nm, that is, violet to red while typical scenarios involve lengths of observation in centimeters and meters. The crucial simplification in geometrical optics relies on light rays — a construct that will soon be elaborated. In the entire chapter, we will only be considering homogeneous and isotropic media. In other words, the media in which light propagates are uniform in all space and in all directions.

1.1 Light Rays

The major simplification in geometrical optics stems from the construction of light rays to approximate the formulation of light as an electromagnetic wave. But first, it is well-established that light exists in quanta, or discrete packages, known as photons. How can the model of an electromagnetic wave be coherent with the inherently quantum nature of light? The answer to this is that there are myriad photons in a normal light beam and the combined system can thus be approximated as a continuous wave. This is similar to how water ripples are formulated as continuous waves, though they comprise individual molecules.

To further streamline the model, a light ray is defined to be a line that is in the direction of energy flow. In a homogeneous and isotropic medium, a light ray will be perpendicular to the wavefronts at each point of intersection. As a result, a point source will "emit" light rays radially outwards. In the case of a plane wave, this formulation leads to immense convenience as all light rays are parallel to one another. Hence, a single light ray is representative of the whole set of light rays.

However, with this definition, we still do not know the shape of light rays in a homogeneous medium. If they were random squiggles, this model would be rendered useless. Well, it is intuitive that the wavefronts of light take on the same general shapes as they propagate in a vacuum. For example, a spherical wavefront becomes a larger spherical wavefront at a later instant. A plane wavefront still remains planar. Hence, a light ray in vacuum is naturally a straight line. However, the fact that a light ray maintains the form of a straight line when propagating in a homogeneous medium that is not a vacuum, is not so obvious.

A multitude of photons impinge on the molecules in a homogeneous medium. However, these molecules cannot be raised into an excited state as their energy gaps do not correspond to photon frequencies in the visible region. Hence, photons are absorbed and simultaneously re-emitted in arbitrary directions. Due to the gargantuan number of incident photons, each molecule effectively re-emits a spherical secondary wave that has a certain phase difference relative to the primary wave instantaneously. This phase difference is constant for all the molecules and arises because the molecules respond as driven dipole oscillators which oscillate at a certain phase difference, relative to the driving force caused by the electric field of the incident EM wave. Due to the constant phase difference relative to the primary wave, the secondary waves constructively interfere in the direction of the primary wave propagation because the secondary waves are emitted as the primary wave "hits" the molecules along its propagation. Furthermore, since the molecules in the medium are densely packed and the wavelength of visible light is much larger than the intermolecular spacings, destructive interference occurs in all other directions, and there is minimal lateral scattering.

As the transmitted wave is the superposition of the primary and secondary waves, the result is that the transmitted wave also follows the same general shape as the primary wave along the same direction of propagation, and a light ray in a homogeneous medium is still a straight line.

Another important phenomenon pertains to the transmitted wave accumulating a progressively larger phase difference relative to the primary wave as more secondary waves are gathered. This manifests itself as a change in the phase velocity of the wave. The primary and secondary light waves both propagate at a phase velocity c, but their superposition produces a transmitted wave of phase velocity $\frac{c}{n}$ where n is the refractive index of the medium. The phase velocity of an electromagnetic wave in a medium is given by

$$v = \frac{1}{\sqrt{\mu\varepsilon}},$$

where μ and ϵ are the magnetic permeability and electric permittivity of the medium, respectively. The speed of light in a vacuum is given by

$$c = \frac{1}{\sqrt{\mu_0 \varepsilon_0}}.$$

Since μ does not deviate much from μ_0 for most materials, the refractive index is given by

$$n = \frac{c}{v} = \sqrt{\frac{\varepsilon}{\varepsilon_0}} = \sqrt{\kappa}$$

$$v = \frac{c}{n}, \tag{1.1}$$

where κ is the dielectric constant, $\varepsilon = \kappa\varepsilon_0$. Stemming from the laws of electromagnetism and the properties of a dipole oscillator, κ in fact depends on the wavelength of the light ray. Hence, the refractive index $n(\lambda)$ is generally a function of the wavelength λ of the light ray it carries. Qualitatively, n decreases as λ increases.[1]

The last theorem that confirms the utility of light rays is a theorem by Malus and Dublin. A pivotal corollary of the theorem is that light rays remain perpendicular to the wavefronts after an arbitrary number of reflections and refractions from various surfaces. Hence, we do not have to worry about light rays varying haphazardly when transiting across different media. Another consequence of this is that we can deduce the orientations of wavefronts by first drawing light rays, which are much more convenient and simpler to visualize.

However, there are a few drawbacks to this model of light rays. Firstly, they do not represent the phase velocity of light waves in a medium, nor do they depict their amplitudes. Furthermore, they cannot incorporate the superposition of light waves, which involves a vector sum of displacements, nor the phenomenon of diffraction which, dominates in the regime where apertures are of sizes comparable to the wavelength of light. Despite these limitations, light rays are still an important construct in geometrical optics, which rarely deals with the above phenomena.

---

[1] Cauchy's equation, which is rather accurate for visible light, states that the refractive index $n = A + \frac{B}{\lambda^2}$, where A and B are constants that depend on the material of the medium (usually determined by graph fitting). This dependence of n on λ is in fact what causes white light to split into a spectrum of colours after transmitting in media such as glass prisms, as waves of different wavelengths are refracted to different extents when transiting across an interface.

1.2   The Law of Reflection

Part of a light wave is scattered back when it impinges on the interface of another medium with a different index of refraction. This phenomenon is known as reflection. If the difference in the refractive indices is large and the transition is sudden, a large proportion of the light wave is reflected.

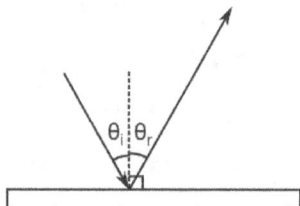

Figure 1.1:   Reflection off a surface

The normal is an imaginary line perpendicular to the instantaneous gradient of a surface. It is usually drawn as a dotted line, as shown in Fig. 1.1. The angle of incidence is the angle between the incident ray and the normal of the point that the incident ray impinges upon. Similarly, the angle of reflection is the angle between the reflected ray and the normal.

The law of reflection states that the angle of incidence and the angle of reflection are equal:

$$\theta_i = \theta_r. \tag{1.2}$$

Furthermore, the incident ray, normal and the reflected ray must all lie on the same plane, known as the plane of incidence. Hence, problems involving reflections can be reduced to effectively two-dimensional problems.

Depending on the smoothness of the surface, the orientation of the reflected rays of a bundle of incident rays will vary. In the case of a smooth flat surface, where all irregularities are small in comparison with the wavelength of light, the reflected rays remain parallel. This is known as specular reflection. However, in the case of a rough surface whose bumps and pits are comparable in size to the wavelength of light, different parallel incident rays will emerge in various directions due to the unevenness of the surface. This is known as diffuse reflection.

Problem: If the brightness of an image depends on the number of light rays entering one's eyes, and the beams from a projector can be assumed to be plane waves, why are projector screens not made of polished mirrors? This is potentially more energy efficient, as a lower intensity of the projected beam would be required to produce an image of the same brightness.

Well, the main limitation is that a purely specular reflection will result in a very limited region of angles where the beams can reach one's eyes. Hence, most reflecting surfaces engender a combination of specular and diffuse reflections.

Plane Mirrors

An ubiquitous application of the law of reflection would be plane mirrors. Based on the law of reflection, these mirrors form sharp images.

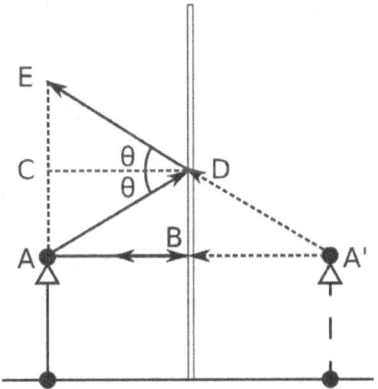

Figure 1.2:   Thin plane mirror

Consider the thin plane mirror in Fig. 1.2; we wish to determine how the image of an object will appear. Note that when determining an image in general, every point on the object acts as a point source of (reflected) light. The location of the image of a particular point is where all final light rays, which are originally emitted from the point, appear to emanate from after a reflection with the mirror. If most of the final light rays appear to converge at a certain point, the image at that point will be sharp or focused. If not, the image will be blurred.

To determine the location of the image of a point, multiple rays in different directions that emanate from the point are drawn and their point of intersection is determined. Note that although all rays do not necessarily coincide at a single point, this should usually be the case in the apparatus that we will consider in this chapter, to a certain degree of accuracy, at least.

Consider point A, the tip of the vertical object. Consider a line AB perpendicular to the plane of the mirror that passes through A. Then, an incident light ray along line AB must also be reflected along line BA. Now, consider another light ray, AD, impinging on the mirror at an arbitrary angle of incidence θ at point D. The angle of reflection is also θ by the law of reflection. Then, let E be the point of intersection of the reflected ray and a line in the plane of incidence that passes through A and is perpendicular to the normal. A, D and E lie in the plane of incidence. Note that the plane of incidence may not necessarily be the plane comprising this page, though it is drawn that way in the figure. Now, let C be the point of intersection of AE with the normal from D.

Evidently, the two reflected rays BA and DE are diverging from the mirror. Their extensions coincide at point A$'$ which is behind the mirror (these are denoted by dotted lines as they are not real rays). Now, triangle ACD is congruent to triangle ECD as they have two equal angles and share the same side CD. Then, C is the midpoint of EA. Furthermore, since the line CD is parallel to AA$'$ as they are both normal to the mirror, D is the midpoint of EA$'$. Hence, by the midpoint theorem, $\overline{AA'} = 2\overline{CD} = 2\overline{AB}$. Hence, the image distance, $\overline{BA'}$, which is the perpendicular distance between the image and the mirror is equal to the object distance, $\overline{AB}$, which is the perpendicular distance between the object and the mirror:

$$\overline{AB} = \overline{BA'}.$$

The vertical position of A$'$ corresponds to the vertical position of A. Lastly, since the angle θ was arbitrary and the location of A$'$ does not depend on θ, all rays from point A will appear to coincide at A$'$. Hence, A$'$ is the location of the image of A.

Then, every point on the vertical object can be correspondingly mapped to a point on the image to obtain the widely-spaced, dashed line in Fig. 1.2 above. The vertical height of the image is the same as that of the object. The image is also described to be upright as its vertical orientation is the same as that of the object (see white arrows). Furthermore, the image is known as a virtual image as light rays do not actually converge behind the mirror. They only appear to do so. If a screen were to be placed at the horizontal position of A$'$, no image will be formed on the screen.

A final quotidian phenomenon that may puzzle some is the apparent left-right reversal of the image in a mirror. If you raise your left hand, your image in the mirror appears to raise its right hand. If your shirt has a letter "S", the image will show a number "2". Does the mirror somehow cause the image to be reversed? Well, the answer is no. The object had already been "reversed" before it was mapped to an image. This can be best illustrated by writing a "S" on a transparent sheet of plastic and holding it in front of yourself, towards a mirror. As expected, the mirror shows an image "2". But if you now look at the sheet of plastic from your perspective, you see that the object also appears as "2"! Hence, what is left or right is only a matter of perspective. A person standing behind the object will see that the left-right orientation of the image is the same as that of the object. This is because, each point on the object is directly mapped via a line normal to the mirror and passing through that particular point to the other side of the mirror.

Problem: As a person moves away from a plane mirror, how does the vertical height of his or her image change (if any)?

The vertical height of the image does not change with the object distance as it is always identical to the height of the object. Images in plane mirrors appear to diminish as we move further away from them in real life as they now cover a smaller angular distance. The same phenomenon occurs when someone moves away from you; he or she appears smaller, though his or her actual height definitely does not shrink. In other words, the images in the mirror appear to shrink visually along with the mirror, while maintaining the relative proportions, but the height of the image definitely does not change.

Mirror Images

A neat trick in determining whether a light ray will impinge on an object after a reflection on a plane mirror is to extend the incident ray beyond the mirror and check if it hits the mirror image of the object.

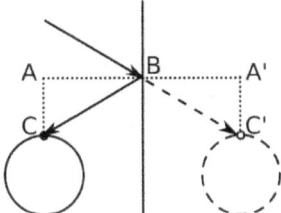

Figure 1.3: Mirror image

The point of intersection of the extended incident light ray and the mirror image corresponds to the point of intersection of the reflected ray and the original object (after a reflection about the mirror). Considering Fig. 1.3, B is the point at which the incident light ray hits the mirror. AB is the normal to the mirror surface. C is the point of intersection of the reflected ray and the object. C' is the point of intersection of the extended incident light ray. A' is obtained from extending the normal AB. It can be seen that

$$\triangle ABC \cong \triangle A'BC'$$

as

$$\angle ABC = \angle A'BC'$$

$$\angle CAB = \angle C'A'B = 90°$$

$$\overline{AB} = \overline{A'B'}.$$

Therefore, if a reflected ray hits an object, the extension of the incident ray also hits the mirror image while preserving all relevant distances.

This method of extending the incident light ray becomes extremely handy when there are multiple mirrors. If an extended incident light ray impinges on the image of a primary image, the light ray will hit the primary image by the above analysis. Consequently, it will also hit the original object. Similarly, if an extended incident ray hits the image of the image of a primary image, it will still hit the original object and so on. The last essential property is that the distance traversed by the light ray in hitting the object is equal to the distance obtained by extending the incident ray to the mirror image as the distances are preserved.

Problem: Consider a cylindrical receiver sandwiched between two plane mirrors separated by a distance $2l$. A point source P lies a distance d away from the cylindrical axis O. Consider the plane depicted in Fig. 1.4. What is the minimum radius of the cylinder R, such that all light rays emitted to the right of P in this plane hits the receiver? (Adapted from Chinese Physics Olympiad)

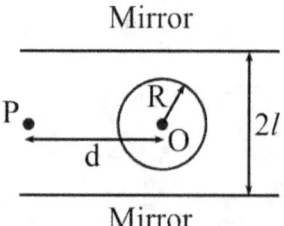

Figure 1.4:   Point source P and receiver

Well, the trick here is to consider the mirror images of the receiver. We will only consider the light rays traveling upwards, as the situation is symmetrical.

The mirror images in this plane form an array of infinite circles whose adjacent centers are separated by a distance $2l$. To ensure that all rays traveling upwards and rightwards reach the receiver, we just have to ensure that all incident rays impinge on an image when extended.

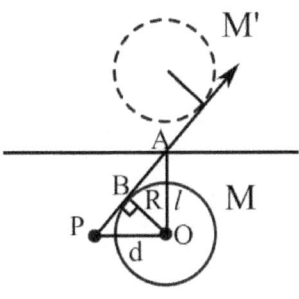

Figure 1.5: Primary image

Referring to Fig. 1.5, notice that if R is not large enough, some light rays can slip through the gap between the object M and primary image M'. Hence, the radius R must be increased until the boundary case, where an extended light ray from P to the point on the top mirror that is at the same horizontal position as O, namely point A, is tangential to both circles. The existence of such a boundary case is evident from the fact that the perpendicular distances from the centers of M and M' to this light ray are always identical. If such a condition is satisfied, all lights rays emitted at an angle $0 \leq \theta \leq \angle APO$, will hit either M or M', where θ is the angle that the light ray subtends with line PO. The minimum R in this case can be determined by observing that triangles $\triangle PBO$ and $\triangle POA$ are similar,

$$\triangle PBO \sim \triangle POA \quad (AA).$$

Hence,

$$\frac{R}{d} = \frac{l}{\sqrt{l^2 + d^2}}$$

$$R = \frac{ld}{\sqrt{l^2 + d^2}}.$$

We are left with showing that once this condition is satisfied, all incident rays emitted an angle $\theta_0 < \theta < 90°$ will eventually hit an image (remember that there are still infinite arrays of images above M' and below M). This is intuitive as it is impossible for a steeper line to pass through the gap between two adjacent circles when a line with a gentler slope, that emanates from the same point, cannot.[2]

---

[2]The reader should try to prove this mathematically.

1.3   Refraction

Refraction occurs when an incident light ray impinges on an interface between two media with different indices of refraction. The transmitted light ray is bent relative to the incident ray. This phenomenon is known as refraction and occurs because the phase velocity of light is different in different media.

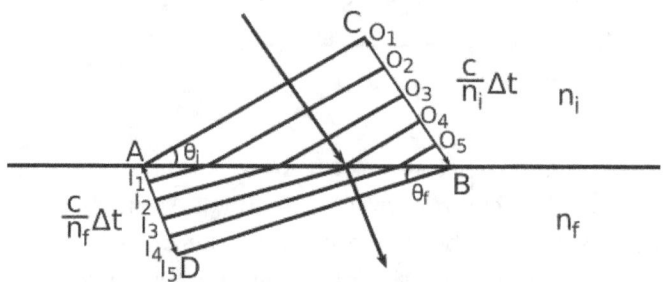

Figure 1.6:   Refraction at an interface

To visualize why a light ray bends, we can return to the more vivid model of waveforms in Fig. 1.6. Consider one incident wavefront out of the myriad described by the incident light ray with an angle of incidence θ_i. Note that the figure above shows a slideshow of the various positions of a single wavefront as it progresses. It is initially at O_1 with its end impinging on the interface at A. Then, it progresses to O_2, O_3, O_4 and O_5 over time. When part of the wave hits the interface, another transmitted wave is re-emitted due to the scattering of the incident waves by the atoms in the second medium.[3] The transmitted wavefront travels and increases in length from I_1 to I_5 as more of the incident wavefront is transmitted, until the entire initial wavefront has been transmitted at I_5. Note that the transmitted light ray must be perpendicular to the transmitted wavefronts. Assuming that this entire process of transmission took a length of time Δt,

$$\overline{CB} = \frac{c}{n_i}\Delta t,$$

$$\overline{AD} = \frac{c}{n_f}\Delta t,$$

where n_i and n_f are the refractive indices of the initial and final media. As triangles $\triangle ABC$ and $\triangle ABD$ share the same side AB and

---

[3]To be precise, this transmitted wavefront is of a constant phase difference relative to the incident wavefront.

$\angle ACB = \angle ADB = 90°$, the sine rule can be applied to obtain

$$\overline{AB} = \frac{\frac{c}{n_i}\Delta t}{\sin \theta_i} = \frac{\frac{c}{n_f}\Delta t}{\sin \theta_f}$$

$$n_i \sin \theta_i = n_f \sin \theta_f, \tag{1.3}$$

or

$$n_{fi} = \frac{n_f}{n_i} = \frac{\sin \theta_i}{\sin \theta_f}, \tag{1.4}$$

where n_{fi} is denoted as the relative refractive index of the final medium to the initial medium. This relationship is known as Snell's law which states additionally that the incident ray, the normal and the transmitted ray all lie in the same plane.

As implied by Snell's law, a light ray entering a medium with a larger refractive index will bend towards the normal. Conversely, a light ray entering a medium with a smaller index of refraction will bend away from the normal.

Apparent Depth

Refraction manifests itself in the perception of depth in a fluid. When one is at the pool, the bodies of the people submerged in the swimming pool appear to have shrunk in height. This can be explained by the ray diagram in Fig. 1.7.

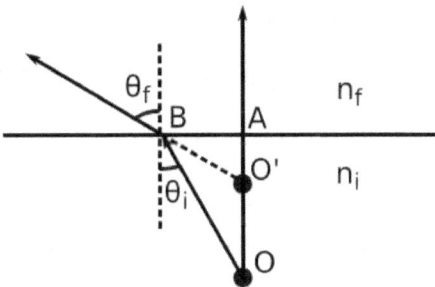

Figure 1.7: Perceived image

Let a point object be at O. Consider a light ray emanating from the object that perpendicularly cuts the interface at point A. Now, consider a second ray that hits the interface at B at an angle of incidence θ_i. The refracted ray will be directed with an angle of refraction θ_f. Now, the virtual image formed by these two rays is at O', which is obtained by extending the refracted ray until it intersects with the extension of the first transmitted ray, as one would perceive light rays to travel in a straight line.[4] Now, the ratio of the apparent depth AO' to the actual depth AO can be computed:

$$\overline{AO'} = \cot \theta_f \overline{AB},$$

$$\overline{AO} = \cot \theta_i \overline{AB},$$

$$\frac{\overline{AO'}}{\overline{AO}} = \frac{\cos \theta_f \sin \theta_i}{\cos \theta_i \sin \theta_f}.$$

From Snell's law,

$$\frac{\sin \theta_i}{\sin \theta_f} = \frac{n_f}{n_i},$$

$$\frac{\overline{AO'}}{\overline{AO}} = \frac{\cos \theta_f n_f}{\cos \theta_i n_i}.$$

At small angles of incidence (i.e. the observer stays near the normal), $\cos \theta_i \approx 1$ and $\cos \theta_f \approx 1$, hence

$$\frac{\overline{AO'}}{\overline{AO}} = \frac{n_f}{n_i}.$$

Since this result is independent of θ_i for small angles of θ_i, all light rays with small θ_i converge at O' — implying that an image is formed there. Hence, the ratio of the apparent depth to the actual depth is

$$\frac{Depth_{app}}{Depth_{act}} = \frac{\overline{AO'}}{\overline{AO}} = \frac{n_f}{n_i}. \tag{1.5}$$

If the final medium is air and the initial medium is water, $n_f = 1$ and $n_i \approx \frac{4}{3}$ and

$$\frac{\overline{AO'}}{\overline{AO}} = \frac{3}{4}.$$

Hence, the perceived height of the bodies of people who are submerged in the pool is $\frac{3}{4}$ of their actual height.

Medium with a Varying Index of Refraction

In certain problems, such as the case of air with a temperature gradient, the refractive index varies with position. In such problems, it is more illuminating

---

[4]Note that even though the transmitted rays are diverging, they are usually captured and focused by the lenses in our eyes.

to express Snell's law as

$$n_i \sin \theta_i = n_f \sin \theta_f.$$

Then, for any general n and θ, which are functions of position,

$$n \sin \theta = c \tag{1.6}$$

for a single light ray, where c is a constant. Consider the following variations of such problems.

Problem: Consider a light ray that emanates from the origin at a certain angle θ_0 with respect to the y-axis. If the index of refraction of the medium of propagation obeys

$$n(y) = \sqrt{1 + ky},$$

determine the trajectory of the light ray, $y(x)$.

Let the angle that the instantaneous slope of the light ray at coordinates (x, y) makes with the y-axis be $\theta(x, y)$. Then, consider an interface at coordinates $y + dy$ in Fig. 1.8.

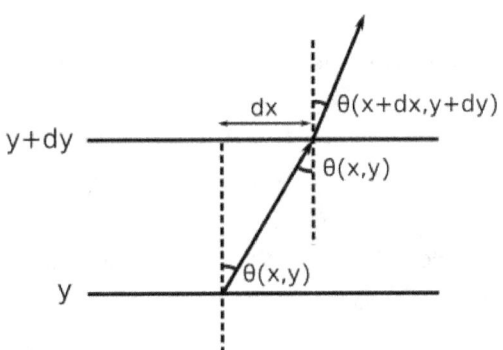

Figure 1.8: Infinitesimally thin slab

By Snell's law,

$$n(y) \sin \theta(x, y) = n(y + dy) \sin \theta(x + dx, y + dy).$$

In other words,

$$n \sin \theta = c$$

for some constant c. The value of c is determined by the initial condition $n(0) = 1$ and $\sin \theta = \sin \theta_0$ at the origin. Hence, $c = \sin \theta_0$.

$$n \sin \theta = \sin \theta_0.$$

Furthermore, $\sin\theta = \dfrac{dx}{\sqrt{(dy)^2+(dx)^2}} = \dfrac{dx}{\sqrt{(\frac{dy}{dx})^2+1}\,dx} = \dfrac{1}{\sqrt{(\frac{dy}{dx})^2+1}}$. Hence,

$$\sqrt{1+ky}\cdot\dfrac{1}{\sqrt{\left(\frac{dy}{dx}\right)^2+1}} = \sin\theta_0$$

$$\frac{dy}{dx} = \sqrt{\frac{ky}{\sin^2\theta_0} - \cot^2\theta_0}$$

$$\int_0^y \frac{1}{\sqrt{\frac{ky}{\sin^2\theta_0} - \cot^2\theta_0}}\,dy = \int_0^x dx$$

$$\frac{2\sin^2\theta_0}{k}\left(\sqrt{\frac{ky}{\sin^2\theta_0} - \cot^2\theta_0} - \cot^2\theta_0\right) = x.$$

Simplifying this equation,

$$y = \frac{k}{4\sin^2\theta_0}\left(x + \frac{2\cos^2\theta_0}{k}\right)^2 + \frac{\cos^2\theta_0}{k}.$$

Therefore, the trajectory of the light ray is a parabola.

Another type of question pertains to the determination of the refractive index of a medium as a function of position, when provided with the trajectory of a light ray. Consider the reverse of the problem above.

Problem: Given that the trajectory of a light ray, beginning at the origin, is $y = kx^2$ in the region $x \geq 0$ where $k > 0$, determine the refractive index as a function of the y-coordinate $n(y)$. It is known that the refractive index is strictly a function of y only.

Well, we can use the fact that

$$n(y)\sin\theta(x,y) = c,$$

where c is a constant and $\theta(x,y)$ is the instantaneous angle that the slope of the trajectory makes with the y-axis at coordinates (x,y). Furthermore, since $k > 0$, $\sin\theta > 0$ as the ray obviously travels in the positive x- and y-directions.

$$\sin\theta = \frac{dx}{\sqrt{(dy)^2+(dx)^2}} = \frac{1}{\sqrt{\left(\frac{dy}{dx}\right)^2+1}} = \frac{1}{\sqrt{4k^2x^2+1}} = \frac{1}{\sqrt{4ky+1}}.$$

Thus,

$$n = \frac{c}{\sin\theta} = c\sqrt{1+4ky}$$

where c is a constant. The physical meaning of c is $n(0)$, which refers to the refractive index of the medium at $y = 0$ as $\sin \theta = 1$ at the origin.

1.4 Total Internal Reflection

Observe that when a light ray, traveling in an optically denser medium, impinges on the interface with an optically less dense medium, Snell's law would require the sine of the refracted angle to be larger than 1 if the angle of incidence is above a certain critical angle, θ_c. Specifically,

$$\sin \theta_c = \frac{n_f}{n_i}.$$

Note that the right-hand side has a value smaller than one, as the light ray attempts to travel from a denser to less dense medium. If the light impinges on the interface at the critical angle, the refracted angle is $\frac{\pi}{2}$ (i.e. the refracted ray is parallel to the interface). If the angle of incidence θ_i is larger than θ_c, the light ray will undergo a phenomenon known as total internal reflection and be entirely reflected in the original medium with an angle of reflection equal to the angle of incidence, so that

$$\theta_i = \theta_r.$$

1.5 Fermat's Principle

Fermat's principle unifies the various laws above. The original principle states that given endpoints A and B, the actual path taken by the light ray to travel from A to B is one that results in the minimum time elapsed. The modern principle now states that the actual path between A and B is such that the time taken by the light to travel takes on a stationary value with respect to all possible small variations. More precisely, any possible variation of the actual path, with fixed end points, will not lead to first order changes in the time elapsed. Referring to Fig. 1.9, given fixed endpoints A and B, we "wiggle" a line connecting A and B until a stationary value for the time traveled by the ray is reached.[5]

Most of the time, the optical path length (OPL) is considered instead of the time taken by the light ray to travel. As light travels at a speed $\frac{c}{n}$ in a medium with an index of refraction n, the time taken for the light to

---

[5]This is similar to extremizing the action (which is analogous to OPL) in Lagrangian mechanics. In fact, one can exploit the fact that the "Hamiltonian" is conserved in the context of OPL to derive the previous laws. See Problem 15 of this chapter.

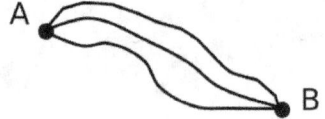

Figure 1.9:   "Wiggling" a path between fixed endpoints A and B

travel a distance s in this medium is n times that required for it to travel the same distance in vacuum. Hence, if we define the infinitesimal OPL to be the refractive index multiplied by an infinitesimal length along the path taken by the ray and the OPL to be the integral of these infinitesimal segments from A to B,

$$d(OPL) = nds$$

$$OPL = \int_A^B nds, \tag{1.7}$$

and the path which adopts a stationary value for the OPL corresponds to a path of stationary time elapsed. Note that the refractive index, in general, may be a function of position in the above expression.

Next, Fermat's principle underscores the reversibility of light rays. If a light ray takes path P from A to B, it will travel along the same path, except in the opposite direction, when originating at B and ending at A.

Now, we can show that Fermat's principle implies all the laws that we have discussed so far. Firstly, it is evident from applying the triangle inequality that the path taken by a light ray from A to B in a homogeneous medium is a straight line connecting A to B, as it is the path with minimum OPL.

Next, Fermat's principle engenders the law of reflection. Considering a horizontal plane mirror in a medium with a uniform refractive index n, we wish to analyze the path taken by a light ray that emanates from A, impinges the mirror and returns to B in Fig. 1.10. Let O be the point of intersection of the incident light ray and the mirror. Its location is variable and our objective is to determine the point O such that the OPL traversed by the light ray between A and B is extremized.

Since we have shown that the paths of light rays are straight lines in homogeneous media, the segments AO and OB must be straight. The total optical path length is correspondingly

$$OPL = n(\overline{AO} + \overline{OB}).$$

We wish to determine an appropriate point O such that this expression takes on a stationary value. Consider the geometrical point B' corresponding to

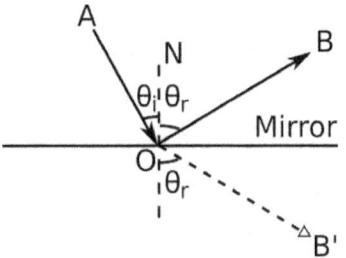

Figure 1.10:   Reflecting off a mirror

the reflection of B about the mirror. Then,

$$\overline{OB'} = \overline{OB}$$

$$\Longrightarrow OPL = n(\overline{AO} + \overline{OB'}).$$

It is evident from the triangle inequality (applied to $\triangle AOB'$) that if A, O and B' are collinear, the OPL will be a minimum. Hence, the actual path taken by the light ray is such that AOB' is a straight line — implying that

$$\theta_i = \theta_r.$$

Furthermore, the condition for collinearity requires A, the normal N and B to lie in the same plane. Hence, the law of reflection is a direct corollary of Fermat's principle.

Lastly, we shall show that Snell's law is also consistent with Fermat's principle.

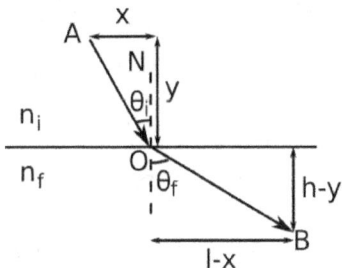

Figure 1.11:   Refraction at an interface between different media

Referring to Fig. 1.11, consider two endpoints A and B in two different media with uniform refractive indices n_i and n_f, respectively. Let the point of intersection between the incident light ray and the horizontal interface be O. Consider the plane containing points A, B and O. Let A and B be separated by horizontal and vertical distances l and h, respectively. Let the horizontal and vertical distances between A and O be x and y, respectively. We wish to find an appropriate point O that produces a stationary OPL — note that y is fixed but x can vary. The OPL is

$$OPL = n_i\sqrt{x^2 + y^2} + n_f\sqrt{(l - x)^2 + (h - y)^2}.$$

The derivative of the OPL with respect to x must be zero:

$$\frac{d(OPL)}{dx} = \frac{n_i x}{\sqrt{x^2 + y^2}} - \frac{n_f(l - x)}{\sqrt{(l - x)^2 + (h - y)^2}} = 0.$$

Notice that this can be rewritten in terms of the angles θ_i and θ_f as

$$n_i \sin\theta_i = n_f \sin\theta_f,$$

which is Snell's law. When θ_i is larger than the critical angle θ_c, no value of θ_f can result in a stationary OPL. Therefore, light does not cross the interface and is instead reflected. All-in-all, we have established that Fermat's principle implies the law of reflection and Snell's law.

Now, if there are multiple paths with the same endpoints A and B that have the same stationary OPL, all of such paths are valid paths that are physically taken by a light ray. Conversely, if we require a light ray to take multiple paths from endpoints A to B, all of these paths must have stationary values of OPL. In practice, these OPL's are usually taken to be identical. This has important consequences in focusing apparatus which redirects various light rays to a single point.

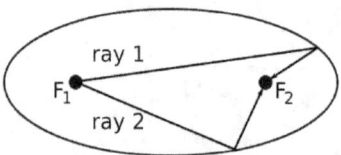

Figure 1.12:   An elliptical room

An instructive example would be the elliptical room whose walls are perfectly reflective, depicted in Fig. 1.12. By the property of an ellipse, all paths that originate from a focus, travel to the wall and finally back to the other focus have the same length. Therefore, by Fermat's principle, all light rays that emanate from one focus will converge at the other focus. If one still has doubts about the applicability of Fermat's principle, one can attempt to verify the following geometrical property of an ellipse: if a line is connected from a focus to a point on the surface and back to the other focus, the angle of incidence is equal to the angle of reflection (i.e. the law of reflection holds). Proving this would show the validity of Fermat's principle in this set-up. Conversely, if you accept the equivalence of the law of reflection and Fermat's principle, the above application of Fermat's principle means that if you strike a billiard ball at a focus of an elliptical table, the ball will always fall into a hole at the other focus (assuming that its collisions are elastic)!

1.6 Optical Apparatus

1.6.1 *Focusing Mirrors*

Suppose that we wish to focus a beam of parallel light rays in a plane such that they coincide at a certain point, by utilising a mirror. What should the shape of the mirror be?

Before we proceed, let us introduce a few definitions. The **optical** or **principal axis** is the axis of symmetry of a mirror or lens. In the case of Fig. 1.13, it is the x-axis. The point of intersection of the optical axis and the optical apparatus is known as the **vertex**, which is point O in this case. The **focal point** is defined as the point on the optical axis where incident light rays, parallel to the principal axis, coincide. The **focal length** is defined as the distance between the vertex and the focal point. If we construct our mirror wisely, rays parallel to the principal axis should converge at the focal point of the mirror.

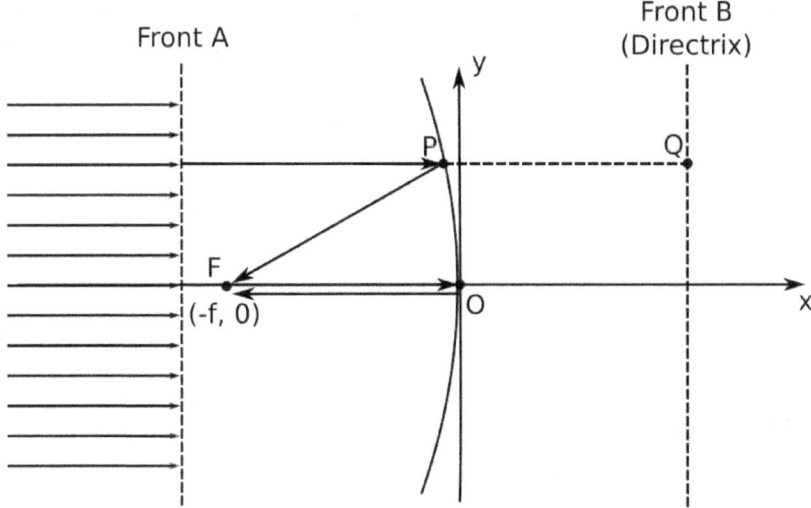

Figure 1.13:   A concave focusing mirror

Consider front A which joins the tips of a bundle of incident light rays, parallel to the principal axis, at this juncture. Now, track the light ray that is along the optical axis. It reflects from O and travels to the focal point F. Since the parallel rays can be assumed to be emitted from a point source infinitely far away, the optical path length traversed by all rays from front A, to the mirror and back to the focal point, must be equal in order for all of the parallel rays to converge at F by Fermat's principle (the rays all emanate from a single point at infinity and are focused at F). From this fact, the shape of the mirror can be determined.

If the rays were not reflected, they would travel to front B, which is described by the equation $x = f$ where f is the focal length $\overline{OF}$. Consider a ray incident at y-coordinate y. Let the point P at which it impinges on the mirror have coordinates (x, y), with O as the origin. Now, if the ray were not reflected, it would have traveled to point Q, of coordinates (f, y). For the OPL of all rays to be equal, $\overline{FP} = \overline{PQ}$ for all possible P's. For those familiar with conic sections, you might recognize that this is the definition of a parabola — F is the focus and front B is the directrix. We can easily prove this.

$$\overline{FP} = \overline{PQ}$$
$$\sqrt{(x + f)^2 + y^2} = f - x$$
$$x^2 + 2fx + f^2 + y^2 = x^2 - 2fx + f^2$$
$$y^2 = -4fx. \tag{1.8}$$

Hence, the shape of the mirror is a parabola described by the equation above. Given an arbitrary parabola of the form $x = ay^2 + by + c$, its equation can be rewritten as

$$\left(y + \frac{b}{2a} \right)^2 = \frac{1}{a} \left(x - c + \frac{b^2}{4a} \right).$$

The focal length can be expressed in terms of the equation of the parabola by comparing the coefficients in front of x, and we obtain

$$|f| = \left| \frac{1}{4a} \right|.$$

The focal length of a concave mirror is defined to be positive as parallel rays converge in front of the mirror.

$$f_{cave} = \left| \frac{1}{4a} \right|. \tag{1.9}$$

This is the reason behind the paraboloid shape of satellite dishes!

The example previously depicted a concave mirror whose focal point is in front of itself. However, let us consider the situation where light rays come from the right of the previous mirror (so that the mirror is now convex with respect to the light rays).

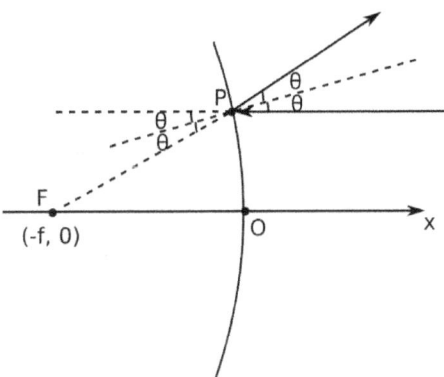

Figure 1.14: A convex focusing mirror

Referring to Fig. 1.14, if we extend any of the incident rays and corresponding reflected rays to the region behind the mirror, we will find that the extensions again intersect at the focal point in the previous section, F. This is due to the extended portions of the rays obeying the law of reflection, as the angles are preserved. Hence, the result from the previous example can be applied directly. It can then be seen that a convex mirror "fictitiously focuses" parallel rays to the focal point behind the mirror (though it causes the rays to diverge in reality). The relationship between the focal length of a convex mirror to its parabolic equation is identical to the previous case of a concave mirror. However, the focal length of a convex mirror is defined to be negative and

$$f_{vex} = -\left|\frac{1}{4a}\right|. \tag{1.10}$$

A Spherical Approximation

In reality, a spherical mirror is much easier to manufacture with precision as one simply has to repeatedly grind two objects together. Hence, we shall generally be analyzing spherical apparatus in this chapter. Consider an arc of the circle described by the equation

$$(x + R)^2 + y^2 = R^2.$$

Figure 1.15 shows the overlay of this circle on a parabola that suitably approximates it near the origin O.

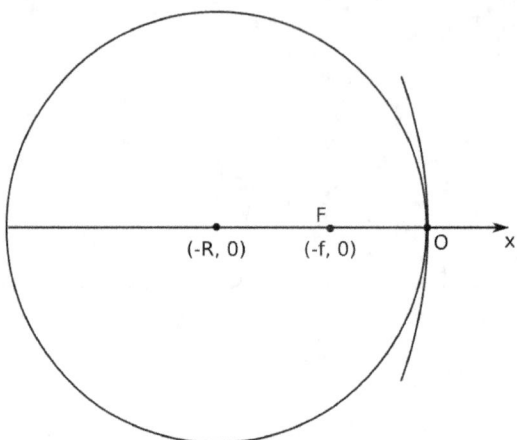

Figure 1.15:   A circle superposed onto a parabola

Note that if we zoom into the region near O, the parabola and circle look roughly identical. Hence, parallel rays in the immediate vicinity of the principal axis, which is known as the **paraxial region**, will still converge at the same focal point after being reflected by the spherical surface. The paraxial region corresponds to points of small y-coordinates and hence, small x-coordinates in this case. Expanding the equation of the circle,

$$x^2 + 2xR + R^2 + y^2 = R^2.$$

The negligible x^2 term is discarded as x is already small, thus

$$y^2 = -2Rx$$

which is the equation that describes the approximating parabola.
Comparing this with Eq. (1.8), it is evident that

$$|f| = \frac{R}{2}$$

for a spherical mirror. The sign of f, again, depends on whether the mirror is convex or concave.

The Mirror Formula

Now that we have determined the focal points of a parabolic and spherical mirror in terms of their geometrical properties, the location of an image of an object with non-negligible height shall be determined. Note that only rays in the paraxial region will be considered — implying that the height of the object should be small relative to the curvature of the mirror. The reason behind this, as we shall soon discover, is that only the rays in the paraxial region converge to form an image.

In Fig. 1.16, let the object be $\overline{BC}$. To identify the location of the image via ray tracing, we first draw Ray 1 which emanates from the top of the object B, travels parallel to the principal axis and is reflected towards the focal point. Ray 2 connects the tip of the object to the focal point and is reflected by the mirror in a direction parallel to the principal axis.[6] The former is due to the property of the focal point and the latter is due to the reversibility of light rays. Let the object and image distances be

$$u = \overline{OC},$$

$$v = \overline{OD},$$

respectively. Both u and v are positive when the object and image are in front of the mirror. Let the object and image heights be

$$h_1 = \overline{BC},$$

$$h_2 = \overline{DE}.$$

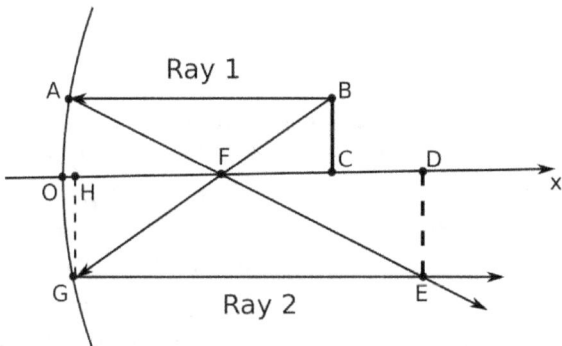

Figure 1.16:   Image due to a mirror

Observe that

$$\triangle BFC \sim \triangle GFH$$

$$\implies \frac{h_1}{h_2} = \frac{u - f}{f - \overline{OH}}.$$

---

[6] As an alternative, one can also draw a ray from B to the vertex O as the law of reflection is easily applicable.

$\overline{OH}$ is negligible as compared to f in the paraxial region, therefore

$$\frac{h_1}{h_2} = \frac{u-f}{f}.$$

Similarly,

$$\triangle AFH \sim \triangle EFD$$

$$\frac{h_1}{h_2} = \frac{\overline{HF}}{\overline{FD}} = \frac{f}{v-f}$$

$$\frac{u-f}{f} = \frac{f}{v-f}$$

$$(u-f)(v-f) = f^2.$$

This is the Newtonian form of the mirror equation which can be expressed as

$$x_o x_i = f^2, \tag{1.11}$$

where x_o and x_i are the distances of the object and image from the focal point. However, the more common form of the mirror equation is obtained from the fact that

$$uv = (u+v)f$$

$$\frac{1}{u} + \frac{1}{v} = \frac{1}{f}. \tag{1.12}$$

Note that u and v are positive if the object and image are located in front of the mirror respectively. u is always positive for real objects by definition (however, the above equations are also valid for $u < 0$ where the incoming light rays seemingly converge at a point behind the mirror[7]). Next, note that a positive v implies a real image in the case of a mirror while the converse implies a virtual image as it is formed behind the mirror (where light rays do not physically converge). f is positive for a concave mirror and negative for a convex mirror. The magnification m, which is the ratio of the height of the image to that of the object, is given by

$$m = -\frac{h_2}{h_1} = -\frac{1}{\frac{u-f}{f}} = -\frac{1}{u\left(\frac{1}{u} + \frac{1}{v}\right) - 1}$$

$$m = -\frac{v}{u}, \tag{1.13}$$

---

[7]This can be proven by abusing the reversibility of light rays and considering the case where the rays emanate from the back of the mirror at object distance v, that is obtained from substituting the negative value of u into the mirror equation (take note that a convex mirror then becomes concave and vice versa).

where a negative sign has been added to indicate the orientation of the image (upright or inverted). If the magnification is negative, the image will be inverted and vice-versa. The expression for magnification also implies that a ray[8] emitted from B to O will converge at E too as the ratio of sides implies that $\triangle BOC \sim \triangle EOD$.

The next technicality pertains to why an object in the form of a straight line is "mapped" to another straight line DE. Notice that we could have chosen any point on the object as point B and the above derivation still follows. The x-coordinate of the imaged point is independent on the y-coordinate of the original point on the object. It then follows from the fact that the object is a continuous line that the image must also be a continuous line with a certain x-coordinate, namely line DE.

Lastly, to be completely rigorous, we can prove that all rays, emitted from B that hits a parabolic mirror with equation $y^2 = 4fx$ at a y-coordinate such that third order and above terms in y are negligible, will converge at E (we have only shown so far that 3 out of myriad rays do so — namely, rays 1, 2 and the ray emitted from B to O). In the following proof, we assume that $\frac{h_1}{u}$ and $\frac{h_2}{v}$ are small as the object must be in the paraxial region and that terms in x^2 are negligible as they are of order four in y. If we wish to show that all light rays from B in the paraxial region converge at E, we just have to show that the optical path length traversed by a light ray from B to any point on the mirror and back to E is the same, at least to second order in y. Consider a light ray that hits O. The OPL_O in this case is

$$OPL_O = \sqrt{h_1^2 + u^2} + \sqrt{h_2^2 + v^2}$$

$$= u\sqrt{1 + \left(\frac{h_1}{u}\right)^2} + v\sqrt{1 + \left(\frac{h_2}{v}\right)^2} \approx u + \frac{h_1^2}{2u} + v + \frac{h_2^2}{2v}.$$

Suppose that a light ray from B hits the mirror at (x, y) and is reflected towards E. The optical path length OPL in this case is

$$OPL = \sqrt{(h_1 - y)^2 + (u - x)^2} + \sqrt{(h_2 + y)^2 + (v - x)^2}.$$

Expanding,

$$OPL = u\sqrt{\frac{h_1^2 - 2h_1y + y^2 + u^2 - 2ux + x^2}{u^2}}$$

$$+ v\sqrt{\frac{h_2^2 + 2h_2y + y^2 + v^2 - 2vx + x^2}{v^2}}.$$

---

[8]See Footnote 6.

Discarding the $\frac{x^2}{u^2}$ and $\frac{x^2}{v^2}$ terms and using the binomial expansion $(1+n)^{\frac{1}{2}} \approx 1 + \frac{1}{2}n$,

$$OPL \approx u + \frac{h_1^2 - 2h_1y + y^2 - 2ux}{2u} + v + \frac{h_2^2 + 2h_2y + y^2 - 2vx}{2v}.$$

Now, consider the difference between OPL and OPL_O.

$$OPL - OPL_O = -\left(\frac{h_1}{u} - \frac{h_2}{v}\right)y + \frac{y^2}{2}\left(\frac{1}{u} + \frac{1}{v}\right) - 2x.$$

The expression in the first bracket is zero as $\frac{h_2}{h_1} = \frac{v}{u}$. The second term in brackets is $\frac{1}{f}$. Lastly, the shape of the mirror provides the relationship $y^2 = 4fx$. Therefore,

$$OPL - OPL_O = \frac{4fx}{2} \cdot \frac{1}{f} - 2x = 0.$$

We have hence proven that all light rays that originate from B and impinge on a parabolic mirror in the vicinity of the optical axis will converge at E.

Problem: By using Eqs. (1.12) and (1.13), what can be deduced about the type (real or virtual), orientation (upright or inverted) and relative size (magnified or diminished) of the image of a real object produced by a convex mirror?

A convex mirror has a negative value of f. Since $u > 0$ for real objects and

$$\frac{1}{v} = \frac{1}{f} - \frac{1}{u}$$

$$\implies |v| < |u|,$$

and $v < 0$. Then,

$$0 < -\frac{v}{u} = m < 1.$$

The image is virtual ($v < 0$), upright ($m > 0$) and diminished ($|m| < 1$).

1.6.2   *Lenses*

A Spherical Refracting Surface

Referring to Fig. 1.17, light rays travel from a point P on the optical axis from medium 1 of refractive index n_1 to medium 2 of refractive index n_2 across a convex spherical interface of radius R. C is the center of the sphere. We claim that all light rays emanating from P and impinging at small angles of incidence on the interface (paraxial region) will be focused to another point P' on the optical axis in medium 2.

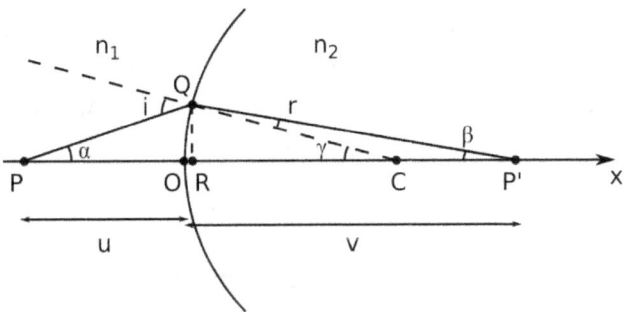

Figure 1.17: Convex spherical refracting surface

Consider a light ray that emanates from P, impinges on the interface with an angle of incidence i and is transmitted with a refracted angle r. As the exterior angle of a triangle is equal to the sum of the two opposite interior angles,

$$i = \alpha + \gamma,$$
$$r = \gamma - \beta.$$

By Snell's law,

$$n_1 \sin i = n_2 \sin r.$$

As i and r are small,

$$n_1 i = n_2 r,$$
$$n_1(\alpha + \gamma) = n_2(\gamma - \beta).$$

Using the small angle approximation $\tan x \approx x$ when x is small,

$$n_1 \left(\frac{\overline{QR}}{\overline{PR}} + \frac{\overline{QR}}{\overline{RC}} \right) = n_2 \left(\frac{\overline{QR}}{\overline{RC}} - \frac{\overline{QR}}{\overline{RP'}} \right).$$

In the paraxial region, $\overline{OR}$ is negligible as compared to the other lengths. Dividing the above by $\overline{QR}$,

$$\frac{n_1}{u} + \frac{n_2}{v} = \frac{n_2 - n_1}{R}.$$

u is the object distance, $u = \overline{PO} \approx \overline{PR}$, and is positive if the object lies in front of the interface while $v = \overline{OP'} \approx \overline{RP'}$ is the image distance and is positive if the image lies behind the interface. Recall that R is the radius of curvature and is by definition, $R = \overline{OC} \approx \overline{RC}$.

It can be seen that the point on the optical axis that a light ray emitted from P in the paraxial region crosses (determined by v) is independent of the angle of incidence. Hence, paraxial light rays from P are focused at P'. By letting the object distance tend to infinity, the light rays from P become a parallel bundle. Then, the value of v in this case, by definition, is the **second** or **image focal length**, f_i. The reason behind this distinction between focal lengths will be elaborated in a moment.

$$\frac{n_2}{f_i} = \frac{n_2 - n_1}{R}$$

$$f_i = \frac{n_2}{n_2 - n_1} R.$$

By letting the image distance v tend to infinity, the value of u becomes the **first** or **object focal length**, f_o (i.e. parallel rays emerging from the right will converge to a point in medium 1, at distance f_o away from the vertex O).

$$\frac{n_1}{f_o} = \frac{n_2 - n_1}{R}$$

$$f_o = \frac{n_1}{n_2 - n_1} R.$$

The focal lengths of a convex spherical refractive surface have hence been determined. For a concave spherical interface, we can leverage the reversibility of light rays and swap all the corresponding quantities (index 1 with 2 and u with v) to obtain

$$\frac{n_1}{u} + \frac{n_2}{v} = \frac{n_1 - n_2}{R}.$$

The corresponding object and image focal lengths are

$$f_i = \frac{n_2}{n_1 - n_2} R,$$

$$f_o = \frac{n_1}{n_1 - n_2} R.$$

The above relationships for convex and concave surfaces can be combined into general equations

$$\frac{n_1}{u} + \frac{n_2}{v} = \frac{n_2 - n_1}{R}, \tag{1.14}$$

$$f_i = \frac{n_2}{n_2 - n_1} R, \tag{1.15}$$

$$f_o = \frac{n_1}{n_2 - n_1} R, \tag{1.16}$$

where R is positive for a convex interface and negative otherwise. Lastly, although not explicitly shown, one can prove that Eq. (1.14) holds for "virtual objects" (light rays in medium 1 that appear to converge at a point in medium 2 before refraction) with negative values of u by exploiting the reversibility of light rays and considering rays that emanate from an object distance v — obtained from substituting the negative value of u into Eq. (1.14) — on the other side of the interface (note that you have to account for the fact that a convex interface becomes concave and vice-versa, and the fact that the refractive indices are swapped).

Spherical Lenses

A simple spherical lens consists of two refracting surfaces enclosing a medium with a refractive index n_l that is usually larger than that of the media that it is immersed in, n_m and n_f. Let us first consider a thin **converging lens**, which is also known as a **bi-convex lens**, in Fig. 1.18.

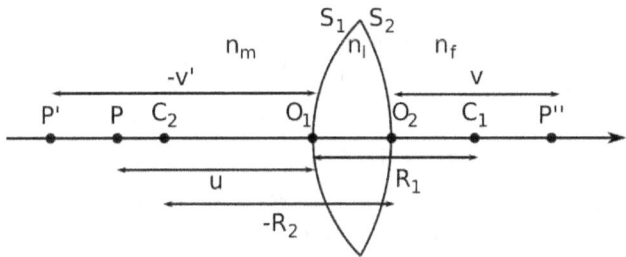

Figure 1.18:   Converging lens

Purely for the sake of illustration, we assume that the image P' of P due to the first surface S_1 is virtual (i.e. in front of the surface). This occurs when the object distance $\overline{PO_1} = u$ is smaller than f_o of S_1. However, the following arguments still hold if the image were real, as Eq. (1.14) holds for negative object distances as well. With respect to the first convex interface S_1,

$$\frac{n_m}{u} + \frac{n_l}{v'} = \frac{n_l - n_m}{R_1},$$

where $v' = \overline{O_1 P'}$ is the image distance with respect to the first spherical surface S_1. With respect to the second concave interface S_2, the light rays effectively emanate from P' from a medium of refractive index n_l (note that even though P' is in medium with a refractive index n_m, the light rays physically travel in the lens) to a medium with refractive index n_f. The object distance of P' with respect to S_2 is $-v' + \overline{O_1 O_2}$. The second term is negligible in the case of a thin lens. Furthermore, since S_2 is concave with respect to the light rays, the radius of curvature R_2 is a negative value. Let v be the distance between the image P'' of P' and the vertex O_2. Then,

$$\frac{n_l}{-v'} + \frac{n_f}{v} = \frac{n_f - n_l}{R_2}.$$

Adding the two previous equations,

$$\frac{n_m}{u} + \frac{n_f}{v} = \frac{n_l - n_m}{R_1} + \frac{n_f - n_l}{R_2}. \tag{1.17}$$

Usually, $n_f = n_m$. Then,

$$\frac{n_m}{u} + \frac{n_m}{v} = (n_l - n_m)\left(\frac{1}{R_1} - \frac{1}{R_2}\right).$$

Dividing both sides by n_m and letting $n = \frac{n_l}{n_m}$ be the relative refractive index of the lens to the medium it is immersed in,

$$\frac{1}{u} + \frac{1}{v} = (n - 1)\left(\frac{1}{R_1} - \frac{1}{R_2}\right). \tag{1.18}$$

This is known as the **Lensmaker's formula**. In the situation above, $R_1 > 0$ and $R_2 < 0$. By letting u and v tend to infinity individually, we discover that the object and image focal lengths are identical. Hence, we drop the prefixes altogether and define the focal length of a spherical lens to be

$$\frac{1}{f} = (n - 1)\left(\frac{1}{R_1} - \frac{1}{R_2}\right). \tag{1.19}$$

Then,

$$\frac{1}{u} + \frac{1}{v} = \frac{1}{f},$$

which is the **Gaussian Lens formula** which also has the following Newtonian form.

$$x_o x_i = f^2$$

where $x_o = u - f$ and $x_i = v - f$. Contrary to the mirror equation, v is now positive if the image lies behind the lens. Note that in the case where a lens is bi-concave (i.e. diverging lens), $R_1 < 0$ and $R_2 > 0$ which causes $f < 0$ (n is always assumed to be greater than 1). For lenses which have a planar surface, their focal lengths can be determined by letting the appropriate radius of curvature tend to infinity (the sign does not matter).

Non-Point Objects

In the previous section, the Gaussian Lens formula was derived for a point source. If the object was not a point source, the above derivation still holds if the object is not large, as we can zoom out of the entire set-up such that it appears as a point source since the above derivation was only valid in the paraxial region in the first place. In this section, the method of ray tracing, used in determining the location of an image, will be illustrated. By convention, a converging lens is depicted by a line with two arrowheads pointing away from the line while a diverging lens is represented by a line with two arrowheads pointing towards the line.

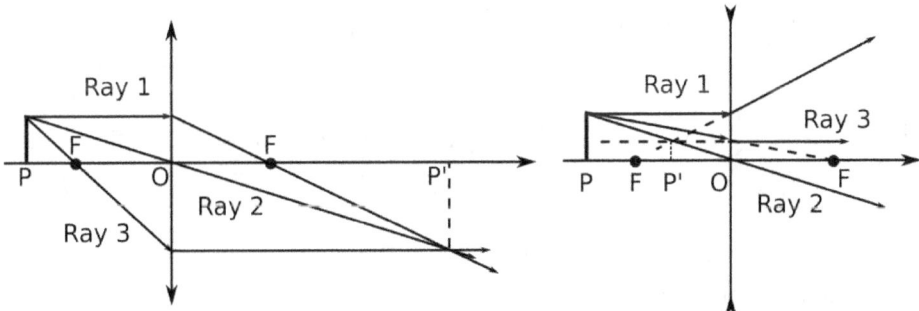

Figure 1.19: Ray diagrams for converging and diverging lenses

Referring to Fig. 1.19, to locate the image of a point produced by a lens, any two of the following three rays from the particular point can be drawn:

- Ray 1: Draw a ray emanating from the point and parallel to the optical axis. After it passes through the lens, either the refracted ray or its extension will pass through the image focal point.
- Ray 2: Draw a ray through the optical center of the lens, O. The ray will pass through the lens without any change in direction.
- Ray 3: Draw a ray emanating from the point such that it or its extension passes through the object focal point. After refracting from the lens, it will travel in a direction parallel to the optical axis.

Note that in the case of a diverging lens, the image and object focal points are in front and behind the lens respectively as it has a negative focal length. Perhaps the only point here that requires further justification is why ray 2 passes through the lens undisturbed. In the vicinity of O, the lens appears like a rectangular block with parallel faces. It is well-known that the transmitted light ray across a rectangular block is parallel to the incident ray, except with a lateral displacement proportional to the thickness of the block (see later section). Since the lens is thin in this case, there will be no deviation of ray 2.

Lastly, the Gaussian Lens formula and the formula for magnification can be easily proven using similar triangles to be

$$\frac{1}{u} + \frac{1}{v} = \frac{1}{f}, \tag{1.20}$$

where u is the image distance $\overline{PO}$, v is the object distance $\overline{OP'}$ and f is the focal length of the lens. Note that v is positive if the image is behind the lens and that f is positive and negative for converging and diverging lenses respectively. The image is real if it lies behind the lens ($v > 0$) and is virtual otherwise ($v < 0$). Its equivalent Newtonian form is

$$x_o x_i = f^2, \tag{1.21}$$

with $x_o = u - f$ and $x_i = v - f$. Finally, the magnification of the image is

$$m = -\frac{v}{u}. \tag{1.22}$$

If the magnification is positive, the image is upright and vice-versa. Hence, a real image is inverted while a virtual image is upright.

Problem: Incident rays that are parallel to the principal axis meet at the image focal point of a lens. Where do incident parallel rays that subtend an angle with the principal axis intersect?

The focal plane is defined as the vertical plane that passes through the focal point. The parallel rays must intersect at one point along the image focal plane as they can be taken to be rays emitted from a point on an object at infinity (the image must be located at the focal plane by the Gaussian Lens formula). To determine the exact point of intersection, simply draw one ray that passes through the optical center of the lens, undeviated — the point of concern is its intersection with the focal plane.

Combination of Lenses

When there are multiple lenses in a system, Eq. (1.20) can be consecutively applied. The image due to a preceding lens will become the object of the following one. Furthermore, the total magnification of the system is the product of the individual magnifications due to each lens. For a two-lenses system with focal lengths f_1 and f_2, let u be the distance between the object and the first lens, v' be the image distance of the first lens and v be the image distance of the second lens. We will first analyze a special case where the two thin lenses are placed together such that the distance between them is negligible. Then, the object distance from the first image to the second lens is $-v'$. Applying Eq. (1.20),

$$\frac{1}{u} + \frac{1}{v'} = \frac{1}{f_1}$$

$$\frac{1}{-v'} + \frac{1}{v} = \frac{1}{f_2}.$$

Adding the two equations together,

$$\frac{1}{u} + \frac{1}{v} = \frac{1}{f_1} + \frac{1}{f_2}.$$

It can be seen that this set-up is equivalent to a thin lens with an effective focal length f given by

$$\frac{1}{f} = \frac{1}{f_1} + \frac{1}{f_2}$$

at the same position. If there are n lenses placed together, the formula above can be repeatedly applied to obtain

$$\frac{1}{f} = \sum_{i=1}^{n} \frac{1}{f_i}. \tag{1.23}$$

The magnification can then be easily determined by dividing v by u. In the case where the two thin lenses are separated by an appreciable distance d, consider a two-lenses system with focal lengths f_1 and f_2 and let u_1 be the distance between the object and the first lens and v_1 be the image distance of the first lens — corresponding definitions hold for the second lens (u_2 and v_2). Then,

$$\frac{1}{u_1} + \frac{1}{v_1} = \frac{1}{f_1} \tag{1.24}$$

$$v_1 = \frac{u_1 f_1}{u_1 - f_1}. \tag{1.25}$$

Since $u_2 = d - v_1$, we can apply the Gaussian Lens formula again to obtain

$$\frac{1}{d - v_1} + \frac{1}{v_2} = \frac{1}{f_2} \tag{1.26}$$

$$v_2 = \frac{(d - v_1)f_2}{d - v_1 - f_2} = \frac{f_2 d(u_1 - f_1) - f_1 f_2 u_1}{(d - f_2)(u_1 - f_1) - u_1 f_1}. \tag{1.27}$$

The magnification is

$$M = -\frac{v_1}{u_1} \cdot -\frac{v_2}{u_2} = \frac{f_1}{u_1 - f_1} \cdot \frac{f_2}{d - v_1 - f_2} = \frac{f_1 f_2}{(u_1 - f_1)(d - f_2) - u_1 f_1}.$$

Even though the above equations completely describe the behavior of a two-lenses system, they are rather cumbersome. Therefore, let us try to adopt a new perspective. Firstly, similar definitions for the object (first) and image (second) focal points hold here for the combined system. The image focal point is the point of intersection between the principal axis and a ray parallel to the principal axis that is incident on the first lens. Conversely, the object focal point is the point of intersection between the principal axis and a ray parallel to the principal axis that is incident on the second lens, in the reverse direction. The latter can be determined as the value of u_1 after setting $v_2 \to \infty$, which causes $u_2 \to f_2$ and $v_1 \to d - f_2$. From Eq. (1.24), the first object distance under these conditions is

$$u_1 = FFL = \frac{f_1(d - f_2)}{d - (f_1 + f_2)}, \tag{1.28}$$

which is known as the front-focal length (FFL). It describes the distance between the object focal point and the first lens. In a similar vein, the cognate back-focal length (BFL) is defined as the distance between the image focal point and the second lens and can be determined as the value of v_2 after setting $u_1 \to \infty$ and $v_1 \to f_1$ in Eq. (1.26). The BFL, which is the second image distance in this case, is

$$v_2 = BFL = \frac{f_2(d - f_1)}{d - (f_1 + f_2)}. \tag{1.29}$$

Now, let us move on to a new formulation. Our goal is to determine an equivalent thin lens system that encapsulates all properties of the image (size and location) produced by this two-lenses system (the intermediate process is not of concern). Let the effective focal length of such a lens be f — this can be determined by imposing the condition that the magnification should be coherent between the set-ups. Defining the object and image distances with respect to this equivalent lens as u and v, the magnification is

$$M = -\frac{v}{u} = \frac{v_1 v_2}{u_1 u_2}.$$

Now as u_1 tends to infinity, $u \to u_1$ as well — enabling us to cancel them in the denominators before they explode. Furthermore, $v_1 \to f_1$, $u_2 \to d - f_1$

and $v_2 \to BFL$. The image distance of the equivalent lens is then

$$v = -\frac{v_1 v_2}{u_2} = -f_1 \cdot \frac{f_2(d - f_1)}{d - f_1 - f_2} \cdot \frac{1}{d - f_1} = \frac{f_1 f_2}{f_1 + f_2 - d},$$

which must be the image focal length of the equivalent lens. Similarly, one can show that the object focal length also takes the form of the above expression by considering $v_2 \to \infty$, $v \to v_2$, $u_2 \to f_2$, $v_1 \to d - f_2$ and $u_1 \to FFL$. Therefore, the common focal length of the equivalent lens is

$$f = \frac{f_1 f_2}{f_1 + f_2 - d}. \tag{1.30}$$

By choosing this particular value of f, we ensure that the magnifications are consistent. We just have to tweak the position of the equivalent lens to locate the image at the correct position. In doing so, we discover that a simple model involving an effective thin lens does not work! The point a distance f away from the object focal point does not correspond to the point a distance f away from the image focal point (i.e. $FFL + d + BFL \neq 2f$) — this should be the case if the system can really be represented by a single thin lens.

To amend this loophole, observe that the intersection of a ray parallel to the principal axis and incident on the first lens and a ray emanating from the second lens in a direction parallel to the principal axis determines the location of the image. We simply have to guarantee that the ends of these rays are correct, while maintaining the magnification. In light of this, instead of having a thin lens where an incident ray on one side immediately emerges from the other side, we can stretch the lens out such that an incident ray on one side is transported to the same vertical position on the other side before proceeding with the same deflection as the case of a thin lens with focal length f. That is, because a system comprising an equivalent lens is lacking some horizontal distance, we artificially supplement it (this does not affect the magnification as it is a mere translation). Referring to Fig. 1.20, the planes forming these two "teleporters" are known as the front and rear principal planes and must be separated by the "missing" distance $FFL + d + BFL - 2f$ (if this is negative, the rear principal plane is actually located in front of the front principal plane). The front principal plane is a distance f on the right of the object focal point while the rear principal plane is a distance f on the left of the image focal point so as to properly concentrate parallel rays at the corresponding focal points and to correctly construct the image from these rays (see first point of this paragraph).

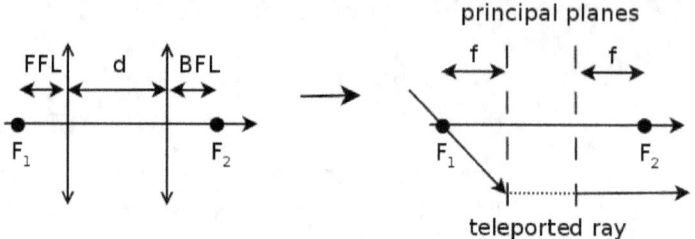

Figure 1.20:　Conversion of lenses into principal planes

Since we have effectively only extended the region inside an equivalent lens of focal length f, the Gaussian Lens formula is valid, as long as u is taken to be the distance between the object and the front principal plane while v denotes the distance between the image and the rear principal plane.

$$\frac{1}{u} + \frac{1}{v} = \frac{1}{f}. \tag{1.31}$$

The Newtonian form holds as well:

$$x_o x_i = f^2 \tag{1.32}$$

where x_o and x_i are still the distances between the object and the object focal point F_1 and between the image and the image focal point F_2. Evidently, the Newtonian form is more useful in this case as it makes no mention of the two principal planes. We can determine the focal points via the BFL and FFL and then apply the above equation to determine the location of the image! Finally, the magnification can be determined through

$$M = -\frac{v}{u} = -\frac{f}{x_o} = -\frac{x_i}{f}. \tag{1.33}$$

To locate the image with the aid of the principal planes in a ray diagram, there are three useful rays to be drawn, as depicted in Fig. 1.21. A ray parallel to the principal axis and incident on the front principal plane will emerge from the rear principal plane and travel towards the image focal point. Similarly, an incident ray passing through the object focal point will emerge from the rear principal plane as a ray parallel to the principal axis. Finally, the two points of intersection between the principal planes and the principal axis are termed as the nodal points, labeled as N_1 and N_2 in Fig. 12.1. A ray crossing the first nodal point will emerge as a parallel ray from the second nodal point (they are effectively the extended version of the optical center).

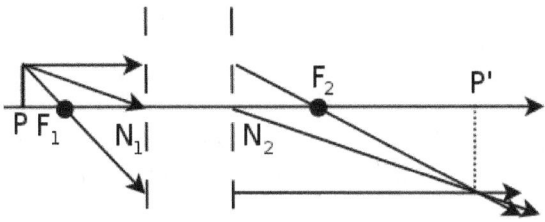

Figure 1.21:   Ray tracing with principal planes

1.6.3   *Applications of Lenses*

Human Eye

The human eye consists of a crystalline lens which focuses light rays at a certain distance to a region at the back of the eye, known as the retina, where light-sensitive cells (rod and cones) are located. Since there can only be a single object distance for which rays are focused at the retina for a single focal length, the eye accommodates to different object distances by varying the focal length of the pliable lens. In the relaxed state, the lens is pretty flat and thus has a large focal length. The eye is then accommodated to objects at infinity — this is why you are advised to gaze at distant trees after staring at the computer for too long. To acclimatize to shorter distances, the ciliary muscles which tug onto the ends of the lens compress the lens and increase the radii of curvature such that the focal length decreases. This process of accommodation can only occur up to a minimum focal length. The object location at which emitted rays can be focused at this juncture is known as the **near point** whose typical distance is 25cm from the lens for a normal eye.

  There are several vision-related conditions which afflict many. The first defect is myopia or near-sightedness. Distant light rays are focused in front of the retina and thus cast a blurred image on the retina. Myopia is usually caused by the radii of curvature of the lens being too large (possibly because it cannot return to its original state) and the eyeball being too long. To occlude the premature convergence of distant rays, a diverging lens can be introduced in front of the eye to correctly focus distant light rays through the spectacle-lens system. Another prevalent defect is hyperopia or long-sightedness. The radii of curvature of the lens are too small, possibly due to the deterioration of the ciliary muscles, or the eyeball is too short such that nearby rays are focused behind the retina. A converging lens can then be introduced to alleviate this symptom.

Refracting Telescope

A telescope is used for astronomical observations — its main purposes are to focus distant rays and to magnify images. A refracting telescope consists of a converging objective lens and an eyepiece which may be converging or diverging. A Keplerian telescope adopts a converging eyepiece while a Galilean telescope uses a diverging eyepiece.

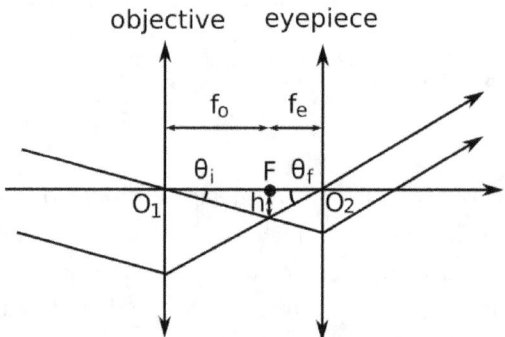

Figure 1.22:   Keplerian telescope

In Fig. 1.22, rays which emanate from a distant object (effectively at infinity) are first focused to a point on the image focal plane of the objective lens. The image of the objective then functions as the object of the eyepiece. To minimize eye strain, the distance between the objective and the eyepiece is adjusted such that the final image is at infinity (as the relaxed eye is accommodated to infinity). If the focal lengths of the objective and eyepiece are f_o and f_e respectively, the separation L between the two lenses is

$$L = f_o + f_e \tag{1.34}$$

so that the image of the objective falls at the object focal length of the eyepiece. Now, a more enlightening measure of image amplification in the case of telescopes, which form images at infinity, is the angular magnification which is defined as the ratio between the angle θ_f that rays from the final image due to an optical apparatus subtend at eye and the angle θ_i that rays from the object, unperturbed by any apparatus, subtend at the unaided eye. Since the object is at infinity, θ_i can be taken to be the angle of parallel rays impinging on the objective instead of those incident on the eye, as the distance between the eye and the objective is relatively negligible. For small angles,[9]

$$\theta_i \approx \tan \theta_i = \frac{h}{f_o},$$

$$\theta_f \approx \tan \theta_f = \frac{h}{f_e},$$

where h is the height of the image of the objective along the focal plane. Thus, the angular magnification M is

$$M = -\frac{\theta_f}{\theta_i} = -\frac{f_o}{f_e}, \tag{1.35}$$

where a negative sign has been added to account for the orientation of the final image, which is located in front of the eyepiece. It can be deduced from the above that the focal length of the objective is usually chosen to be much larger than that of the eyepiece to amplify the angular width of the image.

Problem: What are the possible advantages of the eyepiece being a diverging lens instead of a converging lens?

Firstly, the image is upright which facilitates observations. However, this factor is less significant when observing astronomical objects such as stars which appear as little dots in the sky. Secondly, since $f_e < 0$, the length L of the telescope is reduced such that it is less bulky.

Problem: The objective and eyepiece of a telescope are each bi-convex, with both surfaces having identical radii of curvature (the radii of curvature of the objective and eyepiece may differ though). Suppose that the separation between them is L_0 under normal conditions. If the interior of the telescope is now filled with water, determine the new separation L that the telescope needs to be adjusted to. The refractive indices of the lenses and water are $\frac{3}{2}$ and $\frac{4}{3}$ respectively.

The original focal lengths of the objective and eyepiece are $f_o = R_o$ and $f_e = R_e$ respectively by substituting $n = \frac{3}{2}$ in Eq. (1.19), where R_o and R_e are the respective radii of curvature. Substituting $n_m = 1$, $n_l = \frac{3}{2}$, $n_f = \frac{4}{3}$, $R_1 = -R_2 = R_o$ in Eq. (1.17) and letting u tend to infinity, the image focal

---

[9]This is a reasonable assumption as you usually align your eye with the optical center of the eyepiece and do not want to roll your eyes around much.

length of the objective in water is

$$f'_o = 2R_o.$$

Similarly, substituting $n_m = \frac{4}{3}$, $n_l = \frac{3}{2}$, $n_f = 1$, $R_1 = -R_2 = R_e$ in Eq. (1.17) and letting v tend to infinity, the object focal length of the eyepiece in water is

$$f'_e = 2R_e.$$

Thus, the new separation is

$$L = f'_o + f'_e = 2(R_o + R_e) = 2L_0.$$

1.6.4   *Other Refracting Apparatus*

Rectangular Slab

When an incident light ray travels through a rectangular slab of refractive index n_p immersed in surrounding homogeneous medium of refractive index n_m, the transmitted ray is parallel to the incident light ray with a slight shift, as shown in Fig. 1.23.

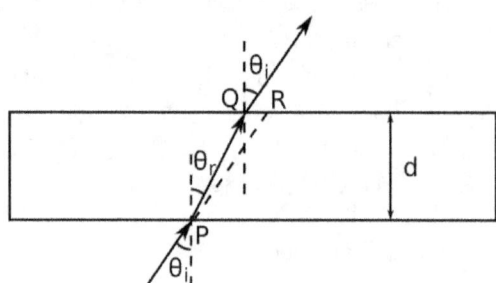

Figure 1.23:   Shift due to a rectangular slab

Point R is the intersection of the second interface with the extension of the incident ray. The parallel shift is

$$\overline{QR} = |d(\tan\theta_i - \tan\theta_r)|,$$

where θ_i and θ_r are related by Snell's law. Now, for rays emanating from a single point and impinging with small angles of incidence, the effect of a rectangular block is to form an image of the point at the same coordinates along the surface of the block but at a different perpendicular distance from the slab. Suppose that the object distance, which is the length of a normal line originating from the first surface and crossing the object, is initially u.

Due to the parallel deviation $\overline{QR}$, the extension of the transmitted ray intersects the same normal line at a perpendicular distance $u - \overline{QR}\cot\theta_i$ from the first surface. The deviation in distance for small angles is

$$\delta = \overline{QR}\cot\theta_i = d\left(1 - \frac{\tan\theta_r}{\tan\theta_i}\right) = d\left(1 - \frac{n_m}{n_p}\right) \tag{1.36}$$

by Snell's law, as $\tan\theta \approx \sin\theta$ for small θ. This expression is independent of θ_i for small angles — implying that an image is formed there. Therefore, a rectangular slab effectively reduces the object distance by $\delta = d(1 - \frac{n_m}{n_p})$, while maintaining the orientation, for rays incident at small angles.

Triangular Prism

Consider a cross section of a triangular prism with an apex angle α. In general, a ray can undergo both refraction and total internal reflection due to a prism. However, let us consider the case of the former only as it is more interesting.

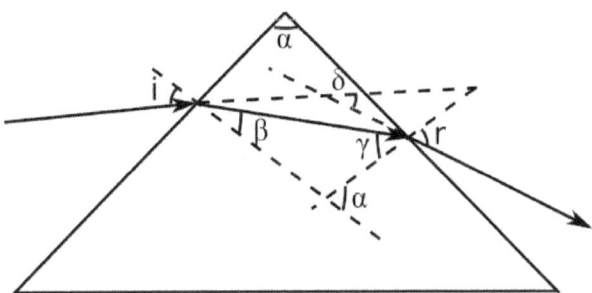

Figure 1.24:   Triangular prism

Referring to Fig. 1.24, given an incident ray with an angle of incidence i, we wish to determine the deviation δ which is the angle subtended by the incident and the transmitted ray. We assume that no total internal reflection occurs. δ is simply the sum of the deviations at each interface, i.e.

$$\delta = i - \beta + r - \gamma.$$

Since $\alpha = \beta + \gamma$,

$$\delta = i + r - \alpha. \tag{1.37}$$

We are left with determining r in terms of i. Assuming that the surrounding medium is air and that the refractive index of the prism is n,

$$\sin r = n\sin\gamma = n\sin(\alpha - \beta) = n(\sin\alpha\cos\beta - \cos\alpha\sin\beta).$$

Since $n \sin \beta = \sin i$ and $\cos \beta = \sqrt{1 - \frac{\sin^2 i}{n^2}}$,

$$\sin r = n \left(\sin \alpha \sqrt{1 - \frac{\sin^2 i}{n^2}} - \cos \alpha \frac{1}{n} \sin i \right) = \sin \alpha \sqrt{n^2 - \sin^2 i} - \cos \alpha \sin i$$

$$r = \sin^{-1} \left(\sin \alpha \sqrt{n^2 - \sin^2 i} - \cos \alpha \sin i \right),$$

$$\delta = i + \sin^{-1} \left(\sin \alpha \sqrt{n^2 - \sin^2 i} - \cos \alpha \sin i \right) - \alpha. \tag{1.38}$$

Equation (1.38) doesn't look very neat, but an interesting problem is to determine the condition on i for the minimal deviation. At this particular value of i, $\frac{d\delta}{di} = 0$. Hence, by implicitly differentiating Eq. (1.37),

$$\frac{dr}{di} = -1.$$

By implicitly differentiating the expression obtained from Snell's law at each interface,

$$\cos i \, di = n \cos \beta d\beta,$$

$$\cos r \, dr = n \cos \gamma d\gamma.$$

Dividing the latter by the former and applying $\frac{dr}{di} = -1$,

$$-\frac{\cos r}{\cos i} = \frac{\cos \gamma}{\cos \beta} \frac{d\gamma}{d\beta}.$$

Differentiating $\alpha = \beta + \gamma$ with respect to β,

$$\frac{d\gamma}{d\beta} = -1$$

$$\implies \frac{\cos r}{\cos i} = \frac{\cos \gamma}{\cos \beta},$$

$$\frac{\cos^2 i}{\cos^2 \beta} = \frac{\cos^2 r}{\cos^2 \gamma}.$$

Applying Snell's law once again ($n \sin \beta = \sin i$ and $n \sin \gamma = \sin r$),

$$\frac{n^2 - n^2 \sin^2 i}{n^2 - \sin^2 i} = \frac{n^2 - n^2 \sin^2 r}{n^2 - \sin^2 r}.$$

Dividing both sides by n^2 and cross multiplying,

$$n^2 - \sin^2 r - n^2 \sin^2 i + \sin^2 i \sin^2 r = n^2 - \sin^2 i - n^2 \sin^2 r + \sin^2 i \sin^2 r$$

$$(n^2 - 1) \sin^2 i = (n^2 - 1) \sin^2 r$$

$$\implies i = r.$$

From Snell's law and $\beta + \gamma = \alpha$, this also implies that

$$\beta = \gamma = \frac{\alpha}{2}.$$

This makes sense in the limiting case of a prism that takes the form of an isosceles triangle, as it implies that the path that this light ray takes is symmetrical — it travels parallel to the base of the prism inside the prism. Moving on, applying Snell's law,

$$\sin i = n \sin \beta = n \sin \frac{\alpha}{2}$$

$$i = \sin^{-1} \left(n \sin \frac{\alpha}{2} \right). \tag{1.39}$$

The minimum deviation is

$$\delta_{min} = i + r - \alpha = 2i - \alpha$$

$$\delta_{min} = 2 \sin^{-1} \left(n \sin \frac{\alpha}{2} \right) - \alpha. \tag{1.40}$$

Conversely, the refractive index of the prism n can be expressed as

$$n = \frac{\sin \frac{\delta_{min} + \alpha}{2}}{\sin \frac{\alpha}{2}}. \tag{1.41}$$

This equation is used in practice to determine n of an arbitrary material. The material is first shaped into a prism, after which δ_{min} and α are experimentally measured. Then, n can be determined.

Problems

Reflection, Refraction and Total Internal Reflection

1. *Optical Fiber**

We model an optical fiber as a cylinder with refractive index n_f surrounded by a cladding of refractive index $n_c < n_f$. The two circular ends of the optical fiber are not covered with cladding. Let θ be the angle of incidence of a light ray impinging at one of the ends of the fibre. Assuming that the surrounding medium has refractive index 1, determine the range of θ for which the light ray is trapped in the optical fiber (i.e. cannot be transmitted to the cladding).

2. *Field of View**

You are at an aquarium with a porthole of radius R, negligible thickness and refractive index $n = \frac{3}{2}$ embedded in an opaque ground. To observe aquatic lifeforms swimming at the bottom of the aquarium (in water with refractive index $\frac{4}{3}$) which is a distance h below the ground, you peek through the porthole. What is the maximum area of the bottom that you can see?

3. *Skewed Mirrors**

Consider two semi-infinite plane mirrors with their finite ends placed together. The angle subtended by the two mirrors is α. An emitter is placed at point P in this two-dimensional plane. Furthermore, a receiver in the form of a circular arc of radius r is sandwiched between the two mirrors. Find the largest angle θ at which a light ray is emitted from P will eventually reach the receiver. How many reflections does this take?

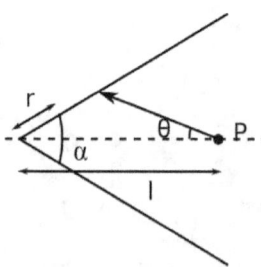

4. *Rebound**

Consider two semi-infinite plane mirrors with their finite ends placed together. The angle subtended by the two mirrors is $2\alpha < \frac{\pi}{2}$. An emitter is placed at point P in this two-dimensional plane. If the perpendicular to the mirror from P is of length a, determine the angle θ at which a ray can be emitted such that it returns to P after a reflection from the top mirror, followed by a reflection by the bottom mirror. What is the distance travelled by the light ray between its emission from and return to point P?

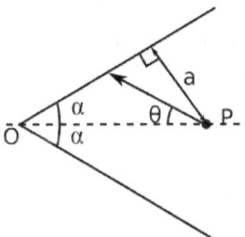

5. *8 Reflections**

Consider two semi-infinite plane mirrors with their finite ends placed together. The angle subtended by the two mirrors is α. A ray, that is parallel to mirror 2, is incident on mirror 1 and after 8 reflections, it emerges parallel to mirror 1 (after a final reflection from mirror 2). Determine α.

6. *Glass Ball**

Half of the surface of a glass sphere with refractive index n is coated with silver. Determine the angle of deflection of a ray that impinges on the non-coated surface of the equator, at an angle of incidence i, after it exits the sphere. Under what conditions can a bundle of parallel light rays — incident on the non-coated half of the equator at small angles of incidence — emerge from the ball, still in a parallel bundle?

7. *Curving Ray***

A medium with a refractive index $n(y)$ fills the region $y > 0$. A light ray traveling along the x-direction in air strikes the medium at a right incidence angle at the origin and begins to propagate within the medium. Determine

$n(y)$ if (a) the ray moves in a circular arc of radius R and (b) the ray moves in a complete sinusoidal curve of amplitude 1m and "wavelength" λ. Given that the largest refractive index is that of diamond with $n = 2.5$ approximately, determine the maximum angular size of the circular arc and the minimum "wavelength" of the sinusoidal curve.

8. Ray in Circle**

A light ray starts from the interior of the circumference of a solid circle of radius R at $\frac{\pi}{4}$ radians with respect to the radial direction, radially inwards. Set the origin to be at the center of the circle. If the refractive index of the circle varies according to the relationship $n(r) = \sqrt{\frac{R^2}{r^2} + 1}$, determine the magnitude of the angular displacement of the light ray when it reaches the center of the circle.

9. Mirage Effect**

On a sweltering afternoon, a man walks along a road. The refractive index of air above the road obeys $n(y) = n_0(1 + \alpha y)$ where α is a constant and y is the height above the road. Firstly, explain qualitatively the reason behind this variation in refractive index and whether α is positive or negative. As a result of this refractive index gradient, the man cannot see the road beyond a certain distance L. If his eyes are a height h above the ground, determine L. Finding the trajectory of a light ray emanating from the road would be a bonus.

10. Trapping Light**

An isotropic point source is placed at the center of a cube of edge length l and refractive index $n > 1$. If the medium surrounding the cube is vacuum and $\sin^{-1}\frac{1}{n} \leq \frac{\pi}{4}$ radians, determine the minimum surface area on the cube that needs to be covered with opaque paint so that no light escapes the cube. Next, for all $n > 1$, determine the minimum painted area if the paint is now perfectly reflective.

11. Emitter in Triangular Room***

A room takes the shape of a right-angled isosceles triangle with base length $8a$. The walls of the room are covered with mirrors and a square receiver of side length a is placed at the right-angled corner of the room. A light ray is emitted at an infinitesimal distance away from the mid-point of the hypotenuse, at an angle θ with respect to the horizontal, such that $\cot\theta = 8$. Determine the distance covered by the light ray before it impinges on the receiver.

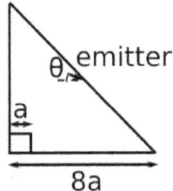

Fermat's Principle

12. *Optimal Path**

Suppose that you wish to cross from one side of a lake (point P) to the other side (point Q). The lake takes the form of a rectangular strip of width d and the vertical distance y between P and Q is much smaller than the horizontal distance L between them. If you move at speed v_1 on shore and $v_2 < v_1$ in water, draw the path that takes the least time from points P to Q. You do not need to calculate the time taken along this path.

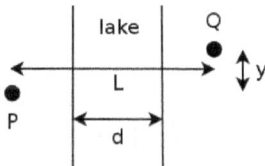

13. *Lensmaker's Formula***

Derive the Lensmaker's formula (Eq. (1.19)) for a thin lens with refractive index n and comprising radii of curvature R_1 and R_2 by considering two rays and applying Fermat's principle. This approach is more direct than that presented in this chapter.

14. *Light in the Atmosphere***

The atmosphere of the Earth can be modeled as an ideal gas with a uniform temperature T and average mass M, wrapped around a uniform spherical Earth of radius r_0. The gravitational field strength in the region of the atmosphere can be taken to be that at the surface of the Earth, g. If the refractive index of a point in the atmosphere is proportional to the density at that point, $n = \alpha\rho$, determine the height h above the surface of the Earth at which a light ray travels in a circle around the Earth. Hint: The ideal gas law is $pV = nRT$ where p, V, n and T are the pressure, volume, moles and temperature of the gas respectively while R is the ideal gas constant.

15. Lagrangian Derivation**

A more rigorous derivation of the laws of reflection and refraction from Fermat's principle uses the Lagrangian formulation. The OPL between two points $y(x_1)$ and $y(x_2)$ visited by a light ray is given by

$$OPL = \int_{x_1}^{x_2} n(x, y)\sqrt{1 + y'^2}\,dx$$

where $n(x, y)$ is the refractive index of the medium and $y(x)$ is the trajectory of the ray between the two fixed endpoints. Use your knowledge of the Lagrangian method to prove the following. Firstly, the path taken by a ray in a homogeneous medium is a straight line. Next, prove the laws of reflection and refraction (hint: under a suitable choice of coordinates, the Hamiltonian is conserved).

Optical Apparatus

16. Minimum Distance*

Determine the minimum distance between a real object and its image produced by a thin converging lens in terms of its focal length f. Ignore the unrealistic case where the object distance is 0.

17. Blurring*

A wire with negligible thickness is placed a distance u in front of a converging lens of unknown focal length and diameter D. When a screen is placed at a distance L behind the lens, a smudge with an appreciable thickness d is formed. Determine the possible focal lengths of the lens.

18. Congealing Lenses*

The flat surfaces of two thin plano-convex lenses of common radius R but different refractive indices n_1 and n_2 are glued together to form a thin converging lens. Determine the focal length of this lens.

19. *Mirror with Liquid* *

A small ball is placed along the axis of a concave mirror of focal length f at an object distance u. The concave surface of the mirror is filled with a thin layer of liquid of refractive index n. If the image of the ball is formed by the rays impinging on the mirror near its vertex, determine the location of the image.

20. *Quarter Prism* **

A glass prism in the shape of a quarter-cylinder rests on a horizontal table. A uniform, horizontal bundle of light impinges perpendicularly on its vertical plane surface as shown in the figure below. Note that all rays are above the surface of the table (though some are infinitesimally close to it). If the radius of the cylinder is $R = 5\text{cm}$ and the refractive index of glass is $n = 1.5$, where on the table beyond the cylinder, will a patch of light be found? A range should be given.

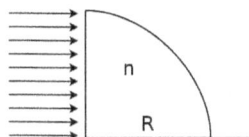

21. *Unique Configuration* **

Two converging lenses of focal lengths f_1 and f_2 are situated between an object and a screen, with the lens with focal length f_1 closer to the object. If we require an image to be produced on a screen which is at a distance l away from the object with $l < 2f_1 + 4f_2$, show that for a given object distance to the first lens, $u > f_1$, there is only one possible position for the second lens.

22. *Moving Image* **

An ant lies along the principal axis of a concave mirror of focal length f. If the ant begins moving, under what conditions will the velocities of the ant and its image be identical? Supposing that the ant travels such that the object distance u increases at the rate $\frac{du}{dt} = \frac{\alpha}{v-f}$ where v is the object distance and α is a constant, starting from an initial object distance u_0, determine $v(t)$.

23. *Prism* **

A right-angled isosceles prism of side length 9cm and refractive index 1.5 is placed 6cm away from a converging lens of focal length $f_1 = 20$cm, followed by a diverging lens of focal length $f_2 = -10$cm a distance 7cm behind it. A 1cm stick is located 8cm above the prism, with one end aligned with the mid-point of the hypotenuse as shown in the figure below. Describe the final image of the stick and its magnification.

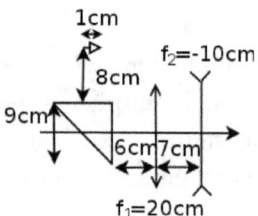

Solutions

1. Optical Fiber*

Let α be the refracted angle at the air-fiber interface. Let β be the angle of incidence of the light ray propagating in the fiber and impinging on the fiber-cladding interface. Then, for the light ray to not escape the fiber,

$$\sin \beta > \frac{n_c}{n_f}$$

for total internal reflection to occur. Since $\beta = \frac{\pi}{2} - \alpha$,

$$\implies \cos \alpha > \frac{n_c}{n_f}.$$

By Snell's law, $\sin \alpha = \frac{\sin \theta}{n_f}$. Then,

$$\sqrt{1 - \frac{\sin^2 \theta}{n_f^2}} > \frac{n_c}{n_f}$$

$$|\sin \theta| < \sqrt{n_f^2 - n_c^2}$$

$$-\sin^{-1} \sqrt{n_f^2 - n_c^2} < \theta < \sin^{-1} \sqrt{n_f^2 - n_c^2}.$$

2. Field of View*

Consider a ray emanating from the bottom of the aquarium that impinges the porthole at an angle of incidence i. The ray is refracted as it enters the porthole. In order to leave the porthole and enter your eyes, the angle of refraction r must be less than the critical angle. That is,

$$\sin r \leq \frac{1}{n} = \frac{2}{3}.$$

By Snell's law,

$$\frac{4}{3} \sin i = \frac{3}{2} \sin r$$

$$\sin i = \frac{9}{8} \sin r \leq \frac{3}{4}.$$

Therefore, the additional radius on the bottom of the aquarium, beyond R, that the observer can see is

$$\Delta R = h \tan i_{max} = h \frac{\frac{3}{4}}{\sqrt{1 - \frac{9}{16}}} = \frac{3}{\sqrt{7}} h.$$

Therefore, the field of view is

$$A = \pi(R + \Delta R)^2 = \pi \left(R + \frac{3}{\sqrt{7}} h \right)^2.$$

3. Skewed Mirrors*

Combining all the mirror images of the receiver, we obtain a full circle, as depicted in Fig. 1.25.

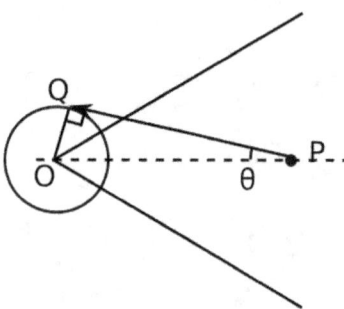

Figure 1.25:   Mirror image

The light ray with the largest θ is tangent to the circle. Hence,

$$\theta = \sin^{-1} \frac{r}{l}.$$

Let the x-axis be along the symmetrical axis, pointing towards the right. Let the origin be at O. If the ray did not undergo any reflection, the angular region on the circle that a ray can reach is $[\frac{\alpha}{2}, -\frac{\alpha}{2}]$. After 1 reflection with either mirror, the region becomes $[\frac{3\alpha}{2}, -\frac{3\alpha}{2}]$. Extending this logic to n reflections, the region increases to $[\frac{(2n+1)\alpha}{2}, -\frac{(2n+1)\alpha}{2}]$. We wish to determine the smallest n for which

$$\left(n + \frac{1}{2} \right) \alpha \geq \angle POQ = \cos^{-1} \frac{r}{l}$$

$$n_{min} = \left\lceil \frac{\cos^{-1} \frac{r}{l}}{\alpha} - \frac{1}{2} \right\rceil.$$

4. Rebound*

In Fig. 1.26, let P' be the mirror image of P after a reflection from the bottom mirror and P'' be the mirror image of P' after a subsequent reflection from the top mirror.

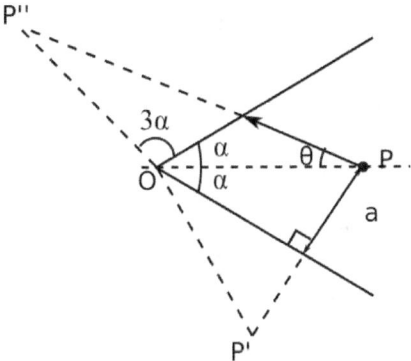

Figure 1.26: Mirror images

Then, the angle θ we wish to find is $\angle OPP''$. Notice that

$$\overline{OP} = \overline{OP'} = \overline{OP''} = \frac{a}{\sin \alpha}.$$

Hence, $\triangle OPP''$ is isosceles and

$$\theta = \frac{\pi - 4\alpha}{2} = \frac{\pi}{2} - 2\alpha.$$

The distance traversed is $\overline{PP''}$ which can be computed via cosine rule:

$$\overline{PP''}^2 = \frac{2a^2}{\sin^2 \alpha}(1 - \cos 4\alpha) = \frac{4a^2 \sin^2 2\alpha}{\sin^2 \alpha} = 16a^2 \cos^2 \alpha$$

$$\implies \overline{PP''} = 4a \cos \theta.$$

5. 8 Reflections*

Due to the symmetrical nature of the set-up and the reversibility of light rays, the path of the light ray between the 4th reflection (with mirror 2) and the 5th reflection (with mirror 1) must be symmetrical about the symmetry axis of the two mirrors as well. That is, the path must be perpendicular to the symmetry axis. Then, the angle subtended by the ray after emerging from the 4th reflection (with mirror 2) and mirror 2 must be $\frac{\pi}{2} - \frac{\alpha}{2}$ (we are referring to the angle closer to the point of connection of the mirrors).

To visualize this angle at the 4th reflection, consider the mirror images in Fig. 1.27.

Image $2'$ is produced by reflecting mirror 2 about mirror 1; image $1'$ is produced by reflecting mirror 1 about image $2'$ while image $2''$ is produced by reflecting image $2'$ about image $1'$. The angle after emerging from the

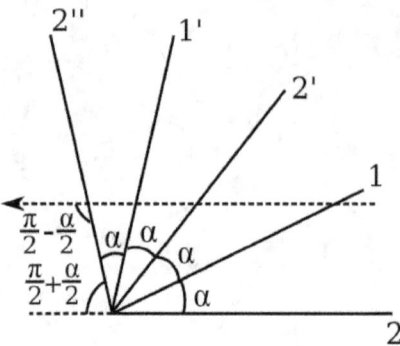

Figure 1.27: Images of mirrors

4th reflection is labeled above. We have

$$4\alpha + \left(\frac{\pi}{2} + \frac{\alpha}{2}\right) = \pi \implies \alpha = \frac{\pi}{9}.$$

6. Glass Ball*

The path of a light ray inside the ball is shown in Fig. 1.28.

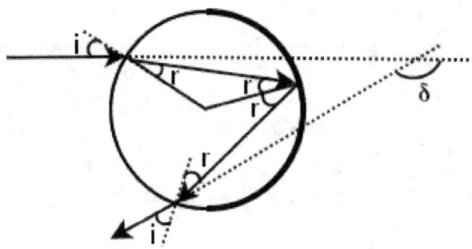

Figure 1.28: Light ray in ball

The angle of deflection is the sum of the individual deviations at each interface.

$$\delta = i - r + (\pi - 2r) + i - r = \pi + 2i - 4r.$$

For small angles of incidence and thus refraction, Snell's law yields

$$r = \frac{i}{n}.$$

The angle of deflection is then

$$\delta = \pi + \left(2 - \frac{4}{n}\right) i.$$

In order for the rays emerging from the ball to remain parallel, δ must be independent of i. This requires

$$n = 2.$$

Incidentally, the angle of deflection in this case is π radians — implying that the reflected bundle is anti-parallel to the incident one (with some losses due to imperfect reflection).

7. Curving Ray**

Referring to Fig. 1.8, define $\theta(x, y)$ as the angle of incidence that the ray makes with the horizontal interface at y-coordinate y. By Snell's law,

$$n(y) \sin \theta = n(0) \sin \theta_0.$$

Since the medium at the origin is air with refractive index $n = 1$ and the angle of incidence θ_0 there is $\frac{\pi}{2}$,

$$n(y) \sin \theta = 1$$

$$n(y) = \csc \theta.$$

We have

$$\sin \theta = \frac{1}{\sqrt{\left(\frac{dy}{dx}\right)^2 + 1}} \implies \csc \theta = \sqrt{\left(\frac{dy}{dx}\right)^2 + 1}.$$

If the path of the ray is a circular arc of radius R, its trajectory is described by

$$x^2 + (y - R)^2 = R^2$$

$$2x + 2(y - R)\frac{dy}{dx} = 0$$

$$\left(\frac{dy}{dx}\right)^2 = \frac{x^2}{(y - R)^2} = \frac{R^2 - (y - R)^2}{(y - R)^2}$$

$$n(y) = \sqrt{\left(\frac{dy}{dx}\right)^2 + 1} = \sqrt{\frac{R^2}{(y - R)^2}}$$

$$n(y) = \frac{R}{R - y}.$$

Note that we only consider $y < R$ as the ray cannot pass by $y = R$ at which the refractive index must tend to infinity. For the sinusoidal arc,

$$y = 1 - \cos(kx)$$

where $k = \frac{2\pi}{\lambda}$. Note that we do not consider other sinusoidal functions, such as $y = \sin(kx)$, because y' at $y = 0$ must be zero. From this trajectory,

$$\frac{dy}{dx} = k\sin(kx) = \frac{2\pi}{\lambda}\sqrt{1 - (1-y)^2}$$

$$n(y) = \sqrt{\left(\frac{dy}{dx}\right)^2 + 1} = \sqrt{1 + \frac{4\pi^2}{\lambda^2}(2y - y^2)}.$$

Given that $n_{max} = 2.5$, since $n(y)$ for a circular arc is increasing with $y \geq 0$,

$$2.5 = \frac{R}{R - y_{max}}$$

$$y_{max} = R\left(1 - \frac{1}{2.5}\right) = 0.6R.$$

The maximum angular size of the circular arc is thus

$$\theta_{max} = \cos^{-1}\left(\frac{R - y_{max}}{R}\right) = \cos^{-1} 0.4 = 66.4° \text{ (3sf)}.$$

Finally, for the sinusoidal trajectory, the maximum refractive index corresponds to $y = 1$ (because $2y - y^2 = 1 - (1-y)^2 \geq 1$ where the equality occurs when $y = 1$). Equating this with $n_{max} = 2.5$,

$$2.5 = \sqrt{1 + \frac{4\pi^2}{\lambda_{min}^2}}$$

$$\lambda_{min} = \frac{2\pi}{\sqrt{5.25}} = 2.74\text{m (3sf)}.$$

8. Ray In Circle**

In Fig. 1.29, let the origin O be located at the center of the circle and define the coordinates of the light ray as (r, θ). Since the refractive index is strictly a function of r, the interfaces are concentric circles. Let ϕ be the angle of incidence of the light ray as it propagates through this circle.

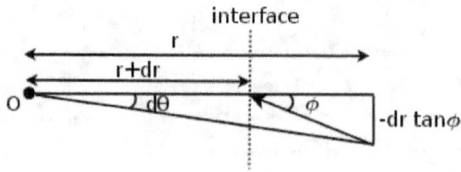

Figure 1.29:   Ray traveling radially inwards from r to $r + dr$

Consider the propagation of the light ray from radial coordinate r to $r + dr$ where $dr < 0$. Evidently, the angular change $d\theta$ is given by

$$r d\theta = -dr \tan \phi.$$

By Snell's law, $n \sin \phi$ is constant.

$$n \sin \phi = n(R) \sin \frac{\pi}{4} = \sqrt{2} \cdot \frac{1}{\sqrt{2}} = 1$$

$$\sin \phi = \frac{1}{n}$$

$$\tan \phi = \frac{\frac{1}{n}}{\sqrt{1 - \frac{1}{n^2}}} = \frac{1}{\sqrt{n^2 - 1}}$$

for $0 \le \phi < \frac{\pi}{2}$. Substituting the expression for $n(r)$,

$$\tan \phi = \frac{r}{R}.$$

Substituting this into the first equation,

$$\int_{\theta_0}^{\theta_0 + \Delta\theta} d\theta = -\int_R^0 \frac{\tan \phi}{r} dr = -\int_R^0 \frac{1}{R} dr$$

$$|\Delta\theta| = 1 \text{ radian.}$$

9. Mirage Effect**

The road absorbs heat from the Sun and transfers heat to its surroundings — establishing a temperature gradient that decreases with height. Since the density of air is inversely proportional to temperature (pressure is approximately fixed), the density of air increases with height, causing the refractive index to increase with height ($\alpha > 0$) as more air molecules are packed into a unit volume such that they scatter light to a greater extent. To analyze this set-up, define the origin at a point on the road and orient the positive x-axis towards the observer. Let the trajectory of a light ray emanating from the origin with an initial angle of incidence θ_0 be $y(x)$. Define $\theta(x, y)$ as the angle of incidence at coordinates (x, y). Referring to Fig. 1.8, Snell's law states that

$$n_0(1 + \alpha y) \sin \theta = n_0 \sin \theta_0.$$

Since $\sin\theta = \dfrac{1}{\sqrt{(\frac{dy}{dx})^2 - 1}}$,

$$\frac{1 + \alpha y}{\sqrt{\left(\frac{dy}{dx}\right)^2 - 1}} = \sin\theta_0$$

$$\implies \frac{dy}{dx} = \sqrt{\frac{(1 + \alpha y)^2}{\sin^2\theta_0} - 1}.$$

Separating variables,

$$\int_0^y \frac{1}{\sqrt{(1 + \alpha y)^2 - \sin^2\theta_0}} dy = \int_0^x \frac{1}{\sin\theta_0} dx.$$

Let us have an intermission at this point. Usually, the integral on the left hand side is solved by introducing a hyperbolic cosine substitution. However, as one may not yet be familiar with this approach, we will introduce another method that circumvents this need after this — the drawback is that though we will be able to find L, we will be unable to determine $y(x)$. Proceeding with the first method, let $1 + \alpha y = \sin\theta_0 \cosh\phi$ for a new variable ϕ such that $\alpha dy = \sin\theta_0 \sinh\phi d\phi$.

$$\int_{\cosh^{-1}\frac{1}{\sin\theta_0}}^{\cosh^{-1}\frac{1+\alpha y}{\sin\theta_0}} \frac{1}{\sqrt{\sin^2\theta_0\left(\cosh^2\phi - 1\right)}} \cdot \frac{\sin\theta_0}{\alpha} \sinh\phi d\phi = \frac{x}{\sin\theta_0}.$$

Applying the identity $\cosh^2\phi - 1 = \sinh^2\phi$,

$$\int_{\cosh^{-1}\frac{1}{\sin\theta_0}}^{\cosh^{-1}\frac{1+\alpha y}{\sin\theta_0}} \frac{1}{\alpha} d\phi = \frac{x}{\sin\theta_0}$$

$$\cosh^{-1}\frac{1 + \alpha y}{\sin\theta_0} - \cosh^{-1}\frac{1}{\sin\theta_0} = \frac{\alpha x}{\sin\theta_0}.$$

Following from this, the trajectory is

$$y = \frac{\sin\theta_0}{\alpha}\cosh\left(\frac{\alpha x}{\sin\theta_0} + \cosh^{-1}\frac{1}{\sin\theta_0}\right) - \frac{1}{\alpha}.$$

When $y = h$, the value of x is

$$x = \frac{\sin\theta_0}{\alpha}\left(\cosh^{-1}\frac{1 + \alpha h}{\sin\theta_0} - \cosh^{-1}\frac{1}{\sin\theta_0}\right).$$

The maximum value of x, which corresponds to L, occurs when $\sin \theta_0 = 1$.

$$L = \frac{1}{\alpha} \cosh^{-1}(1 + \alpha h) = \frac{1}{\alpha} \ln \left(1 + \alpha h + \sqrt{(1 + \alpha h)^2 - 1}\right)$$

in terms of ln. To preclude the need for the cosh substitution, we begin from

$$(1 + \alpha y) \sin \theta = \sin \theta_0.$$

Taking the total derivative of the above with respect to x,

$$\alpha \sin \theta y' + (1 + \alpha y) \cos \theta \theta' = 0$$

$$\cot \theta \theta' = -\frac{\alpha}{1 + \alpha y} y'.$$

Since $\cot \theta = y'$,

$$\theta' = -\frac{\alpha}{1 + \alpha y}.$$

As $1 + \alpha y = \frac{\sin \theta_0}{\sin \theta}$ from Snell's law,

$$\int_{\theta_0}^{\theta_h} - \csc \theta d\theta = \int_0^x \frac{\alpha}{\sin \theta_0} dx$$

where θ_h is the angle of incidence at $y = h$ that can be obtained from Snell's law, as $\theta_h = \sin^{-1} \frac{\sin \theta_0}{1 + \alpha h}$. Then, the x-coordinate of the ray at height h is

$$x = \frac{\sin \theta_0}{\alpha} \ln \frac{\csc \theta_h + \cot \theta_h}{\csc \theta_0 + \cot \theta_0}.$$

The maximum x, which corresponds to L, occurs when $\sin \theta_0 = 1$. Substituting the expression for θ_h,

$$L = \frac{1}{\alpha} \ln (\csc \theta_h + \cot \theta_h)$$

$$= \frac{1}{\alpha} \ln \left(1 + \alpha h + \sqrt{(1 + \alpha h)^2 - 1}\right).$$

10. Trapping Light**

a) Opaque Paint

By symmetrical arguments, the six faces of the cubes should be painted with the exact same pattern. The next astute observation is the incident angle that a light ray makes with a particular face, is always preserved after arbitrary reflections from all faces. To show this, define the x, y and z axes to be perpendicular to the faces of the cube. Then, the direction vector of a ray can be expressed as (v_x, v_y, v_z) such that $v_x^2 + v_y^2 + v_z^2 = \frac{c^2}{n^2}$ where $\frac{c}{n}$ is the phase velocity of light in the cube. Now, label the faces in the xy, yz and xz-planes z, x and y respectively. The angle of incidence that this ray makes with the plane k, θ_k, is related by

$$\cos \theta_k = \begin{pmatrix} v_x \\ v_y \\ v_z \end{pmatrix} \cdot \boldsymbol{n_k} = |v_k|,$$

where $\boldsymbol{n_k}$ is a three-dimensional vector with one or negative one as the kth component (same sign as v_k) and zero as the other components. A reflection can only negate one component of the velocity which does not change the angles of incidence with all planes. Hence, the angle of incidence a light ray makes with a plane remains constant (unless it is absorbed). For a light ray to escape from a face, the angle of incidence must be smaller than the critical angle.

$$\sin \theta_k < \frac{1}{n}$$

$$\implies \cos \theta_k > \frac{\sqrt{n^2 - 1}}{n}$$

$$|v_k| > \frac{\sqrt{n^2 - 1}}{n}.$$

Here comes the crucial observation. We claim that if a light ray is able to escape the cube at all, it is able to escape from the first face it impinges on. This is because, if the light ray is able to escape the cube, $|v_k| > \frac{\sqrt{n^2-1}}{n}$ for some k. Furthermore, the first face it impinges on corresponds to the component of the direction vector with the greatest magnitude. Thus, it must be able to escape from the first face it impinges on. Moreover, the contrapositive of this statement implies that if a light ray cannot escape from the first face it impinges on (i.e. it undergoes total internal reflection), it cannot escape from the cube at all. Following from this, we conclude that we simply need to paint an area on each face to absorb the incoming rays which can directly escape. Since $\sin^{-1} \frac{1}{n} \leq \frac{\pi}{4}$ radians, this corresponds to a circle of radius

$$r = \frac{l}{2} \tan \theta_c = \frac{l}{2\sqrt{n^2 - 1}}$$

where $\theta_c = \sin^{-1} \frac{1}{n}$ is the critical angle. Hence, the total area that needs to be painted is $6\pi r^2 = \frac{3l}{2(n^2-1)}$.

b) Reflective Paint

For all light to be trapped inside the cube, its surfaces must be equivalent to mirrors. Then, we can consider the mirror images of the cube which form an infinite array of cubes. Consider the faces perpendicular to the x-axis (perpendicular to an arbitrary surface) as shown in Fig. 1.30.

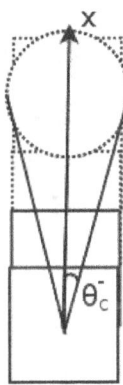

Figure 1.30:   Mirror images of faces perpendicular to x-axis

The argument in the previous case implies that if a ray can escape from one face, it can always escape from that face or the face opposite it after an arbitrary number of reflections. Now consider a light cone with half-angle θ_c^- as shown in the figure above. As light in this cone can always escape from the two faces perpendicular to the x-axis, the intersection of this light cone and these two surfaces and their mirror images must be coated with paint to prevent light from escaping. As the light cone propagates, its base area (a circle) expands and eventually covers the entire face. Therefore, these two faces and thus all faces by symmetry must be completely covered with paint — implying that the answer is $6l^2$.

11. Emitter in Triangular Room***

Drawing the complete mirror images of the receiver would generate the tessellation in Fig. 1.31, where we have defined the origin at O.

We see that the nearest possible horizontal array of images that the light ray can impinge on are those with centers at $y = 0$. Thus, let us see if it will indeed hit the target. The regions spanned by these receivers are

$$(x \in [11a + 16ak, 13a + 16ak], y \in [-a, a] \,|\, k \in \mathbb{Z}).$$

The horizontal distance required for the ray to reach height a is $(4a - a)$ $\cot \theta = 24a$. So the receiver corresponding to $k = 1$ is a likely candidate. At

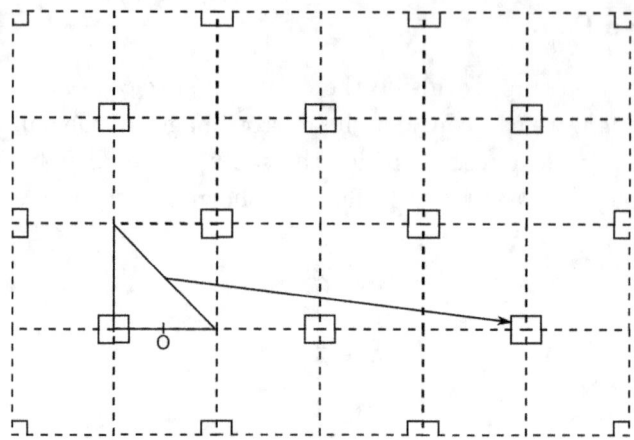

Figure 1.31:   Mirror images

$x = 27a$, the y-coordinate of the light ray is

$$y = 4a - \frac{27a}{\cot\theta} = \frac{5}{8}a.$$

Since $-a \le \frac{5}{8}a \le a$, we see that the light ray hits this receiver. The total length traveled by the ray is then

$$\frac{27a}{\cos\theta} = 27a \cdot \sqrt{\tan^2\theta + 1} = \frac{27\sqrt{65}}{8}a.$$

12. Optimal Path*

The optimal path between P and Q (of least time) is that of light by Fermat's principle. Effectively, the relative "refractive index" between the lake and the shore is $\frac{v_1}{v_2}$. Therefore, a light ray would be refracted at both edges of the lake and be deflected by a vertical distance $d(\tan i - \tan r)$ downwards where i and r are the angles of incidence and refraction respectively. This is because, the lake is effectively a "glass block" of relative refractive index $\frac{v_1}{v_2}$. For the light ray to reach Q from P,

$$L\tan i - d(\tan i - \tan r) = y.$$

Since $y \ll L$, i and r must be small too. Then, we can approximate $\tan i \approx i$ and $\tan r \approx r$. From Snell's law, we also have

$$r = \frac{v_2}{v_1}i.$$

Solving for i,

$$i = \frac{y}{L - \left(1 - \frac{v_2}{v_1}\right)d},$$

$$r = \frac{v_2 y}{v_1 L - (v_1 - v_2)d}.$$

The optimal path is depicted in Fig. 1.32.

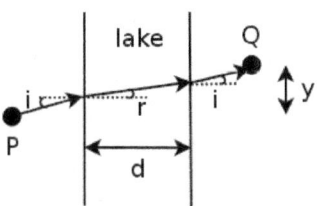

Figure 1.32:   Optimal path

13. Lensmaker's Formula**

Figure 1.33 illustrates a close-up of the lens (R_2 is negative for the right surface). Assume that the tip has a vertical height h. $h \ll R_1$ and $h \ll -R_2$ for a thin lens.

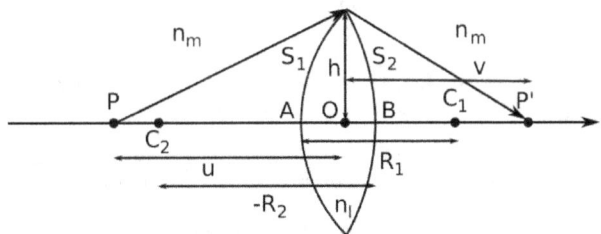

Figure 1.33:   Close-up of lens

Define the optical center O of the lens to be the point of intersection of a vertical line cutting between the two surfaces of the lens and the principal axis. Then, consider an object at point P along the principal axis. We define the object distance u to be that between the object and O (we will take the limit of this to infinity later). This definition differs from the one in the section on lenses afore but we will adopt this for the sake of convenience here. Suppose that the image of this object is formed at point P' along the principal axis (the image distance v is similarly that between the image and O).

Now, consider an axial ray from P to P' (i.e. straight line joining them) and a ray from P passing by the tip of the lens before reaching P'. In order for both light rays to reach P' from P, their optical path length must be identical by Fermat's principle. To compute this, we first determine $\overline{AO}$ and $\overline{OB}$. The equation of a circle of radius R centered at $x = R$ is

$$(x - R)^2 + y^2 = R^2.$$

Therefore, at a particular y-coordinate, x satisfies

$$|R - x| = \sqrt{R^2 - y^2}.$$

For the tip of the first convex surface, $R_1 > x$.

$$R_1 - x = R_1 \sqrt{1 - \frac{h^2}{R_1^2}} \approx R_1 - \frac{h^2}{2R_1}.$$

Therefore,

$$\overline{AO} = x \approx \frac{h^2}{2R_1}.$$

Similarly,

$$\overline{OB} = -\frac{h^2}{2R_2}$$

as R_2 is defined to be negative. Therefore, for the optical path lengths of the two rays to be equal,

$$\sqrt{u^2 + h^2} + \sqrt{v^2 + h^2} = u + v + (n - 1)\left(\overline{AO} + \overline{OB}\right)$$

$$= u + v + \frac{h^2(n - 1)}{2}\left(\frac{1}{R_1} - \frac{1}{R_2}\right).$$

As $u \to \infty$, v tends to the image focal length f. Furthermore, we can perform a Maclaurin expansion of the surds on the left-hand side to obtain

$$u + f + \frac{h^2}{2f} = u + f + \frac{h^2(n - 1)}{2}\left(\frac{1}{R_1} - \frac{1}{R_2}\right),$$

as $u \gg h$ and $f \gg h$. Canceling the similar terms yields

$$\frac{1}{f} = (n - 1)\left(\frac{1}{R_1} - \frac{1}{R_2}\right).$$

Finally, taking the limit $v \to \infty$ and determining the value of u would show that the object and image focal lengths are both f.

14. Light in the Atmosphere**

Define the origin to be at the center of the Earth. Consider the forces on an infinitesimal gas element in spherical coordinates between radial coordinates r and $r+dr$. It experiences pressure $p(r)$ radially outwards, $p(r+dr) = p+dp$ radially inwards and its own weight radially inwards. For it to remain in equilibrium,

$$-\rho g dV = (p+dp)dA - pdA$$

where dV and dA are the volume and area normal to the radial direction of the infinitesimal element respectively. Hence,

$$dp = -\rho g dr.$$

By the ideal gas law,

$$pV = nRT$$

$$p = \frac{\rho}{M}RT.$$

Since T is constant,

$$dp = \frac{d\rho}{M}RT.$$

Substituting this into the other expression for dp,

$$-\rho g dr = \frac{d\rho}{M}RT$$

$$\int_{\rho_0}^{\rho} \frac{1}{\rho}d\rho = \int_{r_0}^{r_0+h} -\frac{Mg}{RT}dr$$

$$\rho = \rho_0 e^{-\frac{Mg}{RT}h}$$

where ρ_0 is the density at the surface of the Earth and h is the altitude above the surface of the Earth. The OPL of a circle at this altitude is

$$OPL = n \cdot 2\pi(r_0 + h) = \alpha \rho_0 e^{-\frac{Mg}{RT}h} \cdot 2\pi(r_0 + h),$$

$$\frac{d(OPL)}{dh} = -\frac{Mg\alpha\rho_0}{RT} \cdot 2\pi(r_0+h)e^{-\frac{Mg}{RT}h} + 2\pi\alpha\rho_0 e^{-\frac{Mg}{RT}h}.$$

This path must have a stationary OPL in order for the light ray to actually travel along it. Then,

$$\frac{d(OPL)}{dh} = 0 \implies \frac{RT}{Mg} = r_0 + h$$

$$h = \frac{RT}{Mg} - r_0.$$

15. Lagrangian Derivation**

Recall that the action of a one-dimensional system (described by its x-coordinate) between times t_1 and t_2 is defined as

$$S = \int_{t_1}^{t_2} \mathcal{L}(x, \dot{x}, t)dt$$

where $\mathcal{L}$ is the Lagrangian. To extremize S, we require

$$\frac{d}{dt}\left(\frac{\partial \mathcal{L}}{\partial \dot{x}}\right) = \frac{\partial \mathcal{L}}{\partial x},$$

which is the Euler-Lagrange equation. In this case, S corresponds to the OPL while the Lagrangian is

$$\mathcal{L}(y, y', x) = n(x, y)\sqrt{1 + y'^2},$$

where y and x have assumed the roles of x and t respectively. To extremize the OPL between two endpoints, we require

$$\frac{d}{dx}\left(\frac{\partial \mathcal{L}}{\partial y'}\right) = \frac{\partial \mathcal{L}}{\partial y}.$$

For a homogeneous medium, n is constant such that

$$\frac{\partial \mathcal{L}}{\partial y} = 0$$

$$\implies \frac{\partial \mathcal{L}}{\partial y'} = c$$

for some constant c. Since $\frac{\partial \mathcal{L}}{\partial y'} = \frac{ny'}{\sqrt{1+y'^2}}$, this implies that

$$y' = C$$

for some constant C. The trajectory of the light ray is thus a straight line. To prove the laws of reflection and refraction, we can orient our coordinate system such that the refractive index is a function of y solely. In this process, we may have to bring the endpoints closer and closer together, but it does not affect our analysis as we can deduce the long-term path of a ray from its immediate response at each juncture. The Lagrangian then becomes

$$\mathcal{L} = n(y)\sqrt{1 + y'^2}.$$

For a Lagrangian that is independent of x, the Hamiltonian H is conserved.

$$H = \frac{\partial \mathcal{L}}{\partial y'} y' - \mathcal{L}$$

$$\frac{ny'}{\sqrt{1 + y'^2}} \cdot y' - n\sqrt{1 + y'^2} = H$$

$$\frac{n}{\sqrt{1 + y'^2}} = -H.$$

To prove the law of reflection, suppose that the light ray now meets an impermeable and perfectly non-absorptive interface. Since it cannot penetrate the interface, it must lie on the same side of the interface before and after it impinges on the interface. Set our endpoints of concern at the locations of the ray directly before and after it impinges on the interface. Since the refractive index is uniform in the thin layer above the interface (i.e. n at these locations are identical), the above result implies that

$$|y'_{after}| = |y'_{before}|$$

$$\implies y'_{after} = -y'_{before}$$

where we reject the option $y'_{after} = y'_{before}$, as the ray cannot pass through the interface and its path cannot be discontinuous — hence proving the law of reflection. To prove the law of refraction, set our endpoints of concern at the locations of the light ray immediately before and after it crosses the relevant interface (which is perpendicular to the y-direction since $n(y)$ is solely a function of y). Observe that if we define θ as the angle subtended by the ray and the normal to the interface,

$$\frac{1}{\sqrt{1 + y'^2}} = \frac{1}{\sqrt{1 + \cot^2 \theta}} = \sin \theta,$$

so we have

$$n(y) \sin \theta = -H$$

for some constant H. This is simply Snell's law!

16. Minimum Distance*

$$\frac{1}{u} + \frac{1}{v} = \frac{1}{f}$$

$$v = \frac{fu}{u - f}.$$

The distance D we wish to minimize is

$$D = u + v = u + f + \frac{f^2}{u - f}$$

$$\frac{dD}{du} = 1 - \frac{f^2}{(u - f)^2} = 0$$

$$\implies u = 2f.$$

To show that this is a minimum point,

$$\frac{d^2 D}{du^2} = \frac{f^2}{2(u - f)^3}$$

$$\left.\frac{d^2 D}{du^2}\right|_{u=2f} = \frac{2}{f} > 0.$$

Substituting $u = 2f$ into D,

$$D_{min} = 4f.$$

17. Blurring*

Let v be the image distance from the lens. There are two possible positions of the screen that lead to a smudge of thickness d, as shown in Fig. 1.34.

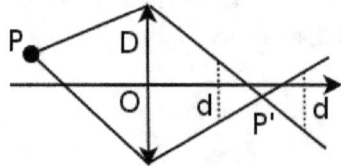

Figure 1.34:   Possible smudges indicated by dotted lines

In either case,

$$\frac{d}{D} = \left| \frac{L}{v} - 1 \right|$$

by similar triangles, where v is the image distance. When $L \geq v$,

$$\frac{1}{v} = \frac{d}{DL} + \frac{1}{L}.$$

Otherwise if $L < v$,

$$\frac{1}{v} = \frac{1}{L} - \frac{d}{DL}.$$

Note that $\frac{d}{D} < 1$ if $L < v$ so this is an entirely valid regime (in ensuring that v is positive so that a real image can be formed). In both cases, one can apply the Gaussian Lens formula to determine the focal length.

$$\frac{1}{f} = \frac{1}{u} + \frac{1}{v}.$$

The two possible focal lengths are

$$f = \frac{1}{\frac{1}{u} + \frac{d}{DL} + \frac{1}{L}},$$

$$f = \frac{1}{\frac{1}{u} - \frac{d}{DL} + \frac{1}{L}}.$$

18. Congealing Lenses*

The individual focal lengths of the two plano-convex lenses are given by the Lensmaker's formula as $\frac{R}{n_1 - 1}$ and $\frac{R}{n_2 - 1}$ respectively (one radius of curvature is infinite). Since these two lenses are thin and are juxtaposed, the effective focal length f obeys

$$\frac{1}{f} = \frac{n_1 - 1}{R} + \frac{n_2 - 1}{R}$$

by Eq. (1.23).

$$f = \frac{R}{n_1 + n_2 - 2}.$$

19. Mirror with Liquid*

The most direct method is to observe that the mirror with a liquid film effectively has a focal length $\frac{f}{n}$. This can be seen by considering an incident bundle of rays parallel and close to the mirror axis. They initially pass by the liquid film undeviated but undergo refraction at the surface of the film after being reflected by the mirror — causing them to converge at a distance $\frac{f}{n}$ from the vertex as the angle that they make with the mirror axis after reflection is increased by a factor of n as compared to the case where the film was absent. Applying the mirror equation with $\frac{f}{n}$ as the effective focal length, the image distance is

$$v = \frac{fu}{nu - f}.$$

A more indirect way is to consider a ray impinging the surface of the liquid film at a small angle of incidence i. The angle of refraction r is then given by

$$r = \frac{i}{n}.$$

This deflected ray intersects with the axis at distance nu from the vertex. All of such rays with small i converge there. Therefore, the effective object distance is nu. Applying the mirror equation,

$$\frac{1}{nu} + \frac{1}{v'} = \frac{1}{f}.$$

Solving for v',

$$v' = \frac{nfu}{nu - f}.$$

Now, we have not accounted for the secondary refraction at the liquid film of a ray after it reflects from the mirror. In this case, the angle is amplified as the ray leaves the liquid into air. Therefore, the effective image distance is decreased by a factor of $\frac{1}{n}$. The real image is located at a distance

$$v = \frac{v'}{n} = \frac{fu}{nu - f}$$

above the vertex.

20. Quarter Prism**

Referring to Fig. 1.35, a ray that impinges the curved surface of the prism at an angular coordinate θ is deflected and hits the table at a distance $x(\theta)$ from the vertical edge of the prism.

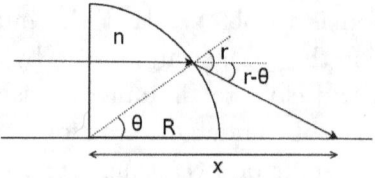

Figure 1.35:   Path of a ray

The angle of refraction r can be computed via Snell's law as

$$r = \sin^{-1}(n \sin \theta),$$

assuming that θ is smaller than the critical angle θ_c. x can thus be computed via simple geometry as

$$x(\theta) = R \sin \theta \cot(\sin^{-1}(n \sin \theta) - \theta) + R \cos \theta.$$

To investigate the behavior of $x(\theta)$ with varying θ, firstly observe that $\cot x$ and $\cos x$ are both strictly decreasing functions of x for $0 < x \leq \frac{\pi}{2}$. Therefore, if we can prove that $\sin^{-1}(n \sin \theta) - \theta$ is a strictly increasing function of θ (for $0 < \theta \leq \theta_c$), we would have shown that $x(\theta)$ is a strictly decreasing function of θ. To this end, consider the derivative of $\sin^{-1}(n \sin \theta) - \theta$ with respect to θ.

$$\frac{d(\sin^{-1}(n \sin \theta) - \theta)}{d\theta} = \frac{1}{\sqrt{1 - n^2 \sin^2 \theta}} \cdot n \cos \theta - 1 > 0$$

as $\sqrt{1 - n^2 \sin^2 \theta} < \sqrt{n^2 - n^2 \sin^2 \theta} = n \cos \theta$. Therefore, $x(\theta)$ is a strictly decreasing function of θ. The minimum value of x occurs at the largest possible of θ in the relevant regime, θ_c. This is because, rays that impinge the curved surface at an angle of incidence above θ_c will undergo total internal reflection and hit the table at a point within the prism (cases that are not of interest).

$$x_{min} = R \sin \theta_c \cot(\sin^{-1}(n \sin \theta_c) - \theta_c) + R \cos \theta_c.$$

Substituting $\theta_c = \sin^{-1} \frac{1}{n} = \sin^{-1} \frac{1}{1.5}$ and $R = 5 \text{cm}$,

$$x = 6.71 \text{cm} \quad (3 \text{sf}).$$

The largest value of x occurs when $\theta \to 0$, and

$$x_{max} = \lim_{\theta \to 0} R \sin \theta \cot(\sin^{-1}(n \sin \theta) - \theta) + \lim_{\theta \to 0} R \cos \theta$$

$$= \lim_{\theta \to 0} R\theta \cot [(n - 1)\theta] + R$$

$$= R\theta \cdot \frac{1}{(n - 1)\theta} + R$$

$$= \frac{R}{n - 1} + R$$

$$= 15 \text{cm}.$$

This value is expected as the curved surface of the prism acts on the ray that corresponds to $\theta = 0$ like a plano-convex lens (since we are dealing with regions close to the right tip of the prism). Applying the Lensmaker's formula, the focal length of this lens obeys

$$\frac{1}{f} = \frac{n-1}{R} \implies f = 10\text{cm}.$$

Therefore, the ray which impinges the curved surface at $\theta = 0$ is focused by the right tip of the prism to the focal point that lies 10cm away from the right tip (since the incident ray is parallel to the "principal axis"). This amounts to a total distance of $10 + R = 15\text{cm}$ from the vertical edge of the prism.

21. Unique Configuration**

Let the image distance to the first lens be v. Then,

$$\frac{1}{f_1} = \frac{1}{u} + \frac{1}{v}$$

$$v = \frac{f_1 u}{u - f_1}.$$

Let the object distance to the second lens be u'. Then, the image distance to the second lens must be $l - u - v - u' = l - u - \frac{f_1 u}{u - f_1} - u'$ for the final image to be formed at the reference point. For the sake of convenience, define $k = l - u - \frac{f_1 u}{u - f_1}$. Then,

$$\frac{1}{f_2} = \frac{1}{u'} + \frac{1}{k - u'}$$

$$u'^2 - ku' + f_2 k = 0$$

$$u' = \frac{k \pm \sqrt{k^2 - 4kf_2}}{2}.$$

For there to be two unique solutions,

$$k^2 - 4kf_2 > 0$$

which implies that $k > 4f_2$ or $k < 0$. However, notice that since the image distance of the second lens must be positive for the image to be real,

$$k - u' > 0$$

for both u''s.

$$\implies \frac{k \mp \sqrt{k^2 - 4kf_2}}{2} > 0$$

which cannot be satisfied if $k < 0$. Hence k must be greater than $4f_2$ to enable the possibility of two u''s.

$$l - u - \frac{f_1 u}{u - f_1} > 4f_2.$$

Multiplying the above by $u - f_1$ and simplifying would yield

$$u^2 + (4f_2 - l)u + f_1 l - 4f_1 f_2 < 0.$$

The inequality above is satisfied when

$$\alpha < u < \beta,$$

where α and β are the roots of the left-hand side (we assume that the discriminant is positive. If this is not the case, we could have concluded that no such u exists already). Furthermore, as a given condition in the question, u must satisfy $u > f_1$. This implies that $\beta > f_1$ and

$$\beta = \frac{l - 4f_2 + \sqrt{16f_2^2 - 8f_2 l + l^2 - 4f_1 l + 16f_1 f_2}}{2} > f_1$$

$$\sqrt{16f_2^2 - 8f_2 l + l^2 - 4f_1 l + 16f_1 f_2} > 2f_1 + 4f_2 - l.$$

Since $2f_1 + 4f_2 > l$, squaring and simplifying would yield

$$4f_1^2 < 0$$

which leads to a contradiction. Hence, there cannot be two values of u' that satisfy the above conditions.

22. Moving Image**

It is more convenient to consider the Newtonian form of the lens equation which states that

$$x_o x_i = f^2$$

where $x_o = u - f$ and $x_i = v - f$. Then,

$$x_o = \frac{f^2}{x_i}.$$

Differentiating the above with respect to time,

$$\frac{dx_o}{dt} = -\frac{f^2}{x_i^2} \cdot \frac{dx_i}{dt}.$$

Since $\frac{dx_o}{dt} = \frac{du}{dt}$ and $\frac{dx_i}{dt} = \frac{dv}{dt}$,

$$\frac{du}{dt} = -\frac{f^2}{x_i^2} \cdot \frac{dv}{dt}.$$

When the velocities of the ant and its image are identical, $\frac{du}{dt} = -\frac{dv}{dt}$. Note that the positive direction of v is opposite to that of u. At this juncture,

$$f^2 = x_i^2$$

which implies that

$$x_i = \pm f,$$

$$x_o = \pm f.$$

That is, the velocities are equal only when $u = v = 2f$ or $u = v = 0$. Substituting $\frac{du}{dt} = \frac{\alpha}{v-f} = \frac{\alpha}{x_i}$ into the previous differential equation,

$$\frac{\alpha}{x_i} = -\frac{f^2}{x_i^2} \cdot \frac{dv}{dt} = -\frac{f^2}{x_i^2} \cdot \frac{dx_i}{dt}$$

$$\int_{x_i^0}^{x_i} \frac{1}{x_i} dx_i = -\int_0^t \frac{\alpha}{f^2} dt$$

$$\ln \left| \frac{x_i}{x_i^0} \right| = -\frac{\alpha}{f^2} t,$$

where x_i^0 is the initial value of x_i. Now, observe that x_i and x_i^0 must always have the same sign as $\frac{dx_i}{dt}$ is proportional to $-x_i$ and thus causes x_i to increase (for negative x_i) or decrease (for positive x_i) until it attains the equilibrium value 0. Thus, we can remove the absolute value brackets and conclude that

$$x_i = x_i^0 e^{-\frac{\alpha}{f^2} t}.$$

Substituting $x_i^0 = \frac{f^2}{x_o^0} = \frac{f^2}{u_0 - f}$ and $x_i = v - f$,

$$v = \frac{f^2}{u_0 - f} e^{-\frac{\alpha}{f^2} t} + f.$$

23. Prism**

Notice that some of the rays impinge on the hypotenuse at a greater angle than the critical angle (e.g. a vertical ray). Therefore, the hypotenuse of the prism acts as a mirror and we can consider the mirror image of the object which is entirely located at 12.5cm left of the mid-point of the hypotenuse.

Now, we have neglected the effect of refraction at the other sides of the prism. To account for this, simply map the mirror images of those edges as well and bend the rays of the mirror image of the object when they cross these virtual edges. The object is then effectively located 8cm on the left of a slab with refractive index 1.5 and thickness 9cm. Since passing through a slab forms another image which is tantamount to reducing the object distance by $t\left(1 - \frac{1}{n}\right)$ where t is the thickness of the slab, the effective object distance to the first lens is

$$8 + 9 - 9\left(1 - \frac{1}{\frac{3}{2}}\right) + 6 = 20\text{cm}$$

which coincides with the front focal point of the first lens. Therefore, the image of the first lens is at infinity. The final image is then formed at

$$v = -10\text{cm}$$

from the second lens. The image is virtual and has a magnification

$$M = -\frac{-10}{20} = \frac{1}{2},$$

where the infinite image distance of the first lens and the infinite object distance of the second lens nullify each other. This positive value implies that the final image is aligned with the first object (obtained after refracting the mirror image of the initial object through the glass slab). Therefore, the final image points downwards. Note that we do not consider rays that are not reflected by the prism (e.g. direct rays from the original object) as they are not paraxial and thus do not converge.

Chapter 2

Thermodynamics and Ideal Gases

In this chapter, we will be looking at thermodynamics — the kinetic theory of heat — from both macroscopic and microscopic perspectives. The zeroth and first law of thermodynamics will be introduced and applied to the specific system of an ideal gas.

2.1 The Zeroth Law

It is common knowledge that if we put a hot object in thermal contact with a cool object, the hot object becomes cooler while the cool object becomes hotter to a certain extent. This is a quotidian phenomenon that occurs until the transfer of "hotness" or "coolness" between the two objects ceases. At this juncture, the two objects have attained thermal equilibrium. Specifically, two objects are said to be in thermal equilibrium if they are in thermal contact and there is no net exchange of heat between them. We will hold off the definition of heat for now and just understand it as a form of energy transfer. Finally, when thermal equilibrium is attained between two objects, they should be similar in a certain respect — if two systems are in thermal equilibrium, they are said to possess the same temperature.

The zeroth law of thermodynamics states that if objects A and B are each at thermal equilibrium with a common object C, objects A and B are at thermal equilibrium with each other. This intuitive concept has vast consequences. Firstly, it standardizes the notion of temperature as the definition of temperature now implies that all objects of the same temperature are in thermal equilibrium. Next, the zeroth law allows us to use object C to determine whether objects A and B will be in thermal equilibrium without physically putting them in thermal contact. This, combined with the fact that object C may experience certain measurable and observable changes when placed in contact with objects A and B, such as a rise in the mercury level due to expansion in a mercury thermometer or the change in the pressure of a gas, allows us to quantify the temperature of an object. For example, we can have two reference points, setting 0°C for ice and 100°C for steam, and divide the interval into 100 equal segments to create the Celsius temperature scale.

2.2   Common Quantities in Thermodynamics

2.2.1   *State Variables*

In thermodynamics, it is important to distinguish between state variables and non-state variables. As its nomenclature implies, state variables — such as pressure, volume and temperature — are functions of the configuration of a system which can be specified by the positions and velocities of all constituents in a system.

Non-state variables are not functions of the configuration of the system and can thus have multiple values at a single state. Consider a car driving from the origin in the xy-plane, stocked with a certain initial amount of fuel. If we define the state of the car to be its position in the xy-plane, the amount of fuel left in the car at a given state is not a state function as the car can traverse different paths to reach the same final state — these paths may consume different amounts of fuel. Most starkly, if the car drives back to the origin, the amount of fuel left is not the same as before! Therefore, the amount of fuel left in the car is definitely not a state variable as we cannot determine its value solely by looking at the car's position.

2.2.2   *Internal Energy*

The internal energy of a system is defined to be the sum of all microscopic forms of energy — energy on the atomic and molecular scale. It is the sum of all microscopic kinetic energy and microscopic potential energy. Crucially, internal energy is uniquely defined for each state of a system and is a state variable.

$$U = \sum \text{K.E}_{mic} + \sum \text{P.E}_{mic}. \tag{2.1}$$

Microscopic kinetic energy results from the possible random motions of individual constituents. For example, molecules may translate, rotate and even vibrate about a common center. The latter two situations only occur in the case of polyatomic molecules. It is important to differentiate microscopic and macroscopic kinetic energies. The former is highly disordered and thus less useful than the latter, to a certain extent. A moving block has a macroscopic kinetic energy associated with the motion of the entire object as a whole but if we zoom into the scale of individual molecules, we may find them jiggling about in random directions and thus can associate a microscopic kinetic energy with that motion.

Microscopic potential energy results from the interactions between the constituents of a system and between the constituents of a system and external factors, on a molecular scale. Chemical bonds between atoms and strong interactions in the nuclei are typical examples of internal interactions. The creation of electric dipoles in atoms due to an external electric field is an example of an interaction with an external entity. A special form of internal interactions, associated with the phase of a system, results in a form of energy known as latent energy. This will be explored in a later chapter.

2.2.3 *Heat and Work*

Heat and work are both energies in transit and are not forms of energy. In the case of closed systems where mass exchange cannot occur, heat and work are the only possible forms of energy transfer. Similar to how the work performed on a particle increases its macroscopic kinetic energy in mechanics, heat and work are just methods of delivering or extracting energy. However, heat can be differentiated from work by observing that its flow requires a temperature gradient. Work, on the other hand, can be performed by a system on another system of the same temperature. For example, two gases, that are separated by a movable wall, may have attained the same temperature but not the same pressure. Then, there is work performed by pushing the wall.

Heat and work done are not state variables as they are just methods of delivering energy to or extracting energy from a system. For the same change in the internal energy of a system which is a state variable, heat and work done can make different contributions to this change, as long as their sum is consistent. Moreover, their final products — namely the change in internal energy of the relevant system — are indistinguishable, so there is no way to deduce their individual contributions by observing the final state of the system alone. This is analogous to how it is impossible to know what your sneaky friend has spent your credit card on by simply analyzing the total amount of money left in your bank account — you have to inspect the bill at the end of the month which details every single purchase (the process of purchasing). Therefore, heat and work done are, most importantly, both process-dependent.

2.3   The First Law of Thermodynamics

The first law of thermodynamics is quintessentially the conservation of energy. Supposing that there is a decrease in the internal energy and macroscopic kinetic energy of a system, this decrease in energy should manifest itself as the physical work done by the system and the heat flowing from the system. Conversely, we can conclude that the increase in the internal energy plus macroscopic kinetic energy of a system is the sum of the heat supplied to and the work done on the system. This is the first law of thermodynamics which can be expressed mathematically as

$$\Delta U + \Delta \mathrm{K.E}_{mac} = W_{on} + Q, \tag{2.2}$$

where W_{on} is the net work done on the system by external agents and Q is the net heat supplied to the system. The conservative work on the right-hand side can be shifted to the left-hand side such that the left-hand side becomes the change in the system's total energy E (internal energy plus macroscopic kinetic and potential energies).

$$\Delta E = W_{on} + Q, \tag{2.3}$$

where W_{on} now only includes the work done by non-conservative forces on the system. Usually, the macroscopic energies are constant such that the above becomes

$$\Delta U = W_{on} + Q. \tag{2.4}$$

In most cases, the heat transferred between systems is prohibitively difficult to determine directly. However, the first law enables us to calculate heat indirectly from measurable quantities such as internal energy and work done.

Sometimes, the first law is expressed in terms of the work done by the system, which is negative of the work done on the system, $W_{by} = -W_{on}$. Then,

$$Q = W_{by} + \Delta U. \tag{2.5}$$

In a certain sense, the heat supplied to the system manifests in terms of the work performed by the system and the increase in internal energy as it stores part of the heat.

2.4   Ideal Gases

Now, we are interested in analyzing the specific system of gases. Microscopically, we can model a gas as a system of molecules that are hard spheres undergoing constant random motion. These particles are assumed to collide elastically, have no interactions with one another (besides collisions) and are small relative to the volume of their container. A gas that exhibits such behavior is known as an ideal gas and is of course, not realistic. When gas particles collide with the walls of the container, they exert a force on the walls of the container. Macroscopically, these collisions manifest as a pressure on the container walls.

A system is in thermodynamic equilibrium when the macroscopic state of every part of the system is not evolving over time. A pressure and temperature can be defined for every part of a gas at equilibrium. However, if the gas were to undergo a sudden change, such as a contraction, it will be in a non-equilibrium state, at least for a short instance. Then, a pressure and temperature are not well-defined at this juncture.

We can define the equilibrium state of an ideal gas using three state variables — temperature, pressure and volume. For a system in general, there will be an equation that relates the different state variables. In the particular case of an ideal gas, its equation of state is known as the ideal gas law. Concretely,

$$pV = nRT \tag{2.6}$$

where p, V and T are the pressure, volume and the temperature of the gas respectively. n is the number of moles of gaseous molecules while R is the ideal gas constant, $R = 8.314\,\text{J mol}^{-1}\text{K}^{-1}$. Note that T is measured in Kelvins, which can easily be calculated from a temperature expressed in degree Celsius with the following conversion formula:

$$T(K) = T(^\circ C) + 273.15.$$

The ideal gas law makes intuitive sense from a microscopic standpoint, it basically states that

$$p \propto \frac{nT}{V}.$$

When the number of moles of molecules increases, more molecules collide with the walls per unit time — increasing the pressure of the gas. When temperature increases, the gaseous molecules become more "excited" and possess a larger average kinetic energy. Thus, they exert a greater force on the container walls per collision and collide with the walls more frequently. Lastly, if the volume of the container is increased, gaseous molecules have to travel a longer distance to collide with the walls, leading to a decrease in the frequency of collision and hence pressure. Another slight technicality is that n strictly refers to the number of moles of gaseous molecules and not the total moles of gaseous particles or atoms. Even if the container encloses k moles of diatomic gaseous molecules, n is still equal to k and not $2k$. The number of elementary entities in a mole is the Avogadro's Constant, $N_A = 6.02 \times 10^{23}$. Thus, we can rewrite the ideal gas equation in terms of the number of gaseous molecules.

$$pV = NkT \tag{2.7}$$

where N is the total number of molecules and $k = \frac{R}{N_A} = 1.38 \times 10^{-23}\,\mathrm{JK}^{-1}$ is the Boltzmann constant.

Ultimately, the ideal gas law encapsulates the following three gas laws which are its predecessors. Firstly, Boyle's law states that the pressure and volume of a gas of fixed mass are inversely proportional when its temperature is held constant. Secondly, Charles' law states that the absolute temperature (Kelvin scale) and volume of a gas of fixed mass are directly proportional when its pressure is held constant. Finally, Gay–Lussac's law asserts that the pressure and absolute temperature of a gas of fixed mass are directly proportional when its volume is held constant.

Problem: A thermally insulated piston of negligible dimensions separates a rectangular container into two regions. The two regions are both filled with ideal gases at an initial temperature of 27°C. The initial configuration of the system is shown in Fig. 2.1, with the piston being initially stationary. The temperature of the gas in region A is now increased to 227°C while the temperature of the gas in the other region is maintained at 27°C. Find x, the distance of the piston from the left end of the container, after the system has equilibrated.

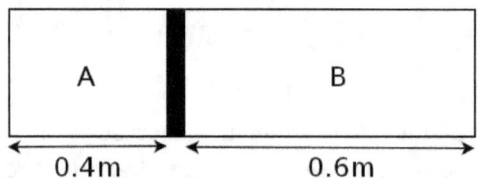

Figure 2.1:   Ideal gases

Let the cross sectional area of the container be A. For the system to be at equilibrium, the pressures due to both gases should be equal. Let the initial and final common pressures be p_1 and p_2 respectively. Since the number of moles of each gas is constant, by the ideal gas equation,

$$p \propto \frac{T}{V}.$$

Applying this relation to gases A and B,

$$\frac{p_1}{p_2} = \frac{T_{1A}V_{2A}}{T_{2A}V_{1A}} = \frac{(27 + 273)x}{(227 + 273) \cdot 0.4}$$

$$\frac{p_1}{p_2} = \frac{T_{1B}V_{2B}}{T_{2B}V_{1B}} = \frac{1 - x}{0.6}.$$

Solving,

$$x = \frac{10}{19}\text{m}.$$

2.4.1 *Internal Energy*

Due to the proposed lack of interactions between ideal gas molecules, the internal energy of an ideal gas stems solely from the microscopic kinetic energy of the moving molecules. By the equipartition theorem in statistical mechanics, energy is shared equally at thermal equilibrium among the modes[1] of a molecule — which arise for each independent contribution to the total energy that is quadratic in a certain variable — such as translational and rotational kinetic energy. Each mode of a molecule contributes an additional $\frac{1}{2}kT$ amount of average energy to a molecule. Due to the lack of internal interactions between molecules, the average energy of a molecule must also be the average kinetic energy. Consequently, the average kinetic energy of a gas molecule is

$$\langle \text{K.E} \rangle = \frac{f}{2}kT, \tag{2.8}$$

where f is the number of degrees of freedom of a particle which is the number of independent forms of motion exhibited by a molecule and is also the number of coordinates required to specify the state of a molecule. Then, the internal energy of an ideal gas is

$$U = N\langle \text{K.E} \rangle = \frac{f}{2}NkT = \frac{f}{2}nRT = \frac{f}{2}pV. \tag{2.9}$$

As expected, the internal energy is a state function as it is only dependent on the temperature of the gas. Ideal gases are usually assumed to be

---

[1] Vibrational degrees of freedom are not included here as the energies associated with them are not quadratic in a certain variable. They are in fact quantized.

monoatomic. Thus, molecules have three degrees of freedom due to possible translations in the x, y and z-directions. This monoatomic property is usually assumed by default unless stated otherwise. For a diatomic gas molecule, there are usually 5 degrees of freedom due to there being three translational and two rotational directions. There are only two rotational degrees of freedom for a diatomic molecule as it is not possible for the diatomic molecule to rotate about the axis joining the two atoms as the atoms are assumed to be small (thus contributing negligible energy due to rotations along this axis). In the general case of polyatomic molecules, vibrational modes may arise, especially at high temperatures. However, we will only be dealing with molecules with no vibrational freedom.

Following from the above discussion, the average translational kinetic energy per molecule of an ideal gas (regardless of the number of atoms per molecule) is

$$\langle \text{K.E}_{trans} \rangle = \frac{3}{2}kT.$$

Then, we can actually relate temperature to the mean square speed[2] of the molecules. Assuming that there are N gaseous molecules which each have mass m,

$$\langle \text{K.E}_{trans} \rangle = \frac{1}{N} \sum_{i=1}^{N} \frac{1}{2} m v_i^2 = \frac{1}{2} m \frac{\sum_{i=1}^{N} v_i^2}{N}.$$

The mean square speed $\langle v^2 \rangle$ and the root-mean-square speed of the molecules v_{rms} are defined as

$$\langle v^2 \rangle = \frac{\sum_{i=1}^{N} v_i^2}{N}$$

$$v_{rms} = \sqrt{\langle v^2 \rangle} = \sqrt{\frac{\sum_{i=1}^{N} v_i^2}{N}}.$$

We can rewrite the expression for the average kinetic energy per molecule as

$$\langle \text{K.E}_{trans} \rangle = \frac{1}{2} m \langle v^2 \rangle = \frac{1}{2} m v_{rms}^2,$$

to conclude that

$$\langle v^2 \rangle = v_{rms}^2 = \frac{3kT}{m}. \tag{2.10}$$

---

[2]The speed in the context of polyatomic molecules would usually refer to the speed of the center of mass.

This implies that temperature can be used as a direct measure of how fast gas molecules are moving and gives a kinetic interpretation of temperature. Lastly, note that the above expressions for the mean square speed and the root-mean-square speed are consistent with the kinetic theory of gases (as we shall show) — a microscopic model of ideal gases that will be introduced later.

2.4.2 General and Reversible Work

Recall that the infinitesimal work done by a force $\boldsymbol{F}$ in moving an object (such as a wall) is

$$\delta W = \boldsymbol{F} \cdot d\boldsymbol{r}$$

where $d\boldsymbol{r}$ is the infinitesimal displacement of the object. The differential in front of W is a small δ which represents an inexact differential, as W is not an actual function and thus does not have a derivative. This is because W is generally not a state function and is dependent on the path that a process takes. Moreover, remember that in the case of a fluid, the force that it exerts can only be perpendicular to its surface as it cannot withstand any shear forces. Then, the work done by an infinitesimal portion of gas near the boundary of our gaseous system on its surroundings across a massless interface of surface area dA can be rewritten as $p_{ex}dAdx$, where dx is the signed displacement of the massless interface in the direction of its area vector (defined to be positive outwards with respect to the gas) such that $dAdx$ is the area swept by the infinitesimal interface. Integrating over the entire boundary surface of the gas, the total work done by the gas on its surroundings after an infinitesimal change is

$$\delta W_{by} = p_{ex}dV$$

where dV is the change in volume of the gas. It is pivotal to understand that p_{ex} refers to the external pressure imposed on the interface by the surroundings and is not the pressure of our gaseous system. It is assumed that the external pressure is well-defined, such as in the case of a force evenly distributed on a massless piston, else the above expression cannot be used either. This dependence on the external pressure can be easily verified in the case where $p_{ex} = 0$ such that even if the gas had a well-defined pressure, it should not perform any work on its surroundings as it will just expand freely. The deeper reason behind is that generally, when the gas pressure initially differs from p_{ex}, the gas pressure will be ill-defined at the next instance as the gas will become inhomogeneous. Since the gas sections immediately adjacent to the interface must balance the force enforced by the external pressure, their volumes are changed slightly such that their pressure (assuming that we consider small enough sections which are approximately homogeneous) are accustomed to the external pressure. However, this information is not instantaneously transmitted to other parts of the gas such that the gas is no longer in equilibrium as a whole. All-in-all, the work done by the gas on its surroundings is that due to the sections surrounding the interface which have pressure p_{ex}.

Observe that δW_{by} is generally an unedifying description of the evolution of our gaseous system as we are unable to relate it to the gas pressure. However, in the case where the initial gas pressure differs from the external pressure by an infinitesimal value, the infinitesimal work done can be written as

$$\delta W_{by} = pdV \qquad (2.11)$$

where p is the pressure of the gas, that is also the external pressure. In order for Eq. (2.11) to be valid, the external pressure can only be varied by infinitesimal amounts (e.g. by carefully placing grains of sand on a piston), such that the system evolves over a series of purely equilibrium states from an initial to final state. Such a slowly-occuring process is known as a quasistatic process and is an idealization. Actually, the condition for the applicability of Eq. (2.11) is much stricter — it requires the process that the gas undergoes to be reversible,[3] under which being quasistatic is a mere prerequisite. For example, when a gas in a container is undergoing quasistatic compression performed by adding grains of sand on top of a gas piston, the gas pressure will generally differ from the external pressure if friction between the piston and the container walls is present. This friction is in fact a form of irreversiblity which renders Eq. (2.11) obsolete. Therefore, we must scrutinize the circumstances in the problems we face to check if we can apply Eq. (2.11) which is only valid for reversible processes.

The total work done during a reversible process is obtained by integrating the infinitesimal work done over the path that a system takes.

$$W_{by} = \int pdV, \qquad (2.12)$$

where the integral indicates that we should track all infinitesimal volume changes as the gas evolves from an initial to final state. When a gas expands, the work done by the gas is positive as the displacement of the interface is

---

[3]This concept of reversibility will be explored in the next chapter.

in the same direction as the force due to gaseous pressure while the work done on the gas is negative. In a similar vein, when a gas contracts, the work done by the gas is negative and the work done on the gas is positive.

Problem: n moles of helium are isolated in a gas piston with initial temperature T_0. If a constant force F is abruptly exerted on the piston for a short period of time such that it contracts the gas by a distance x, determine the final temperature of the gas T_1.

In this scenario, the gas does not undergo a reversible process as the change is sudden — implying that we must use the external pressure $\frac{F}{A}$ where A is the cross sectional area of the piston in computing work done by or on the gas. Since the work done on the gas by the external force is $\frac{F}{A} \cdot A dx$ after it contracts by an infinitesimal distance dx, the total work done on the gas is Fx. Moreover, as the compression is swift, there is negligible heat transfer between the gas and its surroundings such that $Q = 0$. The first law of thermodynamics then implies that

$$\Delta U - \frac{3}{2} nR\Delta T = W_{on} = Fx,$$

$$T_1 = T_0 + \Delta T = T_0 + \frac{2Fx}{3nR}.$$

Microscopic View

Let us adopt a microscopic perspective to better understand the sign of work done by considering a gas in a gas piston again. If the piston is compressing the gas, gas molecules collide with the incoming piston and rebound with a speed larger than that before the collision. Since the mean-square speed of the molecules increases, the internal energy of the gas increases, which means that positive work has been done on the gas. Conversely, if the gas is expanding, gas molecules hit a retreating piston and rebound with a speed smaller than that before the collision. Thus, the internal energy of the gas decreases. This agrees with the macroscopic interpretation that negative work is done on the system when the gas expands.

Work Done

We observe that work done depends on how p varies with V. Hence, there can be different work done by and on the gas for the same final and initial states of the system as there are different paths a process can take. Thus, it is useful to draw Pressure-Volume or PV diagrams to visualize this.

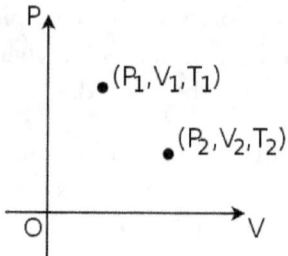

Figure 2.2:   PV diagram

An equilibrium state of an ideal gas can be defined by three macroscopic quantities — namely P, V and T. Due to the ideal equation, these properties are not independent and we can define an equilibrium state of an ideal gas based on two quantities alone if we know the number of moles of gaseous molecules. We usually choose them as P and V so that we can visualize work. Referring to Fig. 2.2, each point on a PV diagram represents a possible equilibrium state consisting of 3 quantities, though it may only be a two-dimensional diagram. The system may undergo a process from one state to another and this is delineated by a line from the initial to final state. The intermediate points on this line correspond to the intermediate equilibrium states of the system as it evolves. Different processes from the same initial state to the same final state will result in different lines. Note that non-quasistatic processes cannot be depicted by a line on a PV diagram as there is no well-defined pressure for the intermediate states.

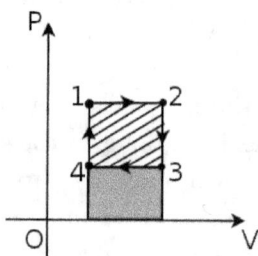

Figure 2.3:   A cyclic process

For example, the PV diagram in Fig. 2.3 shows how the system evolves over four different processes as we consider four specific states of the system. Processes $1 \rightarrow 2$ and $3 \rightarrow 4$ are isobaric processes as the pressure of the system remains constant while processes $2 \rightarrow 3$ and $4 \rightarrow 1$ are isochoric processes as the volume of the system remains constant. The magnitude of work done during a process is simply the area under the curve illustrating the process in the PV graph (remember that work done by the gas is positive if the gas expands and negative otherwise). Thus, the work done by the gas in process $1 \rightarrow 2$ is the sum of the shaded area and filled area, 0 during processes $2 \rightarrow 3$ and $4 \rightarrow 1$ and negative of the filled area during process $3 \rightarrow 4$. Thus, the total work done by the gas during the cycle $1 \rightarrow 2 \rightarrow 3 \rightarrow 4 \rightarrow 1$ is the shaded area and is positive. Note that in general, the magnitude of the work done during a cyclic process, such as above, is the area enclosed by the PV curve. The sign of work done by the gas will depend on the direction of the process. For example, if the process above were to evolve from $1 \rightarrow 4 \rightarrow 3 \rightarrow 2 \rightarrow 1$, the work done by the gas will be negative of the shaded area and hence, positive work is done on the gas. Lastly, the change in internal energy in a cyclic process is zero as the internal energy is a state function and the initial and final states are identical. Then, the area enclosed by a cyclic process in a PV diagram is also directly related to the heat supplied to or extracted from the system.

We are now ready to analyze the work done by a gas during different reversible processes.

Reversible Isochoric Process

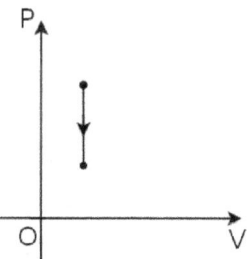

Figure 2.4: Isochoric process

Referring to Fig. 2.4, an isochoric or isovolumetric process is one in which the volume of the system does not change (i.e. $dV = 0$). Then,

$$W_{by} = \int_{V_1}^{V_1} pdV = 0.$$

By the first law of thermodynamics,

$$\Delta U = Q.$$

Reversible Isobaric Process

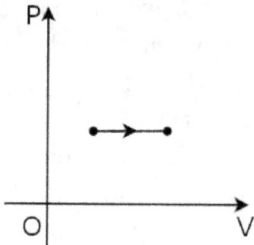

Figure 2.5:   Isobaric process

Referring to Fig. 2.5, an isobaric process is a process in which the pressure of the system remains constant.

$$W_{by} = \int_{V_1}^{V_2} p\,dV = p\Delta V$$

$$\Delta U = Q - p\Delta V.$$

Reversible Isothermal Process

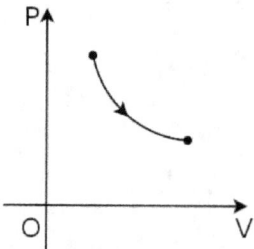

Figure 2.6:   Isothermal process

Referring to Fig. 2.6, an isothermal process is one in which the temperature of the system remains constant.

$$W_{by} = \int_{V_1}^{V_2} p\,dV = \int_{V_1}^{V_2} \frac{nRT}{V}\,dV = nRT\ln\frac{V_2}{V_1}.$$

An example of an isothermal process is the expansion of a cylinder of gas with a thin wall performed by pulling the piston extremely slowly, allowing sufficient time for the gas to gain heat through the container walls to constantly maintain thermal equilibrium with its surroundings. Furthermore, since $\Delta U = \frac{f}{2}nR\Delta T$, $\Delta U = 0$ during an isothermal process. The first law of thermodynamics then implies that

$$Q = W_{by} = nRT \ln \frac{V_2}{V_1}. \tag{2.13}$$

Adiabatic Process

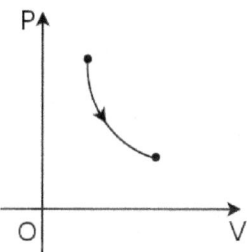

Figure 2.7: Adiabatic process

Referring to Fig. 2.7, an adiabatic process is a reversible (and thus necessarily quasistatic) process in which there is no net heat transfer between a system and its external surroundings (i.e. $Q = 0$).

Adiabatic processes usually involve well-insulated containers. An example would be the gradual increase in external pressure on a thermally insulated gas piston such that the pressure of the gas is always equal to the external pressure as it contracts. An example of a process that involves $Q = 0$ but is not adiabatic would be a sudden compression of a cylinder of gas performed by pushing the piston rapidly such that there is negligible time for heat to escape the system — this is not an adiabatic process as it is non-quasistatic and irreversible. To calculate the work done by the gas in an adiabatic process, we will use the following paramount adiabatic condition. In an adiabatic process,

$$pV^\gamma = c \tag{2.14}$$

where c is an arbitrary constant determined by initial conditions. γ is the adiabatic index and is given by

$$\gamma = \frac{f+2}{f} \tag{2.15}$$

where f is the degrees of freedom of a gas molecule. As a corollary of this condition, adiabats drawn on a PV diagram are steeper than isotherms as $p \propto \frac{1}{V^\gamma}$ for an adiabat with $\gamma > 1$, as compared to $p \propto \frac{1}{V}$ for an isotherm.

Proof: The differential form of the first law of thermodynamics implies that

$$dU = \delta Q + \delta W_{on}.$$

Since $\delta Q = 0$ for an adiabatic process and $\delta W_{on} = -pdV$ for a reversible process,

$$dU + pdV = 0.$$

We know that

$$U = \frac{f}{2} nRT = \frac{f}{2} pV.$$

Taking the total derivative of the above,

$$dU = \frac{f}{2} dpV + \frac{f}{2} pdV$$

$$\frac{f}{2} dpV + \frac{f}{2} pdV + pdV = 0.$$

Dividing the whole equation by $\frac{f}{2} pV$ yields

$$\frac{1}{p} dp + \frac{f+2}{f} \cdot \frac{1}{V} dV = 0.$$

Integrating the above and taking the exponential of both sides,

$$pV^{\frac{f+2}{f}} = c$$

for some constant c. Observe that we have proven the adiabatic condition from the first law of thermodynamics and the ideal gas law. Therefore, if we are interested in analyzing a reversible adiabatic process involving a gas, we can simply use the adiabatic condition, instead of the first law of thermodynamics, in combination with the ideal gas law. The resultant equations are often simpler this way. Now, to determine the work done by the gas in a reversible adiabatic process, we write

$$W_{by} = \int_{V_1}^{V_2} pdV = \int_{V_1}^{V_2} \frac{c}{V^\gamma} dV$$

$$= \left[\frac{cV^{1-\gamma}}{1-\gamma} \right]_{V_1}^{V_2}$$

$$= \frac{1}{1-\gamma}(cV_2^{1-\gamma} - cV_1^{1-\gamma})$$

$$= \frac{1}{1-\gamma}(p_2V_2 - p_1V_1).$$

Substituting $\gamma = \frac{f+2}{f}$,

$$W_{by} = \frac{f}{2}(p_1V_1 - p_2V_2).$$

Alternatively, we could have derived W_{by} from ΔU via the first law of thermodynamics. Since $Q = 0$,

$$W_{by} = -\Delta U.$$

We know that U is given by

$$U = \frac{f}{2}pV$$

$$\implies \Delta U = \frac{f}{2}(p_2V_2 - p_1V_1)$$

$$W_{by} = -\Delta U = \frac{f}{2}(p_1V_1 - p_2V_2).$$

Problem: Burning a piece of wood releases smoke consisting of carbon monoxide (molar mass μ_s) at temperature T_s near the surface of the Earth. If the smoke then rises adiabatically (assume that there is no heat transfer between the atmosphere and the smoke), determine the maximum altitude h that the smoke can attain. The atmosphere can be presumed to have a uniform temperature T_a and molar mass μ_a. Furthermore, $gh \ll \frac{RT_a}{\mu_a}$, where g is the gravitational field strength near the surface of the Earth such that the density of atmospheric air can be assumed to be a constant up till altitude h.

The atmospheric pressure as a function of altitude h is approximately

$$p(h) = p_0 - \rho_a g h$$

where p_0 is the pressure at the surface and $\rho_a = \frac{p_0\mu_a}{RT_a}$ is the uniform mass density of the atmosphere. For the smoke to undergo an adiabatic process, its pressure at all instances must be equal to $p(h)$. By the adiabatic condition, the temperature of the smoke $T(h)$ as a function of altitude obeys

$$p^{1-\gamma}T^\gamma = p_0^{1-\gamma}T_s^\gamma.$$

Since $\gamma = \frac{7}{5}$ for a diatomic gas (carbon monoxide),

$$T(h) = T_s \left(\frac{p}{p_0} \right)^{\frac{2}{7}}.$$

The smoke stops rising when its density is equal to that of the atmosphere ρ_a as the upthrust just balances its weight. This occurs when

$$\frac{p\mu_s}{RT} = \rho_a$$

$$\implies \frac{p\mu_s}{RT_s \left(\frac{p}{p_0} \right)^{\frac{2}{7}}} = \frac{p_0\mu_a}{RT_a}.$$

Substituting the expression for $p(h)$,

$$\left(1 - \frac{\mu_a g h}{RT_a} \right)^{\frac{5}{7}} = \frac{\mu_a T_s}{\mu_s T_a}.$$

Since $gh \ll \frac{RT_a}{\mu_a}$, we can perform a binomial expansion to obtain

$$1 - \frac{5\mu_a g h}{7RT_a} = \frac{\mu_a T_s}{\mu_s T_a}$$

$$h = \frac{7RT_a}{5\mu_a g} - \frac{7RT_s}{5\mu_s g}.$$

2.5   Heat Capacity

If a block of copper and a block of aluminium, that are initially at the same temperature and are of equal masses, are placed into identical beakers of water, the final temperatures of water in the two beakers, at thermal equilibrium, are different. Thus, we conclude that the two blocks must have stored different quantities of heat as internal energy even though they were initially at the same temperature. Therefore, it is natural to define a property that refers to the additional amount of heat required to raise the temperature of a substance by unit temperature as temperature on its own is not a good gauge of the internal energy of a substance. This quantity is known as the heat capacity C of the substance. Concretely,

$$C = \frac{\delta Q}{dT} \qquad (2.16)$$

where Q and T are the heat supplied to the system and the temperature of the system. Applying the first law of thermodynamics,

$$dU = CdT + \delta W_{on} \qquad (2.17)$$

We see that if $\delta W_{on} = 0$ throughout the thermodynamic process, $dU = CdT$ such that C is a good measure of the internal energy stored by the substance. Most notably, we have

$$dU = C_v dT \qquad (2.18)$$

where C_v is the heat capacity of an isochoric process. Since the changes in volume of solids and liquids are often negligible throughout all types of processes (i.e.they are all approximately isochoric), we can simply define a heat capacity C for them that is process-independent (since $dU = CdT$ for all processes and U is a state function). However, the value of C of a gaseous system, on the other hand, depends on the process as W_{on} changes accordingly in Eq. (2.17). Thus, we need to define different values of C for different processes in a gaseous system. Furthermore, C is no longer a measure of the stored internal energy of a gas as part of the heat supplied could have been used as work done by the gas. Before we determine these for isochoric and isobaric processes, it is intuitive that a larger amount of a substance requires a greater quantity of heat for the same change in temperature as a larger system is basically a smaller system duplicated by several parts. It is then natural to define a property for the additional amount of heat required to increase the temperature of a substance by unit temperature, per amount of substance — this is known as the specific heat capacity of the substance. The amount of substance usually refers to the mass of substance m for solids and liquids and the number of moles of gas molecules n for gases. In the case of the latter, the specific heat capacity of gases with respect to the number of moles is known as the molar specific heat capacity. Quantitatively,

$$c = \frac{\delta Q}{mdT}, \qquad \text{(Solids/Liquids)}$$

$$c = \frac{\delta Q}{ndT}, \qquad \text{(Gases)}$$

where c is the specific heat capacity. If c is independent of T,

$$Q = mc\Delta T, \qquad\qquad \text{(Solids/Liquids)}$$

$$Q = nc\Delta T. \qquad\qquad \text{(Gases)}$$

Once again, we emphasize that the molar specific heat capacity of a gas varies across different processes. Therefore, it is convenient to calculate the molar specific heat capacities for specific processes — namely isochoric and isobaric processes. We will derive them for gases with a general number of degrees of freedom.

For an isochoric process, by Eq. (2.18),

$$nc_v dT = dU = \frac{f}{2} nRdT$$

$$\implies c_v = \frac{f}{2} R \qquad\qquad (2.19)$$

In the case of a reversible isobaric process, $\delta W_{on} = -pdV$ and $\delta Q = nc_p dT$ by definition so

$$dU = \frac{f}{2} nRdT = nc_p dT - pdV$$

since $pdV = nRdT$ by the ideal gas law under isobaric conditions,

$$c_p = \frac{f+2}{2} R \qquad\qquad (2.20)$$

where we have also shown that c_v and c_p are independent of temperature for an ideal gas.

We see that the molar specific heat capacity of a gas under constant pressure is larger than that under constant volume as work must be done by the gas (to expand when temperature increases). Quantitatively, $c_p = c_v + R$. Considering these expressions for c_v and c_p, the more general definition of the adiabatic index is in fact

$$\gamma = \frac{c_p}{c_v}. \qquad\qquad (2.21)$$

Problem: When a constant power P is transferred to a solid, its temperature T increases according to

$$T = T_0(1 + \alpha t)^{\frac{1}{4}}$$

where t is the time elapsed and T_0 is the initial temperature. Determine the heat capacity of the solid $C(T)$ as a function of its temperature.

$$\frac{dT}{dt} = \frac{\alpha T_0}{4(1+\alpha t)^{\frac{3}{4}}},$$

$$C = \frac{\delta Q}{dT} = \frac{\delta Q}{dt} \cdot \frac{dt}{dT} = P \cdot \frac{4(1+\alpha t)^{\frac{3}{4}}}{\alpha T_0} = \frac{4PT^3}{\alpha T_0^4}.$$

Enthalpy

It may be noteworthy that a state function known as the enthalpy H of a substance is defined as

$$H = U + pV, \tag{2.22}$$

where U, p and V are its internal energy, pressure and volume respectively. For our purposes, it is merely another state function, derived from other state functions, but chemists prefer to use it for the following reason. Observe that for a substance undergoing a reversible isobaric process,

$$dH = dU + pdV = C_p dT \tag{2.23}$$

where C_p is the heat capacity at constant pressure. Therefore,

$$dH = \delta Q$$

where δQ is the heat absorbed by the substance during the reversible isobaric process. Since experiments on Earth are usually performed under constant pressure (atmospheric pressure), H is a more convenient pathway in specifying the heat absorbed by a substance. Finally, it can be seen that a stronger form of Eq. (2.23) holds for an ideal gas. Since $U = nc_v T$ and $pV = nRT$ for an ideal gas, its enthalpy is

$$H = n(c_v + R)T = nc_p T. \tag{2.24}$$

This implies that the relationship

$$\frac{dH}{dT} = nc_p \tag{2.25}$$

is valid for any general process on an ideal gas of fixed moles, just as $\frac{dU}{dT} = nc_v$.

2.6 Gas Flows

In this section, we will explore how the first law of thermodynamics can be applied to situations where a gas enters or leaves a container. The interpretation of work done in such processes is often more subtle and is dependent of our definition of a system, as the following example shall illustrate.

Problem: In Fig. 2.8, an evacuated chamber is placed on the surface of the Earth where the pressure and temperature of atmospheric air — which can be presumed to be diatomic — are p_0 and T_0. If the cap sealing the chamber is opened, determine the temperature of the gas inside the chamber at the instance where there is no longer any net influx of air molecules into the chamber. The tank is insulated such that there is negligible heat transfer between the inside of the tank and the atmosphere. Assume that no air leaks out of the chamber once it has entered it.

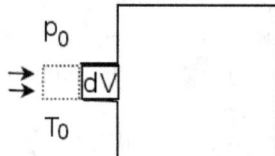

Figure 2.8:   Evacuated chamber

Firstly, note that the relevant final temperature of the gas in the chamber is not necessarily T_0, as the gas may not have attained thermal equilibrium with the atmosphere when a mechanical equilibrium is established (i.e. the final pressure of the gas is p_0). Now, we reach a junction where we have to choose a system to apply the first law of thermodynamics to.

Method 1: Control Mass Just like what we have done in the previous sections, we can pick a set of gas molecules as our system and track them. This method is known as the control mass approach as we fix the constituents of our system. In the context of this problem, we can choose our system as the group of gas particles that will enter the chamber. The change in the macroscopic energies of this system is negligible and there is no heat transfer between the atmosphere and this system. The first law of thermodynamics then states that

$$\Delta U = W_{on}.$$

Now, the origin of W_{on} is rather subtle. Suppose the total volume of our system in the atmosphere is V_0. As our control mass enters the chamber, its posterior experiences a pressure p_0 which is analogous to a piston with pressure p_0. Therefore, we can readily state

$$W_{on} = p_0 V_0 = nRT_0$$

where n is the total number of moles of gas that enters the chamber and T_0 is the atmospheric temperature as the piston pushes volume V_0 of gas into the chamber. Note that the possible work done on the incoming back sections by the front sections which are already in the chamber is excluded precisely because we have defined all gas molecules that will eventually enter the chamber as our system, such that this component of work is not performed by an external agent. In other words, though the work done on the arriving section by the gas already in the chamber increases the internal energy of the arriving section, there is a corresponding decrease in the internal energy of the gas in the chamber and thus no net change in the internal energy of our system due to this factor. With this clarification, we proceed with substituting $\Delta U = \frac{5}{2} nR\Delta T$ for a diatomic gas. Hence,

$$\frac{5}{2} nR\Delta T = nRT_0$$

$$\Delta T = \frac{2}{5} T_0.$$

The final temperature T_f is

$$T_f = T_0 + \Delta T = \frac{7}{5} T_0.$$

Method 2: Control Volume Instead of choosing a predetermined group of particles as our system, we can demarcate a region known as a control volume and analyze the energies entering and leaving this region. In this case, we can define the control volume as the chamber. Let n now denote the instantaneous number of moles stored in the chamber. In a short time interval dt, the only change in energy inside the control volume stems from the dn moles of atmospheric molecules, which occupy volume dV in the atmosphere, entering the chamber. Since their macroscopic energies are negligible, the total energy carried by these molecules is their final internal energy which is their initial internal energy (internal energy in the atmosphere $\frac{5}{2} dnRT_0$) plus the gain in internal energy due to the work done on them by the gas section immediately behind them as they are pushed in.

For purposes of illustration, suppose that the arriving gas section has a cross sectional area dA and a length dx. The force by the gas section at the back of this section on this section is $p_0 dA$ and must have acted over a distance dx to push it into the chamber. Consequently, the work done on the arriving section by its posterior neighbour, which is known as flow work, is $p_0 dV = dn R T_0$. Note that the meaning of this work is slightly different from W_{on} afore. The $p_0 V_0$ term in the previous method arose from the work done on all molecules that will enter the chamber by other atmospheric molecules. However, the $p_0 dV$ term here indicates the work done on incoming gas molecules due to the gas molecules immediately behind them, which may or may not eventually enter the chamber. In a certain sense, we may be including the "internal forces" in our analysis. At this point, you may wonder why we did not consider the work done on the infinitesimal section dV entering the chamber due to the gas already inside the chamber. This is because, the incoming gas section becomes part of the system once it enters the control volume (chamber) — meaning that this does not represent a flow of energy outside of the control volume.

Moving on, the rate of increase of energy, which is manifested solely as internal energy U, inside our control volume is therefore

$$\dot{U} = \frac{5}{2} R T_0 \dot{n} + R T_0 \dot{n} = \frac{7}{2} R T_0 \dot{n}.$$

Integrating and substituting the initial values of U and n as zero,

$$U_f = \frac{7}{2} n_f R T_0$$

where U_f and n_f are the final internal energy and the number of moles of gas inside the chamber respectively. Since $U_f = \frac{5}{2} n_f R T_f$ where T_f is the final temperature,

$$T_f = \frac{7}{5} T_0.$$

Steady Flows

The control volume approach introduced afore presents a neat method of analyzing steady gas flows in which the properties of each point in a system do not vary with time. Recall that a streamline delineates the trajectory of a fluid molecule when the flow is steady. Now, consider the steady flow of a gas along a streamtube which consists of a bundle of adjacent streamlines. Suppose that we wish to relate the flow speeds (v), temperatures (T) and heights h at two points along a streamtube as shown in Fig. 2.9.

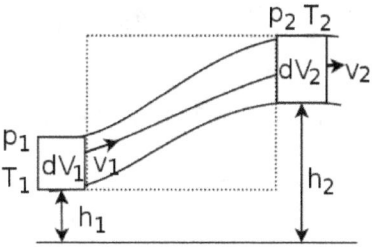

Figure 2.9:   Streamtube

Let the rate of moles of gas molecules flowing through a cross section be $\dot{n}$. This must be uniform through the entire streamtube at steady state and is equivalent to the mass continuity equation. Let the cross sectional areas of the right and left ends next to the demarcated region in the stream tube be A_1 and A_2 respectively. Then, the mass continuity equation is

$$\eta_1 A_1 v_1 = \eta_2 A_2 v_2 = \dot{n}, \tag{2.26}$$

where η represents the number density in a region, which can be computed as $\eta = \frac{p}{RT}$ by the ideal gas law. Thus, the mass continuity equation is equivalent to stating that

$$\frac{pAv}{T} = \text{constant.} \tag{2.27}$$

Moving on, we also can exploit the fact that the energy of the region between these two points should be constant with respect to time at steady state. That is, we can balance the energy influx into and outflow from the demarcated region. In time dt, other than work done and heat transfer into the demarcated region by entities external to the streamtube, there is a change in energy within the region due to dn moles of molecules (with molar mass μ) entering from the left and dn moles of molecules exiting from the right. The net increase in macroscopic kinetic energy is $\frac{1}{2}\mu dn(v_1^2 - v_2^2)$ where v_1 and v_2 are the respective flow speeds while the net increase in gravitational potential energy is $\mu dng(h_1 - h_2)$ (we assume that other forms of potential energy are absent). Meanwhile, the net increase in energy inside the dashed boundary due to the internal energies of the incoming and outgoing molecules is $dnc_v(T_1-T_2)+p_1 dV_1-p_2 dV_2$ where dV_1 and dV_2 are the volumes of the incoming and outgoing molecules at the respective ends. As for the last two terms, remember that we have to include the flow work done by the molecules behind the incoming molecules on the left end (which is positive) and that by the molecules in front of the outgoing molecules on the right end (which is negative as the force due to their pressure opposes the flow velocity v_2). Since $p_1 dV_1 = dn R T_1$ and $p_2 dV_2 = dn R T_2$, the above can be rewritten as

$$dn c_v (T_1 - T_2) + dn R (T_1 - T_2) = dn c_p (T_1 - T_2)$$

where c_p is the molar heat capacity at constant pressure. Another way to see this is that $dn c_v T_1 + p_1 dV_1$ and $dn c_v T_2 + p_2 dV_2$ are simply the enthalpies of the incoming and outgoing molecules, $dn c_p T_1$ and $dn c_p T_2$! All-in-all, the rate of change of energy in the demarcated region is

$$\dot{Q} + \dot{W}_{on} + \frac{1}{2} \mu \dot{n} (v_1^2 - v_2^2) + \mu \dot{n} g (h_1 - h_2) + \dot{n} c_p (T_1 - T_2) = 0. \quad (2.28)$$

Most of the time, the rate of external heat flow $\dot{Q}$ and work done $\dot{W}_{on}$ are zero such that the above becomes

$$\frac{1}{2} \mu v_1^2 + \mu g h_1 + c_p T_1 = \frac{1}{2} \mu v_2^2 + \mu g h_2 + c_p T_2 \quad (2.29)$$

since $\dot{n}$ is uniform. Equivalently,

$$\frac{1}{2} \mu v^2 + \mu g h + c_p T = \text{constant}. \quad (2.30)$$

In words, the sum of the macroscopic kinetic and potential energies, the internal energy of the molecules and flow work performed by posterior molecules at any point along a streamtube is a constant. In cases where the potential energy term is also negligible, the conserved quantity is $\frac{1}{2} \mu v^2 + c_p T$. This quantity divided by c_p is known as the stagnation temperature T_t.

$$T_t = \frac{\mu}{2 c_p} v^2 + T.$$

Its physical meaning is the temperature at the point along the streamline that is stationary. Now, the term "stationary" implies that we need to specify a reference frame for its meaning to be unambiguous. Recall that we have assumed that the flow was steady when deriving the above equations. Therefore, the relevant point must be stationary relative to the frame in which the flow is steady and the streamlines do not move with time. Conversely, we can express the maximum macroscopic speed (when $T = 0$) that the gas can attain with respect to this frame as

$$v_{max} = \sqrt{\frac{2 c_p T_t}{\mu}}.$$

Problem: A rocket in outer space propels itself by burning fuel to release diatomic gas of temperature T_1 in its combustion chamber which has a cross sectional area A_1. The gas then flows adiabatically and is expelled out of the nozzle, which has a cross sectional A_2, at a speed v_2 relative to the rocket and at pressure p_2 and temperature $T_2 < T_1$. If the rocket is designed correctly (i.e. its cross sectional area is varied appropriately) such that steady flow relative to the rocket is achieved, determine the thrust experienced by the rocket.

We will analyze this set-up in the frame of the rocket. Let the pressure of the released gas at the combustion chamber be p_1 and let it have a speed v_1 relative to the rocket. Firstly, the adiabatic condition implies that

$$p_1^{1-\gamma} T_1^{\gamma} = p_2^{1-\gamma} T_2^{\gamma}$$

where $\gamma = \frac{7}{5}$ for a diatomic gas.

$$\implies p_1 = p_2 \left(\frac{T_1}{T_2} \right)^{\frac{7}{2}}.$$

Since the flow is steady in the frame of the rocket, mass continuity (Eq. (2.27)) requires

$$\frac{p_1 A_1 v_1}{T_1} = \frac{p_2 A_2 v_2}{T_2}.$$

Substituting the expression for p_1 in terms of p_2,

$$v_1 = v_2 \frac{A_1}{A_2} \left(\frac{T_2}{T_1} \right)^{\frac{5}{2}}.$$

Applying Eq. (2.29) while neglecting the change in gravitational potential energy,

$$\frac{1}{2} \mu v_1^2 + \frac{7}{2} R T_1 = \frac{1}{2} \mu v_2^2 + \frac{7}{2} R T_2$$

where μ is the molar mass of the diatomic gas. Substituting the expression for v_1 in terms of v_2,

$$v_2^2 = \frac{7R(T_1 - T_2)}{\mu \left(1 - \frac{A_1^2}{A_2^2} \left(\frac{T_2}{T_1} \right)^5 \right)}.$$

The rate of moles of molecules exiting the nozzle is $\eta_2 A_2 v_2 = \frac{p_2 A_2 v_2}{RT_2}$ where η_2 is the number density of gas molecules at the nozzle. As such, after a time interval dt, the momentum of the gas molecules that escape in the frame of the rocket is

$$dp = \frac{p_2 A_2 v_2}{RT_2} dt \cdot v_2 = \frac{7 p_2 A_2 (T_1 - T_2)}{T_2 \left(1 - \frac{A_1^2}{A_2^2} \left(\frac{T_2}{T_1}\right)^5\right)} dt$$

$$\implies \frac{dp}{dt} = \frac{7 p_2 A_2 (T_1 - T_2)}{T_2 \left(1 - \frac{A_1^2}{A_2^2} \left(\frac{T_2}{T_1}\right)^5\right)}.$$

Observe that after this time interval dt, the total momentum of the gas flowing in the rocket increases by dp in the frame of the rocket. Therefore, by the conservation of momentum, the rocket's momentum must have changed by $-dp$. Therefore, the thrust experienced by the rocket is

$$F = -\frac{dp}{dt} = -\frac{7 p_2 A_2 (T_1 - T_2)}{T_2 \left(1 - \frac{A_1^2}{A_2^2} \left(\frac{T_2}{T_1}\right)^5\right)},$$

where the negative sign indicates that the force is opposite in direction to the relative velocity of the ejected gas.

2.7 Kinetic Theory of Gases

This section will discuss the microscopic perspective to ideal gases in classical thermodynamics by modeling a system as a large collection of discrete molecules. Only monoatomic molecules with no rotational and vibrational modes will be considered. In the limit where the volume of the system tends to infinity with a constant density — an ideal known as the thermodynamic limit — thermal fluctuations are smoothed out such that thermodynamic quantities are close to their average values. Quantitatively, taking the average of N independent samples of a variable yields a standard deviation that is $\frac{1}{\sqrt{N}}$ times the standard deviation of a single sample. Since the standard deviation is a natural measure of the spread or uncertainty of a distribution, the decrease in standard deviation with N causes thermal fluctuations to be negligible, as N in this context refers to the number of molecules in a system, which is gargantuan. This notion also sheds light on the statistical nature of thermodynamics which involves probabilistic laws that are accurate in the regime of many constituents.

2.7.1 Distribution Functions

Velocity Distribution Function

A velocity distribution function $f(\boldsymbol{v}) = f(v_x, v_y, v_z)$ is used to describe the fraction of molecules with a velocity in the immediate vicinity of a certain $\boldsymbol{v}$, just like any other probability distribution function. Concretely, it is a three-dimensional probability distribution function (one for each spatial dimension) such that the fraction of molecules with velocities between $\boldsymbol{v} = (v_x, v_y, v_z)$ and $(v_x + dv_x, v_y + dv_y, v_z + dv_z)$ is $f(\boldsymbol{v})dv_x dv_y dv_z$.

Since the motion of gas molecules is proposed to be isotropic, the velocity distribution function should only be dependent on speed and not the direction of velocity.

$$f(\boldsymbol{v}) = f(v).$$

Given this isotropic nature, a common mistake is to assume that the fraction of molecules traveling at speeds between v and $v + dv$ and whose velocities make an angle between θ and $\theta + d\theta$ with a fixed axis, such as the z-axis, is equal for all θ. This confusion is best rectified by considering the velocity space, in Fig. 2.10, which is a sphere that depicts the possible velocities of the molecules as vectors extending from the origin.

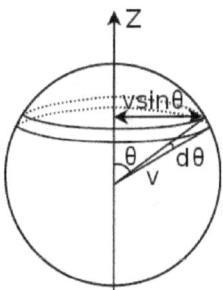

Figure 2.10: Molecules with angles between θ and $\theta + d\theta$

Since every point in velocity space represents a velocity, the velocity distribution function can be ascribed to every point in space to quantify the fraction of molecules possessing that particular velocity per unit volume around that point. Due to the isotropic nature of the distribution, this probability density is uniform over a spherical shell at a constant radius (and thus constant speed) away from the origin. Observe that the fraction of molecules travelling at speeds between v and $v + dv$ that make an angle between θ and $\theta + d\theta$ with respect to the z-axis is an approximately circular hoop of radius

$v \sin \theta$, width $v d\theta$ and thickness dv (spherical coordinates). Then, the relevant fraction is $2\pi v^2 f(v) \sin \theta d\theta dv$ which is non-uniform across different θ for a given speed.

Finally, the velocity distribution function needs to be normalized like any other probability distribution function. This can be evaluated in Cartesian coordinates and also conveniently, in spherical coordinates due to its isotropy.

$$\int_{-\infty}^{\infty} \int_{-\infty}^{\infty} \int_{-\infty}^{\infty} f(v) dv_x dv_y dv_z = 1, \tag{2.31}$$

$$\int_{0}^{\infty} 4\pi v^2 f(v) dv = 1. \tag{2.32}$$

Speed Distribution Function

The distribution of the speeds of molecules can be easily computed from the velocity distribution. Since the velocity distribution is uniform for a constant speed v, the fraction of molecules having a speed between v and $v + dv$ is simply the volume of a spherical shell of radius v and thickness dv, multiplied by $f(v)$. The speed distribution function $f_s(v)$ is then

$$f_s(v) = 4\pi v^2 f(v) \tag{2.33}$$

and is a one-dimension distribution. Then, the fraction of molecules with speeds between v and $v + dv$ and velocities that make an angle between θ and $\theta + d\theta$ with a certain axis can be expressed as

$$\frac{1}{2} f_s(v) \sin \theta d\theta dv. \tag{2.34}$$

Finally, note that if $f(v)$ is normalized, $f_s(v)$ is also normalized as a result of Eq. (2.32).

We have now covered the two important distribution functions in kinetic theory. Do not worry about the exact functions for now as this will be discussed in a later section. Instead, let us focus on how thermodynamic variables can be described in terms of these distributions. However, we will still be using the following results for the mean, mean square and mean cube speeds which are consequences of the speed distribution:

$$\langle v \rangle = \sqrt{\frac{8kT}{\pi m}}, \tag{2.35}$$

$$\langle v^2 \rangle = \frac{3kT}{m}, \tag{2.36}$$

$$\langle v^3 \rangle = \sqrt{\frac{128k^3T^3}{m^3\pi}}, \tag{2.37}$$

where k is the Boltzmann constant, T is the temperature of the gas and m is the mass of a single molecule.

2.7.2 Pressure

Collisions with a Stationary Area

We first analyze the rate of collisions of molecules per unit area with a stationary wall. Consider an infinitesimal area dA and define the positive z-axis to be parallel to its area vector (which is pointing outwards from the container). We will adopt spherical coordinates in this problem. Firstly, we consider molecules that travel at a particular speed v. The volume swept by molecules with velocity v that subtends an angle θ with respect to the z-axis in time dt is of the shape in Fig. 2.11.

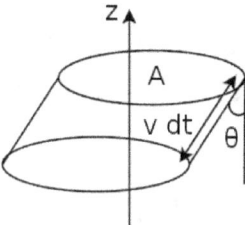

Figure 2.11: Volume of molecules with velocity v that collide with the wall in time dt

The shape has a total volume of

$$dV = v \cos \theta dA dt.$$

By Eq. (2.34), the number of collisions with the area dA in time dt due to this particular class of molecules is thus $\eta dV \cdot \frac{1}{2} f_s(v) \sin \theta d\theta dv = \frac{1}{2}\eta f_s(v) \sin \theta d\theta dv dV$, where η is the number density of molecules which is assumed to be uniform. Therefore, the number of collisions per unit area, per unit time due to molecules that travel at speeds between v and $v + dv$ and angles between θ and $\theta + d\theta$ is

$$\frac{1}{2}\eta v f_s(v) \sin \theta \cos \theta d\theta dv. \tag{2.38}$$

Momentum Transfer Per Collision

When a molecule traveling at speed v and angle θ collides with the stationary wall, it rebounds and effectively reverses its velocity in the z-direction, assuming that the collision is elastic. Therefore, the momentum transferred to the infinitesimal area is $2mv\cos\theta$ in the positive z-direction. The pressure contribution dp due to molecules traveling between speeds v and $v + dv$ and angle θ and $\theta + d\theta$ is then the momentum transferred per collision multiplied by the number of collisions per unit area, per unit time.

$$dp = 2mv\cos\theta \cdot \frac{1}{2}\eta v f_s(v)\sin\theta\cos\theta d\theta dv = \eta m v^2 f_s(v)\sin\theta\cos^2\theta d\theta dv.$$

The total pressure on the wall is then obtained by integrating the above over all relevant v and θ. Note that θ is only integrated from 0 to $\frac{\pi}{2}$ as only molecules travelling in the positive z-direction are germane.

$$p = \int_0^\infty \int_0^{\frac{\pi}{2}} \eta m v^2 f_s(v)\sin\theta\cos^2\theta d\theta dv$$

$$= \frac{1}{3}\eta m \int_0^\infty v^2 f_s(v)dv$$

where $\int_0^{\frac{\pi}{2}}\sin\theta\cos^2\theta d\theta$ can be solved via the substitutions $u = \cos\theta$, $du = -\sin\theta d\theta$. Now, observe that the final integral averages v^2 to produce the mean square speed. Thus,

$$p = \frac{1}{3}\eta m \langle v^2 \rangle \tag{2.39}$$

which is often written as $p = \frac{1}{3}\rho\langle v^2 \rangle$ where $\rho = \eta m$ is the mass density of the gas. Substituting the expression for $\langle v^2 \rangle$ in Eq. (4.7), we can prove the ideal gas equation.

$$p = \eta k T,$$

$$pV = NkT,$$

where N is the total number of molecules.

2.7.3 *Effusion*

Effusion is the process where gas molecules escape from a small hole of area A and a diameter smaller than the mean free path of the molecules — the average distance traveled by the molecules between consecutive collisions. Interesting effusion properties to compute would be the molecular flux out of the hole and the rate of change of internal energy of the gas remaining in the container. The speed distribution of escaped molecules is also intriguing and shall be deferred to a later section. For now, we should understand qualitatively that the speed distribution of effused molecules favors molecules with higher speeds (as compared to the standard speed distribution $f_s(v)$) as these molecules are more energetic and more likely to escape from the hole.

Equation (2.38) is the rate of molecules of speeds between v and $v+dv$ and angles between θ and $\theta + d\theta$ colliding with a stationary wall, per unit area, and is similarly, also the rate of molecules effusing out of a small hole, per unit area. After integration over the relevant range of θ (this does not change the expression's dependence on v), the (instantaneous) speed distribution $f_e(v)$ of escaping molecules is proportional to $vf_s(v)$. It can be seen from the additional factor of v, as compared to $f_s(v)$, that effusion preferentially selects molecules with greater speeds as they are more likely to escape from the hole.

Next, the molecular flux, which is the rate of moles of gas flowing out of the hole, can be calculated by multiplying Eq. (2.38) by A and integrating over the relevant limits.

$$\Phi = \int_0^\infty \int_0^{\frac{\pi}{2}} \frac{1}{2}\eta A v f_s(v) \sin\theta \cos\theta \, d\theta \, dv = \frac{1}{4}\eta A \int_0^\infty v f_s(v) dv$$

$$\Phi = \frac{1}{4}\eta A \langle v \rangle. \tag{2.40}$$

The above can be expressed solely in terms of the thermodynamic properties p and T by substituting $\langle v \rangle = \sqrt{\frac{8kT}{\pi m}}$ and by expressing η in terms of p and T via the ideal gas law:

$$pV = NkT,$$
$$\eta = \frac{N}{V} = \frac{p}{kT}.$$

Therefore,

$$\Phi = \frac{pA}{\sqrt{2\pi mkT}}. \tag{2.41}$$

Since Φ is inversely proportional to $\sqrt{m}$, effusion can be used to separate different gas molecules and isotopes of the same gas. As the lighter molecules effuse at a greater rate, the preponderance of molecules left in the container will be the heavier molecules.

Problem: Effusion is often applied in uranium enrichment processes. Suppose that we have a large sample of two different isotopes of uranium trapped in two different gas molecules of molar masses μ_1 and $\mu_2 > \mu_1$. Initially, the ratio of molecules with molar mass μ_1 to those of molar mass μ_2 is $q < 0.5$. We can purify a sample of homogeneous temperature by allowing it to effuse through a membrane fraught with porous holes that have diameters smaller than the mean free path of the molecules and collecting the molecules that pass through the filter up till a period of time. Backwards effusion is negligible. Determine the number of cycles needed to increase the previous ratio to at least $2q$ by repeatedly applying this procedure.

Suppose that the ratio of molecules with molar masses μ_1 to those with μ_2 is r currently. Since $\Phi \propto \frac{\eta}{\sqrt{m}}$, the ratio of the rates of effusion is

$$\frac{\Phi_1}{\Phi_2} = \frac{r\sqrt{\mu_2}}{\sqrt{\mu_1}}.$$

The new ratio after a single step is evidently

$$r' = \sqrt{\frac{\mu_2}{\mu_1}}r.$$

Therefore, the minimum number of stages required to increase the ratio to at least $2q$ is

$$n = \log_{\sqrt{\frac{\mu_2}{\mu_1}}} 2.$$

Next, it is useful to determine the rate of energy loss engendered by the escaping molecules. Equation (2.38) is the rate of molecules with speed v and angle θ escaping the hole, per unit area. Therefore, the total kinetic energy by this class of particles, that escape the hole, can be determined by multiplying Eq. (2.38) by $\frac{1}{2}mv^2$ (kinetic energy of a molecule of that class) and A, and integrating over the relevant limits.

$$\frac{dE}{dt} = -\int_0^\infty \int_0^{\frac{\pi}{2}} \frac{1}{4}\eta Amv^3 f_s(v) \sin\theta \cos\theta\, d\theta dv$$

$$= -\int_0^\infty \frac{1}{8}\eta Amv^3 f_s(v) dv$$

$$= -\frac{1}{8}\eta Am\langle v^3 \rangle$$

where E is the total internal energy remaining in the container. Substituting $\langle v^3 \rangle = \sqrt{\frac{128k^3 T^3}{m^3 \pi}}$,

$$\frac{dE}{dt} = -\eta A \sqrt{\frac{2k^3 T^3}{m\pi}}. \qquad (2.42)$$

The average energy of an effusing molecule can be determined by dividing the magnitude of the rate of energy lost by the molecular flux.

$$\frac{1}{2} m \langle v_e^2 \rangle = \frac{\left| \frac{dE}{dt} \right|}{\Phi} = 2kT$$

which is evidently more than the average kinetic energy of a molecule originally in the container, $\frac{3}{2}kT$.

2.7.4 Mean Free Path

In this section, we will model the collisions between gas molecules and determine the mean free time and mean free path which are the average time elapsed and distance covered between consecutive collisions of a molecule. Important assumptions in this model are that colliding molecules are scattered elastically in random directions after a collision and that collisions between different time intervals are independent events.

Monoatomic gas molecules are modeled as hard spheres with a radius r. Suppose that we select a particular particle and follow its motion. Then, observe that the tracked molecule can collide with another molecule if the center of the other molecule is within a circular cross section of radius $2r$ from the center of the tracked molecule, as shown in Fig. 2.12.

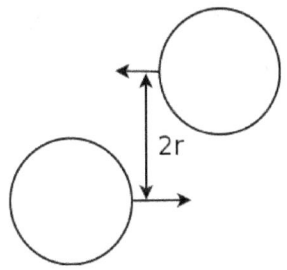

Figure 2.12: Effective collision radius

Therefore, we define the effective collision cross sectional area as

$$\sigma = \pi (2r)^2 = 4\pi r^2.$$

Now, let the tracked particle have a constant velocity v until its next collision and define u to be the velocity of a particular class of other molecules that it could collide with. Then in time dt, the effective collision volume swept by the tracked particle, relative to this class of molecules, is

$$\sigma|\boldsymbol{v} - \boldsymbol{u}|dt.$$

The probability of a collision occurring between the tracked molecule and the particular class of molecules during the time interval, is the above multiplied by the number density of that particular class of molecules, $f(\boldsymbol{u})\eta du_x du_y du_z$.

$$\eta\sigma|\boldsymbol{v} - \boldsymbol{u}|f(\boldsymbol{u})du_x du_y du_z.$$

Integrating the above over all $\boldsymbol{u}$ would yield the probability of the tracked molecule colliding in the time interval dt. Then, we can average the resultant expression over all $\boldsymbol{v}$ (all possible tracked molecules) to determine the probability of a molecule colliding in the time interval dt on average. This probability is

$$\eta\sigma\langle|\boldsymbol{v} - \boldsymbol{u}|\rangle dt = \eta\sigma\langle v_r\rangle dt$$

where v_r is the relative speed between molecules. The average is performed over all possible $\boldsymbol{v}$ and $\boldsymbol{u}$. Now, define $P(t)$ as the probability that a molecule, on average, has not collided from time $t = 0$ to time t. Then, from the first principles of calculus,

$$P(t + dt) = P(t) + \frac{dP}{dt}dt.$$

Since the collision events during different time intervals are independent, the probability of a molecule surviving till $t + dt$ is the product of the probability it survives till t and the probability of it not colliding in the interval between t and $t + dt$. This applies to the average case as well.

$$P(t + dt) = P(t)(1 - \eta\sigma\langle v_r\rangle)dt.$$

Comparing the two expressions for $P(t + dt)$,

$$\frac{dP}{dt} = -\eta\sigma\langle v_r\rangle P$$

$$\int_1^p \frac{1}{P}dP = \int_0^t -\eta\sigma\langle v_r\rangle dt,$$

where the lower limit of P has been set to one as the probability that a molecule, on average, survives till $t = 0$ is one. Therefore,

$$P(t) = e^{-n\sigma\langle v_r\rangle t}.$$

Now, we can use the above to calculate the mean time between collisions. The probability of a molecule surviving till time t and colliding between the time interval from t to $t+dt$, on average, is simply the product of the probability that it survives till time t and the probability of it colliding within the time interval dt.

$$P(t) \cdot (\eta\sigma\langle v_r \rangle)dt = \eta\sigma\langle v_r \rangle e^{-\eta\sigma\langle v_r \rangle t}dt.$$

Therefore, the mean free time is obtained by multiplying the above by t and integrating over all t.

$$\tau = \int_0^\infty t\eta\sigma\langle v_r \rangle e^{-\eta\sigma\langle v_r \rangle t}dt = \frac{1}{\eta\sigma\langle v_r \rangle}. \tag{2.43}$$

It can be shown that the average relative velocity $\langle v_r \rangle$ is exactly $\sqrt{2}\langle v \rangle$ from the velocity distribution of gas molecules. The proof is non-trivial and will not be presented here. Following from this,

$$\tau = \frac{1}{\sqrt{2}\eta\sigma\langle v \rangle}. \tag{2.44}$$

Substituting $\langle v \rangle = \sqrt{\frac{8kT}{\pi m}}$,

$$\tau = \frac{\sqrt{\pi m}}{4\eta\sigma\sqrt{kT}}. \tag{2.45}$$

It can be seen that heavier molecules collide less frequently and that the mean collision interval is shorter for a larger temperature — both properties make intuitive sense. Following from this, the mean free path is

$$\lambda = \langle v \rangle\tau = \frac{1}{\sqrt{2}\eta\sigma}. \tag{2.46}$$

2.7.5 *Statistics of an Ideal Gas*

Macrostates and Microstates

A thermodynamics system can be described in two ways. Firstly, it can be quantified on the whole in terms of the macroscopic properties it exhibits such as temperature and pressure. These are the attributes measured during experiments. A set of such variables is known as a macrostate. Next, we can adopt another perspective by describing a system based on the parameters of all its constituents (e.g. by labeling all particles with their positions and velocities). A configuration consisting of such parameters is known as a microstate. Crucially, several microstates can result in the same macrostate.

For example, suppose that you roll two dice — a possible macrostate may be the sum of the two numbers. Consider the particular sum 4 — it can be formed in three ways: $1 + 3$, $2 + 2$ and $3 + 1$ which are different microstates of the system.

Boltzmann Distribution

Consider a system coupled to another gargantuan system, known as a heat reservoir, such that energy can be exchanged. The reservoir is so large that any heat extracted from or deposited into it does not vary its temperature significantly. If the system is in thermal equilibrium with the reservoir such that the common temperature is T, the probability of the system undertaking a microstate S with a certain energy E is proportional to $e^{-\frac{E}{kT}}$, which is known as the Boltzmann factor.

$$p(S) \propto e^{-\frac{E}{kT}}.$$

Assume that there is only a single microstate corresponding to a single energy such that the probability can be expressed as a function of the energy of the system instead. If there are N microstates with the ith state having energy E_i, the probability of the system adopting the kth microstate with energy E_k is hence

$$p(E_k) = \frac{e^{-\frac{E_k}{kT}}}{\sum_{i=1}^{N} e^{-\frac{E_i}{kT}}}.$$

Let us apply this to the simplest example of a two-state system with energy levels 0 and E. Then, the probability of each microstate is

$$p(0) = \frac{1}{1 + e^{-\frac{E}{kT}}},$$

$$p(E) = \frac{e^{-\frac{E}{kT}}}{1 + e^{-\frac{E}{kT}}}.$$

We can also calculate the average energy as

$$\langle E \rangle = 0 \cdot p(0) + E \cdot p(E) = \frac{E}{e^{\frac{E}{kT}} + 1}.$$

Another intriguing application of the Boltzmann distribution pertains to an isothermal atmosphere with molar mass μ and uniform temperature T. By balancing forces on each gas section, one can obtain from basic mechanics that the pressure $p(h)$ at a small altitude h above the surface of Earth obeys

$$p(h) = p_0 e^{-\frac{\mu g h}{RT}}$$

where p_0 is the pressure at the surface. An alternate perspective can be adopted by considering the distribution of molecules as a function of altitude. Since the gravitational potential energy per molecule at altitude h is mgh where m is the mass of a single molecule (the reference point has been set at the surface of the Earth), the Boltzmann distribution implies that the density $\rho(h)$ of the atmospheric molecules varies with altitude h according to

$$\rho(h) = \rho_0 e^{-\frac{mgh}{kT}}$$

where ρ_0 is the density at the surface of Earth. Multiplying the numerator and denominator of the exponent by the Avogadro's number N_A,

$$\rho(h) = \rho_0 e^{-\frac{\mu g h}{RT}}.$$

Since $\rho = \frac{p\mu}{RT} \implies \rho \propto p$ for an ideal gas,

$$p(h) = p_0 e^{-\frac{\mu g h}{RT}}.$$

Maxwell–Boltzmann Distributions

The Boltzmann distribution can be applied to a single ideal gas molecule by considering all other gas molecules as the heat reservoir. The resultant distributions (for velocity and speed) are known as the Maxwell–Boltzmann distributions. In this process, we are making the assumptions that there are no intermolecular forces and that the intermolecular distances are large as compared to the mean free path (average distance between consecutive collisions) of molecules, such that collisions occur once in a blue moon. These can be satisfied in the case of a very dilute gas. Then, we can approximately say that the system (which is one gas particle) is at equilibrium with a reservoir (all other particles), maintained at a temperature T.

In the case of a monoatomic molecule with only translational freedoms, its total energy (excluding possible macroscopic energies) is given by

$$E = \frac{1}{2}mv^2 = \frac{1}{2}mv_x^2 + \frac{1}{2}mv_y^2 + \frac{1}{2}mv_z^2$$

where the x, y and z-directions are arbitrarily chosen. Then, the probability of a molecule having a velocity v between (v_x, v_y, v_z) and $(v_x + dv_x, v_y +$

$dv_y, v_z + dv_z)$ is proportional to the Boltzmann factor. Since the molecules are assumed to be identical, the distribution of molecules having velocity $\boldsymbol{v}$ is identical to the probability distribution of the velocity of a single molecule. That is, a single molecule is representative of the entire system of molecules as they are identical. Then, the fraction of molecules having velocity $\boldsymbol{v}$, $f(\boldsymbol{v})$, is also proportional to the Boltzmann factor.

$$f(\boldsymbol{v}) = Ae^{-\frac{m(v_x^2 + v_y^2 + v_z^2)}{2kT}}$$

where A is a normalization factor. Note that we have already used the isotropic nature of the distribution to conclude that f is strictly a function of speed and independent of the direction of velocity. Now, we can evaluate A by imposing the condition that

$$\int_{-\infty}^{\infty} \int_{-\infty}^{\infty} \int_{-\infty}^{\infty} f(\boldsymbol{v}) dv_x dv_y dv_z = 1.$$

Before this, let us go through a few integration tricks.

Integration Trick: Differentiating a Parameter

We shall discuss a general method for evaluating integrals of the form $\int_{-\infty}^{\infty} x^{2n} e^{-\alpha x^2} dx$ and $\int_{0}^{\infty} x^{2n+1} e^{-\alpha x^2} dx$ where α is a constant and n is a non-negative integer. Firstly, we begin with the integral

$$I_x = \int_{-\infty}^{\infty} e^{-\alpha x^2} dx.$$

Consider a second integral $I_y = \int_{-\infty}^{\infty} e^{-\alpha y^2} dy$ where y is a variable independent of x. Due to this independence, the product of these integrals can be evaluated by combining their integrands.

$$I_x I_y = \int_{-\infty}^{\infty} \int_{-\infty}^{\infty} e^{-\alpha(x^2 + y^2)} dx dy.$$

These limits of integration are tantamount to the entire xy-plane. Therefore, the above can also be computed in terms of polar coordinates by substituting $x = r \cos \theta$ and $y = r \sin \theta$. Then,

$$I_x I_y = \int_{0}^{\infty} \int_{0}^{2\pi} re^{-\alpha r^2} d\theta dr$$

$$= 2\pi \int_{0}^{\infty} re^{-\alpha r^2} dr$$

$$= 2\pi \cdot \left[-\frac{e^{-\alpha r^2}}{2\alpha} \right]_0^\infty$$

$$= \frac{\pi}{\alpha},$$

where we have also conveniently proven that $\int_0^\infty x e^{-\alpha x^2} = \frac{1}{2\alpha}$. Since $I_x = I_y$,

$$\int_{-\infty}^\infty e^{-\alpha x^2} dx = \sqrt{\frac{\pi}{\alpha}}.$$

Now, notice that the integral above is a function of α.

$$I(\alpha) = \int_{-\infty}^\infty e^{-\alpha x^2} dx.$$

Then, we can take the total derivative of this integral with respect to α.

$$\frac{dI(\alpha)}{d\alpha} = \frac{d}{d\alpha} \left(\int_{-\infty}^\infty e^{-\alpha x^2} dx \right).$$

Since α is independent of x which is the variable that we are integrating with respect to, the derivative can be moved within the integral.

$$\frac{dI(\alpha)}{d\alpha} = \int_{-\infty}^\infty \frac{\partial}{\partial \alpha} \left(e^{-\alpha x^2} \right) dx = \int_{-\infty}^\infty -x^2 e^{-\alpha x^2} dx.$$

Note that the total derivative becomes a partial derivative in the second expression as the integrand is also a function of x. We already know the exact expression for $I(\alpha)$, which is given by $\sqrt{\frac{\pi}{\alpha}}$, such that $\frac{dI(\alpha)}{d\alpha} = -\frac{1}{2}\sqrt{\frac{\pi}{\alpha^3}}$. Then,

$$\int_{-\infty}^\infty x^2 e^{-\alpha x^2} dx = \frac{1}{2}\sqrt{\frac{\pi}{\alpha^3}}.$$

We can repeat this differentiation process to further evaluate expressions of the form $\int_{-\infty}^\infty x^{2n} e^{-\alpha x^2} dx$ in general.

$$\int_{-\infty}^\infty x^{2n} e^{-\alpha x^2} dx = \frac{(2n-1)!}{(n-1)! \cdot 2^{2n-1}} \sqrt{\frac{\pi}{\alpha^{2n+1}}}$$

for $n \geq 1$. Finally, in cases where we wish to compute $\int_0^\infty x^{2n} e^{-\alpha x^2} dx$, observe that the integrand is an even function such that $\int_0^\infty x^{2n} e^{-\alpha x^2} dx = \frac{1}{2}\int_{-\infty}^\infty x^{2n} e^{-\alpha x^2} dx$.

Next, to evaluate[4] $\int_0^\infty x^{2n+1}e^{-\alpha x^2}\,dx$, we start from

$$\int_0^\infty xe^{-\alpha x^2}\,dx = \frac{1}{2\alpha}.$$

In a similar vein, we can differentiate the above with respect to α within the integral to conclude that

$$\int_0^\infty x^3 e^{-\alpha x^2}\,dx = \frac{1}{2\alpha^2}$$

and in general,

$$\int_0^\infty x^{2n+1}e^{-\alpha x^2}\,dx = \frac{n!}{2\alpha^{n+1}}.$$

Normalization

Returning to the previous velocity distribution, we require

$$A\int_{-\infty}^\infty e^{-\frac{mv_x^2}{2kT}}\,dv_x\int_{-\infty}^\infty e^{-\frac{mv_y^2}{2kT}}\,dv_y\int_{-\infty}^\infty e^{-\frac{mv_z^2}{2kT}}\,dv_z = 1.$$

These are integrals of the form $\int_{-\infty}^\infty e^{-\alpha x^2}\,dx$ which can be evaluated to be

$$A\cdot\left(\frac{2\pi kT}{m}\right)^{\frac{3}{2}} = 1$$

$$A = \left(\frac{m}{2\pi kT}\right)^{\frac{3}{2}}.$$

Then, the velocity distribution function is

$$f(v) = \left(\frac{m}{2\pi kT}\right)^{\frac{3}{2}}e^{-\frac{m(v_x^2+v_y^2+v_z^2)}{2kT}} = \left(\frac{m}{2\pi kT}\right)^{\frac{3}{2}}e^{-\frac{mv^2}{2kT}}. \qquad (2.47)$$

It is convenient to express the above in terms of the thermal speed of gas molecules, $v_{th} = \sqrt{\frac{2kT}{m}}$, whose physical meaning is the most probable speed of the gas molecules as we shall prove later.

$$f(v) = \frac{1}{\sqrt{\pi^3 v_{th}^3}}e^{-\frac{v^2}{v_{th}^2}}. \qquad (2.48)$$

---

[4]Note that it is meaningless to determine $\int_{-\infty}^\infty x^{2n+1}e^{-\alpha x^2}\,dx$, which is just zero as the integrand is an odd function.

Distribution of a Component of Velocity

Next, we can derive the one-dimensional distribution of a particular component velocity such as v_x. That is, we are interested in the fraction of molecules with a particular x-component of velocity v_x — molecules with different components in the other directions but the same component in the x-direction still belong to the same class. We argue that the components of velocity of the particles — namely v_x, v_y and v_z — should be independent variables as the different components of velocity should be uncorrelated. Then, the fractional density of the particles attaining a velocity v between (v_x, v_y, v_z) and $(v_x + dv_x, v_y + dv_y, v_z + dv_z)$ is the product of the respective fractional densities.

$$f(v) = f(v_x, v_y, v_z) = g(v_x)g(v_y)g(v_z)$$

where $g(v_i)$ is the distribution along a particular component. Apportioning the different variables (i.e. we put all functions of v_x into $g(v_x)$, functions of v_y into $g(v_y)$ and so on) and normalizing yields

$$g(v_x) = \sqrt{\frac{m}{2\pi kT}} e^{-\frac{mv_x^2}{2kT}} = \frac{1}{\sqrt{\pi}v_{th}} e^{-\frac{v^2}{v_{th}^2}} \tag{2.49}$$

and so on for the other directions.

Speed Distribution

The speed distribution is

$$f_s(v) = 4\pi v^2 f(v) = 4\pi \left(\frac{m}{2\pi kT}\right)^{\frac{3}{2}} v^2 e^{-\frac{mv^2}{2kT}} = \frac{4v^2}{\sqrt{\pi}v_{th}^3} e^{-\frac{v^2}{v_{th}^2}}. \tag{2.50}$$

We shall now prove that v_{th} is the most probable speed (i.e. the maximum of $f_s(v)$). Consider the derivative of $f_s(v)$ with respect to v.

$$\frac{df_s}{dv} = \frac{8v}{\sqrt{\pi^3}v_{th}^3} e^{-\frac{v^2}{v_{th}^2}} - \frac{8v^3}{\sqrt{\pi^3}v_{th}^5} e^{-\frac{v^2}{v_{th}^2}}.$$

For this to be zero,

$$v = v_{th}$$

where the physically incorrect negative solution has been rejected. Finally, one can check that the value of $\frac{df_s}{dv}$ is positive for values of v slightly smaller than v_{th} and negative for values of v slightly larger than v_{th} to show that this corresponds to a maximum. Moving on, $f_s(v)$ is graphed for two values of T in Fig. 2.13.

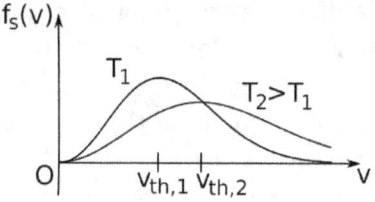

Figure 2.13:   Maxwell–Boltzmann speed distribution

$f_s(v)$ is zero at $v = 0$, has a maximum and tends to zero as v tends to infinity. For larger values of T, the distribution becomes broader but the peak value decreases as the area under the curve must still be unity. The peak also shifts towards the right for larger values of T as v_{th} increases. From the Maxwell–Boltzmann speed distribution, the mean and mean square speeds can be computed as

$$\langle v \rangle = \sqrt{\frac{8kT}{m\pi}},$$

$$\langle v^2 \rangle = \frac{3kT}{m}.$$

This is an important result (but do not overrate its significance) as it relates the temperature of an ideal gas to its mean squared speed. The mean translational kinetic energy is then related to the temperature according to

$$\frac{1}{2}m\langle v^2 \rangle = \frac{3kT}{2}. \tag{2.51}$$

The mean cube speed can also be shown to be

$$\langle v^3 \rangle = \sqrt{\frac{128k^3T^3}{m^3\pi}}.$$

Problem: Determine the speed distribution $f_e(v)$ of molecules effused from a small hole in a compartment given that the distribution of the original gas in the compartment is Maxwellian and that the compartment is maintained at a constant temperature T.

We have previously remarked that $f_e(v)$ is proportional to $vf_s(v)$ and thus $v^3 e^{-\frac{v^2}{v_{th}^2}}$. Therefore,

$$f_e(v) = Av^3 e^{-\frac{v^2}{v_{th}^2}}$$

for some constant A. Normalizing the distribution requires

$$A = \frac{1}{\int_0^\infty v^3 e^{-\frac{v^2}{v_{th}^2}} dv}.$$

Since we have calculated that $\int_0^\infty x^3 e^{-\alpha x^2} dx = \frac{1}{2\alpha^2}$,

$$A = \frac{2}{v_{th}^4},$$

$$f_e(v) = \frac{2v^3}{v_{th}^4} e^{-\frac{v^2}{v_{th}^2}} = \frac{m^2}{2k^2 T^2} v^3 e^{-\frac{v^2}{v_{th}^2}}.$$

Problems

1. Real and Ideal Gas Thermometers*

A constant volume gas thermometer is constructed from connecting a gas chamber of a fixed volume to a manometer. The difference Δh in liquid levels in the manometer reflects the pressure of the gas in the chamber and the temperature T of the gas can then be read off a pre-calibrated linear graph between Δh and T. To measure the temperature of a substance (usually a liquid), the gas chamber is immersed in the substance such that its temperature becomes the temperature of the substance (the heat capacity of the gas is negligible). Now, a certain constant volume gas thermometer contains one mole of a gas whose equation of state is

$$\left(p + \frac{a}{V^2}\right)(V - b) = RT$$

where a and b are characteristic constants of the gas. This is known as the van der Waals equation of state and is commonly used to model real gases. Another constant volume gas thermometer contains one mole of an ideal gas which obeys the ideal gas law, $pV = RT$. The thermometers are calibrated at the ice and steam points to give centigrade scales. Show that the two thermometers will give identical readings when placed in thermal contact with a substance of any temperature.

2. Connected Vessels*

Two thermally insulated vessels of volumes V_1 and V_2 initially contain monoatomic gases of initial pressures and temperatures p_1, T_1 and p_2, T_2. They are then linked by a thermally insulated tube. Determine the final pressure p and temperature T.

3. Isobaric Compression*

A certain amount of helium is cooled at constant pressure p_0. As a result, its volume decreases from V_0 to $\frac{V_0}{2}$. Find the amount of heat lost in this process.

4. Balloon*

A helium balloon is allowed to rise to a height such that the external pressure is half of the ground pressure p_1. Its initial volume and temperature are V_1 and T_1 respectively. Assume that the envelope of the balloon is a perfect insulator and that the process is quasistatic. Calculate the final volume and temperature of the gas and the amount of work done by the gas. (Singapore Physics Olympiad)

5. Cyclic Process*

The current pressure and volume of an ideal gas are p_0 and V_0. It then undergoes a cyclic process as follows. It first expands under the constraint that $p = \frac{p_0}{V_0}V$ to $(2p_0, 2V_0)$. Then, its pressure is reduced isochorically from $2p_0$ to p_0. Finally, it contracts isobarically until its volume returns to V_0. Determine the heat absorbed during this cyclic process.

6. Pushing a Piston*

A thermally insulated container of cross sectional area A is separated into two compartments, A and B, by a frictionless divider which is a perfect insulator. Certain moles of an ideal gas with an adiabatic constant γ fill the two compartments. A massless, thermally insulated piston at one end of compartment B is initially maintained at some pressure p. Initially, the system is at equilibrium such that volumes of A and B are $\frac{2}{3}Al$ and $\frac{1}{3}Al$. The pressure on the piston is then increased so gradually that the system is always at equilibrium, until the combined volume of the two compartments becomes Al'. If the temperature increments in the two compartments are ΔT_A and ΔT_B respectively, determine the number of moles of ideal gas they contain, n_A and n_B.

7. Moving a Division**

A gas-tight, thermally isolated cylinder of total volume V is divided into two compartments A and B by a piston made of a conducting material, which can be controlled by an external agent outside the cylinder. Initially, A and B are of equal volume; they contain respectively 1 and 2 moles of an ideal monoatomic gas, all at temperature T_0 (the external agent holds the piston in place). The external agent then moves the piston to a position such that A and B possess final volumes $\frac{V}{3}$ and $\frac{2V}{3}$ respectively. This is done sufficiently slowly for the temperatures of the two gas samples to remain uniform and equal throughout the process. Find an expression for the final temperature of the system while neglecting the heat capacity of the cylinder and piston.

8. *Pumping a Balloon***

A balloon with surface tension γ (be wary that this is not the adiabatic index) is placed in a vacuum chamber and connected via a small tube to a gas container with a piston. The total number of moles of gas in the balloon and piston is n. The system is allowed to equilibrate such that the pressure of the gas in the combined system is p_0. If the system is maintained at a constant temperature T and the pressure on the piston is quasistatically varied — such that the system is always at thermodynamic equilibrium — until all gas molecules in the piston are transferred to the balloon, determine the amount of work done on the gas by the piston. The final pressure of the gas is p_1. Assume that the balloon constantly maintains a spherical shape.

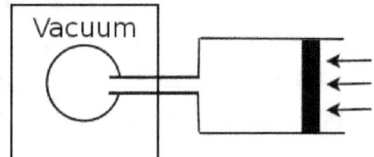

9. *Water Tap***

A container is partially filled with an ideal gas (on top) and incompressible water of density ρ. The initial pressure of the gas is $2p_a$ where p_a is the atmospheric pressure. If the small hole of area A of the bottom of the container is opened such that water begins to flow out of the container, determine the time required for the water to stop flowing if the ideal gas undergoes an isothermal process such that $nRT = k$ where k is a constant. Assume that the flow of water is energy conserving and steady and neglect any difference in pressure due to the height of the water. The velocity of water inside the container is also negligible. Assume that the temperature of the water remains constant as well.

10. *Pumping a Tyre***

A thermally insulated container with a movable, massless piston is connected to a thermally insulated tyre of constant volume V via a thermally insulated tube. During each pumping cycle, the valve in the tube is first closed. Then, the piston is expanded until the pressure and volume of the gas becomes p_a and V_a, by taking in air from the outside. The gas in the piston, which has an adiabatic index γ, is then compressed adiabatically until its volume becomes $\frac{V_a}{2}$. Finally, the valve is opened until equilibrium is reached between the container and the tyre. If the tyre does not contain any gas initially, determine the minimum number of cycles required to increase the pressure in the tyre to $2^{\gamma-1}p_a$.

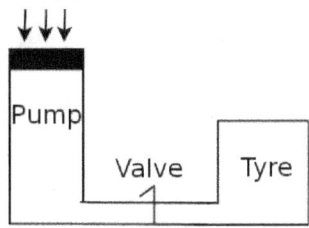

11. *Rotating Gas***

An open container, exposed to the atmosphere, contains water of density ρ_w. An "L-shaped" tube is inserted into it as shown in the figure below. The diameter of the vertical part of the tube is negligible while the horizontal part of the tube has a uniform cross sectional area and length l. Initially, the tube is motionless such that the water level is completely flat at equilibrium. Subsequently, the tube is rotated at a constant angular velocity ω about the vertical column such that the water level in the tube is a height Δh above the water level in the container at equilibrium. If the atmospheric pressure and temperature are p_a and T and if the molar mass of the gas inside the tube is μ, determine Δh. Assume that the gas in the tube undergoes an isothermal process and $l^2\omega^2 \ll \frac{RT}{\mu}$ where R is the ideal gas constant.

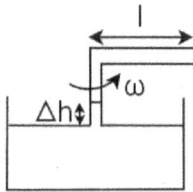

12. *Adiabatic Oscillation***

A small cork of cross sectional area A and mass m blocks the opening of a wine bottle that is filled with an ideal gas with an adiabatic constant γ. If the atmospheric pressure is p_0 and the volume of gas inside the bottle is V_0 at the equilibrium state, determine the angular frequency of small oscillations of the cork about its equilibrium position.

13. *Bouncing Ball***

A thermally insulated container with a constant cross sectional area A is separated into an upper and lower compartment by a divider of mass M. The two compartments are filled with certain moles of ideal gas which can exchange heat with one another as the divider is not thermally insulated. A small ball of a certain mass m is stuck to the bottom face of the divider. Initially, the ratio of the volumes of the upper and lower compartments is $3 : 1$ and the pressure of the gas in the upper compartment is p_1. Then, the ball of mass m falls from the divider and bounces on the bottom of the container, until it eventually comes to rest at the bottom of the lower compartment. If the final ratio of the volumes of the upper and lower compartments is $2 : 1$, determine m.

14. *Dumping Water****

An inverted container with a constant cross sectional area and mass m is floating with its base at the water level as shown in the figure below. The height of the air column is h_0. The plate holding back the water on top is then removed such that water falls down at negligible velocity — causing the instantaneous depth of the container, which is defined to be the distance between the water level and the base of the container, to become h_1. The column of air between the two water sections dissolves and has no impact on the system. Argue qualitatively that the container should sink. If the entire set-up has a constant temperature T and the gas in the container instantaneously attains thermodynamic equilibrium at every depth of the container, determine the velocity of the container at depth h (assume that $\frac{h_0}{h}$ is small). Neglect atmospheric pressure. Now, interpret your results for $h_0 \to 0$.

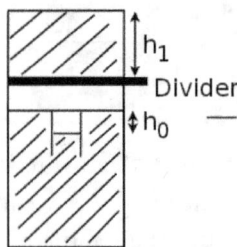

Gas Flows

15. *Combining Flows**

Two tubes carrying an identical ideal gas flowing at pressures p_1, p_2 and temperatures T_1, T_2 merge at a junction into a combined third tube. If the flow velocities at all parts of the tubes are negligible and if the volume flow rate in the first tube is k times that of the second tube, determine the temperature T_3 of the gas exiting from the third tube. The flow is and the tubes are thermally insulated.

16. *Sustaining a Fan**

A fan of cross sectional area A steadily takes in diatomic air molecules of molar mass μ, pressure p_1 and temperature T_1 and expels it at velocity v_2, pressure p_2 and temperature T_2. Determine the electric power needed to sustain the fan (assuming that it is perfectly efficient).

17. *Speed of Sound***

This problem will explore an elegant way of deriving the speed of a one-dimensional sound wave in a gaseous medium: $c = \sqrt{\frac{\gamma p}{\rho}}$ where γ, p and ρ are the adiabatic index, ambient pressure and density of the gaseous medium. Suppose that the sound wave travels adiabatically in the x-direction at velocity c and that the currently oscillating point along the medium travels at a small velocity $-v$ ($v \ll c$) in the lab frame. The density of the currently oscillating section only differs from the ambient pressure by a small amount $\Delta \rho \ll \rho$. Think of a way to apply the equations describing steady flow (mass and energy continuity). Through these two equations and the adiabatic condition, you will obtain two equations that are linear combinations of two variables (one of which is v) that are equated to zero. By exploiting the fact that the determinant must be zero for the two variables to have non-trivial solutions, determine c.

Kinetic Theory of Gases

18. *Pressure**

Prove Eq. (2.39) by considering molecules traveling at a particular z-component of velocity v_z. You will have to relate $\langle v_z^2 \rangle$ to $\langle v^2 \rangle$. (Note that we did not use this simple proof in order to expedite the derivation of Eq. (2.40).)

19. *Equipartition Theorem**

Suppose that the energy of a system in a particular state, quantified by the variable x which can range from $-\infty$ to ∞, is $E = \alpha x^2$ where α is a constant. If the probability of the system adopting a certain state follows the Boltzmann distribution, show that the average energy of the system is $\frac{1}{2}kT$, where k is the Boltzmann constant and T is the temperature of the system. If the energy of the system in a particular state is now $E = \sum_{i=1}^{N} \alpha_i x_i^2$, where the $x_i's$ are independent variables that collectively define a state and each ranges from $-\infty$ to ∞, show that the average energy is given by $\frac{N}{2}kT$.

20. Equilibrating Effusion*

A container is separated into two compartments of volumes V_1 and V_2 by a massive divider. The first compartment initially contains n_0 moles of an ideal gas while the other compartment is empty. If a hole, with a diameter smaller than the mean free path of molecules, is made on the divider and the two compartments are maintained at temperatures T_1 and T_2, determine the pressure in each compartment when the system has equilibrated.

21. Isothermal Leaking**

A hole of area A, whose diameter is smaller than the mean free path of gas molecules, is punctured on the surface of a container of volume V that rests in a vacuum. If the initial number density of ideal gas molecules inside the container is η_0 and the gas is constantly in a state of equilibrium, determine the number density $\eta(t)$ if the gas is maintained at a constant temperature T and if each molecule has mass m. Then, determine the external power supplied to the gas inside the cylinder. Neglect all form of energy loss, other than that due to the escaping molecules.

22. Thermal Conductivity**

This problem concerns estimating the thermal conductivity of an ideal gas via the kinetic theory of gases. By Fourier's law of conduction, the heat flux density, or the power delivered per unit perpendicular area, across an area is proportional to the temperature gradient.

$$\frac{dq}{dt} = -k\frac{dT}{dz}$$

where the z-direction has been set as the direction along which temperature varies. q is the heat flow per unit area — implying that $\frac{dq}{dt}$ is the power per unit area. The negative sign in the above equation implies that heat flows from regions of higher temperature to regions of lower temperature. Finally, k is the thermal conductivity which we aim to determine in this problem.

Now, consider the following set-up. Two large plates parallel to the xy-plane are located at certain z-coordinates. They are maintained at different temperatures such that a steady, position-dependent temperature $T(z)$, that is strictly decreasing with increasing z, is set up in the region between them. An ideal gas with f degrees of freedom fills this region.

(a) Argue qualitatively why there will be power delivered across a plane of a constant z-coordinate based on the varying temperature $T(z)$.

(b) It is known that the gas molecules have a mean free path λ. Now, consider a class of gas molecules with a certain velocity that makes an angle θ with the z-direction. If the gas molecules cut across a plane of z-coordinate z at a particular instance, what is the average kinetic energy carried by them?

(c) Using the previous result, determine the heat flux density and thermal conductivity k across a plane of z-coordinate z, in terms of the degrees of freedom of the gas molecules f, the number density η (assumed to be uniform throughout), λ and the average speed $\langle v \rangle$ of the gas molecules at a plane of z-coordinate z. Assume that λ is small such that second order and above terms in λ are negligible.

23. Adiabatic Condition***

Through the kinetic theory of gases, show that a process involving a monoatomic ideal gas in a thermally insulated container with a slowly moving and thermally insulated piston conserves the quantity $TV^{\frac{2}{3}}$ where T and V are the instantaneous temperature and volume respectively. The speed of the piston is very small as compared to the speed of the gas molecules. Assume that the collisions between the gas molecules and the piston are perfectly elastic.

24. Leaking Container***

A hole of area A, whose diameter is smaller than the mean free path of gas molecules, is made on a thermally insulated container of volume V, that is placed in a large vacuum. If the initial number density of gas molecules inside the container is η_0 and the initial temperature is T_0, show that the number density $\eta(t)$ obeys

$$\eta(t) = \frac{1}{\left(\eta_0^{-\frac{1}{6}} + A\sqrt{\dfrac{kT_0}{72\pi m V^2 \eta_0^{\frac{1}{3}}}}\right)^6}$$

where m is the mass of one molecule. Assume that the gas inside the container constantly attains a homogeneous equilibrium state. Hint: consider the rate of change of number density and the internal energy of the gas inside the container.

Solutions

1. Real and Ideal Gas Thermometers*

Since V is constant, observe that both equations of state imply a linear relationship between p and T. For the van der Waals gas,

$$p = \frac{RT}{V - b} - \frac{a}{V^2}$$

while for the ideal gas,

$$p = \frac{RT}{V}.$$

Since the height difference Δh between the two liquid levels in a manometer is proportional to the difference between the pressure of the gas and the (constant) atmospheric pressure, the above implies that Δh obeys a linear relationship with T for both gases.

$$\Delta h = m_1 T + c_1,$$
$$\Delta h = m_2 T + c_2.$$

Since the calibration itself is used to fit a linear relationship between Δh and T and because we know that the actual relationship between Δh and T is indeed linear for both gases, both thermometers will correctly reflect the real temperature of the substance measured. The readings are then naturally the same.

2. Connected Vessels*

Since the vessels are thermally insulated, the total internal energy must be conserved. $U = \frac{3}{2}nRT = \frac{3}{2}pV$ for an ideal gas. Therefore,

$$\frac{3}{2}p_1 V_1 + \frac{3}{2}p_2 V_2 = \frac{3}{2}p(V_1 + V_2)$$

$$p = \frac{p_1 V_1 + p_2 V_2}{V_1 + V_2}.$$

Next, the total number of moles is

$$n = \frac{p_1 V_1}{RT_1} + \frac{p_2 V_2}{RT_2}.$$

The final temperature is then

$$T = \frac{p(V_1 + V_2)}{nR} = \frac{(p_1 V_1 + p_2 V_2)T_1 T_2}{p_1 V_1 T_2 + p_2 V_2 T_1}.$$

3. Isobaric Compression*

Let the initial and final temperatures of the gas be T_0 and T_1 respectively. By the ideal gas law,

$$\frac{T_1}{T_0} = \frac{\frac{V_0}{2}}{V_0} = \frac{1}{2}$$

$$\implies T_1 = \frac{T_0}{2}.$$

The heat transferred to the gas in the process is then

$$Q = nc_p\Delta T = \frac{5}{2}nR\left(\frac{T_0}{2} - T_0\right) = -\frac{5}{4}nR\frac{p_0V_0}{nR} = -\frac{5}{4}p_0V_0$$

where the negative sign indicates heat loss by the gas.

4. Balloon*

Let the final volume and temperature be V_2 and T_2 respectively. By the adiabatic condition,

$$p_1V_1^\gamma = \frac{p_1}{2}V_2^\gamma$$

$$\frac{V_2}{V_1} = \left(\frac{p_1}{\frac{p_1}{2}}\right)^{\frac{3}{5}} = 2^{\frac{3}{5}}$$

$$V_2 = 2^{\frac{3}{5}}V_1$$

since $\gamma = \frac{5}{3}$ for a monoatomic gas (helium). By the ideal gas law,

$$\frac{T_2}{T_1} = \frac{\frac{p_1}{2}V_2}{p_1V_1} = \frac{1}{2}\cdot 2^{\frac{3}{5}} = 2^{-\frac{2}{5}}$$

$$T_2 = 2^{-\frac{2}{5}}T_1.$$

By the first law of thermodynamics, during an adiabatic process,

$$W_{by} = -\Delta U$$

$$= \frac{3}{2}(p_1V_1 - \frac{p_1}{2}V_2)$$

$$= \frac{3(1 - 2^{-\frac{2}{5}})}{2}p_1V_1.$$

5. Cyclic Process*

The PV curve of the process is a right-angled triangle with side lengths p_0 and V_0. Therefore, the work done by the gas (the reader should check for the sign) is

$$W_{by} = \frac{p_0 V_0}{2}.$$

The internal energy of the gas remains the same after the process as the initial and final states are the same. Then, by the first law of thermodynamics, the heat absorbed by the gas is

$$Q = W_{by} = \frac{p_0 V_0}{2}.$$

6. Pushing a Piston*

At each intermediate stage of the process, the system is in an equilibrium state such that the pressures in the two compartments are equal. Furthermore, since the walls are insulated, the gases in the two compartments undergo adiabatic processes. Let $U_A = \frac{2}{3} Al$, $U_B = \frac{1}{3} Al$, V_A and V_B be the respective initial and final volumes of the gases in the compartments. If the final common pressure is p',

$$p' V_A^\gamma = p U_A^\gamma,$$

$$p' V_B^\gamma = p U_B^\gamma.$$

Dividing the first equation by the second, it can be seen that the ratio of the volumes of the compartments remains the same. That is, $V_A = \frac{2}{3} Al'$ and $V_B = \frac{1}{3} Al'$. By substituting one of these expressions into the corresponding equation above,

$$p' = \frac{p l^\gamma}{l'^\gamma}.$$

Applying the ideal gas law to the gas in compartment A,

$$n_A R \Delta T_A = p' V_A - p U_A = \frac{2}{3} pAl \left(\frac{l^{\gamma-1}}{l'^{\gamma-1}} - 1 \right)$$

$$n_A = \frac{2pAl}{3R \Delta T_A} \left(\frac{l^{\gamma-1}}{l'^{\gamma-1}} - 1 \right).$$

Similarly,

$$n_B = \frac{pAl}{3R \Delta T_B} \left(\frac{l^{\gamma-1}}{l'^{\gamma-1}} - 1 \right).$$

7. Moving a Division**

An important point to note in this problem is that the pressures of the two gases need not be equal at any instance in time (even when thermal equilibrium has been attained) as the forces on the piston can always be balanced by the external agent. Let V_A be the instantaneous volume of compartment A. If the external agent moves the piston such that V_A is changed to $V_A + dV_A$ at this instance, the work done by the external agent on the system comprising the two gases is $(P_B - P_A)dV_A$, where P_A and P_B are the respective pressures of the gases in compartments A and B. In writing this, we have noted that the change in volume of the gas in B must be $-dV_A$. By the work-energy theorem, the work done by the external agent must be equal to the increase in internal energy of the two gases, $3c_v dT = \frac{9}{2}RdT$, where T is the instantaneous common temperature of the gases.

$$(P_B - P_A)dV_A = \frac{9}{2}RdT.$$

Substituting $P_A = \frac{RT}{V_A}$ and $P_B = \frac{2RT}{V-V_A}$,

$$RT\left(\frac{2}{V - V_A} - \frac{1}{V_A}\right)dV_A = \frac{9}{2}RdT$$

$$\int_{\frac{V}{2}}^{\frac{V}{3}}\left(\frac{2}{V - V_A} - \frac{1}{V_A}\right)dV_A = \frac{9}{2}\int_{T_0}^{T}\frac{dT}{T}$$

$$2\ln\frac{\frac{V}{2}}{\frac{2V}{3}} + \ln\frac{\frac{V}{2}}{\frac{V}{3}} = \frac{9}{2}\ln\frac{T}{T_0}$$

$$\ln\frac{27}{32} = \frac{9}{2}\ln\frac{T}{T_0}$$

$$T = T_0\left(\frac{27}{32}\right)^{\frac{2}{9}} = 0.963T_0 \quad (3\text{sf}).$$

8. Pumping a Balloon**

The total work done by a gas in an isothermal process is given by Eq. (2.13). Therefore, the total work done on the gas is

$$W_{on} = -nRT\ln\frac{V_f}{V_i} = nRT\ln\frac{p_1}{p_0}$$

since pV is constant for an isothermal process. This is not work done by the piston on the gas, $W_{piston, on}$, as the balloon also performs work on the gas,

$W_{balloon,\,on}$.

$$W_{piston,\,on} + W_{balloon,\,on} = W_{on}$$

$$W_{piston,\,on} = W_{on} + W_{balloon,\,by}$$

where $W_{balloon,\,by}$ is the work done on the balloon by the gas. This is equal to the negative change in surface energy of the balloon. Recall that the surface energy of a spherical balloon is $4\pi\gamma r^2$ where r is the radius of the balloon. Let the initial and final radii of the balloon be r_0 and r_1. We know that

$$p_0 = \frac{2\gamma}{r_0},$$

$$p_1 = \frac{2\gamma}{r_1},$$

due to the pressure discontinuity caused by surface tension across the surface of a spherical balloon at equilibrium. Solving the above for r_0 and r_1 in terms of the respective pressures, the change in surface energy is

$$W_{balloon,\,by} = 4\pi\gamma r_1^2 - 4\pi\gamma r_2^2 = \frac{16\pi\gamma^3}{p_1^2} - \frac{16\pi\gamma^3}{p_0^2}.$$

Therefore,

$$W_{piston,\,on} = nRT\ln\frac{p_1}{p_0} + \frac{16\pi\gamma^3}{p_1^2} - \frac{16\pi\gamma^3}{p_0^2}.$$

Another method in evaluating the work done by the piston on the gas would be to evaluate $-\int p\,dV$ directly, with V being the volume of gas in the gas piston. Let the instantaneous pressure of the gas and the radius of the balloon be p and r. Since the gas is at equilibrium at every instance,

$$p = \frac{2\gamma}{r}.$$

Furthermore, by the ideal gas law,

$$p = \frac{nRT}{\frac{4}{3}\pi r^3 + V}.$$

Substituting the expression for r in terms of p, obtained from the first equation, into the second equation,

$$p = \frac{nRT}{\frac{32\pi\gamma^3}{3p^3} + V}.$$

Simplifying,

$$V = \frac{nRT}{p} - \frac{32\pi\gamma^3}{3p^3}$$

$$dV = \left(-\frac{nRT}{p^2} + \frac{32\pi\gamma^3}{p^4}\right)dp.$$

The work done on the gas by the piston is then

$$W_{piston,\,on} = -\int p\,dV = \int_{p_0}^{p_1} \left(\frac{nRT}{p} - \frac{32\pi\gamma^3}{p^3}\right)dp$$

$$= nRT\ln\frac{p_1}{p_0} + \frac{16\pi\gamma^3}{p_1^2} - \frac{16\pi\gamma^3}{p_0^2}.$$

9. Water Tap**

Let the instantaneous pressure and volume of the gas be p and V. Then, p and V are related by

$$pV = k.$$

Next, let v be the velocity of water gushing out of the hole. Applying Bernoulli's principle[5] to the water level and the hole,

$$p = p_a + \frac{1}{2}\rho v^2$$

$$v = \sqrt{\frac{2(p - p_a)}{\rho}}.$$

The volume flow rate of water is Av. This is also the rate of increase of the volume of the gas, $\frac{dV}{dt}$.

$$\frac{dV}{dt} = A\sqrt{\frac{2(p - p_a)}{\rho}}.$$

Substituting $p = \frac{k}{V}$,

$$\frac{dV}{dt} = A\sqrt{\frac{2\left(\frac{k}{V} - p_a\right)}{\rho}}.$$

---

[5]The reader may wonder if Bernoulli's principle is valid in this context, especially after perusing the section on gas flows. In our derivation of Bernoulli's principle, the possible change in the internal energy of a fluid was excluded. In the current situation, this does not matter as the temperature of the water is uniform and because water is presumed to be incompressible.

The initial and final volumes are $\frac{k}{2p_a}$ and $\frac{k}{p_a}$. Then,

$$\int_{\frac{k}{2p_a}}^{\frac{k}{p_a}} \frac{V}{\sqrt{kV - p_a V^2}} dV = \int_0^t A\sqrt{\frac{2}{\rho}} dt.$$

To evaluate the left-hand side, use the substitutions $V = \frac{k}{2p_a}\sin\theta + \frac{k}{2p_a}$ and $dV = \frac{k}{2p_a}\cos\theta d\theta$. Then,

$$\int_{\frac{k}{2p_a}}^{\frac{k}{p_a}} \frac{V}{\sqrt{kV - p_a V^2}} dV = \int_{\frac{k}{2p_a}}^{\frac{k}{p_a}} \frac{V}{\sqrt{p_a} \cdot \sqrt{\frac{k^2}{4p_a^2} - \left(V - \frac{k}{2p_a}\right)^2}} dV$$

$$= \int_0^{\frac{\pi}{2}} \frac{k}{2\sqrt{p_a^3}}\sin\theta d\theta + \int_0^{\frac{\pi}{2}} \frac{k}{2\sqrt{p_a^3}} d\theta$$

$$= \frac{k}{2\sqrt{p_a^3}} + \frac{k\pi}{4\sqrt{p_a^3}}$$

$$= \frac{k(2+\pi)}{4\sqrt{p_a^3}}.$$

Then, the time required is

$$A\sqrt{\frac{2}{\rho}}t = \frac{k(2+\pi)}{4\sqrt{p_a^3}}$$

$$t = \frac{k(2+\pi)\sqrt{\rho}}{A\sqrt{32p_a^3}}.$$

10. Pumping a Tyre**

Let the final pressure of the gas after the adiabatic compression be p_a'. Then by the adiabatic condition,

$$p_a V_a^\gamma = p_a'\left(\frac{V_a}{2}\right)^\gamma$$

$$p_a' = 2^\gamma p_a.$$

We have shown in a previous problem that when two thermally insulated vessels of initial pressures and volumes p_1, p_2, V_1 and V_2 are connected, the final pressure is

$$p = \frac{p_1 V_1 + p_2 V_2}{V_1 + V_2}.$$

Let us apply this result to the current problem. Let the pressure inside the tyre after the ith cycle be p_i. Then, the equilibrium pressure after the $(i+1)$th cycle is that obtained by connecting two thermally insulated vessels of initial pressures and volumes $2^\gamma p_a$, p_i, $\frac{V_a}{2}$ and V. Then,

$$p_{i+1} = \frac{2^\gamma p_a V_a + 2 p_i V}{V_a + 2V}.$$

The above can be simplified into

$$p_{i+1} - 2^\gamma p_a = \frac{2V}{V_a + 2V} (p_i - 2^\gamma p_a).$$

It can be seen that the above is a geometric progression with a constant ratio $\frac{2V}{V_a + 2V}$. Using the base case $p_0 = 0$,

$$p_n - 2^\gamma p_a = - \left(\frac{2V}{V_a + 2V} \right)^n 2^\gamma p_a.$$

When $p_n = 2^{\gamma-1} p_a$,

$$\left(\frac{2V}{V_a + 2V} \right)^n = \frac{1}{2}$$

$$n = - \frac{1}{\log_2 \frac{2V}{V_a + 2V}}.$$

The minimum number of cycles is the ceiling of the above value.

11. Rotating Gas**

Firstly, understand that when the tube is rotated, the pressure $p(r)$ in the tube must vary as a function of radial distance r from the axis of rotation to provide the centripetal force required by each gas section to remain at rest relative to the tube. As a consequence of the ideal gas law, the density $\rho(r)$ of the gas must also vary with radial distance. Consider an infinitesimal section of gas between radial distance r and $r + dr$. It has a mass density $\rho(r)$ and we define its cross sectional area to be A. Therefore, its mass is $dm = \rho A dr$. The external forces on this element are pA radially outwards and $(p + dp)A$

radially inwards. The net force must provide the required centripetal force.

$$dpA = dmr\omega_r^2 = \rho r\omega^2 A dr.$$

Then,

$$\frac{dp}{dr} = \rho r\omega^2.$$

Furthermore, we know from the ideal gas law that

$$p = \frac{\rho}{\mu}RT$$

$$\implies \rho = \frac{p\mu}{RT},$$

$$\frac{dp}{dr} = \frac{p\mu r\omega^2}{RT}$$

$$\int_{p_0}^{p} \frac{1}{p}dp = \int_0^r \frac{\mu r\omega^2}{RT}dr$$

$$\ln\frac{p}{p_0} = \frac{\mu r^2\omega^2}{2RT}$$

$$p(r) = p_0 e^{\frac{\mu r^2\omega^2}{2RT}},$$

where p_0 is the pressure at radial distance $r = 0$ (i.e. along the axis of rotation). Now, our objective is to determine p_0 as its difference with the atmospheric pressure enables us to compute Δh via the pressure difference caused by a static column of fluid. To this end, we can exploit the fact that the total mass of gas in the tube must be the same as before. That is,

$$\int_0^l \rho(r)dr = \int_0^l \rho_0 dr$$

where ρ_0 is the uniform density of gas before the tube was rotated.

$$\rho_0 = \frac{\mu p_a}{RT}.$$

Substituting $\rho(r) = \frac{\mu p(r)}{RT}$, the above requires

$$p_a l = \int_0^l p(r)dr$$

$$= \int_0^l p_0 e^{\frac{\mu r^2\omega^2}{2RT}}dr$$

$$\approx \int_0^l p_0 \left(1 + \frac{\mu r^2 \omega^2}{2RT}\right) dr$$

$$= p_0 \left(l + \frac{\mu l^3 \omega^2}{6RT}\right)$$

$$\implies p_0 = \frac{p_a}{1 + \frac{\mu l^2 \omega^2}{6RT}} \approx p_a \left(1 - \frac{\mu l^2 \omega^2}{6RT}\right).$$

This difference in pressure causes the water to rise up the tube until

$$p_a - p_0 = \rho_w g \Delta h$$

$$\Delta h = \frac{\mu l^2 \omega^2 p_a}{6 \rho_w g RT}.$$

12. Adiabatic Oscillation**

At the equilibrium position, the pressure of the gas inside the wine bottle is $p_0 + \frac{mg}{A}$ so that the net force due to pressure balances the weight of the cork. Now, consider a small displacement x upwards, such that the new volume of the gas is

$$V = V_0 + Ax = V_0 \left(1 + \frac{Ax}{V_0}\right).$$

Let the pressure of the gas at this point be p. Applying Newton's second law to the cork,

$$m\ddot{x} = (p - p_0)A - mg.$$

If the oscillations are small and thus slow (by the conservation of energy), the process that the gas in the bottle undergoes is adiabatic. In an adiabatic process, the quantity pV^γ is a constant. Therefore,

$$pV^\gamma = \left(p_0 + \frac{mg}{A}\right) V_0^\gamma$$

$$m\ddot{x} = \frac{(p_0 A + mg)V_0^\gamma}{V^\gamma} - p_0 A - mg = \frac{p_0 A + mg}{\left(1 + \frac{Ax}{V_0}\right)^\gamma} - p_0 A - mg.$$

Performing a binomial expansion on the denominator and discarding second order terms in $\frac{Ax}{V_0}$,

$$m\ddot{x} = (p_0 A + mg)\left(1 - \frac{\gamma Ax}{V_0}\right) - p_0 A - mg = -\frac{\gamma (p_0 A + mg)A}{V_0}x.$$

The angular frequency of oscillations is thus

$$\omega = \sqrt{\frac{\gamma(p_0 A + mg)A}{mV_0}}.$$

13. Bouncing Ball**

Let the initial pressures in the upper and lower compartment be p_1 and p_2. Let the final pressures be p_1' and p_2'. In order for the system to be in mechanical equilibrium, the pressure differences must balance the pressure due to the weight of the piston (and the weight of the ball in the first case).

$$p_2 = p_1 + \frac{(m+M)g}{A},$$

$$p_2' = p_1' + \frac{Mg}{A}.$$

Next, we know that gases must have common initial and final temperatures. Then, the ratio of moles in the two compartments are given by the ideal gas law as

$$\frac{n_1}{n_2} = \frac{p_1 V_1}{p_2 V_2} = \frac{p_1' V_1'}{p_2' V_2'}.$$

Substituting $\frac{V_1}{V_2} = 3$, $\frac{V_1'}{V_2'} = 2$ and the expressions for p_2 and p_2' in terms of p_1 and p_1',

$$p_1' = \frac{3p_1 Mg}{2(m+M)g - p_1 A}.$$

Next, we can apply the conservation of energy to this system. The decrease in gravitational potential energy of the ball must be equal to the increase in the internal energies of the gases and the gravitational potential energy of the divider. Equivalently, the falling ball supplies heat to the system during the collisions. If we let the total volume of the container be V_0,

$$\frac{mgV_0}{4A} = \frac{3}{2}(p_1' V_1' - p_1 V_1 + p_2' V_2' - p_2 V_2) + \frac{MgV_0}{12A}.$$

Simplifying,

$$p_1' = p_1 + \frac{5mg}{12A} - \frac{5Mg}{36A}.$$

Equating the two expressions for p_1' would yield a quadratic equation in m.

$$30m^2 g^2 + (20Mg + 57p_1 A)mg - 36p_1^2 A^2 - 31p_1 AMg - 10M^2 g^2 = 0.$$

Solving for m,

$$m = -\frac{M}{3} - \frac{19p_1 A}{20g} + \sqrt{\frac{4}{9}M^2 + \frac{5p_1 AM}{3g} + \frac{841p_1^2 A^2}{400g^2}}$$

where we have rejected the other solution which is negative.

14. Dumping Water***

Before the water falls onto the container, the upthrust is just enough to balance the weight of the container. However, when water is dumped onto the container, the pressure of the gas inside the container should increase — causing it to contract under isothermal conditions. This results in a shrinking volume of gas and thus a smaller value of upthrust — causing the container to sink further. This propagates a vicious cycle as the more the container sinks, the higher the pressure of the gas and the smaller the upthrust — thus causing it to sink even further.

Let us now try to solve for the velocity of the container $\dot{h}$ as a function of its depth h. Let the density of water be ρ and the cross sectional area of the container be A. Initially, the upthrust must balance the weight of the container.

$$\rho A g h_0 = mg.$$

Under isothermal conditions, the quantity pV is conserved. The initial pressure of the gas is $\rho g h_0 = \frac{mg}{A}$ and the initial volume is $A h_0$. Therefore, the conserved quantity is

$$pV = mgh_0.$$

Now, we aim to calculate the height of the air column x at thermodynamic equilibrium when the depth of the container is h. The pressure and volume of the gas at this juncture are then $\rho g(h + x)$ and Ax. Then,

$$\rho g(h + x)Ax = mgh_0.$$

Since $\rho g = \frac{mg}{A h_0}$,

$$\frac{mg}{h_0}(h + x)x = mgh_0$$

$$x^2 + hx - h_0^2 = 0$$

$$x = \frac{-h + \sqrt{h^2 + 4h_0^2}}{2} = \frac{-h + h\sqrt{1 + 4\frac{h_0^2}{h^2}}}{2}$$

where we have rejected the negative solution. Performing a binomial expansion and neglecting higher order terms in $\frac{h_0^2}{h^2}$,

$$x = \frac{-h + h\left(1 + \frac{2h_0^2}{h^2}\right)}{2} = \frac{h_0^2}{h}.$$

Now, apply Newton's second law to the container — the external forces on it are its weight and the upthrust.

$$m\ddot{h} = mg - \rho A x g = mg - \frac{mg}{h_0}x$$

$$\ddot{h} = g - \frac{gh_0}{h}.$$

Expressing $\ddot{h}$ as $\dot{h}\frac{d\dot{h}}{dh}$ and separating variables,

$$\int_0^{\dot{h}} \dot{h}\,d\dot{h} = \int_{h_1}^h \left(g - \frac{gh_0}{h}\right) dh$$

$$\frac{\dot{h}^2}{2} = g(h - h_1) - gh_0 \ln \frac{h}{h_1},$$

where we have removed the absolute value brackets for the ln term as $h > h_1$. Then,

$$\dot{h} = \sqrt{2g(h - h_1) - 2gh_0 \ln \frac{h}{h_1}}.$$

When $h_0 \to 0$,

$$\dot{h} = \sqrt{2g(h - h_1)}$$

which is just the velocity of a free-falling particle (as there is no upthrust when $h_0 = 0$). Technically, this limit is slightly incorrect as the container should not have been able to stay afloat before the water was dropped.

15. Combining Flows*

Suppose that in time dt, dn_1 and dn_2 moles of gas molecules enter the junction from the first and second tubes respectively. By mass continuity, the number of moles leaving the third tube in this time interval must be $dn_1 + dn_2$. One can now enforce the continuity of energy flow across the junction, similar to the section on gas flows, to show that

$$dn_1 c_p T_1 + dn_2 c_p T_2 - (dn_1 + dn_2) c_p T_3 = 0.$$

Since $pdV = dnRT$, $\frac{dn_1}{dt} = \frac{p_1}{RT_1}\frac{dV_1}{dt}$ and $\frac{dn_2}{dt} = \frac{p_2}{RT_2}\frac{dV_2}{dt}$ where $\frac{dV_1}{dt}$ and $\frac{dV_2}{dt}$ are the volume flow rates in the respective tubes. Dividing the previous equation by dt and substituting these,

$$p_1 \frac{dV_1}{dt} + p_2 \frac{dV_2}{dt} - \left(\frac{p_1}{T_1}\frac{dV_1}{dt} + \frac{p_2}{T_2}\frac{dV_2}{dt} \right) T_3 = 0.$$

Since $\frac{\frac{dV_1}{dt}}{\frac{dV_2}{dt}} = k$,

$$T_3 = \frac{(p_1 + kp_2)T_1 T_2}{p_1 T_2 + kp_2 T_1}.$$

16. Sustaining a Fan*

By Eq. (2.27), mass continuity requires

$$\frac{p_1 v_1}{T_1} = \frac{p_2 v_2}{T_2}$$

where v_1 is the flow velocity entering the fan as the cross sectional area A is common for both sides of the flow.

$$v_1 = \frac{p_2 T_1}{p_1 T_2} v_2.$$

Note that the molar flow rate is

$$\dot{n} = \frac{p_2 A v_2}{RT_2}.$$

Applying Eq. (2.28) with $\dot{Q} = 0$, the rate, work done by the fan on the gas flowing through it is

$$\dot{W}_{on} = \dot{n}\left[c_p(T_2 - T_1) + \frac{1}{2}\mu(v_2^2 - v_1^2) \right].$$

Substituting $c_p = \frac{7}{2}R$ for a diatomic gas and $\dot{n} = \frac{p_2 A v_2}{RT_2}$,

$$\dot{W}_{on} = \frac{p_2 A v_2}{2RT_2}\left[7R(T_2 - T_1) + \mu\left(1 - \left(\frac{p_2 T_1}{p_1 T_2} \right)^2 \right) v_2^2 \right]$$

which is also the power required to sustain the fan.

17. Speed of Sound**

Since the sound wave propagates at velocity c in the x-direction, the pressure and density should be constant with time in a frame that travels at c in the x-direction relative to the lab frame — implying that the flow is steady. In this new frame, the speed of the oscillating section is $c + v$ while the speed of the sections that are not oscillating is c. Enforcing mass continuity,

$$(\rho + \Delta\rho)(c + v) = \rho c.$$

Furthermore, by Eq. (2.29) and neglecting the gravitational potential energy terms,

$$\frac{1}{2}\mu(c + v)^2 + c_p(T + \Delta T) = \frac{1}{2}\mu c^2 + c_p T$$

where μ and c_p are the molar mass and isobaric molar heat capacity of the medium. T is the ambient temperature and $T + \Delta T$ is the temperature of the oscillating section. Discarding terms that are second order in v or $\Delta\rho$ in the above equations,

$$\rho v + \Delta\rho c = 0,$$

$$\mu c v + c_p \Delta T = 0.$$

ΔT can be related to $\Delta\rho$ through the adiabatic condition. Since $p^{1-\gamma}T^\gamma =$ constant and $\rho \propto \frac{p}{T}$ by the ideal gas law,

$$\rho^{1-\gamma}T = c$$

for some constant c. Taking the total derivative of the above,

$$(1 - \gamma)\rho^{-\gamma}T d\rho + \rho^{1-\gamma}dT = 0$$

$$dT = \frac{(\gamma - 1)T}{\rho}d\rho.$$

Since $\Delta\rho$ and ΔT are small,

$$\Delta T \approx \frac{(\gamma - 1)T}{\rho}\Delta\rho.$$

Substituting this expression for ΔT and summarizing our equations,

$$\mu c v + \frac{c_p(\gamma - 1)T}{\rho}\Delta\rho = 0$$

$$\rho v + c\Delta\rho = 0.$$

The above set of equations can be written in matrix form as

$$\begin{pmatrix} \mu c & \frac{c_p(\gamma-1)T}{\rho} \\ \rho & c \end{pmatrix} \begin{pmatrix} v \\ \Delta\rho \end{pmatrix} = \begin{pmatrix} 0 \\ 0 \end{pmatrix}.$$

For non-trivial solutions to exist for v and $\Delta\rho$, the determinant of the first matrix must be zero.

$$\mu c^2 - c_p(\gamma-1)T = 0$$

$$c = \sqrt{\frac{c_p(\gamma-1)T}{\mu}}.$$

Notice that $c_p(\gamma-1) = c_p \cdot \frac{c_p - c_v}{c_v} = \frac{c_p}{c_v} \cdot R = \gamma R$ and $\frac{\mu}{RT} = \frac{\rho}{p}$ by the ideal gas law such that the above becomes

$$c = \sqrt{\frac{\gamma p}{\rho}}.$$

18. Pressure*

Consider an infinitesimal area dA and define the z-axis to be parallel to its area vector. Consider a class of molecules that are travelling at z-component of velocity v_z. In time dt, the volume of such molecules colliding with the area is

$$v_z dA dt.$$

The number of such molecules colliding the infinitesimal area, per unit time and area is then

$$\eta v_z g(v_z) dv_z$$

where η is the number density of molecules and $g(v_z)dv_z$ is the fraction of molecules with z-components of velocity between v_z and $v_z + dv_z$. The elastic collision of one of such molecules with the wall results in $2mv_z$ amount of momentum transferred to the wall. The pressure on the wall due to this class of molecules is then the rate of such molecules colliding with the wall, per unit area, multiplied by the momentum transferred per molecule. The total pressure is then obtained by integrating the above over all classes of molecules (i.e. v_z from 0 to ∞).

$$p = \int_0^\infty 2\eta m v_z^2 g(v_z) dv_z.$$

Note that the integral $\int_{-\infty}^\infty v_z^2 g(v_z) dv_z = \langle v_z^2 \rangle$ — implying that $\int_0^\infty v_z^2 g(v_z) dv_z = \frac{1}{2}\langle v_z^2 \rangle$. Then,

$$p = \eta m \langle v_z^2 \rangle.$$

Now, we need to relate $\langle v_z^2 \rangle$ to $\langle v^2 \rangle$.

$$\langle v^2 \rangle = \langle v_x^2 + v_y^2 + v_z^2 \rangle.$$

Since the different components of velocities are independent,

$$\langle v^2 \rangle = \langle v_x^2 \rangle + \langle v_y^2 \rangle + \langle v_z^2 \rangle.$$

Moreover, the three directions are symmetrical such that $\langle v_x^2 \rangle = \langle v_y^2 \rangle = \langle v_z^2 \rangle$.

$$\langle v_z^2 \rangle = \frac{1}{3}\langle v^2 \rangle,$$

$$p = \frac{1}{3}\eta m \langle v^2 \rangle.$$

19. Equipartition Theorem*

By the Boltzmann distribution, the probability of attaining a state x which has energy αx^2 obeys the relationship

$$p(x) \propto e^{-\frac{\alpha x^2}{kT}}.$$

Therefore, the average energy is

$$\langle E \rangle = \frac{\int_{-\infty}^\infty \alpha x^2 e^{-\frac{\alpha x^2}{kT}} dx}{\int_{-\infty}^\infty e^{-\frac{\alpha x^2}{kT}} dx} = \frac{\frac{1}{2}\sqrt{\frac{\pi k^3 T^3}{\alpha^3}} \cdot \alpha}{\sqrt{\frac{\pi kT}{\alpha}}} = \frac{1}{2}kT$$

where integrals of the form $\int_{-\infty}^\infty e^{-\alpha x^2} dx$ and $\int_{-\infty}^\infty x^2 e^{-\alpha x^2} dx$ have been computed previously. In the second scenario, the probability of attaining a state

$(x_1, x_2, ..., x_N)$ which has energy $E = \sum_{i=1}^{N} \alpha_i x_i^2$ is

$$p(x_1, x_2, \ldots, x_N) \propto e^{-\frac{\sum_{i=1}^{N} \alpha_i x_i^2}{kT}}.$$

The average energy is therefore

$$\langle E \rangle = \frac{\int_{-\infty}^{\infty} \int_{-\infty}^{\infty} \cdots \int_{-\infty}^{\infty} \left(\sum_{i=1}^{N} \alpha_i x_i^2 \right) e^{-\frac{\sum_{i=1}^{N} \alpha_i x_i^2}{kT}} dx_1 dx_2 ... dx_N}{\int_{-\infty}^{\infty} \int_{-\infty}^{\infty} \cdots \int_{-\infty}^{\infty} e^{-\frac{\sum_{i=1}^{N} \alpha_i x_i^2}{kT}} dx_1 dx_2 ... dx_N}$$

$$= \sum_{j=1}^{N} \frac{\int_{-\infty}^{\infty} \int_{-\infty}^{\infty} \cdots \int_{-\infty}^{\infty} \alpha_j x_j^2 e^{-\frac{\sum_{i=1}^{N} \alpha_i x_i^2}{kT}} dx_1 dx_2 ... dx_N}{\int_{-\infty}^{\infty} \int_{-\infty}^{\infty} \cdots \int_{-\infty}^{\infty} e^{-\frac{\sum_{i=1}^{N} \alpha_i x_i^2}{kT}} dx_1 dx_2 ... dx_N}$$

$$= \sum_{j=1}^{N} \frac{\left(\int_{-\infty}^{\infty} \alpha_j x_j^2 e^{-\frac{\alpha_j x_j^2}{kT}} dx_j \right) \cdot \prod_{i \neq j} \left(\int_{-\infty}^{\infty} e^{-\frac{\alpha_i x_i^2}{kT}} dx_i \right)}{\prod_{i=1}^{N} \left(\int_{-\infty}^{\infty} e^{-\frac{\alpha_i x_i^2}{kT}} dx_i \right)}$$

$$= \sum_{j=1}^{N} \frac{\int_{-\infty}^{\infty} \alpha_j x_j^2 e^{-\frac{\alpha_j x_j^2}{kT}} dx_j}{\int_{-\infty}^{\infty} e^{-\frac{\alpha_j x_j^2}{kT}} dx_j}$$

$$= \sum_{j=1}^{N} \frac{1}{2} kT$$

$$= \frac{N}{2} kT.$$

20. Equilibrating Effusion*

The effusion rate is proportional to $\frac{p}{\sqrt{T}}$ where p and T are the pressure and temperature of the gas. Let the pressures of the compartments at equilibrium be p_1 and p_2. Then,

$$\frac{p_1}{\sqrt{T_1}} = \frac{p_2}{\sqrt{T_2}}.$$

Note that the pressure on both sides are not necessarily equal for an equilibrium to be attained as we just have to ensure that there is no net transfer of molecules. Moving on, the total number of molecules must be conserved.

$$\frac{p_1 V_1}{T_1 R} + \frac{p_2 V_2}{T_2 R} = n_0.$$

Solving the two equations above,

$$p_1 = \frac{n_0 R T_1 T_2}{V_1 T_2 + V_2 \sqrt{T_1 T_2}},$$

$$p_2 = \frac{n_0 R T_1 T_2}{V_2 T_1 + V_1 \sqrt{T_1 T_2}}.$$

21. Isothermal Leaking**

From the effusion equation,

$$V \frac{dn}{dt} = -\frac{1}{4} n A \langle v \rangle$$

$$\int_{\eta_0}^{\eta} \frac{1}{\eta} d\eta = -\int_0^t \frac{A}{V} \sqrt{\frac{kT}{2\pi m}} dt$$

$$\eta = \eta_0 e^{-\frac{A}{V} \sqrt{\frac{kT}{2\pi m}} t}.$$

To compute the power supplied to the container, we can subtract the total rate of change of internal energy of the gas by the rate of kinetic energy lost by the escaped molecules. The latter is given by Eq. (2.42) as

$$\frac{dE}{dt} = -\eta A \sqrt{\frac{2k^3 T^3}{m\pi}}.$$

The former can be obtained by differentiating $U = \frac{3}{2} \eta V k T$. Since V and T are constant,

$$\frac{dU}{dt} = \frac{3}{2} \cdot \frac{d\eta}{dt} V k T.$$

Therefore, the external power is

$$P = \left| \frac{dU}{dt} - \frac{dE}{dt} \right| = A \sqrt{\frac{k^3 T^3}{8\pi m}} \eta_0 e^{-\frac{A}{V} \sqrt{\frac{kT}{2\pi m}} t}.$$

22. Thermal Conductivity**

(a) Consider a plane of a certain z-coordinate z. In time dt, some molecules on the bottom and top of this plane crosses the plane. Since the bottom region possesses a higher temperature, more molecules on the bottom cross the plane than those on the top and they carry a larger kinetic energy with them. Then, there will be a net energy transfer from the bottom to the top, across the plane at z-coordinate z.

(b) The molecules crossing the plane at coordinate z would have, on average, traveled a distance λ since their last collision. Therefore, the molecules traveling at an angle θ with respect to the z-axis would have traveled a distance $\lambda \cos \theta$ in the z-direction on average and are representative of the temperature $T(z - \lambda \cos \theta)$ as their kinetic energies do not change until their next collisions. The average kinetic energy of such molecules is then

$$\frac{f}{2} k_b T(z - \lambda \cos \theta)$$

where we have used k_b to denote the Boltzmann constant to avoid confusion with the thermal conductivity k.

(c) Now, consider the net energy change due to one molecule with speed v and angle θ leaving the plane at z and due to one molecule arriving with speed v and angle θ, with temperature $T(z - \lambda \cos \theta)$. The net energy change due to the exchange of one such pair of molecules is

$$-\frac{f}{2} k_b T(z) + \frac{f}{2} k_b T(z - \lambda \cos \theta) \approx -\frac{f}{2} k_b \lambda \cos \theta \frac{dT}{dz}.$$

Next, we know from Eq. (2.38) that the fraction of molecules with speed v and angle θ crossing the plane, per unit area and time, is $\frac{1}{2} \eta v f_s(v) \sin \theta \cos \theta d\theta dv$. Therefore, the heat flux density is obtained by multiplying this by $-\frac{f}{2} k_b \lambda \cos \theta \frac{dT}{dz}$ and integrating over all relevant v and θ. Note that the limits of θ are from 0 to π as we want to encompass molecules from both above and below the plane of z-coordinate z. However, by combining these into a single integral, we are assuming that the temperature variation is small across distances in orders of λ as $f_s(v)$ would change across different z-coordinates. Including such variations would result in second order terms in λ in the expression for the heat flux density which will be discarded anyway. Therefore, we can integrate over all relevant limits with a constant $f_s(v)$, taken to be the speed distribution at coordinate z.

$$\frac{dq}{dt} = -\frac{f}{4} \eta k_b \lambda \frac{dT}{dz} \int_0^\pi \sin \theta \cos^2 \theta d\theta \int_0^\infty v f_s(v) dv$$

$$= -\frac{f}{6} \eta k_b \lambda \langle v \rangle \frac{dT}{dz}.$$

Therefore, the thermal conductivity is approximately

$$k = \frac{f}{6}\eta k_b \lambda \langle v \rangle$$

where $\langle v \rangle$ is the average speed at z-coordinate z.

23. Adiabatic Condition***

Consider an infinitesimal area dA on the piston and define the positive x-direction to be parallel to its area vector. Let the velocity of this area be u. The number of molecules with an x-component of velocity v_x colliding with this area in time dt is

$$\eta g(v_x)dv_x(v_x - u)dAdt$$

where η is the number density of molecules and $g(v_x)dv_x$ is the fraction of molecules with an x-component of velocity between v_x and $v_x + dv_x$. The energy change in the gas due to collisions with the piston can be computed by observing that the final x-component of velocity of a gas molecule is $(v_x - 2u)$ in the reverse direction after a collision. Therefore, if the mass of a molecule is m, the change in energy due to one collision is

$$\frac{1}{2}m(v_x - 2u)^2 - \frac{1}{2}mv_x^2 = -2muv_x$$

where we have discarded the second order term in u. Therefore, the change in internal energy of the ideal gas due to the collision between this class of molecules with the infinitesimal area dA is

$$-2m\eta u g(v_x)v_x^2 dv_x dAdt.$$

Then, the total change in internal energy is obtained by integrating the above over all classes of molecules and all areas on the piston. In the case of the latter, we are essentially integrating $udAdt$ over the surface of the piston which results in an infinitesimal change in volume dV. Thus, the total change in energy is

$$dE = -2m\eta dV \int_0^\infty g(v_x)v_x^2 dv_x.$$

Since $\int_0^\infty g(v_x)v_x^2 dv_x = \frac{1}{2}\langle v_x^2 \rangle$,

$$dE = -m\eta\langle v_x^2 \rangle dV = -\frac{1}{3}m\eta\langle v^2 \rangle dV$$

where v is the speed of a molecule and the angle brackets represent taking the mean of. Next, since the internal energy E of a gas is simply the total microscopic kinetic energy,

$$E = \frac{N}{2}m\langle v^2 \rangle = \frac{\eta V}{2}m\langle v^2 \rangle$$

$$\implies m\eta\langle v^2 \rangle = \frac{2E}{V}.$$

Substituting this into the expression for dE,

$$dE = -\frac{2E}{3V}dV$$

$$\frac{1}{E}dE = -\frac{2}{3V}dV$$

$$\ln E = -\frac{2}{3}\ln V + c$$

$$EV^{\frac{2}{3}} = C$$

for some constant C. Next, E is proportional to $\langle v^2 \rangle$ and thus T (by the Boltzmann distribution). We can also state that E is proportional to T directly by the equipartition theorem which is actually a consequence of the Boltzmann distribution. Exploiting $E \propto T$, the quantity $TV^{\frac{2}{3}}$ must be conserved.

24. Leaking Container***

From the effusion equation, we know that

$$V\frac{d\eta}{dt} = -\frac{1}{4}\eta A\langle v \rangle$$

$$\frac{d\eta}{dt} = -\frac{A\eta}{V}\sqrt{\frac{kT}{2\pi m}}.$$

From Eq. (2.42), the total rate of change of internal energy is

$$\frac{dE}{dt} = -A\eta\sqrt{\frac{2k^3 T^3}{m\pi}}.$$

Now, we need to solve the system of equations comprising $\frac{dE}{dt}$ and $\frac{d\eta}{dt}$. In this process, we note that T is a variable as the more energetic molecules are favored in escaping the container — causing the average energy of the molecules remaining in the container to decrease with time. Hence, we first express everything in terms of E and η to eliminate T. Since $E = \frac{3}{2}\eta V kT$,

$$T = \frac{2E}{3\eta V k}$$

$$\frac{d\eta}{dt} = -A\sqrt{\frac{E\eta}{3\pi m V^3}},$$

$$\frac{dE}{dt} = -A\sqrt{\frac{16E^3}{27m\eta\pi V^3}}.$$

From the two equations above,

$$\frac{dE}{dt} = \frac{4E}{3\eta} \cdot \frac{d\eta}{dt}$$

$$\frac{1}{E}dE = \frac{4}{3\eta}d\eta$$

$$\implies \frac{E}{\eta^{\frac{4}{3}}} = \frac{E_0}{\eta_0^{\frac{4}{3}}} = c$$

where $E_0 = \frac{3}{2}\eta_0 V kT_0$ is the initial energy. We let the right-hand side be c for the sake of convenience. Since $E = \frac{3}{2}\eta V kT = c\eta^{\frac{4}{3}}$,

$$T = \frac{2c}{3kV}\eta^{\frac{1}{3}}.$$

Substituting this expression for T into $\frac{d\eta}{dt}$,

$$\frac{d\eta}{dt} = -A\sqrt{\frac{c}{3\pi m V^3}}\eta^{\frac{7}{6}}.$$

Solving this differential equation by separating variables would yield the desired result.

$$\eta(t) = \frac{1}{\left(\eta_0^{-\frac{1}{6}} + A\sqrt{\frac{kT_0}{72\pi m V^2 \eta_0^{\frac{1}{3}}}}\right)^6}.$$

Chapter 3

The Second Law and Heat Engines

The first law of thermodynamics, which was the main focus of the previous chapter, is basically the principle of the conservation of energy. It asserts that energy should be conserved in physical processes but it does not delineate a direction for physical processes. Hence, this chapter will discuss the second law of thermodynamics which concisely sets a particular direction for all processes, and examine its implications on heat engines and related systems.

3.1 Kelvin-Planck's and Clausius' Statements

In nature, certain processes are observed to only proceed in a single direction spontaneously, though other processes that are consistent with the conservation of energy are seemingly possible. A cup of hot coffee will lose heat to its cool surroundings but never gain heat from it, without any external work, though the latter is perfectly coherent with the first law of thermodynamics. The very notion of temperature does not help either. The zeroth law of thermodynamics only implies that when two objects are in thermal equilibrium, they have the same temperature. It does not dictate the direction of heat conduction.

More generally, the classical laws so far are temporally reversible. That is, if you take a video of an egg that is dropped onto the ground, so precise that you can track the motion of individual atoms, and reverse the video such that a cracked egg reverts to a complete egg, the system in reverse will still obey all the classical laws that have been introduced. Therefore, this replay is permissible. However, we know from common experience that this never seems to occur. Therefore, a new law — known as the second law of thermodynamics — is needed to prescribe the direction of evolution of a system.

The second law of thermodynamics can be stated in various, equivalent forms which are remarkably succinct. The two most intuitive ones are Kelvin-Planck's statement and Clausius' statement. Kelvin-Planck's statement asserts that it is impossible to construct a perpetual motion machine of the second kind — a cyclic engine whose sole effect is to absorb heat from a heat reservoir and produce an equivalent amount of work done. Note that a heat reservoir is defined to be a large repository of internal energy (as compared to the system it is connected to) such that its temperature stays approximately constant throughout heat transfer. As a concomitant of Kelvin-Planck's statement, an engine that is 100% thermally efficient is precluded. Clausius' statement, on the other hand, purports that it is impossible to construct a device that solely transfers heat from a body of lower temperature to one of higher temperature. Note that the word "solely" in the context of the two statements implies that there should not be any changes imposed on the external environment.

These seemingly disparate statements are actually equivalent, as we shall show later, but for now, let us examine their ramifications on the feasibility of various processes. Due to these axioms, some processes are deemed to be impossible. Consequently, processes can be categorized as reversible or irreversible. A reversible process is an evolution of a system from an initial state to a final state such that there is a process that allows the system to return to its initial state without leaving any changes to its surroundings. An irreversible process does not satisfy this requirement. An important fact to understand is that a process is deemed irreversible only if you try every path from the final state to the initial state (the path is not necessarily the original movie played in reverse) and the above criterion is still not fulfilled.

Lastly, note that a system can always be reverted from a final state to its initial state, regardless of whether the original process is reversible or irreversible. However, this reversion may involve changes to the external surroundings of the system if the original process was irreversible. Ultimately, the reversibility of a process is a completely different concept from whether a system can be restored to its original state — the former concerns whether a system can be reverted without leaving any traces of the occurrence of the original process.

Reversible processes are idealizations and do not exist in reality, both because of inherent irreversibilities, such as friction, in practical processes and the infeasibility in meaningfully using a reversible process, as we shall see. As a result of Kelvin-Planck's and Clausius' statements, the following processes are irreversible.

It is impossible to reverse any process in which heat due to friction is produced. A fraction of work done is inevitably converted to heat by friction. In order to reverse the process, this heat must be converted back to an equivalent amount of work done which violates the Kelvin-Planck's statement. In a similar vein, it is impossible to reverse any process that occurs too quickly. If the intermediate states of a system are not in thermodynamic equilibrium, frictional losses and turbulence will arise and these are irreversible based on the previous argument. Therefore, a reversible process must first be quasistatic — implying that it would take eons for a reversible process to be completed.

Direct heat transfer from a high-temperature body to a low-temperature body with a finite temperature difference is also irreversible due to Clausius' statement. Since heat transfer can only occur across a temperature gradient between two bodies, a reversible heat transfer process is physically impossible. However, we can "cheat" for theoretical purposes by putting two bodies with an infinitesimal temperature difference in thermal contact to approximate a reversible heat transfer process. Such a conceptual process is infeasible in real life as it would take an eternity for a significant amount of heat to be spontaneously transferred.

Finally, it is also impossible to reverse a process in which a gas expands or contracts without performing work or absorbing heat. An example of such a phenomenon is a gas escaping a ruptured membrane to fill up the evacuated portion of a thermally insulated container — this process is known as a Joule expansion and will be analyzed later. We can prove this by contradiction. Suppose that there were a reverse process for Joule expansion. Then, one could first run an isothermal expansion from an initial state to a final state to absorb a certain amount of heat and produce an equivalent amount of work (as internal energy does not vary in an isothermal process). Afterwards, running the reverse Joule expansion process[1] would yield a cyclic system which absorbs a certain amount of heat to produce the same amount of work — hence violating Kelvin-Planck's statement. As another corollary, the process of mixing different gases is also irreversible as it is effectively two Joule expansions of two gases.

---

[1] In order for a reverse Joule expansion process to be the reverse process of an isothermal process, the final and initial temperatures of a Joule expansion must first be identical. This is indeed the case as the gas does no work (there is no external pressure) and does not receive any heat during a Joule expansion.

3.2   Heat Engines and Refrigerators

A heat engine works by receiving heat from a heat source and producing work. Its process is cyclic so that it can produce a steady output. As the operation cannot be perfectly efficient — as forbidden by Kelvin-Planck's statement — some exhaust heat must be dumped into a heat sink so that the system can return to its initial state. The schematic in Fig. 3.1 summarizes the design of a heat engine.

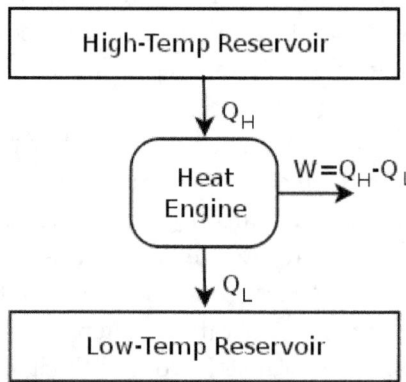

Figure 3.1:   Heat engine

In a single cycle, the heat engine draws Q_H amount of heat from the high-temperature reservoir and deposits Q_L amount of leftover heat while producing $W = Q_H - Q_L$ amount of work. There must be zero net heat or work flowing into the heat engine during a cycle as the internal energy of the heat engine must remain unchanged after a single cycle.

Next, a refrigerator works in a different way — its essential function is to extract heat from a low-temperature reservoir and to deposit it in a high-temperature reservoir. In practice, heat is constantly transferred from the lower-temperature refrigerant to the higher-temperature room in order to keep the refrigerator cold. However, this cannot occur spontaneously, as precluded by Clausius' statement. Therefore, a refrigerator is a cyclic system that operates by receiving a certain amount of work to bring some heat from a low-temperature reservoir to one of higher temperature.

Referring to Fig. 3.2, in every cycle, the refrigerator extracts Q_L amount of heat from a low-temperature reservoir and transfers Q_H amount of heat to the high-temperature reservoir, requiring $W = Q_H - Q_L$ amount of external work in the process.

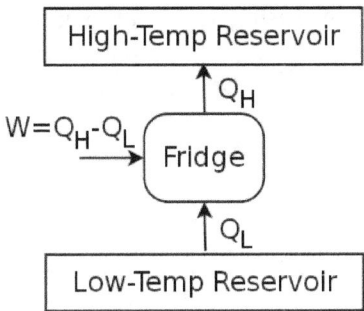

Figure 3.2: Refrigerator

3.2.1 Equivalence of Kelvin-Planck's and Clausius' Statements

With an understanding of how heat engines and refrigerators work, we shall now prove the equivalence of Kelvin-Planck's and Clausius' statements by contradiction. Suppose there exists a device that violates Kelvin-Planck's statement. Then, we can use it as a heat engine and connect its output to a refrigerator such that it supplies the necessary external power to the refrigerator. Both devices are connected to the same low-temperature and high-temperature reservoirs.

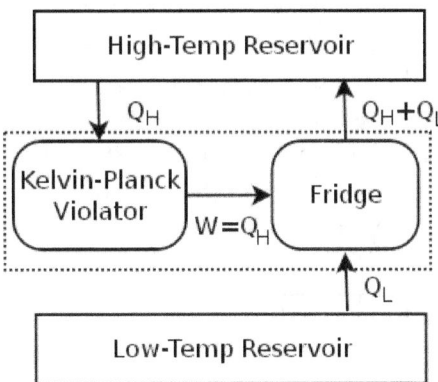

Figure 3.3: Conceptual set-up

As shown in Fig. 3.3, the hypothetical engine receives Q_H heat from the high-temperature reservoir and delivers $W = Q_H$ amount of work to the refrigerator. The refrigerator then draws Q_L heat from the low-temperature reservoir and stores $Q_H + Q_L$ heat in the high-temperature reservoir. Now, closely observe the part of the system that is enclosed by the dotted lines (the combined system comprising the heat engine and refrigerator). If we were to place it inside a black box (no peeking!) and observe its effects from the outside, we would obtain the equivalent system in Fig. 3.4 which acts as a refrigerator.

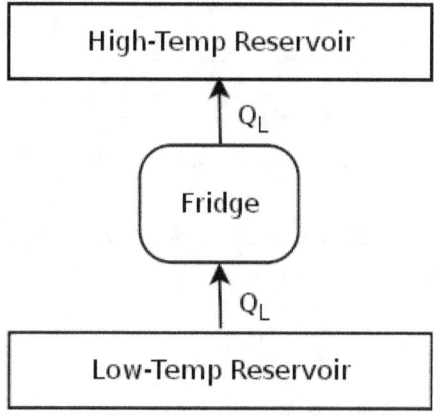

Figure 3.4:   Equivalent system

The equivalent system effectively draws Q_L amount of heat from the low-temperature reservoir and deposits it completely into a high-temperature reservoir — a phenomenon that is forbidden by Clausius' statement. Thus, we have proven that if Kelvin-Planck's statement is violated, Clausius' statement would be violated as well. To prove the converse, we consider a similar set-up. Suppose there exists a device whose sole effect is to deliver Q_L amount of heat from a low-temperature reservoir to a high-temperature one. Then, we can use this system as a refrigerator to continuously pump the heat, deposited during a heat engine cycle, back into the heat source.

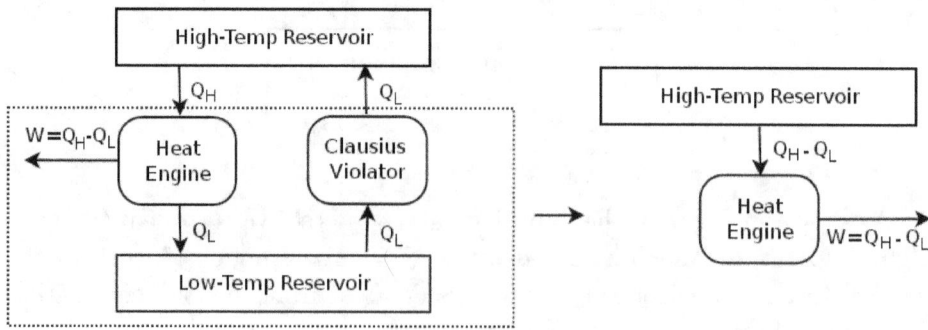

Figure 3.5:   Conceptual set-up 2

Referring to Fig. 3.5, a heat engine absorbs Q_H heat from a high-temperature reservoir and releases Q_L heat back into a low-temperature reservoir, producing $W = Q_H - Q_L$ amount of work to an external system in the process. The hypothetical refrigerator then pumps Q_L heat from the low-temperature reservoir back into the heat source. By considering the heat engine, refrigerator and the heat sink as a whole, one would obtain a heat engine which absorbs $Q_H - Q_L$ amount of heat from the heat source and produces $W = Q_H - Q_L$ amount of work — a perfectly thermally efficient device which is forbidden by Kelvin-Planck's statement. Having proven the converse, we have shown that Kelvin-Planck's and Clausius' statements are in fact equivalent.

3.2.2 *Carnot's Principles*

Kelvin-Planck's and Clausius' statements prescribe a theoretical limit on the efficiencies of heat engines and refrigerators — measures that we shall now quantify. Since the purpose of a heat engine is to produce useful work, the efficiency η of a heat engine is defined as the work produced W divided by the total amount of heat input Q_H.

$$\eta = \frac{W}{Q_H} = \frac{Q_H - Q_L}{Q_H} = 1 - \frac{Q_L}{Q_H} \tag{3.1}$$

where Q_L is the total heat deposited into heat sinks. In general, Q_H is not necessarily extracted from a single heat source and can be accumulated across different heat sources at different junctures in a heat engine cycle. Similarly, heat can also be deposited into various heat sinks at different instances. However, for this section, we will solely be studying systems operating between two reservoirs which means that these systems can only exchange heat with this fixed pair of reservoirs and nothing else.

Since the purpose of a refrigerator is to extract heat from a low-temperature reservoir, the efficiency of a refrigerator is quantified by the ratio of the total amount of heat extracted to the external work received. This measure is known as the coefficient of performance (COP).

$$\text{COP} = \frac{Q_L}{W} = \frac{Q_L}{Q_H - Q_L}. \tag{3.2}$$

It should be noted that the COP can be greater than unity (this is why the misleading word "efficiency" is not used for refrigerators). Proceeding to the main topic, Carnot deduced certain principles concerning these efficiencies from Kelvin-Planck's and Clausius' statements. Carnot's principles state that

(1) An irreversible heat engine is less efficient than a reversible heat engine when operating between the same two heat reservoirs. The COP of an irreversible refrigerator is smaller than that of a reversible refrigerator when operating between the same two reservoirs.
(2) The efficiencies of all reversible heat engines operating between identical pairs of heat reservoirs are the same. Accompanying this, the coefficients of performance of all reversible refrigerators operating between identical pairs of heat reservoirs are the same.

We can prove the two claims by contradiction. A crucial component of our proofs would entail the fact that a reversible heat engine, unsurprisingly, can be reversed to operate as a refrigerator with the same Q_H, Q_L and W (except that they are in the opposite directions) and vice-versa. Beginning with the first principle, we assume that an irreversible heat engine is more efficient than a reversible heat engine. We then connect this irreversible heat engine and the reverse of the reversible heat engine (a refrigerator) to the same pair of heat reservoirs in Fig. 3.6.

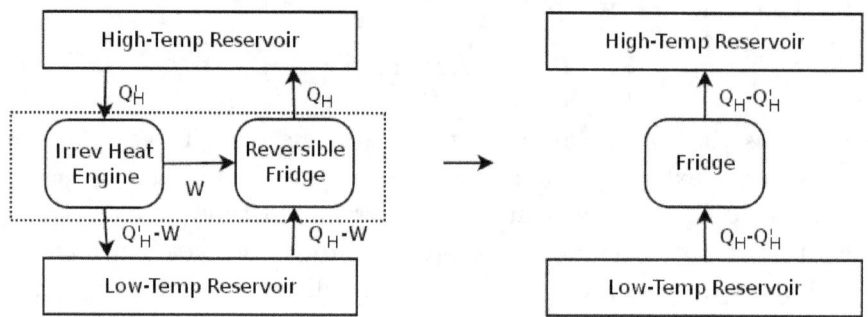

Figure 3.6:   Conceptual set-up 3

The irreversible heat engine extracts Q'_H heat from the source and produces W amount of work which is used to power the refrigerator. The exhaust heat is then $Q'_H - W$. The refrigerator then extracts $Q_H - W$ heat from the low-temperature reservoir and deposits Q_H amount of heat into the high-temperature one. Since the efficiency of the irreversible heat engine is greater than that of the reversible heat engine (which is currently running in reverse),

$$\frac{W}{Q'_H} > \frac{W}{Q_H}$$

$$\implies Q_H > Q'_H$$

and

$$Q_H - W > Q'_H - W.$$

Now, consider the combined system of the heat engine and refrigerator. Its sole effect is to deliver $Q_H - Q'_H$ amount of heat from the low-temperature reservoir to the high-temperature reservoir, in contravention of Clausius' statement. Therefore, the efficiency of an irreversible engine must be less than or equal to that of a reversible engine. However, the equality case cannot hold — if not, the fridge will be the reverse process for the irreversible heat engine (as the combined system of the heat engine and fridge results in no heat and work everywhere) and thus violate the premise. The efficiency of an irreversible engine must then be less than that of a reversible engine. In the case of refrigerators, one can assume that an irreversible refrigerator has larger COP than a reversible refrigerator and consider a similar set-up by running the reversible refrigerator in reverse as a heat engine that supplies work to the irreversible refrigerator. The combined set-up would then violate Clausius' statement — leading to the conclusion that the COP of an irreversible refrigerator is smaller than that of a reversible refrigerator.

To prove the second claim, compare two reversible engines A and B. Suppose that A is more efficient than B. Then, one can use A as a heat engine, in replacement of the irreversible heat engine, and run B in reverse as the refrigerator in the previous set-up. Then, one would conclude that A must be less efficient or equally efficient as B. Afterwards, one can reverse the roles of A and B to conclude that B must also be less efficient or equally efficient as A. Therefore, A and B must have the same efficiency. Since a reversible refrigerator is its equivalent reversible heat engine in reverse, the coefficient of performance of the refrigerator and the efficiency of its equivalent heat engine can be related, if the heat reservoirs remain unchanged. From the definitions of efficiency and COP and the facts that the magnitude of the heat transfers and work are the same,

$$\text{COP} = \frac{1}{\eta} - 1.$$

Therefore, if all reversible heat engines have the same η, all reversible refrigerators should also have the same COP. Lastly, take note that these arguments did not mention the working substance — the medium facilitating the process — of the heat engines or refrigerators. Then, the efficiencies and COPs of reversible heat engines and refrigerators across a constant pair of reservoirs are independent of their working substance. This means that the efficiencies of a reversible heat engine that uses an ideal gas as its medium and one that uses a real gas are the same across the same two reservoirs!

Thermodynamic Temperature

In this section, we shall study how the efficiency of reversible heat engines can be used to formalize the definition of temperature through the Kelvin scale. We have just concluded that the efficiency of a reversible heat engine is independent of the engine process and the working substance. Then, the efficiency η, and thus $\frac{Q_L}{Q_H}$, can only be functions of the temperatures of the high-temperature and low-temperature heat reservoirs, T_L and T_H.

$$\frac{Q_L}{Q_H} = f(T_L, T_H).$$

Now, observe that we can choose the reversible heat engine that operates between a high-temperature reservoir at temperature T_H and a low-temperature reservoir at temperature T_L as one that consists of the two reversible heat engines and a middle-temperature reservoir at temperature T_M in Fig. 3.7.

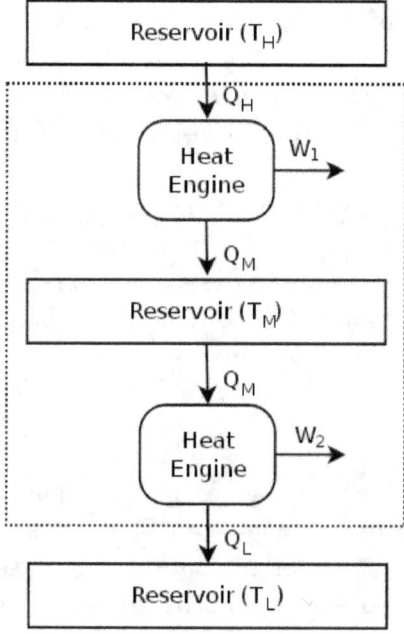

Figure 3.7:   Engine comprising two reversible heat engines

The combined system, enclosed in dotted lines, is still a reversible heat engine operating between the two reservoirs. Now, we know that

$$\frac{Q_M}{Q_H} = f(T_M, T_H),$$

$$\frac{Q_L}{Q_M} = f(T_L, T_M),$$

where Q_M is the heat delivered via the middle reservoir. Then,

$$\frac{Q_L}{Q_H} = \frac{Q_L}{Q_M} \cdot \frac{Q_M}{Q_H}$$

$$f(T_L, T_H) = f(T_L, T_M) \cdot f(T_M, T_H).$$

Observe that the left-hand side is independent of T_M. The only way for this equation to be satisfied is for

$$f(T_L, T_M) = \frac{g(T_L)}{g(T_M)},$$

$$f(T_M, T_H) = \frac{g(T_M)}{g(T_H)},$$

for some function $g(T)$ so that the terms in T_M cancel. That is, in general,

$$f(T_x, T_y) = \frac{g(T_x)}{g(T_y)}.$$

Therefore, for a heat engine receiving Q_H from a heat source at temperature T_H and depositing Q_L into a heat sink of temperature T_L,

$$\frac{Q_L}{Q_H} = \frac{g(T_L)}{g(T_H)}.$$

We can choose any monotonic function $g(T)$ — the exact function will determine how the temperature scale is defined. Lord Kelvin chose the simplest function $g(T) = T$ and established his Kelvin scale. On this scale, the ratio between two temperatures is equal to the ratio between the heat transferred to and from a reversible heat engine connected to reservoirs of those two temperatures. This definition of temperature is independent of any thermometric, physical property — such as the expansion of mercury — and is thus known as a thermodynamic scale. Lastly, we need to have a reference temperature in order to define all other temperatures. In accordance with international standards, 273.16K is defined to be the triple point of water (the unique temperature at which the solid, liquid and gaseous phases of water co-exist). Then, all other temperatures can be defined by setting

T_L or T_H as 273.16K. For example, suppose that a heat engine connected between a heat source at temperature T and a heat sink at 273.16K draws Q_H heat from the source and deposits Q_{ref} heat into the sink. Then the temperature T in Kelvins, is given by

$$\frac{Q_{ref}}{Q_H} = \frac{273.16K}{T}$$

$$\implies T = \frac{Q_H}{Q_{ref}} \cdot 273.16K.$$

Thankfully, this novel thermodynamic scale does not differ much from the previously pervasive Celsius scale. The magnitude of an additional Kelvin is in fact equal to the magnitude of an additional degree Celsius and the conversion formula between Kelvins and degree Celsius is

$$T(K) = T(°C) + 273.15.$$

As a result of the Kelvin temperature scale, the efficiency of a reversible heat engine operating between a heat source and sink at respective temperatures T_H and T_L is by definition

$$\eta = 1 - \frac{Q_L}{Q_H} = 1 - \frac{T_L}{T_H} \tag{3.3}$$

where T_L and T_H are measured in Kelvins.

3.2.3   *What does it take to be Reversible?*
— The Carnot Engine

Now that we have established that a reversible engine is the most efficient when operating between a pair of heat reservoirs and have determined its efficiency, let us analyze the criteria required for such an engine to be reversible.

Firstly, the operation of the engine must be frictionless and quasistatic, as we have previously shown that processes involving friction are irreversible. Secondly, when the engine exchanges heat with a reservoir, the engine's temperature must be identical to the reservoir's temperature. This point is less obvious so we shall provide a formal proof. Referring to Fig. 3.8, use a reversible heat engine to deliver work to its refrigerator counterpart (which exists due to its reversible nature), while operating between the same pair of heat reservoirs.

Observe that if we isolate the engine, refrigerator and heat sink, this combined system withdraws Q_H from the heat source and returns it back — all while leaving no traces in itself and external entities. Therefore, the transfers of Q_H between the engine and the heat source as well as between the

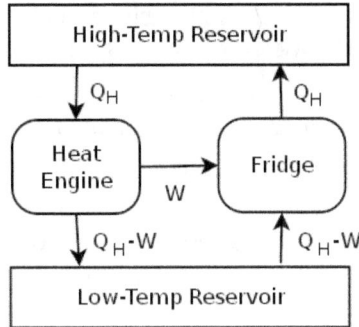

Figure 3.8: Reversible heat engine and refrigerator (reverse engine)

refrigerator and the heat source must be reversible! The only way for this to occur is when the engine and refrigerator temperatures are equal to that of the heat source during the process of heat exchange. By considering the engine, refrigerator and heat source as a combined system, we can similarly deduce that the engine and refrigerator temperatures must be identical to that of the heat sink during their interactions.

With this knowledge, we can now construct a reversible heat engine. There can only be two states of the engine — when it is exchanging heat with a reservoir and when it is not. In the case of the former, its temperature, must be equal to the reservoir's temperature, so the only way for it to exchange significant heat is to undergo an isothermal process (at the reservoir's temperature). In the case of the latter, since the engine cannot interact with other entities beside the reservoirs by assumption, it can only undergo an adiabatic process. As such, we have drastically narrowed down the possible processes (absent of frictional losses) of a reversible engine that operates between a pair of reservoirs.

Now, the total number of reversible engines with the maximum efficiency $1 - \frac{T_L}{T_H}$ actually depends on a slight technicality in the definition of efficiency. When we say that the engine withdraws Q_H amount of heat from the heat source, one perspective is that the engine can only receive heat from the heat source at each juncture and cannot lose any heat to it (i.e. it is a one-way heat flow of Q_H). Another perspective is that the engine can receive and return heat from and to the heat source, for a net heat intake of Q_H. Evidently, the former definition is more restrictive, but it is in fact the more pervasive one. This is because in a more general cycle which interacts with a multitude of reservoirs, the heat sources and sinks are not labeled for us. We then usually take the reservoirs that the system gains or loses heat from and to as the heat sources and sinks respectively. That is, when we are computing the total heat input to a system (for efficiency calculations), we can simply add the values of heat influxes and ignore the heat outflows.

Adopting the former perspective, there is in fact only one set of processes that a reversible system operating between two reservoirs can undergo. Visualizing the PV diagram of our engine, we have two enforced isotherms (at the temperatures of the reservoirs) and only adiabats to connect them. A reversible heat engine operating between two reservoirs must thus entail two isotherms and two adiabats, as depicted in Fig. 3.9.

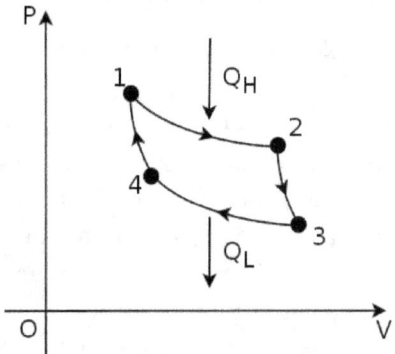

Figure 3.9:   Carnot cycle

(1) $1 \to 2$: An isothermal expansion
(2) $2 \to 3$: An adiabatic expansion
(3) $3 \to 4$: An isothermal compression
(4) $4 \to 1$: An adiabatic compression

In particular, the Carnot engine follows this set of operations and uses an ideal gas as its working fluid. Now, let us practice deriving the efficiency of a heat engine by verifying Eq. (3.3) for the Carnot engine. In this derivation, it is important to note that the engine only receives heat during process $1 \to 2$ from a heat source of temperature T_H and deposits heat during process $3 \to 4$ to a heat sink of temperature T_L. Let the pressure, volume and temperature of the ith state of the ideal gas system (i ranges from 1 to 4) in the Carnot engine be p_i, V_i and T_i. Note that $T_1 = T_2 = T_H$ and $T_3 = T_4 = T_L$ as there cannot be a finite temperature difference during a reversible heat transfer. By the ideal gas equation,

$$p_1 V_1 = p_2 V_2 \implies \frac{p_2}{p_1} = \frac{V_1}{V_2}, \qquad \text{(Isothermal)}$$

$$p_3 V_3 = p_4 V_4 \implies \frac{p_3}{p_4} = \frac{V_4}{V_3}. \qquad \text{(Isothermal)}$$

Based on the adiabatic condition,

$$p_2 V_2^\gamma = p_3 V_3^\gamma,$$
$$p_1 V_1^\gamma = p_4 V_4^\gamma.$$

Dividing the former equation by the latter and substituting the results we obtained from the two equations at the top,

$$\left(\frac{V_2}{V_1}\right)^{\gamma-1} = \left(\frac{V_3}{V_4}\right)^{\gamma-1} \implies \frac{V_2}{V_1} = \frac{V_3}{V_4}.$$

The heat absorbed during process $1 \to 2$, Q_H, and the heat ejected during $3 \to 4$, Q_L, can be calculated by the first law of thermodynamics since the internal energy of a gas does not change during an isothermal process.

$$Q_H = W_{12by} = \int_{V_1}^{V_2} \frac{nRT_H}{V} dV = nRT_H \ln \frac{V_2}{V_1},$$

$$Q_L = -W_{34by} = -\int_{V_3}^{V_4} \frac{nRT_L}{V} dV = nRT_L \ln \frac{V_3}{V_4}.$$

There is a negative sign in front of Q_L as it is defined to be the heat ejected from the system during process $3 \to 4$ (and not heat supplied to the ideal gas). Thus, the efficiency of a Carnot engine is

$$\eta = 1 - \frac{Q_L}{Q_H} = 1 - \frac{nRT_L \ln \frac{V_3}{V_4}}{nRT_H \ln \frac{V_2}{V_1}} = 1 - \frac{T_L}{T_H}$$

where we have used the fact that the ratios of the volumes are equal. This is to be expected as we have precisely defined temperature according to the Kelvin scale to ensure this!

3.3 Clausius' Inequality and Entropy

Scrutinizing the efficiency of a reversible heat engine that is operating between two heat reservoirs at temperatures T_L and T_H, we observe that

$$\frac{Q_L}{Q_H} = \frac{T_L}{T_H}. \tag{3.4}$$

If we now standardize the sign of heat flow such that heat supplied to a system is positive while heat extracted from a system is negative, we have for a reversible engine

$$-\frac{Q_L}{Q_H} = \frac{T_L}{T_H}$$

as Q_L was previously defined to be a positive quantity that represents the heat extracted from the system into a heat sink. Rearranging,

$$\frac{Q_L}{T_L} + \frac{Q_H}{T_H} = 0.$$

Note that Q_L and Q_H are the only heat exchanges of the system with an external environment and these exchanges must occur when the system is at temperature T_L and T_H respectively (as there cannot be a finite temperature gradient during a reversible heat transfer). Therefore, we can write the above sum as a path integral along the cycle that the system takes

$$\oint \frac{\delta Q}{T} = 0,$$

where δQ is an infinitesimal heat transfer (positive if it is a heat input) and T is the instantaneous temperature of the system along the path it takes. The loop superimposed on the integral underscores the fact that this path is a complete cycle (i.e. starts and ends at the same point). Note that the integrand is only non-zero when the system is exchanging heat with the two reservoirs and hence evaluates to the previous sum above. We see that the integral of $\frac{\delta Q}{T}$ over a reversible engine cycle operating between two reservoirs must be zero! This brings us to the question of determining this integral along general cycles, both reversible and irreversible. The answer to this is Clausius' inequality.

Clausius' Inequality: For any arbitrary cyclic process that a closed system undergoes,

$$\oint \frac{\delta Q}{T} \leq 0, \tag{3.5}$$

where δQ is an infinitesimal heat transfer into the system and T is the instantaneous temperature of the system at each juncture along its path. A closed system refers to a system in which no mass exchange with an external environment occurs. The equality case holds if and only if the cycle is internally reversible.

Now, it is important to make a distinction between internal and external irreversibilities (we have held this off till now). We will only consider closed systems. An internally reversible process is one in which no irreversibilities, such as friction and heat transfer between components of a system with a finite temperature gradient, occur within a system. Due to this definition, an internally irreversible process that a system undergoes must be quasistatic. However, irreversibilities at the boundaries of the system, such as heat transfer between the system and its external environment across a finite temperature difference, is still allowed. A process involving a system is externally reversible if no irreversibilities occur during the interaction of the system and its environment at their boundaries. Lastly, a process is totally reversible if it is both internally and externally reversible. This total reversibility is what we have been considering up till now.

Proof of Clausius' inequality: A closed system undergoing a general cyclic process may be connected to various reservoirs of different temperatures at different junctures along the cycle. Then, let the ith reservoir in the process[2] have a temperature T_i in Fig. 3.10. Its temperature is matched to the instantaneous temperature of the system during their point of interaction. Now, consider the following hypothetical set-up where all of the reservoirs, that are in direct correspondence with the system, are connected to a common principal reservoir of temperature T_p via reversible engines. Each engine may function as a heat engine or a refrigerator.

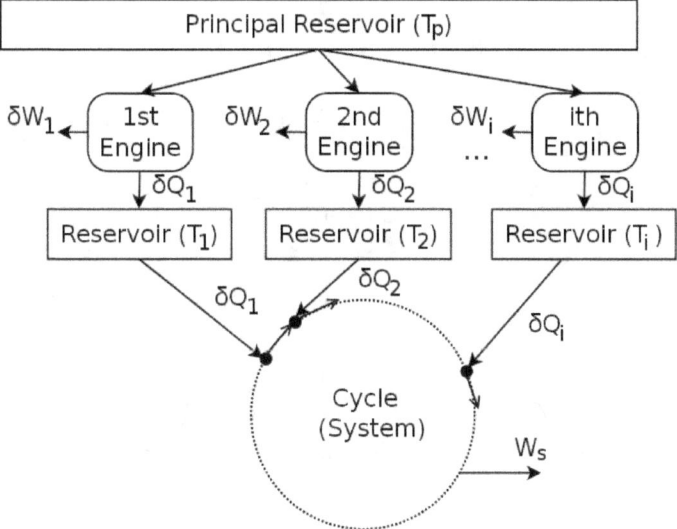

Figure 3.10: System connected to an array of heat reservoirs

During the ith infinitesimal step along the cycle, the principal reservoir supplies a certain amount of heat to the ith reversible engine which produces

---

[2]We will consider the case of discrete reservoirs first, for the sake of clarity.

δW_i amount of work in the process. The ith engine also transfers δQ_i heat to a reservoir of temperature T_i, which in turn delivers δQ_i heat to the system. We can in fact relate δW_i to δQ_i as the facilitator is a reversible engine which obeys

$$\frac{\delta Q_L}{T_L} = \frac{\delta Q_H}{T_H}$$

$$\delta W = \left(\frac{T_H}{T_L} - 1\right)\delta Q_L$$

$$\implies \delta W_i = \left(\frac{T_p}{T_i} - 1\right)\delta Q_i.$$

In a single complete cycle, which comprises myriad such infinitesimal steps, the system takes in a net heat of $\sum \delta Q_i$ and hence produces $W_s = \sum \delta Q_i$ amount of work. It cannot keep any heat as internal energy because it returns to its original state. Note that the system does not directly contravene Kelvin-Planck's statement as some δQ_i may be negative — implying that some heat flows out of the system too. With that out of the way, consider all auxiliary reservoirs and the system as a combined system in Fig. 3.11.

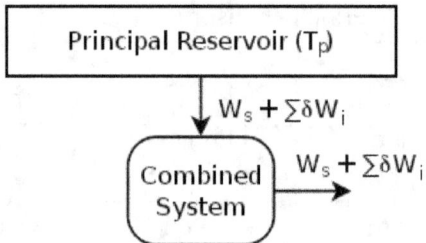

Figure 3.11:   Equivalent system

The net effect of the combined system is seemingly to absorb $W_s + \sum \delta W_i$ amount of heat from the principal reservoir and produce an equivalent amount of work. Since this is forbidden by Kelvin-Planck's statement,

$$W_s + \sum \delta W_i \leq 0.$$

Substituting $W_s = \sum \delta Q_i$ and $\delta W_i = (\frac{T_p}{T_i} - 1)\delta Q_i$,

$$\sum \delta Q_i + \sum \left(\frac{T_p}{T_i} - 1\right)\delta Q_i \leq 0$$

$$T_p \sum \frac{\delta Q_i}{T_i} \leq 0.$$

Changing the above from a discrete sum to a closed loop integral (by imagining the hypothetical set-up as a continuous, infinite series of reservoirs) and dividing by T_p which is a positive quantity,

$$\oint \frac{\delta Q}{T} \leq 0$$

for an arbitrary cyclic process taken by the system. Since we have matched the temperatures of the auxiliary reservoirs to the instantaneous temperature of the system, T here represents the temperature of the system so this inequality strictly involves only quantities of the cyclic system.

To show that the equality case must hold if the cycle is internally reversible, we first realize that an internally reversible cycle of the system is fully reversible in this context as it only interacts with heat reservoirs with infinitesimal temperature gradients. Now, suppose that the integral is negative such that $W_s + \sum \delta W_i$ is negative in the previous set-up. Then, reversing the system and all the reversible heat engines would yield a combined system with a positive $W_s + \sum \delta W_i$ — an evident contradiction as it violates Kelvin-Planck's statement.

To show that an irreversible cycle must result in a negative value of the closed loop integral, suppose that the integral results in zero such that $W_s + \sum \delta W_i$ is zero. Then, the combined system, comprising the set of reservoirs and the system, draws no heat from the principal reservoir and performs zero net work on the external environment. The states of the interior of the combined system — the system and the reversible engines — have not changed either as all processes are cyclic. The combined system thus results in no change to the environment and itself — showing that each part of it must be totally reversible and contradicting the premise of the irreversibility of the original system. The equality case then occurs if and only if the cycle is internally reversible.

3.3.1 *Entropy*

Clausius' inequality implies that

$$\oint \frac{\delta Q_{rev}}{T} = 0$$

for any internally reversible, cyclic process. A subscript has been added to highlight the fact that this cycle must be internally reversible. As this integral is zero for all such cycles, the following integral from an initial state 1 to a final state 2 via an internally reversible route is solely dependent on the initial and final states.

$$\Delta S = \int_1^2 \frac{\delta Q_{rev}}{T}. \tag{3.6}$$

The above integral is defined as the change in entropy from state 1 to 2, ΔS. Though the calculation of ΔS is performed over an internally reversible path, the entropy changes between an initial and final state along all paths are the same — the system can even undergo an irreversible process. ΔS is only dependent on the initial and final states of a system and the absolute entropy S, assuming that it exists,[3] is a state function that is defined uniquely for each state of the system.

General Path

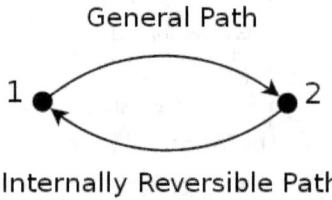

1 $\qquad\qquad\qquad$ 2

Internally Reversible Path

Figure 3.12:   Cycle

Referring to Fig. 3.12, consider a cyclic process where a system takes a general route, internally reversible or irreversible, from an initial state 1 to a final state 2 and takes an internally reversible route back to the initial state. Applying Clausius' inequality,

$$\int_1^2 \frac{\delta Q}{T} + \int_2^1 \frac{\delta Q_{rev}}{T} \leq 0$$

where the first integral is along the general path and the second integral is along the internally reversible route. Rearranging,

$$\int_1^2 \frac{\delta Q_{rev}}{T} \geq \int_1^2 \frac{\delta Q}{T}$$

$$\Delta S \geq \int_1^2 \frac{\delta Q}{T}.$$

Now for an isolated system which involves no mass and heat transfer, δQ is zero such that

$$\Delta S \geq 0. \tag{3.7}$$

---

[3]It actually does but this is beyond the scope of this book and is not germane here.

The entropy of an isolated system can only increase! This is another common statement of the second law of thermodynamics. The equality case only occurs when the process in the system is internally reversible. Applying this idea to the entire universe (the set of all particles) which is considered as an isolated system, the entropy of the universe can only increase!

We can use this notion of entropy to determine whether a process in a system A is totally reversible. Since the entire universe contains all possible systems, including those that system A interacts with, the process in A is totally reversible if and only if this process, with respect to the entire universe, is internally reversible (since external reversibilities of the process with respect to A also become internal irreversibilities of the universe). Then, the process in a system is totally reversible if and only if it does not result in a net entropy change of the universe. At this juncture, we may be confused by internal and external irreversibilities. The distinction between them is entirely dependent on the choice of the system as the following example shall illustrate.

Problem: Consider two bodies of different temperatures T_1 and T_2 that are brought into thermal contact with $T_1 > T_2$. An infinitesimal amount of heat, dQ, is transferred from the high-temperature body to the low-temperature body. The two bodies only interact with each other and not other parts of their surroundings. Is the high-temperature body undergoing an internally reversible process? What about the low-temperature body? What about the combined system? Which of the above systems are isolated systems? Find the entropy changes of each body and the entire universe.

The one-body systems are both undergoing an internally reversible process as heat transfer with an external system is not counted as an irreversibility within the system. The combined system comprising both objects does not undergo an internally reversible process as there is heat transfer within components of the system. Only the system comprising both objects is an isolated system as there is no heat transfer between this system and its surroundings. The infinitesimal changes in entropy of the high-temperature and low-temperature bodies are

$$dS_H = -\frac{dQ}{T_1},$$
$$dS_L = \frac{dQ}{T_2}.$$

The net change in entropy of the entire universe is the sum of these changes.

$$dS_{total} = \frac{dQ}{T_2} - \frac{dQ}{T_1} > 0$$

as dQ is a positive quantity and $T_1 > T_2$. As expected of an irreversible process, the entropy change of the universe is positive. Incidentally, this also shows that the entropy version of the second law of thermodynamics implies Clausius' statement. If $T_1 < T_2$, the entropy change of the universe is negative — a result that is forbidden by $\Delta S \geq 0$. Since we have already shown that Kelvin-Planck's and Clausius' statements result in $\Delta S \geq 0$ and because we have just proven the converse, these three versions of the second law of thermodynamics are equivalent.

3.3.2   *Entropy Calculations*

The key to evaluating entropy changes is to actually ignore the process that a system actually undergoes as entropy is a state function. However, we must keep track of the system's initial and final states and then identify a reversible path along which entropy changes can be computed.

Entropy Change of a Heat Reservoir

If a heat reservoir, maintained at constant temperature T, receives Q amount of heat, its entropy change is

$$\Delta S = \frac{Q}{T}. \tag{3.8}$$

This is the simplest case of a system receiving heat as its temperature does not vary. However, there are still a few subtleties in writing the above expression as we must ensure that the process that we use in computing the entropy change is indeed internally reversible. The Q amount of heat may not be dumped into the reservoir as one lump sum as the reservoir may not go through a series of equilibrium states (though it is extremely large). Therefore, it is best to inject heat sparingly on many different occasions and to sum all the individual entropy changes to compute the total entropy change. Since the total sum of these interspersed heat inputs is still Q and the temperature is always T, the above expression is valid.

Entropy Change of a System with a Constant Heat Capacity

A body of a constant heat capacity C receives or loses heat to an external system such that its temperature changes from T_i to T_f. To determine its change in entropy, consider the quasistatic process where heat is injected or ejected in infinitesimal amounts on each occasion, over infinitely many occasions. This could be attained theoretically by connecting the body to a series of reservoirs that establish an infinitesimal temperature difference with the current temperature of the body. The body is always in thermodynamic equilibrium and the process is internally reversible. Each infinitesimal amount of heat dQ received results in an infinitesimal change in temperature $dT = \frac{dQ}{C}$. Therefore, the entropy change of the body is

$$\Delta S = \int_{T_i}^{T_f} \frac{dQ}{T} = \int_{T_i}^{T_f} C \cdot \frac{dT}{T} = C \ln \frac{T_f}{T_i}. \tag{3.9}$$

Entropy Change of Ideal Gases

If n moles of an ideal gas with a specific heat capacity at constant volume c_v undergoes a process such that its temperature and volume changes from T_i and V_i to T_f and V_f respectively, its entropy change is

$$\begin{aligned}
\Delta S &= \int \frac{\delta Q}{T} \\
&= \int \frac{dU + pdV}{T} \\
&= \int_{T_i}^{T_f} \frac{nc_v}{T} dT + \int_{V_i}^{V_f} \frac{nR}{V} dV \\
\Delta S &= nc_v \ln \frac{T_f}{T_i} + nR \ln \frac{V_f}{V_i},
\end{aligned} \tag{3.10}$$

since $\delta Q = dU + \delta W_{by}$ by the first law of thermodynamics and $\delta W_{by} = pdV$ for an internally reversible process that we have chosen for the integration. We reiterate the fact that the above expression is valid regardless of the actual process that the gas undergoes. It does not matter if the process is non-quasistatic — such that the gas is not always in equilibrium — as the entropy change is dependent only on the initial and final states.

Joule Expansion

A thermally insulated rectangular container of total volume V_f is initially separated into two regions by a membrane. The left side of the divider contains an ideal gas of volume V_i and temperature T while the right side is empty. A large hole is then punctured on the membrane such that the gas begins to expand freely and finally attains equilibrium. What is the entropy change of this process?

As the container is thermally insulated, there is no heat transfer between the gas and the container. Since there is no work done on or by the gas, as there is nothing to resist the gas' expansion, the internal energy of the gas remains constant. Since the internal energy of a gas is strictly only dependent on the gas' temperature, the temperature of the gas stays constant throughout. Therefore, the entropy change of this process is that of a reversible isothermal expansion of a gas from volume V_i to V_f (even though the actual process is a non-equilibrium process). Substituting $T_i = T_f = T$ into Eq. (3.10),

$$\Delta S = nR \ln \frac{V_f}{V_i}. \tag{3.11}$$

This example emphasizes how the entropy change of an irreversible process should still be calculated via a reversible path between the same initial and final states.

3.4   Fundamental Relation of Thermodynamics

Based on the definition of entropy, we know that for an internally reversible process,

$$\delta Q = T dS,$$

where δQ is the infinitesimal heat supplied to a system at temperature T and dS is the infinitesimal change in entropy of the system (which is a state function and thus has an actual derivative). On another note, the work done during an internally reversible process on a system by an external agent is

$$\delta W_{on} = -p dV.$$

Therefore, the first law of thermodynamics yields

$$dU = T dS - p dV \tag{3.12}$$

for an internally reversible process. However, observe that the equation above is expressed entirely in terms of state variables. Then, Eq. (3.12) must hold for all processes, reversible or irreversible! Eq. (3.12) is known as the fundamental relation of thermodynamics and relates the changes in the state variables of a system that has a uniform pressure and temperature. To further convince yourself that the above equation is valid for all processes, consider the following. The first law of thermodynamics states that

$$dU = \delta Q + \delta W_{on}$$

and this is valid for all processes, both reversible and irreversible. In an irreversible process,[4] $\delta W_{on} > -pdV$ or $\delta W_{by} < pdV$. For example, some possible work by the system could have been converted to heat due to friction. and is thus rendered useless. However, in an irreversible process, $\delta Q < TdS$ as implied by Clausius' inequality. Equation (3.12) just implies that the differences in δW_{on} and δQ in reversible and irreversible processes exactly cancel out!

3.4.1 *Spontaneous Reactions*

Usually, the external environment of a system imposes constraints on the evolution of a system. Suppose that the relevant system only interacts with its surroundings which has instantaneous temperature T_{ext} and pressure p_{ext}. In an infinitesimal process, the system gains δQ heat from its environment — resulting in an entropy change of $-\frac{\delta Q}{T_{ext}}$ in the environment. The entropy change of the system dS must thus satisfy

$$dS - \frac{\delta Q}{T_{ext}} \geq 0$$

for the total entropy change of the universe to be non-negative. Applying the first law of thermodynamics $\delta Q = dU + \delta W_{by}$, where dU is the change in internal energy of the system and δW_{by} is the work done by the system,

$$dU + \delta W_{by} - T_{ext}dS \leq 0.$$

If the work done by the system is only that against its external environment, $\delta W_{by} = p_{ext}dV$ where dV is the change in volume of the system. Thus,

$$dU + p_{ext}dV - T_{ext}dS \leq 0 \tag{3.13}$$

where the equality case only holds for a totally reversible process of the system.

A spontaneous process is defined as one that **can** (not **will**) occur in one direction without any external intervention (defined as inputting work or heat through media besides the surroundings of the system). Since a spontaneous process is foremost irreversible — else, its reverse process can

---

[4] Actually, this result is really proven from the fundamental relation of thermodynamics and Clausius' inequality but intuition guides us to the result as well. Since $dU = \delta Q + \delta W_{on} = TdS - pdV$ and $\delta Q < TdS$, $\delta W_{on} > -pdV$ for irreversible processes.

also occur — we can investigate the conditions for spontaneity under different constraints via the inequality case of the above relationship. However, keep in mind that these are not the only criteria to spark off a reaction as they simply govern whether a process is possible and not whether it will actually take place.[5]

(a) **Isolated System:** Since $\delta Q = 0$ for an isolated system, Eq. (3.13) implies

$$dS > 0$$

for a spontaneous process. The equilibrium state is thus the state of maximum entropy.

(b) **A Solid (Classical Physics):** Since the volume of a solid is approximately constant and its molecules are fixed (such that its entropy cannot vary significantly), $dV = 0$ and $dS = 0$. We then require

$$dU < 0$$

for a spontaneous process — a familiar property in classical mechanics. The equilibrium state is thus the state of minimum internal energy.

(c) **System at Constant Entropy and External Pressure:** In such cases where there is work performed, we are in a slight conundrum as we ideally wish to express everything in terms of properties of the system but δW_{by} is only related to p_{ext}. To circumvent this, we can simply look at states where the system has attained mechanical equilibrium with its environment such that its pressure is $p = p_{ext}$. The process between a pair of such states is then most generally a finite process. As $dS = 0$ and p_{ext} is constant, Eq. (3.13) for a finite step requires

$$\Delta U + p_{ext}\Delta V \leq 0.$$

Since the system's pressure is $p = p_{ext}$ in its initial and final states,

$$\Delta(U + pV) = \Delta H < 0$$

for a spontaneous process, where H is the enthalpy of the system. Notice that the possibility of the intermediate states possessing pressures that differ from p_{ext} does not affect this result since H is a state function.

(d) **System at Constant Volume and External Temperature:** In this case, we only consider states of the system that have attained thermal

---

[5]We also have to take into account the kinetics of the process in general (whether the process will proceed readily).

equilibrium with the external environment (i.e. the system's temperature is $T = T_{ext}$). Since $dV = 0$ and T_{ext} is constant, Eq. (3.13) for a finite step is

$$\Delta U - T_{ext}\Delta S \leq 0.$$

As the system's temperature is $T = T_{ext}$ in its initial and final states,

$$\Delta(U - TS) = \Delta A < 0$$

for a spontaneous process, where $A = U - TS$ is the Helmholtz free energy of the system.

(e) **System at Constant External Temperature and Pressure:** Since p_{ext} and T_{ext} are constant, Eq. (3.13) for finite processes is

$$\Delta U + p_{ext}\Delta V - T_{ext}\Delta S \leq 0.$$

Using a similar procedure as above, we only consider equilibrium states of the system whose pressures and temperatures are equal to those of the external environment. Then, since $p = p_{ext}$ and $T = T_{ext}$ in the initial and final states of the system,

$$\Delta(U + pV - TS) = \Delta(H - TS) = \Delta G < 0$$

for a spontaneous process, where $G = H - TS$ is the Gibbs free energy of the system. Chemists are the most familiar with this condition as their experiments are often conducted under standard laboratory environments (fixed pressure and temperature).

Problems

Heat Engines and Refrigerators

1. *Heat Pump***

Besides the heat engine and refrigerator, there is a third common appliance known as the heat pump. The main objective of a heat pump is to deliver heat to a high-temperature system. It operates by receiving a certain amount of external work $W = Q_H - Q_L$ to withdraw heat Q_L from a low-temperature reservoir of temperature T_L and depositing heat Q_H into the high-temperature reservoir (the system) of temperature $T_H > T_L$. Suggest a measure for the performance of a heat pump (call it the coefficient of performance COP_{HP}) and express the maximum COP_{HP} in terms of T_L and T_H.

   As a concrete example, a building at temperature T is heated by a heat pump which uses a river at temperature T_0 as a heat source. The heat pump, which has the ideal performance, consumes a constant power W while the building loses heat to its surroundings at a rate $\alpha(T - T_0)$, where α is a constant. Show that the equilibrium temperature of the building, $T_e > T_0$, is given by

$$T_e = T_0 + \frac{W}{2\alpha} \left(1 + \sqrt{1 + \frac{4\alpha T_0}{W}} \right).$$

2. *Brayton Cycle***

A Brayton cycle uses an ideal gas as its working substance and consists of the four following reversible steps. The ideal gas is first compressed adiabatically and then expanded isobarically. Afterwards, it is further expanded adiabatically and finally compressed isobarically to its initial state. Determine the efficiency of this engine in terms of the temperatures of the first and second states, T_1 and T_2.

3. *Otto Cycle***

An Otto cycle uses an ideal gas with an adiabatic index γ as its working substance and consists of the four following reversible steps. The ideal gas is first compressed adiabatically and its pressure is then increased isochorically. Afterwards, it is expanded adiabatically and its pressure is finally decreased isochorically to its initial value. Determine the efficiency of this engine in terms of the volumes of the first and second states, V_1 and V_2, and γ.

4. *Stirling Cycle***

A Stirling cycle uses an ideal gas with an adiabatic index γ as its working substance and consists of the four following reversible steps. The ideal gas is first expanded isothermally and its pressure is then decreased isochorically. Afterwards, it is compressed isothermally and its pressure is isochorically increased to its initial value. Determine the efficiency of this engine in terms of the volumes of the first and third states and γ.

5. *Is Carnot Still the Most Efficient?***

We have shown that the Carnot engine is the most efficient engine when operating between two heat reservoirs with temperatures T_H and T_L ($T_H > T_L$), whereby its efficiency is given by $\eta_{Carnot} = 1 - \frac{T_L}{T_H}$. Now, suppose that you are given a series of reservoirs with temperatures ranging between T_L and T_H and are asked to construct a heat engine that interacts with any subset of these reservoirs. Show that the efficiency of an engine with internal irreversibilities (e.g. friction) but a well-defined temperature at every juncture (i.e. its process is quasistatic such that it is always in an equilibrium state) is smaller than that of an internally reversible engine following the same cycle of equilibrium states (hint: use Clausius' inequality). Next, prove that the efficiency of an internally reversible engine constructed with the given reservoirs is no larger than the Carnot efficiency $\eta_{Carnot} = 1 - \frac{T_L}{T_H}$. This shows that a Carnot engine operating between the highest and lowest temperature reservoirs is still the most efficient when a series of reservoirs with intermediate temperatures is available.

The Second Law and Entropy

6. *Gaseous Processes**

Determine the entropy change of n moles of a gas with an adiabatic constant γ in expanding from an initial volume V to a final volume kV under isothermal and isobaric conditions.

7. *Mixing**

A thermally insulated container of total volume V is separated by a frictionless divider into two compartments A and B that have volumes αV and $(1 - \alpha)V$ respectively. n moles of a certain gas fills compartment A and a certain amount of a different gas fills compartment B such that the system is in equilibrium. Determine the entropy change, of the system comprising the two gases, associated with mixing the two gases by removing the divider and waiting till the system attains thermodynamic equilibrium once again.

8. Connected Vessels*

Two thermally insulated vessels of volumes V_1 and V_2 initially contain n_1 and n_2 moles of different monoatomic gases that are at pressures p_1 and p_2. These vessels are then connected by a thermally insulated tube. After the system of vessels equilibrates, determine the change in entropy of the universe.

9. Transferring via a Carnot Engine**

A small body with constant heat capacity C is placed in direct thermal contact with a large reservoir of temperature T_2 such that its temperature is changed from T_1 to T_2. Determine the entropy changes of the body, reservoir and the universe. Show that the entropy change of the universe is non-negative regardless of the relative magnitudes of T_1 and T_2. Now if the heat is delivered to or extracted from the small body via a Carnot engine operating between the large reservoir and the small body, determine the entropy changes of the body, reservoir and the universe.

10. Verifying the Second Law**

Two substances of heat capacities C_1, C_2 and initial temperatures T_1 and T_2 are placed in thermal contact. They are isolated from their surroundings. When thermal equilibrium is subsequently achieved, determine the entropy change of the universe and show that it must be non-negative.

11. Maximum Work Done**

Determine the maximum work obtainable from a heat engine connected to two reservoirs of constant heat capacities C_H and C_L at initial temperatures T_H and $T_L < T_H$.

12. Pushing a Piston**

A cylindrical container is separated by a fixed divider with a valve and a frictionless piston is attached to its open right end. The walls of the cylinder, divider and piston are perfect thermal insulators. The cylinder is filled with 12g of helium in the left compartment and 2g of helium in the right. Initially, the pressures, volumes and temperatures of the gases are respectively, 6atm,

11.2L and 273K in the left side, and 1atm, 11.2L and 273K in the right side. The specific heat capacity (note that this is per unit mass) of helium at constant pressure is $c_p = 5.25\text{J/g K}$. The piston is pushed towards the divider by a reversible compression until the pressure on the right side equals 6atm. At this juncture, the valve opens and the whole system is allowed to reach equilibrium. What is the final equilibrium temperature? Find the total entropy change of the whole process. (Singapore Physics Olympiad)

13. *Reversible Heat Transfer***

A body of constant heat capacity C is heated up from a temperature T_1 to T_2 by bringing it into thermal contact and waiting till it establishes thermal equilibrium with N large reservoirs of temperatures $T_1 + \Delta T$, $T_1 + 2\Delta T, \ldots, T_1 + (N-1)\Delta T$, T_2 in ascending order of temperature, where $\Delta T = \frac{T_2 - T_1}{N}$. Determine the net entropy change of the universe due to this process. Then, take the limit of $N \to \infty$ and $\Delta T \to dT$ where dT is an infinitesimal change in temperature and show that this process is reversible.

Solutions

1. Heat Pump**

Since the primary aim of a heat pump is to deliver heat Q_H via external work W, its coefficient of performance is

$$COP_{HP} = \frac{Q_H}{W} = \frac{Q_H}{Q_H - Q_L} = \frac{1}{1 - \frac{Q_L}{Q_H}}.$$

There are various ways to show that the maximum COP_{HP} occurs when the heat pump is a Carnot cycle in reverse. The most direct way to do this is to observe that $COP_{HP} = COP_{FR} + 1$ where COP_{FR} is the coefficient of performance if we used the heat pump as a refrigerator. Since COP_{FR} is maximized when the refrigerator is reversible (reverse Carnot cycle), COP_{HP} is maximized when the heat pump is a reverse Carnot cycle. Another method is to exploit the fact that the entropy change of the universe must be non-negative due to a heat pump cycle. Then,

$$\frac{Q_H}{T_H} - \frac{Q_L}{T_L} \geq 0$$

$$\frac{Q_L}{Q_H} \leq \frac{T_L}{T_H}$$

$$COP_{HP} \leq \frac{1}{1 - \frac{T_L}{T_H}} = \frac{T_H}{T_H - T_L}.$$

Next, when the building is at its equilibrium temperature T_e, its rate of heat loss to its surroundings must be equal to its rate of heat gain from the heat pump.

$$\alpha(T_e - T_0) = W\frac{T_e}{T_e - T_0}$$

$$\alpha T_e^2 - (2\alpha T_0 + W)T_e + \alpha T_0^2 = 0$$

$$T_e = \frac{2\alpha T_0 + W + \sqrt{(2\alpha T_0 + W)^2 - 4\alpha^2 T_0^2}}{2\alpha} = T_0 + \frac{W}{2\alpha}\left(1 + \sqrt{1 + \frac{4\alpha T_0}{W}}\right),$$

where we have chosen the root that is greater than T_0.

Note: In our following solutions for heat engines and refrigerators, the ith state is defined to have pressure p_i, volume V_i and temperature T_i.

2. Brayton Cycle**

The system takes in heat during process $2 \to 3$ and releases heat during process $4 \to 1$.

$$Q_{23} = nc_p\Delta T = \frac{c_p}{R}(p_3 V_3 - p_2 V_2) = \frac{c_p}{R}p_2(V_3 - V_2)$$

$$Q_{41} = \frac{c_p}{R}p_1(V_1 - V_4)$$

where c_p is the isobaric molar heat capacity of the gas medium. The total work done by the gas in a single cycle is the net heat supplied to the ideal gas as its internal energy remains constant.

$$W = Q_{23} + Q_{41} = \frac{c_p}{R}[p_2(V_3 - V_2) + p_1(V_1 - V_4)].$$

The efficiency is then

$$\eta = \frac{W}{Q_{in}} = \frac{W}{Q_{23}} = \frac{p_2(V_3 - V_2) + p_1(V_1 - V_4)}{p_2(V_3 - V_2)}.$$

To simplify the above expression, we know from the adiabatic condition, applied to processes $1 \to 2$ and $3 \to 4$, that

$$p_1 V_1^\gamma = p_2 V_2^\gamma$$

$$p_4 V_4^\gamma = p_3 V_3^\gamma \implies p_1 V_4^\gamma = p_2 V_3^\gamma,$$

$$\frac{V_1}{V_4} = \frac{V_2}{V_3}.$$

The efficiency can be expressed as

$$\eta = 1 + \frac{p_1(V_1 - V_4)}{p_2(V_3 - V_2)}$$

$$= 1 + \frac{p_1 V_1 \left(1 - \frac{V_4}{V_1}\right)}{p_2 V_2 \left(\frac{V_3}{V_2} - 1\right)}$$

$$= 1 - \frac{p_1 V_1}{p_2 V_2}$$

$$= 1 - \frac{T_1}{T_2}.$$

3. Otto Cycle**

The engine absorbs heat during process $2 \to 3$ and ejects heat during process $4 \to 1$.

$$Q_{23} = nc_v(T_3 - T_2) = \frac{c_v}{R}(p_3V_3 - p_2V_2) = \frac{c_v}{R}V_2(p_3 - p_2),$$

$$Q_{41} = \frac{c_v}{R}(p_1V_1 - p_4V_4) = \frac{c_v}{R}V_1(p_1 - p_4),$$

where c_v is the molar heat capacity of the gas at constant volume. The total work done by the gas in a single cycle is the net heat supplied.

$$W = Q_{23} + Q_{41}.$$

The efficiency is then

$$\eta = \frac{W}{Q_{23}} = 1 + \frac{V_1(p_1 - p_4)}{V_2(p_3 - p_2)}.$$

To simplify the above expression, apply the adiabatic condition to processes $1 \to 2$ and $3 \to 4$.

$$p_1 V_1^\gamma = p_2 V_2^\gamma$$

$$p_4 V_1^\gamma = p_3 V_2^\gamma$$

$$\implies \frac{p_1}{p_4} = \frac{p_2}{p_3}.$$

Then,

$$\eta = 1 + \frac{p_1 V_1 \left(1 - \frac{p_4}{p_1}\right)}{p_2 V_2 \left(\frac{p_3}{p_2} - 1\right)} = 1 - \frac{p_1 V_1}{p_2 V_2}.$$

Since $p_1 V_1^\gamma = p_2 V_2^\gamma$, $\frac{p_1 V_1}{p_2 V_2} = \frac{V_2^{\gamma-1}}{V_1^{\gamma-1}}$.

$$\eta = 1 - \left(\frac{V_2}{V_1}\right)^{\gamma-1}.$$

4. Stirling Cycle**

The work done by an ideal gas in a reversible isothermal process in which its volume changes from V_i to V_f is

$$\int pdV = \int \frac{nRT}{V} dV = nRT \ln \frac{V_f}{V_i}.$$

The work done by the ideal gas during processes $1 \to 2$ and $3 \to 4$ are

$$W_{12} = nRT_1 \ln \frac{V_3}{V_1},$$

$$W_{34} = nRT_3 \ln \frac{V_1}{V_3}.$$

The total work done by the gas is then

$$W = W_{12} + W_{34} = nR(T_1 - T_3) \ln \frac{V_3}{V_1}.$$

The ideal gas absorbs heat during processes $4 \to 1$ and $1 \to 2$. The former heat absorbed is $nc_v \Delta T = \frac{1}{\gamma - 1} nR(T_1 - T_3)$ while the latter is the work done by the gas from $1 \to 2$, W_{12}, as its internal energy remains constant during an isothermal process.

$$Q = \frac{1}{\gamma - 1} nR(T_1 - T_3) + nRT_1 \ln \frac{V_3}{V_1}.$$

The efficiency of the engine is then

$$\eta = \frac{W}{Q} = \frac{nR(T_1 - T_3) \ln \frac{V_3}{V_1}}{\frac{nR(T_1 - T_3)}{\gamma - 1} + nR(T_1 - T_3) \ln \frac{V_3}{V_1}} = \frac{T_1 - T_3}{\frac{T_1 - T_3}{(\gamma - 1) \ln \frac{V_3}{V_1}} + T_1 - T_3}.$$

5. Is Carnot Still the Most Efficient?**

Since the internal energy of an engine does not change after a cycle, the efficiency of a heat engine is

$$\eta = \frac{W}{Q_{in}} = \frac{Q_{in} - Q_{out}}{Q_{in}} = 1 - \frac{Q_{out}}{Q_{in}}$$

where Q_{in} is the sum of the positive values of heat influxes while Q_{out} is the sum of the positive values of heat outflows. By Clausius' inequality,

$$\delta Q \leq T dS$$

for any infinitesimal process (the equality case holds when the process is internally reversible). Now, we shall classify the various δQ and dS's into positive quantities and negative quantities. The positive quantities are labeled as δQ_{pos} and dS_{pos} while the absolute values of the negative quantities are labeled as δQ_{neg} and dS_{neg}. Then,

$$Q_{in} = \int \delta Q_{pos},$$

$$Q_{out} = \int \delta Q_{neg},$$

while Clausius' inequality implies that

$$\delta Q_{pos} \leq TdS_{pos}$$

$$-\delta Q_{neg} \leq -TdS_{neg} \implies \delta Q_{neg} \geq TdS_{neg}.$$

Therefore,

$$\eta = 1 - \frac{\int \delta Q_{neg}}{\int \delta Q_{pos}} \leq 1 - \frac{\int TdS_{neg}}{\int TdS_{pos}}$$

where the equality case only holds for internally reversible engines. We have hence proven the first claim. To prove the second one, observe that

$$\int TdS_{neg} \geq \int T_L dS_{neg} = T_L \int dS_{neg},$$

$$\int TdS_{pos} \leq \int T_H dS_{pos} = T_H \int dS_{pos}.$$

Furthermore,

$$\int dS_{neg} = \int dS_{pos}$$

as $\int dS_{pos} - \int dS_{neg} = 0$ for the cycle to return to its original state (since entropy is a state function). The efficiency of an internally reversible engine is

$$\eta = 1 - \frac{\int TdS_{neg}}{\int TdS_{pos}} \leq 1 - \frac{T_L \int dS_{neg}}{T_H \int dS_{pos}} = 1 - \frac{T_L}{T_H} = \eta_{Carnot}.$$

6. Gaseous Processes*

We will be using Eq. (3.10). During an isothermal expansion,

$$\Delta S = nR \ln \frac{kV}{V} = nR \ln k.$$

During an isobaric expansion, the pressure is constant. Since the volume of the gas expands by k times, its temperature must also increase by a factor of k by the ideal gas law. Then,

$$\Delta S = nc_v \ln k + nR \ln k = \frac{\gamma}{\gamma - 1} nR \ln k.$$

7. Mixing*

The two gases initially have a common temperature and pressure. In order for the pressures to be balanced, the number of moles of gas molecules in compartment B must be

$$\frac{1 - \alpha}{\alpha} n.$$

After the removal of the divider, the two gases undergo free expansion, just like the Joule expansion. Their final common temperature must be the same as their initial common temperature as energy is conserved (the walls of the container are insulated). Therefore, the total change in entropy is the sum of the two changes in entropy associated with the free expansions of the two gases. By Eq. (3.11),

$$\Delta S = nR \ln \frac{V_f}{V_A} + \frac{1 - \alpha}{\alpha} nR \ln \frac{V_f}{V_B} = -nR \ln \alpha - \frac{1 - \alpha}{\alpha} nR \ln (1 - \alpha).$$

8. Connected Vessels*

We need to determine the final pressure p of the combined set-up. Since the system is thermally insulated, its total energy must be the same. Its initial energy is $\frac{3}{2} p_1 V_1 + \frac{3}{2} p_2 V_2$. Then,

$$\frac{3}{2} p(V_1 + V_2) = \frac{3}{2} p_1 V_1 + \frac{3}{2} p_2 V_2$$

$$p = \frac{p_1 V_1 + p_2 V_2}{V_1 + V_2}.$$

Next, we will need to calculate the final temperature of the combined system T_f. By the ideal gas equation,

$$p(V_1 + V_2) = (n_1 + n_2) R T_f$$

$$T_f = \frac{p_1 V_1 + p_2 V_2}{(n_1 + n_2) R}.$$

Applying Eq. (3.10) to the two gases, the total change in entropy of the universe is

$$\Delta S = \frac{3}{2}n_1 R \ln \frac{T_f}{T_1} + n_1 R \ln \frac{V_1 + V_2}{V_1} + \frac{3}{2}n_2 R \ln \frac{T_f}{T_2} + n_2 R \ln \frac{V_1 + V_2}{V_2}$$

$$= \frac{3}{2}n_1 R \ln \frac{n_1(p_1 V_1 + p_2 V_2)}{(n_1 + n_2)p_1 V_1} + n_1 R \ln \frac{V_1 + V_2}{V_1}$$

$$+ \frac{3}{2}n_2 R \ln \frac{n_2(p_1 V_1 + p_2 V_2)}{(n_1 + n_2)p_2 V_2} + n_2 R \ln \frac{V_1 + V_2}{V_2},$$

where T_1 and T_2 are the initial temperatures of the respective vessels.

9. Transferring via a Carnot Engine**

In the first case, the body receives $C(T_2 - T_1)$ amount of heat. Therefore, the entropy change of the large reservoir is

$$\Delta S_{res} = -\frac{C(T_2 - T_1)}{T_2}$$

as the reservoir correspondingly loses $C(T_2 - T_1)$ amount of heat (which is possibly negative though). The entropy change of the body is given by Eq. (3.9),

$$\Delta S_{body} = C \ln \frac{T_2}{T_1}.$$

The entropy change of the universe is the sum of the two above entropies.

$$\Delta S_{universe} = C \ln \frac{T_2}{T_1} - \frac{C(T_2 - T_1)}{T_2}.$$

Defining a new variable $x = \frac{T_1}{T_2}$,

$$\Delta S_{universe} = C(x - 1) - C \ln x.$$

Substituting $x = 1$, we obtain $\Delta S_{universe} = 0$ which is expected. Now consider

$$\frac{d\Delta S_{universe}}{dx} = C - \frac{C}{x}$$

which is negative when $x < 1$ and positive when $x > 1$. Coupled with the fact that $\frac{d\Delta S_{universe}}{dx} = 0$ when $x = 1$, $x = 1$ is a local minimum and $\Delta S_{universe} \geq 0$ for all x. In the second case, the entropy change of the reservoir is different as it needs to supply more heat (as some of the heat is converted to work by the Carnot engine). Then, let t be the instantaneous temperature of the body, δQ_H be the infinitesimal heat extracted from the reservoir and δQ_L be the infinitesimal heat delivered to the body in a single infinitesimal cycle (note that δQ_H and δQ_L are possibly negative). The infinitesimal change in entropy of the reservoir due to this infinitesimal Carnot engine cycle is

$$dS_{res} = -\frac{\delta Q_H}{T_2}.$$

We also know that for a Carnot engine,

$$\frac{\delta Q_H}{T_2} = \frac{\delta Q_L}{t}.$$

Then,

$$\Delta S_{res} = \int -\frac{\delta Q_H}{T_2} = -\int_{T_1}^{T_2} \frac{\delta Q_L}{t} = -\int_{T_1}^{T_2} C \cdot \frac{dt}{t} = -C \ln \frac{T_2}{T_1}.$$

The entropy change of the body remains the same as its initial and final states are identical to those in the first case, $\Delta S_{body} = C \ln \frac{T_2}{T_1}$. The total entropy change of the universe is then

$$\Delta S_{universe} = 0,$$

which is expected of a Carnot engine as it is reversible.

10. Verifying the Second Law**

Denoting the final equilibrium temperature of the two substances as T_f, the conservation of internal energy implies that

$$T_f = \frac{C_1 T_1 + C_2 T_2}{C_1 + C_2}.$$

The total entropy change of the universe is

$$\Delta S = \int_{T_1}^{T_f} \frac{C_1 dT}{T} + \int_{T_2}^{T_f} \frac{C_2 dT}{T}$$

$$= C_1 \ln \frac{T_f}{T_1} + C_2 \ln \frac{T_f}{T_2} = C_1 \ln \left(\frac{C_1 + C_2 \frac{T_2}{T_1}}{C_1 + C_2} \right) + C_2 \ln \left(\frac{C_1 \frac{T_1}{T_2} + C_2}{C_1 + C_2} \right).$$

Defining $x = \frac{T_2}{T_1}$,

$$\Delta S = C_1 \ln \left(\frac{C_1 + C_2 x}{C_1 + C_2} \right) + C_2 \ln \left(\frac{\frac{C_1}{x} + C_2}{C_1 + C_2} \right).$$

When $x = 1$, $\Delta S = 0$ as expected. Furthermore,

$$\frac{d\Delta S}{dx} = C_1\frac{C_2}{C_1 + C_2 x} - C_2\frac{\frac{C_1}{x}}{C_1 + C_2 x} = \frac{C_1 C_2}{C_1 + C_2 x}\left(1 - \frac{1}{x}\right)$$

which is negative when $x < 1$ and positive when $x > 1$. Since $\frac{d\Delta S}{dx} = 0$ when $x = 1$, $x = 1$ is a local minimum — implying that $\Delta S \geq 0$.

Alternatively, we can apply the AM-GM inequality as follows:

$$\frac{T_1 + T_1 + \cdots + T_2 + T_2 + \cdots}{kC_1 + kC_2} \geq \sqrt[kC_1 + kC_2]{T_1 \times T_1 \times \cdots \times T_2 \times T_2 \times \cdots}$$

where k is a real number such that kC_1 and kC_2 are integers and where T_1 and T_2 are included kC_1 and kC_2 times respectively. Then,

$$\left(\frac{C_1 T_1 + C_2 T_2}{C_1 + C_2}\right)^{k(C_1 + C_2)} \geq T_1^{kC_1} T_2^{kC_2}$$

$$\implies T_f^{C_1 + C_2} \geq T_1^{C_1} T_2^{C_2},$$

$$(C_1 + C_2)\ln T_f \geq C_1 \ln T_1 + C_2 \ln T_2$$

$$\Delta S = C_1 \ln \frac{T_f}{T_1} + C_2 \ln \frac{T_f}{T_2} \geq 0.$$

The equality only holds if $T_1 = T_2$.

11. Maximum Work Done**

Let the final common temperature of the reservoirs be T_f. Then, the total work done by the heat engine is

$$W = C_H(T_H - T_f) + C_L(T_L - T_f)$$

by the first law of thermodynamics. We then seek to minimize T_f. Let the instantaneous temperatures of the two reservoirs be t_H and t_L respectively. Applying the second law of thermodynamics to an infinitesimal heat engine cycle,

$$\frac{\delta Q_H}{t_H} + \frac{\delta Q_L}{t_L} \geq 0$$

where δQ_H and δQ_L are the infinitesimal heat supplied to the heat source and sink, respectively. Since $\delta Q_H = C_H dt_H$ and $\delta Q_L = C_L dt_L$, integrating over the relevant limits would yield

$$\int_{T_L}^{T_f} C_L \frac{dt_L}{t_L} \geq \int_{T_H}^{T_f} -\frac{C_H dt_H}{t_H}$$

$$C_L \ln \frac{T_f}{T_L} \geq C_H \ln \frac{T_H}{T_f}$$

$$T_f \geq T_H^{\frac{C_H}{C_H+C_L}} T_L^{\frac{C_L}{C_H+C_L}}.$$

Therefore,

$$W \leq C_H T_H + C_L T_L - (C_H + C_L) T_H^{\frac{C_H}{C_H+C_L}} T_L^{\frac{C_L}{C_H+C_L}}.$$

12. Pushing a Piston**

When the piston is pushed towards the divider, the gas on the right undergoes an adiabatic compression. When the valve is opened subsequently, mixing occurs and an equilibrium is attained.

Let T_i, T_f, P_i and P_f denote the initial and final temperatures and pressures of the right gas before and after the compression (immediately before the mixing). By the adiabatic condition,

$$T_f = T_i \left(\frac{P_f}{P_i}\right)^{1-\frac{1}{\gamma}} = 273 \times 6^{\frac{2}{5}} = 559K,$$

where we have substituted $T_i = 273K$, $P_i = 1\text{atm}$, $P_f = 6\text{atm}$ and $\gamma = \frac{5}{3}$ for a monoatomic gas (helium). When the gases are mixed, the total internal energy in the cylinder must remain constant. Denoting n_l and n_r as the number of moles of helium in the left and right compartments respectively,

$$\frac{3}{2}n_l R \cdot 273 + \frac{3}{2}n_r R \cdot 559 = \frac{3}{2}(n_l + n_r)R \cdot T$$

where T is the final equilibrium temperature. Since $n_l = 6n_r$,

$$T = 314K.$$

There is no entropy change during the adiabatic compression. To study the entropy change due to the mixing, observe that the final equilibrium pressure is still 6atm. This is because the total internal energy of the set-up is conserved during the mixing and the initial total internal energy is (by the ideal gas law) $\frac{3}{2}pV_l + \frac{3}{2}pV_r = \frac{3}{2}p(V_l + V_r)$ where V_l and V_r are the volumes of the left and right gases before mixing while $p = 6\text{atm}$. The final total internal energy is $\frac{3}{2}p_f(V_l + V_r)$ where p_f is the final equilibrium pressure. Therefore,

$p_f = p$. Since the final equilibrium pressure is still 6atm, the entropy change of each gas is equivalent to that of it undergoing a reversible isobaric process from its initial to final temperature.

$$dS = \frac{\delta Q_{rev}}{T} = \frac{mc_p dT}{T}$$

$$\Delta S = c_p m_l \int_{273}^{314} \frac{dT}{T} + c_p m_r \int_{559}^{314} \frac{dT}{T}$$

$$= 5.25 \left(12 \ln \frac{314}{273} + 2 \ln \frac{314}{559} \right) = 2.76 \text{J/K}$$

where m_l and m_r are the masses of the gases originally in the left and right compartments respectively.

13. Reversible Heat Transfer**

By considering the initial and final states of the body and applying Eq. (3.9), the total entropy change of the body is

$$\Delta S_{body} = C \ln \frac{T_2}{T_1}.$$

Moving on, observe that the ith reservoir of temperature $T_1 + i\Delta T$ is responsible for increasing the body's temperature from $T_1 + (i-1)\Delta T$ to $T_1 + i\Delta T$. The heat transferred from the reservoir to the body in this process is $C\Delta T$. Then, the change in entropy of the ith reservoir is evidently $-\frac{C\Delta T}{T_i + i\Delta T}$ where the negative sign indicates that it loses heat. The total entropy change of the universe due to the entire procedure in the question is the sum of the entropy changes of the body and the reservoirs.

$$\Delta S_{universe} = C \ln \frac{T_2}{T_1} - C\Delta T \sum_{i=1}^{N} \frac{1}{T_1 + i\Delta T}.$$

As $N \to \infty$ and $\Delta T \to dT$, the latter sum becomes an integral. The total entropy change of the universe is then

$$\Delta S_{universe} = C \ln \frac{T_2}{T_1} - C \int_{T_1}^{T_2} \frac{1}{T} dT$$

$$= C \ln \frac{T_2}{T_1} - C \ln \frac{T_2}{T_1}$$

$$= 0$$

which shows that this process is reversible!

Chapter 4

Heat Transfer and Phase Transitions

This chapter will analyze the common forms of heat transfer — convection, conduction and radiation — and their accompanying effects such as expansion and phase changes.

4.1 Convection

Convective heat transfers are difficult to analyze rigorously but a rule of thumb adequate for small temperature differences is Newton's law of cooling. It states that the net heat flux density $\dot{q}$ — the net power transmitted per unit perpendicular area — between a small area on a liquid or solid surface and the surrounding air (which convects heat away) is proportional to the temperature difference between them for small differences. Concretely,

$$\dot{q} = -h(T_s - T_a)$$

where T_s is the temperature of the small area on the surface while T_a is the temperature of the air shrouding our set-up. The negative sign hinges on the fact that a surface of higher temperature loses heat to its surroundings. h is a constant of proportionality that must be determined empirically (as this is only an approximate relationship) and is commonly referred to as the heat transfer coefficient. The total net power $\dot{Q}$ transferred between a surface with uniform temperature and its surroundings is then $\dot{q}$ multiplied by its surface area A.

$$\dot{Q} = -hA(T_s - T_a).$$

Problem: Assuming that Newton's law of cooling holds with a heat transfer coefficient h, determine the instantaneous temperature $T(t)$ of a cup of coffee with constant heat capacity C, whose interface with air has a constant surface area A, as a function of time. The temperature of air in the room is approximately a constant T_a as air is vast and the initial temperature of the coffee is $T_0 > T_a$. Assume that the coffee is homogeneous at all times.

Let the instantaneous temperature of the coffee be T. By Newton's law of cooling, its net heat flux with its environment is

$$\dot{Q} = -hA(T - T_a).$$

Since $\dot{Q} = C\dot{T}$,

$$C\dot{T} = -hA(T - T_a)$$

$$\int_{T_0}^{T} \frac{1}{T - T_a} dT = \int_0^t -\frac{hA}{C} dt$$

$$\ln \left| \frac{T - T_a}{T_0 - T_a} \right| = -\frac{hA}{C} t.$$

Observing that $T \geq T_a$ at all times since $T_0 > T_a$ (more specifically, $\dot{T}$ is negative only when $T > T_a$ and becomes zero when $T = T_a$), we can remove the absolute value brackets.

$$T = (T_0 - T_a)e^{-\frac{hA}{C}t} + T_a.$$

4.2   Conduction

Conduction occurs within a substance due to collisions between its constituent particles and the diffusion of particles. The collisions between excited particles and less energetic particles and the net diffusion of more energetic particles result in the transfer of energy from regions of higher temperature to regions of lower temperature. Quantitatively, Fourier's law of conduction states that the heat flux density is proportional to the temperature gradient. In a one-dimensional heat flow along the x-direction,

$$\dot{q} = -k\frac{dT}{dx} \tag{4.1}$$

where $\dot{q}$ is the heat flux density and $\frac{dT}{dx}$ is the temperature gradient. k is known as the thermal conductivity and is dependent on the various properties of the conducting medium. The negative sign stems from the fact that the heat flux density points in the direction of decreasing temperature. Since a one-dimensional flow is assumed, the total heat flux $\dot{Q}$ across a surface of area A normal to the x-direction is

$$\dot{Q} = -kA\frac{dT}{dx}. \tag{4.2}$$

Usually, we will be analyzing steady state systems where there is no longer any change in the temperature of any point on the substance with respect to time. There can, however, still be heat conducted throughout the substance as long as the heat influx is equal to the heat outflow for each point on the substance if heat is not generated anywhere in the substance. This condition is known as the continuity of heat flux which ensures that no net heat is stored anywhere in the substance. If there is heat generated by portions of the substance itself, the outflow must be greater than the influx for equilibrium to be maintained.

Problem: Consider a slab of thickness l and uniform cross sectional area A, in Fig. 4.1. Its ends are maintained at T_0 and T_1. Assuming that the system has reached steady state, find the heat flux through the cross section of the slab and the temperature of a layer at a distance x from the end at T_0 as a function of x.

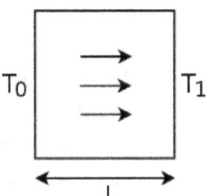

Figure 4.1:   Slab

From Fourier's law of conduction,

$$\dot{Q} = -kA\frac{dT}{dx}$$

where the heat flux $\dot{Q}$ is defined to be positive rightwards. Since the system is at equilibrium, we can leverage the fact that $\dot{Q}$ is uniform throughout all cross sections to determine its value.

$$\int_0^l \dot{Q}dx = \int_{T_0}^{T_1} -kAdT$$

$$\dot{Q} = \frac{kA(T_0 - T_1)}{l}. \tag{4.3}$$

To determine the temperature $T(x)$ of a layer at a distance x from the left end, we integrate the expression with more general limits.

$$\int_0^x \dot{Q}dx = \int_{T_0}^T -kAdT$$

$$T - T_0 = -\frac{\dot{Q}}{kA}x$$

$$T = \frac{T_1 - T_0}{l}x + T_0.$$

Observe that from Eq. (4.3), we can relate the temperature difference across the two ends of the slab and the heat flux in the following manner.

$$T_0 - T_1 = \Delta T = \dot{Q} \cdot \frac{l}{kA}.$$

This holds for all substances with a uniform thermal conductivity and cross sectional area (slabs in general). Scrutinizing the above, one may notice that it is completely analogous to Ohm's law for a resistor,

$$V = IR,$$

where V is the voltage, I is the current and R is the electrical resistance. Temperature and heat flux are then analogous to voltage and current in a circuit. $\frac{l}{kA}$ is the thermal equivalent of electrical resistance, which we shall refer to as thermal resistance. Observe that the expression for thermal resistance is also completely analogous to that for electrical resistance for a resistor with a constant cross section. In the case of the latter, for a resistor with conductivity σ, length l and a constant cross sectional area A,

$$R = \frac{l}{\sigma A}.$$

With that said, the following two equations can be written down. For a steady state one-dimensional heat conduction with no heat generation at any point in the system, the temperature difference between two surfaces is directly proportional to the heat flux across them.

$$\Delta T = \dot{Q}R \tag{4.4}$$

where R is the thermal resistance. Its value for a substance of uniform thermal conductivity k, length l and cross sectional area A is

$$R = \frac{l}{kA}. \tag{4.5}$$

The thermal resistances of more general configurations need to be calculated in other ways. Besides the similarity of resistances, analogies can be drawn between Kirchhoff's laws and certain properties in a thermal circuit. Kirchhoff's loop rule, which states that the sum of voltages along a loop is zero, is superficial in this context as its thermal counterpart basically asserts that the sum of temperature differences along a loop is zero. However, the analogous version of Kirchhoff's junction rule, which enforces the condition that the net current flowing out of a junction is zero at steady state, is rather crucial. This is in fact the continuity of heat flux which asserts that the net heat flux emanating from each point in a set-up with no heat generated must be zero at steady state (else its temperature will vary). We will delve further into the ramifications of these analogies right after the following example.

Varying Contact Area

For certain geometries of substances, the contact area may vary with x. However, the heat flux should still be continuous throughout layers of the substance in the steady state regime, as long as the substance does not generate any heat by itself. Then, we may need to perform an integration to calculate the rate of heat conduction.

Problem: In Fig. 4.2, a long cylindrical shell has an inner radius r_0 and outer radius r_1, length l ($l \gg r_1$) and a uniform thermal conductivity k. The temperatures of the inner and outer surfaces are maintained at T_0 and T_1 respectively. When the system has attained steady state, determine the heat flux across cylindrical shells and thus the thermal resistance of this set-up. Neglect any edge effects.

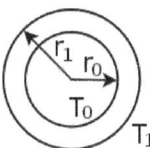

Figure 4.2: Cylindrical shell

Due to the axial symmetry and comparatively large length of this set-up, heat purely flows in the radial direction, perpendicular to the cylindrical axis, while temperature is purely a function of radial distance from the axis. Consider a cylindrical shell that is of radius r, length l and thickness dr from the center of the original cylindrical shell. The heat flux density should be uniform across this surface of area $2\pi r l$ due to symmetry. By Fourier's law of conduction, the total heat flux through this shell is

$$\dot{Q} = -k2\pi rl\frac{dT}{dr}.$$

For the system to be in steady state, the heat flux across all cylindrical shells must be the same so that the net heat flux into each layer is zero. Then,

$$\dot{Q}\int_{r_0}^{r_1}\frac{1}{r}dr = -2\pi kl\int_{T_0}^{T_1}dT$$

$$\dot{Q} = \frac{2\pi kl(T_0 - T_1)}{\ln\frac{r_1}{r_0}}.$$

It can also be seen that the thermal resistance is

$$R = \frac{\ln\frac{r_1}{r_0}}{2\pi kl}. \tag{4.6}$$

4.2.1   *Equivalent Resistance*

The analogy between thermal resistors and electrical resistors extends beyond a single resistor. We can determine the effective thermal resistance for parallel and series configurations of various materials with different thermal conductivity due to the analogy between continuities of heat and current fluxes in steady state systems. When no heat is generated by a substance, there must be no net heat flux entering or leaving each surface as this would lead to an accumulation or deficit in internal energy — implying that the system has not reached steady state yet. As remarked previously, this is similar to Kirchhoff's junction rule in circuitry.

Series Configuration

Before we derive an expression for the general case, consider the following auxiliary problem. Two slabs of different thicknesses, l_1 and l_2, uniform cross sectional area A and thermal conductivities k_1 and k_2 are connected in series as shown in Fig. 4.3. The two ends are maintained at temperatures T_0 and T_2 respectively. Find the heat flux, the effective thermal resistance of the combined system and the temperature of the interface, T_1, at steady state.

Again, it is important to note the continuity of heat fluxes at the two sides of the middle interface. If the heat flux between the left side and the interface is $\dot{Q}$, the heat flux between the interface and the right surface must also be $\dot{Q}$. Next, an equivalent thermal circuit can be drawn as shown above.

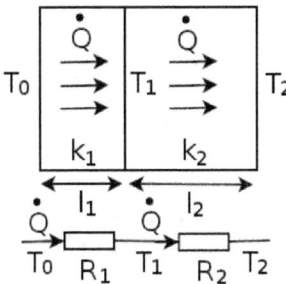

Figure 4.3:   Slabs in series

Recalling our definition of R previously, we can define

$$R_1 = \frac{l_1}{k_1 A},$$

$$R_2 = \frac{l_2}{k_2 A}.$$

Then applying the results derived previously,

$$\dot{Q} = \frac{T_0 - T_1}{R_1},$$

$$\dot{Q} = \frac{T_1 - T_2}{R_2}.$$

Eliminating T_1, we get

$$T_0 - T_2 = (R_1 + R_2)\dot{Q}$$

$$\dot{Q} = \frac{T_0 - T_2}{R_1 + R_2} = \frac{(T_0 - T_2)A}{\frac{l_1}{k_1} + \frac{l_2}{k_2}}.$$

The equivalent resistance is defined such that

$$T_0 - T_2 = R_{eq}\dot{Q}$$

$$\implies R_{eq} = R_1 + R_2 = \frac{l_1}{k_1 A} + \frac{l_2}{k_2 A}.$$

T_1 can be solved for by eliminating $\dot{Q}$ in our original simultaneous equations.

$$\left(\frac{1}{R_1} + \frac{1}{R_2}\right) T_1 = \frac{T_0}{R_1} + \frac{T_2}{R_2}$$

$$T_1 = \frac{R_2 T_0 + R_1 T_2}{R_1 + R_2} = \frac{k_1 l_2 T_0 + k_2 l_1 T_2}{k_1 l_2 + k_2 l_1}.$$

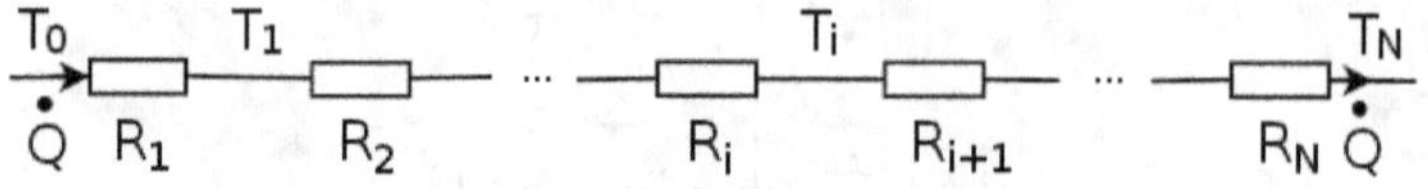

Figure 4.4:   Equivalent circuit

In general, for a thermal circuit constructed from an array of thermal resistors arranged in series, an equivalent thermal resistance can be derived. Referring to Fig. 4.4, let there be a total of N thermal resistors with resistances $R_1, R_2, \ldots, R_N$ and let T_0 and T_N be the temperatures of the ends of the circuit (maintained at constant temperature). Then for $0 < i < N$, let T_i be the temperature of the interface between the ith and $(i+1)$th thermal resistors. We would like to find the equivalent resistance of this circuit R_{eq} and the various T_i's. For a system in equilibrium, $\dot{Q}$ must be constant throughout. Thus,

$$R_{i+1}\dot{Q} = T_i - T_{i+1}$$

for all $0 \le i < N$. Summing the above for all i, we get

$$\sum_{i=1}^{N} R_i \dot{Q} = T_0 - T_N$$

$$\dot{Q} = \frac{T_0 - T_N}{\sum_{i=1}^{N} R_i},$$

$$R_{eq} = \sum_{i=1}^{N} R_i. \tag{4.7}$$

To calculate T_i we can spilt the circuit into two components, one containing resistors R_1 to R_i and the other containing R_{i+1} to R_N. Then, we can compute the equivalent resistances for these two parts.

$$R_{1i} = \sum_{j=1}^{i} R_j,$$

$$R_{(i+1)N} = \sum_{j=i+1}^{N} R_j.$$

Then,

$$\dot{Q} = \frac{T_0 - T_i}{R_{1i}},$$

$$\dot{Q} = \frac{T_i - T_N}{R_{(i+1)N}}.$$

Eliminating $\dot{Q}$,

$$T_i = \frac{R_{(i+1)N}T_0 + R_{1i}T_N}{R_{1i} + R_{(i+1)N}}.$$

Parallel Configuration

Now consider another auxiliary problem of two slabs, of surface areas A_1 and A_2, equal thickness l and thermal conductivities k_1 and k_2 connected in parallel, as depicted in Fig. 4.5. Let the ends be maintained at temperatures T_0 and T_1.

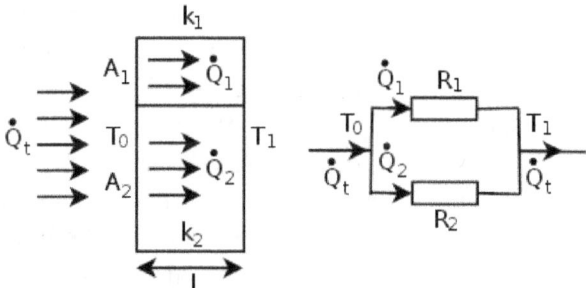

Figure 4.5: Slabs in parallel

The definition of the equivalent thermal resistance R_{eq} is such that

$$(T_0 - T_1) = \dot{Q}_t R_{eq},$$

where $\dot{Q}_t$ is the total heat flux from the left end to the right end. Next, the total heat flux is simply the sum of the individual heat fluxes across the slabs as there cannot be any accumulation of energy anywhere in the system.

$$\dot{Q}_t = \dot{Q}_1 + \dot{Q}_2 = \frac{T_0 - T_1}{R_1} + \frac{T_0 - T_1}{R_2} = \frac{T_0 - T_1}{R_{eq}}$$

$$\implies \frac{1}{R_{eq}} = \frac{1}{R_1} + \frac{1}{R_2}.$$

The total heat flux is then

$$\dot{Q}_t = \frac{T_0 - T_1}{R_{eq}} = \frac{k_1 A_1 + k_2 A_2}{l}(T_0 - T_1).$$

In general, if we have N thermal resistors in parallel that are of resistances $R_1, R_2, \ldots, R_N$,

$$\dot{Q}_t = \sum_{i=1}^{N} Q_i = \sum_{i=1}^{N} \frac{T_0 - T_1}{R_i}.$$

By definition of the equivalent resistance,

$$\dot{Q}_t = \frac{T_0 - T_1}{R_{eq}}$$

$$\frac{1}{R_{eq}} = \sum_{i=1}^{N} \frac{1}{R_i}. \tag{4.8}$$

With these equivalent resistances, various thermal conduction problems can be solved as if they were simple circuit problems.

Incidentally, the notion of an equivalent resistance provides an alternative derivation of the thermal resistance of a cylindrical shell given by Eq. (4.6). Due to axial symmetry, the entire cylindrical shell can be divided into many shells of varying radius and infinitesimal thickness whose inner and outer surfaces individually possess uniform temperature. Furthermore, since the heat flux must be continuous across all layers, the thermal resistance of the cylindrical shell can be deemed as summing those of the infinitesimal shells in series. Lastly, because the heat flow is radial and perpendicular to each shell, the thermal resistance of a shell of radius r and thickness dr follows the format of the thermal resistance of a slab.

$$\frac{dr}{kA} = \frac{dr}{2\pi k l r},$$

where the contact area in this context is now the cylindrical surface of radius r and length l, $A = 2\pi r l$. For readers who are not yet convinced that we can do this, we can further divide the shell into strips of length l, thickness dr and width $r d\theta$ (cylindrical coordinates). These strips are effectively slabs and hence have thermal resistance $\frac{dr}{klrd\theta}$. The previous shell is composed of myriad such strips placed side-by-side or connected in parallel as the heat fluxes across these strips are along different "branches" of a circuit. The effective resistance of a shell is then obtained from integrating the reciprocal of the thermal resistance of a strip $\frac{klrd\theta}{dr}$ from $\theta = 0$ to $\theta = 2\pi$ and subsequently taking the inverse of this result which yields $\frac{dr}{2\pi klr}$. With this clarification, we can proceed with determining the equivalent resistance of the entire cylindrical shell. Since the equivalent resistance of resistors in series is the sum of all the individual resistances, the equivalent resistance of the cylindrical shells at different radii is tantamount to the integral of $\frac{dr}{2\pi klr}$ from r_0 to r_1.

$$R = \int_{r_0}^{r_1} \frac{dr}{2\pi klr} = \frac{\ln \frac{r_1}{r_0}}{2\pi kl}$$

which is consistent with Eq. (4.6). Now, there is a pivotal warning to be made here. When we claim that a resistor is composed of different components connected in series or parallel, we must first check that the surfaces of the components are each of uniform temperature. This is because resistance is foremost, only defined for objects with surfaces of uniform temperature (e.g. the slab whose two ends have uniform, albeit different, temperatures). In the above example of a cylindrical shell, the uniform temperatures of the inner and outer surfaces of each infinitesimal shell enable us to add the infinitesimal shells in series. In general, caution must be taken in slicing a resistor into surfaces with uniform temperature if one wants to apply the technique of adding resistors.

4.3 Radiation

Thermal radiation is the energy emitted in the form of electromagnetic waves. Unlike conduction and convection, these electromagnetic waves do not require any medium to propagate and in fact travel at the theoretically maximum speed. Thermal radiation is emitted by every object with a non-zero absolute temperature (i.e. measured with respect to the Kelvin scale). The Stefan-Boltzmann law states that the total heat flux density $\dot{q}$ radiated by a surface across all wavelengths due to a black body can be computed as

$$\dot{q} = \sigma T^4, \tag{4.9}$$

where T is the temperature of the surface on the black body and σ is the Stefan-Boltzmann constant whose numerical value is $5.670 \times 10^{-8}\, Wm^{-2}K^{-4}$. A black body is an idealized physical entity that absorbs all incident electromagnetic radiation and emits the maximum amount of radiation for a given temperature and surface area.

For realistic bodies, the heat flux density $\dot{q}$ radiated from a surface is less than that of a black body and is calculated as

$$\dot{q} = \varepsilon\sigma T^4 \tag{4.10}$$

where $0 \leq \varepsilon \leq 1$ is known as the emissivity of the body. It measures the ability of a body to emit thermal radiation in comparison to a black body counterpart. Next, the luminosity L of a body is defined to be the total power emitted via radiation by a body. For a body with a uniform surface temperature T, it is simply $\dot{q}$ multiplied by the exposed surface area A of the body.

$$L = \varepsilon \sigma A T^4. \tag{4.11}$$

4.3.1   Wien's Displacement Law

In general, each wavelength of light contributes a different proportion to the total power radiated by a black body. Wien's displacement law relates the peak wavelength, which makes the largest contribution to this radiated power, to the temperature of the black body.

$$\lambda_{peak} T = b \tag{4.12}$$

where b is Wien's constant which has a numerical value 2.898×10^{-3} mK. The inversely proportional nature of temperature relative to the peak wavelength provides a rough explanation of why blue stars are actually hotter than red stars.

4.3.2   Radiation at a Surface

In general, when radiation strikes a surface, a fraction of it may be absorbed, reflected or transmitted in accordance with the absorptivity (α), reflectivity (ρ) and transmissivity (t) of the surface. Since these are the only effects possible, the sum of these coefficients should be unity.

$$\alpha + \rho + t = 1.$$

For a black body, $\alpha = 1$, $\rho = 0$ and $t = 0$. That is, a black body is a perfect absorber as well. In general, these coefficients are dependent on the wavelength of radiation and the temperature of the surface. However, due to the prevalent insensitivity of these coefficients to temperature and wavelength variations in real materials, the absorptivity, reflectivity and transmissivity are assumed to be uniform across all wavelengths and temperatures. Such an ideal object is known as a gray body. Finally, it may be helpful to note that in some cases, the surfaces are thick enough such that they can be assumed to be opaque — causing $t = 0$ and further simplifying the set-up.

Kirchhoff's Law of Radiation

Kirchhoff's law states that the absorptivity and emissivity of a gray body are equal[1] when the body is at thermodynamic equilibrium with its surroundings.

$$\alpha = \varepsilon. \tag{4.13}$$

That is, a good emitter is also a good absorber. Kirchhoff's law can be proven by leveraging the impossibility of heat transfer between two bodies which are at the same temperature. Consider an arbitrary gray body of an arbitrary shape and size, with an absorptivity α and emissivity ε. Now, imagine enclosing this body with a slightly larger black body replica of a similar shape and size. The arbitrary body and the black body are both at temperature T and both have surface area A. Evidently, all radiation that is emitted by the arbitrary body impinges on the surface of the black body. The luminosity of the arbitrary body is $\varepsilon\sigma AT^4$. The power it absorbs, due to the radiation by the black body, is $\alpha\sigma AT^4$. As there can be no net heat flux between these two bodies, these expressions must be equal.

$$\implies \alpha = \varepsilon.$$

Even though we have considered a particular set-up in our proof of Kirchhoff's law, the absorptivity and emissivity of a gray body are properties that are independent of the external environment and hence identical across all types of surroundings. Furthermore, since α and ε are uniform for a gray body across all wavelengths and temperature by proposition, Kirchhoff's law of radiation is often applied, even in the case where the gray body has not attained thermodynamic equilibrium with its surroundings. That is, $\alpha = \varepsilon$ is assumed to hold in all cases for a gray body.

Problem: Two large, thin plates of area A are oriented parallel to each other in a vacuum in Fig. 4.6. The left plate is a black body while the right plate is an opaque gray body with an emissivity ε. If the left plate is maintained at a temperature T_1, determine the equilibrium temperature of the right plate T_2. Note that each plate has two surfaces.

---

[1] Actually, a stronger version states that the absorptivity and emissivity of a body for all wavelengths of radiation are equal at thermodynamic equilibrium. However, this is not particularly enlightening as it implies that we would need to define an absorptivity and emissivity for each wavelength. Thus, we shall just consider all wavelengths of radiation as a whole by adopting the gray body assumption.

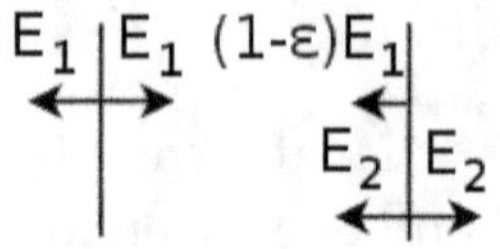

$$E_1 \big| E_1 \ (1\text{-}\varepsilon)E_1 \big|$$
$$E_2 \big| E_2$$

Figure 4.6:   Radiating plates

The incident power on the right plate, due to the left, is $E_1 = \sigma A T_1^4$. By Kirchhoff's law, $\varepsilon \sigma A T_1^4$ amount of power is absorbed by the right plate and $(1 - \varepsilon)\sigma A T_1^4$ amount of power is reflected back to the left plate. Next, the right plate also emits $E_2 = \varepsilon \sigma A T_2^4$ amount of power on each side. Therefore, the net heat flux between the left and right plates is $E_1 - (1 - \varepsilon)E_1 - E_2 = \varepsilon \sigma A(T_1^4 - T_2^4)$. For the right plate to be at equilibrium, this must also be equal to the heat flux on its right side, $E_2 = \varepsilon \sigma A T_2^4$. Therefore,

$$\varepsilon \sigma A(T_1^4 - T_2^4) = \varepsilon \sigma A T_2^4$$

$$T_2 = \frac{T_1}{\sqrt[4]{2}}.$$

Notice that in the above example, we imposed the condition that the heat flux must be continuous instead of enforcing the fact that the power emitted by the second plate must equal the power absorbed. Both methods will work fine but the former is often simpler and less messy in more complicated set-ups.

Problem: Two large black plates of area A are oriented parallel to each other and are maintained at temperatures T_i and T_f. Now, if N identical black plates are slotted between them — numbered from 1 to N from left to right, with the first plate being the closest to the plate with temperature T_i — determine the net heat flux transferred between adjacent plates at steady state and the temperature of the jth plate in the N intermediate plates.

Let $\dot{Q}$ be the common net heat flux between adjacent plates, positive rightwards. By the continuity of heat flux,

$$\sigma A(T_i^4 - T_1^4) = \dot{Q}$$

$$\sigma A(T_1^4 - T_2^4) = \dot{Q}$$

$$\vdots$$

$$\sigma A(T_N^4 - T_f^4) = \dot{Q}.$$

Summing all of the above equations,

$$\sigma A(T_i^4 - T_f^4) = (N+1)\dot{Q}$$

$$\dot{Q} = \frac{\sigma A(T_i^4 - T_f^4)}{N+1}.$$

From the previous series of equations, we have the "arithmetic progression"

$$T_j^4 = T_{j-1}^4 - \frac{\dot{Q}}{\sigma A}$$

with T_0 being T_i. Therefore,

$$T_j^4 = T_i^4 - \frac{j\dot{Q}}{\sigma A} = \frac{(N+1-j)T_i^4 + jT_f^4}{N+1}$$

$$T_j = \sqrt[4]{\frac{(N+1-j)T_i^4 + jT_f^4}{N+1}}.$$

View Factor

In the previous problems, the emitted radiation by a plate, in the direction of another, was completely projected on the other plate. However, this is not necessarily true in general. Consider the case where there are two radiating bodies, A and B, at temperatures T_0 and T_1. They possess surface areas A_A and A_B respectively. Each body emits thermal radiation and also receives thermal radiation from the other body. The view factor, F_{AB}, is defined as the fraction of radiation emitted by A that strikes the surface of B (note that this is not the fraction absorbed by B and that radiation reflected by A is not counted). The view factor is a purely geometric property that is dependent on many factors such as the orientations of the bodies. The reciprocity theorem states that

$$F_{AB}A_A = F_{BA}A_B. \tag{4.14}$$

This can be proven, again, by imposing the condition that there cannot be a net heat flux between two objects of the same temperature. Suppose that A and B were black bodies at the same temperature T, such that there is no reflected radiation. Then, the power incident on B due to radiation by A is $F_{AB}\sigma A_A T^4$. Similarly, the power incident on A due to B is $F_{BA}\sigma A_B T^4$. For the net heat flux to be zero, these quantities must be equal — implying that

$$F_{AB}A_A = F_{BA}A_B.$$

Since the view factors and surface areas are purely geometric properties, the above result must hold for non-black bodies as well — hence proving the reciprocity theorem.

Moving on, we wish to compute the net heat flux between A and B in general. We first start off with the simplest case where A and B are both black bodies. The amount of power absorbed by B due to the thermal radiation by A is

$$\dot{Q}_{A \to B} = F_{AB}\sigma A_A T_0^4.$$

Similarly, the amount of power absorbed by A due to B is

$$\dot{Q}_{B \to A} = F_{BA}\sigma A_B T_1^4.$$

The net heat flux between A and B is

$$\dot{Q}_{AB} = \dot{Q}_{A \to B} - \dot{Q}_{B \to A} = F_{AB}\sigma A_A T_0^4 - F_{BA}\sigma A_B T_1^4.$$

Employing the reciprocity theorem,

$$\dot{Q}_{AB} = F_{AB}\sigma A_A (T_0^4 - T_1^4). \tag{4.15}$$

Proceeding with a new set-up, consider the special case where A is a small object with emissivity ε_A in a room whose surrounding temperature is T_1. B in this case is the surroundings of A and acts as a black body such that $\varepsilon_B = 1$. The general system of a gray body A and black body B cannot be solved with just their view factors as F_{AB} is not generally representative of the radiation reflected from the surface of the gray body that is incident on the black body. One would expect the distribution of reflected light to differ from that of light emitted by A. However, in this case, we know that both the emitted and reflected forms of radiation by A are completely received by B due to its all-encapsulating nature — we can thus circumvent this loophole. The power emitted by B and incident on A is

$$F_{BA}\sigma A_B T_1^4.$$

$\varepsilon_A F_{BA}\sigma A_B T_1^4$ amount of power is absorbed by A and the rest is reflected back to B. The surface area A_B of the surroundings is not well-defined at the moment but we will apply the reciprocity theorem later to circumvent this muddy point. Moving on, the power emitted by A and absorbed by B is

$$\varepsilon_A F_{AB} \sigma A_A T_0^4.$$

The net heat flux between the object and its surroundings is

$$Q = \varepsilon_A F_{AB} \sigma A_A T_0^4 - \varepsilon_A F_{BA} \sigma A_B T_1^4.$$

Applying the reciprocity theorem yields

$$Q = \varepsilon_A \sigma A_A (T_0^4 - T_1^4) \qquad (4.16)$$

since $F_{AB} = 1$ as the surroundings B receives all radiation by A. This is an extremely useful result that is expressed solely in terms of the properties of object A.

Problem: A spherical black body of absolute temperature T_0 and radius r is covered by a thin, concentric, and black spherical shell of radius R. Let the temperature of the surroundings far away be T_2. What is the equilibrium temperature of the shell, T_1?

The view factor of the sphere to the shell is 1 since all radiation emitted from the sphere reaches the shell. The net heat flux from the sphere to the shell, $\dot{Q}$, is then given by Eq. (4.15) as

$$\dot{Q} = \sigma 4\pi r^2 (T_0^4 - T_1^4).$$

The net heat flux from the shell to the surroundings, $\dot{Q}'$ is

$$\dot{Q}' = \sigma 4\pi R^2 (T_1^4 - T_2^4)$$

by Eq. (4.15) again as the view factor of the shell to the surroundings is also 1. Lastly, for the shell to be at equilibrium, the heat fluxes must be equal.

$$\dot{Q} = \dot{Q}'.$$

Solving,

$$T_1 = \sqrt[4]{\frac{r^2 T_0^4 + R^2 T_2^4}{r^2 + R^2}}.$$

4.3.3 System of Gray Bodies

Having discussed a few special systems, this section will try to analyze a more general system of gray bodies. But first, we define the following quantities for the sake of convenience. The radial exitance M of an object is the total power **emitted** by the surface of an object per unit area. We emphasize the fact that this is the power emitted which implies that reflected radiation is not counted. The irradiance E on an object is the total power incident (not absorbed by!) on the surface of an object per unit area. Finally, the radiosity J of an object is the total power leaving the surface of an object per unit area. Power emitted, reflected and transmitted by a surface all contribute to the radiosity.

In a system of gray bodies, radiation may be reflected back and forth between gray bodies. Then, relevant quantities such as the net heat flux between two bodies may be determined by summing an infinite series, or better yet, by solving simultaneous equations involving the quantities we have just defined. Consider the following problem.

Problem: Two large opaque plates with area A and emissivities ε_1 and ε_2 are parallel to each other. If the two plates are maintained at temperatures T_1 and T_2 respectively, determine the net heat flux between them.

By Kirchhoff's law, the absorptivities of the two plates are ε_1 and ε_2 respectively. Observe that when plate 1 emits a certain amount of exitance M_1, a fraction ε_2 of it is absorbed by plate 2 and the rest is reflected back to plate 1. Plate 1 then absorbs a fraction ε_1 of the power again and reflects the rest and so on. A similar process occurs for the exitance emitted by plate 2. The net power per unit area emitted from plate 1, in the direction of the plate 2, is then

$$\dot{q} = M_1 - \varepsilon_1(1 - \varepsilon_2)M_1 - \varepsilon_1(1 - \varepsilon_1)(1 - \varepsilon_2)^2 M_1$$

$$- \varepsilon_1(1 - \varepsilon_1)^2(1 - \varepsilon_2)^3 M_1 - \cdots$$

$$- \varepsilon_1 M_2 - \varepsilon_1(1 - \varepsilon_1)(1 - \varepsilon_2)M_2 - \varepsilon_1(1 - \varepsilon_1)^2(1 - \varepsilon_2)^2 M_2 - \cdots$$

$$= M_1 - \frac{\varepsilon_1(1 - \varepsilon_2)M_1}{1 - (1 - \varepsilon_1)(1 - \varepsilon_2)} - \frac{\varepsilon_1 M_2}{1 - (1 - \varepsilon_1)(1 - \varepsilon_2)},$$

where the negative terms involving M_1 are due to the reflected portions of M_1 that plate 1 absorbs back and the negative terms involving M_2 stem from plate 1 absorbing part of the radiation emitted by plate 2. The heat flux between the two plates is then the above multiplied by the area of plate 1.

$$\dot{Q} = A\left(M_1 - \frac{\varepsilon_1(1 - \varepsilon_2)M_1}{1 - (1 - \varepsilon_1)(1 - \varepsilon_2)} - \frac{\varepsilon_1 M_2}{1 - (1 - \varepsilon_1)(1 - \varepsilon_2)} \right).$$

Substituting $M_1 = \varepsilon_1 \sigma T_1^4$ and $M_2 = \varepsilon_2 \sigma T_2^4$,

$$\dot{Q} = \frac{\sigma A(T_1^4 - T_2^4)}{\frac{1}{\varepsilon_1} + \frac{1}{\varepsilon_2} - 1}.$$

A more elegant method employs the definition of irradiance and radiosity. The radiosity J of each plate, in the direction towards the other, is only the sum of its radial exitance M and the reflected power per unit area as the opaque plates do not transmit any power. The reflected power per unit area of a plate is simply one minus its absorptivity (which is equal to its emissivity) multiplied by the irradiance on the plate E.

$$J_1 = M_1 + (1 - \varepsilon_1)E_1,$$

$$J_2 = M_2 + (1 - \varepsilon_2)E_2.$$

However, we know that the irradiance on a particular plate is simply the radiosity of the other plate. Then,

$$J_1 = M_1 + (1 - \varepsilon_1)J_2,$$

$$J_2 = M_2 + (1 - \varepsilon_2)J_1.$$

Solving these equations simultaneously would yield

$$J_1 = \frac{M_1 + (1 - \varepsilon_1)M_2}{1 - (1 - \varepsilon_1)(1 - \varepsilon_2)},$$

$$J_2 = \frac{M_2 + (1 - \varepsilon_2)M_1}{1 - (1 - \varepsilon_1)(1 - \varepsilon_2)}.$$

The net heat flux density $\dot{q}$ emanating from plate 1, in the direction towards plate 2, is the radiosity of plate 1 minus the irradiance on plate 1 which is the radiosity of the plate 2. Thus,

$$\dot{q} = J_1 - J_2 = \frac{\varepsilon_2 M_1 - \varepsilon_1 M_2}{1 - (1 - \varepsilon_1)(1 - \varepsilon_2)}.$$

The net heat flux between the plates is then the net heat flux density from plate 1 multiplied by the area of plate 1 as all of the net heat flux density emerging from plate 1 is incident on plate 2.

$$\dot{Q} = \dot{q}A = \frac{\sigma A(T_1^4 - T_2^4)}{\frac{1}{\varepsilon_1} + \frac{1}{\varepsilon_2} - 1}.$$

Opaque gray Systems with Partial Capturing of Radiosity

In the previous problem, the radiosity of a plate completely impinged on the other. However, the irradiance on a component in a system due to another component is only a portion of the latter's radiosity in general due to the relative orientations of the components. In an attempt to rectify this, one might immediately think of the view factor F_{ji} which was defined as the fraction of radiation emitted by a component j that is projected on another component i. However, we cannot directly say that the irradiance on component i due to component j is $\frac{F_{ji}J_jA_j}{A_i}$ where J_j is the radiosity of component j. This is due to the fact that the direction of reflected light from component j will most probably be different from that of its emitted light (whose direction is arbitrary). However, if we assume that reflections off component j are diffuse — such that light is scattered off the surface of component j haphazardly — we can indeed say that the irradiance on component i due to component j is $\frac{F_{ji}J_jA_j}{A_i}$. This assumption of diffuse reflections is very common.

Now, consider a system of N opaque components with the ith component possessing a surface area A_i, radiosity J_i, emissivity ε_i and exitance M_i. The irradiance on component i is $\sum_{j=1}^{N} \frac{F_{ji}J_jA_j}{A_i}$. Then, the various J_i's can be related by

$$J_i = M_i + (1 - \varepsilon_i) \sum_{j=1}^{N} \frac{F_{ji}J_jA_j}{A_i}.$$

The summation includes component i, as in general, a portion of its own radiosity may be incident on itself. By the reciprocity relation of view factors,

$$F_{ji}A_j = F_{ij}A_i.$$

Therefore,

$$J_i = M_i + (1 - \varepsilon_i) \sum_{j=1}^{N} F_{ij}J_j. \tag{4.17}$$

Following from this, we have a system of N variables (the various J_i's) and N equations. Therefore, the radiosity of each surface can be solved for, in principle. Afterwards, we can compute the net heat flux $\dot{Q}_i$ (defined to be positive when emitted) emanating from component i by taking the product of its area A_i and by its radiosity subtracted by the total irradiance on it.

$$\dot{Q}_i = A_i \left(J_i - \sum_{j=1}^{N} F_{ij}J_j \right).$$

This can be further simplified by employing Eq. (4.17).

$$\dot{Q}_i = A_i \left(J_i - \frac{J_i - M_i}{1 - \varepsilon_i} \right) = \frac{A_i}{1 - \varepsilon_i}(M_i - \varepsilon_i J_i).$$

Substituting $M_i = \varepsilon_i \sigma T_i^4$ where T_i is the temperature of the ith surface,

$$\dot{Q}_i = \frac{A_i \varepsilon_i}{1 - \varepsilon_i} (\sigma T_i^4 - J_i). \tag{4.18}$$

Problem: Two long concentric cylinders of radii r_1 and r_2, with $r_1 < r_2$ and emissivities ε_1 and ε_2, are maintained at temperatures T_1 and T_2 respectively. Determine the net heat flux between the cylinders if they have length l. Ignore any edge effects and assume that reflections off the cylinders are diffuse.

Let the exitances and radiosities of the cylinders be M_1, M_2, J_1 and J_2 respectively. Then,

$$J_1 = M_1 + (1 - \varepsilon_1) F_{11} J_1 + (1 - \varepsilon_1) F_{12} J_2,$$

$$J_2 = M_2 + (1 - \varepsilon_2) F_{21} J_1 + (1 - \varepsilon_2) F_{22} J_2.$$

Evidently, all radiation emitted by cylinder 1 is received by cylinder 2. Then, $F_{12} = 1$ and $F_{11} = 0$. By the reciprocity theorem,

$$A_1 = F_{21} A_2$$

$$2\pi r_1 l = F_{21} 2\pi r_2 l$$

$$F_{21} = \frac{r_1}{r_2} \implies F_{22} = 1 - \frac{r_1}{r_2},$$

as the leftover radiosity from the larger cylinder that is not incident on the smaller one must be redirected to itself. Substituting these values into the radiosities,

$$J_1 = M_1 + (1 - \varepsilon_1) J_2,$$

$$J_2 = M_2 + (1 - \varepsilon_2) \frac{r_1}{r_2} J_1 + (1 - \varepsilon_2) \left(1 - \frac{r_1}{r_2} \right) J_2.$$

Solving these (with $M_1 = \varepsilon_1 \sigma T_1^4$ and $M_2 = \varepsilon_2 \sigma T_2^4$) would yield

$$J_1 = \frac{\left(\varepsilon_1 \varepsilon_2 + \varepsilon_1 \frac{r_1}{r_2} - \varepsilon_1 \varepsilon_2 \frac{r_1}{r_2} \right) \sigma T_1^4 + (1 - \varepsilon_1) \varepsilon_2 \sigma T_2^4}{\varepsilon_2 + \varepsilon_1 \frac{r_1}{r_2} - \varepsilon_1 \varepsilon_2 \frac{r_1}{r_2}}.$$

Finally, the net heat flux between the cylinders is also the net heat flux emanating from cylinder 1.

$$\dot{Q} = \frac{A_1 \varepsilon_1}{1 - \varepsilon_1} (\sigma T_1^4 - J_1) = \frac{2\pi r_1 l \sigma (T_1^4 - T_2^4)}{\frac{1}{\varepsilon_1} + \frac{r_1}{r_2} \left(\frac{1}{\varepsilon_2} - 1 \right)}.$$

4.4   Thermal Expansion

Objects usually expand when heated because their molecules vibrate and move about faster, causing intermolecular distances to increase. Similarly, objects usually contract when cooled. Empirically, it is found that for small changes in temperature, the fractional change in length along a single dimension is proportional to the change in temperature.

$$\frac{\Delta L}{L_0} = \alpha \Delta T \qquad (4.19)$$

where ΔL is the change in length and L_0 is the original length before the temperature change. α is known as the coefficient of linear expansion which varies across different objects and ΔT refers to the change in temperature of the object (usually in Kelvins or degree Celsius). This equation is valid for small fractional changes, $\frac{\Delta L}{L_0} \ll 1$. An equivalent form of the above equation is

$$L = L_0(1 + \alpha \Delta T) \qquad (4.20)$$

where L is the final length of the object.[2] Similarly, we can define the coefficient of expansion for area and volume.

$$\frac{\Delta A}{A_0} = \alpha_A \Delta T,$$

$$\frac{\Delta V}{V_0} = \alpha_V \Delta T,$$

where A and V refer to area and volume respectively.

$$A = A_0(1 + \alpha_A \Delta T),$$

$$V = V_0(1 + \alpha_V \Delta T).$$

For objects that expands isotropically (the same percentage in all directions) and for small fractional changes,

$$\alpha_A \approx 2\alpha, \qquad (4.21)$$

$$\alpha_V \approx 3\alpha. \qquad (4.22)$$

To show these, let the initial lengths of an object along three perpendicular directions be x_1, y_1 and z_1 respectively. Let the final lengths be x_2, y_2 and z_2.

---

[2]You may worry that the above expression gives different results for the same rise in temperature if we intersperse the heating of the object, as compared to the case where its temperature is increased on only one occasion. However, such disparities are second order and can be neglected.

We only consider the lengths in the x and y-directions in the case of area for the sake of illustration. For an isotropic expansion,

$$A \propto xy,$$

$$V \propto xyz,$$

where x, y and z are the object's dimensions. Therefore,

$$\frac{A}{A_0} = \frac{x_2}{x_1} \cdot \frac{y_2}{y_1} = (1 + \alpha \Delta T)^2 \approx 1 + 2\alpha \Delta T,$$

$$\frac{V}{V_0} = \frac{x_2}{x_1} \cdot \frac{y_2}{y_1} \cdot \frac{z_2}{z_1} = (1 + \alpha \Delta T)^3 \approx 1 + 3\alpha \Delta T,$$

where second order and above terms in $\alpha \Delta T$ have been discarded. Comparing the different expressions for $\frac{A}{A_0}$ and $\frac{V}{V_0}$, it can be seen that

$$a_A \approx 2\alpha,$$

$$a_V \approx 3\alpha.$$

Problem: Find the mean radius of curvature r when an initially straight bimetallic strip consisting of two metal strips, with coefficients of linear expansion α_1 and α_2 ($\alpha_2 > \alpha_1$) and a small common thickness x, is heated such that its temperature increases by ΔT.

Let the initial length of the bimetallic strip be L_0. Let the final length of the strips with coefficients of α_1 and α_2 be L_1 and L_2 respectively. Then,

$$L_1 = (1 + \alpha_1 \Delta T) L_0,$$

$$L_2 = (1 + \alpha_2 \Delta T) L_0.$$

Evidently, the first strip should occupy the inner part of the arc while the second strip occupies the outer part. The mean radius of curvature is the distance between the interface of the two strips and the center of the circle. If the arc produced by the bimetallic strip subtends an angle θ,

$$L_1 = \theta \left(r - \frac{x}{2} \right),$$

$$L_2 = \theta \left(r + \frac{x}{2} \right).$$

Then,

$$\frac{r - \frac{x}{2}}{r + \frac{x}{2}} = \frac{L_1}{L_2} = \frac{(1 + \alpha_1 \Delta T)}{(1 + \alpha_2 \Delta T)}$$

$$r = \frac{x[2 + (\alpha_1 + \alpha_2)\Delta T]}{2(\alpha_2 - \alpha_1)\Delta T}.$$

Problem: A straight line is drawn using a marker on a uniform circular plate. It takes the form of a chord that lies a perpendicular distance h from the center of the circle. If the coefficient of linear expansion of the plate is α, determine the final shape of the line after the plate is heated such that its temperature is increased by ΔT. The expansion of the plate is isotropic. Tom claims that the line will now be bent. Is he correct?

Intuitively, an isotropic expansion is akin to us taking a photo of the plate and then enlarging the image. Therefore, we would expect that the final line takes the form of a chord that lies a perpendicular distance $(1 + \alpha \Delta T)h$ from the center of the circle and that Tom is wrong. If one is not satisfied with this argument, one can consider the following more quantitative proof. Define the origin at the center of the circle, x-axis to be parallel to the chord and the y-axis to be perpendicular to the chord. Define θ to be the clockwise angular coordinate of a point on the line from the y-axis. The radial coordinate of a point on the initial straight line as a function of θ, $r(\theta)$, is

$$r(\theta) = \frac{h}{\cos \theta}.$$

The new radial coordinate of a point on the line as a function of θ, $r'(\theta)$, after the isotropic expansion is simply $r(\theta)$ scaled by a factor of $(1 + \alpha \Delta T)$ as the circular disk is stretched radially.

$$r'(\theta) = \frac{h(1 + \alpha \Delta T)}{\cos \theta}.$$

This equation takes the same form as the previous equation, except that h is replaced by $h(1 + \alpha \Delta T)$. Therefore, the new curve represents a chord that lies a perpendicular distance $h(1 + \alpha \Delta T)$ from the center of the circle.

4.5   Phase Transitions

A phase is defined as a physically distinct state of matter that is homogeneous. Common phases[3] include the solid, liquid and vapour (gaseous)

---

[3]There are in fact other exotic phases but we shall only consider the three common ones.

phase. The process involving a pure substance — whose chemical composition is uniform across all molecules — that evolves from one phase to another is known as a phase transition. Consider the following phase transitions of water.

When a block of ice is heated at atmospheric pressure, one would find that its temperature rises until its melting point. At this juncture, ice begins to melt into water. However, the temperatures of ice and water stagnate at the melting point, though heat is continuously supplied, until the ice completely melts. Similarly, heating the water further would increase its temperature until its boiling point, at which water begins its transition to its vapour state (steam). Again, the heat supplied during this transition is not embodied as rises in the temperatures of the water and steam, until all water has boiled off. Afterwards, the temperature of steam continues to increase as it absorbs more heat.

There is a common trend where the temperature of a substance remains constant during such phase transitions. There must be some explanation for this seemingly missing heat that is not manifested as an increase in temperature of the substance. We name the dormant heat supplied to facilitate solid-liquid and liquid-gas transitions the latent heats of fusion and vaporization respectively.

To understand why a latent heat is necessary, we consider the first law of thermodynamics. During a phase transition, there is a change in the potential energy of the substance. During melting, the substance is transformed from an ordered lattice into a disordered liquid whose particles are further apart. Energy is required for the molecules to overcome the attractive bonding between them so that they can escape from their rigid structure. From another perspective of energy, the potential energy in a liquid is larger than that in a solid (less negative as the potential energy between two molecules is usually negative due to the attractive nature of their interactions) as molecules are further apart. In a similar vein, vapor molecules are essentially liberated during boiling and the intermolecular forces between them become negligible. Energy is required to help them overcome the attractive bonding in the liquid state. The potential energy of a vapor is virtually zero and, thus, is larger than the potential energy of a liquid.

Besides a change in potential energy, work is also performed by the substance during a phase transition between solid, liquid and vapor phases due to discontinuities in densities. Specifically, work must be performed by the substance in overcoming the external pressure when expanding or contracting during a phase transition.

Therefore, latent heat plays the roles of changing the microscopic potential energy of a substance and enabling it to perform work as it changes phase. Since the potential energy and volume changes of a liquid-vapor transition often outstrip those of a solid-liquid transition, the latent heat of vaporization is much larger than the latent heat of fusion. Moreover, the latent heat supplied does not lead to an increase in the microscopic kinetic energy of the substance — implying that its temperature remains constant.

Quantitatively, it is convenient to define the specific latent heat of fusion and vaporization, which is the latent heat per unit mass of substance, as the latent heat required to completely boil or melt a substance often scales with mass. In general, we define L as the specific latent heat of a substance during a particular phase transition. L is different for different states at which phase transition occurs as the work done by the substance varies. The amount of heat, Q, that needs to be supplied to facilitate the particular phase transition of mass m of a substance from a phase of lower internal energy to one of higher internal energy is

$$Q = mL. \tag{4.23}$$

Since a phase transition is an internally reversible process as the different phases must coexist at the same temperature (such that there is no heat transfer between constituents of different temperatures),

$$Q = T\Delta S = Tm\Delta s$$

where T is the temperature at which the phase transition occurs and Δs is the entropy change per unit mass of the substance (specific entropy change), in completely converting from one state to another. Then,

$$L = T\Delta s, \tag{4.24}$$

where the phase with a larger internal energy also possesses greater entropy.

4.5.1   *Phase Diagrams*

To visualize the phases of a substance at different equilibrium states, a phase diagram can be drawn. Each state of a substance can be ascribed a unique pressure, volume and temperature, which are in fact connected by an equation of state (such as the ideal gas law in the case of ideal gases). Therefore, only two properties are needed to specify a state of a system. In light of this clarification, a phase diagram is usually plotted as a pressure-temperature diagram, exemplified by Fig. 4.7.

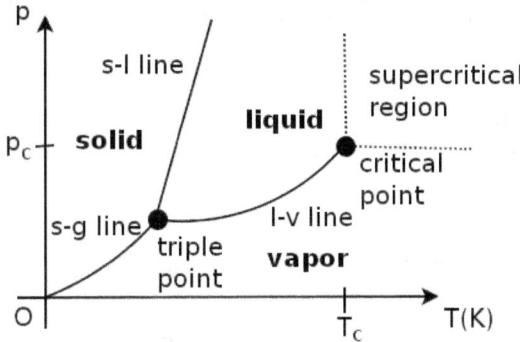

Figure 4.7:   Phase diagram of common substances

The experimental phase diagram of a typical substance[4] is depicted above. There are three lines that demarcate boundaries between phases in the P-T diagram. These lines are the respective coexistence lines where different phases of the substance can coexist at a single equilibrium pressure and temperature. A phase transition occurs when the state of the substance is on a coexistence line (as one phase is progressively converted to another) and is completed when the state crosses over this line. To navigate over a coexistence line from a phase with lower internal energy to one with higher internal energy, a latent heat needs to be supplied to the substance — this latent heat is dependent on the point on the coexistence line and the direction in which the state of the substance traverses. On another note, an interesting observation is that the substance can actually directly transition from a solid to vapor without passing by the liquid state at low pressures. Such a phase transition is known as sublimation and the reverse process is known as deposition.

At low temperatures and high pressures, the substance takes the form of a solid as expected. At high temperatures and low pressures, the substance is a vapor. At intermediate temperatures and pressures, the substance is a liquid.

There are a few interesting properties of the phase diagram. Firstly, there is a single temperature and pressure at which the three phases can coexist — this is known as the triple point. At pressures below that at the triple point, the substance can sublime. Furthermore, the pressure at the triple point is the lowest pressure at which a liquid can exist for all substances while the

---

[4]Water is an atypical substance. Its solid-liquid coexistence line has a negative gradient for reasons that will be elaborated later.

temperature at the triple point is the lowest temperature at which a liquid can exist for typical substances (not water).

Problem: Determine the specific latent heat of sublimation L_s at the triple point if the specific latent heats of fusion and vaporization are L_f and L_v at the triple point respectively.

Let the specific entropies of the solid, liquid and vapor states of the substance at the triple point be s_s, s_l and s_v respectively. Let the temperature at the triple point be T. Then,

$$L_f = T(s_l - s_s),$$
$$L_v = T(s_g - s_l),$$
$$L_s = T(s_g - s_s) = L_v - L_f.$$

Moving on, interesting observations regarding the coexistence lines can be made. The solid-vapor line originates at absolute zero ($0\,\mathrm{K}$) and zero pressure and ends at the triple point. The solid-liquid line extends from the triple point to infinity. However, the liquid-vapor line starts from the triple point and terminates at a certain juncture! This state is known as the critical point and the temperature and pressure at this point are termed the critical temperature T_c and the critical pressure p_c respectively.

So what actually occurs in the supercritical region, at states with temperatures and pressures larger than the corresponding critical values? The liquid and vapor phases become indistinguishable and the substance morphs into a homogeneous fluid (which is neither liquid or gaseous and is simply referred to as a fluid). Surface tension vanishes such that the meniscus dividing the two phases disappears. The density of the substance also evolves continuously — a stark contrast with the previously discontinuous densities of the liquid and gaseous states. Therefore, if you change the state of a substance from the liquid region to the supercritical region and back to the vapor region, you won't actually observe a phase transition! These properties are rather counter-intuitive as the supercritical region is rather exotic. For example, the critical pressure of water is roughly 218 atm which is enormous and hard to achieve.

To better illustrate the prevalent abrupt jump in density during a liquid-vapor phase change and the seamless transition in the supercritical region, consider the temperature against volume graphs of a pure substance heated at constant pressure from a liquid state, for different values of pressure in Fig. 4.8. Note that we usually analyze heating at constant pressure as it is a decent representation of the processes on Earth which are commonly conducted under atmospheric pressure.

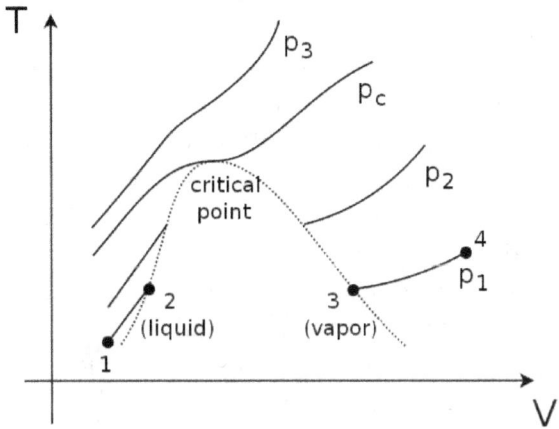

Figure 4.8:   Isobars on T-V diagram

Let us focus on the curve describing the particular pressure p_1 (which is the smallest out of all the pressures that we will consider and is smaller than the critical pressure p_c) which is reflected by the bottom-most graph. As the liquid is heated, its temperature and volume increase from state 1 to state 2. At this juncture, the graph becomes disjoint between segments 12 and 34. The substance attains an equilibrium state where its liquid and vapor phases, which respectively correspond to states 2 and 3, coexist with different volumes. Collectively, this pair of disjoint points correspond to a single point on the liquid-vapor coexistence line on the P-T diagram and thus have the same temperature. The coexistence of phases will persist until sufficient heat (commensurate with the latent heat of vaporization at constant pressure p_1) is supplied to completely vaporize the substance. To visualize this on the above diagram, we can instead define V as the average volume of the substance such that V increases from state 2 (where the substance is completely liquid) to state 3 (where the substance is completely vapor) along a horizontal line. Subsequent heating beyond this point would cause the temperature and volume of the vapor to increase indefinitely (e.g. from states 3 to 4).

At a slightly larger pressure $p_2 > p_1$, which is still smaller than p_c, a similar trend of discontinuous lines occurs. However, the graph is shifted upwards and the horizontal gap between the disjoint points is reduced. Plotting the locus of the pairs of disjoint points at different pressures, we obtain the bell-shaped curve depicted in dotted lines. The portion on the left of the peak corresponds to the liquid phases while that on the right corresponds to the vapor phases when the two phases coexist. Notice that there is a certain minimum pressure (defined as the critical pressure p_c) where the states of the substance is a continuous curve — its point of inflexion produces the peak of the bell-shaped curve. That is, at this pressure, the supposedly disjoint pair of points converge to form a single point which is a point of inflexion. At pressures above p_c, such as $p_3 > p_c$, the states of the substance during heating under constant pressure is a continuous curve such that there is no volume discontinuity as the liquid and vapor phases become indistinguishable and an "integrated fluid."

The above analysis suggests that if we are given an equation of state that models the liquid or vapor phases of a substance, we can determine the critical pressure p_c by finding the pressure that produces a point of inflexion on the T-V diagram when the substance is heated or cooled at constant pressure. Afterwards, the critical temperature T_c can also be determined.

Problem: In light of the ineptness of the ideal gas law in describing phase transitions, the van der Waals model was developed and proposes that the equation of state of a real gas is

$$\left(p + \frac{a}{v^2}\right)(v - b) = RT,$$

where p and T are the pressure and temperature of the gas and v is the volume of the gas per unit mole. a and b are known constants. Determine the critical pressure and temperature predicted by this model.

To determine the pressure p at which there is a point of inflexion in the T-V diagram when the gas is cooled under constant pressure, we need to determine

$$\left(\frac{\partial T}{\partial V}\right)_p = 0,$$

$$\left(\frac{\partial^2 T}{\partial V^2}\right)_p = 0,$$

where the subscript p underscores the fact that we treat p as a constant in computing the partial derivative. Since the number of moles n is fixed and $V = nv$, the above is equivalent to finding

$$\left(\frac{\partial T}{\partial v}\right)_p = 0,$$

$$\left(\frac{\partial^2 T}{\partial v^2}\right)_p = 0.$$

From the van der Waals equation of state,

$$\left(\frac{\partial T}{\partial v}\right)_p = \frac{1}{R}\left[-\frac{2a}{v^3}(v-b) + p + \frac{a}{v^2}\right] = 0,$$

$$\left(\frac{\partial^2 T}{\partial v^2}\right)_p = \frac{1}{R}\left[\frac{6a}{v^4}(v-b) - \frac{4a}{v^3}\right] = 0.$$

From the second equation, we obtain the specific molar volume at this juncture.

$$3(v-b) = 2v$$

$$\implies v = 3b.$$

Substituting this into the first equation, the critical pressure is

$$-\frac{2a \cdot 2b}{27b^3} + p_c + \frac{a}{9b^2} = 0$$

$$\implies p_c = \frac{a}{27b^2}.$$

Finally, substituting p_c and v into the equation of state, the critical temperature is

$$T_c = \frac{8a}{27Rb}.$$

4.5.2 Coexistence of Phases

This section will analyze the coexistence lines in greater detail. As an introduction, consider a closed system containing the liquid and vapor phases of a substance that has not yet established an equilibrium.

If the pressure on the liquid due to the vapor is too low, liquid molecules will escape the liquid (evaporate) at a greater rate than gas molecules entering the liquid (condensing). Thus, the liquid will vaporize to produce more gas molecules — causing the vapor pressure to increase. Conversely, if the vapor pressure is too high, there will be a net influx of molecules into the liquid — condensing the gas and reducing the vapor pressure.

Therefore, there is a tendency for the system to equilibrate until there is no net exchange of molecules between the phases. A dynamic equilibrium is established such that the rate of molecules evaporating from the liquid phase is equal to the rate of molecules condensing from the vapor phase. When such an equilibrium has been established, the liquid and vapor are referred to as a saturated liquid and vapor respectively. The vapor pressure at this juncture, for a given common temperature T between the liquid and vapor, is known as the saturated vapor pressure $p_s(T)$. Similarly, the common temperature, for a given vapor pressure p, is known as the saturation temperature $T_s(p)$. The liquid-gas coexistence line represents the saturation pressure at various temperatures or equivalently, the saturation temperature at various pressures. Observe from the coexistence line that in general, there is a one-to-one mapping between the equilibrium pressure and temperature when two or more phases coexist. Therefore, in a certain sense, there is only a single independent variable when phases coexist.

In light of the above discussion, another important point to understand is that a substance generally does not exist as a purely liquid or a purely vapor phase at equilibrium due to evaporation and condensation. Evaporation always occurs, because the energy distribution of surface molecules in a liquid is Boltzmann-like such that some highly energetic molecules will definitely leave the liquid over time. Therefore, a purely liquid phase cannot be at equilibrium. However, there is still a slight chance for a purely vapor phase to attain an equilibrium as a vapor will in fact not condense in empty space to form small droplets. This is because, when a liquid phase has yet to form, the intermolecular forces are too weak to cause molecules to congregate together to produce a liquid. From the perspective of energy, the molecules need to provide the surface energy required to build the liquid surface — a difficult barrier to overcome. A nucleation center, such as a dust particle, is in fact required to keep the molecules together and to spark off condensation. It reduces the interface of the liquid with its vapor (as part of the surface is stuck to the nucleation center) such that the energy barrier is lowered. Therefore, in the case of extremely clean vapors, it is possible for them to attain an equilibrium. Such vapors which exist in the vapor region of the phase diagram and lie outside of the coexistence lines are known as supersaturated vapors. They exist in a state of unstable equilibrium as the presence of a nucleation center will immediately trigger condensation.

Clausius–Clapeyron Equation

The equation of a coexistence line $p(T)$ is modeled by the Clausius–Clapeyron equation which states that

$$\frac{dp}{dT} = \frac{L}{T\Delta v},$$

where L is the specific latent heat during the transition at the current state (T, p) on the coexistence line and Δv is the specific change in volume (volume change per unit mass) across the two phases, from one of lower internal energy to one of higher internal energy, at the particular (T, p) state.

Proof: There is a delightful proof of the Clausius–Clapeyron equation that is based on conjuring a hypothetical reversible heat engine, that utilizes two coexisting phases of a pure substance as its working substance, and imposing the efficiency of a reversible heat engine dictated by the second law of thermodynamics.

For purposes of illustration, let our working substance be a combination of a saturated liquid and vapor stored in a container (this proof also works for other coexisting phases). Now, consider the following four processes of an infinitesimal Carnot cycle performed by this working substance.

(1) The working substance is expanded isothermally at temperature T when it is put in thermal contact with a reservoir of temperature T. Mass m of the liquid is vaporized in this process such that the total volume per unit mass of the working substance changes from v_1 to v_2. Since this change in total volume is much steeper than the change in pressure of the working substance (this extends to all pairs of coexisting phases as well but holds especially in the case where one phase is a vapour), in the limit where $v_2 - v_1 \ll v_1$, this process on the PV diagram of the working substance is approximately depicted by a straight line at a constant pressure p.

(2) The working substance is expanded adiabatically such that its temperature and pressure decrease to $(T + dT)$ and $(p + dp)$ respectively, where $dT < 0$ and $dp < 0$.

(3) The working substance is compressed isothermally at temperature $(T + dT)$ when it is put in thermal contact with a reservoir of temperature $(T + dT)$. Mass m of the vapour is condensed in this process. Again, the pressure of this process is constant at $(p + dp)$.

(4) Finally, the working substance is compressed adiabatically such that its pressure and temperature reverts from $(p + dp)$, $(T + dT)$ to p, T.

Since the graph depicting this cycle on a PV diagram is approximately a parallelogram with edge length $m(v_2 - v_1) = m\Delta v$ and height $-dp$, the work done by this infinitesimal Carnot cycle is $-m\Delta v dp$. Furthermore, the working substance only receives heat during the first process which is of amount mL where L is the specific latent heat of vaporisation. Consequently, the efficiency of this cycle is

$$\eta = \frac{W}{Q_{in}} = -\frac{\Delta v dp}{L}.$$

As all processes are reversible, this must be equal to the Carnot efficiency $1 - \frac{T+dT}{T} = -\frac{dT}{T}$.

$$-\frac{\Delta v dp}{L} = -\frac{dT}{T}$$

$$\implies \frac{dp}{dT} = \frac{L}{T\Delta v}.$$

Scrutinizing the Clausius–Clapeyron equation, one can see that the gradients of the coexistence lines are usually positive as Δv is positive for most transitions. Furthermore, the solid-liquid coexistence line should be extremely steep at Δv is small. That said, water is an anomaly as the density of ice is actually smaller than that of water — causing Δv to be negative. Then, the solid-liquid line for water has a steep, negative gradient.

Unfortunately, the Clausius–Clapeyron equation is hard to solve for in general as L and Δv are both functions of state that are difficult to model. For example, L for a liquid-gas transition decreases as temperature increases, attaining zero at the critical temperature and causing the liquid-gas line to terminate.

However, we can determine approximate solutions for solid-gas and liquid-gas transitions at temperatures much lower than T_c for a small temperature change. The specific latent heat L remains approximately constant and Δv can be taken to be the specific volume of the gaseous state, v_g, which is much larger than the specific volumes of the other phases. Then,

$$\frac{dp}{dT} = \frac{L}{Tv_g}.$$

Assuming that the ideal gas law holds,

$$pv_g = \frac{RT}{\mu}$$

where μ is the molar mass of the gas molecules. Then,

$$\frac{dp}{dT} = \frac{Lp\mu}{RT^2}$$

$$\int_{p_0}^{p} \frac{1}{p} dp = \int_{T_0}^{T} \frac{L\mu}{RT^2} dT$$

$$\ln \frac{p}{p_0} = \frac{L\mu}{RT_0} - \frac{L\mu}{RT}$$

$$p = p_0 e^{-\frac{L\mu}{R}\left(\frac{1}{T} - \frac{1}{T_0}\right)}, \tag{4.25}$$

where (T_0, p_0) is a reference point on the coexistence line.

4.5.3 *Mixture of Gases*

In this section, we will study the coexistence of the liquid and vapor phases of a substance when its vapor phase is mixed with a disparate gas. This is a ubiquitous phenomenon as Earth is brimming with air (which is a mixture of different gases) that envelopes all other substances. First and foremost, a pivotal assumption regarding gas mixtures is the Dalton model. It states that the total pressure p of a mixture of N gases that occupy a certain volume at thermal equilibrium is simply the sum of the individual pressures, referred to as the partial pressures p_i, that each gas would have caused in that volume.

$$p = \sum_{i=1}^{N} p_i$$

where p_i is the partial pressure of the ith gas. This relationship is evident from the kinetic theory of gases as the different gases would simply engender their own pressures if they do not interact with each other. That is, the Dalton model is simply stating that each gas operates as if it is the only gas in that particular volume and is unaffected by the presence of other gases (assuming that there are no interactions). The partial pressure p_i can be expressed as a fraction of p via the ideal gas law. If there are n_i moles of the ith gas occupying the common volume V at temperature T,

$$p_i V = n_i RT$$

for all $1 \leq i \leq N$. Summing the above for all i,

$$pV = nRT$$

where

$$n = \sum_{i=1}^{N} n_i$$

is the total number of moles of gas molecules. Dividing the ideal gas law of the ith gas by the previous equation,

$$p_i = \frac{n_i}{n} p.$$

That is, the partial pressure generated by the ith gas is simply its mole fraction relative to the entire mixture multiplied by the total pressure.

Next, we proceed with our main topic — analyzing the coexistence of the liquid phase of a substance with a mixture of gases at equilibrium. For the sake of illustration, consider a set-up involving liquid water and its vapor (water vapor) mixed with atmospheric air. Since atmospheric air normally contains a portion of water vapor, we shall explicitly exclude this component in referring to air and instead, treat it as part of the vapor. Air, with water vapor removed from its constituents, is often referred to as dry air. Now, there are two common assumptions made in this context. Firstly, the coexistence of liquid water and its vapor is presumed to be unaffected by the presence of dry air. That is, when a dynamic equilibrium has been established such that the amount of liquid water remains constant, the partial pressure of water vapor must be the saturated vapor pressure corresponding to temperature of the mixture T, $p_s(T)$. Secondly, the experiment is often conducted at constant pressure, which is atmospheric pressure p_0, such that the sum of the partial pressures of water vapor and dry air must be equal to p_0 at all instances.

Let us first consider a set-up where the liquid phase has yet to form — there is solely a mixture of supersaturated water vapor and dry air in a container. Now, as this mixture is cooled at constant pressure, the partial pressure of water vapor remains constant as its mole fraction relative to the mixture remains constant. Eventually, the mixture attains a temperature at which water vapor first begins to condense — this temperature is known as the dew point. In other words, the dew point of a mixture of water vapor and air is the temperature at which liquid water first begins to form when the mixture is cooled at constant pressure. Since the partial pressure p_w of water vapor remains constant during this process, the dew point is simply the saturation temperature $T_s(p_w)$ of water vapor at that constant partial pressure p_w.

The above discussion implies that we need to know the partial pressure of water vapor p_w in order to calculate the dew point. However, a more common measure which enables the indirect calculation of p_w is the relative humidity ϕ of an air-water vapor mixture. Firstly, an air-water vapor mixture is defined to be saturated when a dynamic equilibrium has been established between the water vapor and liquid water (i.e. the state of water is on the liquid-vapor coexistence line). The relative humidity is then defined as the ratio of the mole fraction of water vapor in the current mixture to the mole fraction of water vapor in a saturated mixture at the same temperature and total pressure. This is equivalent to the ratio of the partial pressure of water vapor in the current mixture p_w to the saturation pressure of water vapor at the same temperature $p_s(T)$.

$$\phi = \frac{p_w}{p_s(T)}.$$

Problem: Determine the dew point of a mixture of dry air and water vapor with relative humidity ϕ and current temperature T_0. Suppose that you have a P-T graph of the liquid-vapor coexistence line of water.

Firstly, we need to compute the partial pressure of water vapor in the mixture which is

$$p_w = \phi p_s(T_0).$$

$p_s(T_0)$ can be determined from the P-T graph by drawing a vertical line at T-coordinate T_0 and finding the pressure of the point of intersection of this line and the liquid-vapor coexistence line of water. Moving on, the dew point is the saturation temperature $T_s(p_w)$ at vapor pressure p_w. This can be identified by drawing a horizontal line on the P-T diagram at P-coordinate p_w and finding the temperature of the point of intersection of this line and the liquid-vapor coexistence line.

Next, what occurs if we continue to cool the previous mixture at constant pressure after the dew point has been reached and wait for an equilibrium to be established (assume that its final temperature is still greater than the triple point temperature)? Firstly, note that water cannot solely exist in its liquid phase at the end of this process as we have already remarked that a purely liquid phase cannot be at equilibrium. Instead of completely condensing into liquid water, what actually occurs is that water vapor partially condenses such that its partial pressure decreases as its mole fraction decreases. Its final partial pressure must correspond to the saturation pressure at the final temperature of the set-up in order for liquid water and water vapor to attain a dynamic equilibrium. Since the pressure of the mixture is immutable, this also implies that the partial pressure of air increases.

Problem: A mixture of supersaturated water vapor, with initial partial pressure p_{w0}, and n moles of dry air molecules is cooled at constant pressure p_0 from an initial temperature T_0 to a smaller final temperature T_1 that is below the temperature of the dew point but above the temperature of the triple point of water. Describe the evolution of the state of water during this process on a P-T diagram and determine the number of moles of water vapor Δn that is condensed. Assume that you know the saturation pressure of water vapor as a function of temperature, $p_s(T)$.

On a phase diagram, the state of water begins as a supersaturated vapor at (T_0, p_{w0}) and travels along a horizontal line (at constant partial pressure

p_{w0}) until it intersects the liquid-vapor coexistence line (the temperature at this point of intersection is the dew point). As the mixture is further cooled, the state of water travels along the coexistence line, towards decreasing temperature, until temperature T_1.

The mole ratio between water vapor and air is simply the ratio between their partial pressures. Therefore, the initial number of moles of water vapor is

$$n_{w0} = \frac{p_{w0}}{p_0 - p_{w0}} n.$$

The final partial pressure of water vapor is the saturation pressure at temperature T_1, $p_s(T_1)$. Therefore, the final number of moles of water vapor is

$$n_{w1} = \frac{p_s(T_1)}{p_0 - p_s(T_1)} n.$$

The moles of water vapor condensed is then

$$\Delta n = n_{w0} - n_{w1} = \left(\frac{p_{w0}}{p_0 - p_{w0}} - \frac{p_s(T_1)}{p_0 - p_s(T_1)} \right) n.$$

Finally, let us consider the reverse process of the previous set-up. Suppose that we start with liquid water and an air-water vapor mixture and heat it at constant pressure p_0. Liquid water will first begin to vaporize and increase the partial pressure of water vapor as its mole ratio increases. The equilibrium state of water initially moves along the liquid-vapor coexistence line towards increasing temperature. However, as the partial pressure of water increases, the partial pressure of air must decrease for the pressure of the mixture to remain constant. This insinuates that there is a certain limiting temperature where it is no longer possible for water to have an equilibrium state with coexisting liquid and vapor phases as the partial pressure of air decreases below zero. At this juncture, liquid water is said to boil as the bubbles formed by evaporation can no longer be restrained by the external pressure p_0. Since liquid water just begins to boil when the partial pressure of air is zero and when the partial pressure of its saturated vapor in a bubble is at least as large as the external pressure for the bubble to continue expanding, the boiling temperature T_b corresponds to the temperature at which the saturation pressure of water is the constant external pressure p_0.

$$p_s(T_b) = p_0.$$

Note that p_0 refers to the atmospheric pressure p_{atm} in most situations. During boiling, like any other phase transition, the temperature of water remains constant until sufficient latent heat has been supplied to completely vaporize liquid water.

Problem: The immiscible liquid phases of two substances A and B are stored together in an open container at atmospheric pressure p_{atm}. Given their saturation pressures as functions of temperature $p_{sA}(T)$ and $p_{sB}(T)$, determine the condition on the temperature T at which boiling first occurs. Afterwards, explain how one can determine which substance has a higher molar rate of boiling throughout this boiling process. Assume that the heights of the liquids are small such that the pressure is uniform throughout the set-up.

Observe that at the interface between A and B, a bubble comprising the saturated vapors of both A and B can form. If we assume that Dalton's law holds, the total pressure of the bubble is the sum of the partial pressures of the vapors of A and B. Therefore, boiling first occurs at this interface, rather than the possible interfaces of A and B with air. The boiling temperature T satisfies

$$p_{sA}(T) + p_{sB}(T) = p_{atm}.$$

To identify the substance that boils at a greater rate, observe that the temperatures of the two substances remain constant during boiling — implying that the partial pressures of their saturated vapors contained in the bubbles remain constant too as there is a one-to-one mapping between saturation pressure and temperature. Since the molar ratio of the saturated vapors of A and B contained in a bubble is equal to the ratio of their partial pressures, the substance with a higher saturation pressure at this boiling temperature T will boil at a greater molar rate, throughout the entire boiling process since this molar ratio is constant.

Problems

Conduction

1. Concentric Spheres **

Two concentric, hollow spheres have radii r_1 and r_2 respectively with $r_1 < r_2$. Denote their instantaneous temperatures as T_1 and T_2. If the space between them is filled with a material with thermal conductivity k and negligible heat capacity, determine the instantaneous heat flux between the two spheres. Using the previous result, find $T_1(t)$ and $T_2(t)$ if the heat capacities of the spheres are C_1 and C_2 and if their initial temperatures are T_{10} and T_{20}.

2. Cylindrical Shell with Felt **

Suppose that the cylindrical shell in Section 4.2 is now covered with felt that has a uniform thermal conductivity k_2 and an outer radius r_2. Let the thermal conductivity of the cylindrical shell, with inner and outer radii r_0 and r_1, be k_1. The inner surface of the cylindrical shell is maintained at T_0 while the outer surface of the felt is maintained at T_1. Determine the heat flux across cylindrical layers.

3. Current in a Wire **

Consider a long cylindrical wire with a radius R and thermal conductivity k. A current runs through it such that each unit volume of the wire produces p amount of heat per unit time. If the temperature of the cylindrical surface of the wire is maintained at T_0, determine the temperature distribution in the wire $T(r)$ as a function of its radial coordinate r.

4. Conducting Gas **

n_0 moles of an ideal gas fill a container of constant cross sectional area A and length l. It is known that the thermal conductivity of a section of ideal gas is proportional to the square root of its temperature $k = c\sqrt{T}$. If the ends of the container are maintained at temperatures T_1 and T_2 respectively, determine the pressure of the gas at steady state. Assume one-dimensional heat flow in the direction perpendicular to the cross section of the container.

5. Truncated Cone **

A truncated cone has two circular surfaces of radii r_0 and r_1, $r_0 < r_1$, which are maintained at temperatures T_0 and T_1 respectively. The perpendicular distance between these two surfaces is h. Find the heat flux in the direction of the axis. Assume that $r_1 - r_0 \ll h$ such that the half-angle of the cone is small. Where is this assumption necessary in your working?

6. *Regular Polygon*****

The N vertices of a homogeneous regular N-gon are maintained at temperatures $T_1, T_2, \ldots, T_N$ respectively by an external agency. Determine the steady state temperature of the centroid.

7. *Slabs and Gases*****

Three slabs (filled with black) have thermal conductivities k_1, k_2 and k_3, cross sectional areas $A_1 = A_2 = A$ and A_3 and lengths l_1, l_2 and l_3. They are connected by tubes filled with gases (shaded gray) of heat transfer coefficient h as shown in the figure below. The cross sectional areas of the gas tubes are not given and are irrelevant. If the left end of the left slab and the right end of the right slab are maintained at temperatures T_{1l} and T_{2r}, determine the condition for the middle slab to have a uniform temperature at steady state.

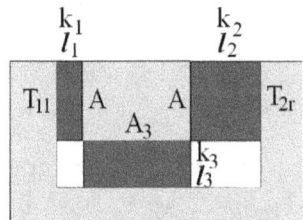

Radiation

8. *Radiation Pressure**

A small, black plate of area A is stationary at a large distance away from the Sun which is a spherical black body with radius r_s, mass M and constant temperature T_s. Determine the mass of the plate, m. Neglect all other gravitational effects and assume that the surface of the plate is perpendicular to the line joining the center of the Sun to it.

9. *Spherical Space Station**

A space station takes the form of a black sphere in outer space with surroundings at zero absolute temperature. Due to the operation of the space station, its internal appliances produce a certain amount of power that is conducted isotropically within the sphere. If the equilibrium temperature of the space station under such circumstances is T, determine the new equilibrium temperature T' of the space station after a black spherical shell, of a slightly larger radius than the space station, is used to envelope the space station. What if N thin black shields are used? What if a single thin shield, made of an opaque gray material of emissivity ε, is used?

10. *Transmitting Plate* **

Two plates of emissivities ε_1 and ε_2 are oriented parallel to each other. The first plate is opaque while the second plate has a reflectivity r. If the two plates are maintained at temperatures T_1 and T_2 respectively, determine the heat flux transmitted across the second plate.

11. *Three Gray Plates* **

Two large, gray and opaque plates with emissivities ε_1 and ε_3 are oriented parallel to each other and are maintained at temperatures T_1 and T_3 respectively. Now, another plate of equal emissivity, absorptivity and transmittivity is placed between the two plates. Determine the equilibrium temperature of this plate, T_2.

12. *Earth's Atmosphere* ***

In this problem, we will model the effect of an atmosphere on Earth. Suppose that the Sun is a black body with temperature T_1 and radius r_1. The Earth is a sphere that is located at a distance R from the Sun and has a radius r_3. The emissivity of the Earth is ε_3.

(1) If there is no atmosphere on Earth, determine the temperature of the Earth at equilibrium, T_3.

(2) Now, we consider the effects of an atmosphere. Model the atmosphere as a spherical shell of gas, with an emissivity ε_2 and outer radius $r_2 > r_3$, surrounding the Earth. At thermal equilibrium, its absorptivity for both ultraviolet and infrared light is ε_2. The atmosphere transmits a fraction t of ultraviolet light but is completely opaque to infrared. Assuming that the Sun emits ultraviolet light while the Earth emits and re-emits infrared, determine the temperature of the atmosphere T_2 and the Earth, T_3, at thermodynamic equilibrium. Assume that the atmosphere is a perfect thermal conductor such that all incident radiation is instantaneously evenly distributed across it.

Thermal Expansion

13. *Ring**

A flat, circular ring has an inner radius and outer radius. If the ring is now heated such that it undergoes isotropic expansion, does the area of the hole in the middle increase or decrease?

14. *Spherical Balls***

Spherical ball A is hung down from a massless, inextensible string that is connected to a wall. Spherical ball B lies motionless on a horizontal floor. The same quantity of heat Q is supplied to both balls. Assuming no heat losses, are the final temperatures of the balls the same? If not, estimate the difference in the final temperatures in terms of parameters of your choice. (International Physics Olympiad)

Phase Transitions

15. *Latent Heat**

Consider a container with a piston that contains a certain amount of gaseous and liquid states of the same substance. The piston is first fixed and the system is at equilibrium at temperature T. The latent heat of vaporization per mole of gas of this configuration is determined to be L. Now, consider the case where the massive piston is not fixed and is instead, balanced by the difference between the interior pressure and atmospheric pressure. The system is initially at equilibrium at temperature T. Determine the latent heat of vaporization per mole of gas of this new configuration in terms of L and T. Assume that the gaseous form of the substance is ideal and attains thermodynamic equilibrium at every instance.

16. *Gas in Rocket**

A motionless cylindrical vessel of cross sectional area A in outer space initially contains an ideal gas of total mass M and initial pressure $p \ll p_s$ where p_s is the saturation pressure at its current temperature (which is above the triple point temperature but below the critical temperature). The vessel is then given a constant acceleration a along its cylindrical axis while its temperature is maintained. Determine the mass of liquid m formed by condensation due to this motion after the system has equilibrated. Hint: you have to consider different regimes of a.

17. Melting Ice**

1kg of ice at 0°C floats in 5kg of water at 50°C. The whole system is thermally isolated. Determine the change in entropy of the whole system when thermal equilibrium has been reached. The specific heat capacity of water is $4.2 \text{kJkg}^{-1}\text{K}^{-1}$ and the latent heat of fusion of ice is 333kJkg^{-1}.

18. Boiling Point**

Model the atmosphere as a spherical shell of uniform gas with molar mass μ and uniform temperature T_a that envelopes the spherical, uniform Earth. If the atmospheric pressure and boiling point of water at the surface of the Earth is p_0 and T_0 respectively, determine the boiling point of water at a small height z from the surface of the Earth. Assume that the latent heat of vaporization is a constant L in this regime and that the specific volume of water vapor is much larger than that of water.

19. Heating a Container**

A closed container of constant volume currently contains certain amount of gaseous and liquid states of the same substance at equilibrium. If the current temperature of the system is T and the specific latent heat of vaporization in the current state is L, determine the fractional change in the moles of gaseous molecules due to evaporation if the equilibrium temperature of the system is slightly increased by $\Delta T \ll T$. The specific volume of the gas can be assumed to be much greater than that of the liquid. The gas molecules have molar mass μ.

Solutions

1. Concentric Spheres**

Similar to the case of a cylindrical shell, the isotropic nature of the set-up implies that the heat flux is purely radial. Let the heat flux between adjacent spherical shells be $\dot{Q}$ (positive towards the outer shells). Then, consider the heat flux across a spherical shell of radius r.

$$\dot{Q} = -k4\pi r^2 \frac{dT}{dr}.$$

Since $\dot{Q}$ is uniform throughout all spherical shells (as the material cannot absorb any heat else it will experience an infinite temperature change),

$$\int_{r_1}^{r_2} \frac{\dot{Q}}{r^2} dr = -\int_{T_1}^{T_2} 4\pi k dT$$

$$\dot{Q}\left(\frac{1}{r_1} - \frac{1}{r_2}\right) = 4\pi k(T_1 - T_2)$$

$$\dot{Q} = \frac{4\pi k(T_1 - T_2)}{\frac{1}{r_1} - \frac{1}{r_2}}.$$

Another method is to calculate the thermal resistance between the two shells as the sum of the thermal resistances of many infinitesimal shells of varying radius in series. Because the heat flux density is perpendicular at every point on a spherical shell, the thermal resistance of an infinitesimal shell of radius r and thickness dr is analogous to that of a slab, $\frac{dr}{kA(r)} = \frac{dr}{4\pi kr^2}$. The total resistance between the two shells is then

$$R = \int_{r_1}^{r_2} \frac{dr}{4\pi kr^2} = \frac{1}{4\pi k}\left(\frac{1}{r_1} - \frac{1}{r_2}\right)$$

so the heat flux is

$$\dot{Q} = \frac{T_1 - T_2}{R} = \frac{4\pi k(T_1 - T_2)}{\frac{1}{r_1} - \frac{1}{r_2}}.$$

In the second part of the problem, we have

$$C_1 \frac{dT_1}{dt} = -\dot{Q} = -A(T_1 - T_2),$$

$$C_2 \frac{dT_2}{dt} = \dot{Q} = A(T_1 - T_2),$$

where $A = \frac{4\pi k}{\frac{1}{r_1} - \frac{1}{r_2}}$. To decouple this pair of equations, multiply the first equation by C_2 and subtract the second equation, multiplied by C_1, from it.

$$C_1 C_2 \frac{d(T_1 - T_2)}{dt} = -(C_1 + C_2)A(T_1 - T_2),$$

$$\frac{d(T_1 - T_2)}{dt} = -\frac{(C_1 + C_2)A}{C_1 C_2}(T_1 - T_2).$$

The solution to this differential equation in variable $(T_1 - T_2)$, after substituting the initial conditions, is

$$T_1 - T_2 = (T_{10} - T_{20})e^{-\frac{(C_1 + C_2)A}{C_1 C_2}t}.$$

On the other hand, since the total internal energy of the spheres must be conserved,

$$C_1 T_1 + C_2 T_2 = C_1 T_{10} + C_2 T_{20}.$$

Solving these equations simultaneously,

$$T_1 = \frac{C_2}{C_1 + C_2}(T_{10} - T_{20})e^{-\frac{(C_1 + C_2)A}{C_1 C_2}t} + \frac{C_1}{C_1 + C_2}T_{10} + \frac{C_2}{C_1 + C_2}T_{20},$$

$$T_2 = -\frac{C_1}{C_1 + C_2}(T_{10} - T_{20})e^{-\frac{(C_1 + C_2)A}{C_1 C_2}t} + \frac{C_1}{C_1 + C_2}T_{10} + \frac{C_2}{C_1 + C_2}T_{20}.$$

2. Cylindrical Shell with Felt**

Recall that we derived the thermal resistance of a cylindrical shell as

$$R = \frac{\ln \frac{r_1}{r_0}}{2\pi k}.$$

The shells and the felt are connected in series. Thus, the equivalent thermal resistance is

$$R_{eq} = R_1 + R_2 = \frac{\ln \frac{r_1}{r_0}}{2\pi k_1} + \frac{\ln \frac{r_2}{r_1}}{2\pi k_2}.$$

Thus, the rate of heat conduction is

$$\dot{Q} = \frac{T_0 - T_1}{R_{eq}} = \frac{2\pi(T_0 - T_1)}{\frac{\ln \frac{r_1}{r_0}}{k_1} + \frac{\ln \frac{r_2}{r_1}}{k_2}}.$$

3. Current in a Wire**

Consider a cylindrical shell between radii r and $r+dr$. The heat flux entering this shell is $\dot{Q}(r)$ while the heat flux emanating from it is $\dot{Q}(r+dr)$. Let the length of the wire be l. Then, the volume of this shell is $2\pi rldr$ which implies that the heat generated per unit time is $2\pi prldr$. At equilibrium, the net heat flow through this shell must be equal to the heat generated.

$$\dot{Q}(r+dr) - \dot{Q}(r) = 2\pi plrdr.$$

Shifting dr to the left-hand side and applying the first principles of calculus,

$$\frac{d\dot{Q}}{dr} = 2\pi plr.$$

Separating variables and integrating, we can determine the heat flux through a cylindrical shell between radial distances r and $r+dr$.

$$\int_0^{\dot{Q}} d\dot{Q} = \int_0^r 2\pi plrdr$$

$$\dot{Q} = \pi plr^2,$$

where we have used the fact that the heat flux must be zero at $r=0$ as the shell at $r=0$ has negligible volume and thus generates negligible heat. Now, to determine the temperature distribution, we apply Fourier's law of conduction.

$$\dot{Q} = kA\frac{dT}{dr} = k2\pi rl\frac{dT}{dr}.$$

Then,

$$\frac{dT}{dr} = \frac{pr}{2k}.$$

Integrating this and imposing the limit $T = T_0$ at $r = R$, the temperature $T(r)$ is

$$\int_{T_0}^T dT = \int_R^r \frac{pr}{2k}dr$$

$$T = T_0 + \frac{pr^2}{4k} - \frac{pR^2}{4k}.$$

4. Conducting Gas**

At steady state, the heat flux and pressure must be continuous throughout the gas. By Fourier's law of conduction, the heat flux across a cross section is

$$\dot{Q} = -cA\sqrt{T}\frac{dT}{dx}$$

where the x-direction has been set to be the direction perpendicular to the cross section, pointing from the end at T_1 to the end at T_2. Shifting dx to the left and integrating,

$$\int_0^l \dot{Q}dx = \int_{T_1}^{T_2} -cA\sqrt{T}dT.$$

Since $\dot{Q}$ is a constant,

$$\dot{Q} = \frac{2cA(\sqrt{T_1^3} - \sqrt{T_2^3})}{3l}.$$

To determine the temperature at a distance x from the end at temperature T_1, we perform the previous integration over more general limits.

$$\int_0^x \dot{Q}dx = \int_{T_1}^{T} -cA\sqrt{T}dT.$$

Substituting the expression for $\dot{Q}$,

$$T = \left(\sqrt{T_1^3} - \frac{(\sqrt{T_1^3} - \sqrt{T_2^3})x}{l}\right)^{\frac{2}{3}}.$$

Now, we need to ensure that the total number of moles is n_0. We know from the ideal gas law that

$$p = \eta RT$$

where η is the molar density of molecules. Then,

$$\eta = \frac{p}{RT} = \frac{p}{R\left(\sqrt{T_1^3} - \frac{(\sqrt{T_1^3} - \sqrt{T_2^3})x}{l}\right)^{\frac{2}{3}}}.$$

Now, we need to ensure that

$$n_0 = A\int_0^l \eta dx.$$

In this integration, remember that p must be uniform at steady state. Then,

$$\int_0^l \eta dx = \int_0^l \frac{p}{R\left(\sqrt{T_1^3} - \frac{\left(\sqrt{T_1^3} - \sqrt{T_2^3}\right)x}{l}\right)^{\frac{2}{3}}} dx$$

$$= \left[-\frac{p}{R}\left(\sqrt{T_1^3} - \frac{\left(\sqrt{T_1^3} - \sqrt{T_2^3}\right)x}{l}\right)^{\frac{1}{3}} \cdot \frac{3l}{\sqrt{T_1^3} - \sqrt{T_2^3}} \right]_0^l$$

$$= \frac{3pl\left(\sqrt{T_1} - \sqrt{T_2}\right)}{R(\sqrt{T_1^3} - \sqrt{T_2^3})}.$$

Substituting $n_0 = A \int_0^l \eta dx$,

$$n_0 = \frac{3pAl\left(\sqrt{T_1} - \sqrt{T_2}\right)}{R\left(\sqrt{T_1^3} - \sqrt{T_2^3}\right)},$$

$$p = \frac{n_0 R\left(\sqrt{T_1^3} - \sqrt{T_2^3}\right)}{3Al(\sqrt{T_1} - \sqrt{T_2})} = \frac{n_0 R(T_1 + \sqrt{T_1 T_2} + T_2)}{3Al}.$$

5. Truncated Cone**

Let h_2 be the height of the truncated part of the cone. Let the origin be at the center of the circular surface with radius r_0 and let the positive x-axis be directed perpendicular towards the other surface. We can calculate h_2 by using similar triangles,

$$\frac{h_2}{r_0} = \frac{h_2 + h}{r_1}$$

$$h_2 = \frac{hr_0}{r_1 - r_0}.$$

The half-angle of the cone, θ, is

$$\theta = \tan^{-1}\frac{r_0}{h_2} = \tan^{-1}\frac{r_1 - r_0}{h}.$$

Now we consider a circular surface with thickness dx that is at x-coordinate x from the origin. The radius of this circular surface is

$$r = (x + h_2)\tan\theta.$$

By Fourier's law of conduction, the heat flux through this surface is

$$\dot{Q} = -kA\frac{dT}{dx} = -k\pi r^2\frac{dT}{dx} = -k\pi(x+h_2)^2\tan^2\theta\frac{dT}{dx}.$$

Since the system is in thermal equilibrium, $\dot{Q}$ is constant.

$$\dot{Q}\int_0^h \frac{1}{(x+h_2)^2}dx = -\int_{T_0}^{T_1} k\pi\tan^2\theta\, dT$$

$$\dot{Q} = \frac{k\pi\tan^2\theta(T_0-T_1)}{\frac{1}{h_2}-\frac{1}{h+h_2}} = \frac{\pi r_0 r_1 k(T_0-T_1)}{h}.$$

The assumption of a small half-angle is necessary in ensuring that all circular surfaces possess uniform temperatures such that we can approximate the heat flow to be solely one-dimensional (perpendicular to the bases of the truncated cone) in the above working.

An alternative method is to deem the truncated cone as many circular disks of infinitesimal thickness dx connected in series. Firstly, the uniform temperatures of the two surfaces of each infinitesimal disk cause the resistance of an infinitesimal disk to be well-defined. Next, because the heat flux density is perpendicular at all points on each disk, the thermal resistance of an infinitesimal disk at x-coordinate x is analogous to that of a slab.

$$\frac{1}{kA(x)}dx = \frac{1}{k\pi(x+h_2)^2\tan^2\theta}dx.$$

Integrating the above from $x=0$ to $x=h$, the total thermal resistance of the truncated cone is

$$R = \int_0^h \frac{1}{k\pi(x+h_2)^2\tan^2\theta}dx = \frac{1}{k\pi\tan^2\theta}\left(\frac{1}{h_2}-\frac{1}{h+h_2}\right)$$

$$\implies \dot{Q} = \frac{T_0-T_1}{R} = \frac{k\pi\tan^2\theta(T_0-T_1)}{\frac{1}{h_2}-\frac{1}{h+h_2}} = \frac{\pi r_0 r_1 k(T_0-T_1)}{h}.$$

6. Regular Polygon**

Let the external power supplied to the ith vertex be $\dot{Q}_i$ and the temperature at the centroid be T_c when an equilibrium has been established. At steady state, the net power received by the polygon must be zero. This implies that

$$\sum_{i=1}^N \dot{Q}_i = 0.$$

Now, rotate the entire set-up by $\frac{2\pi}{N}$ radians $(N-1)$ times to obtain $(N-1)$ rotationally symmetric set-ups. Superposing these $(N-1)$ set-ups with the original set-up, $\sum_{i=1}^{N} \dot{Q}_i$ power flows into each vertex to establish a vertex temperature $\sum_{i=1}^{N} T_i$ and centroid temperature NT_c at steady state. Since $\sum_{i=1}^{N} \dot{Q}_i = 0$, no external power is delivered to any point on the polygon — implying that its temperature must be uniform at steady state. Equating the temperature of the centroid and the temperature of a vertex in this set-up,

$$NT_c = \sum_{i=1}^{N} T_i$$

$$T_c = \frac{\sum_{i=1}^{N} T_i}{N}.$$

7. Slabs and Gases**

Let the steady state temperatures of the right end of the first slab, left end of the second slab and the top and bottom ends of the third slab be T_{1r}, T_{2r}, T_{3t} and T_{3b} respectively. Furthermore, let the steady state temperatures of the gases in the top and bottom sections be T_{g1} and T_{g2} respectively. Drawing the thermal circuit of the set-up, we obtain Fig. 4.9.

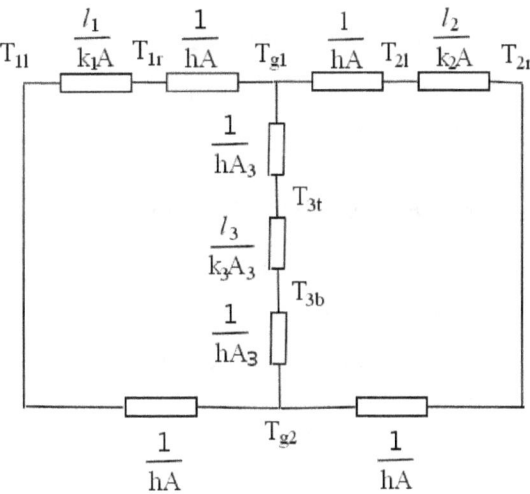

Figure 4.9:   Thermal circuit

The thermal resistance between the surface of a slab of area A' and a gas of heat transfer coefficient h is $\frac{1}{hA'}$ as Newton's law of cooling states that $\dot{Q} = -hA'\Delta T$. For the temperature of the middle (third) slab to be uniform, $T_{3t} = T_{3b}$ which implies that there is no heat flux in the middle branch. Then,

T_{g1} and T_{g2} must be equal. For this to occur, the ratio between the left and right portions of the bottom and top horizontal segments must be equal (since they are connected in parallel between predetermined temperatures T_{1l} and T_{2r}). That is,

$$\frac{\frac{1}{hA}}{\frac{1}{hA}} = \frac{\frac{1}{hA} + \frac{l_1}{k_1 A}}{\frac{1}{hA} + \frac{l_2}{k_2 A}}$$

$$\implies \frac{l_1}{k_1} = \frac{l_2}{k_2}.$$

8. Radiation Pressure*

The luminosity of the Sun is

$$L = 4\pi r_s^2 \sigma T_s^4.$$

Let the plate be located at a distance r from the center of the Sun. The irradiance on the plate at a distance r from the center of the Sun, due to the Sun's radiation, is

$$E = \frac{4\pi r_s^2 \sigma T_s^4}{4\pi r^2} = \frac{r_s^2 \sigma T_s^4}{r^2}.$$

Since the plate is a black body, all incident radiation is absorbed. This implies that its absorbed power is

$$P = \frac{r_s^2 A \sigma T_s^4}{r^2}.$$

For a photon, its energy E is related to its momentum p by

$$E = pc$$

where c is the speed of light. Since P is the average energy per photon multiplied by the rate of photons incident on the plate and because the force on the plate is the average momentum per photon multiplied by the rate of impinging photons,

$$F = \frac{P}{c} = \frac{r_s^2 A}{r^2 c} \sigma T_s^4.$$

This force must balance the gravitational force on the plate due to the sun.

$$\frac{GMm}{r^2} = \frac{r_s^2 A}{r^2 c} \sigma T_s^4$$

$$m = \frac{r_s^2 A}{GMc} \sigma T_s^4.$$

9. Spherical Space Station*

Let the heat flux density produced by the internal appliances of the space station be $\dot{q}$. Since the original equilibrium temperature of the black space station was T, by the continuity of heat flux,

$$\dot{q} = \sigma T^4.$$

Now, observe that when a black shield is used to cover the space station, the heat flux density throughout the space station and the shield must still be $\dot{q}$ by the continuity of heat flux as the internal appliances still operate in the same manner. This directly implies that the equilibrium temperature of the shield is T (as the shield is akin to the original space station) and that the shield emits an exitance (heat flux density) $\dot{q}$, both radially inwards and outwards. The space station therefore receives $\dot{q}$ heat flux density from the internal appliances and $\dot{q}$ heat flux density from the shield — for a total of $2\dot{q}$ which must also be the heat flux density radiated by it at steady state. Consequently,

$$\sigma T'^4 = 2\dot{q} = 2\sigma T^4$$

$$T' = \sqrt[4]{2}T.$$

When N black shields are used, we can repeat the above arguments to show that the exitance of the outer-most shield is $\dot{q}$, which causes the exitance of the second outer-most shield to be $2\dot{q}$, which causes the exitance of the third outer-most shield to be $3\dot{q}$ and so on as the interior of a shield always transmits a heat flux density $\dot{q}$ to that shield by the continuity of heat flux. The space station therefore receives $\dot{q}$ heat flux density from the internal appliances and $N\dot{q}$ heat flux density from the inner-most shield — for a total of $(N+1)\dot{q}$.

$$\sigma T'^4 = (N+1)\dot{q} = (N+1)\sigma T^4$$

$$T'' = \sqrt[4]{N+1}T.$$

In the second scenario, the equilibrium temperature of the shield is not T but it still emits an exitance $\dot{q}$, both radially inwards and outwards. The space station then receives $2\dot{q}$ from the heat produced by its internal appliances and from the exitance of the shield. However, even though the space station emits exitance $\sigma T''^4$ (where T'' is its equilibrium temperature), a fraction $(1 - \varepsilon)$ is reflected back by the shield and reabsorbed by the space station. Therefore, the space station effectively only radiates a heat flux density

$\varepsilon\sigma T''^4$. Balancing the heat flux densities received by and emanating from the space station,

$$\varepsilon\sigma T''^4 = 2\dot{q} = 2\sigma T^4$$

$$T'' = \sqrt[4]{\frac{2}{\varepsilon}}T.$$

10. Transmitting Plate**

Let M_1, J_1 and M_2, J_2 be the exitances and radiosities of the first and second plates in the directions towards each other respectively. Then,

$$J_1 = M_1 + (1 - \varepsilon_1)J_2,$$

$$J_2 = M_2 + rJ_1.$$

Note that the coefficient of J_1 in the second equation is not $(1 - \varepsilon_2)$ as some radiation is transmitted as well. Solving the two equations above simultaneously,

$$J_1 = \frac{M_1 + (1 - \varepsilon_1)M_2}{1 - (1 - \varepsilon_1)r},$$

$$J_2 = \frac{M_2 + rM_1}{1 - (1 - \varepsilon_1)r}.$$

The net heat flux from the first plate to the second is then

$$\dot{Q} = (J_1 - J_2)A = \frac{[(1 - r)M_1 - \varepsilon_1 M_2]A}{1 - (1 - \varepsilon_1)r}.$$

Part of this heat flux is transmitted and the rest is absorbed by the second plate (note that the reflected portion has already been excluded). The portion of radiation transmitted is then

$$Q_t = \frac{t}{\varepsilon_2 + t}\dot{Q} = \frac{1 - \varepsilon_2 - r}{1 - r}\dot{Q} = \frac{A(1 - \varepsilon_2 - r)\left[(1 - r)\varepsilon_1\sigma T_1^4 - \varepsilon_1\varepsilon_2\sigma T_2^4\right]}{(1 - r)(1 - r + \varepsilon_1 r)},$$

where t is the transmissivity of the second plate and where we have substituted $M_1 = \varepsilon_1\sigma T_1^4$ and $M_2 = \varepsilon_2\sigma T_2^4$.

11. Three Gray Plates**

The emissitivity, absorptivity and transmittivity of the middle plate are each $\frac{1}{3}$. Define J_1 and J_3 to be the radiosities of the first and third plate, towards the middle plate. Define J_{21} and J_{23} to be the radiosities of the middle plate in the directions of the first and third plates respectively. Finally, let M_1, M_2 and M_3 be the respective exitances of the plates. Then,

$$J_1 = M_1 + (1 - \varepsilon_1)J_{21},$$

$$J_3 = M_3 + (1 - \varepsilon_3)J_{23}.$$

In relating the radiosities of the middle plate, we must take extra care.

$$J_{21} = M_2 + \frac{1}{3}J_3 + \frac{1}{3}J_1.$$

The second term $\frac{1}{3}J_3$ is the contribution to the radiosity towards the first plate, due to the transmitted radiation from the third plate. The third term $\frac{1}{3}J_1$ stems from the reflected irradiance on the second plate, due to the first plate. Similarly,

$$J_{23} = M_2 + \frac{1}{3}J_1 + \frac{1}{3}J_3.$$

Now, $\frac{1}{3}J_1$ is the transmitted portion and $\frac{1}{3}J_3$ is the reflected portion. We immediately realize that

$$J_{21} = J_{23}.$$

When the second plate has attained thermodynamic equilibrium, the net heat fluxes on both sides of this plate must be equal. Then,

$$J_1 - J_{21} = J_{23} - J_3,$$

$$J_1 + J_3 = 2J_{21}.$$

Substituting this expression into the previous equation in J_{21} yields

$$J_{21} = M_2 + \frac{2}{3}J_{21}$$

$$J_{21} = J_{23} = 3M_2 = \sigma T_2^4.$$

Then,

$$J_1 = M_1 + (1 - \varepsilon_1)\sigma T_2^4,$$

$$J_3 = M_3 + (1 - \varepsilon_3)\sigma T_2^4.$$

Adding these equations together and using $J_1 + J_3 = 2J_{21} = 2\sigma T_2^4$,

$$M_1 + M_3 = (\varepsilon_1 + \varepsilon_3)\sigma T_2^4.$$

Substituting $M_1 = \varepsilon_1 \sigma T_1^4$ and $M_3 = \varepsilon_3 \sigma T_3^4$,

$$T_2 = \sqrt[4]{\frac{\varepsilon_1 T_1^4 + \varepsilon_3 T_3^4}{\varepsilon_1 + \varepsilon_3}}.$$

12. Earth's Atmosphere***

When there is no atmosphere, there must be no net heat flux between the Sun and the Earth. The Sun emits a total power of $4\pi r_1^2 \sigma T_1^4$. The intensity of this radiation at a distance R from the sun is then

$$\frac{4\pi r_1^2 \sigma T_1^4}{4\pi R^2} = \frac{r_1^2 \sigma T_1^4}{R^2}.$$

The power absorbed by the Earth at thermodynamic equilibrium is then the intensity multiplied by the cross sectional area of the Earth πr_3^2 and its absorptivity ε_3.

$$P = \frac{\varepsilon_3 \pi r_1^2 r_3^2 \sigma T_1^4}{R^2}.$$

This must be equal to the power radiated by the Earth which is $\varepsilon_3 4\pi r_3^2 \sigma T_3^4$. Equating these powers,

$$\frac{\varepsilon_3 \pi r_1^2 r_3^2 \sigma T_1^4}{R^2} = \varepsilon_3 4\pi r_3^2 \sigma T_3^4$$

$$T_3 = \sqrt[4]{\frac{r_1^2}{4R^2}} T_1.$$

In the presence of an atmosphere, we can use a similar process to first determine the equilibrium temperature of the atmosphere T_2.

Referring to Fig. 4.10, we first define the power, radiated by the Sun and incident on the atmosphere, spread per unit area of the atmosphere, as

$$E = \frac{\frac{\pi r_1^2 r_2^2 \sigma T_1^4}{R^2}}{4\pi r_2^2} = \frac{\sigma T_1^4 r_1^2}{4R^2}.$$

Now, note that the net heat flux between the atmosphere and the Earth must be zero at equilibrium. Then, the net heat flux between the atmosphere and exterior surroundings (including the Sun) must also be zero for the heat flux to be continuous. The atmosphere reflects a fraction $1 - \varepsilon_2 - t$ of the incident radiation by the Sun. Therefore, the net heat flux between the atmosphere

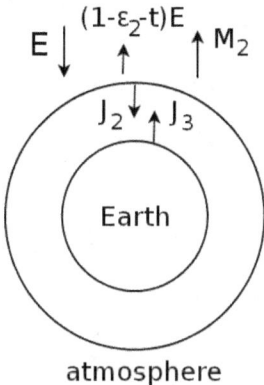

atmosphere

Figure 4.10: Earth with atmosphere

and its exterior surroundings is

$$4\pi r_2^2(E - (1 - \varepsilon_2 - t)E - M_2)$$

where $M_2 = \varepsilon_2 \sigma T_2^4$. Imposing the condition that the above expression is zero and substituting the expression for E,

$$(\varepsilon_2 + t)E = M_2,$$

$$T_2 = \sqrt[4]{\left(1 + \frac{t}{\varepsilon_2}\right)\frac{r_1^2}{4R^2}}\, T_1.$$

Now, we move onto the condition that there must be no net heat flux between the atmosphere and the Earth. When an atmosphere is present, there will be infrared light repeatedly bouncing between the atmosphere and the Earth. However, no infrared light that is emitted or reflected by the Earth escapes the combined system of the atmosphere and Earth due to the opacity of the atmosphere to infrared. That said, the atmosphere still permits a portion of the Sun's ultraviolet radiation to be transmitted. Define J_2, M_2, J_3 and M_3 to be the respective radiosities and exitances of the atmosphere and the Earth, in the direction towards each other. Then,

$$J_2 = M_2 + tE + (1 - \varepsilon_2)F_{23}J_3 + (1 - \varepsilon_2)F_{22}J_2,$$

$$J_3 = M_3 + (1 - \varepsilon_3)F_{32}J_2 + (1 - \varepsilon_3)F_{33}J_3.$$

Since the spherical Earth is enclosed by the atmosphere, it is evident that $F_{32} = 1$ and $F_{33} = 0$. Then, F_{23} and F_{22} can also be determined. By the reciprocity theorem,

$$F_{32}A_3 = F_{23}A_2.$$

Substituting $F_{32} = 1$, $A_3 = 4\pi r_3^2$ and $A_2 = 4\pi r_2^2$,

$$F_{23} = \frac{r_3^2}{r_2^2}.$$

The fraction of radiation emitted by the atmosphere that is not received by the Earth must be returned to the atmosphere. Thus,

$$F_{22} = 1 - F_{23} = \left(1 - \frac{r_3^2}{r_2^2}\right).$$

Then,

$$J_2 = M_2 + tE + (1 - \varepsilon_2)\frac{r_3^2}{r_2^2}J_3 + (1 - \varepsilon_2)\left(1 - \frac{r_3^2}{r_2^2}\right)J_2,$$

$$J_3 = M_3 + (1 - \varepsilon_3)J_2.$$

Solving these equations simultaneously and sparing the gory details would yield

$$J_2 = \frac{tE + M_2 + (1 - \varepsilon_2)\frac{r_3^2}{r_2^2}M_3}{\varepsilon_2 + \varepsilon_3\frac{r_3^2}{r_2^2} - \varepsilon_2\varepsilon_3\frac{r_3^2}{r_2^2}},$$

$$J_3 = \frac{(1 - \varepsilon_3)tE + (1 - \varepsilon_3)M_2 + \left(\varepsilon_2 + \frac{r_3^2}{r_2^2} - \varepsilon_2\frac{r_3^2}{r_2^2}\right)M_3}{\varepsilon_2 + \varepsilon_3\frac{r_3^2}{r_2^2} - \varepsilon_2\varepsilon_3\frac{r_3^2}{r_2^2}}.$$

The net heat flux between the atmosphere and the Earth is then by Eq. (4.18),

$$\dot{Q} = \frac{A_3\varepsilon_3}{1 - \varepsilon_3}\left(\sigma T_3^4 - J_3\right) = \frac{A_3\varepsilon_3}{1 - \varepsilon_3} \cdot \frac{(\varepsilon_3 - 1)tE + (\varepsilon_3 - 1)M_2 + \varepsilon_2(1 - \varepsilon_3)\sigma T_3^4}{\varepsilon_2 + \varepsilon_3\frac{r_3^2}{r_2^2} - \varepsilon_2\varepsilon_3\frac{r_3^2}{r_2^2}}$$

$$= \frac{A_3\varepsilon_3\left(-tE - M_2 + \varepsilon_2\sigma T_3^4\right)}{\varepsilon_2 + \varepsilon_3\frac{r_3^2}{r_2^2} - \varepsilon_2\varepsilon_3\frac{r_3^2}{r_2^2}}.$$

Since $\dot{Q}$ must be zero,

$$-tE - M_2 + \varepsilon_2\sigma T_3^4 = 0.$$

Substituting $M_2 = (\varepsilon_2 + t)E$ (this was previously derived when computing the temperature of the atmosphere),

$$\sigma T_3^4 = \frac{\varepsilon_2 + 2t}{\varepsilon_2} E.$$

Substituting the expression for E would yield

$$T_3 = \sqrt[4]{\frac{(\varepsilon_2 + 2t)r_1^2}{4\varepsilon_2 R^2}} T_1.$$

This temperature is $\sqrt[4]{\frac{\varepsilon_2 + 2t}{\varepsilon_2}} > 1$ times of the equilibrium temperature of the Earth without an atmosphere — the model in this problem then implies that the presence of an atmosphere warms the Earth.

13. Ring*

Since the ring undergoes isotropic expansion, the entire ring, including the hole, is scaled radially by a certain factor. Therefore, the size of the hole increases. An alternate perspective to this scaling is that we are enlarging an image of the ring such that the hole becomes larger.

14. Spherical Balls**

An important observation is that the balls expand when heated. However, the center of mass of the ball connected by a string drops while the center of mass of the ball on the horizontal ground rises when the balls expand. Thus there is positive work done on the ball by gravity in the former case while there is negative work done on the ball by gravity in the latter case. Let r be the initial radii of the spheres and Δr_1 and Δr_2 be the respective changes in radii of the spheres due to thermal expansion. Applying the first law of thermodynamics to both balls (we use the subscript 1 to denote the ball hung by a string and 2 for the other ball),

$$Q = \Delta U_1 - W_{on1} = mc\Delta T_1 - mg\Delta r_1,$$
$$Q = \Delta U_2 - W_{on2} = mc\Delta T_2 + mg\Delta r_2,$$

where ΔT_1 and ΔT_2 are the changes in the temperatures of the balls. c is the standard specific heat capacity of the balls which accounts for the slight work done against the atmosphere due to thermal expansion of the balls when they are heated, in addition to their changes in internal energy. Lastly,

$$\Delta r = r\alpha\Delta T$$

where α is the coefficient of linear expansion. Thus,

$$(mc - mgr\alpha)\Delta T_1 = Q,$$

$$(mc + mgr\alpha)\Delta T_2 = Q,$$

$$\Delta T_1 - \Delta T_2 = \frac{Q}{mc}\left(\frac{1}{1 - \frac{gr\alpha}{c}} - \frac{1}{1 + \frac{gr\alpha}{c}}\right) \approx \frac{2Qgr\alpha}{mc^2},$$

where we have used the Maclaurin expansion ($\frac{1}{1+x} = 1 - x + \cdots$) and discarded second order and above terms in α.

15. Latent Heat*

In the first case, the heat supplied to the system is entirely embodied in the increase in potential energy of the molecules of the substance as they transition from the liquid phase into the solid phase. Assuming that Δn moles of liquid become gas,

$$\Delta U = \Delta n L$$

where ΔU is the increase in potential energy of the n moles of molecules. In the second case, part of the heat supplied to the system is also manifested as work performed by the expanding gas molecules.

$$Q = \Delta U + W_{by}.$$

Since the gas is always at thermodynamic equilibrium, its pressure must be constant so that the piston does not move. Stemming from this constant pressure, the temperature of the gas must also be maintained at T in order for it to attain equilibrium with the liquid state (remember that there is a one-to-one mapping between equilibrium pressure and temperature along a coexistence line). The first law of thermodynamics then yields

$$Q = \Delta U + p\Delta V = \Delta n L + p\Delta V.$$

Furthermore, from the ideal gas law,

$$pV = nRT$$

$$p\Delta V = \Delta n RT$$

as p and T are constant. Then,

$$Q = \Delta n(L + RT).$$

The latent heat of vaporization per mole of gas in this configuration is thus $L + RT$.

16. Gas in Rocket*

After the system has equilibrated, the gas at the rear of the rocket (relative to its direction of acceleration) must have the largest pressure p_{rear} while the front of the rocket has the lowest pressure p_{front} as the force engendered by the pressure at the back of any gas section exceeding the pressure at its front causes that gas section to accelerate forwards at a at equilibrium. As we will be comparing p_{rear} to p_s, p_{front} is negligible when computing the net force on the entire gas remaining in the vessel for the following reason. The initial pressure $p = \frac{\rho_i RT}{\mu} \ll p_s$ where ρ_i, T and μ are the initial density, temperature and molar mass of the gas. After the system has equilibrated, there will be a new density distribution $\rho(x)$ of the ideal gas where x is the x-coordinate along the cylindrical axis of the vessel. However, as some gas molecules are condensed, $\int \frac{\rho(x)RT}{\mu} dx < \int \frac{\rho_i RT}{\mu} dx$ where the integral is performed over the length of the cylindrical axis. Since $\frac{\rho(x)RT}{\mu} = p'(x)$ where $p'(x)$ is the pressure distribution after the system has equilibrated, this further implies that $\int p'(x)dx < \int pdx \ll \int p_s dx$. Since $p'(x)$ is monotonic and $p'(x)$ at the rear end of the cylinder (p_{rear}) is possibly comparable to p_s, $p'(x)$ at the front end (p_{front}) must be much smaller than p_s to satisfy the above inequality. Therefore, only the rear pressure p_{rear} effectively produces the force required to accelerate the remaining gas of mass M' at a.

$$p_{rear} = \frac{M'a}{A}.$$

If none of the gas condenses, p_{rear} must be smaller than the saturation pressure p_s while $M' = M$.

$$p_{rear} = \frac{Ma}{A} < p_s.$$

Therefore, the gas does not condense if $a < \frac{p_s A}{M}$. If a does not satisfy this bound, a dynamic equilibrium will be established between the liquid and vapor phases of the substance. Then, p_{rear} must correspond to the saturation pressure p_s.

$$p_{rear} = \frac{M'a}{A} = p_s,$$

$$M' = \frac{p_s A}{a}.$$

The number of moles condensed is thus

$$M - M' = M - \frac{p_s A}{a}.$$

17. Melting Ice**

We must first do a preliminary check on whether all the ice melts. Assuming that the ice does not completely melt, the final equilibrium temperature of the system will be 0°C. The amount of heat lost by the water is then

$$Q_{wat} = m_{wat}c\Delta T = 5 \cdot 4200 \cdot (0 - 50) = -1050000J.$$

The amount of heat required for the ice to completely melt is

$$Q_{melt} = m_{ice}L = 1 \cdot 333000 = 333000J.$$

Since $|Q_{water}| > Q_{melt}$, the ice completely melts. Now we can calculate the final equilibrium temperature of the system, T, using the fact that

$$Q_{ice} + Q_{water} = 0$$

$$333000 + 4200 \cdot T + 5 \cdot 4200(T - 50) = 0$$

$$T = 28.452°C = 301.602K.$$

The total change in entropy of the system is the sum of the changes in entropy due to the melting of ice, heating of the melted ice and the cooling of water respectively.

$$\Delta S_{total} = \Delta S_{melt} + \Delta S_{melted\,ice} + \Delta S_{wat}$$

$$= \int_{melt} \frac{dQ}{T} + \int_{melted\,ice} \frac{dQ}{T} + \int_{wat} \frac{dQ}{T}$$

$$= \frac{m_{ice}L}{273} + \int_{273}^{301.602} \frac{m_{ice}c}{T}dT + \int_{323}^{301.602} \frac{m_{wat}c}{T}dT$$

$$= \frac{333000}{273} + 4200\ln\frac{301.602}{273} + 5 \cdot 4200\ln\frac{301.602}{323}$$

$$= 199JK^{-1}.$$

18. Boiling Point**

From balancing forces on individual sections of the atmosphere or from the Boltzmann distribution, the pressure of an isothermal atmosphere of temperature T and molar mass μ at an altitude z above the surface of the Earth can be shown to be

$$p(z) = p_0 e^{-\frac{\mu g z}{RT_a}},$$

where p_0 is the pressure at the surface of the Earth. When water starts to boil, the saturated vapor pressure must be equal to the external pressure due to the atmosphere. We then need to determine the temperature corresponding to $p(z)$ on the coexistence line. Applying the Clausius–Clapeyron equation and neglecting the specific volume of water,

$$\frac{dp}{dT} = \frac{L}{Tv_g} = \frac{Lp\mu}{RT^2}$$

where v_g is the specific volume of the gas that can be expressed as $\frac{RT}{p\mu}$ via the ideal gas law. Then,

$$\int_{p_0}^{p} \frac{1}{p} dp = \int_{T_0}^{T} \frac{L\mu}{RT^2} dT$$

$$p = p_0 e^{-\frac{L\mu}{R}\left(\frac{1}{T} - \frac{1}{T_0}\right)}.$$

Equating the saturation vapor pressure and the external atmospheric pressure,

$$p_0 e^{-\frac{\mu g z}{RT_a}} = p_0 e^{-\frac{L\mu}{R}\left(\frac{1}{T} - \frac{1}{T_0}\right)}.$$

Solving for T,

$$T = \frac{LT_a T_0}{gzT_0 + LT_a}.$$

19. Heating a Container**

The Clausius–Clapeyron equation yields the following case when the specific volume of the gas v_g is dominant.

$$\frac{dp}{dT} = \frac{L}{Tv_g}.$$

By the ideal gas law,

$$pv_g = \frac{1}{\mu}RT$$

$$\frac{dp}{dT} = \frac{Lp\mu}{RT^2}$$

$$\implies \frac{dp}{p} = \frac{L\mu}{RT^2}dT.$$

We also obtain from the ideal gas law that

$$n = \frac{pV}{RT}.$$

Taking the total derivative of both sides,

$$dn = \frac{dpV}{RT} - \frac{pV}{RT^2}dT$$

where V has been taken to be a constant as the volume of the gas does not change significantly (as the volume of the liquid is negligible and the volume of the container remains constant). Dividing the above by $n = \frac{pV}{RT}$,

$$\frac{dn}{n} = \frac{dp}{p} - \frac{dT}{T}.$$

Substituting the previously derived expression for $\frac{dp}{p}$,

$$\frac{dn}{n} = \left(\frac{L\mu}{RT^2} - \frac{1}{T}\right)dT.$$

Since ΔT is small, the fractional change $\frac{\Delta n}{n}$ can be approximated by substituting ΔT for dT and taking T to be the initial temperature in the above equation.

$$\frac{\Delta n}{n} = \left(\frac{L\mu}{RT^2} - \frac{1}{T}\right)\Delta T.$$

Chapter 5

Electrostatics

This chapter will study stationary systems of charge particles, referred to as electrostatic systems, and useful quantities associated with them such as charge, electric fields and electric potential.

5.1 Electric Charges

It is observed that if an "electrical" object A attracts another "electrical" object B and if another "electrical" object C attracts A, then C will definitely repel B. This resulted in the development of the concept of electric charges and the existence of two classes of charges — namely, positive and negative charges. It is then said that like charges repel while unlike charges attract. Note that the way through which the signs of charges are assigned is completely arbitrary. For all we know, the charges that are currently deemed positive could have been defined as negative and vice-versa — the laws of electromagnetism will work just as well.

Charges are measured in terms of the unit Coulombs (C) where one Coulomb is defined as the amount of charge transported by one Ampere[1] ($1A$) of current across a cross section in one second. One Coulomb of unbalanced charge is an extremely large amount of charge. To put things into proportion, the surface charge density of the Earth's surface (net charge per unit area) is only of the order of nano-Coulombs per square meter ($10^{-9}Cm^{-2}$).

Finally, an important property of charges is that the net amount of charge in a closed system is conserved. This does not necessarily mean that electric charges carried by subatomic particles cannot be created or destroyed. Rather, it means that these particles are created or destroyed in a specific ratio such that the net change in the total amount of charge is zero.

---

[1]The SI unit for current is the Ampere.

5.2 Coulomb's Law

Coulomb's law quantifies the "electrical" force, $\boldsymbol{F}_{t1}$, on a point, test charge q_t due to another point charge q_1.

$$\boldsymbol{F}_{t1} = \frac{1}{4\pi\varepsilon_0} \frac{q_1 q_t}{r_{1t}^2} \hat{\boldsymbol{r}}_{1t} = \frac{1}{4\pi\varepsilon_0} \frac{q_1 q_t}{r_{1t}^3} \boldsymbol{r}_{1t}. \tag{5.1}$$

$\boldsymbol{r}_{1t}$ is the vector pointing from q_1 to q_t and is equal to $\boldsymbol{r}_t - \boldsymbol{r}_1$ where $\boldsymbol{r}_t$ and $\boldsymbol{r}_1$ are the respective position vectors of q_t and q_1. ε_0 is a constant known as the permittivity of free space and has a numerical value of $8.854 \times 10^{-12} m^{-3} kg^{-1} A^2 s^4$ in SI units. Firstly, observe that Coulomb's law naturally satisfies Newton's third law as $\boldsymbol{F}_{t1} = -\boldsymbol{F}_{1t}$; this had better be the case. Next, we see that the Coulomb force is, again, an inverse-square central force similar to the gravitational force. A direct ramification is that the angular momentum of the system of two charges is conserved about any point.

The total force on a test charge due to an array of charges is given by the vector sum of the forces on that charge due to each individual charge by the principle of superposition. The principle of superposition is by no means trivial and should not be taken for granted as it cannot be derived from the other axioms. It basically states the total effect due to multiple sources is the sum of the effects produced by each individual source. The net Coulomb force $\boldsymbol{F}_t$ on a test charge q_t placed in a system of N charges is

$$\boldsymbol{F}_t = \sum_{i=1}^{N} \frac{1}{4\pi\varepsilon_0} \frac{q_i q_t}{r_{it}^2} \hat{\boldsymbol{r}}_{it}, \tag{5.2}$$

where $\hat{\boldsymbol{r}}_{it}$ is the unit vector along the vector pointing from q_i to q_t. For continuous charge distributions,

$$\boldsymbol{F}_t = \int \frac{1}{4\pi\varepsilon_0} \frac{q_t}{r^2} \hat{\boldsymbol{r}} dq, \tag{5.3}$$

where $\hat{\boldsymbol{r}}$ is the unit vector along the vector pointing from an infinitesimal charge element dq on the charge distribution to the test charge q_t. The integration is performed over the entire charge distribution.

Problem: In Fig. 5.1, two charges are hung from identical strings of length l from the same point on the ceiling. They then attain static equilibrium when the two strings are mutually perpendicular. If the charges have masses m_1 and m_2 respectively, determine the product of the magnitude of the charges, divided by the product of their masses.

Label the charges from left to right as q_1 and q_2 respectively. Let the string connecting q_1 subtend an angle θ with the vertical. We balance the

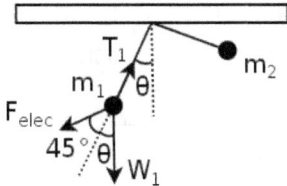

Figure 5.1: Hanging charges

forces on each charge, in the directions perpendicular to the string holding it to avoid the need to consider the tensions in the strings. Since the distance between the two charges is $\sqrt{2}l$,

$$m_1 g \sin \theta = \frac{q_1 q_2}{4\pi\varepsilon_0 \cdot 2l^2} \cdot \sin 45°,$$

$$m_2 g \cos \theta = \frac{q_1 q_2}{4\pi\varepsilon_0 \cdot 2l^2} \cdot \cos 45°.$$

Since $\sin^2 \theta + \cos^2 \theta = 1$,

$$\left(\frac{q_1 q_2}{8\sqrt{2}\pi\varepsilon_0 g l^2} \right)^2 \left(\frac{1}{m_1^2} + \frac{1}{m_2^2} \right) = 1$$

$$\frac{q_1 q_2}{m_1 m_2} = \frac{8\sqrt{2}\pi\varepsilon_0 g l^2}{\sqrt{m_1^2 + m_2^2}}.$$

Problem: There are two fixed positive point charges of charges q_1 and q_2 with position vectors r_1 and r_2 respectively. Find the position vector, r_3, of the point at which a third charge q_3 should be placed such that it will be at equilibrium. For a particular value of q_3, the entire system can remain at equilibrium without being held by any external force. Determine this particular value.

It is obvious that the three charges must be collinear, with the third charge sandwiched between the others, for the third charge to be at equilibrium. The magnitude of the forces on charge q_3 due to q_1 and q_2 are

$$F_{31} = \frac{1}{4\pi\varepsilon_0} \frac{q_1 q_3}{r_{13}^2},$$

$$F_{32} = \frac{1}{4\pi\varepsilon_0} \frac{q_2 q_3}{r_{23}^2},$$

where r_{13} and r_{23} are the distances between q_1 and q_3 and q_2 and q_3 respectively. For the third charge to remain stationary,

$$F_{31} = F_{32} \implies \frac{r_{23}}{r_{13}} = \sqrt{\frac{q_2}{q_1}}.$$

q_3 is located at a point that divides the line joining q_1 and q_2 into two segments with the above ratio. By the ratio theorem in vectors,

$$r_3 = \frac{\sqrt{q_2}r_1 + \sqrt{q_1}r_2}{\sqrt{q_1} + \sqrt{q_2}}.$$

Proceeding with the second part of the problem, since we know the ratio between r_{13} and r_{23}, balancing forces on charge q_1 yields

$$\frac{q_1 q_3}{4\pi\varepsilon_0 r_{13}^2} + \frac{q_1 q_2}{4\pi\varepsilon_0 (r_{13} + r_{23})^2} = 0$$

$$q_3 = -\frac{r_{13}^2}{(r_{13} + r_{12})^2} q_2 = -\frac{q_1 q_2}{(\sqrt{q_1} + \sqrt{q_2})^2}.$$

We can check that this value of q_3 results in no net force on charge q_2 as well. The simplest way of doing so is to observe that the above expression of q_3 is symmetric in q_1 and q_2 — implying that if we swapped the positions of q_1 and q_2 such that q_2 becomes the current q_1, the force balance condition on q_2 still produces the above expression for q_3.

5.3   Electric Field

In light of Coulomb's law and the principle of superposition, it is convenient to formulate a construct known as the electric field to describe a system of charges. A field basically ascribes each point in space a local quantity. The electric field, in this case, is a vector field which assigns each point in space the force per unit charge that will be exerted on a charge that is placed at that point. With this definition, the electric field at a point in space due to a single stationary point charge q is

$$E = \frac{1}{4\pi\varepsilon_0} \frac{q}{r^2} \hat{r}, \tag{5.4}$$

where r is the vector pointing from q to the point of concern. Note that r is neither the position vector of q nor that of the point of concern — it is the vector separating them. The magnitude of the electric field, E, is known as the electric field strength. Next, the electric field at a point in space due to a system of charges is given by the principle of superposition as

$$E = \sum_{i=1}^{N} \frac{1}{4\pi\varepsilon_0} \frac{q_i}{r_i^2} \hat{r}_i = \int \frac{1}{4\pi\varepsilon_0 r^2} \hat{r} \, dq \tag{5.5}$$

for a system of N discrete charges and a continuous charge distribution, respectively. r_i is the vector pointing from the ith charge to the point of concern while r in the integrand represents the vector pointing from the infinitesimal charge dq under consideration to the point of concern. These expressions again follow from the principle of superposition. Having defined the electric field, the force on a point test charge q_t placed at a point in an electric field is then

$$F_t = q_t E. \tag{5.6}$$

The electric field is an extremely useful construct as it reveals the force per unit charge on any charge placed at a point without further information of the surrounding charge distributions. Two different charge distributions can possibly produce the same electric field at a point in space and a charge at that point will still respond in the same manner in both cases. Let us now evaluate the electric fields at certain points in the following examples.

Problem: Find the electric field at the centroid of the equilateral triangle in the given charge configuration in Fig. 5.2.

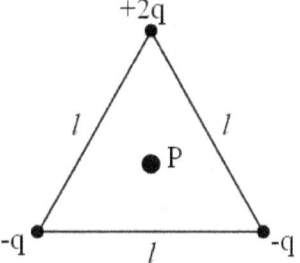

Figure 5.2: Charges arranged in an equilateral triangle

The electric field in the horizontal direction is zero due to symmetry. Since the distance between a vertex and the centroid of an equilateral triangle is $\frac{\sqrt{3}}{3}l$, the electric field in the y-direction (defined to be positive upwards) is given by

$$E_y = -\frac{2q}{4\pi\varepsilon_0 \left(\frac{\sqrt{3}}{3}l\right)^2} - \frac{q}{4\pi\varepsilon_0 \left(\frac{\sqrt{3}}{3}l\right)^2} \cdot \sin 30° \cdot 2 = -\frac{9q}{4\pi\varepsilon_0 l^2}.$$

Problem: Consider a rod of uniform linear charge density, λ. Find the electric field at point P that is a perpendicular distance h away from the rod, as shown in Fig. 5.3. The two anti-clockwise angles that the lines joining P and the ends of the rod subtend with the vertical are θ_0 and θ_1, respectively.

We define the origin at the foot of the perpendicular from the point of concern, P. Consider an infinitesimal length element dx on the rod with ends

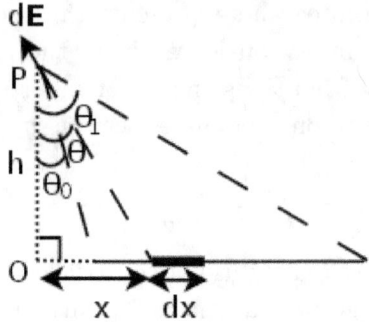

Figure 5.3:   Rod

at x and $x + dx$. Its contribution to the electric field at the point P is

$$d\boldsymbol{E} = \frac{dq}{4\pi\varepsilon_0(x^2 + h^2)} \begin{pmatrix} -\dfrac{x}{\sqrt{x^2 + h^2}} \\ \dfrac{h}{\sqrt{x^2 + h^2}} \end{pmatrix}.$$

Thus, the electric field at point P can be obtained by integrating this expression over the whole rod. Using $dq = \lambda dx$,

$$E_x = \int_{h\tan\theta_0}^{h\tan\theta_1} -\frac{\lambda x}{4\pi\varepsilon_0(x^2 + h^2)^{\frac{3}{2}}} dx = \left[\frac{\lambda}{4\pi\varepsilon_0\sqrt{x^2 + h^2}} \right]_{h\tan\theta_0}^{h\tan\theta_1}$$

$$= \frac{\lambda}{4\pi\varepsilon_0 h}(\cos\theta_1 - \cos\theta_0).$$

To integrate the expression for the electric field in the y-direction, polar coordinates should be used. We can label each point on the rod with a coordinate θ which is the anti-clockwise angle subtended by the vector pointing from P to the particular point on the rod and the vertical. Then, the x-coordinate of an infinitesimal segment can be expressed as

$$x = h\tan\theta$$

$$dx = h\sec^2\theta\, d\theta,$$

$$E_y = \int_{h\tan\theta_0}^{h\tan\theta_1} \frac{\lambda}{4\pi\varepsilon_0(x^2 + h^2)} \cdot \frac{h}{\sqrt{x^2 + h^2}} dx$$

$$= \int_{\theta_0}^{\theta_1} \frac{\lambda}{4\pi\varepsilon_0 h^2 \sec^2\theta} \cdot \cos\theta \cdot h\sec^2\theta\, d\theta$$

$$= \left[\frac{\lambda}{4\pi\varepsilon_0 h} \sin\theta \right]_{\theta_0}^{\theta_1}$$

$$= \frac{\lambda}{4\pi\varepsilon_0 h} (\sin\theta_1 - \sin\theta_0).$$

As the concept of an electric field is rather abstract, electric field lines — which are generally continuous curves, except at certain singularities (such as a point charge) — can be drawn to visualize the direction of the electric field at various locations. The direction of the electric field at a point corresponds to the direction of the electric field line at that point. Like any other field lines, electric field lines cannot cross each other as that would imply that there are two possible directions for the electric fields at a single point. Interestingly, electric field lines can only begin from positive charges and terminate[2] at negative charges. The facts that positive and negative charges are the sources and sinks of field lines are quite intuitive but our assertion that they are the **only** ones is rather astonishing. The reader should ponder the reason behind this after he or she has learnt Gauss' law. Next, another intriguing property of electric field lines is that they can never form closed loops — this is a direct corollary of the fact that the line integral of an arbitrary electrostatic field over a closed loop is zero, which we shall show later. If an electric field line in the shape of a closed loop indeed exists, we can perform a line integral along this loop to yield a non-zero result, contradicting the previous sacrosanct property of electrostatic fields.

Note that it is generally difficult to infer the magnitude of the electric field strength at a given point in space from looking at the field lines alone. We can at most infer that the electric field is generally stronger or weaker in a region by looking at the density of field lines cutting a given surface located in that region. The following are some examples of the electric field lines of isolated charges and pairs of charges in a single plane.

---

[2]If a field line reaches a point where $\boldsymbol{E} = \boldsymbol{0}$ exactly, we do not say that the field line is cut off as we can append it to a nearby point where the electric field is non-zero — the existence of such a point is guaranteed by the fact that a field line can only end at a negative charge. On another note, a field line can also extend indefinitely, such as in the case of a system with solely positive charges, but whether we classify that as terminating or not is a matter of semantics.

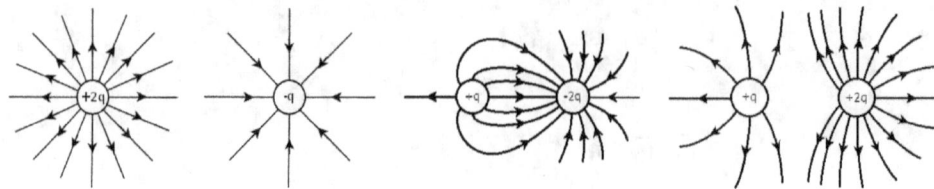

As verified by the above diagrams, electric field lines only emanate from positive charges and flow into negative charges. No closed loops are formed either. Moreover, charges of larger magnitude have a larger density of field lines surrounding them — indicating that they produce stronger electric fields. Actually, the field lines of the two diagrams on the left involving single charges are good indicators of the electric field strength as a function of radial distance from a single charge. The density of field lines cutting through an infinitesimal surface on a sphere of radius r centered about the charge decreases with $\frac{1}{r^2}$ (note that the diagrams are really three-dimensional such that the two left diagrams represent isotropic field lines in the radial direction) — in accordance with Coulomb's law.

Problem: Two charges $q_1 > 0$ and $q_2 < 0$ are located along the x-axis with the charge q_2 having the larger positive x-coordinate. A field line emanates from q_1 at an angle α with the positive x-axis. Determine if this field line will terminate at charge q_2. If so, determine the angle β that the field line makes with the negative x-axis as it is received by q_2.

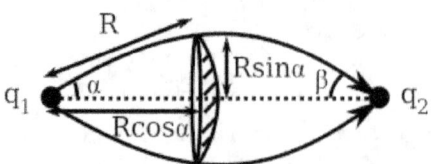

Figure 5.4:   Field lines and spherical cap

Firstly, suppose that a field line exits from q_1 and enters q_2 as shown in Fig. 5.4 (the positive x-axis is taken to be rightwards). As this set-up is symmetric about the x-axis, we can rotate the field line about the x-axis for a complete revolution to generate other axial-symmetric field lines. Notice that by doing so, we have enclosed a volume with these field lines. Since field lines cannot cross, field lines emanating from q_1 in its neighboring region, bounded by this volume, cannot escape this volume. Furthermore, since field lines cannot form closed loops and must terminate at a negative charge, all the previously mentioned field lines must eventually be received by q_2. Therefore, the number of field lines emanating from q_1 and entering q_2 are identical within this volume. Now, we need to determine the total fraction of field lines emitted by q_1 that is captured in this region. Observe that in the immediate vicinity of q_1, the electric field of q_2 is negligible as compared to that of q_1. Therefore, the electric field lines are isotropic about the neighborhood of q_1 (similar to the case where q_1 is the only charge in the system) — we then have to determine the proportion of the total number of isotropic field lines that are bounded by a cone of half-angle α. This fraction can be computed by dividing the surface area of the spherical cap, of a small radius R, corresponding to the angle α by the total surface area of a sphere of radius R, $4\pi R^2$. Since the surface area of a spherical cap is $\pi(a^2 + h^2)$ where $a = R\sin\alpha$ is the radius of its base circle and $h = R - R\cos\alpha$ is the altitude between the vertex and the center of its base, the fraction of the total number of field lines emitted by q_1 that is captured within this volume is

$$\frac{\pi\left[R^2\sin^2\alpha + (R - R\cos\alpha)^2\right]}{4\pi R^2} = \frac{1 - \cos\alpha}{2} = \sin^2\frac{\alpha}{2}.$$

Applying a similar argument to q_2, the fraction of the total number of field lines that it receives, enclosed within this volume, is $\sin^2\frac{\beta}{2}$. Since the number of field lines emitted by a positive charge or received by a negative charge is proportional to the magnitude of the charge, in order for the number of field lines emerging from q_1 and terminating at q_2 to be equal within this region,

$$q_1\sin^2\frac{\alpha}{2} = -q_2\sin^2\frac{\beta}{2}$$

$$\sin\frac{\beta}{2} = \sqrt{-\frac{q_1}{q_2}}\sin\frac{\alpha}{2}.$$

We do not consider $\sin\frac{\beta}{2} = -\sqrt{\frac{q_1}{q_2}}\sin\frac{\alpha}{2}$ as a field line can only propagate in a single plane in this case, such that a field line emitted at positive α and ending at negative β (or vice-versa) would have to cross the field line joining the two charges (but field lines cannot intersect). Observe that a solution for β only exists if $\sqrt{-\frac{q_1}{q_2}}|\sin\frac{\alpha}{2}| < 1$ else the field line would extend to infinity. If this condition if fulfilled,

$$\beta = 2\sin^{-1}\left(\sqrt{-\frac{q_1}{q_2}}\sin\frac{\alpha}{2}\right).$$

5.4   Gauss' Law

To supersede the cumbersome integrations that have to be performed for charge distributions, we can apply Gauss' law, which is a mathematical equivalent to Coulomb's law, to efficiently calculate the electric field of symmetrical objects. Before Gauss' law is analyzed, we first introduce a few related quantities. As a recap, the area vector of an infinitesimal surface with area dA is a vector pointing normally from the surface that possesses magnitude dA. There are two possible choices for the direction of the area vector and the exact choice is arbitrary if the entire surface, comprising all the infinitesimal surfaces, is open (it does not enclose any volume). For closed surfaces which enclose volume, the area vector of each infinitesimal surface is defined to be outwards.

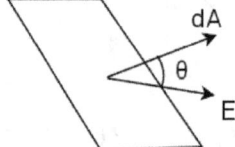

Figure 5.5:   Infinitesimal surface

Referring to Fig. 5.5, the electric flux $d\Phi$ through an infinitesimal surface of area $d\boldsymbol{A}$ is given by the dot product between the electric field at this surface and its area vector $d\boldsymbol{A}$. Note that the electric field is assumed to be uniform throughout the infinitesimal surface due to its minuscule size.

$$d\Phi = \boldsymbol{E} \cdot d\boldsymbol{A} = EdA\cos\theta. \tag{5.7}$$

The total flux through a closed surface S is

$$\Phi = \oiint_S \boldsymbol{E} \cdot d\boldsymbol{A},$$

where we are integrating the electric flux contributions by infinitesimal areas over the whole closed surface S. The loop on the integral represents the fact that the integration is performed over a (imaginary) closed surface which has no distinct boundaries and a definitive "inside" and "outside".

Moving on, Gauss' law states that the total electric flux cutting through a closed surface S is directly proportional to the charge that the surface S encloses, q_{enc}.

$$\oiint_S \boldsymbol{E} \cdot d\boldsymbol{A} = \frac{q_{enc}}{\varepsilon_0}. \tag{5.8}$$

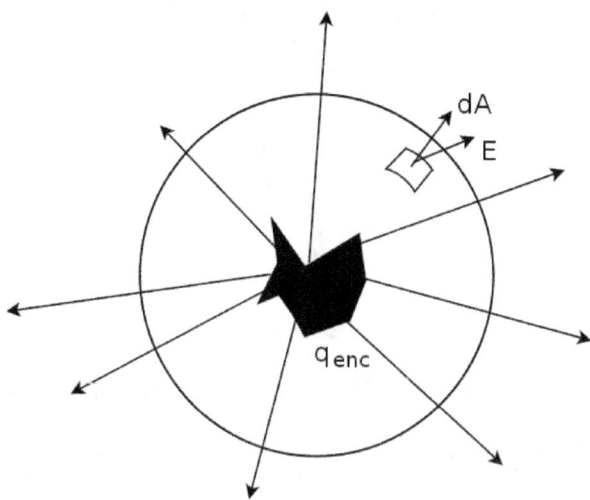

Figure 5.6: Spherical Gaussian surface

The closed surface S can be arbitrarily chosen and is known as a Gaussian surface. A spherical Gaussian surface is depicted in Fig. 5.6. Since the infinitesimal area vectors are defined to be pointing outwards, the electric flux at an infinitesimal area is positive if electric field lines are crossing out of the surface.

Proof: We first consider a single point charge q. The total electric flux through a spherical Gaussian surface of radius r, centered at the charge q, is

$$E \cdot 4\pi r^2 = \frac{q}{4\pi\varepsilon_0 r^2} \cdot 4\pi r^2 = \frac{q}{\varepsilon_0},$$

as the electric field at each point on the spherical surface is radially outwards (normal to the infinitesimal surface at that point) and of uniform magnitude $\frac{q}{4\pi\varepsilon_0 r^2}$. Evidently, Gauss' law is valid for a spherical Gaussian surface encapsulating a single charge. Now, consider an arbitrary closed surface encapsulating the charge and the projections of all infinitesimal area elements on the surface of a sphere of radius r onto the arbitrary surface. We first assume that the arbitrary surface has no folds such that each area element of the sphere is mapped to a single area element on the arbitrary surface.

Consider an area element on the sphere and its projection in Fig. 5.7. Assume that the base of the projected area (the base makes an angle θ with the projected area and is parallel to the original area element on the sphere) is a distance R away from the charge. If the area of the element on the sphere is dA, the area of the base is then $dA\frac{R^2}{r^2}$ by similarity arguments. Therefore,

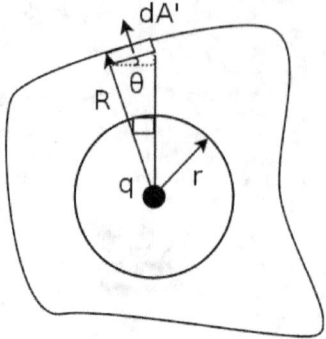

Figure 5.7:   Projection of element on sphere onto arbitrary surface

the area of the projected area is $dA' = dA\frac{R^2}{r^2\cos\theta}$. Next, the electric field strength at the projected area is $\frac{q}{4\pi\varepsilon_0 R^2}$. However, note that the electric field subtends an angle θ with the area vector of the projected area as the electric field is radially outwards. The electric flux through the projected area is then

$$d\Phi = \frac{q}{4\pi\varepsilon_0 R^2} \cdot dA\frac{R^2}{r^2\cos\theta} \cdot \cos\theta = \frac{qdA}{4\pi\varepsilon_0 r^2},$$

which is equal to the electric flux cutting across the infinitesimal area dA on the sphere. This argument can be applied to all area elements on the sphere and their projections. Therefore, the electric flux across the arbitrary surface is equal to that across the sphere which was previously derived to be $\frac{q}{\varepsilon_0}$. In the case where the arbitrary surface has folds such that a certain projection corresponds to multiple surfaces on the arbitrary surface, simply observe that the electric field lines will be emitted from some of these surfaces (labeled "out") and received by some surfaces (labeled "in") as shown in Fig. 5.8.

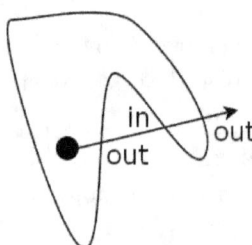

Figure 5.8:   Electric field line cutting multiple surfaces

The electric flux through one "in" element exactly cancels one "out element", as we have proven that the magnitudes of the electric fluxes across all possible projections must be equal to the magnitude of the electric flux across the original element on the sphere, and because these fluxes are of different signs. Observe that there will always only be a single net "out" surface in every direction as the Gaussian surface encloses the charge. Therefore, the electric flux through an arbitrary closed surface with folds, enclosing a single point charge q, must still be $\frac{q}{\varepsilon_0}$.

Finally, we need to show that a charge outside of a Gaussian surface does not result in a net electric flux through it.

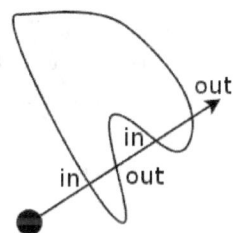

Figure 5.9: Gaussian surface enclosing zero charge

In a similar vein, the projections of a sphere on to the Gaussian surface can be considered and possible projection candidates can be labeled with "in" and "out" in Fig. 5.9. There will always be the same number of elements labeled "in" and elements labeled "out" in every direction as the charge is outside of the Gaussian surface. Therefore, the net electric flux through a Gaussian surface due to a charge outside is zero. We have officially shown that the total electric flux across an arbitrary Gaussian surface enclosing a single point charge q is $\frac{q}{\varepsilon_0}$, regardless of the charges outside the surface. Since a general charge distribution can be seen as a collection of point charges, the total electric flux across an arbitrary Gaussian surface enclosing a net charge q_{enc} is $\frac{q_{enc}}{\varepsilon_0}$, regardless of the charges outside the surface.

An intuitive understanding of Gauss' law can be obtained by imagining the isotropic emission of water (perhaps, in a region where there is no gravity) from each charge particle enclosed in a Gaussian surface at a steady rate. Consider a single point charge q for now. The water emerging from it at a certain instance propagates as a spherical wavefront that travels at a constant velocity (uniform across all charges) such that the surface density of water decreases with $\frac{1}{r^2}$, where r is the radial distance from the source. If we set the rate of water flowing out of this source to be proportional to q, the surface density of water at a point will be analogous to the electric field strength at that point. Moreover, the direction of water flow (radial) is also aligned with the direction of the electric field due to the charge. Evidently, the electric flux through an infinitesimal surface is then analogous to the rate of water flowing through it. When we compute the total electric flux crossing a Gaussian surface S, that encloses q, we are simply determining the net rate of water flowing out of S which must equal the rate of water leaking out of q, as the total volume of water that S can contain is fixed! This fact is independent of the position of q in S or the shape of S. By the same logic, if q is external to S, the net rate of water flowing out of S must be zero. Considering the superposition of similar set-ups involving different charges, the net rate of water flowing out of a Gaussian surface S must be identical to the net rate of water emitted by the sources enclosed by S. Since the counterpart of the former is the total electric flux Φ crossing S while the latter is proportional to the charge q_{enc} enclosed by S, Φ must be proportional to q_{enc} and is independent of the configuration of charges within S or the shape of S (as long as it encapsulates the same amount of charge). To summarize, Gauss' law is basically stating that instead of summing the rate of water flowing through small windows over an entire closed surface, we can adopt an alternate perspective and determine the total rate of water leaking out of the sources within the closed surface instead!

5.4.1   *Applications of Gauss' Law*

Gauss' law is extremely effective in determining the electric fields of certain distributions. Generally, the electric field deviates from point to point which causes the electric flux across a general surface to be tedious to determine. However, for symmetrical objects, we can choose a convenient surface such that the electric field strength is always constant throughout certain area elements or such that the electric field strength at some area elements is zero. Then, the electric field can be integrated trivially over the Gaussian surface to obtain the electric flux. The application of Gauss' law to symmetric systems shall be illustrated through the following examples.

Problem: Find the electric field at a perpendicular distance r away from an infinitely long line with a uniform linear charge density λ.

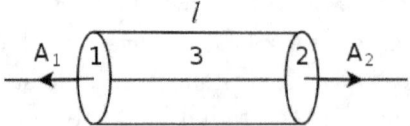

Figure 5.10:   Infinite rod

We define a cylindrical Gaussian surface with an arbitrary length l as shown in Fig. 5.10. The electric fluxes through bases 1 and 2 are zero as the electric field in the direction of the area vectors of the bases (along the line) is zero due to symmetry (there is no reason to prefer left over right or vice-versa). Furthermore, the electric field strength is uniform over curved surface 3. The electric field strengths of points that are on the same circular cross section are equal due to the axial symmetry of the line. Moreover, the electric field of the points that are on different cross sections are the same as the line is infinitely long and each cross section is the same in that respect. Besides having a uniform magnitude, the electric fields on curved surface 3 are also directed radially so that they are perpendicular to the portion of the surface they cut through by symmetry. By Gauss' law,

$$\oiint \boldsymbol{E} \cdot d\boldsymbol{A} = \Phi_1 + \Phi_2 + \Phi_3 = \frac{q_{enc}}{\varepsilon_0}$$

$$0 + 0 + 2\pi r l \cdot E = \frac{\lambda l}{\varepsilon_0}$$

$$E = \frac{\lambda}{2\pi \varepsilon_0 r},$$

as the charge enclosed by the cylindrical Gaussian surface is λl in this case. E is the electric field strength at a perpendicular distance r from the line. The electric field is directed radially outwards, in the direction perpendicular to the line.

Problem: Find the electric field in all space due to an infinitely large sheet of charge with a surface charge density σ, and negligible thickness.

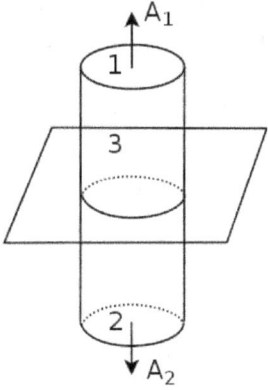

Figure 5.11:   Infinite plane

Similarly, we define a cylindrical Gaussian surface as illustrated in Fig. 5.11 (actually any closed surface with a uniform cross sectional area will suffice). In this case, the electric flux through surface 3 is zero as the electric field in the plane parallel to the sheet is zero due to symmetry. There should only be an electric field perpendicular to the plane that is uniform throughout all points that are the same perpendicular distance away from the plane due to the infinite nature of the plane. In light of this, the electric field strengths should be identical through bases 1 and 2 due to symmetry if we define them to be at the same distance above and below the plane (however, the electric fields are opposite in direction so that they are aligned with the area vectors of the bases). Let E be the electric field strength at the bases. By Gauss' law,

$$\oiint \boldsymbol{E} \cdot d\boldsymbol{A} = \Phi_1 + \Phi_2 + \Phi_3 = \frac{q_{enc}}{\varepsilon_0}$$

$$EA + EA + 0 = \frac{\sigma A}{\varepsilon_0}$$

$$E = \frac{\sigma}{2\varepsilon_0},$$

as the charge enclosed (the circle in the middle) by the proposed Gaussian surface is σA. The electric field at every point in space points in the direction normal to and emanating from the plane. If σ is positive, the electric field at surface 1 will be directed upwards and the electric field at surface 2 will be directed downwards. Observe that E is in fact independent of the length of the Gaussian cylinder! This implies that the electric field is uniform in the regions above and below the plane.

Problem: Find the electric field due to a spherical charge distribution of radius R and uniform charge density ρ in all space. Consider both regions within the sphere and outside the sphere.

The electric field of a sphere can only be in the radial direction, with a uniform electric field strength across a concentric spherical surface, due to the radial symmetry imposed by the sphere. Applying Gauss' law to a concentric spherical Gaussian surface of radius $r \geq R$ outside the sphere,

$$E_{r \geq R} \cdot 4\pi r^2 = \frac{q_{enc}}{\varepsilon_0},$$

where $E_{r \geq R}$ is the electric field strength of a point on the Gaussian surface. Then,

$$E_{r \geq R} = \frac{q_{enc}}{4\pi \varepsilon_0 r^2},$$

where q_{enc} is the total charge of the sphere $q_{enc} = \frac{4}{3}\rho\pi R^3$. It can be seen that the electric field due to a spherical charge distribution outside the sphere is tantamount to that due to a concentrated point charge q_{enc} at the center of the sphere. Next, to determine the electric field within the sphere, apply Gauss' law to a concentric spherical Gaussian surface of radius $r < R$ inside the sphere.

$$E_{r<R} \cdot 4\pi r^2 = \frac{q_{enc}}{\varepsilon_0}.$$

This time, the charge enclosed is not the total charge of the spherical charge distribution. q_{enc} is the amount of charge captured by the Gaussian surface.

$$q_{enc} = \frac{4}{3}\rho\pi r^3.$$

Then,

$$E_{r<R} = \frac{\rho r}{3\varepsilon_0}.$$

Plotting the graph of the electric field strength $E(r)$ against the radial distance r from the center, we will obtain a linear graph that increases from $E = 0$ at $r = 0$ to $E = \frac{\rho R}{3\varepsilon_0}$ at $r = R$, after which it decays with $\frac{1}{r^2}$.

A Neat "Trick" in Computing Electric Fields

Often times, we may be obsessed with the formula of the electric field or the elegance of Gauss' law and forget that the electric field at a point is quintessentially the force per unit charge exerted on a charge placed there. This physical meaning can in fact be leveraged to determine the electric field at a point due to a charge distribution by placing an imaginary unit charge there. Instead of computing the force exerted on the test charge by the set-up, we can determine the negative of the force exerted on the set-up by the test charge as a consequence of Newton's third law — a feat that can sometimes be simpler.

Problem: Determine the electric field at a point $\frac{l}{2}$ above the center of a square with a uniform surface charge density σ and edge length l.

Referring to Fig. 5.12, we shall determine the magnitude of electric field (the direction is obvious) at the required point by placing an imaginary unit charge there and determining the magnitude of the force experienced by the square due to this unit charge. The force experienced by an infinitesimal area dA (with area vector pointing downwards) on the square is $\sigma E dA$ where E is the electric field due to the unit charge at that area. Since the net force on the plane can only be in the vertical direction, we can simply sum the vertical components of the forces experienced by each infinitesimal area element over

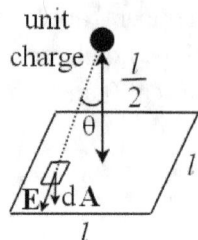

Figure 5.12:   Square and imaginary charge

the entire square. This is equivalent to computing $\iint_{square} \sigma E \cos \theta dA$ with the angle θ defined in the figure above, over the entire square. Making the astute observation that $\iint_{square} \sigma E \cos \theta dA = \sigma \iint_{square} \boldsymbol{E} \cdot d\boldsymbol{A}$ where $d\boldsymbol{A}$ is the area vector of the infinitesimal area element under consideration, the last integral is simply the total electric flux due to the unit charge cutting across the square! Now, imagine that the unit charge is located at the center of a cube with edge length l. The total electric flux across the cube is $\frac{1}{\varepsilon_0}$ by Gauss' law — implying that $\frac{1}{6\varepsilon_0}$ flux leaks out of each face of the cube (which is exactly our current set-up)! Thus, the force that the imaginary unit charge exerts on the square and the electric field at the required point due to the square are $\frac{\sigma}{6\varepsilon_0}$ downwards and upwards respectively.

Finally, it is not difficult to extend this argument to show that more generally, the component of electric field at a location — in the direction normal to a planar surface of uniform surface charge density σ — is simply σ multiplied by the electric flux, due to an imaginary unit charge placed at that particular location, cutting across the planar surface.

Earnshaw's Theorem

Gauss' law has direct ramifications on the stability of the equilibrium established by an electrostatic system. The following statement is known as Earnshaw's Theorem.

Problem: Show that a collection of point charges cannot be maintained in a stable equilibrium solely by their electrostatic interactions.

Firstly, we have to recall the meaning of a stable equilibrium — it requires that once a particle is displaced from its equilibrium position, there is a correcting force exerted on it that tends to return it to its equilibrium position. In the current context, if a certain charge rests in a stable equilibrium state, the electric field due to all other charges in the region around this equilibrium position must point towards (if this particular charge is positive) and away from (if the charge is negative) the equilibrium position. However, applying Gauss' law to an infinitesimal surface (such as a small sphere) surrounding the mentioned equilibrium position, the total electric flux through this surface would be non-zero (as the fields at all points on the surface either all point inwards or outwards) even though the enclosed charge is zero (note that the charge in the equilibrium position does not count as we are looking at the electric field due to all other charges) — leading to a contradiction. Therefore, a stable equilibrium cannot exist in a system purely held by electrostatic forces. Only an unstable equilibrium, along at least one direction, or a neutral equilibrium can occur.

Some problems, such as the following example, exploit Earnshaw's Theorem in an elegant fashion.

Problem: Place a positive charge q at an arbitrary position within an arbitrary configuration of fixed positive charges. Now, string the charge q along a wire and hold it at rest (it may be at rest even if you gently released it). Show that you can always design a path for the wire such that the charge can escape to infinity once you give it a small velocity along the wire, at its initial position.

If a net electric field is present at the charge's initial position, we are done as we can just arrange for our wire to follow the electric field line (due to the fixed positive charges) there, along the direction of the electric field. Since there are no negative charges for the electric field line to terminate at, it must extend to infinity. Moving on, we consider the case where the charge's initial position is an equilibrium state. Observe that only the following two scenarios can occur at an equilibrium position as a consequence of Earnshaw's Theorem: (1) an unstable equilibrium exists along at least one direction which implies that a non-zero electric field exists at a neighboring point. (2) a neutral equilibrium exists such that the electric field is zero at all neighboring points. In light of this, we can construct an escape route for q using the following algorithm with the initial state set at q's initial location and a fixed search direction emanating from q's initial location (i.e. proposed direction of initial velocity). Given a current state, the first step is to check if situation (1) or (2) is met. If situation (1) is fulfilled, erase all previously proposed wire segments and record the position of the point that we have detected to possess a non-zero electric field. Direct the initial segment of the wire from the initial position of q to this recorded position via a straight line, after which the wire can be adjusted to follow the electric field line crossing the recorded position indefinitely. Terminate the algorithm as we have constructed an escape route. If situation (2) is met, construct a wire segment joining the current state to a new neighboring state along the fixed search direction. Update the current state and repeat from the first step.

This algorithm will either stop — after encountering a state with scenario (1) and constructing an escape route — or run indefinitely to build an infinite straight wire that emanates from the initial position along the fixed search direction. The second case is also an escape route as it only occurs when all encountered states are neutral equilibria such that there is nothing to change the initial velocity of q and to prevent it from escaping to infinity (e.g. when q is placed in a region absent of other charges).

5.5 Line Integral of Electrostatic Field

Path Independence of the Line Integral of Electrostatic Field

Being a central force, the Coulomb force is conservative. We shall prove this fact once and for all now by showing that the line integral of the electrostatic[3] field $\int_{P_1}^{P_2} \boldsymbol{E} \cdot d\boldsymbol{s}$ between two points P_1 and P_2 is path-independent as this property will be used prodigiously. As a recap, a line integral means that we integrate the dot product of the electric field $\boldsymbol{E}$ at a point along a predetermined path with the infinitesimal displacement vector $d\boldsymbol{s}$ pointing from that particular point to a neighboring point along the path, over the entire path.

Referring to Fig. 5.13, we first consider the system of an isolated stationary charge q and compute the line integrals of its electric field from a point P_1 to another point P_2 along paths A and B. Defining the origin at q, P_1 and P_2 are at radial distances r_1 and r_2 from the origin, respectively.

We see that the line integral along path A can be easily computed as it consists of purely radial and tangential displacements. The line integral of the electric field in the tangential direction along path A is simply zero as the electric field is always perpendicular to the displacement of the charge. The line integral in the radial direction of path A is

$$\int_{r_1}^{r_2} E_r \cdot dr = \int_{r_1}^{r_2} \frac{q}{4\pi\varepsilon_0 r^2} dr = \frac{q}{4\pi\varepsilon_0} \left(\frac{1}{r_1} - \frac{1}{r_2} \right), \tag{5.9}$$

which is the total line integral along path A. Now, consider an arbitrary path such as path B. Regardless of the actual path taken, it must still pass through a shell of radius r and thickness dr (for all $r_1 \leq r \leq r_2$) as shown

---

[3]This refers to the steady fields that do not change with time and are produced by stationary charges. We will later discover that moving charges or rather, changing magnetic fields, engender another form of electric field that is non-conservative.

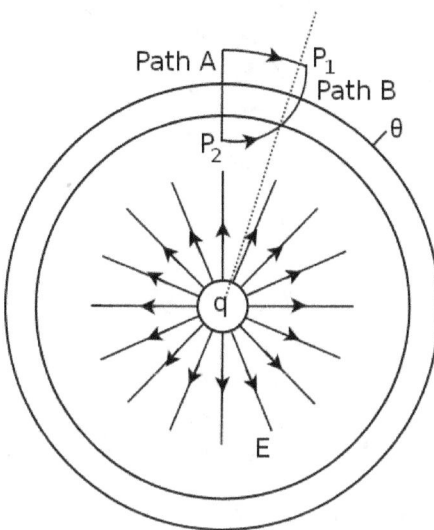

Figure 5.13: Two different paths

in the figure above. Let the path taken through the shell subtend an angle θ with respect to the radial direction. The infinitesimal line integral due to this path crossing this shell is

$$\boldsymbol{E} \cdot d\boldsymbol{s} = E_r \cdot \frac{dr}{\cos \theta} \cdot \cos \theta = E_r dr$$

as the length of the path inside the shell is $\frac{dr}{\cos \theta}$, but an additional factor of $\cos \theta$ is included in the dot product between the electric field (which is radial) and the infinitesimal displacement. Observe that this is equal to the line integral incurred by the displacement of the corresponding segment of path A in the same shell. Adding the contributions from the segments in all shells, the line integral over path B must be identical to that of path A, given by the first expression. The last technicality is that the actual path taken by the particle may cross a shell multiple times. However, it can also be concluded from the above that the magnitudes of the line integral when a path enters or leaves the shell are the same, but opposite in sign. Thus, they cancel out — leaving only one "net path" crossing the shell. Since every system of charges can be thought to consist of only point charges, the line integral of an electrostatic field is always path-independent.

An extremely important corollary of the above result is that the closed loop integral of an electric field is always equal to zero.

Consider the loop in Fig. 5.14, the line integral of the electric field along path A is equal in magnitude to that along path B but of different signs.

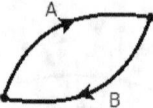

Figure 5.14:   An arbitrary loop

Thus,

$$\oint \boldsymbol{E} \cdot d\boldsymbol{s} = 0 \tag{5.10}$$

along the loop.

Change in Electric Field across Thin Charged Surfaces

Leveraging Gauss' law and the nullity of the closed loop integral of an electric field, we can determine the relationship between the electric fields at different sides of a charged boundary. For any infinitesimally thin surface of charge with a surface charge density σ, we can draw a cylindrical Gaussian surface of infinitesimal length (the two horizontal segments along this page) and base area (perpendicular to this page), as shown in Fig. 5.15.

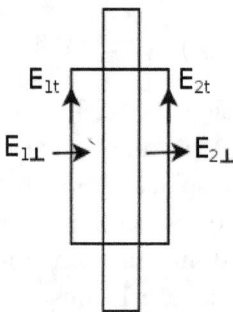

Figure 5.15:   Side view of charged surface

With an infinitesimal length, the electric flux through the curved surface of the cylindrical Gaussian surface is zero. The electric fields at the two bases (on the left and right) are not necessarily the same but they do not vary along their respective surfaces as the surfaces are assumed to be small. Then, if we denote $E_{1\perp}$ and $E_{2\perp}$ as the electric field strengths normal to the surface at the left and the right of the surface respectively (rightwards is defined to be positive), we can apply Gauss' law to determine the change in the normal component of the electric field while noting that the enclosed charge is σdA, where dA is the cross sectional area of the Gaussian cylinder.

$$(E_{2\perp} - E_{1\perp})dA = \frac{\sigma dA}{\varepsilon_0}$$

$$E_{2\perp} - E_{1\perp} = \frac{\sigma}{\varepsilon_0}.$$

Next, to determine the change in components of electric field parallel to the surface, draw a rectangular loop of width w (lying in the plane of the surface) and infinitesimal length (normal to the surface). For the sake of illustration, consider the cross section of the cylindrical Gaussian surface that we had considered previously, depicted in the figure above, such that the two vertical segments have length w but the two horizontal segments are infinitesimal. The path integral of the electric field along this loop must be zero by Eq. (5.10). Since the lengths are infinitesimal, the contributions to the path integral along the two horizontal segments are negligible. Then,

$$(E_{2t} - E_{1t})w = 0$$

$$E_{2t} = E_{1t}.$$

We have only shown that the component of the electric field along one particular line lying on the surface is continuous across the surface. By "rotating" the loop and considering all other lines, we can show that the entire component of electric field lying in the plane of the surface is continuous. Therefore, the change in electric field across the surface is purely in the normal direction.

$$\Delta \boldsymbol{E} = (E_{2\perp} - E_{1\perp})\hat{\boldsymbol{n}} = \frac{\sigma}{\varepsilon_0}\hat{\boldsymbol{n}} \tag{5.11}$$

where $\Delta \boldsymbol{E}$ represents the change in the electric field vector and $\hat{\boldsymbol{n}}$ is a unit vector perpendicular to the plane and in the positive direction (rightwards in this case).

Next, another interesting question to ask is that, given $E_{1\perp}$ and $E_{2\perp}$, what is the normal component of force, per unit area, that the charged surface experiences? This seems perplexing at first as the normal component of the electric field is inherently discontinuous. However, the correct answer is rather simple: we take the average of the two fields and multiply it by the surface charge density. That said, the reasoning is rather subtle.

The important observation is that $E_{1\perp}$ and $E_{2\perp}$ include both the external field and the field produced by the charged surface. The force that the surface feels is only due to the external field. If we let the normal components of the fields of the surface and external entities be E_s outwards and E_{ext} rightwards,

$$E_{1\perp} = E_{ext} - E_s,$$
$$E_{2\perp} = E_{ext} + E_s.$$

The normal component of the force on the surface, per unit area, is thus

$$P = \sigma E_{ext} = \frac{\sigma(E_{1\perp} + E_{2\perp})}{2}. \tag{5.12}$$

This expression is sometimes known as the electrostatic pressure on a charged surface.

Problem: A soap bubble with a total charge Q uniformly distributed over its spherical surface is currently at equilibrium. If the pressure inside the soap bubble is identical to that outside, determine its equilibrium radius given that the surface tension of the bubble is γ.

Applying Gauss' law to a spherical Gaussian surface inside and outside the bubble, one can conclude that the electric field inside the bubble is zero as there is no charge enclosed. The electric field immediately outside the bubble is

$$E \cdot 4\pi R^2 = \frac{Q}{\varepsilon_0}$$

where R is the equilibrium radius of the bubble.

$$E = \frac{Q}{4\pi\varepsilon_0 R^2}.$$

Therefore, the pressure on the surface of the bubble is given by Eq. (5.12) as

$$P = \frac{Q^2}{32\pi^2\varepsilon_0 R^4}.$$

Now, cut the bubble into two hemispheres. The electrostatic pressure tends to push the two hemispheres apart while surface tension tends to keep them together. The force on one hemisphere due to the electrostatic pressure is the electrostatic pressure multiplied by the cross sectional area of the hemisphere, $P \cdot \pi R^2$. More rigorously, one can integrate the component of force, in the direction of the line joining the center to the vertex of a hemisphere (direction of net force), due to the electrostatic pressure over infinitesimal areas on the hemisphere, but one can also show that this is equivalent to taking the pressure multiplied by the area of the projection of the hemisphere onto its equatorial plane[4] (which is the equatorial circle with area πR^2). The surface tension force on this hemisphere is $2\pi\gamma R$. Balancing the forces,

$$\frac{Q^2}{32\pi^2\varepsilon_0 R^4} \cdot \pi R^2 = 2\pi\gamma R$$

$$R = \sqrt[3]{\frac{Q^2}{64\pi^2\varepsilon_0\gamma}}.$$

Similar to how we investigated the pressure discontinuity across a spherical bubble in the chapter on fluids, we can also solve this problem via the principle of virtual work. However, we shall hold this off for now as a potential energy, associated with the electrostatic interactions of a system, can in fact be defined. Then, we can find the radius that extremizes the total potential energy (surface energy and electric potential energy) which is a shortcut to applying the principle of virtual work. This notion of electric potential energy and other related ideas will be explored in the next few sections.

5.6   Electric Potential Energy

Due to the conservative nature[5] of electrostatic interactions, we can define a potential energy for a charge under the sole influence of Coulomb forces or equivalently, in an electrostatic field. The potential energy of a point charge, q_t, at a point in space, with position vector $\boldsymbol{r}$, is defined as the work done by an external force in bringing it from infinity to that point without a change in kinetic energy. In this definition, we have adopted the convention of defining the potential energy at infinity to be zero. This is often implicitly assumed but does not have to be the case in general.[6]

$$U = W_{ext} = -W_{Coulomb} = -\int_\infty^r q_t \boldsymbol{E} \cdot d\boldsymbol{s}, \tag{5.13}$$

where $\boldsymbol{s}$ is the position vector of the charge when it was in the midst of being moved via a certain path from infinity to $\boldsymbol{r}$. By definition of a conservative force, the integral can be evaluated along any path from infinity to $\boldsymbol{r}$ as all paths would yield the same result.

---

[4]See solution to Problem 20 of Chapter 5.

[5]We have already shown that the line integral of an electrostatic field is path-independent.

[6]There are some cases where the zero reference point should be set somewhere else, lest certain quantities diverge.

Problem: Derive the potential energy of a point charge q placed at a distance r away from the center of a uniform spherical charge distribution of total charge Q. q is outside of the sphere.

We have shown that the electric field due to a uniformly charged sphere is radially outwards and of magnitude

$$E = \frac{Q}{4\pi\varepsilon_0 s^2}$$

at a radial distance s from the center, for regions exterior to the sphere. We can then find the electric potential energy of this charge by bringing it along a path that lies strictly in the radial direction, with the origin defined at the center of the sphere.

$$U = -\int_\infty^r \frac{qQ}{4\pi\varepsilon_0 s^2} \hat{s} \cdot d\boldsymbol{s} = -\int_\infty^r \frac{qQ}{4\pi\varepsilon_0 s^2} ds = \frac{qQ}{4\pi\varepsilon_0 r}.$$

Note that the choice of a radial path (which doesn't affect the value of U which is path-independent) enables us to conclude that $\hat{s} \cdot d\boldsymbol{s} = ds$ — greatly simplifying our calculations. The above expression is in fact identical to the electric potential energy of a point charge q due to another point charge Q at the center of the sphere, whose expression shall soon be derived. This is intuitive as q cannot differentiate between the electric fields of a point charge Q located at the center of the sphere and a uniformly charged sphere of total charge Q as long as it remains outside the sphere. This is the beauty of the formulation of an electric field as it is only a local property! Finally, the discussion here also implies that the above expression is only valid when the charge is outside of the sphere — the reader should try to find U for regions within the sphere.

Moving on, Eq. (5.13) enables us to determine the potential energy of a test charge in a steady electric field (i.e. all other charges must be fixed) if we know the electric field in all space. More commonly, we only know the charge distribution of the system — implying that we have to take an additional intermediate step in determining the electric field everywhere to calculate the potential energy of a charge. To circumvent this sinuous route, we can instead directly determine the potential energy of a test charge from the distribution of relevant charges. We first consider the case where the external electric field on a test charge q_t is solely due to a fixed point charge q_1 when their current separation is r_{1t}. Since the line integral of the electric field due to a single charge has been computed in the previous section, the potential energy of q_t is simply the negative of Eq. (5.9) (after substituting

$r_2 \to \infty$ and $r_1 = r_{1t}$), multiplied by q_t.

$$U_{t1} = \frac{q_t q_1}{4\pi\varepsilon_0 r_{1t}}. \tag{5.14}$$

The total electric potential energy of a single point test charge q_t due to N other discrete fixed point charges is then obtained by the principle of superposition.

$$U_t = \sum_{i=1}^{N} \frac{q_t q_i}{4\pi\varepsilon_0 r_{it}} \tag{5.15}$$

where r_{it} is the current distance between the ith charge q_i and the test charge q_t.

Lastly, based on the definition of the electric potential energy, we can obtain the net Coulomb force on a test charge q_t due to the other fixed charges by taking negative of the potential energy gradient if we know the potential energy as a function of spatial coordinates. In the prevalent Cartesian, spherical and cylindrical coordinates,

$$\boldsymbol{F}(x, y, z) = -\frac{\partial U}{\partial x}\hat{\boldsymbol{i}} - \frac{\partial U}{\partial y}\hat{\boldsymbol{j}} - \frac{\partial U}{\partial z}\hat{\boldsymbol{k}},$$

$$\boldsymbol{F}(r, \theta, \phi) = -\frac{\partial U}{\partial r}\hat{\boldsymbol{r}} - \frac{\partial U}{r\partial\theta}\hat{\boldsymbol{\theta}} - \frac{\partial U}{r\sin\theta\partial\phi}\hat{\boldsymbol{\phi}},$$

$$\boldsymbol{F}(r, \theta, z) = -\frac{\partial U}{\partial r}\hat{\boldsymbol{r}} - \frac{\partial U}{r\partial\theta}\hat{\boldsymbol{\theta}} - \frac{\partial U}{\partial z}\hat{\boldsymbol{z}}.$$

In vector calculus notation, $\boldsymbol{F} = -\nabla U$. As always, an easy way to remember the gradient in a certain coordinate system is to note that the denominators are the infinitesimal length segments in that coordinate system.

Problem: A positive point charge Q of mass m lies at the midpoint of the line joining two fixed positive point charges, both of charge q. The two charges q are a distance $2l$ apart. Find the angular frequency of small oscillations of charge Q about its equilibrium position along the line joining the charges via two methods — forces and potential energy. Thus, state whether charge Q is at a stable equilibrium in the direction of the line joining them. What happens if the charge had $-Q$ charge instead?

We shall take rightwards to be positive in Fig. 5.16. Firstly, observe that charge Q is at equilibrium only if it is at the midpoint of the line joining the two other charges.

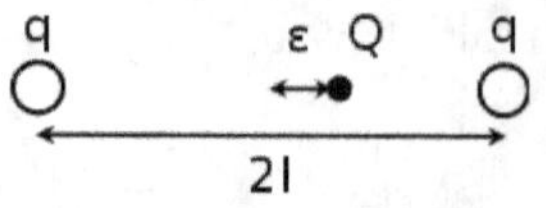

Figure 5.16:   3 Charges

Method 1: Forces

Let the displacement of charge Q from its equilibrium position be ϵ. The force on Q is

$$
\begin{aligned}
F &= \frac{qQ}{4\pi\varepsilon_0(l+\epsilon)^2} - \frac{qQ}{4\pi\varepsilon_0(l-\epsilon)^2} \\
&= \frac{qQ}{4\pi\varepsilon_0 l^2}\left(\frac{1}{\left(1+\frac{\epsilon}{l}\right)^2} - \frac{1}{\left(1-\frac{\epsilon}{l}\right)^2}\right) \\
&\approx \frac{qQ}{4\pi\varepsilon_0 l^2}\left(1 - 2\frac{\epsilon}{l} - 1 - 2\frac{\epsilon}{l}\right) \\
&= -\frac{qQ\epsilon}{\pi\varepsilon_0 l^3}.
\end{aligned}
$$

By Newton's second law, $F = m\ddot{\epsilon}$.

$$
\ddot{\epsilon} = -\frac{qQ}{\pi m\varepsilon_0 l^3}\epsilon.
$$

This equation of motion indicates a simple harmonic motion of angular frequency

$$
\omega = \sqrt{\frac{qQ}{\pi m\varepsilon_0 l^3}}.
$$

Method 2: Potential Energy

The potential energy of charge Q when it is at a distance r from the left charge, along the horizontal, is

$$
U = \frac{qQ}{4\pi\varepsilon_0 r} + \frac{qQ}{4\pi\varepsilon_0(2l-r)},
$$

$$
\frac{dU}{dr} = -\frac{qQ}{4\pi\varepsilon_0 r^2} + \frac{qQ}{4\pi\varepsilon_0(2l-r)^2},
$$

$$
\frac{d^2U}{dr^2} = \frac{qQ}{2\pi\varepsilon_0}\left[\frac{1}{r^3} + \frac{1}{(2l-r)^3}\right].
$$

Since the charge is at equilibrium when $r = l$,

$$\omega = \sqrt{\frac{U''(l)}{m}} = \sqrt{\frac{qQ}{\pi m \varepsilon_0 l^3}}.$$

Revisit the chapter on oscillations for a review of this potential energy method if necessary. Charge Q is at a stable equilibrium along horizontal direction, as any deviation of its position from the equilibrium position tends to be corrected by a restoring force (evident from its oscillation). If the charge had a negative charge, $-Q$, it will be at an unstable equilibrium along the horizontal direction, as any deviation of its position from the equilibrium position tends to be amplified by the net force due to the two other charges. This can also be directly concluded from the fact that $\frac{d^2 U}{dr^2}$ when $r = l$ is negative if the charge were to be negative — meaning that the potential energy of the negative charge at that equilibrium position is a local maximum in the horizontal direction. Incidentally, observe that the charge at the center has different stabilities of equilibrium in different directions — a charge that is stable in the horizontal direction will be unstable in directions perpendicular to the horizontal and vice versa (verifying Earnshaw's Theorem).

5.7 Electric Potential

The electric potential of a point in space, with position vector $\boldsymbol{r}$, is defined as

$$V = -\int_{\infty}^{\boldsymbol{r}} \boldsymbol{E} \cdot d\boldsymbol{s} \tag{5.16}$$

which is the line integral of the electric field along a path from infinity to $\boldsymbol{r}$. Again, the convention of defining the zero reference point at infinity has been adopted. $\boldsymbol{s}$ is the position vector of an intermediate point on the path. Physically, the electric potential at a point in space represents the work done per unit charge by an external force in bringing a charge from infinity to that point, without a change in kinetic energy. The potential is usually measured in Volts (V) which is $\text{kgm}^2\text{s}^{-3}\text{A}^{-1}$ in SI units or one Joule divided by one Coulomb JC^{-1}.

Problem: Determine the potential due to a spherical charge distribution, of radius R and uniform charge density ρ, in all space.

Define the origin at the center of the sphere. We have shown that the electric fields at a radial distance s from the center of the sphere, outside and inside the sphere, are

$$E_{s \geq R} = \frac{q_{enc}}{4\pi\varepsilon_0 s^2} = \frac{\rho R^3}{3\varepsilon_0 s^2},$$

$$E_{s < R} = \frac{\rho s}{3\varepsilon_0}.$$

Therefore, the potential at a point P with a radial distance $r \geq R$, outside the sphere, is

$$V_{r \geq R} = -\int_{\infty}^{r} \frac{\rho R^3}{3\varepsilon_0 s^2} \hat{s} \cdot d\boldsymbol{s}$$

$$= -\int_{\infty}^{r} \frac{\rho R^3}{3\varepsilon_0 s^2} ds$$

$$= \frac{\rho R^3}{3\varepsilon_0 r}$$

where we have once again integrated along a strictly radial path to establish $\hat{s} \cdot d\boldsymbol{s} = ds$ for the sake of convenience. The potential at a point P at a radial distance $r < R$ is the sum of the change in potential from infinity to a radial distance R and that from a radial distance R to r. We have to split the integration into these parts as the electric field is incongruous.

$$V_{r < R} = -\int_{\infty}^{R} \frac{\rho R^3}{3\varepsilon_0 r^2} dr - \int_{R}^{r} \frac{\rho r}{3\varepsilon_0} dr$$

$$= \frac{\rho R^2}{3\varepsilon_0} - \frac{\rho r^2}{6\varepsilon_0} + \frac{\rho R^2}{6\varepsilon_0}$$

$$= \frac{\rho(3R^2 - r^2)}{6\varepsilon_0}$$

where we have deliberately and cautiously integrated along a path in the radial direction again since the path integral should be independent of the path taken. Similar to Eq. (5.13), Eq. (5.16) relates the electric potential to the electric field. To directly determine the electric potential due to a charge distribution, we can easily show that the potential at a point P at a distance r from a single, isolated charge q is

$$V = \frac{q}{4\pi\varepsilon_0 r},$$

by dividing Eq. (5.14) by q_t and substituting q for q_1 and r for r_{1t}, since V is basically U_{t1} divided by q_t. The potential of a point P due to a system of

N discrete point charges is then

$$V = \sum_{i=1}^{N} \frac{q_i}{4\pi\varepsilon_0 r_i} \tag{5.17}$$

by the principle of superposition, where r_i is the distance between the ith charge, q_i, and point P. For a continuous charge distribution, the potential at a point P can be obtained by summing the contributions to the potential by each infinitesimal charge.

$$V = \int \frac{1}{4\pi\varepsilon_0 r} dq \tag{5.18}$$

where r is the distance between point P and the infinitesimal charge dq under consideration.

Following from our definition of the potential, we can also rewrite the following expressions. The potential energy of a test charge due to a fixed system of charges that result in a potential V at the test charge's location is

$$U = -\int_{\infty}^{r} q\boldsymbol{E} \cdot d\boldsymbol{s} = qV. \tag{5.19}$$

With regard to the problem in the previous section on the electric potential energy of a point charge q due to a fixed, uniform sphere of charge Q and radius R, this implies that the electric potential energy of q at a radial distance $r < R$ from the center of the sphere is $V_{r<R}$ in the previous problem, multiplied by q ($\frac{qQ(3R^2-r^2)}{8\pi\varepsilon_0 R^3}$).

Next, based on the definition of the electric potential, the electric field at a point is the negative potential gradient at that point.

$$\boldsymbol{E} = -\nabla V. \tag{5.20}$$

In light of this, we now have an alternative method of determining the electric field in all space. We can first calculate the electric potential everywhere from the given charge distribution directly — a task that is often simpler due to the scalar nature of the potential — and subsequently apply Eq. (5.20) to obtain the electric field everywhere.

Another consequence of Eq. (5.20) is that electric field lines always point from a region of higher electric potential to a region of lower electric potential. Furthermore, it may be useful to define equipotential lines and surfaces which are basically contour plots delineating adjacent points of the same electric potential. There is no component of the electric field along equipotential lines and surfaces as there is no potential difference between points on them. Thus, the electric field of a point along an equipotential line or surface is always perpendicular to the tangent of the equipotential line or the equipotential surface at that particular point.

5.8   Potential Energy of a System

Electric Potential Energy of a System

Now that we have determined the electric potential energy of a point charge due a distribution of fixed charges, we would like to find the electric potential energy of a system of charges that are all unfettered. We have shown in the chapter on energy that the potential energy of a system of N particles undergoing central force interactions is simply the sum of the potential energies due to the interactions between each pair of particles, with no repeats due to the permutations of a pair.

$$U_{sys} = \sum_{i<j} U_{ij}$$

where U_{ij} is the potential energy associated with the interactions between the ith and jth particles. Its value is given by the potential energy of one of those particles, given that the other particle is fixed. Applying the above to N charges with the ith charge having charge q_i,

$$U_{ij} = \frac{q_i q_j}{4\pi r_{ij}}$$

where r_{ij} is the distance between the ith and jth charges. Therefore,

$$U_{sys} = \sum_{i<j} \frac{q_i q_j}{4\pi r_{ij}}. \tag{5.21}$$

Now, we present another perspective to the above expression. The electric potential energy of a system is also the work done by an external force in assembling it and is the energy stored in the system of charges. Let us imagine assembling the system by bringing individual point charges into the system one at a time, while holding the other charges that are already in the system fixed. We begin with an empty system.

1st Charge: The work done by an external force in bringing this charge to its final position is 0 as the system is initially empty and there is no Coulomb force on this charge.

2nd Charge: There is a Coulomb force on q_2 due to q_1. Thus, the work done by an external force in bringing q_2 to its final position is

$$W_2 = \frac{q_1 q_2}{4\pi\varepsilon_0 r_{12}}.$$

3rd Charge: There are Coulomb forces on q_3 due to q_1 and q_2. Thus, the work done by an external force is

$$W_3 = \frac{q_1 q_3}{4\pi\varepsilon_0 r_{13}} + \frac{q_2 q_3}{4\pi\varepsilon_0 r_{23}}.$$

ith Charge: There are Coulomb forces on q_i due to $q_1, q_2 \ldots q_{i-1}$. Thus, the work done by an external force is

$$W_i = \sum_{j=1}^{i-1} \frac{q_i q_j}{4\pi\varepsilon_0 r_{ji}}.$$

Thus, we see that the total work done in assembling a system of N charges, in this particular way, is simply the sum of the potential energy contributions between all pairs of charges with no repeats. Since the potential energy of a system should be independent of the process through which the charges got there, this must be the potential energy of the system.

$$U_{sys} = \sum_{i<j} \frac{q_i q_j}{4\pi r_{ij}}.$$

If we include the repeats, we can write the total potential energy of the system as

$$U_{sys} = \frac{1}{2} \sum_{i,j\, i\neq j} \frac{q_i q_j}{4\pi\varepsilon_0 r_{ij}} = \frac{1}{2} \sum_{i=1}^{N} q_i \sum_{j\neq i} \frac{q_j}{4\pi\varepsilon_0 r_{ij}} = \frac{1}{2} \sum_{i=1}^{N} q_i V_i, \qquad (5.22)$$

where V_i is the potential at the position of the ith charge due to all other charges. In the case of continuous charge distributions, an analogous version of the above is

$$U_{sys} = \frac{1}{2} \int V \, dq \qquad (5.23)$$

where V is the potential at the position of an infinitesimal charge dq due to all other charges. The integration is performed over the entire charge distribution. Since the contribution of an infinitesimal surface or volume charge to the potential at its own location is negligible,[7] V can in fact be taken to

---

[7]The electric potential is proportional to charge and inversely proportional to distance. Meanwhile, surface and volume charges are proportional to the square and cube of their

be the potential due to the entire charge distribution! The above expression is hence extremely useful in the case of continuous charge distributions.

Now, the reader may be confused by the fact that the electric potential energy of a charge q at a position with external potential V was asserted in the previous section to be qV, yet there is an additional factor of $\frac{1}{2}$ here for the potential energy of a system. Why can't we simply argue that $U_{sys} = \int V dq$ since the potential energy of a charge dq under a steady external potential V is $V dq$? The difference here is that qV is the formula for the potential energy of q, given that all other charges are fixed, but in this section, we are computing the potential energy of a system of charges that are all free to move. If we claim that $U = \int V dq$ (i.e. sum the potential energy of each charge, given that all other charges are fixed) instead of $U = \frac{1}{2} \int V dq$, we have double-counted[8] the potential energy associated with each pair of charges. This is similar to the chapter on energy where we discussed the total energy of two masses connected by a spring with spring constant k and a current extension x. The elastic potential energy of each mass, given that the other mass is fixed, is $\frac{1}{2}kx^2$ but we cannot add these individual energies together to argue that the elastic potential energy of the system of two masses is kx^2. It is obvious that the elastic potential energy of the combined system is still $\frac{1}{2}kx^2$ in this case, as we can visualize that the spring is only stretched by that x amount. In the case of electric potential energy, we cannot directly perceive the manifestation of the potential energy as a physical change in the system, but the fallacy of double-counting still exists.

Problem: Find the energy stored by the system of charges in Fig. 5.2.

$$U_{sys} = -2 \times \frac{2q^2}{4\pi\varepsilon_0 l^2} + \frac{q^2}{4\pi\varepsilon_0 l^2} = -\frac{3q^2}{4\pi\varepsilon_0 l^2}.$$

Problem: A soap bubble with a total charge Q uniformly distributed over its spherical surface is currently at equilibrium. If the pressure inside the soap bubble is identical to that outside, determine its equilibrium radius, given that the surface tension of the bubble is γ, by considering the total potential energy of the bubble.

---

length dimensions respectively. Therefore, the contributions of surface and volume charges to the potentials at their own locations are proportional to their length dimension and squared length dimension respectively — yielding an insignificant result when their length dimensions are negligible.

[8]This double-counting is easy to see in the discrete case, as we will be computing $U_{sys} = \sum_{i \neq j} \frac{q_i q_j}{4\pi r_{ij}} = 2\sum_{i<j} \frac{q_i q_j}{4\pi r_{ij}}$ instead of the correct $U_{sys} = \sum_{i<j} \frac{q_i q_j}{4\pi r_{ij}}$.

Let the radius of the bubble be a variable r. The total potential energy comprises the surface energy, which is $U_{surface} = 4\pi\gamma r^2$, and the electric potential energy of the soap bubble, which is essentially a spherical shell of charge Q. Applying Gauss' law to a concentric, spherical Gaussian surface outside of a spherical shell, one can show that as a consequence of radial symmetry, the spherical shell is akin to a charge Q placed at its center in the region outside of the shell. Therefore, the potential V due to the shell at the surface of the shell is akin to that at a distance r from a point charge Q.

$$V = \frac{Q}{4\pi\varepsilon_0 r}.$$

Applying Eq. (5.23) over the surface of the shell while noting that V is uniform over it, the electric potential energy of the bubble is

$$U_{elec} = \frac{1}{2} \cdot \frac{Q}{4\pi\varepsilon_0 r} \cdot Q = \frac{Q^2}{8\pi\varepsilon_0 r}.$$

The total potential energy of the bubble is thus

$$U_{tot} = U_{surface} + U_{elec} = 4\pi\gamma r^2 + \frac{Q^2}{8\pi\varepsilon_0 r}.$$

The equilibrium radius $r = R$ extremizes U_{tot} by the principle of virtual work as all forces on the bubble are conservative.

$$\frac{dU_{tot}}{dr}\bigg|_{r=R} = 8\pi\gamma R - \frac{Q^2}{8\pi\varepsilon_0 R^2} = 0$$

$$\implies R = \sqrt[3]{\frac{Q^2}{64\pi^2\varepsilon_0\gamma}}.$$

Incidentally, an apparent paradox, that can deepen our understanding of the principle of virtual work, can be constructed from this problem. Suppose that we now constrain the bubble to have a fixed uniform surface charge density σ, instead of a fixed total charge Q; intuition should hint that the equilibrium radius can be obtained from substituting $Q = 4\pi\sigma R^2$ from the start and repeating the above derivations as nothing in the set-up has essentially changed (we have merely replaced a word for another). However, one will find that the approach based on balancing forces (shown previously) and the approach here yield different answers! Why is this so?

It turns out that the answer based on balancing forces will be correct if we simply repeat both procedures. You will actually obtain a negative equilibrium radius, which is absurd, from the principle of virtual work. The problem with applying the principle of virtual work in this context lies in the origin of the change in electric potential energy. In the original problem, the change solely arose from the interactions between charges on the bubble. However, in the modified problem, the change arises from both the inter-actions between charges already on the bubble and the work done by an external force in packing more charge on the bubble (as the total amount of charge now varies). Only the first portion is relevant in inducing forces on the charges already on the bubble (electrostatic pressure) and it is hence the factor that affects whether the bubble is currently in equilibrium. To rectify this loophole, we can either subtract the work done by an external force in adding more charge in applying the principle of virtual work, or better yet, directly apply the principle of virtual work with a fixed amount of total charge Q equal to that at the equilibrium radius. The latter approach works because whether the bubble remains in equilibrium should only depend on the current configuration of charges, which produces the forces on it at this instance, and not the future configuration. After all, how would the bubble know if we are going to add some charge or not? It would only dare to think that it carries a constant amount of charge!

Problem: Find the energy stored in a uniform, spherical charge distribution of radius R and total charge Q. Do this by both applying Eq. (5.23) and assembling the charges.

Let $\rho = \frac{Q}{\frac{4}{3}\pi R^3}$ be the uniform volume charge density of the sphere. We have previously shown that the potential inside a uniformly charged sphere is

$$V = \frac{\rho(3R^2 - r^2)}{6\varepsilon_0}.$$

Applying Eq. (5.23),

$$U_{sys} = \frac{1}{2}\int_0^R\int_0^\pi\int_0^{2\pi} \frac{\rho^2(3R^2 - r^2)}{6\varepsilon_0} \cdot r^2 \sin\theta \, d\phi d\theta dr$$

$$= \frac{1}{2}\int_0^R \left(\frac{2\pi\rho^2 R^2 r^2}{\varepsilon_0} - \frac{2\pi\rho^2 r^4}{3\varepsilon_0}\right) dr$$

$$= \frac{1}{2}\left(\frac{2\pi\rho^2 R^5}{3\varepsilon_0} - \frac{2\pi\rho^2 R^5}{15\varepsilon_0}\right)$$

$$= \frac{4\pi\rho^2 R^5}{15\varepsilon_0}$$

$$= \frac{3Q^2}{20\pi\varepsilon_0 R}.$$

Next, we can consider assembling the sphere by repeatedly bringing in spherical shells of thickness dr and radius r to an already assembled sphere of radius r until r eventually increases to R. We have derived that the electric potential energy of a point charge, dq, at a radius r (outside) from the center of a spherical charge distribution of charge Q_0 is

$$dU = \frac{Q_0}{4\pi\varepsilon_0 r}dq.$$

Applied to the current context, Q_0 is due to the already established spherical distribution of radius r, as we bring in the next shell of charge.

$$Q_0 = \rho\frac{4}{3}\pi r^3.$$

The infinitesimal external work done in bringing dq charge to the surface of the already assembled sphere of radius r is then

$$dU = \frac{\rho r^2}{3\varepsilon_0}dq.$$

When assembling a given shell of charge of thickness dr, the external work done in bringing each infinitesimal charge is the same, as seen from the above expression. Thus, we can just integrate the above expression by summing up the contributions due to each spherical shell. We can directly write $dq = \rho 4\pi r^2 dr$.

$$U_{sys} = \int_0^R \frac{4\pi\rho^2}{3\varepsilon_0}r^4 dr = \frac{4\pi\rho^2 R^5}{15\varepsilon_0} = \frac{3Q^2}{20\pi\varepsilon_0 R}.$$

Energy Density of an Electric Field

An alternate perspective to the electric potential energy of a system can be embraced by associating an energy density function with the electric field. It is convenient to invent the existence of such a construct as it is intuitive to think that the work done by an external force to assemble the system should be "stored" somewhere, which in this case, is the electric field. The energy associated with an infinitesimal volume, dV, with an electric field $\boldsymbol{E}$ is in fact (we will state this without proof)

$$dU = \frac{1}{2}\varepsilon_0 E^2 dV$$

where E^2 can be computed as $\boldsymbol{E}\cdot\boldsymbol{E}$. The total energy associated with an electric field is then

$$U = \int_{allspace} \frac{1}{2}\varepsilon_0 E^2 dV, \tag{5.24}$$

where the integral is performed over all space. Ultimately, note that the above is equivalent to the work done approach and is in fact derived from it.

That said, a major pitfall lurks in this formulation — it leads to singularities in a system of point charges as the electric field diverges at the positions of the point charges. This is because, this method assumes that there is some energy required to assemble the point charge itself. It is technically correct as it takes an infinite amount of energy to pack non-zero charge into zero volume, but this is not particularly edifying. Thus, in such cases, this method is not favored and the work done approach should be applied instead.

Problem: Find the energy stored in a uniform, spherical charge distribution of radius R and total charge Q by considering the energy density of the electric field.

We know that the electric field strengths at a radial distance r outside and inside the sphere are

$$E_{r \geq R} = \frac{Q}{4\pi\varepsilon_0 r^2},$$

$$E_{r < R} = \frac{\rho r}{3\varepsilon_0},$$

where ρ is the charge density of the sphere. Then, the total energy associated with the electric field can be integrated over spherical shells spanning all space as each infinitesimal volume element on a shell possesses the same amount of energy.

$$U = \int_0^R \frac{1}{2}\varepsilon_0 E_{r<R}^2 dV + \int_R^\infty \frac{1}{2}\varepsilon_0 E_{r\geq R}^2 dV$$

$$= \int_0^R \frac{1}{2}\varepsilon_0 \left(\frac{\rho r}{3\varepsilon_0}\right)^2 4\pi r^2 dr + \int_R^\infty \frac{1}{2}\varepsilon_0 \left(\frac{Q}{4\pi\varepsilon_0 r^2}\right)^2 4\pi r^2 dr$$

$$= \left[\frac{2\pi\rho^2 r^5}{45\varepsilon_0}\right]_0^R + \left[-\frac{Q^2}{8\pi\varepsilon_0 r}\right]_R^\infty$$

$$= \frac{3Q^2}{20\pi\varepsilon_0 R},$$

where we have used the fact that $Q = \frac{4}{3}\pi\rho R^3$.

Electric Potential Energy of a System of Charges in Equilibrium

The electric potential energy of a system of charges in equilibrium actually takes on a special value. For example, we can compute the total electric potential of the system in the second part of the second problem in Section 5.2 as

$$\frac{q_1 q_2}{4\pi\varepsilon_0 r_{12}} + \frac{q_1 q_3}{4\pi\varepsilon_0 r_{13}} + \frac{q_2 q_3}{4\pi\varepsilon_0 r_{23}} = \frac{q_1 q_2}{4\pi\varepsilon_0 (r_{13} + r_{12})}$$

$$-\frac{q_1^2 q_2}{4\pi\varepsilon_0 (\sqrt{q_1} + \sqrt{q_2})^2 r_{13}}$$

$$-\frac{q_1 q_2^2}{4\pi\varepsilon_0 (\sqrt{q_1} + \sqrt{q_2})^2 r_{23}} = 0$$

since $q_3 = -\frac{q_1 q_2}{(\sqrt{q_1} + \sqrt{q_2})^2}$ and $\frac{r_{23}}{r_{13}} = \sqrt{\frac{q_2}{q_1}}$. Therefore, we may surmise that the total electric potential energy of a general system of charges, held in equilibrium solely by their electrostatic interactions, is zero.

To prove this, we can show that the total work done by an external force in assembling the system is zero or equivalently, that the total work done by an external force in disassembling the system (i.e. bringing all charges back to infinity) is zero. Constructing a single method of doing so would prove our claim as the total potential energy of a system must be independent of how we constructed the system. To this end, exploiting the inverse-square nature of the Coulomb force, observe that if we scale the relative distances between all charges in a system by a factor k, the Coulomb force between each pair of charges will decrease by a factor of $\frac{1}{k^2}$. This implies that if the system was originally in equilibrium such that the net force on each charge is zero, the system will still be in equilibrium after the scaling as the net force on each charge is simply scaled by a factor of $\frac{1}{k^2}$ but $\frac{0}{k^2}$ is still **0**! Therefore, we can disintegrate the system without performing any external work by progressively expanding the system by a small scale factor until all relative distances tend to infinity!

Observe that the inverse-square nature of the Coulomb force was not crucial in our derivation — any force between two particles that depended on the distance between them to a certain power would work. Furthermore, by making the astute observation that the gravitational force is also an inverse-square law such that the net electrostatic and gravitational forces on a charge in a system governed solely by electrostatic and gravitational interactions are scaled by the same amount when the relative distances in the system are scaled, we can strengthen our theorem to state that

Theorem: The total electric and gravitational potential energy of a general system of particles, held in equilibrium solely by their electrostatic and gravitational interactions, is zero.

That said, the reverse of the statement is not true — a system of particles with zero total potential energy is not necessarily in equilibrium. However, this theorem can be applied to swiftly identify equilibrium configurations of a system when we are guaranteed that they exist or when we can easily verify if they are indeed equilibrium states.

Problem: Determine the charge Q that needs to be placed at the center of four charges q that form the four vertices of a square, to keep the entire system in equilibrium.

Let l be the edge length of the square. Instead of balancing forces, we can apply the previous theorem and assert that if an equilibrium can indeed be established, the total potential energy of the four charges q and Q must be zero. This is less tedious as we do not need to consider vectors.

$$4 \cdot \frac{q^2}{4\pi\varepsilon_0 l} + 2 \cdot \frac{q^2}{4\pi\varepsilon_0 \cdot \sqrt{2}l} + 4 \cdot \frac{qQ}{4\pi\varepsilon_0 \cdot \frac{l}{\sqrt{2}}} = 0$$

$$Q = -\frac{2\sqrt{2}+1}{4}q.$$

Problems

Discrete Charges

1. *Two Charges**

Two point charges, which are not necessarily of the same charge, are placed along the positive x-axis. The electric potential is found to tend to positive infinity when approaching $x = 0$. Next, it is known that the electric potential is zero at two points on the positive x-axis, where x_0 is the larger x-coordinate of the two. Lastly, the electric potential at $x = \alpha x_0$ is a local minimum along the x-direction where α is a positive constant. What can you say about the signs and relative magnitude of the charges? Determine the distance d between the charges.

2. *Triangle of Charges**

Three identical charges of mass m and charge q are initially positioned such that they form the vertices of an equilateral triangle with sides l_0. If they were initially stationary, determine the velocities of the charges when their relative separation becomes l afterwards.

3. *Exploding Charge***

A negative point charge $-q$ of mass m is currently orbiting a fixed, positive charge Q at a radius of rotation r_0. Now, the orbiting charge suddenly disintegrates such that it ejects half of its mass in the radial direction at a negligible velocity, relative to the rest of the charge. However, the leftover mass still retains the entire charge $-q$. Determine the minimum distance between the leftover charge and Q, r, in the motion thereafter while neglecting any gravitational effects.

Charge Distributions

4. *Cube Potential**

A cube of length l possesses a uniform volume charge density ρ. Find the ratio of the electric potential at one of its vertices to that at its center. **Hint:** Use scaling arguments.

5. Flux at Corner of Cube*

A point charge q is placed at one of the vertices of an imaginary cubic Gaussian surface. Find the electric flux through one of the faces of the cube that is non-zero.

6. Flux Through Spherical Cap*

A charge q is placed off-center in an imaginary sphere. Determine the total electric flux cutting the spherical cap (in bold) depicted in the figure below, that is characterized by the distances a and h, for $h = 0$. Next, solve for the electric flux across the spherical cap for general a and h and check that your result yields the right answer for the previous limiting case.

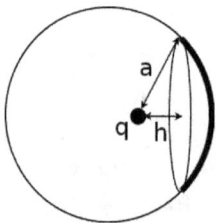

7. Force on Cube Face*

The six faces of an insulating cube of edge length l are coated with a uniform surface charge density σ. Determine the force experienced by one face of the cube.

8. Electric Field of Equilateral Triangle*

Determine the electric field at a point P located at a height $\frac{l}{2\sqrt{6}}$ above the centroid of an equilateral triangle with edge length l and uniform surface charge density σ.

9. Chessboard*

Every black tile of an insulating chessboard of dimensions $l \times l$ is painted with a uniform surface charge density σ while every white tile is entirely neutral. Determine the electric field at a point $\frac{l}{2}$ above the center of the chessboard without any integration.

10. *Finite Line Charge Revisited**

Prove that the direction of the electric field at an arbitrary point P due to a finite line charge, with uniform linear charge density λ and ends at A and B, always bisects $\angle APB$. Furthermore, show that the magnitude of the electric field at P is

$$E = \frac{\lambda}{2\pi\varepsilon_0 h}\sin\alpha,$$

where h is the perpendicular distance between P and the line and $\alpha = \frac{1}{2}\angle APB$. Even though we have previously derived the electric field of a line charge, do not overlook this part but rather, search for a different method that directly proves the above properties. Next, determine the electric field vector at (a, a) in the xy-plane due to two line charges, with uniform linear charge density λ, that lie from $(a, 0)$ to $(+\infty, 0)$ and from $(0, a)$ to $(0, +\infty)$.

11. *Hydrogen Atom**

A hydrogen atom is made up of a proton and an electron. The proton may be regarded as a point charge q at $r = 0$, the center of the atom. Meanwhile, the motion of the electron causes its charge to be "smeared out" into a spherically symmetric distribution around the proton, such that the electron is equivalent to a charge density

$$\rho(r) = -\frac{q}{\pi a_0^3}e^{-\frac{2r}{a_0}},$$

where a_0 is a constant known as the Bohr radius and r is the radial distance from the center. Note that e is Euler's number and not the charge of the electron.

(a) Find the total amount of the hydrogen atom's charge that is enclosed within a sphere of radius r, centered about the proton. Check your answer for the limit $r \to \infty$ and explain why it makes sense.

(b) Find the expression for the electric field strength E as a function of r.
(c) Find the expression for the electric potential V as a function of r.

12. *Field Line of Two Opposite Charges***

Two charges $q > 0$ and $-q$ are located along the x-axis at $(-d, 0, 0)$ and $(d, 0, 0)$. Determine the radial distance from the origin that a field line emanating from q at an angle α with respect to the positive x-axis intersects with the yz-plane.

13. *Average Values***

(a) Consider the isolated system of a single point charge q. Determine the average electric field vector over the surface of an imaginary sphere, centered about an arbitrary point and possessing an arbitrary radius r, that encloses the charge q.
(b) Using the same system as above, determine the average electric field vector over the surface of an imaginary sphere, centered about an arbitrary point and possessing an arbitrary radius r, that does not enclose the charge q. Let the vector pointing from q to the center of the sphere be $\boldsymbol{R}$.
(c) Show that for an arbitrary system of point charges, the average potential over the surface of an imaginary sphere, centered about an arbitrary point and possessing an arbitrary radius r, that does not enclose any charge is identical to the potential at the center of the sphere; $V_{avg} = V_{center}$. Explain why this proves Earnshaw's Theorem.
(d) More generally, show that for an arbitrary system of point charges, the average potential over the surface of an imaginary sphere, centered about an arbitrary point and possessing an arbitrary radius r, is $V_{avg} = V_{center} + \frac{q_{enc}}{4\pi\varepsilon_0 r}$ where V_{center} is the potential at the center of the sphere and q_{enc} is the total charge enclosed by the sphere.

14. *Charged Disk***

Determine the electric field at a point of height h above the center of a thin, circular disk that has a uniform surface charge density σ and radius R. The point of concern is along the axis of the disk.

In light of the above result, determine the electric field, due to the truncated cone with a uniform volume charge density ρ shown in the figure on the next page, at the vertex of the original cone.

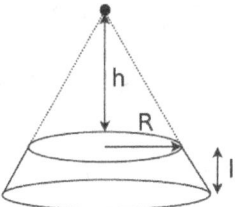

15. *Infinite Plane with Hole***

A hole of radius R is carved out of a thin infinite plane with a positive surface charge density σ that is uniform. Place a charge q at the center of the hole. Neglecting all gravitational effects, show that the center of the hole corresponds to an equilibrium position. Determine the stability of the equilibrium of charge q with mass m when it is slightly displaced in the direction normal to the plane (you have to consider different values of q). If the equilibrium is stable, determine the angular frequency of small oscillations that q exhibits. You may find the result of the previous problem to be useful.

16. *Cylinder***

The cylindrical axis of a cylinder with length l and radius R is aligned with the z-axis. The centers of the bases of the cylinder are at $z = 0$ and $z = l$. If the cylinder has a uniform charge density ρ, determine the electric field everywhere along the cylindrical axis, both inside and outside of the cylinder.

17. *Sphere with Cavity***

A solid sphere of radius R is made up of an insulating material and has a volume charge density ρ. A spherical cavity of radius a is removed from the sphere. The center of the cavity is at a position $\mathbf{d}$ with respect to the center of the sphere. Determine the electric field everywhere within the cavity.

18. *Two Rods***

Two identical thin rods of length l have equal uniform linear charge density λ. They both lie along the x-axis with their centers separated by a distance $d > l$. Determine the magnitude of the Coulomb force exerted on the right rod by the left rod at this instant. Check if your answer returns the correct limit when $d \gg l$.

19. Parallel Square Plate Capacitor**

A square conducting plate of side length $2a$, centered about the origin in the xy-plane, is charged with a uniform surface charge density σ.

(a) Prove that following integral where z is independent of x and y:

$$\int_0^a \int_0^a \frac{dxdy}{(x^2 + y^2 + z^2)^{\frac{3}{2}}} = \frac{1}{z} \tan^{-1} \frac{a^2}{z\sqrt{2a^2 + z^2}}.$$

(b) Determine the electric field $\boldsymbol{E}$ at $(0, 0, z)$ due to this surface charge distribution. Find the limits where $z \to 0^+$ and $z \to 0^-$ and explain why they make sense.

Now, in addition to the previous charged square plate, there is another square plate of the same size, parallel to the xy-plane and centered at $(0, 0, d)$. This additional plate is uniformly charged with a surface charge density $-\sigma$.

(c) Determine the electric field at $(0, 0, z)$ for all z due to the uniform charge distributions on both plates (we assume, albeit incorrectly, that they stay uniform in the presence of each other).

(d) Assuming $d \ll a$, find the asymptotic solution to the previous field for all z. As an aside, the charge distributions on both plates indeed remain uniform in this limit as the plates are effectively infinitely large as compared to the separation between them.

(e) Based on your previous answer, determine the potential difference V between the two plates. The capacitance of the two plates is defined as

$$C = \left| \frac{Q}{V} \right|$$

where Q is the total charge on either plate. Find C and state the variables it depends on.

20. Electric Field due to Spherical Sections**

An insulating sphere of radius R is coated with a uniform surface charge density σ on its exterior surface. Suppose that we cut off a spherical cap corresponding to a half-angle α from a certain axis (i.e. the cap has base radius $R \sin \alpha$) and remove the rest of the sphere. Determine the electric field due to the cap at the center of the original sphere. Thus, state the electric field at the center of an insulating hemisphere with surface charge density σ coated over its curved surface. Finally, using this result for a hemisphere, determine the electric field at the center of the original sphere if a spherical wedge of half-angle α (a slice of watermelon) is extracted from the sphere and the rest of the sphere is removed instead.

21. *Equilateral Triangle* **

Three line charges, each of length L, are arranged in the form of an equilateral triangle. The line charges carry uniform charge densities 2λ, λ and λ.

(a) Determine the electric field at the centroid of the triangle.
(b) Determine the electric potential at the centroid of the triangle.
(c) Find a point inside the triangle where the electric field is zero.

22. *Potential at Rim of Disk* **

Determine the electric potential at the rim of an insulating disk of radius R and uniform surface charge density σ. Hint: adopt polar coordinates about a point on the rim for the integration. In light of your result, derive the electric potential energy stored in the disk.

23. *Potential at Vertex of Square* **

Firstly, determine the potential at the vertex which is sandwiched between the two equal edges (which subtend an angle 2α) of an isosceles triangle with uniform surface charge density σ, if the height of the triangle from this vertex is h. Using this result, determine the potential at the center and thus the potential of a corner of a square with edge length l and a uniform surface charge density σ.

24. *Force on Hemispheres* **

Consider a charged sphere of radius R whose northern and southern hemispheres carry volume charge densities ρ_1 and ρ_2 respectively. Determine the force between these hemispheres.

25. *Two Hemispherical Shells* ***

Determine the force between two uniform and concentric hemispherical shells of radii R, $r < R$ and charges Q, q respectively. The common center and the apexes of the shells are collinear.

Solutions

1. Two Charges*

Since the potential can attain a value of zero, the two charges must be of opposite signs. This is due to the fact that a negative charge will lead to a negative potential at all points while a positive charge will result in the converse. Since the electric potential tends to positive infinity as one approaches $x = 0$, the positive charge must be positioned at $x = 0$. Let this charge be q_1 and the other negative charge be $-q_2$ at x-coordinate $x = d$. The two points on the positive x-axis which correspond to zero potential are located between q_1 and $-q_2$ and on the right of $-q_2$ ($x_0 > d$). Since a zero potential point can occur on the right of $-q_2$ such that the distance between this point and q_1 is larger than that to $-q_2$, $q_1 > q_2$. Enforcing the potential at $x = x_0$ to be zero,

$$\frac{q_1}{x_0} - \frac{q_2}{x_0 - d} = 0$$

where we have utilized $x_0 > d$. Furthermore, since the potential at $x = \alpha x_0$ is a minimum along the x-direction, the electric field in the x-direction at this point must be zero as $E_x = -\frac{\partial V}{\partial x}$ and $\frac{\partial V}{\partial x} = 0$ at this minimum.

$$\frac{q_1}{\alpha^2 x_0^2} - \frac{q_2}{(\alpha x_0 - d)^2} = 0.$$

In writing the above, we have asserted that the minimum lies outside of the region between the charges. Otherwise, the negative sign in the equation above would be a positive sign — resulting in no solutions for d (the field between two opposite charges cannot be zero). Solving,

$$d = \alpha(2 - \alpha)x_0.$$

2. Triangle of Charges*

The total energy of the system is conserved. The total energy is the potential energy of the initial configuration.

$$E = \frac{3q^2}{4\pi\varepsilon_0 l_0}.$$

The total energy of the final configuration is

$$\frac{3q^2}{4\pi\varepsilon_0 l} + \frac{3}{2}mv^2$$

where we have exploited the symmetry of the system to conclude that their relative separations must be identical for all pairs and that their speeds must be equal. Since energy must be conserved,

$$\frac{3q^2}{4\pi\varepsilon_0 l} + \frac{3}{2}mv^2 = \frac{3q^2}{4\pi\varepsilon_0 l_0}$$

$$v = \sqrt{\frac{q^2}{2\pi\varepsilon_0}\left(\frac{1}{l_0} - \frac{1}{l}\right)}.$$

3. Exploding Charge**

Let the initial speed of the charge undergoing circular motion be v. Since the Coulomb force on $-q$ must provide the centripetal force in sustaining its circular motion,

$$\frac{qQ}{4\pi\varepsilon_0 r_0^2} = \frac{mv^2}{r_0}$$

$$v = \sqrt{\frac{qQ}{4\pi m\varepsilon_0 r_0}}.$$

This must also be speed (purely tangential) of the leftover mass immediately after the explosion, by the conservation of momentum, as $\frac{m}{2}$ is ejected in the radial direction relative to the leftover mass at negligible velocity. The total mechanical energy of the leftover mass is

$$E = \frac{1}{2}\cdot\frac{m}{2}\cdot v^2 - \frac{qQ}{4\pi\varepsilon_0 r_0} = -\frac{3qQ}{16\pi\varepsilon_0 r_0}.$$

The total angular momentum of the leftover mass, with respect to Q, is

$$L = \frac{mr_0 v}{2}.$$

The total mechanical energy and angular momentum of the leftover mass are conserved. When the leftover mass reaches the point of minimum distance, it must have no radial velocity. Therefore, we let its purely tangential velocity be v_t. By the conservation of angular momentum and energy,

$$r_0 v = r v_t$$

$$\frac{mv_t^2}{4} - \frac{qQ}{4\pi\varepsilon_0 r} = -\frac{3qQ}{16\pi\varepsilon_0 r_0}.$$

Eliminating v_t in the above set of equations would yield

$$3r^2 - 4r_0 r + r_0^2 = 0$$

$$r = \frac{4 \pm \sqrt{16 - 12}}{6}r_0.$$

The minimum distance is obtained by taking the expression with the negative sign.

$$r_{min} = \frac{r_0}{3}.$$

4. Cube Potential**

Observe that the electric potential is proportional to charge and inversely proportional to a length dimension. The potential at the center of a cube of side length $2l$ is then 4 times the potential at the center of a cube of side length l as its volume and thus charge are larger by a factor of 8 while its length is augmented by a factor of 2. Furthermore, the potential at the center of a cube of side length $2l$ is equal to 8 times the potential at the corner of a cube of side length l by the principle of superposition (as the large cube is made up of 8 smaller cubes). Equating these expressions for the potential at the center of a cube of side length $2l$, one can see that the ratio of the potential at the center of a cube of length l to that at the corner is $2 : 1$.

5. Flux at Corner of Cube*

Let the original cube under consideration have side length l. Piece eight of such cubes together to form a larger cube of side length $2l$ with the charge q at its center. The total electric flux cutting through the large cube is $\frac{q}{\varepsilon_0}$ by Gauss' law. By symmetry, the electric flux across a single face of a cube of length l (with non-zero flux) is $\frac{q}{24\varepsilon_0}$ as there are 24 such faces on the larger cube of length $2l$. Another (slightly dubious) method is to visualize the point charge as a small sphere, of which only $\frac{1}{8}$ lies inside the original cube. Since the charge "enclosed" by the original cube is $\frac{q}{8}$ and there are three symmetrical faces with non-zero flux, the flux across a single face with non-zero flux is $\frac{q}{24\varepsilon_0}$.

6. Flux Through Spherical Cap*

When $h = 0$, the charge can be symmetrically enclosed by two identical spherical caps. The electric flux through a single spherical cap is then given by Gauss' law as $\frac{q}{2\varepsilon_0}$. In the general case, the crucial observation is that any arbitrary surface with the same circular boundary as the spherical cap must possess the same magnitude of electric flux, as long as the arbitrary surface and the spherical cap, glued together, do not enclose the charge. This is a direct consequence of Gauss' law applied to the closed surface formed by combining the arbitrary surface and the spherical cap. If there is no charge enclosed, the total electric flux cutting through the closed surface must be zero — implying that the electric fluxes across the two surfaces must be equal in magnitude and opposite in direction.

The electric flux through the desired spherical cap is then identical to that through a spherical cap of radius a, centered about the charge. Then, we simply have to determine the proportion of the sphere of radius a, centered about the charge, that this spherical cap occupies — this can be done by determining the surface area of the curved surface of the spherical cap. The surface area of a spherical cap with radius R and altitude H is $2\pi RH$. Therefore, the surface area of this cap with radius a and altitude $a - h$ is $2\pi a(a - h)$. Dividing this by the surface area of a complete sphere of radius a, $4\pi a^2$, the proportion of the sphere that the spherical cap occupies is $\frac{1-\frac{h}{a}}{2}$. The total electric flux cutting through the desired spherical cap is then this fraction multiplied by the total flux emitted by the charge, $\frac{q}{\varepsilon_0}$.

$$\frac{\left(1 - \frac{h}{a}\right) q}{2\varepsilon_0}$$

which correctly becomes $\frac{q}{2\varepsilon_0}$ in the limiting case where $h = 0$.

7. Force on Cube Face*

Let the electric field at the surface of one face, due to the five other faces, be $\boldsymbol{E}$. Due to symmetry, the net force experienced by the particular face can only be normal to it. The normal component of force on an infinitesimal surface element on that particular face with area vector $d\boldsymbol{A}$ (pointing outwards) is $\sigma \boldsymbol{E} \cdot d\boldsymbol{A}$ where $\boldsymbol{E}$ is the electric field due to the five other faces at the infinitesimal element. The total force experienced by the face is then

$$F = \sigma \iint \boldsymbol{E} \cdot d\boldsymbol{A} = \sigma \Phi$$

where the integral is performed over the entire face. The integral is simply equal to the electric flux due to the five other faces cutting across this particular face. The electric flux crossing this particular face due to the entire cube is one sixth of the electric flux of the entire cube, $\frac{1}{6} \cdot \frac{6\sigma l^2}{\varepsilon_0} = \frac{\sigma l^2}{\varepsilon_0}$, outwards of the face. Meanwhile, the electric flux emanating outwards from this particular face, due to the charges on the face itself, is $\frac{\sigma l^2}{2\varepsilon_0}$, as half of its total flux is emitted on each side of the face. The outwards electric flux Φ crossing this particular face due to the other five faces is then

$$\Phi = \frac{\sigma l^2}{\varepsilon_0} - \frac{\sigma l^2}{2\varepsilon_0} = \frac{\sigma l^2}{2\varepsilon_0}.$$

The force experienced by this face is thus

$$F = \frac{\sigma^2 l^2}{2\varepsilon_0}$$

and is directed in the outwards direction, normal to the face.

8. Electric Field of Equilateral Triangle*

Place an imaginary unit charge at point P. The electric field due to the equilateral triangle at point P is the force on this unit charge exerted by it and is equal to the negative of the force exerted by the unit charge on the equilateral triangle. Similar to the fourth problem in Section 5.4.1, the normal component of the force (which is the only component present due to the symmetrical nature of the equilateral triangle) exerted by the unit charge on the equilateral triangle is equal to σ multiplied by the electric flux, due to the unit charge, cutting across the equilateral triangle. Now, observe that the imaginary unit charge can be enclosed by four of such equilateral triangles, which form a tetrahedron. Then, the electric flux due to the unit charge crossing one face of the tetrahedron is $\frac{1}{4}$ of the total electric flux that it emits (which is $\frac{1}{\varepsilon_0}$ by Gauss' law). Therefore, the magnitude of the force on the equilateral triangle due to the unit charge, and hence the electric field strength at point P due to the equilateral triangle, is $\frac{\sigma}{4\varepsilon_0}$. The electric field is directed normally outwards from the centroid of the equilateral triangle towards point P.

9. Chessboard*

Since the chessboard is symmetrical about both diagonals, the electric field at points directly above its center can only be normal to the plane of the chessboard. Now, observe that if we rotate the chessboard by 90° and superpose the original chessboard, we obtain a square plate of edge length l that is completely covered by a uniform surface charge density σ. We have determined that the electric field of such a set-up at a point $\frac{l}{2}$ above its center is $\frac{\sigma}{6\varepsilon_0}$ in the final example of Section 5.4.1. The electric field at that point due to the chessboard is thus half of this value

$$E = \frac{\sigma}{12\varepsilon_0}$$

and is normally outwards from the board.

10. Finite Line Charge Revisited*

Referring to the definition of the angle θ in Fig. 5.3, the magnitude of the electric field at P due to an infinitesimal segment between θ and $\theta + d\theta$ can be computed as follows.

$$dE = \frac{\lambda dx}{4\pi\varepsilon_0 r^2},$$

$$x = h\tan\theta \implies dx = h\sec^2\theta d\theta,$$

$$dE = \frac{\lambda}{4\pi\varepsilon_0 \frac{h^2}{\cos^2\theta}} \cdot h\sec^2\theta d\theta = \frac{\lambda}{4\pi\varepsilon_0 h}d\theta.$$

That is, the magnitude of the electric field does not depend on the angular coordinate θ, but only the angular "width" $d\theta$ of the segment.

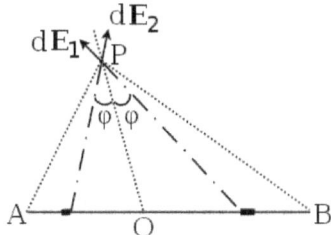

Figure 5.17: Electric fields due to two segments which subtend equal angles from angle bisector

Consider two infinitesimal line segments that subtend angle ϕ with the angle bisector OP and possess equal angular width $d\phi$ (this also ensures that they have equal angular widths in terms of θ as both ϕ and θ are measured about P) in Fig. 5.17. The electric fields at P due to the right and left segments are $d\boldsymbol{E}_1$ and $d\boldsymbol{E}_2$ respectively. Since the magnitudes of these vectors are equal, there is no contribution to the electric field perpendicular to the angle bisector due to this pair of segments. Applying this argument to all pairs for $0 < \phi \le \alpha$, the net electric field at P due to the line charge must bisect $\angle APB$. To determine the magnitude of the net electric field, observe that one pair of segments corresponding to angle ϕ contributes to a value $\frac{\lambda\cos\phi d\phi}{2\pi\varepsilon_0 h}$ to the component of electric field parallel to the angle bisector. Integrating ϕ from 0 to α,

$$E = \frac{\lambda}{2\pi\varepsilon_0 h}\int_0^\alpha \cos\phi d\phi = \frac{\lambda}{2\pi\varepsilon_0 h}\sin\alpha.$$

For the next question, notice that if we apply the above result to a semi-infinite line charge, the electric field vector at a point — at which a line joining this point to the finite end of the line charge is perpendicular to the line charge — subtends an angle $45°$ with the direction along the line charge. Therefore, the net electric field at (a, a) due to the two semi-infinite line charges in the problem will be zero as their individual electric fields nullify each other there.

11. Hydrogen Atom*

(a) Integrating over spherical shells of infinitesimal thickness dr, the total amount of charge within a sphere of radius r (remember to include the proton) is

$$Q = q + \int_0^r -\frac{q}{\pi a_0^3} e^{-\frac{2r}{a_0}} \cdot 4\pi r^2 dr$$

$$= q\left(1 - \frac{4}{a_0^3}\int_0^r r^2 e^{-\frac{2r}{a_0}} dr\right)$$

$$= q\left(1 + \frac{2}{a_0^2}\left[r^2 e^{-\frac{2r}{a_0}} - 2\int_0^r r e^{-\frac{2r}{a_0}} dr\right]\right)$$

$$= q\left(1 + \frac{2}{a_0^2}r^2 e^{-\frac{2r}{a_0}} + \frac{2}{a_0}\left[r e^{-\frac{2r}{a_0}} - \int_0^r e^{-\frac{2r}{a_0}} dr\right]\right)$$

$$= q\left(1 + \frac{2}{a_0^2}r^2 e^{-\frac{2r}{a_0}} + \frac{2}{a_0}r e^{-\frac{2r}{a_0}} + e^{-\frac{2r}{a_0}} - 1\right)$$

$$= q\left(1 + \frac{2}{a_0^2}r^2 + \frac{2}{a_0}r\right)e^{-\frac{2r}{a_0}}.$$

As $r \to \infty$, $Q \to 0$ which makes sense as the hydrogen atom is neutral overall.

(b) Exploiting the spherical symmetry of this charge distribution, Gauss' law gives

$$E \cdot 4\pi r^2 = \frac{Q}{\varepsilon_0}$$

$$E = \frac{q}{4\pi\varepsilon_0}\left(\frac{1}{r^2} + \frac{2}{a_0^2} + \frac{2}{a_0 r}\right)e^{-\frac{2r}{a_0}}.$$

(c) The electric potential is

$$V = -\int_\infty^r \boldsymbol{E} \cdot d\boldsymbol{r} = -\int_\infty^r E\,dr$$

$$V = -\frac{q}{4\pi\varepsilon_0}\int_\infty^r \left(\frac{1}{r^2} + \frac{2}{a_0^2} + \frac{2}{a_0 r}\right)e^{-\frac{2r}{a_0}}\,dr.$$

To evaluate this, observe that by performing integration-by-parts,

$$\int_\infty^r \frac{1}{r^2}e^{-\frac{2r}{a_0}}\,dr = -\frac{1}{r}e^{-\frac{2r}{a_0}} - \int_\infty^r \frac{2}{a_0 r}e^{-\frac{2r}{a_0}}\,dr.$$

Therefore,

$$V = -\frac{q}{4\pi\varepsilon_0}\cdot-\frac{1}{r}e^{-\frac{2r}{a_0}} - \frac{q}{4\pi\varepsilon_0}\int_\infty^r \frac{2}{a_0^2}e^{-\frac{2r}{a_0}}\,dr = \frac{q}{4\pi\varepsilon_0}\left(\frac{1}{a_0}+\frac{1}{r}\right)e^{-\frac{2r}{a_0}}.$$

12. Field Line of Two Opposite Charges**

Trace a field line that emanates from q at an angle α with the positive x-axis — suppose that it intersects the yz-plane at a radial distance R from the origin. Since this set-up is symmetric about the x-axis, we can rotate this field line about the x-axis for a complete revolution to generate other axial-symmetric field lines. This set of field lines and the circular base (of radius R) spanned by their intersections with the yz-plane form a closed surface. Applying Gauss's law to this closed surface, the relevant electric flux Φ only stems from the circular base as the electric field is always tangential along the other curved surface (by definition of an electric field line). The x-component of the electric field due to this set-up at $x = 0$ and a radial distance r in the yz-plane is

$$E_x = \frac{qd}{2\pi\varepsilon_0(r^2 + d^2)^{\frac{3}{2}}}.$$

Therefore, the electric flux through this Gaussian surface (obtained by integrating over circular shells of perimeter $2\pi r$ and thicknes dr) is

$$\Phi = \int_0^R \frac{qd}{2\pi\varepsilon_0(r^2 + d^2)^{\frac{3}{2}}}\cdot 2\pi r\,dr = \frac{qd}{\varepsilon_0}\left(\frac{1}{d} - \frac{1}{\sqrt{R^2 + d^2}}\right).$$

Now, the fraction captured in a cone with half-angle α, out of the total electric flux emitted by q (which is $\frac{q}{\varepsilon_0}$ by Gauss' law), was previously calculated in Section 5.1 as $\sin^2\frac{\alpha}{2}$. Therefore, the above flux must be equal to

$$\Phi = \frac{q}{\varepsilon_0} \cdot \sin^2\frac{\alpha}{2}.$$

Solving for R,

$$R = d\sqrt{\sec^4\frac{\alpha}{2} - 1}.$$

13. Average Values**

(a) Distribute a charge Q evenly over the surface of the imaginary sphere. The average electric field due to q over the imaginary spherical surface is simply the force exerted by q on this spherical shell of charge Q, divided by Q. This is equal to the negative of the force exerted by Q on q, divided by Q, by Newton's third law. However, by applying Gauss' law to a concentric spherical Gaussian surface within the spherical shell of charge Q, we know that it generates zero electric field within itself. Therefore, the force exerted by Q on q must be zero — implying that the average electric field over the imaginary spherical surface is $\boldsymbol{E}_{avg} = \boldsymbol{0}$.

(b) Repeating the same process as above, we compute the force exerted by Q on q. Applying Gauss' law to a concentric spherical Gaussian surface outside of the spherical shell of charge Q, the electric field due to Q at a radial distance $R > r$ from the center of the shell is $\frac{Q}{4\pi\varepsilon_0 R^2}$ (akin to a point charge Q placed at the center). Therefore, the force exerted by Q on q, divided by Q, is $-\frac{q}{4\pi\varepsilon_0 R^2}\hat{\boldsymbol{R}}$ where $\hat{\boldsymbol{R}}$ is the unit vector that points from q to the center of the imaginary sphere. The average electric field due to q over the imaginary spherical surface is the negative of this and is equal to $\frac{q}{4\pi\varepsilon_0 R^2}\hat{\boldsymbol{R}}$.

(c) Similar to the previous sections, distribute a charge Q evenly over the surface of the imaginary sphere. Consider the special case where there is a single charge q outside the imaginary sphere. The average potential due to q over the spherical surface is simply the electric potential energy of the spherical shell of charge Q when q is fixed, divided by Q. The electric potential energy of the spherical shell of charge Q when q is fixed is equal to the electric potential energy of q when the spherical shell is fixed. This is because for any two point charges q_i and q_j, the electric potential energy of q_i when q_j is fixed is $\frac{q_i q_j}{4\pi\varepsilon_0 r_{ij}}$ where r_{ij} is the distance separating them — this is evidently equal to the electric potential energy of q_j when q_i is fixed. The electric potential energy of q when the spherical shell of charge Q is fixed is basically $\frac{qQ}{4\pi\varepsilon_0 R}$, where R is the distance between q and the center of the shell, as the shell is essentially a point charge Q located at its center, for regions outside of the shell. The average potential due to q over the imaginary spherical surface is thus $\frac{q}{4\pi\varepsilon_0 R}$ which is equal to the potential engendered by it at the center of the imaginary sphere. Therefore, $V_{avg} = V_{center}$ for a single charge q. By the principle of superposition, this must also be the case for an arbitrary distribution of charges outside of the imaginary sphere. This result proves Earnshaw's Theorem as it shows that a potential maximum or minimum cannot be located at a point in space with no net charge. Suppose that a maximum existed there — the average potential of its neighbours (over a spherical surface with an infinitesimal radius centered about the point of concern) will then be smaller than the potential at that point (the center of the sphere), contradicting our recently established theorem. A similar logic holds for the existence of a minimum. Since a maximum or minimum cannot be located at a particular point in free space, there is at least one neighboring point with a lower potential than that particular point (which produces an unstable equilibrium in that direction for a positive charge placed at that particular point) and one with a higher potential than that particular point (which produces an unstable equilibrium for a negative charge) since the electric field is the negative potential gradient.

(d) In a manner analogous to the previous part, distribute a charge Q evenly over the surface of the imaginary sphere of radius r and consider the special case where a single charge q is enclosed in an arbitrary location within the imaginary sphere. The electric potential energy of q when the spherical shell Q is fixed is simply q multiplied by the potential due to Q at the location of q (within the shell) which is simply the potential at the surface of the shell since the electric field within the shell is zero. The latter is $\frac{Q}{4\pi\varepsilon_0 r}$ since a spherical shell is akin to a point charge Q placed at its center for regions outside of it. The average potential due to q over the spherical surface can then be computed as $\frac{q}{4\pi\varepsilon_0 r}$ by dividing the electric potential energy of q when the spherical shell Q is fixed, by Q. Proceeding with a general charge distribution involving an arbitrary configuration of charges outside of the imaginary sphere and q_{enc} amount of charge enclosed within it, the q_{enc} charge contributes a value of $\frac{q_{enc}}{4\pi\varepsilon_0 r}$ to the average potential over the spherical surface by the principle of superposition. Combining this with the result of the previous part — which states that the charge configuration outside of the sphere contributes a value of V_{center} to the average potential — the average potential over the imaginary spherical surface is $V_{avg} = V_{center} + \frac{q_{enc}}{4\pi\varepsilon_0 r}$.

14. Charged Disk**

We will consider the disk in polar coordinates along the plane of disk. Define the origin at the center of the disk and the z-axis to be normal to the disk. Consider an infinitesimal area element on the disk at a radial distance r and polar angle θ with sides $rd\theta$ and dr. The distance between this element and P is r' given by

$$r'^2 = h^2 + r^2.$$

Therefore, the magnitude of the electric field at P due to this element is

$$\frac{dq}{4\pi\varepsilon_0 r'^2} = \frac{\sigma r dr d\theta}{4\pi\varepsilon_0 (h^2 + r^2)}.$$

Due to the symmetry of the disk, the electric field at P can only be in the z-direction. Therefore, we simply need to integrate the z-component of the electric field at P due to all infinitesimal elements over the entire disk. Thus, the electric field at P is

$$
\begin{aligned}
E_z &= \int_0^R \int_0^{2\pi} \frac{\sigma r d\theta dr}{4\pi\varepsilon_0 (h^2 + r^2)} \cdot \frac{h}{\sqrt{h^2 + r^2}} \\
&= \int_0^R \frac{\sigma h r dr}{2\varepsilon_0 (h^2 + r^2)^{\frac{3}{2}}} \\
&= \left[-\frac{\sigma h}{2\varepsilon_0 \sqrt{h^2 + r^2}} \right]_0^R \\
&= \frac{\sigma}{2\varepsilon_0} - \frac{\sigma h}{2\sqrt{h^2 + R^2}}.
\end{aligned}
$$

In the next problem, slice the truncated cone into disks of infinitesimal thickness in the direction perpendicular to the axis of the cone. Now, we argue that these disks must make the same contribution to the electric field at the vertex of the original cone. Let the perpendicular distances between two of such disks to the vertex be h_1 and h_2. Observe that the electric field at the vertex due to one disk must be proportional to its charge and inversely proportional to the squared distance between it and the vertex. As compared to the first disk, the distance between the second disk and the vertex is $\frac{h_2}{h_1}$ times that of the first disk. However, its charge is also larger by a factor of $\frac{h_2^2}{h_1^2}$ due to the differences in area. The two effects cancel out — causing the contributions by the two disks to be equal. Therefore, the electric field due to the entire truncated cone is equivalent to the disk at the top, with a surface charge density of ρl. The electric field at the vertex is then

$$E_z = \frac{\rho l}{2\varepsilon_0} - \frac{\rho l h}{2\sqrt{h^2 + R^2}}.$$

15. Infinite Plane with Hole**

We have proven that the electric field due to an infinite plane of charge density σ is $\frac{\sigma}{2\varepsilon_0}$ everywhere, directed outwards from the plane, and that the electric field at a height h above the center of a circular disk of charge density σ is $\frac{\sigma}{2\varepsilon_0} - \frac{\sigma h}{2\sqrt{h^2+R^2}}$, directed outwards from the disk. Therefore, the electric field at a height h above the center of the hole in the infinite plane is

$$E(h) = \frac{\sigma}{2\varepsilon_0} - \left(\frac{\sigma}{2\varepsilon_0} - \frac{\sigma h}{2\sqrt{h^2 + R^2}} \right) = \frac{\sigma h}{2\sqrt{h^2 + R^2}}$$

and is directed normally outwards from the center of the hole. When $h = 0$, $E(0) = 0$ which shows that the center of the hole corresponds to an equilibrium position. Furthermore, since $E(h)$ is directed normally outwards, the equilibrium for $q > 0$ is unstable (as the Coulomb force tends to push it further away from the hole) while the equilibrium for $q < 0$ is stable (as the Coulomb force tends to correct its deviation from the hole). For small values of $h << R$, the electric field becomes

$$E(h) = \frac{\sigma h}{2R}.$$

Therefore, Newton's law yields, for the charge $q < 0$,

$$m\ddot{h} = \frac{q\sigma h}{2R}$$

$$\implies \ddot{h} = -\frac{-q\sigma}{2Rm}h,$$

where $-q$ is a positive quantity. The above equation of motion describes a simple harmonic motion with angular frequency

$$\omega = \sqrt{\frac{-q\sigma}{2Rm}}.$$

16. Cylinder**

We will determine the electric field at a point P with coordinates $(0, 0, h)$ by integrating the contributions due to each infinitesimal element in cylindrical coordinates over the entire cylinder. Consider an infinitesimal element at coordinates (r, ϕ, z). It has volume $r d\phi dr dz$ and thus has charge $\rho r d\phi dr dz$. The electric field strength at point P due to this element is

$$\frac{dq}{4\pi\varepsilon_0 [r^2 + (h-z)^2]} = \frac{\rho r d\phi dr dz}{4\pi\varepsilon_0 [r^2 + (h-z)^2]}.$$

However, by symmetrical arguments, only the z-component of this electric field will survive after an integration over the entire cylinder. Therefore, we just have to take the z-component of the above which is

$$\frac{\rho r d\phi dr dz}{4\pi\varepsilon_0 [r^2 + (h-z)^2]} \cdot \frac{h-z}{\sqrt{r^2 + (h-z)^2}}.$$

Integrating the above over the entire cylinder,

$$\int_0^l \int_0^R \int_0^{2\pi} \frac{\rho r (h-z)}{[r^2 + (h-z)^2]^{\frac{3}{2}}} d\theta dr dz$$

$$= \int_0^l \int_0^R \frac{2\pi \rho r (h-z)}{[r^2 + (h-z)^2]^{\frac{3}{2}}} dr dz$$

$$= \int_0^l \left[-\frac{2\pi\rho(h-z)}{\sqrt{r^2 + (h-z)^2}} \right]_0^R dz$$

$$= \int_0^l \left(\frac{2\pi\rho(h-z)}{\sqrt{(h-z)^2}} - \frac{2\pi\rho(h-z)}{\sqrt{R^2 + (h-z)^2}} \right) dz$$

where we have adopted Gaussian units $\left(\frac{1}{4\pi\varepsilon_0} = 1 \right)$ for the sake of convenience.

At this point we cannot rashly conclude whether $\sqrt{(h-z)^2} = h - z$ or $z - h$ and must instead consider the relative values of z and h. If $h \geq l$ (i.e. the point P is above the cylinder), $\sqrt{(h-z)^2} = h - z$ for the entire regime of integration.

$$E_{h \geq l} = \int_0^l \left(2\pi\rho - \frac{2\pi\rho(h-z)}{\sqrt{R^2 + (h-z)^2}} \right) dz$$

$$= 2\pi\rho l + 2\pi\rho\sqrt{R^2 + (h-l)^2} - 2\pi\rho\sqrt{R^2 + h^2}.$$

If $h \leq 0$ (i.e. the point P is below the cylinder), $\sqrt{(h-z)^2} = z - h$ for the entire regime of integration.

$$E_{h \leq 0} = \int_0^l \left(-2\pi\rho - \frac{2\pi\rho(h-z)}{\sqrt{R^2 + (h-z)^2}} \right) dz$$

$$= -2\pi\rho l + 2\pi\rho\sqrt{R^2 + (h-l)^2} - 2\pi\rho\sqrt{R^2 + h^2}.$$

Now if $0 < h < l$ (i.e. the point P is within the cylinder), $\sqrt{(h-z)^2} = h - z$ for $0 \leq z \leq h$ and $\sqrt{(h-z)^2} = z - h$ for $h \leq z \leq l$. Then, we have to divide the first term in the integral into two regimes of integration.

$$E_{0 < h < l} = \int_0^h 2\pi\rho dz + \int_h^l -2\pi\rho dz + \int_0^l -\frac{2\pi\rho(h-z)}{\sqrt{R^2 + (h-z)^2}} dz$$

$$= 2\pi\rho(2h - l) + 2\pi\rho\sqrt{R^2 + (h-l)^2} - 2\pi\rho\sqrt{R^2 + h^2}.$$

17. Sphere with Cavity**

Recall that the electric field within a sphere with a uniform charge density ρ at a radial vector r from the center was derived to be

$$E = \frac{\rho}{3\varepsilon_0} r.$$

Moving back to the original question, we can first fill up the cavity with the same uniform charge density ρ and then superpose another sphere of density $-\rho$, corresponding to the original cavity, to remove this additional charge. Now, let d be the vector from the center of the large sphere to the center of the cavity, r be the vector from the center of the large sphere to the point in the cavity at which the electric field is of concern and R be the vector from the center of the cavity to the point of concern. Then, the electric field at the point of concern is the superposition of the fields due to the two spheres.

$$E = \frac{\rho}{3\varepsilon_0}(r - R) = \frac{\rho d}{3\varepsilon_0}.$$

This implies that the electric field is uniform within the cavity. Note that this expression is only valid within the cavity as the electric field used in this derivation was computed **inside** a spherical charge distribution.

18. Two Rods**

Define the origin such that the two rods span from $x = 0$ to $x = l$ and $x = d$ to $x = l + d$. We first determine the electric field at points along the x-axis due to the rod between $x = 0$ and $x = l$. The electric field at a point along the x-axis at x-coordinate $y > l$ is obtained by integrating the contributions due to infinitesimal segments between coordinates x and $x + dx$ from $x = 0$ to $x = l$.

$$E(y) = \int_0^l \frac{\lambda dx}{4\pi\varepsilon_0 (x - y)^2}$$

$$= \left[-\frac{\lambda}{4\pi\varepsilon_0 (x - y)} \right]_0^l$$

$$= \frac{\lambda}{4\pi\varepsilon_0 (y - l)} - \frac{\lambda}{4\pi\varepsilon_0 y}.$$

Now, we consider the effects of this field on the second rod. The force on a segment of the second rod between x-coordinates y and $y + dy$ is

$$dqE(y) = \lambda E(y) dy.$$

We integrate this force over the entire second rod which ranges from $y = d$ to $y = d + l$ to obtain the total force on the second rod due to the first rod.

$$\int_d^{l+d} \lambda E(y) dy = \int_d^{l+d} \left(\frac{\lambda^2}{4\pi\varepsilon_0 (y - l)} - \frac{\lambda^2}{4\pi\varepsilon_0 y} \right) dy$$

$$= \frac{\lambda^2}{4\pi\varepsilon_0} \ln \frac{d}{d - l} - \frac{\lambda^2}{4\pi\varepsilon_0} \ln \frac{l + d}{d}$$

$$= \frac{\lambda^2}{4\pi\varepsilon_0} \ln \frac{d^2}{d^2 - l^2}.$$

When $d \gg l$, the two rods are effectively two point charges λl separated by a distance d. Therefore, we expect the force between them to be $\frac{\lambda^2 l^2}{4\pi\varepsilon_0 d^2}$. This is indeed the case as

$$\frac{\lambda^2}{4\pi\varepsilon_0} \ln \frac{d^2}{d^2 - l^2} = -\frac{\lambda^2}{4\pi\varepsilon_0} \ln \left(1 - \frac{l^2}{d^2} \right) \approx -\frac{\lambda^2}{4\pi\varepsilon_0} \cdot -\frac{l^2}{d^2} = \frac{\lambda^2 l^2}{4\pi\varepsilon_0 d^2},$$

where we have used the first order Maclaurin expansion $\ln(1 + x) \approx x$.

19. Parallel Square Plate Capacitor**

(a) Let $x = \sqrt{y^2 + z^2} \tan\theta$ and $dx = \sqrt{y^2 + z^2} \sec^2\theta \, d\theta$.

$$\int_0^a \int_0^a \frac{dx\,dy}{(x^2 + y^2 + z^2)^{\frac{3}{2}}} = \int_0^a \int_0^{\tan^{-1}\frac{a}{\sqrt{y^2+z^2}}} \frac{\cos\theta \, d\theta \, dy}{y^2 + z^2}$$

$$= \int_0^a \frac{a}{(y^2 + z^2)\sqrt{y^2 + a^2 + z^2}} dy.$$

Let $y = \sqrt{a^2 + z^2} \tan\phi$ and $dy = \sqrt{a^2 + z^2} \sec^2\phi \, d\phi$.

$$\int_0^a \frac{a}{(y^2 + z^2)\sqrt{y^2 + a^2 + z^2}} dy$$

$$= \int_0^{\tan^{-1}\frac{a}{\sqrt{a^2+z^2}}} \frac{a}{(z^2 \sec^2\phi + a^2 \tan^2\phi)\sqrt{a^2 + z^2} \sec\phi}$$

$$\cdot \sqrt{a^2 + z^2} \sec^2\phi \, d\phi$$

$$= \int_0^{\tan^{-1}\frac{a}{\sqrt{a^2+z^2}}} \frac{a\cos\phi}{(z^2 + a^2 \sin^2\phi)} d\phi$$

$$= \int_0^{\frac{a^2}{\sqrt{2a^2+z^2}}} \frac{du}{(z^2 + u^2)}$$

where $u = a\sin\phi$. Using the standard result $\int \frac{du}{z^2+u^2} = \frac{1}{z}\tan^{-1}\frac{u}{z} + c$, The integral becomes

$$\int_0^a \int_0^a \frac{dx\,dy}{(x^2 + y^2 + z^2)^{\frac{3}{2}}} = \frac{1}{z}\tan^{-1}\frac{a^2}{z\sqrt{2a^2 + z^2}}.$$

(b) Due to the symmetry of the square plate about the x and y-directions, the electric field at $(0, 0, z)$ should only be in the z-direction.

$$E_z = \int_{-a}^a \int_{-a}^a \frac{\sigma z}{4\pi\varepsilon_0(x^2 + y^z + z^2)^{\frac{3}{2}}} dx\,dy = \frac{\sigma z}{\pi\varepsilon_0}\int_0^a \int_0^a \frac{dx\,dy}{(x^2 + y^2 + z^2)^{\frac{3}{2}}}$$

$$E = \frac{\sigma}{\pi\varepsilon_0}\tan^{-1}\left(\frac{a^2}{z\sqrt{2a^2 + z^2}}\right)\hat{k}.$$

When $z \to 0^+$, $\tan^{-1}(\frac{a^2}{z\sqrt{2a^2+z^2}}) \to \frac{\pi}{2}$.

$$E = \frac{\sigma}{\pi\varepsilon_0} \cdot \frac{\pi}{2} = \frac{\sigma}{2\varepsilon_0}\hat{k}.$$

When $z \to 0^-$, $\tan^{-1}(\frac{a^2}{z\sqrt{2a^2+z^2}}) \to -\frac{\pi}{2}$.

$$E = -\frac{\sigma}{2\varepsilon_0}\hat{k}.$$

These values make sense as the only infinitesimal area on the plate that contributes to a z-component of electric field at $(0,0,0^+)$ or $(0,0,0^-)$ is the infinitesimal area at $(0,0,0)$. The electric fields of the other infinitesimal patches are along the plane of the plate. Therefore, the net electric fields in such limits are simply those of a small area with a surface charge density $\sigma - \frac{\sigma}{2\varepsilon_0}$ emanating outwards from the area.

(c) The resultant electric field due to the two plates is

$$E = \frac{\sigma}{\pi\varepsilon_0} \tan^{-1}\left(\frac{a^2}{z\sqrt{2a^2 + z^2}}\right)\hat{k} - \frac{\sigma}{\pi\varepsilon_0} \tan^{-1}\left(\frac{a^2}{(z-d)\sqrt{2a^2 + (z-d)^2}}\right)\hat{k}.$$

(d) Since $d << a$, $d << z$ for regions outside the plates.

$$E = \frac{\sigma}{\pi\varepsilon_0} \tan^{-1}\left(\frac{a^2}{z\sqrt{2a^2 + z^2}}\right)\hat{k} - \frac{\sigma}{\pi\varepsilon_0} \tan^{-1}\left(\frac{a^2}{(z\sqrt{2a^2 + z^2}}\right)\hat{k} = 0$$

outside the plates. Between the two plates, $z < d$ so the electric field is better expressed as

$$E = \frac{\sigma}{\pi\varepsilon_0} \tan^{-1}\left(\frac{a^2}{z\sqrt{2a^2 + z^2}}\right)\hat{k} + \frac{\sigma}{\pi\varepsilon_0} \tan^{-1}\left(\frac{a^2}{(d-z)\sqrt{2a^2 + (d-z)^2}}\right)\hat{k}.$$

Since $z < d$ in between the two plates, $z << a$ and $d - z << a$. Thus,

$$E = \frac{\sigma}{\pi\varepsilon_0} \cdot \frac{\pi}{2}\hat{k} + \frac{\sigma}{\pi\varepsilon_0} \cdot \frac{\pi}{2}\hat{k} = \frac{\sigma}{\varepsilon_0}\hat{k}.$$

(e) The potential difference between the two plates is

$$V = -\int_{\text{top plate}}^{\text{bottom plate}} E \cdot dr = \int_0^d \frac{\sigma}{\varepsilon_0}dz = \frac{\sigma d}{\varepsilon_0}.$$

(f) Based on the previous result,

$$C = \left|\frac{4\sigma a^2}{V}\right| = \frac{\varepsilon_0(4a^2)}{d}.$$

The capacitance only depends on the dimensions of the plates (e.g. $4a^2$ is the area of a plate and d is the separation between the plates). It is independent of the total charge on each plate and the potential difference between them.

20. Electric Field due to Spherical Sections**

To determine the component of electric field at the center of the original sphere along an arbitrary z-direction, due to the charges distributed over the curved surface of a spherical section, place an imaginary unit charge at the center. The required quantity is simply the negative of the z-component of the force that this unit charge exerts on the spherical section. Observe that since the unit charge is located at the center, the force that it exerts on each infinitesimal surface element on the spherical section is equal and radial in direction, the unit charge effectively engenders a pressure $p = \frac{\sigma}{4\pi\varepsilon_0 R^2}$ over the surface of the spherical section, where R is the radius of the sphere such that $\frac{1}{4\pi\varepsilon_0 R^2}$ is the electric field due to the unit charge at the surface of the sphere.

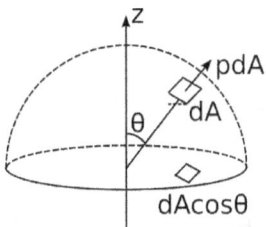

Figure 5.18: Force on an infinitesimal area

Now, consider the force that the unit charge exerts on an infinitesimal area located at an angular coordinate θ from the positive z-axis in Fig. 5.18. The z-component of this force is $pdA\cos\theta$ but $dA\cos\theta$ is simply the projection of the infinitesimal area onto the equatorial plane perpendicular to the z-axis! Therefore, the total z-component of the force exerted by the unit charge on the surface of a spherical section is p multiplied by the projection of the surface onto the equatorial plane.

For a spherical cap, define the positive z-axis to point from the center of the sphere to its vertex. The projection of the spherical cap onto the equatorial plane perpendicular to the z-axis (the xy-plane) is simply a circle of radius a, where $a = R\sin\alpha$ is the base radius of the spherical cap. Therefore, the z-component of the force that the unit charge exerts on the cap (which actually constitutes the net force due to symmetry) is

$$F_z = p \cdot \pi R^2 \sin^2 \alpha = \frac{\sigma \sin^2 \alpha}{4\varepsilon_0},$$

which implies that the electric field at the center of the sphere due to the spherical cap is $\frac{\sigma \sin^2 \alpha}{4\varepsilon_0}$ in the negative z-direction. The electric field due to a hemispherical surface charge is then obtained from substituting $\alpha = \frac{\pi}{2}$ into the above expression.

$$E_{hemi} = \frac{\sigma}{4\varepsilon_0}.$$

To determine the electric field at the center due to a spherical wedge, notice that a hemisphere can be formed by congealing two spherical wedges of half-angles α and $\frac{\pi}{2} - \alpha$, as shown in Fig. 5.19.

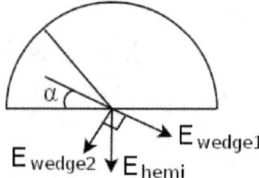

Figure 5.19:   Electric field due to hemisphere composed of two wedges

The electric fields produced by each spherical wedge must be along its line of symmetry. Therefore, the individual electric fields of the two wedges at the center must be mutually perpendicular. Since, their vector sum yields the electric field due to a hemisphere, the electric field strength at the center due to the wedge of half-angle α must be

$$E_{wedge1} = E_{hemi} \sin \alpha = \frac{\sigma \sin \alpha}{4\varepsilon_0}$$

and is directed along the symmetrical axis of the wedge, towards the center.

21. Equilateral Triangle**

(a) Applying the result that we have previously derived for a finite line charge, the electric field due to a line charge of length L and charge density λ along a symmetry axis perpendicular to the line is given by

$$\frac{\lambda L}{4\pi\varepsilon_0 r \sqrt{\frac{L^2}{4} + r^2}},$$

where r is the perpendicular distance between the point of concern on the symmetry axis and the center of the line. Returning to the original problem, suppose that the 2λ line charge is oriented along the y-axis (centered about the origin) while the other two rods are in the positive x-region. Since this set-up is symmetric about the x-axis, the net electric field at the centroid is solely along the x-direction. Noting that the perpendicular distance between an edge of the triangle and its centroid is $\frac{\sqrt{3}L}{6}$, the electric field at the centroid is

$$E = \frac{2\lambda L}{4\pi\varepsilon_0 \cdot \frac{\sqrt{3}L}{6}\sqrt{\frac{L^2}{4} + \frac{L^2}{12}}} - 2 \cdot \frac{\lambda L}{4\pi\varepsilon_0 \cdot \frac{\sqrt{3}L}{6}\sqrt{\frac{L^2}{4} + \frac{L^2}{12}}} \cos 60° = \frac{3\lambda}{2\pi\varepsilon_0 L}.$$

(b) Performing the negative line integral of the electric field along the symmetry axis perpendicular to the line charge from infinity to r, the electric potential at a point on the symmetry axis a perpendicular distance r from the line charge is

$$V = -\int_\infty^r \frac{\lambda L}{4\pi\varepsilon_0 r\sqrt{\frac{L^2}{4} + r^2}} dr.$$

Adopting the substitution $r = \frac{L}{2}\tan\theta$ and $dr = \frac{L}{2}\sec^2\theta d\theta$,

$$V = -\int_{\frac{\pi}{2}}^{\tan^{-1}\frac{2r}{L}} \frac{\lambda L}{4\pi\varepsilon_0 \cdot \frac{L}{2}\tan\theta \cdot \frac{L}{2}\sec\theta} \cdot \frac{L}{2}\sec^2\theta d\theta$$

$$= -\int_{\frac{\pi}{2}}^{\tan^{-1}\frac{2r}{L}} \frac{\lambda}{2\pi\varepsilon_0 \sin\theta} d\theta$$

$$= \left[\frac{\lambda}{2\pi\varepsilon_0}\ln|\csc\theta + \cot\theta|\right]_{\frac{\pi}{2}}^{\tan^{-1}\frac{2r}{L}}$$

$$= \frac{\lambda}{2\pi\varepsilon_0}\ln\left(\sqrt{1 + \frac{L^2}{4r^2}} + \frac{L}{2r}\right).$$

Therefore, the net potential at the centroid due to the three line charges is

$$V = 4 \cdot \frac{\lambda}{2\pi\varepsilon_0}\ln\left(\sqrt{1 + \frac{L^2}{4 \cdot \frac{L^2}{12}}} + \frac{L}{2 \cdot \frac{\sqrt{3}L}{6}}\right) = \frac{2\lambda}{\pi\varepsilon_0}\ln(2 + \sqrt{3}).$$

(c) Recall from Problem 10 of this chapter that the electric field at an arbitrary point P due to a finite line charge with uniform linear charge density λ bisects the angle formed by joining the ends of the line charge to P. It is of magnitude

$$E = \frac{\lambda}{2\pi\varepsilon_0 h}\sin\theta,$$

where θ is half of the angle formed by joining the ends of the line charge to P and h is the perpendicular distance between P and the line charge. It is natural to check if the net electric field is zero at a point on the symmetry axis of the charge distribution.

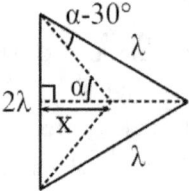

Figure 5.20:   Point corresponding to angle α

Referring to Fig. 5.20, for the electric field at a point corresponding to angle α to be zero,

$$\frac{2\lambda}{2\pi\varepsilon_0 \cdot \frac{L\cos\alpha}{2\sin\alpha}}\sin\alpha = 2 \cdot \frac{\lambda}{2\pi\varepsilon_0 \cdot \frac{L\sin(\alpha-30°)}{2\sin\alpha}}\sin\frac{\alpha}{2}\cos\frac{\alpha}{2}$$

$$\implies \frac{\sin^2\alpha}{\cos\alpha} = \frac{\sin^2\alpha}{2\sin(\alpha-30°)},$$

$$\cos\alpha = 2\sin(\alpha - 30°) = \sqrt{3}\sin\alpha - \cos\alpha$$

$$\implies \cot\alpha = \frac{\sqrt{3}}{2}.$$

Observe that the perpendicular distance between the 2λ line charge and this point is

$$x = \frac{L}{2}\cot\alpha = \frac{\sqrt{3}L}{4},$$

which is exactly halfway along the perpendicular bisector of the 2λ line charge. Another perspective is as follows: suppose we center our origin at the midpoint of the perpendicular bisector of the 2λ line charge.

Observe that the perpendicular distances to the 2λ and λ line charges are $x = \frac{\sqrt{3}L}{4}$ and $\frac{x}{2} = \frac{\sqrt{3}L}{8}$ in Fig. 5.21. Recall that in our solution to Problem 5, we showed that a segment along a line charge with uniform linear charge

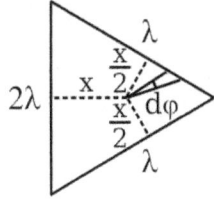

Figure 5.21: Segment with angular width $d\phi$

density λ that spans an angular width $d\phi$ produces an electric field at the origin of magnitude

$$dE = \frac{\lambda}{4\pi\varepsilon_0 h} d\phi$$

where h is the perpendicular distance between the line charge and the origin. Most notably, this value is independent of the angular coordinate ϕ of the infinitesimal segment. Therefore, the electric field strengths of all infinitesimal segments with angular width $d\phi$ on the three line charges at the origin are identical (for the 2λ line, the doubling of charge density is perfectly canceled by the doubling of the perpendicular distance from the origin). As the angular coordinates of the infinitesimal segments on the three line charges span all 2π radians, the net electric field at the origin is the superposition of a vector of constant magnitude rotated over 2π radians — resulting in zero net electric field at the origin. Incidentally, this analysis also paves a way to extend our analysis for the general case where one side of the triangle is now k times the linear charge densities of the other sides ($k > 0$ so the signs of the charges are identical). We simply have to find the point whose perpendicular distance x to this particular side is k times the perpendicular distance to the other two sides.

$$\frac{x}{\left(\frac{\sqrt{3}L}{2} - x\right) \cdot \sin 30^\circ} = k$$

$$x = \frac{\sqrt{3}k}{2(k+2)}L.$$

Finally, one can even apply this procedure to a general triangle formed with sides of equal linear charge densities. The point with zero net electric field is in fact the incenter of the triangle (the center of an inscribed circle) as the perpendicular distances between the incenter and the three sides of the triangle are identical (since they are equal to the radius of the inscribed circle).

22. Potential at Rim of Disk**

We will determine the potential at the top of the disk as shown in Fig. 5.22, by adopting polar coordinates about this point (defined as the origin).

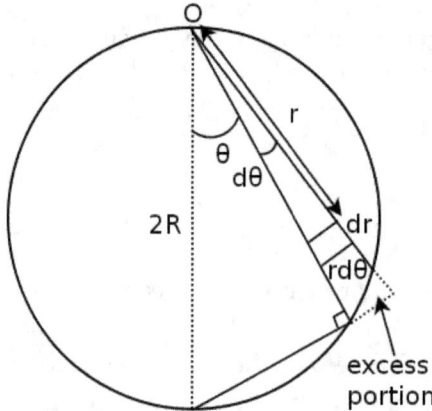

Figure 5.22:   Polar coordinates about point on rim

Define θ as the anti-clockwise angle between the position vector of a point on the disk and the vertical and r as the radial distance from the origin. Observe that since a triangle with a diameter as its hypotenuse inscribed in a circle has a right-angle, r ranges from $r = 0$ to $r = 2R\cos\theta$ across the disk for a given θ. An infinitesimal rectangle of sides $rd\theta$ and dr at (r, θ) contributes a potential $\frac{\sigma r d\theta dr}{4\pi\varepsilon_0 r} = \frac{\sigma}{4\pi\varepsilon_0}dr d\theta$ at the origin. Integrating over the entire disk, the potential at the origin is

$$V = \int_{-\frac{\pi}{2}}^{\frac{\pi}{2}} \int_0^{2R\cos\theta} \frac{\sigma}{4\pi\varepsilon_0} dr d\theta$$

$$= \int_{-\frac{\pi}{2}}^{\frac{\pi}{2}} \frac{\sigma R}{2\pi\varepsilon_0} \cos\theta d\theta$$

$$= \frac{\sigma R}{\pi\varepsilon_0}.$$

As a word of precaution, we cannot adopt limits of integration, similar to those used in this problem, for any general distribution. This is because the limits of integration may not be exact. For example, in this problem, there is an excess portion when considering an infinitesimal rectangle near the edge of the disk (shown in the figure above), as the perpendicular direction to a line joining the origin to an edge is not aligned with the tangent of the circle there. This did not invalidate our answer as such deviations contribute to negligible potential at the origin — however, this is not always true. For example, suppose that we applied a similar method to compute the potential at an arbitrary point P within a charged ring (one-dimensional distribution). The above method would suggest that each segment of the ring, between angular coordinates θ and $\theta + d\theta$ about the origin at P, contributes to the same amount to the potential at P as the length, and hence charge, of a segment is proportional to its radial distance from P such that the charge over radial distance is uniform across all segments. This ludicrous answer stems from the fact that the inaccuracy in representing a ring in polar coordinates about P is no longer negligible (because the charge of an infinitesimal segment is proportional to a single infinitesimal length as opposed to the product of two infinitesimal lengths in a disk).

Moving on, to determine the electric potential energy stored in the disk, we can compute the external work done in assembling the disk by bringing in progressively larger circular shells of radius r and thickness dr from infinity, beginning from $r = 0$ to $r = R$. Since the potential at the rim of a disk of radius r is $\frac{\sigma r}{\pi \varepsilon_0}$, the external work done in bringing a shell of radius r and thickness dr (the next layer) to an already assembled disk of radius r is

$$dW = \frac{\sigma r}{\pi \varepsilon_0} \cdot 2\pi r dr = \frac{2\sigma r^2}{\varepsilon_0} dr.$$

The total external work done in assembling the disk this way, which is the stored potential energy, is then

$$W = \int_0^R \frac{2\sigma r^2}{\varepsilon_0} dr = \frac{2\sigma R^3}{3\varepsilon_0}.$$

23. Potential at Vertex of Square**

Adopt polar coordinates (similar to the previous problem) about the tip of the triangle of concern, as shown in Fig. 5.23.

Let θ denote the anti-clockwise angular coordinate of a point on the triangle from the vertical and r denote its radial distance from the origin. For a given θ, r ranges from $r = 0$ to $r = \frac{h}{\cos \theta}$ along the triangle. Since the contribution to the potential at the origin (the tip of concern) due to an infinitesimal rectangle of sides $rd\theta$ and dr at (r, θ) is $\frac{\sigma dr d\theta}{4\pi\varepsilon_0}$, the total

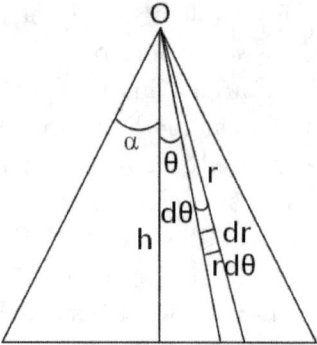

Figure 5.23:   Polar coordinates about tip

potential at the tip is

$$V = \int_{-\alpha}^{\alpha} \int_{0}^{\frac{h}{\cos\theta}} \frac{\sigma}{4\pi\varepsilon_0} dr\, d\theta$$

$$= \int_{-\alpha}^{\alpha} \frac{\sigma h}{4\pi\varepsilon_0 \cos\theta} d\theta$$

$$= \frac{\sigma h}{2\pi\varepsilon_0} \ln(\sec\alpha + \tan\alpha).$$

With regard to the second part of the problem, we first compute the potential at the center of a square of edge length l and uniform surface charge density σ. Since the square is composed of four isosceles triangles with height $\frac{l}{2}$ and $\alpha = \frac{\pi}{4}$, the potential at its center is given by the principle of superposition as

$$V_{center} = 4 \cdot \frac{\sigma l}{4\pi\varepsilon_0} \ln(\sqrt{2}+1) = \frac{\sigma l}{\pi\varepsilon_0} \ln(\sqrt{2}+1).$$

To determine the potential at a corner, we can use scaling arguments. Firstly, the potential at the center of a square of edge length $2l$ and the same charge density σ must be twice that of a square of edge length l since its charge is larger by a factor of four while its length dimension is increased by a factor of two (the potential is proportional to charge over a length dimension). Next, the potential at the center of a square of edge length $2l$ is composed of the potentials of four squares of edge length l at a corner. Therefore, the potential at the corner of a square of length l must be half of that at its center.

$$V_{corner} = \frac{V_{center}}{2} = \frac{\sigma l}{2\pi\varepsilon_0} \ln(\sqrt{2}+1).$$

24. Force on Hemispheres**

Firstly, consider the special case where the sphere has a uniform charge density ρ_1. The electric field at a radius r from the center of the sphere, due to the entire sphere of charge, is

$$E = \frac{\rho_1 r}{3\varepsilon_0}$$

by Gauss' law. Though this electric field is due to both hemispheres, we can simply integrate this electric field over the charge of one hemisphere since a hemisphere evidently cannot exert a net force on itself. Define the z-axis (in spherical coordinates) to pass through the vertex of the hemisphere that we are considering. The magnitude of the force on an infinitesimal volume element $dV = r^2 \sin\theta d\phi d\theta dr$ at coordinates (r, ϕ, θ) in spherical coordinates about the center of the sphere is

$$dq \cdot E = \rho_1 \cdot \frac{\rho_1 r}{3\varepsilon_0} \cdot r^2 \sin\theta d\phi d\theta dr = \frac{\rho_1^2 r^3 \sin\theta}{3\varepsilon_0} d\phi d\theta dr.$$

The net force on this hemisphere should only be in the z-direction. Therefore, we can simply integrate the z-component of the force on each infinitesimal volume element. Multiplying the previous expression by $\cos\theta$ (to retrieve the z-component) and integrating over the entire hemisphere,

$$F = \int_0^R \int_0^{\frac{\pi}{2}} \int_0^{2\pi} \frac{\rho_1^2 r^3 \sin\theta \cos\theta}{3\varepsilon_0} d\phi d\theta dr$$

$$= \int_0^R \int_0^{\frac{\pi}{2}} \frac{2\pi \rho_1^2 r^3 \sin\theta \cos\theta}{3\varepsilon_0} d\theta dr$$

$$= \int_0^R \int_0^{\frac{\pi}{2}} \frac{\pi \rho_1^2 r^3 \sin 2\theta}{3\varepsilon_0} d\theta dr$$

$$= \int_0^R \left[-\frac{\pi \rho_1^2 r^3}{6\varepsilon_0} \cos 2\theta \right]_0^{\frac{\pi}{2}} dr$$

$$= \int_0^R \frac{\pi \rho_1^2 r^3}{3\varepsilon_0} dr$$

$$= \frac{\pi \rho_1^2 R^4}{12\varepsilon_0}.$$

We claim that in the original situation where the southern hemisphere has charge density ρ_2 instead, the force is simply scaled by a factor of $\frac{\rho_2}{\rho_1}$. The net force between the northern and southern hemispheres is the sum of the forces between a charge on the northern hemisphere and a charge on the southern hemisphere, performed over all possible pairs. Therefore, if we scale each charge on the southern hemisphere shell by a factor of $\frac{\rho_2}{\rho_1}$ while preserving all relative distances in the system, the force between the hemispheres should be scaled by a factor of $\frac{\rho_2}{\rho_1}$ since the force between two charges is proportional to the product of their charges. Therefore, the force between the two hemispheres given in the problem is

$$F = \frac{\pi \rho_1 \rho_2 R^4}{12\varepsilon_0}.$$

25. Two Hemispherical Shells***

Firstly, observe that there are actually two possible configurations where the common center and the apexes of the shells are collinear. Let the forces between the shells when they lie on the same and different sides of the common center be F_1 and F_2 respectively. Now, consider the special case where $q = Q$ and duplicate the set-up to form two complete spherical shells depicted in Fig. 5.24.

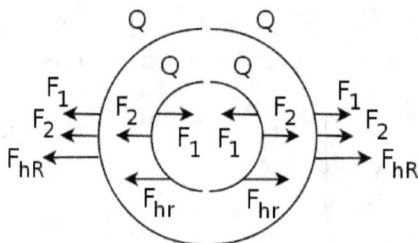

Figure 5.24:   Forces on hemispherical shells

If we let F_{hR} and F_{hr} be the forces between the hemispherical shells of radii R and r respectively, the forces on each hemispherical shell are labeled above. All forces must be directed along the horizontal direction due to the symmetry of the shells. Firstly, observe that since the outer spherical shell produces no net electric field inside itself, the net force that one inner hemispherical shell experiences must be only due to the other inner hemispherical shell and is hence F_{hr}. From this, we directly obtain

$$F_1 = F_2,$$

which shows that there was in fact no ambiguity in our question (even though we are considering the special case for now, the above two forces will be scaled by the same amount when the inner hemispherical shell has charge q instead). Next, the discrepancy between F_{hR} and the net force experienced by an outer hemispherical shell must be due to the electric field produced by the inner spherical shell, which has a value of $\frac{Q}{2\pi\varepsilon_0 R^2}$ at the surface of the outer spherical shell. This electric field produces an electrostatic pressure of $\frac{Q}{2\pi\varepsilon_0 R^2} \cdot \frac{Q}{2\pi R^2} = \frac{Q^2}{4\pi^2\varepsilon_0 R^4}$ on an outer hemispherical shell, where $\frac{Q}{2\pi R^2}$ is its uniform surface charge density. This pressure results in a horizontal force of $\frac{Q^2}{4\pi^2\varepsilon_0 R^4} \cdot \pi R^2 = \frac{Q^2}{4\pi\varepsilon_0 R^2}$ on an outer hemispherical shell. Thus,

$$F_1 + F_2 = \frac{Q^2}{4\pi\varepsilon_0 R^2}.$$

Combining the two equations,

$$F_1 = F_2 = \frac{Q^2}{8\pi\varepsilon_0 R^2}.$$

To revert to the original situation where the inner hemispherical shell carries charge q, simply observe that the force between two charges is proportional to the product of the two charges. The net force between the outer and inner hemispherical shells is the sum of the forces between a charge on the outer shell and a charge on the inner shell, performed over all possible pairs. Therefore, if we scale each charge on the inner shell by a factor of $\frac{q}{Q}$ while preserving all relative distances in the system, the force between the hemispherical shells should be scaled by a factor of $\frac{q}{Q}$ as well. Therefore, returning to the configuration given in the problem,

$$F_1 = F_2 = \frac{Qq}{8\pi\varepsilon_0 R^2}.$$

This result is surprising in that not only are the forces in the two possible configurations equal, their magnitudes are independent of the radius r of the inner hemispherical shell!

Chapter 6

Conductors and Dielectrics

The previous chapter analyzed configurations of charges in vacuum. Proceeding with our study of electrostatics, this chapter will explore how the laws of electrostatics can be applied to conducting and dielectric media.

6.1 Properties of Conductors

Conductors are generally a difficult class of problems to study. Charge carriers, such as ions and electrons, are free to move within a conductor. Consequently, when a conductor is placed in the presence of an external electric field, the charge carriers will redistribute themselves. The redistribution of these charge carriers in the conductor will then produce their own electric fields which then, again, affect the positions of the charge carriers. Ultimately, we will need to know the positions of the charge carriers to determine the net electric field everywhere, but we also need to know the net electric field in order to identify the positions of the mobile charges within the conductor — rendering this problem seemingly intractable.

Important Properties

Thankfully, we have the following properties of conductors which slightly simplify such problems. When the charges of a conductor, that are under the sole influence of electrostatic forces, have attained equilibrium,

(1) The net electric field $E = 0$ everywhere inside a conductor. Note that the net electric field E comprises both the external field and the field due to the conductor.
(2) The net charge density, ρ, inside a conductor is zero. $\rho = 0$. Thus, any net charge on a conductor resides on its surface.
(3) At any point immediately outside the conductor, the electric field E is perpendicular to the surface of the conductor there.

(4) The surface of a conductor is an equipotential surface. Since $E = 0$ inside the conductor by property 1, the electric potentials of the surface and of regions inside the conductor are the same as the line integral of the electric field between any two points of the conductor is zero. Then, the potential is uniform over the entirety of the conductor and is equal to some value ϕ (we shall use ϕ to denote the potential of a conductor instead of V, which will be used more generally).

Even when there are external charges, the charge carriers in the conductor will always redistribute themselves until the four conditions above are satisfied. Of course, this is assuming that the charge carriers do not experience any other non-electrostatic force, such as gravity (which is negligible). In this chapter, conductors are assumed to be ideal such that their charges instantaneously redistribute themselves to attain equilibrium. Furthermore, it is usually implicitly assumed that an electrostatic situation is achievable such that charges of conductors are perpetually in equilibrium.

Proof:

(1) If $E \neq 0$ inside a conductor, charge carriers will continue to move and thus contradict the premise that a static situation has been reached. This occurs even if there is already no net charge density everywhere in the conductor as the positive and negative charges in a conductor will diverge in different directions. Thus, the net electric field $E = 0$ inside a conductor.

(2) Drawing an infinitesimal Gaussian surface around a point inside the conductor, the electric flux through this surface is zero as $E = 0$ inside a conductor. Thus, the charge density $\rho = 0$ at that point by Gauss' law. Repeating this procedure for all points within a conductor, $\rho = 0$ at all points inside a conductor (but not necessarily at the surface as a portion of the Gaussian surface would lie outside the conductor).

(3) Similar to (1), if a tangential component of electric field exists immediately outside the surface of a conductor, surface charges will begin to move tangentially, in contravention of the electrostatic assumption. Therefore, the electric field immediately beyond the surface of a conductor can only be perpendicular to it. Note that a perpendicular component of electric field is permissible since the conductor is usually assumed to be surrounded by a perfectly insulating material (e.g. vacuum) such that no charge flows normally to it.

(4) Since the tangential component of the electric field is zero at the surface of a conductor, the surface of a conductor must be an equipotential

surface as a line integral of the electric field in the tangential direction yields zero. Let the electric potential of the surface be ϕ_0. Then, the electric potential of the entire region inside the conductor is also ϕ_0 as $\boldsymbol{E} = 0$ inside a conductor.

Problem: Two conducting spheres of radii r and $3r$ that are infinitely far apart possess charges q and $2q$ initially. Their surfaces are then connected by a conducting wire. When the system has reached an equilibrium, find the final charge on each sphere.

Since the two conducting spheres are far away such that they are essentially individual isolated systems, the charges on each sphere will be uniformly distributed over its surface (recall that there must be no net charge anywhere inside a conductor by property 2). Since a uniform shell of charge is identical to a point charge, commensurate with the total charge carried by the shell, placed at its center for regions outside of the shell, the potential at the surface of a uniform shell of radius R and total charge Q is equal to the potential of a point located at a distance R from a point charge Q.

$$\phi = \frac{Q}{4\pi\varepsilon_0 R}.$$

When the system eventually reaches its final equilibrium state, the electric potential at the surfaces of both conducting spheres must be the same by property 4 as the wire is a conductor. If we let the final charge on the sphere of radius r be x, the final charge on the other sphere is $3q - x$ by the conservation of charge. Equating the two electric potentials,

$$\frac{x}{4\pi\varepsilon_0 r} = \frac{3q - x}{4\pi\varepsilon_0 \cdot 3r}$$

$$x = \frac{3}{4}q$$

$$3q - x = \frac{9}{4}q.$$

Surface Charge and Pressure on a Conductor

Suppose that we know the electric field strength E directly outside an infinitesimal surface of a conductor at equilibrium ($\boldsymbol{E}$ must only have a normal component). Then, applying Eq. (5.11) which governs the change in electric field across an interface, yields the surface density σ of the infinitesimal surface.

$$\sigma = \varepsilon_0 E \qquad (6.1)$$

since the electric field strength inside a conductor at equilibrium is zero. The electrostatic pressure on the surface of a conductor is then given by Eq. (5.12) as

$$P = \frac{\sigma(E+0)}{2} = \frac{\sigma^2}{2\varepsilon_0} = \frac{1}{2}\varepsilon_0 E^2. \tag{6.2}$$

The last expression bears a stark resemblance to the energy density associated with an electrostatic field, but this is merely a natural consequence. Suppose that an infinitesimal surface dA, at which the surface electric field was originally E, was pushed outwards by the electrostatic pressure for a distance dx such that the volume that the conductor occupies increases by dV. The work done by the conductor is PdV and must be equal to the decrease in potential energy of the conductor. The latter is just $\frac{1}{2}\varepsilon_0 E^2 dV$ as the electric field E, which was originally present in the volume dV, vanishes after the conductor expands (remember that the electric field must be zero within a conductor). Equating the two, $P = \frac{1}{2}\varepsilon_0 E^2$.

6.2   The Uniqueness Theorems

It is beneficial to ponder the clues that one requires to uniquely define a potential function or electric field in a region of space. The rationale behind this is that if we are able to identify such conditions, we can guess a potential function or electric field that satisfies them and conclude that we have the right answer (since the answer is unique), instead of tackling the seemingly intractable problems with conductors head-on. In other words, we are looking for an avenue that enables us to solve a problem through trial-and-error.

A uniqueness theorem specifies the required constraints that guarantee a unique solution to the potential function or electric field in a volume, given that the electric field is consistent with the charge distribution inside the region of interest (i.e. Gauss' law holds for the proposed potential function or electric field). The pivotal components, that our ensuing development of two uniqueness theorems is predicated upon, are the following two theorems.

Theorem: In a region of space free of net charges, the electric potential has no local maximum or minimum. Then, maxima and minima can only exist at the boundaries of this region.

This theorem is basically another way of expressing Earnshaw's Theorem. If a local maximum or minimum exists respectively, a negative or positive test charge placed there will lie in a state of stable equilibrium (as the electric field is the negative potential gradient). Therefore, Earnshaw's Theorem forbids the existence of local maxima and minima inside a region empty of charges. The converse is also easy to prove as the lack of local maxima or minima precludes the existence of a stable equilibrium. This theorem is therefore equivalent to Earnshaw's Theorem.

The second important theorem is known as Green's reciprocity theorem.

Theorem: Let the volume charge densities and potential functions of two arbitrary, separate charge distributions (that do not even need to lie in the same region and may come from two completely different problems) be $\rho_1(x, y, z)$, $\rho_2(x, y, z)$ and $V_1(x, y, z)$, $V_2(x, y, z)$, under the same coordinate system. Then,

$$\iiint \rho_1 V_2 d\Omega_1 = \iiint \rho_2 V_1 d\Omega_2 \qquad (6.3)$$

where $d\Omega_1$ and $d\Omega_2$ represent infinitesimal volume elements in the first and second set-ups respectively. The integrals are performed over the entirety of the charge distributions of the first and second configurations.

Proof: The proof of this seemingly complex theorem is remarkably simple! Suppose that we place the two charge distributions in the same region. Then, $\iiint \rho_1 V_2 d\Omega_1$ is the total electric potential energy of the charges in the first set-up due to those in the second set-up, when the charges in the second set-up are fixed. Note that $\iiint \rho_1 V_2 d\Omega_1$ does not include the potential energy due to the internal interactions between the constituents of the first set-up. Similarly, $\iiint \rho_2 V_1 d\Omega_2$ is the total electric potential energy of the charges in the second set-up due to those in the first set-up, when the charges in the first set-up are fixed. These two expressions must obviously be equal as the potential energy of a charge q_i located at a distance r_{ij} away from another fixed charge q_j is $\frac{q_i q_j}{4\pi\varepsilon_0 r_{ij}}$ — this is identical to the potential energy of q_j when q_i is fixed as well, since the expression is symmetric in q_i and q_j.

Below is a trivial application of Green's reciprocity theorem that yields an astonishing result!

Problem: Suppose we have two arbitrary conductors that are initially neutral, placed at arbitrary locations. If we put q amount of charge on conductor 1, the potential of conductor 2 is ϕ_{21}. If we instead put q amount of charge on conductor 2, the potential of conductor 1 is ϕ_{12}. Show that $\phi_{12} = \phi_{21}$.

Let the set-ups 1 and 2, that we will apply Green's reciprocity theorem to, be those when q amount of charge resides on conductors 1 and 2 respectively (while the other neutral conductor is also present). Since the potential of a conductor is uniform throughout it, applying Green's reciprocity theorem to these set-ups implies that

$$q\phi_{12} = q\phi_{21}$$

as $\iiint \rho_1 V_2 d\Omega_1$ produces $q\phi_{12}$ over the volume of conductor 1 in set-up 1 and zero over the volume of conductor 2 in set-up 1 due to its neutrality. A similar logic holds for set-up 2. Thus,

$$\phi_{12} = \phi_{21}.$$

This result is truly astounding as we have not specified anything about the shapes or locations of the conductors! Returning to our main topic, we have the following two uniqueness theorems.

6.2.1   *First Uniqueness Theorem*

Theorem: Specifying the potential at the boundaries of a volume Ω and the charge density $\rho(x, y, z)$ inside Ω uniquely determines the electric field $E(x, y, z)$ **and** potential $V(x, y, z)$ inside Ω. Certain boundaries can also be taken to be at infinity where the potential is 0 by definition.

Proof: To start off, we identify the imposed constraints on the potential function $V(x, y, z)$ (we only consider the potential function for now as the electric field can be later computed as $E = -\nabla V$) in the volume Ω. Let the imposed potential at the boundaries of Ω be $V_b(x, y, z)$. Firstly, the electric field associated with $V(x, y, z)$ must be consistent with $\rho(x, y, z)$ (i.e. Gauss' law is valid for all possible Gaussian surfaces within Ω). Next, $V(x, y, z)$ must correspond to the potential $V_b(x, y, z)$ at the boundaries of Ω. Now, suppose that $V(x, y, z)$ is not unique under such circumstances such that another function $V'(x, y, z)$ also satisfies the above constraints.

   Observe that $-V'(x, y, z)$ must be a valid solution[1] in the volume Ω with charge density $-\rho(x, y, z)$ and boundary potential $-V_b(x, y, z)$. Superposing the set-ups involving $V(x, y, z)$ and $-V'(x, y, z)$, $V(x, y, z) - V'(x, y, z)$ must be a valid solution in the volume Ω with zero charge density everywhere and a uniform zero potential at its boundaries. Since the boundaries of the volume Ω have zero potential and since the only maxima and minima can be located at the boundaries of Ω by the previous theorem (there is now no net charge inside Ω after the superposition), the potential inside Ω must be

---

[1]We can only claim that it is **a** solution but not **the** solution.

zero everywhere. Therefore,

$$V(x, y, z) - V'(x, y, z) = 0$$
$$V(x, y, z) = V'(x, y, z)$$

which shows that the solution $V(x, y, z)$, and thus the electric field $-\nabla V$ associated with it, is unique.

Corollary: In a region, devoid of net charges, whose boundaries have a uniform potential V_0, the electric field is zero everywhere inside the region.

The potential function $V = V_0$ everywhere within the region satisfies the boundary condition and is consistent with Gauss' law (as there is no net charge everywhere). Therefore, this must be the correct solution as guaranteed by the first uniqueness theorem. This uniform potential function then implies that the electric field is zero everywhere in the region.

6.2.2   *Second Uniqueness Theorem*

Theorem: In a volume Ω surrounded by conductors, specifying the total charges (and not the distribution of charges) on the conductors, and the charge density $\rho(x, y, z)$ inside Ω uniquely determines the electric field $E(x, y, z)$ (and not necessarily the potential) inside Ω. The outer boundary of Ω can be taken to be infinity (where the total charge is zero) if Ω is not enclosed by a conductor. That said, even though the potentials of the conductors are not specified, the conductors must still be individually equipotential.

Proof: Suppose that we have a solution for the electric field $E(x, y, z)$. Let there be N boundary conductors and let the total charge on the ith conductor be specified as Q_i. The imposed conditions are the facts that $E(x, y, z)$ must satisfy Gauss' law applied to any Gaussian surface surrounding the charge distribution $\rho(x, y, z)$ in Ω, produce the correct total charge on each conductor Q_i (when Gauss' law is applied to the surface of each conductor) and be normal to the surfaces of the conductors (by property 3 of conductors). Due to the last requirement, the potential of the ith conductor, associated with the solution $E(x, y, z)$, must be some ϕ_i uniform across the entire conductor.

Now, suppose that another electric field $E'(x, y, z)$ satisfies the afore constraints and produces a uniform potential ϕ_i' on the ith conductor. Note that ϕ_i' is not necessarily equal to ϕ_i as the imposed condition only requires the conductors to be separately equipotential (the potential on each conductor is not specified such that any uniform potential would do). Then,

$-\boldsymbol{E}'(x, y, z)$ would be a valid solution for the electric field inside Ω with charge density $-\rho(x, y, z)$, when the ith conductor possesses total charge $-Q_i$. The field $-\boldsymbol{E}'(x, y, z)$ also causes the ith conductor to have potential $-\phi_i'$. Superposing this with the original set-up, $\boldsymbol{E}(x, y, z) - \boldsymbol{E}'(x, y, z)$ must be a valid solution for the electric field inside Ω with zero net charge density everywhere and when the conductors each carry zero total charge. Note that each conductor may not be neutral at all points on its surface as $\sigma = \varepsilon_0(\boldsymbol{E} - \boldsymbol{E}')$ and the whole point of our exercise is to accommodate the possibility of $\boldsymbol{E}(x, y, z) \neq \boldsymbol{E}'(x, y, z)$. However, each conductor must still be neutral overall as $\boldsymbol{E}(x, y, z)$ and $\boldsymbol{E}'(x, y, z)$ produce the same total charge on each conductor by proposition. Due to this superposition, the ith conductor now has a uniform potential $\phi_i - \phi_i'$.

We shall now prove that $\phi_i - \phi_i' = 0$ for all conductors by applying Green's reciprocity theorem to two wisely-concocted set-ups. Set-up 1 shall be the superposed set-up highlighted in the previous paragraph. As for set-up 2, we choose the configuration of the same N conductors when the jth conductor has a certain non-zero total charge q_j'' and when all other conductors have zero total charge. Applying Green's reciprocity theorem to these set-ups, the left-hand side yields

$$\iiint \rho_1 V_2 d\Omega_1 = 0.$$

The reason behind this is that when performing the above integral over the ith conductor, its contribution to the final result is simply its total charge in set-up 1 multiplied by its potential in set-up 2 since the potential of a conductor must be uniform (such that the integral becomes trivial). Since the total charges of all conductors are zero in set-up 1, the above integral must yield zero. Now, we proceed with the right-hand side of Green's reciprocity theorem.

$$\iiint \rho_2 V_1 d\Omega_2 = q_j''(\phi_j - \phi_j').$$

The only contribution to the above integral is the jth conductor as all other conductors have zero total charge in set-up 2. Since the jth conductor has total charge q_j'' in set-up 2 and potential $\phi_j - \phi_j'$ in set-up 1, its contribution is $q_j''(\phi_j - \phi_j')$. Equating the two sides of Green's reciprocity theorem, we must have

$$\phi_j - \phi_j' = 0,$$

as $q_j'' \neq 0$. Repeating this argument for all conductors, we must have $\phi_i - \phi_i' = 0$ for all $1 \leq i \leq N$. This result dictates that for the super-posed set-up, the boundary of Ω has zero potential while Ω contains no net charge everywhere. Applying the previous corollary established from the first uniqueness theorem (or the fact that the potential maxima and minima can only occur at the boundaries of a region free of net charges), the electric field in the superposed set-up must be zero everywhere! That is,

$$\boldsymbol{E}(x, y, z) = \boldsymbol{E}'(x, y, z).$$

Now, in our above proof, we implicitly adopted the convention that the zero reference point for the potential function is set at infinity (this was assumed in the derivation of Green's reciprocity theorem). However, the above result $\boldsymbol{E}(x, y, z) = \boldsymbol{E}'(x, y, z)$ should be valid even when the zero reference point is defined at some other location, such that the potential at infinity becomes some constant V_0, as adding a constant value V_0 to the potential everywhere will not change the potential gradient and hence the electric field. Therefore, in the general case of an arbitrary zero reference point, it is the electric field in Ω that is uniquely defined and not the potential function. However, once a zero reference point has been decided, the potential function will be uniquely defined too.

An Example Application of the Uniqueness Theorems

To catch a glimpse of the potency of these uniqueness theorems, consider a system comprising a spherical conductor with an arbitrary cavity in it. The conductor is neutral and electrically isolated. Inside the cavity lies a fixed charge q which is not necessarily at the center of the sphere, O. We would like to find the electric field at a point P that is a radius r from the center of the sphere, outside the sphere.

q is assumed to be positive in Fig. 6.1 for illustration purposes, but it can be either positive or negative in the general case. Because of the charge q in the cavity, we would intuitively expect a net charge[2] of the opposite sign to be attracted to the inner surface of the conductor. We can prove that a total charge of $-q$ resides on the inner surface of the conductor by drawing a Gaussian surface as shown above in the dotted lines. The electric flux through it is zero as $\boldsymbol{E} = 0$ inside a conductor. Thus, the total charge

---

[2]We cannot say anything about the charge distribution on the inner surface (e.g. whether the surface charge density is negative everywhere) but we would anticipate that the total charge on the inner surface is of opposite sign.

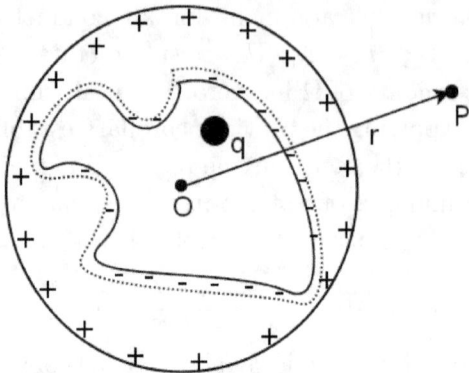

Figure 6.1:   Spherical conductor with cavity

enclosed by the Gaussian surface must be zero — implying that $-q$ net charge resides on the inner surface. As the conductor is neutral, there must be q amount of net charge deposited at the outer surface of the sphere by the conservation of charge.

We now claim that this q amount of net charge must be evenly distributed over the outer surface and shall prove this by construction. Suppose that we remove this q amount of net charge on the exterior surface and allow the system to equilibrate. In the new equilibrium state, the electric field must still be zero inside the conductor. $-q$ net charge resides on the inner surface while the outer surface is neutral overall. We further assert that the electric field is also zero everywhere outside the conductor. Well, this satisfies the boundary condition that the total charge on the outer surface of the conductor is zero and that the outer surface is equipotential. Therefore, this must be the correct electric field by the second uniqueness theorem! Applying Eq. (6.1) to the electric field at the outer surface, we conclude that not only must the outer surface be neutral overall, all points on the outer surface must be neutral!

Following from this, we can now slowly return the q amount of charge that we had extracted, in small amounts at a time. Since these charges experience no net electric field, due to the charge q in the cavity and the $-q$ net charge residing on the inner surface, they will distribute themselves in a spherically symmetric manner. Another way of reaching this conclusion is to assert that this distribution (superposed on the previously neutral outer surface) satisfies the boundary condition that the outer surface is equipotential and has total charge q, after which the second uniqueness theorem and Eq. (6.1) can be applied again.

Since the charges on the inner surface and the cavity produce no net electric field outside of the cavity, the electric field outside the conductor is only due to the uniformly distributed charge q on the exterior surface of the conductor. This is effectively the electric field of a uniform spherical shell, centered about O and with charge q, which is

$$E = \frac{q}{4\pi\varepsilon_0 r^2}$$

at a radial distance r from O, by Gauss' law. Therefore, the electric field at point P due to the entire set-up is given by the above expression and is directed radially outwards from the center of the spherical conductor, O.

To recapitulate, notice that in this particular case, the $-q$ charge on the inner surface redistributes itself such that the net effect of the fields due to the induced charge $-q$ and the enclosed charge q (this does not include the field due to the charges on the outer surface) cancel everywhere beyond the inner surface! Furthermore, the distribution of the total charge q on the outer surface is independent of what is going on inside the conductor! Overall, the electric field due to this set-up is equivalent to the electric field due to a concrete conducting sphere of charge q, in the region beyond the cavity!

6.2.3 *Electrostatic Shielding*

The above results lead us to very general conclusions. In the most general configuration, a conductor can have an arbitrary number of cavities that each enclose an arbitrary charge distribution. The conductor may be electrically isolated (similar to the previous problem) or maintained at a constant potential relative to infinity. Moreover, certain fixed external charges can also be placed outside of the conductor. Even under such broad circumstances, we claim that for a cavity C in a conductor which encloses a certain amount of total charge q, the combined electric field generated by the induced charges[3] $-q$ on the surface of the cavity and the enclosed charges is zero in the region outside of the cavity. We can prove this by construction.

Electrically Isolated Conductor

For an electrically isolated conductor, suppose that we remove the induced and enclosed charges (of total charges $-q$ and q) in cavity C and patch up cavity C with conducting material while maintaining the charges at the surfaces of the rest of the conductor (there could be other cavities as well) and the fixed external charges. Then, the remaining charges will redistribute

---

[3]Its total charge must be $-q$ by Gauss' law applied to a Gaussian surface straddling the cavity.

themselves until the electric fields inside the original conductor and the region where cavity C was originally located are zero. However, the total charges induced on the other cavities and the exterior surface of the conductor must remain the same (by Gauss' law applied to a surface straddling each of the cavities, and the conservation of charge[4] for the exterior surface).

Now, consider an isolated, separate set-up consisting of a thin conducting shell of the shape of cavity C, with a net charge $-q$, that encloses the q amount of charge afore, with the same distribution. By Gauss' law, for the electric field within this shell to be zero, the inner surface must carry $-q$ amount of charge while the outer surface is neutral. Then, the electric field in the region outside this shell is also zero by the second uniqueness theorem as a null electric field satisfies the boundary conditions that the total charge on the outer surface is zero and that the outer surface is equipotential (this is essentially the same argument as the previous problem). Finally, the superposition of these two set-ups (after carving cavity C again) guarantees that the electric field in the conductor is zero everywhere in the original conductor such that all boundaries of the conductor are equipotential and carry the correct amount of charge. The second uniqueness theorem then guarantees that this superposition produces the correct electric field everywhere (inside cavity C, in the other cavities, inside the conductor and outside the conductor). Applying Eq. (6.1) to cavity C, the correct charge distribution induced on cavity C must be the charge distribution on the inner surface of the second set-up that we had considered! This is because the electric field in the interior of cavity C is only due to the second set-up, even after the superposition, as the first set-up produces no net electric field in cavity C (as we had patched it up).

Therefore, the enclosed and induced charges of an arbitrary cavity produce no net electric field outside the cavity (since the second set-up didn't). Furthermore, the distribution of induced charges on the surface of the cavity is independent of whatever is outside of the cavity — it is given by the second set-up which evidently does not consider charges beyond the cavity. In a certain sense, the exterior of the cavity is "shielded" from the interior while the interior is also "shielded" from the external surroundings.

Conductor Maintained at Constant Potential

For a conductor maintained at a constant potential ϕ_0 relative to infinity, suppose that we remove the induced and enclosed charges of cavity C while

---

[4]In applying the conservation of charge, note that the net charge of the conductor increases by q after we remove the induced charge $-q$ in this process.

maintaining the potential of the rest of the conductor (there could be other cavities as well) and retaining the fixed external charges. Then, the remaining charges in the conductor will redistribute themselves and possibly flow out of the conductor to infinity until the potential is ϕ_0 everywhere within the conductor. We do not patch cavity C as it could change[5] the total amount of charge stored in this case (as charges may now flow into the conductor). However, this does not change the fact that this first set-up produces zero net electric field inside cavity C as a consequence of the corollary established from the first uniqueness theorem (cavity C has a uniform potential ϕ_0 in this first set-up and encloses zero net charge everywhere as we had removed all of it).

Next, consider a separate set-up identical to the second set-up in the previous section. Since there is no electric field outside of the shell, the potential everywhere outside of the shell, due to the shell, is zero. Therefore, the superposition of the two mentioned set-ups produce potential ϕ_0 everywhere within the conductor, outside the cavity C. This satisfies the boundary condition that the potential is ϕ_0 over all surfaces of the conductor — implying that this superposition must yield the correct potential and electric field everywhere (inside cavity C, in the other cavities, inside the conductor and outside the conductor) by the first uniqueness theorem.

Similar to the previous scenario, the electric field within cavity C is only due to the second set-up — implying that the distribution of the induced charges on cavity C is also given by the second set-up (independent of the charges exterior to C). Consequently, the induced and enclosed charges of an arbitrary cavity must produce zero net electric field outside the cavity (like the second set-up).

Exterior Surface of the Conductor

With these facts in mind, what can we say about the distribution of charges on the exterior surface (the surface that is not a cavity) of both an electrically isolated conductor and one maintained at a constant potential relative to infinity?

For an electrically isolated conductor, suppose that the entire conductor possesses a total charge Q_{tot} and encloses Q amount of total charge, dispersed over an arbitrary number of cavities. A total of $-Q$ amount of charge will

---

[5]Intuitively, a conductor of a larger volume should be able to contain more charge for the same potential as it can spread the charges stored by a smaller conductor into the excess volume and hence reduce the overall potential — allowing for more charges to flow in.

be induced on the surfaces of the cavities which leaves $Q_{tot} + Q$ amount of charge for the exterior surface. As the electric field produced by the charges enclosed in the cavities and the induced charges is zero, the distribution of the $Q_{tot} + Q$ amount of charge on the exterior surface is independent of the positions of the charges inside the cavities and the shapes and sizes of the cavities. The distribution is equivalent to that due to a concrete conductor with no cavities that carries $Q_{tot} + Q$ amount of charge on its surface. Therefore, the distribution of charges on the exterior surface, compounded with the external charges, also leads to zero electric field inside the volume bounded by the exterior surface (including the cavities) by themselves. The exterior surface would think that the conductor is filled up while the $Q_{tot} + Q$ charges, that it carries, redistribute themselves. This redistribution occurs until these charges produce an electric field that cancels the electric field due to the external charges outside of the conductor, everywhere in the volume bounded by the exterior surface.

Similarly, for a conductor maintained at a constant potential ϕ_0, the amount and distribution of charges on the exterior surface are independent of the cavities and the charges enclosed by them, as the induced and enclosed charges of the cavities produce zero electric field outside them (and thus no contribution to potential). The charge distribution on the exterior surface is equivalent to that due to a concrete conductor with no cavities that is maintained at ϕ_0 in the same external environment. Therefore, the distribution of charges on the exterior surface, compounded with the external charges, also leads to zero electric field inside the volume bounded by the exterior surface (including the cavities) by themselves. The charges on the exterior surface will vary in amount and location until the conductor acquires a uniform potential ϕ_0, while taking into account the influence of the external charges.

Properties of Electrostatic Shielding

We summarize the important results above as follows:

(1) The total induced charges on the surface of a cavity is the negative of the charge that it encloses by Gauss' law. The total charge on the exterior surface of a conductor can then be determined by the conservation of charge if the conductor is electrically isolated.

(2) The enclosed and induced charges in a cavity produce no net electric field outside the cavity.

(3) The distribution of induced charges on the surface of the cavity is independent of the charges outside the cavity.

(4) The combination of the charges on the exterior surface of a conductor and the external charges produces zero net electric field within the region bounded by the exterior surface, including the cavities.

(5) The distribution of the exterior charges of a conductor is independent of the positions of the charges inside the cavities and the sizes and shapes of the cavities. It is only dependent on the external charge distribution outside the conductor.

(6) As a corollary of Property 2 and 4, the electric field in a cavity is solely caused by the induced and enclosed charges of that particular cavity.

(7) As a corollary of Property 2, the potential of a conductor is only caused by and dependent on the charges on the exterior surface and the charges external to the conductor.

(8) As a corollary of Property 4, the potential difference, between a point in a cavity and the conductor that contains it, is only due to the induced and enclosed charges of that particular cavity.

These concepts are intricately weaved into the following problem.

Problem: In Fig. 6.2, Q amount of total charge is placed into arbitrary cavities inside an electrically isolated, neutral spherical conductor of radius R. It is known that a particular cavity is spherical with a radius r and that its center is a distance d away from the center of the conductor. A point charge q_1 lies at the center of this cavity. If an external charge Q' lies at a distance l from the center of the sphere, determine the electric potential at the point charge q_1 due to all other charges at equilibrium. All enclosed charges and Q' are fixed and the center O of the conductor is not enclosed by a cavity.

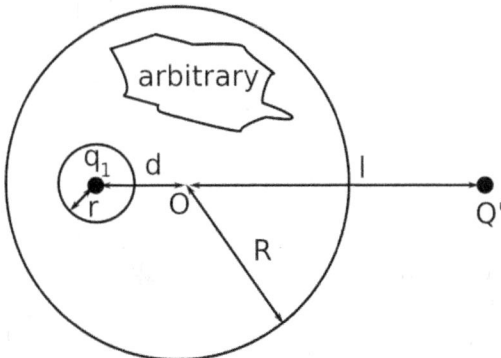

Figure 6.2: Spherical conductor with arbitrary cavities

A total charge $-Q$ is induced on the cavities — implying that a total charge Q resides on the exterior surface of the conductor. Similarly, $-q_1$ charge is induced on the surface of the particular cavity highlighted in the problem. The potential at q_1 is the sum of the potential of the conductor and the potential difference between the center of the cavity and the conductor, both due to all charges besides q_1.

The former can be computed by applying Property 7. Though we do not know the exact distribution of Q over the exterior surface of the conductor, we can consider the potential that they induce at the center O of the sphere. Since all charges on the exterior surface are equidistant from the center, they contribute $\frac{Q}{4\pi\varepsilon_0 R}$ to the potential at O. This, in combination, with the potential caused by Q' causes the potential at O to be

$$\phi_{cond} = \frac{Q}{4\pi\varepsilon_0 R} + \frac{Q'}{4\pi\varepsilon_0 l}.$$

This is the potential of the entire conductor as the conductor must be equipotential at equilibrium. That said, this is the potential of the conductor due to all charges, including q_1 — we have to subtract this contribution. Consider the potential at the surface of the particular cavity that encloses q_1. The contribution to the potential on this surface due to q_1 is $\frac{q_1}{4\pi\varepsilon_0 r}$. Therefore, the potential on this surface due to all other charges is

$$V = \frac{Q}{4\pi\varepsilon_0 R} + \frac{Q'}{4\pi\varepsilon_0 l} - \frac{q_1}{4\pi\varepsilon_0 r}.$$

Moving on, by Property 8, the potential difference between the center of the cavity and the conductor due to all charges besides q_1 is simply that due to the induced charges on the surface of that cavity. By Property 3, the $-q_1$ charges on the surface should be uniformly distributed — they then result in zero electric field within the cavity and thus zero potential difference. As such, we conclude that the potential at q_1 due to all other charges is

$$V = \frac{Q}{4\pi\varepsilon_0 R} + \frac{Q'}{4\pi\varepsilon_0 l} - \frac{q_1}{4\pi\varepsilon_0 r}.$$

6.2.4   *Direct Construction of Solutions*

The uniqueness theorems also enable us to directly construct a solution which satisfies the prescribed boundary conditions, after which we can claim with confidence that this must be the only and correct solution. A classic example of this approach is provided below.

Problem: Determine the charge distribution on a thin conducting disk of radius R that possesses a total charge Q.

Firstly, we have to identify the appropriate boundary conditions and choose the corresponding uniqueness theorem to apply. In this case, we can exploit the fact — that the volume outside the disk is bounded by a conductor (the disk) of total charge Q and infinity — to subsequently use the second uniqueness theorem. Now, we simply have to construct an electric field outside the disk that causes the disk to be equipotential (i.e. no component of electric field in the plane of the disk) and possess total charge Q. To this end, let us first consider a spherical shell of radius R and uniform surface charge density $\sigma_0 = \frac{Q}{4\pi\varepsilon_0 R^2}$.

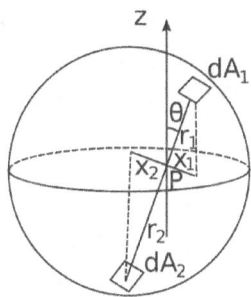

Figure 6.3: Infinitesimal areas at angle θ from z-axis

Consider spherical coordinates about an arbitrary point P with an arbitrarily defined z-axis. Notice that the electric fields due to two infinitesimal areas (in the same plane) between angles θ and $\theta + d\theta$ anti-clockwise from the positive and negative z-axes nullify each other. For example, consider the two infinitesimal areas dA_1 and dA_2 with distances r_1 and r_2 from P, as labeled in Fig. 6.3. The electric fields due to these areas are opposite in direction and are of magnitudes $\frac{\sigma_0 dA_1}{4\pi\varepsilon_0 r_1^2}$ and $\frac{\sigma_0 dA_2}{4\pi\varepsilon_0 r_2^2}$. Since $\frac{dA_1}{dA_2} = \frac{r_1^2}{r_2^2}$, these electric fields cancel out. This argument works for all infinitesimal areas on the spherical shell about P and for all possible point P's within the shell — this is a geometric argument of why the electric field within a spherical shell should be zero.

Now, consider the projection of areas dA_1 and dA_2 onto the equatorial plane (which we define as the disk) perpendicular to the z-axis while maintaining their charges $\sigma_0 dA_1$ and $\sigma_0 dA_2$. Observe that these projections must also produce zero electric field at P along the line joining them (and hence zero component in the equatorial plane) as the charges that they carry are still proportional to the squares of the projected distances $\frac{dA_1}{dA_2} = \frac{x_1^2}{x_2^2}$. Repeating this argument for all pairs of such areas on the spherical shell, the component of the electric field at P in the equatorial plane must be zero. Since point P was chosen randomly, the electric field due to the charge distribution obtained from the projection of the spherical shell onto the disk ensures that the disk is equipotential and results in the right amount of charge on the disk. Note that even though we performed the above arguments with diametrically opposite areas, the upper hemisphere is projected onto the upper surface of the disk and vice-versa for the lower hemisphere (the above argument would still work if we consider corresponding areas on the same hemisphere). Now, the second uniqueness theorem guarantees that the electric field produced by this configuration is correct. Applying Eq. (6.1), we can also assert that the charge distribution obtained from this projection, which we shall now determine, is correct. Redefine the origin at the center of the sphere and the z-axis to be perpendicular to the disk. Let θ denote the angular coordinate that an area on the spherical shell makes with the z-axis such that areas axially symmetric about the z-axis have the same θ coordinate. When an infinitesimal area at coordinate θ is projected onto the plane, the corresponding area on the plane is scaled by a factor of $\cos \theta$ but it carries the same charge. Therefore, the charge density on one surface of the disk, at a radial distance r from its center, is

$$\sigma(r) = \frac{\sigma_0}{\cos \theta} = \frac{Q}{4\pi R^2} \cdot \frac{R}{\sqrt{R^2 - r^2}} = \frac{Q}{4\pi R \sqrt{R^2 - r^2}},$$

which is valid for both upper and lower surfaces of the disk.

6.2.5 *Image Charges*

As a consequence of the uniqueness theorems, we can conjure an elegant method that drastically simplifies set-ups with conductors whose charge distributions are tedious to determine. Specifically, we can invent imaginary charges such that the electric field or potential of the system comprising the original charges and these image charges fulfils the boundary conditions in a region of interest imposed by the original system. This is known as the method of image charges. By an appropriate uniqueness theorem, the electric field or potential of the original system in the chosen regions are identical to that due to the new system that includes the image charges (since the solution is unique)! Note that image charges cannot be inside the region of interest as their presence will change the charge density inside the region that we wish to apply the uniqueness theorems to. The following common examples will elucidate this technique.

Problem: A point charge q lies at a height h above an infinitely large, grounded conducting plane in Fig. 6.4. Find the electric field at the surface of the plane and the potential in the region above the plane. Then, determine the surface charge density of the conducting plane, the total induced charge residing on the plane and the potential energy stored in this system.

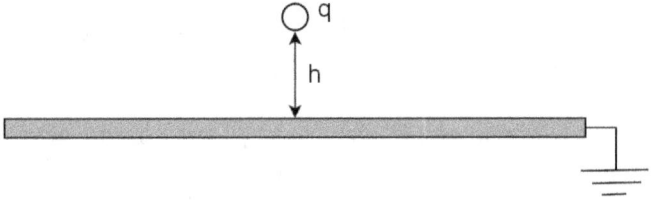

Figure 6.4: Charge above grounded conducting plate

This problem seems impossible to solve as it is difficult to directly determine the charge distribution on the surface of the conducting plane. However, the method of image charges can be applied. We define the z-axis to be positive vertically upwards and choose the volume $z \geq 0$ as our region of interest. The boundary conditions for the first uniqueness theorem include the facts that the electric potentials at the entire surface of the plane and infinity are zero. We observe that these boundary conditions are easily satisfied if we imagine a mirror charge of charge $-q$ at a distance $2h$ directly below the original charge. This reduces the original system to the "equivalent" system in Fig. 6.5, for the regime $z \geq 0$.

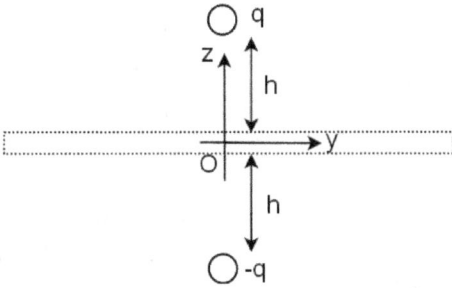

Figure 6.5: Equivalent system with image charge

The electric field and potential in the regime $z \geq 0$ can then be computed. The electric potential at a point (x, y, z) with $z \geq 0$ is then

$$V = \frac{q}{4\pi\varepsilon_0} \left(\frac{1}{\sqrt{x^2 + y^2 + (z - h)^2}} - \frac{1}{\sqrt{x^2 + y^2 + (z + h)^2}} \right).$$

We emphasize that this is invalid for the region $z \leq 0$ as we can only apply the uniqueness theorem to $z \geq 0$ (where the image charge is absent). The electric field at the upper surface of the conductor is

$$E_z(x,y) = 2 \cdot \frac{-q}{4\pi\varepsilon_0(x^2+y^2+h^2)} \cdot \frac{h}{\sqrt{x^2+y^2+h^2}}$$

$$= -\frac{qh}{2\pi\varepsilon_0(x^2+y^2+h^2)^{\frac{3}{2}}},$$

which means that the charge density $\sigma(x,y)$ of the induced charges on the upper surface of the conductor is

$$\sigma(x,y) = \varepsilon_0 E_z = -\frac{qh}{2\pi(x^2+y^2+h^2)^{\frac{3}{2}}}$$

by Eq. (6.1). The total charge Q on the plane should be $-q$ by intuition as all of the field lines emitted by q should reach the upper surface of the infinite plane. This can be proven rigorously by integrating the surface charge density over the entire upper surface. The integral can be performed easily in polar coordinates, centered about the origin O, such that $x^2 + y^2 = r^2$ where r is the radial distance from O.

$$Q = \int_0^\infty \int_0^{2\pi} \sigma r\, d\theta dr = -\int_0^\infty \frac{qhr}{(r^2+h^2)^{\frac{3}{2}}} dr = \left[\frac{qh}{\sqrt{r^2+h^2}}\right]_0^\infty = -q.$$

The potential energy of this system is trickier. One may be tempted to directly say that the potential energy of this system is that between a pair of charges q and $-q$ separated by a distance $2h$. This yields

$$U_{sys} = -\frac{q^2}{8\pi\varepsilon_0 h}$$

which is wrong. The correct answer is in fact half of the above. It is best to compute this by directly integrating the work done by an external force on charge q in bringing it from infinity to its current position, without a change in kinetic energy. Assuming that the conducting plane has reached steady state at each juncture, the image charge is at a z-coordinate $-z$ when the real charge is at coordinate z. The electrostatic force on charge q due to the induced charges in the conductor is

$$F_{elec} = -\frac{q^2}{16\pi\varepsilon_0 z^2}.$$

Therefore, the work done by an external force in bringing q from infinity to its current position, along a path aligned with the z-axis, without a change in kinetic energy is

$$W_{ext} = -W_{elec} = \int_\infty^h \frac{q^2}{16\pi\varepsilon_0 z^2} dz = -\frac{q^2}{16\pi\varepsilon_0 h}.$$

This method also reveals the flaw in the previous reasoning. When the real charge is brought closer to the plane, the image charge also moves closer to the plane, but for free. It is analogous to how work must be done on you to move you closer to the mirror but your image in a real-life mirror moves at no expense of energy. However, the previous reasoning assumed that work is also performed by an external agent in moving the image charge to its final location — a phenomenon which is not true, as the physical charges are already present on the surface of the conductor and redistribute themselves spontaneously. An alternative perspective to this factor of half can be obtained by applying the method of image charges to the region $z \leq 0$. The boundary conditions are simply that the potential is zero at $z = 0$ and infinity. These can be readily satisfied by introducing an image charge $-q$ to nullify the original charge q such that the electric field is zero in $z < 0$. This must be the correct solution by the first uniqueness theorem. Then the factor of half can be reasoned from the energy density of the electric field. When we say that the potential energy is that between a pair of charges q and $-q$ separated by a distance $2h$, we are assuming that the electric field spans both $z \geq 0$ and $z \leq 0$. However, in reality, the electric field is present in $z \geq 0$ and zero in $z \leq 0$ — leading to a factor of half. Incidentally, the zero electric field in $z \leq 0$ also implies that the surface charge density on the lower surface of the plane is zero everywhere.

Problem: A charge q_1 is held at a distance r from the center of a grounded conducting spherical shell of radius R. Find the potential in the region exterior to the conducting shell.

The appropriate image charge system here is not as obvious. However, it never hurts to guess. If we coincidentally find a valid solution, it must be the unique solution, assuming that we can apply a uniqueness theorem. Firstly, we choose the region of interest to be the entire space, excluding the shell. The boundary conditions are that the potentials at the shell and infinity must be zero. It is then intuitive to guess an image charge along the line joining the center of the shell to q_1, exemplified by Fig. 6.6, as the system is symmetric about this axis. Let the image charge have charge q_2 and be located at a distance a away from the center of the shell, in the direction towards q_1.

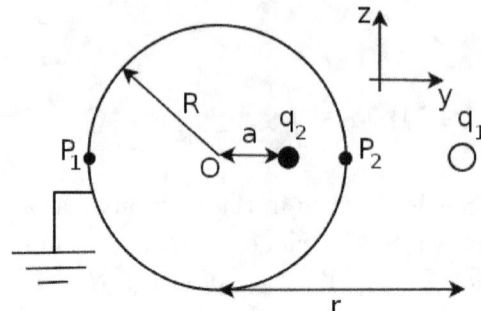

Figure 6.6:   Image charge in the sphere

For the new system, which includes the image charge, to satisfy the constraint that the surface of the shell is at zero potential, the new system must first satisfy the condition that the potential at points P_1 and P_2 are zero. This will enable us to easily determine q_2 and a. Considering the electric potential at point P_1,

$$\frac{q_2}{R+a} + \frac{q_1}{R+r} = 0.$$

Considering that at P_2,

$$\frac{q_2}{R-a} + \frac{q_1}{r-R} = 0.$$

Solving,

$$q_2 = -\frac{R}{r}q_1, \tag{6.4}$$

$$a = \frac{R^2}{r}. \tag{6.5}$$

Now we need to verify if these particular values of q_2 and a satisfy the boundary condition at all points on the surface of the shell. If this is true, our initial guess regarding the geometry and orientation of the image charge must be correct. Define the origin O at the center of the shell and consider a point (x, y, z) on the surface of the shell. Its electric potential is given by

$$V = -\frac{R}{r}q_1 \cdot \frac{1}{4\pi\varepsilon_0\sqrt{x^2 + \left(y - \frac{R^2}{r}\right)^2 + z^2}}$$

$$+ q_1 \cdot \frac{1}{4\pi\varepsilon_0\sqrt{x^2 + (y - r)^2 + z^2}}$$

$$= -\frac{q_1 R}{4\pi\varepsilon_0 r\sqrt{R^2 - \frac{2R^2}{r}y + \frac{R^4}{r^2}}} + \frac{q_1}{4\pi\varepsilon_0\sqrt{R^2 - 2ry + r^2}}$$

$$= -\frac{q_1}{4\pi\varepsilon_0\sqrt{R^2 - 2ry + r^2}} + \frac{q_1}{4\pi\varepsilon_0\sqrt{R^2 - 2ry + r^2}}$$

$$= 0$$

where we have used the fact that $x^2 + y^2 + z^2 = R^2$. Thus, our guess was in fact correct! The potential of a point (x, y, z) outside of the shell is then

$$V = -\cdot\frac{q_1 R}{4\pi\varepsilon_0 r\sqrt{x^2 + (y - \frac{R^2}{r})^2 + z^2}} + \frac{q_1}{4\pi\varepsilon_0\sqrt{x^2 + (y - r)^2 + z^2}}.$$

Note that this expression is again invalid in the region containing the mirror charge (inside the shell).

There are myriad other variations of such problems involving image charges. The only limit is our own imagination! Consider the following as another example.

Problem: A grounded infinite conducting plane has a hemispherical bulge of radius R, which is also made of conducting material, as shown in Fig. 6.7. A charge q is placed a distance $r > R$ above the center of the hemisphere. Determine the force that q experiences.

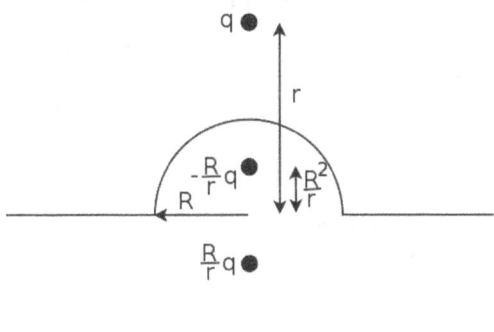

Figure 6.7: Infinite plane with bulge

In this case, we have to conjure an image charge configuration that produces zero potential on the infinite plane with the bulge, in combination with the original charge q. Suppose that we simply put an image charge $-\frac{R}{r}q$ at a distance $\frac{R^2}{r}$ above the center of the hemisphere. The hemispherical surface now has zero potential but the infinite plane does not. To rectify this loophole, we further "reflect" the original charge q and image charge $-\frac{R}{r}q$ about the infinite conducting plane to form two more image charges $-q$ and $\frac{R}{r}q$ at distances r and $\frac{R^2}{r}$ below the center of the hemisphere. Then, all boundary conditions are fulfilled and the electric field due to this configuration above the plane must be correct. The force on the original charge q (defined to be positive towards the plane) is

$$F = \frac{q \cdot \frac{R}{r}q}{4\pi\varepsilon_0 \left(r - \frac{R^2}{r}\right)^2} + \frac{q^2}{4\pi\varepsilon_0(2r)^2} - \frac{q \cdot \frac{R}{r}q}{4\pi\varepsilon_0 \left(r + \frac{R^2}{r}\right)^2}$$

$$= \frac{q^2 R^3 r}{2\pi\varepsilon_0(r^4 - R^4)} + \frac{q^2}{16\pi\varepsilon_0 r^2}.$$

6.3 Capacitors

For a single, isolated conductor, its uniform potential relative to infinity is proportional to the total charge residing on its surface. If the surface charge distribution on a conductor is doubled everywhere, the electric field due to the conductor will double in the region outside itself. Note that the additional charges will not redistribute themselves, as twice the original electric field satisfies the boundary conditions required to apply the second uniqueness theorem and must hence be the correct electric field. Since the electric field is doubled everywhere, the potential of the conductor is also doubled. An argument in the reverse direction can also be made to conclude that if the potential of the conductor is doubled, the total charge that it carries is also doubled. It is therefore natural to define a characteristic of a conductor that is independent of the total charge that it stores and its potential. This description is known as the capacitance and is defined as the total charge residing on a conductor per unit potential (relative to infinity).

$$C = \frac{q}{\phi} \tag{6.6}$$

where q is the total charge carried by the conductor and ϕ is its potential relative to infinity. Capacitance is an intrinsic quantity that is only dependent on the geometry of the conductor and not its total charge or potential. The capacitance is often measured in Farads (F) where one farad is one Coulomb over one Volt (qV^{-1}).

Problem: Determine the capacitance of a metal sphere of radius R.

To compute the capacitance of a set-up, we inject q amount of charge to the conductor and find its resultant potential ϕ relative to infinity. Afterwards, we can calculate the capacitance C as $\frac{q}{\phi}$. In this case, when q amount of charge is added to the metal sphere, it is distributed evenly over its surface due to symmetry. Since the electric field of a spherical shell of charge q is identical to that of a point charge q placed at its center for regions outside of the shell, the potential at the surface of the metal sphere is the potential of a point located a distance R away from a point charge q.

$$\phi = \frac{q}{4\pi\varepsilon_0 R}.$$

The capacitance of the metal sphere is thus

$$C = \frac{q}{\phi} = 4\pi\varepsilon_0 R$$

which, as expected, is independent of q and ϕ.

Now, this notion of capacitance can be extended to a system of two conductors — known as a capacitor. A capacitor is a component that stores charge and electric potential energy when a potential difference is applied across two ends. Conversely, a capacitor with stored charge can also act as a battery that produces a potential difference.

For the rest of the chapter, we will assume that the two conductors in a capacitor have equal amounts of charge of opposite signs. Firstly, this is often what occurs in reality (some charge from one initially neutral conductor is transferred to another via a circuit). Secondly, we will often consider capacitors where a conductor A encloses another conductor B (i.e. B lies in a cavity of A). Suppose that the conductor B has a total charge q, $-q$ net charge must be induced on the inner surface of conductor A by Gauss' law. Even if some net charge Q resides on the exterior surface of conductor A, it does not generate a potential difference between two points within conductor A as a consequence of electrostatic shielding (Property 4). Therefore, the potential difference between conductor B and the inner surface of conductor A is only dependent on the charge stored on either surface (which must be equal in magnitude). In light of the above discussion, when we refer to the charge of a capacitor, we actually mean the magnitude of charge on either of the surfaces of the conductors that are adjacent.

Now, the definition of the capacitance C of a system of two conductors is the amount of charge stored per unit potential difference between their adjacent surfaces and is a positive quantity.

$$C = \left| \frac{q}{\Delta V} \right|. \qquad (6.7)$$

Note that since conductors at equilibrium are guaranteed to be equipotential, the potential difference can be evaluated unambiguously across any pair of points, each lying on a conductor in the capacitor system. The capacitance measures the ability of a system to store charge and thus electric potential energy. It is emphasized once again that the capacitance is solely an intrinsic geometric quantity of the capacitor.

The most pervasive capacitor would be the parallel-plate capacitor illustrated in Fig. 6.8. Two large plates of area A are separated with a small distance d in between them. Since the plate separation d is small, we can regard the plates as infinite planes and assume that there are no fringe effects (effects at the ends of the plates which cause the electric field to bend instead of following a straight path from one plate to another).

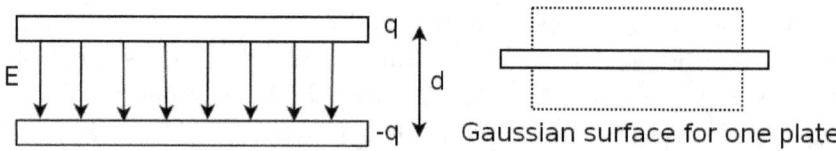

Figure 6.8:   Side view of a parallel-plate capacitor

To determine the capacitance of the parallel-plate system, place q charge on one plate and $-q$ charge on the other. The distribution of charges should be uniform throughout each of the plates due to the "infinite nature" of the plane. The plates then carry uniform surface charge densities $\sigma = \frac{q}{A}$ and $-\sigma = -\frac{q}{A}$. Each of the plates is basically equivalent to the infinite plane of charge in Section 5.4.1. By Gauss' law, the electric field due to one plate in the region between the plates is

$$2E_{plate} \cdot A = \frac{\sigma A}{\varepsilon_0}$$

$$E_{plate} = \frac{\sigma}{2\varepsilon_0}$$

where σ is the surface charge density on one plate. Thus, the total electric field in the region between the two plates is a uniform value

$$E = 2E_{plate} = \frac{\sigma}{\varepsilon_0}$$

as the electric fields of individual plates are mutually reinforced. In regions outside of the gap, the electric field is zero as the effects of the two plates nullify each other. Incidentally, this uniform field also implies that the outer surfaces of the two plates are entirely neutral by Eq. (6.1) when the two plates carry equal magnitudes of opposite charges. The charge distribution when this condition is not satisfied is the subject of Problem 2 in this chapter. Moving on, the potential difference between the two plates is

$$|\Delta V| = Ed = \frac{\sigma d}{\varepsilon_0}.$$

Thus, the capacitance of the system is

$$C = \left|\frac{q}{\Delta V}\right| = \frac{\sigma A}{\frac{\sigma d}{\varepsilon_0}} = \varepsilon_0 \frac{A}{d}. \tag{6.8}$$

Problem: Find the capacitance per unit length of a system comprising a long conducting cylinder of radius r_1 and length $L \gg r_1$ surrounded by a concentric long cylindrical shell of radius r_2 and equal length, depicted in Fig. 6.9.

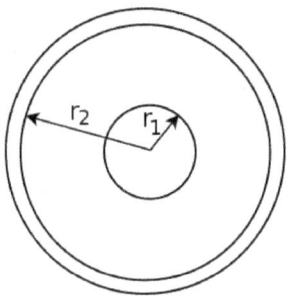

Figure 6.9: Top view of system

Let the cylinder possess charge q which is distributed uniformly across its curved surface due to its axial symmetry and infinite nature. Then, we can draw a concentric cylindrical Gaussian surface of an arbitrary length l and an arbitrary radius r, $r_1 \leq r \leq r_2$, around the cylinder. Let the linear charge density of the cylinder be $\lambda = \frac{q}{L}$. By Gauss' law, the electric field at a radius r from the central axis of the system is

$$E \cdot 2\pi rl = \frac{\lambda l}{\varepsilon_0}$$

$$\implies E = \frac{\lambda}{2\pi \varepsilon_0 r}.$$

The potential difference between the shell and the cylinder is then

$$\Delta V = -\int_{r_2}^{r_1} \frac{\lambda}{2\pi\varepsilon_0 r} dr = \left[-\frac{\lambda}{2\pi\varepsilon_0} \ln r \right]_{r_2}^{r_1}$$

$$= \frac{\lambda}{2\pi\varepsilon_0} \ln \frac{r_2}{r_1}.$$

The capacitance per unit length of this system is thus

$$\frac{C}{L} = \left| \frac{\lambda}{\Delta V} \right| = \frac{2\pi\varepsilon_0}{\ln \frac{r_2}{r_1}}.$$

Energy Stored in a Capacitor

Work is required to move a charge from one conductor to another against a potential difference. This work is often indirectly supplied by an electromotive force. The work done by an external force in moving an infinitesimal charge, dq, across a potential difference ΔV is

$$dW_{ext} = dq\Delta V.$$

Thus, if we define the potential energy of a system of two neutral conductors as zero, the potential energy of a system with charges q and $-q$ on the conductors is equal to the work done by an external force in moving q amount of charge from one conductor to another. Interspersing this transfer of charge over many intervals such that an infinitesimal amount of charge dq is transferred during each event,

$$U = W_{ext} = \int_0^q \Delta V dq = \int_0^q \frac{q}{C} dq = \frac{1}{2}\frac{q^2}{C} = \frac{1}{2}q\Delta V = \frac{1}{2}C\Delta V^2 \qquad (6.9)$$

where we have used the fact that the potential difference between two conductors is $\Delta V = \frac{q}{C}$, where C is the capacitance between them. The final three expressions yield equivalent formulae for the energy stored in a capacitor.

Equivalent Capacitance

Capacitors in Series

The equivalent capacitance of N capacitors in series is

$$\frac{1}{C_{eq}} = \sum_{i=1}^{N} \frac{1}{C_i}. \qquad (6.10)$$

Capacitors in Parallel

The equivalent capacitance of N capacitors in parallel is

$$C_{eq} = \sum_{i=1}^{N} C_i. \tag{6.11}$$

These can be proven via the facts that the charges on the capacitors must be identical for initially neutral capacitors in series and that the potential differences must be identical for capacitors in parallel.

Breakdown Potential and Electric Field

As a form of non-ideal behavior, the originally insulating material between the two conductors in a capacitor may actually become conductive when the electric field at any point in the medium is large enough to ionize its atoms. The capacitor then fails as it is unable to store charge when given a potential difference, since charges begin to flow from one conductor to the other via the conducting medium — discharging the capacitor in the process. This is analogous to how lightning strikes when the potential difference between a cloud and the ground is large enough.

The minimum electric field at any point within an insulating material, for which it becomes conductive, is known as the breakdown electric field E_b. Since the electric field produced by a capacitor is proportional to the potential difference between its two constituent conductors, an analogous quantity known as the breakdown potential V_b is defined as the minimum potential difference between the two conductors for the capacitor to fail.

An interesting question to ask is that if N initially neutral capacitors, with the ith capacitor having capacitance C_i and breakdown potential V_{bi}, are connected in parallel and series respectively, what is the maximum external potential V_{max} that can be applied between the ends of this array of capacitors such that no capacitor fails?

The answer is trivial in the case of parallel connections as the potential difference across each capacitor is simply the external potential V. Therefore,

$$V_{max} = \min_{1 \leq i \leq N} V_{bi}. \tag{6.12}$$

Given an increasing external potential, the capacitor with the minimum V_{bi} breaks down first. In the case of series connections, the common charge q stored by each capacitor when an external potential V is applied to the ends of the array is

$$q = VC_{eq} = \frac{V}{\sum_{1 \le j \le N} \frac{1}{C_j}}.$$

The potential difference across the ith capacitor is thus

$$\frac{q}{C_i} = \frac{V}{C_i \sum_{1 \le j \le N} \frac{1}{C_j}}.$$

The required external potential for the ith capacitor to break down is

$$V = V_{bi} C_i \sum_{1 \le j \le N} \frac{1}{C_j}.$$

Thus,

$$V_{max} = \min_{1 \le i \le N} \left(V_{bi} C_i \sum_{1 \le j \le N} \frac{1}{C_j} \right) = \sum_{1 \le j \le N} \frac{1}{C_j} \cdot \min_{1 \le i \le N} (V_{bi} C_i). \qquad (6.13)$$

Given an increasing external potential, the capacitor with the minimum $V_{bi} C_i$ breaks down first. In the special case where $V_{bi} = V_b$ for all $1 \le i \le N$,

$$V_{max} = V_b \sum_{1 \le j \le N} \frac{1}{C_j} \cdot \min_{1 \le i \le N} C_i.$$

Problem: In the circuit shown in Fig. 6.10, capacitors (which are depicted by two identical, parallel lines) $C_1 = 4.00\mu F$, $C_2 = 12.00\mu F$, $C_3 = 5.00\mu F$, $C_4 = 6.00\mu F$ are initially neutral and an external potential difference is applied across terminals a and b. If the breakdown potential of each capacitor is 12V, determine the maximum charge residing on each capacitor for which no capacitor fails.

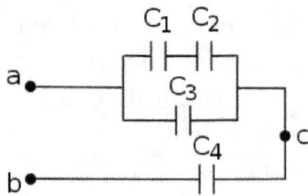

Figure 6.10:   Circuit of capacitors

Applying the formulae that we have just derived, the maximum external potential across the branch containing C_1 and C_2 is

$$12 \left(\frac{1}{4} + \frac{1}{12} \right) \cdot \min(4, 12) = 16V.$$

Since the branch containing C_1 and C_2 and the branch containing C_3 are connected in parallel, the maximum external potential across branch ac is

$$\min(16, 12) = 12\text{V}.$$

Now, the equivalent capacitance of branch ac is

$$\frac{C_1 C_2}{C_1 + C_2} + C_3 = 8\mu\text{F}.$$

Therefore, the capacitors in branch ac are equivalent to a capacitor 8μF with a breakdown potential 12V. As branches ac and cb and connected in series, the maximum external potential across ab is

$$V_{max} = 12\left(\frac{1}{8} + \frac{1}{6}\right) \cdot \min(8, 6) = 21\text{V}.$$

Because the charge residing on each capacitor increases monotonically with the external potential difference applied between terminals a and b, the maximum charge stored by each capacitor occurs when $V_{ab} = V_{max}$. At this juncture, since C_4 breaks down first, the potential difference across branch cb must be equal to its maximum external potential difference (the breakdown potential of C_4).

$$V_{cb} = 12\text{V}.$$

The maximum charge stored by C_4 is

$$q_4 = C_4 V_{cb} = 72\mu\text{C}.$$

The potential difference across branch bc is

$$V_{cb} = V_{max} - V_{cb} = 9\text{V}.$$

The maximum charge stored by C_3 is

$$q_3 = C_3 V_{ab} = 45\mu\text{C}.$$

The maximum charge stored by C_1 and C_2 is

$$q_1 = q_2 = \frac{C_1 C_2}{C_1 + C_2} V_{ab} = 27\mu\text{C}.$$

Another way of seeing this is that since the system comprising the right plates of C_2, C_3 and C_4 is electrically isolated, its total charge — which was originally zero — must be conserved. Therefore,

$$q_2 + q_3 = q_4 \implies q_1 = q_2 = q_4 - q_3 = 27\mu\text{C}.$$

6.3.1 *An Elegant Method for Determining Total Charges*

A common class of conductor problems entails determining the total charge that resides on each conductor (and not the charge distributions) in the presence of external charges. Surprisingly, such problems can be solved with rather elementary methods. One crucial idea in such problems is to exploit scaling arguments and the principle of superposition to construct fictitious capacitors.

Problem: Two parallel, thin and grounded infinite conducting plates are separated by a perpendicular distance d. A charge q is sandwiched between the two plates and is located at a perpendicular distance $x < d$ from one plate. Determine the total charges induced on the two plates.

Firstly, observe that if we scale the charge q by a factor of k, the total charges induced on each of the two plates will also be scaled by a factor k. This implies that if we spread the charge q such that it becomes a uniform infinite plane of total charge q, located at the same perpendicular distance from the plates, the total charges induced on each of the two plates will remain the same, as the infinite plane of charge can be seen as the superposition of many infinitesimal charges dq which each induce $\frac{dq}{q}$ of the total charges, caused by the original charge q, on the two plates.[6] The distributions of charges on the plates in these set-ups will be different but the total charges must be identical. In fact, in the set-up that we have constructed, charges should be evenly distributed over each plate due to the infinite nature of the set-up. Now, let the total charge induced on the plate located at a perpendicular distance d from the plane of charge be q_1 and that on the other plate be q_2. Firstly, we can draw a Gaussian cylinder straddling the two conducting plates and the plane of charge sandwiched between them, with its axis perpendicular to the plates. The electric field in this set-up is solely normal to the plates such that electric flux cutting the curved surface of this Gaussian cylinder is zero — resulting in a net electric flux of zero, as the electric field at its ends are also zero since they are located inside the conducting plates. This implies that

$$q_1 + q_2 + q = 0 \implies q_1 + q_2 = -q,$$

as the charge enclosed by the Gaussian cylinder must be zero by Gauss' law. Now, this infinite plane of charge q can be seen as two planes of charges

---

[6]That the charge distribution induced by the array of infinitesimal charges is the superposition of the individual charge distributions induced by each infinitesimal charge is guaranteed by the first uniqueness theorem.

$-q_1$ and $-q_2$. Then, the entire set-up becomes two parallel-plate capacitors storing charges q_1 and $-q_2$ that are connected in series. Let the capacitance of the respective plates be C_1 and C_2 respectively. Observe that since both conducting planes are grounded, there must be no potential difference between the exterior ends of the two capacitors. That is,

$$\frac{q_1}{C_1} - \frac{q_2}{C_2} = 0.$$

Furthermore, we know that the capacitance of a parallel-plate capacitor is inversely proportional to the plate separation such that $\frac{C_1}{C_2} = \frac{d-x}{x}$.

$$\frac{q_1}{d-x} = \frac{q_2}{x}.$$

Solving for q_1 and q_2,

$$q_1 = -\left(1 - \frac{x}{d}\right)q,$$

$$q_2 = -\frac{x}{d}q.$$

Actually, another way of solving such problems involves a clever application of Green's reciprocity theorem. Define the plate that is a distance x away from q as plate 1 and the other plate as plate 2. We can carefully concoct two set-ups to apply Green's reciprocity theorem to. We will choose set-up 1 as the system mentioned in the problem. For set-up 2, we can consider the case where the plate 1 has potential ϕ_0 while plate 2 is grounded (i.e. has potential 0) — there is no longer a charge q between the plates. In set-up 2, the potential at a perpendicular distance x from plate 1 is evidently $\frac{d-x}{d}\phi_0$. The charges on these two plates in set-up 2 are unknown but this does not hinder the application of Green's reciprocity theorem as the potentials of both plates in set-up 1 are zero anyway. Applying Green's reciprocity theorem to these set-ups,

$$\iiint \rho_1 V_2 d\Omega_1 = \iiint \rho_2 V_1 d\Omega_2.$$

The left-hand side is $q_1\phi_0 + \frac{d-x}{d}q\phi_0$ where the contributions stem from plate 1 and the charge q in set-up 1 (even though plate 2 also possesses a charge q_2 in set-up 1, its potential is zero in set-up 2 as we had deliberately designed it to be so). The right-hand side is zero as the potentials of the two grounded plates in set-up 1 are zero.

$$q_1\phi_0 + \frac{d-x}{d}q\phi_0 = 0.$$

We directly obtain

$$q_1 = -\left(1 - \frac{x}{d}\right)q.$$

Considering a separate set-up 2 where the potentials of plates 1 and 2 are now reversed, we can similarly show that

$$q_2 = -\frac{x}{d}q.$$

6.4   Electric Fields in Matter

The electric fields in the previous sections only propagate in vacuum which is characterized by a permittivity ε_0. This section will analyze electric fields in matter which is a more realistic situation.

6.4.1   *Electric Dipoles*

A physical dipole is a pair of opposite charges of the same magnitude that somehow has a fixed separation d. For a dipole with a common charge magnitude q, we can define the dipole moment as

$$\boldsymbol{p} = q\boldsymbol{d}, \tag{6.14}$$

where $\boldsymbol{d}$ is the vector pointing from the negative charge to the positive charge. An ideal dipole has a minuscule separation ($d \to 0$) and a titanic charge ($q \to \infty$) such that the product of qd is a finite value p. Now, what is the field due to a dipole at a point P far away[7] from the dipole? Our approach is to first compute the potential due to the dipole and then take the negative gradient of it to determine the electric field. Let $\boldsymbol{r}$ be the vector from the middle of the dipole to the point of concern, P. Then, we define angle θ as the clockwise angle subtended by the vector joining the center of the dipole to P and $\boldsymbol{d}$. The potential and electric field at the point of concern P are then,

$$V = \frac{p\cos\theta}{4\pi\varepsilon_0 r^2} = \frac{\boldsymbol{p}\cdot\hat{\boldsymbol{r}}}{4\pi\varepsilon_0 r^2}, \tag{6.15}$$

$$\boldsymbol{E} = \frac{p\cos\theta}{2\pi\varepsilon_0 r^3}\hat{\boldsymbol{r}} + \frac{p\sin\theta}{4\pi\varepsilon_0 r^3}\hat{\boldsymbol{\theta}} = \frac{1}{4\pi\varepsilon_0 r^3}[3(\boldsymbol{p}\cdot\hat{\boldsymbol{r}})\hat{\boldsymbol{r}} - \boldsymbol{p}]. \tag{6.16}$$

---

[7]By far away, we mean that the distance between the dipole and P is much larger than d.

Derivation:

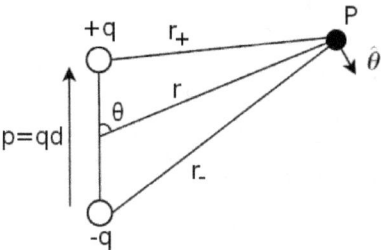

Figure 6.11: Dipole

Referring to Fig. 6.11, the potential at point P is the sum of the individual potentials due to each charge.

$$V = \frac{q}{4\pi\varepsilon_0 r_+} - \frac{q}{4\pi\varepsilon_0 r_-}.$$

By the cosine rule,

$$r_+ = \sqrt{r^2 - rd\cos\theta + \frac{d^2}{4}} \approx r\sqrt{1 - \frac{d}{r}\cos\theta},$$

$$r_- = \sqrt{r^2 + rd\cos\theta + \frac{d^2}{4}} \approx r\sqrt{1 + \frac{d}{r}\cos\theta}$$

as $d \ll r$.

$$V = \frac{q}{4\pi\varepsilon_0 r}\left(\frac{1}{\sqrt{1 - \frac{d}{r}\cos\theta}} - \frac{1}{\sqrt{1 + \frac{d}{r}\cos\theta}}\right)$$

$$\approx \frac{q}{4\pi\varepsilon_0 r}\left[1 + \frac{d}{2r}\cos\theta - \left(1 - \frac{d}{2r}\cos\theta\right)\right]$$

$$= \frac{qd\cos\theta}{4\pi\varepsilon_0 r^2}$$

$$= \frac{p\cos\theta}{4\pi\varepsilon_0 r^2}$$

where we have used the Maclaurin expansion $\frac{1}{1+x} = 1 - x + \cdots$ and discarded second order and above terms in $\frac{d}{r}$. Since θ is the angle between $\boldsymbol{r}$ and $\boldsymbol{p}$, the above can be rewritten as

$$V = \frac{\boldsymbol{p} \cdot \hat{\boldsymbol{r}}}{4\pi\varepsilon_0 r^2}.$$

To calculate the electric field, we can find the negative potential gradient in cylindrical coordinates. We define the z-axis to be perpendicular to the plane containing the dipole and P.

$$E = -\frac{\partial V}{\partial r}\hat{r} - \frac{1}{r}\frac{\partial V}{\partial \theta}\hat{\theta} - \frac{\partial V}{\partial z}\hat{k}$$

$$= \frac{p\cos\theta}{2\pi\varepsilon_0 r^3}\hat{r} + \frac{p\sin\theta}{4\pi\varepsilon_0 r^3}\hat{\theta}$$

$$= \frac{1}{4\pi\varepsilon_0 r^3}[3p\cos\theta\hat{r} - (p\cos\theta\hat{r} - p\sin\theta\hat{\theta})]$$

$$= \frac{1}{4\pi\varepsilon_0 r^3}[3(\boldsymbol{p}\cdot\hat{r})\hat{r} - \boldsymbol{p}].$$

Torque due to Uniform External Electric Field

When placed in an external electric field that is locally uniform, a physical dipole tends to orient its axis along the direction of the electric field.

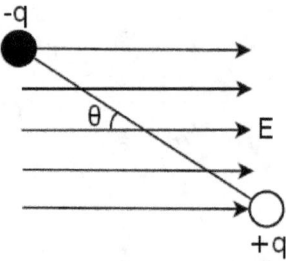

Figure 6.12:   Dipole in external electric field

Referring to Fig. 6.12, when the dipole moment $\boldsymbol{p}$ subtends an angle θ with respect to the external electric field $\boldsymbol{E}$, the torque on the dipole due to the external electric field is

$$\tau = -qdE\sin\theta = -pE\sin\theta,$$

$$\boldsymbol{\tau} = \boldsymbol{p}\times\boldsymbol{E}. \tag{6.17}$$

Note that we do not need to mention the origin that the torque is calculated with respect to, as there is no net force on the dipole due to the external electric field — implying that the torques about all pivots are the same.[8]

---

[8]We proved this in the chapter on statics.

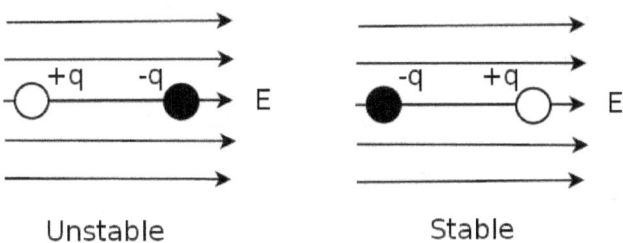

Unstable   Stable

Figure 6.13:   Anti-parallel and parallel configurations

Next, Fig. 6.13 illustrates the two possible equilibrium orientations. The parallel configuration is a stable equilibrium while the anti-parallel configuration is an unstable equilibrium as shown above. The equilibria are stable and unstable respectively as a slight angular displacement tends to be minimized and amplified by the external field respectively.

Lastly, we can define the potential energy of a dipole, which depends on its current angular orientation, as the work by an external agent (in opposition to the external electric field) required in bringing the dipole from infinity to its current configuration, without a change in kinetic energy. Exploiting the path independence of the potential energy, we can do this in the dipole in the following manner. We first bring in the two charges from infinity to the current region, in the orientation where the dipole moment is perpendicular to the external electric field. There is no external work done in this process even though the external electric field may be non-uniform in general (we only require it to be uniform in the vicinity of the final position of the dipole). This is because, the difference in electric potentials at the locations of the two charges after this procedure is zero as the line integral of the electric field along a straight line connecting the two charges is zero (the electric field is always perpendicular to the line). This implies that the final electric potential energies of the two opposite charges of the dipole nullify each other — insinuating that no net external work has been performed. Subsequently, we exert an external torque in rotating the dipole into its current orientation, against a local uniform electric field. The change in potential energy due to the work done by an external force in orienting the dipole at an angle θ with respect to the external electric field $\boldsymbol{E}$, while beginning from $\theta = \frac{\pi}{2}$, is

$$\Delta U = \int_{\frac{\pi}{2}}^{\theta} \tau_{ext}\, d\theta = -\int_{\frac{\pi}{2}}^{\theta} \tau_{elec} d\theta = \int_{\frac{\pi}{2}}^{\theta} pE \sin\theta d\theta = -pE \cos\theta,$$

where we have used the fact that the torque by the external agent must be negative of the torque due to the electric field for the kinetic energy of the dipole to remain constant, $\tau_{ext} = -\tau_{elec}$. This change in potential energy is the total potential energy of the dipole.

$$U(\theta) = -pE\cos\theta = -\boldsymbol{p}\cdot\boldsymbol{E}. \tag{6.18}$$

Force due to Non-Uniform External Electrostatic Field

A physical dipole does not experience a net force in a locally uniform external electric field as the forces on the opposite charges cancel out. However, it will indeed feel a net force from a non-uniform field due to the discrepancy of the external electric field at the location of the two charges. If the position vector of the $-q$ charge is $\boldsymbol{r}$, the net force on the dipole is

$$\boldsymbol{F} = q[\boldsymbol{E}(\boldsymbol{r}+\boldsymbol{d}) - \boldsymbol{E}(\boldsymbol{r})],$$

where $\boldsymbol{E}$ refers to the external electric field and $\boldsymbol{d}$ is the vector pointing from $-q$ to q. The x-component of force is

$$F_x = q([E_x(\boldsymbol{r}+\boldsymbol{d}) - E_x(\boldsymbol{r})].$$

If we use d_x, d_y and d_z to denote the components of $\boldsymbol{d}$ and as $d \to 0$,

$$E_x(\boldsymbol{r}+\boldsymbol{d}) = E_x(\boldsymbol{r}) + \frac{\partial E_x}{\partial x}d_x + \frac{\partial E_x}{\partial y}d_y + \frac{\partial E_z}{\partial z}d_z$$

$$= E_x(\boldsymbol{r}) + \boldsymbol{d}\cdot\nabla E_x$$

$$\implies F_x = q\boldsymbol{d}\cdot\nabla E_x = \boldsymbol{p}\cdot\nabla E_x.$$

Similarly,

$$F_y = \boldsymbol{p}\cdot\nabla E_y,$$

$$F_z = \boldsymbol{p}\cdot\nabla E_z,$$

so we can condense them into the single expression

$$\boldsymbol{F} = (\boldsymbol{p}\cdot\nabla)\,\boldsymbol{E}, \tag{6.19}$$

where the $(\boldsymbol{p}\cdot\nabla)$ term in brackets should be read as an operator that acts on each component of $\boldsymbol{E}$ to produce the corresponding component of $\boldsymbol{F}$. It turns out that we can write

$$\nabla(\boldsymbol{p}\cdot\boldsymbol{E}) = (\boldsymbol{p}\cdot\nabla)\boldsymbol{E} + \boldsymbol{p}\times(\nabla\times\boldsymbol{E}),$$

where $\nabla \times \boldsymbol{E}$ is the curl of $\boldsymbol{E}$ given by

$$\nabla \times \boldsymbol{E} = \begin{pmatrix} \frac{\partial}{\partial x} \\ \frac{\partial}{\partial y} \\ \frac{\partial}{\partial z} \end{pmatrix} \times \begin{pmatrix} E_x \\ E_y \\ E_z \end{pmatrix} = \begin{pmatrix} \frac{\partial E_z}{\partial y} - \frac{\partial E_y}{\partial z} \\ \frac{\partial E_x}{\partial z} - \frac{\partial E_z}{\partial x} \\ \frac{\partial E_y}{\partial x} - \frac{\partial E_x}{\partial y} \end{pmatrix}$$

in Cartesian coordinates. Recall that we once remarked in the chapter on energy that the curl of a vector field must vanish at all points for it to be conservative. Since the electrostatic field is conservative (we have shown that its line integral is path-independent), we must have $\nabla \times \boldsymbol{E} = \boldsymbol{0}$. Thus,

$$\boldsymbol{F} = (\boldsymbol{p} \cdot \nabla)\boldsymbol{E} = \nabla(\boldsymbol{p} \cdot \boldsymbol{E})$$

which makes sense in retrospect as the right-hand side is simply the negative gradient of the potential energy of the dipole $(-\boldsymbol{p} \cdot \boldsymbol{E})$!

6.4.2 *Multipole Expansion*

Let us now invent an analogous notion for the dipole moment of a continuous charge distribution.

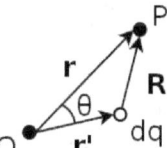

Figure 6.14: Relevant vectors

Referring to Fig. 6.14, suppose we are interested in the potential at a distant point P which has a position vector $\boldsymbol{r}$ relative to the origin O. Let $\boldsymbol{r}'$ denote the position vector of an infinitesimal charge dq on the charge distribution and $\boldsymbol{R}$ denote the vector pointing from dq to P. The potential at P is

$$V(\boldsymbol{r}) = \int \frac{1}{4\pi\varepsilon_0 R} dq.$$

We can relate $\boldsymbol{R}$, $\boldsymbol{r'}$ and $\boldsymbol{r}$ via the cosine rule. Using the definition of θ in the figure above,

$$R^2 = r^2 + r'^2 - 2rr' \cos \theta$$

$$R = \sqrt{r^2 + r'^2 - 2rr' \cos \theta},$$

$$V(\boldsymbol{r}) = \int \frac{1}{4\pi\varepsilon_0 \sqrt{r^2 + r'^2 - 2rr' \cos \theta}} dq.$$

Since

$$(r^2 + r'^2 - 2rr' \cos \theta)^{-\frac{1}{2}} = \frac{1}{r}\left(1 + \frac{r'^2}{r^2} - \frac{2r'}{r} \cos \theta\right)^{-\frac{1}{2}},$$

if we let $\varepsilon = \frac{r'^2}{r^2} - \frac{2r'}{r} \cos \theta$,

$$(r^2 + r'^2 - 2rr' \cos \theta)^{-\frac{1}{2}}$$

$$= \frac{1}{r}(1 + \varepsilon)^{-\frac{1}{2}}$$

$$= \frac{1}{r}\left(1 - \frac{\varepsilon}{2} + \frac{3\varepsilon^2}{8} - \cdots\right)$$

$$= \frac{1}{r}\left(1 - \frac{r'^2}{2r^2} + \frac{r'}{r} \cos \theta + \frac{3}{8}\left(\frac{r'^2}{r^2} - \frac{2r'}{r} \cos \theta\right)^2 - \cdots\right)$$

$$= \frac{1}{r}\left(1 + \frac{r'}{r} \cos \theta + \frac{3 \cos^2 \theta - 1}{2} \frac{r'^2}{r^2} + O\left(\frac{r'^3}{r^3}\right)\right),$$

where third order and higher terms in $\frac{r'}{r}$ are included in $O(\frac{r'^3}{r^3})$ and are relatively negligible when $r \gg r'$ (assuming that the larger terms are non-zero). Substituting this expansion into the electric potential,

$$V(\boldsymbol{r}) = \int \frac{1}{4\pi\varepsilon_0 r} dq + \int \frac{r' \cos \theta}{4\pi\varepsilon_0 r^2} dq + \int \frac{(3 \cos \theta - 1)r'^2}{8\pi\varepsilon_0 r^3} dq + \cdots .$$

Observe that for large $r \gg r'$, the primary dominant term is

$$\int \frac{1}{4\pi\varepsilon_0 r} dq = \frac{Q}{4\pi\varepsilon_0 r}$$

where Q is the total charge of the distribution which is also known as the monopole moment. This makes sense as any charge distribution should look like a point charge commensurate with the total charge if we zoom out of the picture far enough. However, for a charge distribution with zero charge, the first term is zero such that the next term in line is

$$\int \frac{r' \cos\theta}{4\pi\varepsilon_0 r^2} dq = \int \frac{r' \cdot \hat{r}}{4\pi\varepsilon_0 r^2} dq$$

since $r' \cdot \hat{r} = r' \cos\theta$. As r is constant in the context of integrating over the charge distribution, the above is equivalent to

$$\frac{\left(\int r' dq\right) \cdot \hat{r}}{4\pi\varepsilon_0 r^2} = \frac{p \cdot \hat{r}}{4\pi\varepsilon_0 r^2}$$

where the dipole moment p is defined as

$$p = \int r' dq. \tag{6.20}$$

Now, an important property of p is its independence of the choice of origin if the total charge of the distribution is zero (which is precisely when we want to examine it). To prove this, consider the dipole moment p' with respect to a separate origin O' that is at a position vector a relative to O.

$$p' = \int (r' - a) dq = \int r' dq - a \int dq = p$$

since $\int dq = Q = 0$ under our assumption of a collectively neutral distribution. This property also enables us to move the origin close to the charge distribution such that $r \gg r'$ is satisfied.

As seen from the above, the dipole moment of a charge distribution is merely a term involved in the expansion of the potential function (known as the multipole expansion) and is significant in describing the distant electric field of a distribution that is neutral overall (assuming that the dipole moment is non-zero). If the dipole moment too vanishes, we have to look at the higher-order moments, like quadrupole and octopole moments, to describe the far-field behaviour of the charge distribution. For our purposes, the dipole moment usually suffices such that the distant potential of a charge distribution with dipole moment p is

$$V(r) = \frac{p \cdot \hat{r}}{4\pi\varepsilon_0 r^2} \tag{6.21}$$

which is identical to that generated by a physical dipole. Taking the negative gradient in cylindrical coordinates, the distant electric field is also akin to that of a physical dipole.

$$E(r) = \frac{1}{4\pi\varepsilon_0 r^3} [3(p \cdot \hat{r})\hat{r} - p]. \tag{6.22}$$

6.4.3 *Dielectrics*

In insulators, charge carriers are not free to move and can only deviate slightly from their equilibrium positions. A dielectric is an insulating material whose molecules are polarized in the presence of an external electric field, such that molecular dipoles arise. There are mainly two phenomena that may occur. If permanent dipoles exist within the internal structures of the molecules of the dielectric (e.g. water), the polar molecules will experience a torque due to the external field that tends to rotate them into the stable equilibrium configuration. Otherwise, even if the molecules are non-polar and neutral, there may also be induced net charges in certain regions of the molecules as the external field produces forces on positive and negative charges in opposite directions — the nucleus and electron cloud in an atom thus tend to be "torn apart" and their centers no longer coincide. Either way, the molecules in a dielectric will look something like Fig. 6.15 in the presence of an external electric field. The exact cause of polarization is not of concern.

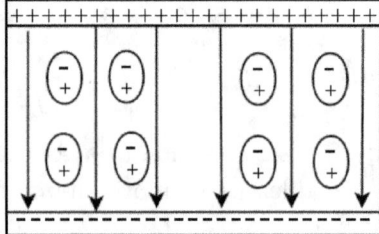

Figure 6.15:   Dielectric in an external electric field

The effect of a dielectric is to reduce the net electric field in it by generating its own electric field that opposes the external field, via induced dipoles.

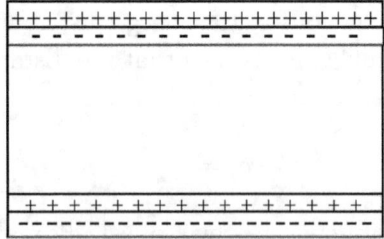

Figure 6.16:   Charges due to dielectric under uniform polarization

The equivalent system of charges due to molecules in a dielectric under uniform polarization is shown in Fig. 6.16. The positive charge on the bottom end of one molecule "cancels" the negative charge on the top end of another molecule below it. Hence, there are only net charges left at the ends of the dielectric. These charges evidently generate an electric field that opposes the original electric field.

6.4.4 Bound Charges

With the physical scenario in mind, let us proceed with a more quantitative analysis. Firstly, we can define the polarization density P of a dielectric as the dipole moment per unit volume, induced by the external electric field. We do not understand the exact mechanism behind the creation of such dipoles but this is perfectly fine as it turns out that P alone characterizes the effects of the dielectric to a large extent. Actually, the expression for P is very difficult to be determined directly from the external field in general as the external field induces some dipoles which then produce their own fields that spur the creation, destruction or realignment of other dipoles and so on. Therefore, we will instead examine the ramifications of a polarization density P while assuming that it is a given. In this section, keep in mind that when we refer to the electric field, we mean the electric field due to the polarization P only (there could be fields due to external charges added to the dielectric and charges outside of the dielectric which are excluded).

Firstly, consider an infinitesimal cuboid of edge lengths dx, dy and dz with a polarization P in the z-direction in Fig. 6.17.

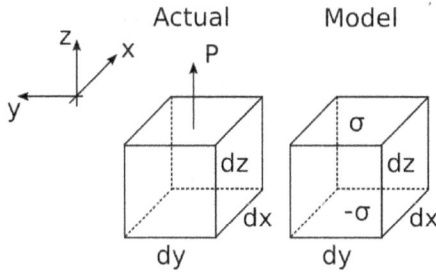

Figure 6.17: Actual cuboid and equivalent charge distribution

Its dipole moment is

$$p = Pdxdydz\hat{k}.$$

Since we have shown that the distant field of an arbitrary charge distribution with zero total charge is only dependent on its dipole moment and independent of its inner workings, we can assert that the distant field due to this cuboid is equivalent to a cuboid with uniform surface charge densities $\sigma = P$ and $-\sigma = -P$ painted on the upper and lower surfaces that are perpendicular to the z-axis. To show this, we can exploit the fact that dipole moments add. That is, we can sum the dipole moment of a portion of charges and the dipole moment of the leftovers to obtain the total dipole moment. Applied to this situation, we can sum the dipole moments of pairs of infinitesimal charges dq and $-dq$ that lie on the upper and lower surfaces, with the same (x, y) coordinates. Observing that each pair forms a physical dipole with dipole moment $dq dz \hat{k}$, the total dipole moment of the right cuboid must be

$$\boldsymbol{p} = (\sigma dx dy) dz \hat{\boldsymbol{k}}$$

after summing over all pairs. Hence, we require

$$\sigma = P$$

for the dipole moments to match. A pivotal caveat in this analysis is that the electric fields due to the two different cuboids in the regions within them are most probably dissimilar as they depend on the internal structure of the dipoles (but it turns out that we are not interested in the exact internal field anyway). We are only claiming that the right cuboid is merely a model for the left cuboid for the exterior electric field at points at distances, much larger than the length scale of the cuboids, from the cuboids.

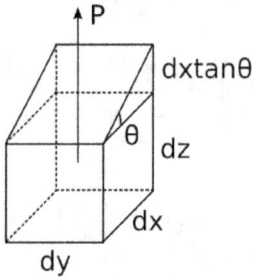

Figure 6.18:   Cuboid with inclined surface

Now, what should be the equivalent charge distribution if, as in Fig. 6.18, one face of the original cuboid is now inclined such that its normal vector now subtends an angle θ with the z-direction ($\boldsymbol{P}$ is still in the z-direction)? Well, if we slice off the additional wedge on top of the cuboid and calculate its equivalent surface charge distribution, we can then superpose the equivalent distributions of the wedge and the cuboid to obtain our desired answer. To this end, consider the wedge in Fig. 6.19.

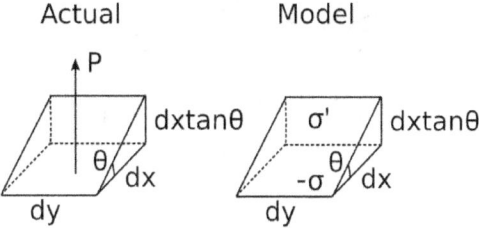

Figure 6.19: Actual wedge and equivalent charge distribution

Its dipole moment is

$$p = \frac{1}{2}dx^2 dy \tan\theta P \hat{k}$$

which suggests that we should place surface charge densities σ' and $-\sigma$ on the "hypotenuse" and the bottom base. Assume that the surface charge densities are uniform over the respective surfaces as we wish to propose a simple equivalent charge distribution. Notice that if we still hope to consider charges on the two surfaces in pairs (i.e. physical dipoles), we must have

$$\sigma' = \sigma \cos\theta,$$

as the area of the "hypotenuse" is $\frac{1}{\cos\theta}$ times larger than the bottom base. Summing the dipole moments of all corresponding pairs, the dipole moment of the wedge is the total magnitude of charge on either surface, multiplied by the average distance between a pair of corresponding charges $\left(\frac{dx \tan\theta}{2}\right)$.

$$p = \sigma dx dy \cdot \frac{dx \tan\theta}{2} \hat{k}$$

where we have accounted for the direction as well. Matching our proposed model with the true dipole moment, we must have

$$\sigma = P,$$

$$\sigma' = P \cos\theta.$$

Therefore, our original trapezoidal element (left in Fig. 6.20) has surface charge density $P \cos\theta$ on its upper slanted surface and $-P$ on its lower surface.

Observe that the surface charge density on each surface of the element can be neatly summarized by

$$\sigma = \boldsymbol{P} \cdot \hat{n},$$

where $\hat{n}$ is the normal unit vector (pointing outwards of the surface). Next, notice that we can apply a similar argument for elements with other inclined

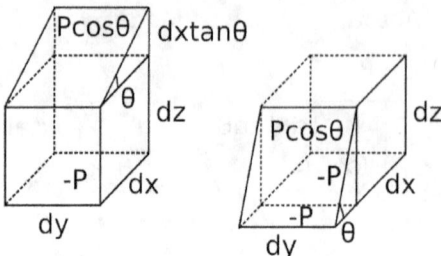

Figure 6.20:   Cuboid with inclined surfaces

faces (slice off a wedge) such as that on the right of Fig. 6.20. This shows that the relationship is valid, regardless of how many inclined surfaces there are and where they are located. To extend this result to a general polarization direction, we simply have to split the polarization into its components and apply the principle of superposition. All in all, the equivalent surface charge distribution on an element constructed from attaching wedges to a cuboid is

$$\sigma_b = \boldsymbol{P} \cdot \hat{\boldsymbol{n}}.$$

We have added an additional subscript b to emphasize that this surface charge distribution is merely a model for the actual cuboid (which is specified by its polarization due to an external field). Such equivalent charge distributions for polarized matter are known as bound charges as they are induced by polarization and do not include the charges we add to the dielectric (known as free charges) to spark off the polarization in the first place.

Armed with the previous relationship, we can now proceed with dissecting a dielectric with significant volume. First and foremost, for fields outside of a relevant region of dielectric, we can always cut the dielectric into many cuboids (in the interior) and cuboid with inclined faces (at the exterior surface of the volume). Therefore, the electric field outside of this dielectric region is equivalent to that produced by a surface charge density

$$\sigma_b = \boldsymbol{P} \cdot \hat{\boldsymbol{n}} \tag{6.23}$$

plastered on the exterior surface of the region and certain charges in the interior which shall now be examined. Within the interior, $\boldsymbol{P}$ is generally a function of position and the equivalent bound charge distribution is obtained from conjoining many infinitesimal cubes together. After such a surgical procedure, consider the infinitesimal volume element in Fig. 6.21 with a polarization that varies over its faces.

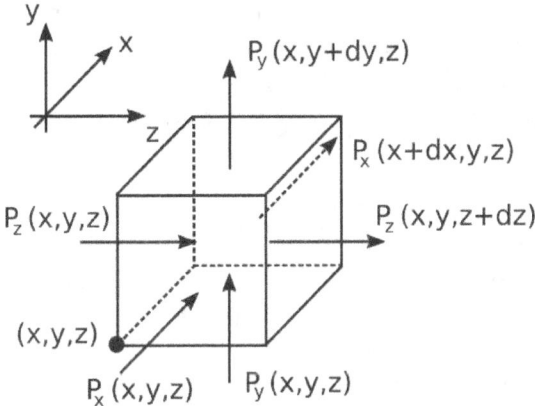

Figure 6.21: Surface and volume charges

By Eq. (6.23), the total net charge on the surface of the cube is

$$P_x(x+dx,y,z)dydz - P_x(x,y,z)dydz + P_y(x,y+dy,z)dxdz$$

$$- P_y(x,y,z)dxdz + \cdots$$

$$= \frac{\partial P_x}{\partial x}dxdydz + \frac{\partial P_y}{\partial y}dxdydz + \frac{\partial P_z}{\partial z}dxdydz,$$

by the definition of the partial derivative. Since our model of bound surface charges on infinitesimal cuboids is originally neutral overall, the cube above must also have zero net bound charge. Actually, this is ordained by the fact that a dielectric is initially neutral[9] overall (before any free charges are added to it) such that our model of equivalent bound charges must also have zero net bound charge, as we must lack the monopole term (total charge) in the multipole expansion for our model to be coherent with the true field of the polarized matter. Therefore, the negative of the above expression must be the total charge contained within the cube — implying that a bound volume charge density

$$\rho_b = -\frac{\partial P_x}{\partial x} - \frac{\partial P_y}{\partial y} - \frac{\partial P_z}{\partial z} = -\nabla \cdot \boldsymbol{P} \tag{6.24}$$

resides within the cube, where $\nabla \cdot \boldsymbol{P}$ is a short form for the preceding expression that is known as the divergence of $\boldsymbol{P}$. To recapitulate, the electric field

---

[9] All atoms, which are the building blocks of matter, are neutral precisely because of the large electrostatic forces that tend to pull charges of opposite signs together.

outside of a dielectric due to a polarization $\boldsymbol{P}$ is akin to that produced by a surface bound charge density σ_b on its exterior surface and a volume bound charge density ρ_b in its interior.

An important point to keep in mind is that our model only produces the correct electric field outside the dielectric because the electric field due to an infinitesimal cuboid is only valid at distances much larger than the size of the cuboid. Thus, the electric field produced by this model does not reflect the true field within the dielectric (as interior points are surrounded by infinitesimal cuboids) but correctly captures the field even directly outside the dielectric (note that we are comparing length scales with the size of an infinitesimal cuboid and not the entire dielectric).

Macroscopic Field Within Dielectric

Finally, what can we say about the electric field (due to polarization) within the dielectric? Well, it must be incredibly complicated as we could suddenly run into an electron, which causes the electric field to diverge towards negative infinity in its vicinity, at one location and then into a positive nucleus at the next location. Precisely because of this intricacy, it is not edifying to study the microscopic field within the dielectric. Instead, we usually consider the macroscopic field which is the electric field averaged over a small volume that encapsulates many dielectric molecules. Ideally, we would hope that our model of bound charges correctly reflects the macroscopic field within the dielectric. Though it seems like a far stretch currently, this claim is in fact true, as we shall now show!

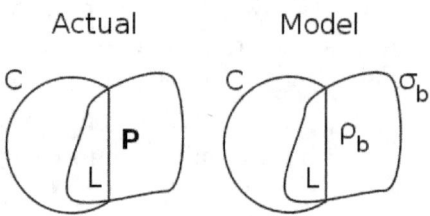

Figure 6.22:   Contour C on actual polarized matter and bound charges model

Figure 6.22 depicts a small region around a point within the dielectric (the rest of the dielectric is not shown). Consider the line integrals of the electric field over a loop C that protrudes out of this region in both the original system (which has a certain polarization $\boldsymbol{P}$) and our model of bound charges (σ_b and ρ_b). Denoting the electric fields of the original system and our model as E and E' respectively, we have

$$\oint_C E \cdot ds = 0,$$

$$\oint_C E' \cdot ds = 0,$$

over the contour C as the closed loop integral of the electrostatic field must be zero. Since $E = E'$ outside the dielectric,

$$\int_L E \cdot ds = \int_L E' \cdot ds$$

along the part of the contour C that lies within the region. Therefore, the line integrals of the electric fields E and E' between any two points on the surface of the region, along a path L entirely within the region, must be identical!

After some scrutiny, observe that this paves a way for comparing the average electric fields within the selected region of the two set-ups! Firstly, notice that the net electric field (due to polarization) at a point within the region has two causes — due to the polarized matter outside of this region (which can be assumed to be sufficiently distant from this region for our bound charge model to accurately reflect the true field) and the dielectric within this region (where the previous assumption is no longer true). By choosing a parallel bundle of lines (such as an array of vertical lines) throughout the dielectric and equating the line integrals, we can argue that the average fields within the two set-ups in the figure above, due to the dielectric within the demarcated region, must be identical! Thus, our model of bound charges, surprisingly, gives the correct macroscopic field within the dielectric. Henceforth, it shall be understood that when we refer to the field inside a dielectric, we mean the macroscopic field.

Problem: Determine the electric fields outside and inside of a dielectric sphere with radius R and a uniform polarization density P.

Define the origin at the center of the sphere and the positive z-axis to be along the direction of P. Let θ be the angle subtended by the position vector of a point on the surface of the sphere and the positive z-axis. Since the bound surface charge density at a point on the surface of the sphere is

$\sigma_b = \boldsymbol{P} \cdot \hat{\boldsymbol{n}}$ where $\hat{\boldsymbol{n}}$ is the normal vector of an infinitesimal area element there,

$$\sigma_b(\theta) = P \cos \theta.$$

Furthermore, since $\boldsymbol{P}$ is uniform, there are no bound volume charges induced within the sphere. We are left with finding the electric field engendered by this charge distribution. Now, we can make the astute observation that this charge density that varies with $\cos \theta$ on the surface and has zero volume charge density can actually be constructed by superposing two spheres of the same radius R and uniform volume charge densities ρ and $-\rho$ in Fig. 6.23. Their centers are separated by a small z-distance d.

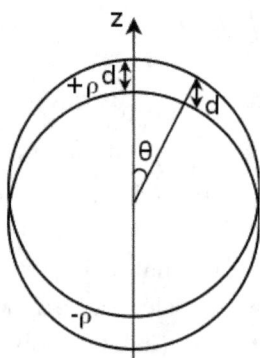

Figure 6.23:   Superposition of spheres with volume charge densities $-\rho$ and ρ

Evidently, the volume charge density is zero in the overlapping region such that as we take $d \to 0$, the two spheres converge to produce zero volume charge density and solely a surface charge density. To see why the surface charge density of this set-up varies with $\cos \theta$, observe that the vertical height of the non-overlapping region is always d, but the radial distance is $d \cos \theta$ at an angle θ from the z-axis. As the two spheres converge, the volume charges are compressed into surface charges and the surface charge density as a function of θ is determined by the radial distance of the non-overlapping portion multiplied by the volume charge density. Therefore, the surface charge density of the sphere formed by bringing these two spherical charge distributions together is

$$\sigma(\theta) = \rho d \cos \theta.$$

For this set-up involving the two spherical distributions to properly reflect the bound surface charge density, we must set

$$\rho d = P.$$

That is, as we take $d \to 0$, we must also take $\rho \to \infty$ such that the product ρd is maintained at P. Since we know that a uniform sphere of charge is akin to a commensurate point charge located at its center for regions outside the sphere, the two spheres can be reduced to two point charges $\frac{4}{3}\pi\rho R^3$ and $-\frac{4}{3}\pi\rho R^3$ that are separated by a vertical distance d when calculating the electric field outside them. As d tends to zero, what we obtain is an ideal dipole at the origin (center of the original sphere) with the following dipole moment!

$$p = \frac{4}{3}\pi\rho R^3 \cdot d = \frac{4}{3}\pi R^3 P.$$

In vector notation,

$$\boldsymbol{p} = \frac{4}{3}\pi R^3 \boldsymbol{P}.$$

The electric field at a position vector $\boldsymbol{r}$ from the origin is then given by Eq. (6.16) as

$$\boldsymbol{E} = \frac{2R^3 P \cos\theta}{3\varepsilon_0 r^3}\hat{\boldsymbol{r}} + \frac{R^3 P \sin\theta}{3\varepsilon_0 r^3}\hat{\boldsymbol{\theta}}$$

in spherical coordinates — an incredibly elegant result! To determine the interior field, we first specify the potential on the spherical surface which is simply that due to an ideal dipole (Eq. (6.15)).

$$V(\theta) = \frac{p\cos\theta}{4\pi\varepsilon_0 R^2} = \frac{Pz}{3\varepsilon_0}$$

where z represents the z-coordinate of a point on the surface. Now that we have specified the potential on the boundary of the interior volume and the interior charge distribution (zero everywhere), we are ensured that the electric potential and electric field within the sphere are unique by the first uniqueness theorem. We can easily guess a solution for the general potential as $V(z) = \frac{Pz}{3\varepsilon_0}$ where z is the z-coordinate of an interior point. Since this obviously satisfies the boundary potential, let us check if it is coherent with the charge distribution within the sphere. The electric field associated with this proposed potential is

$$\boldsymbol{E}(z) = -\nabla V = -\frac{P}{3\varepsilon_0}\hat{\boldsymbol{k}}$$

which is uniform — implying that there is no net charge anywhere within the sphere as the closed surface integral of the electric field is zero about any surface within the sphere. Having met the conditions of the first uniqueness theorem, we can now confidently claim that the electric field within the sphere is given by the above expression.

6.4.5   *Electric Displacement*

Now that we have determined the bound charges which describe the electric field solely due to polarization, we can apply Gauss' law to an arbitrary Gaussian surface S that bounds a volume V in a homogeneous dielectric. In this section, we are interested in the net electric field which is generated by both polarized matter and free charges. Then,

$$\oiint_S \boldsymbol{E} \cdot d\boldsymbol{A} = \frac{q_f + q_b}{\varepsilon_0}.$$

The left-hand side refers to the electric flux cutting across S while q_f and q_b refer to the total free and bound charges enclosed by S. q_b is the volume integral of ρ_b performed over the volume V.

$$q_b = \iiint_V \rho_b dV.$$

However, we also know that the dielectric is initially neutral (neglecting the free charges) such that q_b must be the negative of the total bound surface charge on the Gaussian surface S! Since the bound surface charge on an infinitesimal area dA on the Gaussian surface is $\boldsymbol{P} \cdot d\boldsymbol{A}$,

$$q_b = -\oiint_S \boldsymbol{P} \cdot d\boldsymbol{A}.$$

Substituting this back into the first equation,

$$\oiint_S (\varepsilon_0 \boldsymbol{E} + \boldsymbol{P}) \cdot d\boldsymbol{A} = q_f.$$

This expression is extremely convenient as it directly relates the closed surface integral of a certain quantity to the free charges within a dielectric, without mentioning the bound charges which are difficult to determine directly. The quantity in the brackets on the left-hand side is defined as the electric displacement $\boldsymbol{D}$.

$$\boldsymbol{D} = \varepsilon_0 \boldsymbol{E} + \boldsymbol{P}. \tag{6.25}$$

The previous equation then becomes

$$\oiint_S \boldsymbol{D} \cdot d\boldsymbol{A} = q_f. \tag{6.26}$$

That is, the surface integral of $\boldsymbol{D}$ over a closed surface S is the total free charge enclosed in S. We thus have a "Gauss' law" for $\boldsymbol{D}$.

Boundary Conditions

At the interface between two inhomogeneous dielectric media 1 and 2, we have the following boundary conditions. Firstly, the component of electric field in the plane of the interface must be continuous to ensure that the closed loop integral of the electric field over an infinitesimal loop around a point along the boundary is zero.

$$E_{\|2} = E_{\|1}. \tag{6.27}$$

Next, applying Gauss' law for D to an infinitesimal pillbox with negligible height and bases with area vectors dA normal to the interface, we must have a discontinuity in the normal component of D at the interface.

$$D_{\perp 2}dA - D_{\perp 1}dA = \sigma_f dA$$

$$D_{\perp 2} = D_{\perp 1} + \sigma_f, \tag{6.28}$$

where σ_f is the free surface charge density at the point of concern on the interface and where the positive directions of $D_{\perp 1}$ and $D_{\perp 2}$ have been set to be along the normal direction pointing from medium 1 to medium 2.

6.4.6 Linear Dielectrics

Up till now, we have not substituted an expression for P which is incredibly difficult to determine. However, for linear dielectrics, the polarization density P at a particular point is directly proportional to the net electric field E (due to both bound and free charges) at that point.

$$P = \varepsilon_0 \chi_e E \tag{6.29}$$

where χ_e is a constant known as the electric susceptibility of the linear dielectric. Concomitantly, the electric displacement D at every point is also proportional to E.

$$D = \varepsilon_0 E + P = \varepsilon_0(1 + \chi_e)E.$$

The constant of proportionality is often written as ε which is known as the permittivity of the dielectric.

$$D = \varepsilon E \tag{6.30}$$

where $\varepsilon = (1 + \chi_e)\varepsilon_0$. The permittivity of the dielectric is also commonly written as

$$\varepsilon = \kappa\varepsilon_0, \tag{6.31}$$

where κ is termed the dielectric constant and is related to the electric susceptibility by

$$\kappa = 1 + \chi_e. \tag{6.32}$$

With all these related definitions out of the way, Gauss' law for the electric displacement yields

$$\oiint_S \varepsilon \boldsymbol{E} \cdot d\boldsymbol{A} = q_f$$

$$\oiint_S \boldsymbol{E} \cdot d\boldsymbol{A} = \frac{q_f}{\varepsilon} \tag{6.33}$$

for a closed surface S drawn entirely in a homogeneous[10] dielectric. We see that the effect of a homogeneous linear dielectric is to "modify" the permittivity of the region it occupies though its exact mechanism operates by inducing bound charges to oppose an external field. The electric field in a homogeneous dielectric is then that if the medium were vacuum, decreased by a factor of $\frac{1}{\kappa}$.

$$\boldsymbol{E} = \frac{\varepsilon_0}{\varepsilon} \boldsymbol{E}_{vac} = \frac{\boldsymbol{E}_{vac}}{\kappa}.$$

Another perspective is to simply replace ε_0 in $\boldsymbol{E}_{vac}$ by ε. One main application of a dielectric is to enhance the capacitance of a capacitor. If the region between the two conductors of a capacitor is completely filled with a homogeneous dielectric, the electric field and hence potential difference between the conductors is decreased by a factor of $\frac{1}{\kappa}$, which implies that the capacitance is increased by a factor of κ.

$$C = \kappa C_{vac}. \tag{6.34}$$

For a parallel-plate capacitor consisting of two plates of area A, separation d and a dielectric with dielectric constant κ that occupies the gap between them,

$$C = \kappa \varepsilon_0 \frac{A}{d} = \varepsilon \frac{A}{d}. \tag{6.35}$$

The volume exterior to the plates does not need to be filled with the dielectric as $E_{vac} = 0$ outside of the gap — implying that there will be no polarization there anyway.

---

[10] A prerequisite of homogeneity is that no interfaces are included in the closed surface S. If the surface contains boundaries between different dielectrics, Eq. (6.33) may not be true as ε varies across regions, and because bound surface charges at the boundary have to be considered.

Solving Linear Dielectric Problems

A general problem in linear dielectrics is to solve for a certain quantity (net electric field, potential, bound charge distribution, etc) due to a set-up involving a dielectric placed under certain external conditions (free charges, exterior charges or external electric field). Unfortunately, such problems are extremely difficult and our solutions will mostly rely entirely on guesswork. However, recall that there is something that goes hand-in-hand with guessing — uniqueness theorems! Thankfully, our tools developed in the section on conductors can be transferred to linear dielectrics.

Recall that uniqueness theorems require us to specify the volume charge distribution within a given volume of interest. In the case of dielectrics, this comes from both free and bound charges. Therefore, we ideally want to show that we can determine the bound volume charge density when given the free volume charge density. This can be performed as follows. In a homogeneous linear dielectric,

$$\rho_b = -\nabla \cdot \boldsymbol{P} = -\nabla \cdot \chi_e \varepsilon_0 \boldsymbol{E} = -\frac{\chi_e \varepsilon_0}{\varepsilon} \nabla \cdot \boldsymbol{D} = -\frac{\kappa - 1}{\kappa} \rho_f, \tag{6.36}$$

where κ is the dielectric constant and where we have applied Gauss' law for $\boldsymbol{D}$ to an infinitesimal cuboid in writing $\nabla \cdot \boldsymbol{D} = \rho_f$. Furthermore, there are also boundary conditions at the interface of different dielectric media. Firstly, the parallel component of electric field must be continuous, i.e.

$$\boldsymbol{E}_{\|2} = \boldsymbol{E}_{\|1}.$$

Next, we can rewrite the second boundary condition (Eq. (6.28)) between two linear dielectrics with permittivities ε_1 and ε_2 as

$$\varepsilon_2 E_{\perp 2} = \varepsilon_1 E_{\perp 1} + \sigma_f, \tag{6.37}$$

or in terms of potentials V_1 and V_2 in the two dielectrics,

$$\varepsilon_2 \frac{\partial V_2}{\partial n} = \varepsilon_1 \frac{\partial V_1}{\partial n} - \sigma_f, \tag{6.38}$$

where $\frac{\partial}{\partial n}$ indicates the partial derivative in the normal direction (pointing from media 1 to 2). However, note that the potentials V_1 and V_2 must still be continuous along the interface.

$$V_1 = V_2 \quad \text{(along interface)}. \tag{6.39}$$

It turns out that if we specify the free charge distribution (volume and surface) everywhere in a volume Ω and the potential everywhere on the surface S of Ω, there will only be a unique electric field in Ω that is coherent with the volume free charge density and the boundary conditions above (at dielectric interfaces within Ω). The proof of this analogous first uniqueness theorem shall be omitted.

Problem: A dielectric sphere of radius R and dielectric constant κ, which is initially neutral everywhere, is placed in a region of uniform external field $\boldsymbol{E}_0$. Determine the resultant electric field in all space.

Firstly, since there are no free volume charges in the sphere, there must not be any net volume charges anywhere within the sphere in the final configuration. Furthermore, we can exploit the ambiguous origin of the external field $\boldsymbol{E}_0$ to say that it is produced by two opposite charges (of appropriate magnitudes) placed at diametrically opposite points with respect to the sphere, at distances much greater than R from the sphere. We need not state the explicit charge magnitude and distance from the sphere, as our objective is only to show that it is possible to specify an external charge configuration that produces $\boldsymbol{E}_0$. Having specified the charge distribution in all space, we assert that since the potential at infinity must tend to zero, the first uniqueness theorem guarantees that the electric field in all space is unique.

Now, we can safely guess a solution to the net electric field within the sphere. It is always beneficial to start from the field within the dielectric so that we can first solve for the polarization and thus the field due to polarization (which extends outside of the sphere). A natural instinct is to seek for the simplest solution — which in this case is a uniformly polarized sphere. Suppose that the resultant polarization of the sphere is a uniform $\boldsymbol{P}$ along the direction of $\boldsymbol{E}_0$. If we let the net electric field within the sphere be $\boldsymbol{E}_{in}$, which should be uniform to engender a uniform polarization, we have

$$\boldsymbol{P} = (\kappa - 1)\varepsilon_0 \boldsymbol{E}_{in}.$$

However, we also know from the example problem about a uniformly polarized sphere that the electric field due to polarization, within the sphere, is

$$\boldsymbol{E}_{in}^{pol} = -\frac{\boldsymbol{P}}{3\varepsilon_0}.$$

Since we have $\boldsymbol{E}_{in} = \boldsymbol{E}_{in}^{pol} + \boldsymbol{E}_0$,

$$\boldsymbol{P} = (\kappa - 1)\varepsilon_0(\boldsymbol{E}_{in}^{pol} + \boldsymbol{E}_0) = (\kappa - 1)\varepsilon_0\left(-\frac{\boldsymbol{P}}{3\varepsilon_0} + \boldsymbol{E}_0\right)$$

$$\implies \boldsymbol{P} = \frac{3(\kappa - 1)}{\kappa + 2}\varepsilon_0 \boldsymbol{E}_0,$$

$$E_{in}^{pol} = -\frac{\kappa - 1}{\kappa + 2}E_0,$$

$$E_{in} = \frac{3}{\kappa + 2}E_0.$$

We see that our assumption has worked as the polarization and net electric field within the sphere are indeed uniform! In making this guess, we should have already anticipated this self-consistency to a certain extent as a uniform polarization requires a uniform net electric field, and yet a uniformly polarized sphere produces a uniform electric field due to polarization within itself, which, combined with the uniform external field, results in a uniform net electric field!

The net electric field outside the sphere is simply the superposition of E_0 and the field due to a sphere with uniform polarization P (which we have previously derived). Adopting spherical coordinates about the center of the sphere,

$$E_{out}(r) = E_0\hat{k} + \frac{2R^3 P \cos\theta}{3\varepsilon_0 r^3}\hat{r} + \frac{R^3 P \sin\theta}{3\varepsilon_0 r^3}\hat{\theta}$$

$$= E_0\cos\theta\hat{r} - E_0\sin\theta\hat{\theta} + \frac{2(\kappa - 1)R^3 E_0 \cos\theta}{(\kappa + 2)r^3}\hat{r}$$

$$+ \frac{(\kappa - 1)R^3 E_0 \sin\theta}{(\kappa + 2)r^3}\hat{\theta}$$

$$= \left(\frac{2(\kappa - 1)R^3}{(\kappa + 2)r^3} + 1\right)E_0\cos\theta\hat{r} + \left(\frac{(\kappa - 1)R^3}{(\kappa + 2)r^3} - 1\right)E_0\sin\theta\hat{\theta}$$

where the positive z-axis has been defined to be along E_0. Finally, note that we did not need to check if the boundary conditions are satisfied at the dielectric-vacuum interface as we proposed that the electric fields outside and inside the sphere are caused by the same charge distribution such that the boundary conditions are naturally satisfied. If we suggested different charge distributions for different regions (e.g. disparate image charge configurations), we have to check if the boundary conditions are fulfilled.

6.4.7   Force on Dielectrics

Due to the formation of dipoles in molecules and atoms, the electric fields at the locations of positive and negative charges of a dipole are generally different — resulting in a net force on the dielectric. However, this effect is often entirely due to the fringe fields which are notoriously difficult to solve for. That said, a slick way of accurately evaluating the force on a dielectric, based on the principle of virtual work, exists. Consider the following example.

Problem: A parallel-plate capacitor of length l, width w and plate separation d is partially filled with a dielectric with width w, thickness d and permittivity ε. If the current potential difference between the plates is V, determine the force on the dielectric due to the capacitor when length x of the dielectric is sandwiched between the plates. The width of the plates and the dielectric are aligned.

Well, one might expect no force on the dielectric, despite the separation of charges, as the field between the plates is uniform! However, this is not true for the fields at the edges of the plates which curve away from the plate and are the true cause behind the force on the dielectric. To evaluate the force on the dielectric due to the capacitor, we take an indirect route by first determining the external force F_{ext} required to maintain the dielectric at static equilibrium. Suppose that the dielectric is further driven into the plates by a virtual distance δx, the total virtual work performed by the entire system (the external force, capacitor plates and other external entities) must be zero by the principle of virtual work.

$$\sum W = 0.$$

The exact forms of work performed depend on the actual set-up. We will consider the two cases where the parallel plates are electrically isolated such that the charges on the plate remains constant and where the parallel plates are connected by a battery such that the potential difference between them is constant. In the first case, the only forms of work performed on the system are that by the external force and that by the capacitor. Then,

$$\delta W_{ext} + \delta W_{cap} = 0.$$

Next, we can exploit the fact that $W_{cap} = -\Delta U_{cap}$ and $U_{cap} = \frac{q^2}{2C}$ where q is the charge on the capacitor and C is the capacitance. Then,

$$F_{ext}\delta x = dU_{cap}$$

$$F_{ext} = \frac{dU_{cap}}{dx} = -\frac{q^2}{2C^2}\frac{dC}{dx}.$$

To compute $\frac{dC}{dx}$, observe that the equivalent capacitance, when length x of the dielectric lies within the plates, is that of two capacitors of capacitances $\varepsilon_0\frac{(l-x)w}{d}$ and $\varepsilon\frac{xw}{d}$ connected in parallel as the potential difference across both regions must be identical at equilibrium in light of the conducting plates.

$$C = \varepsilon_0 \frac{(l - x)w}{d} + \varepsilon \frac{xw}{d},$$

$$\frac{dC}{dx} = \frac{(\varepsilon - \varepsilon_0)w}{d},$$

$$F_{ext} = -\frac{q^2}{2C^2} \cdot \frac{(\varepsilon - \varepsilon_0)w}{d} = -\frac{(\varepsilon - \varepsilon_0)V^2 w}{2d}.$$

Since F_{ext} is the external force required to maintain the dielectric in static equilibrium, the force on the dielectric due to the capacitor must be $-F_{ext}$.

$$F = -F_{ext} = \frac{(\varepsilon - \varepsilon_0)V^2 w}{2d}$$

which is independent of the length of dielectric inside the capacitor, x. Since this quantity is positive (towards increasing x), the capacitor tends to pull the dielectric into itself. Proceeding with the second set-up where the plates are maintained at a constant potential difference by a battery, one might think that the forms of work involved are exactly the same as the previous case. However, observe that the total charges on the plates are now variable — implying that work W_{bat} must be done by the battery in driving charge over a constant potential difference. The principle of virtual work in this case yields

$$\delta W_{ext} + \delta W_{bat} = \delta U_{cap}.$$

Suppose that the positive plate gains charge δq while the negative plates loses the same amount of charge when the length of the dielectric inside the capacitor is increased by δx. The work done by the battery in transferring this charge across the potential difference V is then

$$\delta W_{bat} = V \delta q$$

$$F_{ext} = \frac{dU_{cap}}{dx} - V \frac{dq}{dx}.$$

To evaluate $\frac{dU_{cap}}{dx}$, it is more convenient to express U_{cap} as $\frac{1}{2}CV^2$ in light of the constancy of V.

$$\frac{dU_{cap}}{dx} = \frac{1}{2}V^2 \frac{dC}{dx}.$$

We then express $\frac{dq}{dx}$ in terms of $\frac{dC}{dx}$ by substituting $q = CV$.

$$\frac{dq}{dx} = V\frac{dC}{dx}$$

$$F_{ext} = \frac{1}{2}V^2\frac{dC}{dx} - V^2\frac{dC}{dx} = -\frac{1}{2}V^2\frac{dC}{dx},$$

which is the same expression as before. Substituting the expression for $\frac{dC}{dx}$ and taking the negative of the result produces the force on the dielectric due to the capacitor.

$$F = -F_{ext} = \frac{(\varepsilon - \varepsilon_0)V^2 w}{2d}.$$

It makes sense that the results of the two set-ups are coherent. The current force on the dielectric should only depend on the charge distributions in the capacitor and the dielectric at the current instance — this is characterized by the current potential difference V and the length of the dielectric within the plates, x. The different external connections only serve to affect the charge distributions in the future. Finally, this discussion also suggests that we have the liberty of devising our own set-ups in determining the force on a certain object via the principle of virtual work as all set-ups should yield the same answer.

Problems

Conductors

1. *Connecting Conductors**

A conducting sphere that is currently carrying a total charge q is enclosed by a neutral spherical shell of inner radius r_1 and outer radius r_2. Both the sphere and the shell are electrically isolated from the rest of the world. A thin wire is then used to connect the sphere and the inner surface of the shell. Determine the final distribution of charge everywhere.

2. *Capacitor Surfaces**

A parallel-plate capacitor is formed by two identical conducting plates, of a large common area, separated by a small distance. One plate has total charge Q_1 and the other has total charge Q_2. Though the plates are thin, they each still have two surfaces. Neglecting edge effects, calculate the surface charge density on all four surfaces at equilibrium.

3. *Collision with Infinite Conducting Plane**

A charge q of mass m is placed at a height h above a grounded, infinite conducting plane. If it is then released from rest, determine the time taken for the charge to collide with the plane.

4. *Flux of Conducting Cube***

An electrically isolated conducting cube of side length l is neutral. It contains an arbitrary number of cavities that enclose a total amount of charge Q. Determine the electric flux cutting across a $l \times l$ plane parallel to and directly above a face of the cube. There are no charges outside the cube.

5. *Induced Charges***

A spherical conducting sphere of radius R is maintained at a potential V_0 relative to infinity. It contains an arbitrary number of cavities that enclose arbitrary amounts of charge but not the center of the sphere. If a point charge Q is placed at a distance d from the center of the sphere, outside the sphere, determine the total charge residing on the exterior surface of the sphere at equilibrium.

6. Spherical Shell and Charge Revisited**

Considering the problem of a charge q_1 that is brought to a distance r from the center of a grounded spherical shell of radius R, what is the appropriate image charge configuration if the shell was not grounded and had a net charge q_3 instead? What is the force on q_1 due to the shell?

7. Semi-Infinite Conducting Plates**

A semi-infinite conducting plate covers the entire yz-plane in the region $z \geq 0$ while another semi-infinite plate covers the entire xy-plane in the region $x \geq 0$. A charge q is placed at $(l, 0, l)$. What is the corresponding image charge configuration and the force on q due to the conducting plates?

8. Sphere in Electric Field**

Determine the charge distribution on an electrically isolated, neutral conducting sphere placed in a region with a uniform external electric field E.

9. Dipole above Sphere**

Define the origin at the center of a neutral conducting sphere of radius R and set up a Cartesian coordinate system. An idealized dipole with dipole moment $p = p\hat{k}$ lies at a positive z-coordinate $r \gg R$. Determine the approximate force experienced by this dipole.

10. Cylinder in Uniform Field**

In this problem, we shall directly construct the solution to the charge distribution induced on an infinite conducting cylinder of radius R placed in an electric field E perpendicular to its cylindrical axis. The conducting cylinder is neutral and electrically isolated. Firstly, prove that the electric field is uniform within the overlapping region of two infinite, parallel cylinders of radius R and volume charge densities ρ and $-\rho$. By tweaking this set-up to satisfy the boundary conditions imposed by a particular uniqueness theorem, find the charge distribution on the surface of the conducting cylinder in the original problem.

11. Third Uniqueness Theorem**

Prove that if you specify the charge density $\rho(x, y, z)$ in a volume Ω and the normal derivative of the potential $\frac{\partial V}{\partial n}$ everywhere on the surface S bounding Ω, the electric field within Ω is uniquely determined. You do not need to be completely rigorous — an intuitive explanation is fine (e.g. by considering electric field lines).

12. *Cylindrical Conductors****

Our objective is to determine the capacitances per unit length of the following two systems: (1) two long cylindrical conductors and (2) a long cylindrical conductor and an infinite conducting plane.

(a) In three-dimensional Cartesian coordinates, two infinitely long lines at $x = -a$ and $x = a$ with $y = 0$ carry linear charge densities λ and $-\lambda$ respectively. Show that the equipotential surfaces of this set-up are infinitely long cylinders. Define the potential at the origin as zero.

(b) Consider a set-up with two infinitely long cylindrical conductors with radius R and their cylindrical axis along lines $x = -b$ and $x = b$ $(b > R)$ with $y = 0$. By maintaining the conductors at $x = -b$ and $x = b$ at potentials V_0 and $-V_0$ relative to the origin, determine the capacitance per unit length of this system of conductors in light of the previous result.

(c) An infinitely long cylindrical conductor of radius R is now placed with its center a distance $b > R$ above an infinite conducting plane that is grounded. The cylinder is parallel to the plane. Determine the capacitance of this system, per unit length of the cylinder.

Capacitors and Dielectrics

13. *Joining Capacitors**

A capacitor with capacitance C_1 is initially charged with q amount of charge. Then, its ends are connected via long conducting wires to an initially neutral conductor of capacitance C_2. Calculate the energy loss in connecting the capacitors when the system has equilibrated.

14. *Cylindrical Breakdown***

A cylindrical capacitor has an inner conductor of variable radius $a > 0$ and an outer conductor of fixed radius b. If the breakdown electric field is given by E_b (a fixed value), determine the relation between a and b such that the capacitor is able to store the most energy. What if we wish to maximize the potential difference between the two conductors?

15. *Work on Capacitor Plates***

Two large, electrically isolated capacitor plates of area A and charge densities σ and $-\sigma$ are oriented parallel to each other. Suppose that one of the plates is fixed. Determine the work done by an external force in increasing the plate separation by x, by pushing away the other plate without increasing its kinetic energy. Do this in three ways: by directly calculating work from force, considering the energy density of the electric field and the potential energy stored in a capacitor.

16. *Charge in Spherical Shell***

A conducting sphere of radius r_1 is enclosed by a thin concentric conducting shell of radius $r_2 > r_1$. Both conductors are grounded. A charge q is now placed between the sphere and the shell at a distance d $(r_1 < d < r_2)$ from the common center. Determine the total charges induced on the sphere and the shell with and without Green's reciprocity theorem.

17. *Spherical Capacitor***

Consider two concentric spherical shells of radii r_1 and r_2 with $r_1 < r_2$. If q and $-q$ amounts of charge are uniformly spread on the inner and outer shells respectively, determine the energy stored in this capacitor.

Now, consider a separate problem where a point charge $-q$ is located at the center of a spherical, conducting shell of inner and outer radii r_1 and r_2. The conducting shell is initially neutral and is electrically isolated. Determine the external work done in moving the point charge $-q$ through a narrow hole drilled in the shell to infinity, without a change in kinetic energy. Try to use the previous result.

18. *Tilted Plate***

Find the resultant capacitance of a "parallel"-plate capacitor if one of the plates were to be tilted slightly at an angle $\theta \ll 1$ with respect to the horizontal. Each place has surface area A, horizontal length l and width w. The smallest vertical distance between the plates is d.

19. *Half-Filled Capacitor***

Determine the equivalent capacitance of a parallel-plate capacitor, with plates of length l, width w (directed into the page) and plate separation

h ($h \ll l$ and $h \ll w$), that is half-filled with a triangular dielectric of permittivity ε_1.

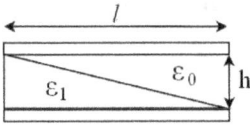

Electric Fields in Matter

20. *Rising Water**

A square parallel-plate capacitor with dimensions $a \times a$ and a separation d is placed inside a beaker of water with density ρ and dielectric constant κ. Two edges of each plate are aligned with the vertical. A battery is connected to the capacitor such that a constant potential difference V is maintained across its plates. The water between the plates rises up to a height h above the water level in the beaker. Neglecting capillary effects, determine h.

21. *Parallel Plates with Dielectrics**

Two large parallel plates with a narrow separation carry fixed, uniform surface charge densities σ and $-\sigma$ respectively. Two large slabs with identical surface areas as the plates and permittivities ε_1 and ε_2 are then slotted between the plates such that the gap within the plates is filled completely. If the slab with permittivity ε_1 is closer to the plate with surface charge density σ, determine the electric displacements, electric fields and polarizations in the two slabs. Finally, determine the bound charges everywhere.

22. *Dielectric with Cavity**

A dielectric with dielectric constant κ, that is initially neutral everywhere, fills all space. If a spherical cavity of radius R is carved and a uniform external electric field $\boldsymbol{E}_0$ permeates all space, determine the net electric field everywhere.

23. *Dipole in Dielectric Sphere***

An ideal dipole with dipole moment $\boldsymbol{p}$ is embedded at the center of a spherical dielectric with permittivity κ and radius R. Determine the resultant electric field in all space. In solving this problem, model the ideal dipole as two opposite point charges with a small separation.[11]

24. *Spherical Conductor in Dielectric**

An electrically isolated, spherical conductor, of radius R and carrying a total charge Q, is centered about the origin. The region $z \geq 0$ (excluding the conductor) is vacuum while the region $z < 0$ (excluding the conductor) is filled with a dielectric with permittivity ε. By trying a solution of the form $V(r) = \frac{A}{r}$ (where A is a constant to be determined) for the potential outside the conductor as a function of radial distance from the origin, determine the potential in all space. Assume that the conditions of this problem are set up such that the potential is unique.

25. *Charge Above Infinite Dielectric Plane**

A dielectric medium of dielectric constant κ fills the entire region $z \leq 0$. A point charge q is placed at a positive z-coordinate $z = d$ (the region $z > 0$ is in vacuum). Determine the force experienced by q.

---

[11]It turns out that the result will vary according to how we model an ideal dipole. For example, imagining it as a small sphere with a uniform polarization would lead to a radically different result!

Solutions

1. Connecting Conductors*

After the sphere and the shell are connected, they must become equipotential. Applying the corollary of the first uniqueness theorem, because the region between the inner surface of the shell and the sphere is free of net charge and bounded by a uniform potential, it must be completely devoid of electric field. Then, by applying Eq. (6.1), there must be no net charge anywhere on the inner surface of the shell and the sphere. Therefore, all of the q charge must reside on the outer surface of the shell. We claim that this q charge is evenly distributed over the outer surface. Well, the electric field produced by this configuration satisfies the condition that the outer surface is equipotential and that a total amount of q charge lies on the outer surface. Applying the second uniqueness theorem to the region outside of the shell, this electric field must be correct and hence the distribution of charge with uniform surface density $\frac{q}{4\pi\varepsilon_0 r_2^2}$ on the outer surface must also be correct by Eq. (6.1).

2. Capacitor Surfaces*

Consider the side view of the capacitors shown in Fig. 6.24.

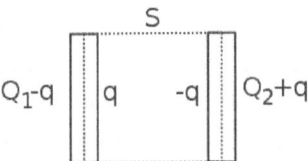

Figure 6.24: Capacitor surfaces

Let the charges on the surfaces from the left to right be $Q_1 - q$, q, $-q$ and $Q_2 + q$ respectively. These must be evenly distributed over the surfaces due to the infinite nature of the set-up. The inner surfaces must have equal magnitudes of charge of the opposite sign — this assertion can be proven by drawing the Gaussian surface S (dotted lines) shown in the figure above (its width extends in the direction normal to the page). The electric field inside the capacitors is zero — thus yielding zero net electric flux. The electric fields at the other surfaces are in the plane of the surfaces — implying that the total electric flux through this Gaussian surface is zero. Then, there must be no total net charge in the volume enclosed — causing the surfaces to have q and $-q$ charge respectively.

Now, the electric field everywhere is the superposition of that of four individual plates. The electric field due to a single plate with charge density σ is $\frac{\sigma}{2\varepsilon_0}$, emanating normally outwards from the plate. To determine q, we just need to enforce the condition that the electric field within the left capacitor is zero. If the area of the plates is A,

$$\frac{Q_1 - q}{2A\varepsilon_0} - \frac{q}{2A\varepsilon_0} + \frac{q}{2A\varepsilon_0} - \frac{Q_2 + q}{2A\varepsilon_0} = 0$$

$$q = \frac{Q_1 - Q_2}{2}.$$

Thus, the charges on the surfaces from left to right are $\frac{Q_1+Q_2}{2}$, $\frac{Q_1-Q_2}{2}$, $\frac{Q_2-Q_1}{2}$ and $\frac{Q_1+Q_2}{2}$.

Another way of solving this problem, which is arguably more insightful, is to introduce $-\frac{Q_1+Q_2}{2}$ amount of charge to both conducting plates such that their total charges become $\frac{Q_1-Q_2}{2}$ and $\frac{Q_2-Q_1}{2}$. Since these are of equal magnitude, we know that these charges must reside solely on the inner surfaces of the plates while the outer surfaces must be neutral everywhere in this new set-up (see parallel-plate capacitor in Section 6.3). The potential difference between the two plates is then $\frac{Q_1-Q_2}{2C}$ where C is the capacitance between them. The crucial observation here is that introducing the same amount of charge to both plates does not change the potential difference between them as the electric fields produced by these additional charges cancel out in the region outside of the plates. Therefore, the potential difference between the two plates in our original set-up must also be $\frac{Q_1-Q_2}{2C}$. Now, the next important observation is that the potential difference between the two plates in our original set-up is only due to the charges q and $-q$ on the inner surfaces. This can be concluded from applying Gauss' law to a Gaussian cylinder with one end inside a plate and the other end in the region between the two plates such that the only charge enclosed is on the inner surface of a plate. Therefore, the potential difference between the plates in the original set-up is $\frac{q}{C}$. Equating the two expressions for the potential difference, we obtain $q = \frac{Q_1-Q_2}{2}$ and the charges on the outer surfaces can then be computed from the conservation of charge.

3. Collision with Infinite Conducting Plane*

Using the method of image charges, the force on the charge when it is at a distance x from the plane is

$$F = -\frac{q^2}{16\pi\varepsilon_0 x^2}$$

where the negative sign reflects the fact that the force is attractive towards the plane and tends to reduce x. By Newton's second law,

$$\ddot{x} = -\frac{q^2}{16\pi\varepsilon_0 m x^2}.$$

It is important to note that if we defined x as the separation between the physical charge and the image charge instead, the force would be $F = -\frac{q^2}{4\pi\varepsilon_0 x^2}$ but $\ddot{x} = \frac{2F}{m}$ as the image charge also "accelerates" towards the physical charge. Using our current definition and expressing $\ddot{x} = \dot{x}\frac{d\dot{x}}{dx}$,

$$\int_0^{\dot{x}} \dot{x}\, d\dot{x} = -\frac{q^2}{16\pi\varepsilon_0 m}\int_h^x \frac{1}{x^2}\, dx$$

$$\dot{x}^2 = \frac{q^2(h-x)}{8\pi\varepsilon_0 m h x}$$

$$\dot{x} = -\sqrt{\frac{q^2(h-x)}{8\pi\varepsilon_0 m h x}}$$

where we have chosen the negative root, since $\dot{x} = 0$ initially and $\ddot{x} < 0$ for positive x. Adopting the substitutions $x = h\sin^2\theta$ and $dx = 2h\sin\theta\cos\theta\, d\theta$,

$$\dot{x} = 2h\sin\theta\cos\theta\frac{d\theta}{dt} = -\sqrt{\frac{q^2}{8\pi\varepsilon_0 m h}}\frac{\cos\theta}{\sin\theta}$$

$$\sqrt{\frac{q^2}{32\pi\varepsilon_0 m h^3}}\int_0^\tau dt = -\int_{\frac{\pi}{2}}^0 \sin^2\theta\, d\theta = \frac{\pi}{4}$$

$$\tau = \pi\sqrt{\frac{2\pi\varepsilon_0 m h^3}{q^2}}.$$

A simpler solution is to observe that the equation of motion can be expressed as

$$\ddot{x} = -\frac{\mu}{x^2}$$

where $\mu = \frac{q^2}{16\pi\varepsilon_0 m}$. Therefore, the analogous Kepler's third law is

$$T^2 = \frac{4\pi^2 a^3}{\mu}$$

where T is the "period" and a is the "semi-major axis" of the charge's elliptical orbit. Now, we simply have to tailor a to fit the context of this question. We can see the collision between the charge and the plane as (half) an elliptical orbit with an eccentricity $e \to 1$. Then, $a = \frac{h}{2}$ as the focus, where the origin of the inverse-square force on the charge is located, is situated at the periapsis when $e \to 1$, while the distance between the focus and the apoapsis is h based on the conditions in the problem. Therefore,

$$T^2 = \frac{4\pi^2 \cdot \left(\frac{h}{2}\right)^3}{\frac{q^2}{16\pi\varepsilon_0 m}} = \frac{8\pi^3\varepsilon_0 m h^3}{q^2}.$$

The time taken for the charge to collide with the plane is half the "period."

$$\tau = \frac{T}{2} = \pi\sqrt{\frac{2\pi\varepsilon_0 m h^3}{q^2}}.$$

4. Flux of Conducting Cube**

By Gauss' law, a total of $-Q$ charge must be induced on the surfaces of the cavities — implying that Q amount of charge resides on the surface of the cube. By Property 5 of electrostatic shielding and because there are no external charges, $\frac{Q}{6}$ amount of charge must be deposited on each face of the cube by symmetry. Next, draw a rectangular Gaussian box of infinitesimal thickness, width l and length l, parallel to a single face of the cube. One $l \times l$ surface lies directly above the cube while the other lies in the interior of the cube. We now apply Gauss' law to this Gaussian box. The electric flux through the surfaces with infinitesimal thickness is negligible. The electric flux through the $l \times l$ surface inside the cube is zero as the electric field is zero inside a conductor at equilibrium. Therefore, the total electric flux through this Gaussian box only stems from the electric flux Φ through the $l \times l$ surface above the cube (this is our objective). Gauss' law then implies

$$\Phi = \frac{Q}{6\varepsilon_0}.$$

5. Induced Charges**

By Property 7 in Section 6.2.3, the potential V_0 of the conductor is only caused by the charges on the exterior surface and the external charge Q. Suppose that a total charge q resides on the exterior surface (we do not know this distribution). Then, consider the center of the sphere. The potential here due to the charges on the exterior surface is

$$\frac{q}{4\pi\varepsilon_0 R}$$

as the charges are all equidistant from the center. This, coupled with the contribution from Q, results in the potential at the center of the sphere which must be V_0.

$$\frac{q}{4\pi\varepsilon_0 R} + \frac{Q}{4\pi\varepsilon_0 d} = V_0.$$

Then,

$$q = 4\pi\varepsilon_0 R V_0 - \frac{QR}{d}.$$

6. Spherical Shell and Charge Revisited**

The image charge configuration in this case is that in Fig. 6.6, with an additional image charge $q_3 - q_2$ at the center of the shell where q_2 is the image charge in Fig. 6.6. This will ensure that the shell is an equipotential surface with total charge q_3. Then, the electric field in the region outside the shell due to this image charge configuration must be the correct solution, as guaranteed by the second uniqueness theorem. The force on q_1 at this instance is then

$$F = \frac{q_1(q_3 - q_2)}{4\pi\varepsilon_0 r^2} + \frac{q_1 q_2}{4\pi\varepsilon_0 (r - a)^2}.$$

Substituting $q_2 = -\frac{R}{r}q_1$ and $a = \frac{R^2}{r}$ that were derived previously,

$$F = \frac{q_1(rq_3 + Rq_1)}{4\pi\varepsilon_0 r^3} - \frac{q_1^2 Rr}{4\pi\varepsilon_0 (r^2 - R^2)^2}$$

in the direction away from the shell.

7. Semi-Infinite Conducting Plates**

There are three image charges: $-q$ at $(-l, 0, l)$, $-q$ at $(l, 0, -l)$ and q at $(-l, 0, -l)$. The force on the real charge due to these image charges are then

$$\frac{q^2}{4\pi\varepsilon_0 \left(2\sqrt{2l}\right)^2} \begin{pmatrix} \frac{\sqrt{2}}{2} \\ 0 \\ \frac{\sqrt{2}}{2} \end{pmatrix} - \frac{q^2}{4\pi\varepsilon_0 \cdot (2l)^2} \begin{pmatrix} 1 \\ 0 \\ 0 \end{pmatrix} - \frac{q^2}{4\pi\varepsilon_0 \cdot (2l)^2} \begin{pmatrix} 0 \\ 0 \\ 1 \end{pmatrix}$$

$$= \frac{(\sqrt{2} - 4)q^2}{64\pi\varepsilon_0 l^2} \begin{pmatrix} 1 \\ 0 \\ 1 \end{pmatrix}.$$

8. Sphere in Electric Field**

The trick in this question pertains to how we interpret this uniform external electric field $\boldsymbol{E}$. Define the positive z-axis to point in the direction of $\boldsymbol{E}$ and the origin at the center of the sphere. We can construct $\boldsymbol{E}$ by placing two charges q and $-q$ at $(0,0,-r)$ and $(0,0,r)$ where $r \gg R$. The field in the region of the sphere is then uniform (especially after we take the limit of r to infinity later) and of magnitude

$$E = \frac{q}{2\pi\varepsilon_0 r^2}.$$

The above shows that in order for our constructed pair of charges to properly reflect the electric field $\boldsymbol{E}$ given in the problem, they must satisfy

$$\frac{q}{r^2} = 2\pi\varepsilon_0 E.$$

Moving on, this pair of charges produces an image pair of charges $-\frac{R}{r}q$ and $\frac{R}{r}q$ at $(0,0,-\frac{R^2}{r})$ and $(0,0,\frac{R^2}{r})$ respectively. Now, let us take the limit of $r \to \infty$ while maintaining the ratio of $\frac{q}{r^2}$ at $2\pi\varepsilon_0 E$. Observe that through such a maneuver, the distance between the two image charges tends to zero while their charges tend to infinity. Therefore, what we obtain is an idealized dipole with a dipole moment

$$\boldsymbol{p} = \frac{R}{r}q \cdot \frac{2R^2}{r}\hat{\boldsymbol{k}} = 4\pi\varepsilon_0 R^3 E\hat{\boldsymbol{k}}.$$

Now, adopt spherical coordinates about the center of the sphere. Applying Eq. (6.16), the electric field due to the image dipole at a point corresponding to polar angle θ in spherical coordinates, immediately outside the surface of the sphere, is

$$\boldsymbol{E}_p(\theta) = \frac{p\cos\theta}{2\pi\varepsilon_0 R^3}\hat{\boldsymbol{R}} + \frac{p\sin\theta}{4\pi\varepsilon_0 R^3}\hat{\boldsymbol{\theta}} = 2E\cos\theta\hat{\boldsymbol{R}} + E\sin\theta\hat{\boldsymbol{\theta}},$$

where $\hat{\boldsymbol{R}}$ is the radial unit vector pointing from the origin to the point of concern on the surface of the sphere and $\hat{\boldsymbol{\theta}}$ is the polar unit vector in spherical coordinates. The external electric field expressed in terms of these coordinates at the point with angular coordinate θ is

$$\boldsymbol{E} = E\cos\theta\hat{\boldsymbol{R}} - E\sin\theta\hat{\boldsymbol{\theta}}.$$

Their superposition yields the net electric field $\boldsymbol{E}_{tot}$ immediately outside of the sphere, at a point that subtends an angle θ with the positive z-axis, as

$$\boldsymbol{E}_{tot} = 3E\cos\theta\hat{\boldsymbol{R}}$$

which only has a normal component and lacks a tangential component as expected (so that the surface of the sphere is an equipotential). The surface charge density of a point on the surface of the sphere that makes an angle θ

with the positive z-axis is then

$$\sigma(\theta) = 3\varepsilon_0 E \cos\theta$$

by Eq. (6.1). Another way of solving this problem is to directly construct the surface charge configuration in a manner similar to Problem 10.

9. Dipole above Sphere**

Let the dipole be constituted by a negative charge $-q$ located at $(0,0,r-\frac{d}{2})$ and a positive charge q located at $(0,0,r+\frac{d}{2})$ such that $p = qd$. We will take two limits in our solution at two different junctures — firstly, we will take $d \to 0$ and then $r \gg R$. The image charges of this charge pair are $\frac{R}{r-\frac{d}{2}}q$ at $(0,0,\frac{R^2}{r-\frac{d}{2}})$ and $-\frac{R}{r+\frac{d}{2}}q$ at $(0,0,\frac{R^2}{r+\frac{d}{2}})$. Performing binomial expansions and discarding second order and above terms in $\frac{d}{r}$, these become $\frac{R}{r}q + \frac{dR}{2r^2}q$ at $(0,0,\frac{R^2}{r} + \frac{dR^2}{2r^2})$ and $-\frac{R}{r}q + \frac{dR}{2r^2}q$ at $(0,0,\frac{R^2}{r} - \frac{dR^2}{2r^2})$. Observe that the $\frac{R}{r}q$ and $-\frac{R}{r}q$ charges effectively form a dipole of dipole moment $\boldsymbol{p'} = \frac{R}{r}q \cdot \frac{dR^2}{r^2}\boldsymbol{\hat{k}} = \frac{R^3}{r^3}p\boldsymbol{\hat{k}}$ at $(0,0,\frac{R^2}{r})$. The remaining $\frac{dR}{2r^2}q$'s at $(0,0,\frac{R^2}{r} + \frac{dR^2}{2r^2})$ and $(0,0,\frac{R}{r}q - \frac{dR}{2r^2}q)$ effectively converge to form a charge $\frac{dR}{r^2}q = \frac{R}{r^2}p$ at $(0,0,\frac{R^2}{r})$ as $d \to 0$. Finally, remember that this is not the full image charge configuration as the sphere must be neutral overall — there must be another compensating image charge $-\frac{R}{r^2}p$ that is located at the center of the sphere (to ensure that it is equipotential). Therefore, we have another pair of opposite charges with magnitude $\frac{R}{r^2}p$ separated by a distance $\frac{R^2}{r}$ which converge to form a dipole with dipole moment $\boldsymbol{p''} = \frac{R^3}{r^3}p\boldsymbol{\hat{k}}$ at the origin when $r \gg R$. Therefore, the overall image charge configuration, when we take $r \gg R$, is an idealized dipole with total dipole moment

$$\boldsymbol{p}_{image} = \boldsymbol{p'} + \boldsymbol{p''} = \frac{2R^3}{r^3}\boldsymbol{p}$$

located at the origin. Applying Eq. (6.16), the electric field produced by this effective image dipole at a z-coordinate z along the z-axis is

$$\boldsymbol{E}_{image} = \frac{\boldsymbol{p}_{image}}{2\pi\varepsilon_0 r^3} = \frac{R^3\boldsymbol{p}}{\pi\varepsilon_0 r^6}.$$

The net force exerted by this image dipole on the original dipole (defined to be positive in the positive z-direction) is thus

$$F = -\frac{R^3 qp}{\pi\varepsilon_0 \left(r - \frac{d}{2}\right)^6} + \frac{R^3 qp}{\pi\varepsilon_0 \left(r + \frac{d}{2}\right)^6}$$

$$\approx -\frac{R^3 qp}{\pi\varepsilon_0 r^6}\left(1 + \frac{3d}{r}\right) + \frac{R^3 qp}{\pi\varepsilon_0 r^6}\left(1 - \frac{3d}{r}\right)$$

$$= -\frac{6R^3 p^2}{\pi\varepsilon_0 r^7}.$$

Note that another way of deriving this involves applying the result $F = (p \cdot \nabla)E_{image}$, where p refers to the dipole moment of the real dipole.

10. Cylinder in Uniform Field**

Firstly, let us determine the electric field $E(r)$ within an infinitely long cylinder of volume charge density ρ, at a radial distance r from the cylindrical axis. Draw a cylindrical Gaussian surface of radius r and length l whose axis is aligned with the cylinder's. Due to the symmetry and infinite nature of the set-up, the electric field is of uniform magnitude across the curved surface of the Gaussian cylinder and is solely in the radial direction. Therefore,

$$E \cdot 2\pi r l = \frac{\rho \cdot \pi r^2 l}{\varepsilon_0}$$

as the electric flux $E \cdot 2\pi r l$ through this Gaussian surface only comes from its curved surface while the charge that it enclosed is $\rho\pi r^2 l$. Therefore, the electric field at a radial distance r is

$$E(r) = \frac{\rho r}{2\varepsilon_0},$$

in vector notation where r is the radial vector, perpendicular to the cylindrical axis, pointing from the cylindrical axis to the point where the electric field is of concern. Now, consider two overlapping cylindrical distributions of charge with volume charge densities ρ and $-\rho$. Let d be the vector pointing from the center of the $-\rho$ distribution to the center of the other distribution and r_1 and r_2 be the radial vectors pointing from the centers of the cylindrical distributions of charge densities ρ and $-\rho$ to a point in the overlapping region (in the cross section containing that point). The electric field at that point is

$$\frac{\rho(r_1 - r_2)}{2\varepsilon_0} = -\frac{\rho d}{2\varepsilon_0}.$$

This shows that the electric field is uniform in the entire overlapping region! Now, when a cylindrical conductor is placed in a uniform electric field, charges will redistribute themselves along its surface until their electric field cancels the uniform external field inside the conductor. This can be achieved with our two cylindrical distributions of charges as we take $d \to 0$ if

$$E - \frac{\rho d}{2\varepsilon_0} = 0$$

$$\implies \rho d = 2\varepsilon_0 E,$$

where E is the uniform external electric field. That is, even though $d \to 0$, ρ must tend to infinity such that the value of ρd is maintained at $2\varepsilon_0 E$. As the distance d between the two cylindrical distributions tends to zero, they essentially converge to form a cylinder with zero volume charge distribution and solely a surface charge distribution

$$\sigma(\theta) = \rho d \cos\theta = 2\varepsilon_0 E \cos\theta,$$

where θ is the angle subtended by a vector pointing from the center of the cylinder to a point on the surface (in the same cross section) and the external electric field E. Refer to Fig. 6.23 for a pictorial representation to see why the $\cos\theta$ factor arises. Let us now check if the electric field produced by this charge distribution satisfies the boundary conditions for the second uniqueness theorem. Firstly, the total charge on the surface of the cylinder produced by this distribution is zero and corresponds to the total charge on the conducting cylinder (which is neutral). Secondly, the electric field on the surface of the cylinder is purely radial because the electric field is zero within the cylinder and only surface charges, whose patches produce electric fields normal to themselves, exist. Therefore, the constraint for the cylinder to be equipotential is also satisfied. The second uniqueness theorem then guarantees that the electric field produced by this charge distribution outside the cylinder is correct. Applying Eq. (6.1) to the surface of the cylinder, we can conclude that the surface charge distribution is indeed

$$\sigma(\theta) = \rho d \cos\theta = 2\varepsilon_0 E \cos\theta.$$

11. Third Uniqueness Theorem**

An intuitive proof goes as follows: suppose you had two electric fields, $E(x, y, z)$ and $E'(x, y, z)$, that satisfy the conditions given in the problem. Then, $E(x, y, z) - E'(x, y, z)$ is a solution to the electric field within the volume Ω when there is no net charge anywhere and when the normal component of electric field is zero everywhere along the surface of S (boundary of Ω). The second condition stems from the fact that specifying $\frac{\partial V}{\partial n}$ along S ensures that the normal components of E and E' are identical along S.

Since there are no net charges (which produce electric field lines) and no field lines entering the volume Ω from the outside, the entire volume Ω must be completely devoid of field lines! That is,

$$\boldsymbol{E}(x, y, z) - \boldsymbol{E}'(x, y, z) = \boldsymbol{0}$$

$$\boldsymbol{E}(x, y, z) = \boldsymbol{E}'(x, y, z),$$

which shows that the electric field is unique within Ω. For the sake of rigor, we shall now present a general proof that ties the various uniqueness theorems together. Suppose that $\boldsymbol{E}_1(x, y, z)$ and $\boldsymbol{E}_2(x, y, z)$ are two solutions to the electric field within a volume Ω with a stipulated charge density $\rho(x, y, z)$. Then, $\boldsymbol{E}_3 = \boldsymbol{E}_1 - \boldsymbol{E}_2$ is a solution to the electric field within Ω with zero charge density everywhere. By Gauss' law (you can see this as applying the integral form of Gauss' law to an infinitesimal cuboid),

$$\nabla \cdot \boldsymbol{E}_3 = \frac{\rho}{\varepsilon_0} = 0,$$

since $\rho = 0$ within Ω. If we let $V_3(x, y, z)$ denote the potential associated with $\boldsymbol{E}_3$, as $\boldsymbol{E}_3 = -\nabla V_3$,

$$\nabla \cdot \nabla V_3 = 0.$$

$\nabla \cdot \nabla V_3$ is usually written as $\nabla^2 V_3$ where ∇^2 is the Laplacian. Thus, we have

$$\nabla^2 V_3 = 0.$$

Now, consider the quantity $\nabla \cdot (V_3 \nabla V_3)$. This can be rewritten via vector calculus identities as

$$\nabla \cdot (V_3 \nabla V_3) = V_3 \nabla^2 V_3 + |\nabla V_3|^2.$$

Since $\nabla^2 V_3 = 0$ within Ω,

$$\nabla \cdot (V_3 \nabla V_3) = |\nabla V_3|^2$$

within Ω. Integrating both sides over the volume Ω,

$$\iiint_\Omega \nabla \cdot (V_3 \nabla V_3) d\Omega = \iiint_\Omega |\nabla V_3|^2 d\Omega.$$

From the divergence theorem, $\iiint_\Omega \nabla \cdot (V_3 \nabla V_3) d\Omega = \oiint_S (V_3 \nabla V_3) \cdot d\boldsymbol{A}$ where S is the surface that bounds Ω. Thus,

$$\oiint_S (V_3 \nabla V_3) \cdot d\boldsymbol{A} = \iiint_\Omega |\nabla V_3|^2 d\Omega.$$

Now, observe that if we specify sufficient boundary conditions to make the left-hand side zero, the right-hand side will be the integral of something squared that yields zero — implying that the integrand must be zero throughout Ω and we will thus have a unique electric field within Ω (as $\nabla V_3 = E_2 - E_1$). So what are the possible boundary conditions that can be imposed? For example, if we stipulate the potential on S originally, the potentials $V_1(x, y, z)$ and $V_2(x, y, z)$ associated with E_1 and E_2 must have the same value on S — indicating that $V_3 = V_1 - V_2 = 0$ throughout S! Then, the left-hand side is zero and we retrieve our first uniqueness theorem.

To see how the second and third uniqueness theorems can be deduced from the above equation, we can rewrite $\nabla V_3 = -E_3$ such that $\oiint_S (V_3 \nabla V_3) \cdot dA = - \oiint_S V_3 E_3 \cdot dA = - \oiint_S V_3 E_{3\perp} dA$, where $E_{3\perp}$ is the normal component of E_3 on S. If the entire surface S solely comprises the surfaces of conductors, let the surface of the ith conductor that is part of S be S_i. Then, V_3 is a constant over each surface (as a conductor must be equipotential). If we let the constant V_3 over the ith conductor surface be V_{3i},

$$- \oiint_S V_3 E_{3\perp} dA = - \sum_i V_{3i} \oiint_{S_i} E_{3\perp} dA = - \sum_i V_{3i} \frac{Q_i}{\varepsilon_0}$$

where Q_i is the total charge on the ith conductor in the solution E_3 (the last equality comes from Gauss' law). Therefore, if we had specified the total charge on each conductor surface beforehand, we must have $Q_i = 0$ for all i (the crucial condition in the second uniqueness theorem) and hence a unique electric field within Ω.

For the third uniqueness theorem, since we have $-E_{3\perp} = \frac{\partial V_3}{\partial n}$ where the latter is the partial derivative of V_3 in the normal direction on the surface S, we have

$$\oiint_S V_3 \frac{\partial V_3}{\partial n} dA = \iiint_\Omega |\nabla V_3|^2 d\Omega.$$

Thus, if we had indicated the partial derivative of the potential on S in the normal direction beforehand, $\frac{\partial V_1}{\partial n}$ and $\frac{\partial V_2}{\partial n}$ must have the same value on S — implying that $\frac{\partial V_3}{\partial n} = \frac{\partial V_1}{\partial n} - \frac{\partial V_2}{\partial n} = 0$. Correspondingly, the electric field within Ω must be unique and this proves our third uniqueness theorem. Incidentally, the above procedure also shows that we can mix and match boundary conditions. For example, we could specify the potential on some portions of S, the total charges on some conducting surfaces that are part of S and $\frac{\partial V}{\partial n}$ on the rest of S such that the electric field within Ω would still be uniquely determined!

12. Cylindrical Conductors***

We have shown via Gauss' law that the electric field strength at a perpendicular distance r from an infinitely long line of linear charge density λ is

$$E = \frac{\lambda}{2\pi\varepsilon_0 r}.$$

The electric field vector is directed radially outwards from the cylindrical axis. The potential at a perpendicular distance r from the line at $x = -a$, relative to the origin, is then

$$V = -\int_a^r \frac{\lambda}{2\pi\varepsilon_0 r} dr = \frac{\lambda}{2\pi\varepsilon_0} \ln\frac{a}{r}.$$

A similar statement can be made for the other line. The potential at a point (x, y, z) in space is then

$$V(x, y, z) = \frac{\lambda}{2\pi\varepsilon_0} \ln\frac{a}{\sqrt{(x+a)^2 + y^2}} - \frac{\lambda}{2\pi\varepsilon_0} \ln\frac{a}{\sqrt{(x-a)^2 + y^2}}$$

$$= \frac{\lambda}{4\pi\varepsilon_0} \ln\frac{(x-a)^2 + y^2}{(x+a)^2 + y^2}.$$

Now, we wish to determine the points that correspond to a fixed potential V.

$$\frac{(x-a)^2 + y^2}{(x+a)^2 + y^2} = e^{\frac{4\pi\varepsilon_0 V}{\lambda}}.$$

Defining the constant on the right-hand side as c for the sake of convenience,

$$(x-a)^2 + y^2 = c[(x+a)^2 + y^2]$$

$$(1-c)x^2 - 2a(1+c)x + (1-c)y^2 + a^2(1-c) = 0.$$

Simplifying the above yields

$$\left[x - \frac{a(1+c)}{1-c}\right]^2 + y^2 = \frac{a^2(1+c)^2}{(1-c)^2} - a^2$$

which is the equation of an infinitely long cylinder with its cylindrical axis at $x = \frac{a(1+c)}{1-c}$, $y = 0$ and squared radius $\frac{a^2(1+c)^2}{(1-c)^2} - a^2$. We can then apply the result of this auxiliary problem to determine the image charges in the set-up involving the two conductors of potential V_0 and $-V_0$ at $x = -b$ and $x = b$. We first consider the conductor of potential V_0. The idea is to choose the pair of infinitely long lines such that the equipotential cylinder of potential V_0 coincides with the conductor with potential V_0. Then, we have to precisely choose λ and a for the lines to achieve this. Since the axis of the conductor with potential V_0 is at $x = -b$ and possesses radius R,

$$\frac{a(1+c)}{1-c} = -b$$

$$\frac{a^2(1+c)^2}{(1-c)^2} - a^2 = R^2$$

$$a^2 = b^2 - R^2$$

$$a = \sqrt{b^2 - R^2}.$$

The positive root is chosen as one can show that the negative root would result in an invalid solution for c. To solve for λ, we substitute this expression for a into the first equation.

$$\frac{1+c}{1-c} = -\frac{b}{\sqrt{b^2 - R^2}}$$

$$c = \frac{b + \sqrt{b^2 - R^2}}{b - \sqrt{b^2 - R^2}}.$$

Substituting $c = e^{\frac{4\pi\varepsilon_0 V_0}{\lambda}}$,

$$\lambda = \frac{4\pi\varepsilon_0 V_0}{\ln \frac{b+\sqrt{b^2-R^2}}{b-\sqrt{b^2-R^2}}}.$$

Therefore, the image charges that satisfy the boundary conditions of the conductor at potential V_0 are lines of linear charge densities $\lambda = \frac{4\pi\varepsilon_0 V_0}{\ln \frac{b+\sqrt{b^2-R^2}}{b-\sqrt{b^2-R^2}}}$ and $-\lambda = -\frac{4\pi\varepsilon_0 V_0}{\ln \frac{b+\sqrt{b^2-R^2}}{b-\sqrt{b^2-R^2}}}$ at $x = -\sqrt{b^2 - R^2}$ and $x = \sqrt{b^2 - R^2}$. One can show that these are the same image charges that satisfy the boundary condition of the conductor with potential $-V_0$ as well.

Now, we simply need to determine the total induced charges on the surfaces of the conductors to compute the capacitance of this system (we already know that the potential difference is $2V_0$). Firstly, observe that the image charges are completely enclosed in the volumes of the conductors. Then, one can draw Gaussian surfaces encapsulating each cylinder and apply Gauss' law to conclude, based on the electric flux through the Gaussian surfaces, that the total induced charge on the surface of each cylinder is that of its image charge. Therefore, the capacitance per unit length of this system of conductors is

$$\frac{C}{l} = \frac{\lambda}{2V_0} = \frac{2\pi\varepsilon_0}{\ln\frac{b+\sqrt{b^2-R^2}}{b-\sqrt{b^2-R^2}}}.$$

In the final problem, drawing the image charges would yield the exact same set-up as the previous problem. However, the potential difference in this case is half of the previous case as the potential difference is taken between the plane and the cylinder, rather than between the two cylinders. The capacitance per unit length is then twice the previous answer.

$$\frac{C'}{l} = \frac{4\pi\varepsilon_0}{\ln\frac{b+\sqrt{b^2-R^2}}{b-\sqrt{b^2-R^2}}}.$$

13. Joining Capacitors*

The potential differences across the two capacitors must be identical as they are connected by conducting wires in parallel. The system is then equal to two capacitors connected in parallel with a total charge q and equivalent capacitance $C_1 + C_2$. The energy loss is then

$$\Delta U = \frac{q^2}{2(C_1 + C_2)} - \frac{q^2}{2C_1} = -\frac{C_2 q^2}{2C_1(C_1 + C_2)}.$$

14. Cylindrical Breakdown*

Suppose that the inner conductor carries a linear charge density λ while the outer conductor carries $-\lambda$. The electric field at a perpendicular distance r from the axis of the capacitor, $a \leq r \leq b$, is given by Gauss' law as

$$E = \frac{\lambda}{2\pi\varepsilon_0 r},$$

directed radially outwards from the inner cylinder. As λ is increased, the breakdown electric field will occur at $r = a$.

$$E_b = \frac{\lambda}{2\pi\varepsilon_0 a}.$$

The potential difference between the two conductors is

$$V = -\int_b^a E\,dr = \frac{\lambda}{2\pi\varepsilon_0}\ln\frac{b}{a} = E_b a \ln\frac{b}{a}.$$

The capacitance per unit length is

$$c = \frac{\lambda}{V} = \frac{2\pi\varepsilon_0}{\ln\frac{b}{a}}.$$

The stored energy per unit length is thus

$$u = \frac{1}{2}cV^2 = \frac{1}{2} \cdot \frac{2\pi\varepsilon_0}{\ln\frac{b}{a}} \cdot \left(E_b a \ln\frac{b}{a}\right)^2 = \pi\varepsilon_0 E_b^2 a^2 \ln\frac{b}{a}.$$

To maximize the energy per unit length,

$$\frac{du}{da} = 0$$

$$2a \ln\frac{b}{a} + a^2 \cdot -\frac{1}{a} = 0$$

$$\frac{b}{a} = \sqrt{e}$$

as $a \neq 0$. To maximize the potential difference between the two conductors,

$$\frac{dV}{da} = 0$$

$$\ln\frac{b}{a} + a \cdot -\frac{1}{a} = 0$$

$$\frac{b}{a} = e.$$

15. Work on Capacitor Plates**

The electric field strength at the location of one infinitely large plate, due to the other plate, is $\frac{\sigma}{2\varepsilon_0}$ by Gauss' law. Therefore, the electrostatic force on the movable plate is

$$F_{elec} = -\sigma A \cdot \frac{\sigma}{2\varepsilon_0} = -\frac{\sigma^2 A}{2\varepsilon_0},$$

where the negative sign indicates that the electrostatic force acts along the direction that tends to reduce the distance between the two plates. The external force applied in moving the plate must then be negative of this to not result any change in kinetic energy of the plate. The work done by the external force in increasing the plate separation by x is then

$$W = F_{ext}x = \frac{\sigma^2 A x}{2\varepsilon_0}.$$

Now, consider the energy density of the field. The field between the two plates has a uniform magnitude $\frac{\sigma}{\varepsilon_0}$ and direction. The field outside the gap is zero. The energy density between the plates is then

$$\frac{1}{2}\varepsilon_0 E^2 = \frac{\sigma^2}{2\varepsilon_0}.$$

The change in the potential energy associated with the electric field in increasing the plate separation by x is the change in volume, Ax, multiplied by the energy density above. This must also be equal to the work done by the external force.

$$W = \Delta U = \frac{\sigma^2 Ax}{2\varepsilon_0}.$$

The change in potential energy can also be determined via the potential energy stored in a capacitor $U = \frac{q^2}{2C}$. Then,

$$W = \Delta U = \frac{q^2}{2} \cdot \left(\frac{1}{C'} - \frac{1}{C}\right) = \frac{\sigma^2 A^2}{2}\left(\frac{d+x}{\varepsilon_0 A} - \frac{d}{\varepsilon_0 A}\right) = \frac{\sigma^2 Ax}{2\varepsilon_0}.$$

16. Charge in Spherical Shell**

First and foremost, note that the outer surface of the shell is neutral everywhere. The region outside the shell has boundaries with zero potential (the shell and infinity) and contains no net charge. Therefore, a zero potential and electric field in this entire region is a valid solution to these boundary conditions and must be the correct solution by the first uniqueness theorem. Therefore, applying Eq. (6.1) to the outer surface of the shell, we conclude that it is neutral everywhere.

Now, let the total charges induced on the sphere and the inner surface of the shell be q_1 and q_2 respectively. Firstly, applying Gauss' law to a concentric spherical Gaussian surface that is within the shell and noting that the total electric flux through it is zero, we can directly conclude that $q_1 + q_2 = -q$ since the total enclosed charge must be zero. Let set-up 1, that we will apply Green's reciprocity theorem to, be the original set-up in the problem. Let set-up 2 be the configuration where the charge q is removed and the inner sphere is maintained at a constant potential ϕ_0. Due to the symmetry of this system, when the inner sphere is maintained at potential ϕ_0, the charge distributions on all surfaces must be uniform. The shell must produce no electric field within itself. Therefore, the potential in the region between the sphere and the shell must decrease with the inverse of the radial distance from the center as the inner sphere, which is the sole contributor to the potential difference in the region between the sphere and the shell, is akin to a point charge at its center. Consequently, the potentials at radial distances

d and r_2 are $\frac{\phi_0 r_1}{d}$ and $\frac{V_0 r_1}{r_2}$ in set-up 2. Applying Green's reciprocity theorem to our chosen set-ups,

$$q_1 \phi_0 + q \cdot \frac{\phi_0 r_1}{d} + q_2 \cdot \frac{\phi_0 r_1}{r_2} = 0.$$

Solving this equation simultaneously with $q_1 + q_2 = -q$,

$$q_1 = -\frac{(r_2 - d)r_1}{(r_2 - r_1)d}q,$$

$$q_2 = -\frac{(d - r_1)r_2}{(r_2 - r_1)d}q.$$

To solve this problem without Green's reciprocity theorem, we can spread q into a uniform spherical shell of total charge q and this would not change the total charges induced on the two relevant surfaces (see argument in Section 6.3.1). Now, as an intermediate step, let us determine the capacitance between two shells of radii a and $b > a$. Place q amount of charge on the inner shell which will be distributed evenly over it due to spherical symmetry. Since a spherical shell is akin to a point charge at its center in the region outside it, the inner shell produces a potential $\frac{q}{4\pi\varepsilon_0 r}$ at a radial distance r that satisfies $a \leq r \leq b$. Therefore, the potential difference between the two shells is $\Delta V = \frac{q}{4\pi\varepsilon_0}\left(\frac{1}{a} - \frac{1}{b}\right)$. The capacitance is thus

$$C = \frac{q}{\Delta V} = \frac{4\pi\varepsilon_0}{\frac{1}{a} - \frac{1}{b}}.$$

Now, after dispersing the charge q into a spherical shell of charge, we can further deem it as two shells of total charges $-q_1$ and $-q_2$. Then, the system of the sphere, the spherical shell of charge and the outer conducting shell is akin to two spherical-shell capacitors (between the sphere of charge q_1 and the shell of charge $-q_1$ and between the shell of charge $-q_2$ and the outer shell of charge q_2) with total charges q_1 and $-q_2$ connected in series. The capacitances of the inner and outer capacitors are $C_1 = \frac{4\pi\varepsilon_0}{\frac{1}{r_1} - \frac{1}{d}}$ and $C_2 = \frac{4\pi\varepsilon_0}{\frac{1}{d} - \frac{1}{r_2}}$. Imposing the condition that there is no potential difference across this capacitor system (as the sphere and the conducting shell are grounded),

$$\frac{q_1}{C_1} - \frac{q_2}{C_2} = 0.$$

Solving this equation simultaneously with $q_1 + q_2 = -q$, we retrieve

$$q_1 = -\frac{(r_2 - d)r_1}{(r_2 - r_1)d}q,$$

$$q_2 = -\frac{(d - r_1)r_2}{(r_2 - r_1)d}q.$$

17. Spherical Capacitor**

The electric field at a radial distance r from the center of the spherical shell, $r_1 \leq r \leq r_2$, is given by Gauss' law as

$$E = \frac{q}{4\pi\varepsilon_0 r^2}.$$

The potential difference between the inner and outer shells is given by

$$\Delta V = -\int_{r_2}^{r_1} \frac{q}{4\pi\varepsilon_0 r^2}dr = \frac{q}{4\pi\varepsilon_0}\left(\frac{1}{r_1} - \frac{1}{r_2}\right).$$

Since the potential energy stored in a capacitor is $U = \frac{1}{2}q\Delta V$, the current potential energy stored in the spherical shell is given by

$$U = \frac{1}{2}q\Delta V = \frac{q^2}{8\pi\varepsilon_0}\left(\frac{1}{r_1} - \frac{1}{r_2}\right).$$

Integrating the energy density of the electric field would also yield the same result. In the next problem, observe that q amount of charge will be uniformly induced on the inner surface of the shell so that the electric field within the shell is zero. This leaves $-q$ amount of charge uniformly distributed on the outer surface. We can determine the external work done by determining the change in potential energy of the system. There are several methods to go about this. Firstly, we can sum the potential energy of the charge due to the two shells of charges and the potential energy between the shells of charges (which was the previous result). To compute the former, simply observe that the potential at the center of a spherical shell of charge Q and radius R is simply

$$\frac{Q}{4\pi\varepsilon_0 R},$$

as all charges are equidistant from the center. The potential at the center due to the two spherical shells is then

$$V_{center} = \frac{q}{4\pi\varepsilon_0 r_1} - \frac{q}{4\pi\varepsilon_0 r_2}.$$

The potential energy of the charge due to the shells is then

$$U_{charge} = -qV_{center} = \frac{q^2}{4\pi\varepsilon_0 r_2} - \frac{q^2}{4\pi\varepsilon_0 r_1}.$$

The total potential energy of the initial set-up is then

$$U = U_{charge} + U_{shells} = \frac{q^2}{4\pi\varepsilon_0 r_2} - \frac{q^2}{4\pi\varepsilon_0 r_1} + \frac{q^2}{8\pi\varepsilon_0}\left(\frac{1}{r_1} - \frac{1}{r_2}\right)$$

$$= \frac{q^2}{8\pi\varepsilon_0 r_2} - \frac{q^2}{8\pi\varepsilon_0 r_1}.$$

The potential energy of the set-up after the charge has been extracted is zero. Therefore, the external work done is

$$W = \Delta U = 0 - U = \frac{q^2}{8\pi\varepsilon_0}\left(\frac{1}{r_1} - \frac{1}{r_2}\right).$$

There is another perspective to the change in potential energy involving the electric fields. The electric field of the initial set-up is that of a point charge, excluding the portion in the spherical shell of inner radius r_1 and outer radius r_2 while the electric field of the final set-up is simply that of a point charge. Therefore, the increase in potential energy is that carried by the field in the spherical shell — this is essentially the previous result. Thus,

$$W = \frac{q^2}{8\pi\varepsilon_0}\left(\frac{1}{r_1} - \frac{1}{r_2}\right).$$

18. Tilted Plate**

We define the origin to be at the left edge where the distance between the plates is the smallest.

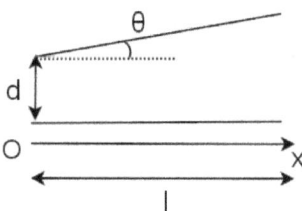

Figure 6.25: Tilted plate

Define the x-axis to be positive rightwards in Fig. 6.25. The separation between the plates at a coordinate x is $d + x\sin\theta$. The set-up can be taken to be many capacitors of infinitesimal thickness, obtained from making myriad vertical cuts, connected in parallel since the potential differences across them are identical due to the equipotential nature of both plates. Each of these infinitesimal plates has length dx and hence area $w dx$. The plate separation of an infinitesimal plate at coordinate x is again $d + x \sin \theta$. Since θ is small, the electric field between the plates is approximately vertical so the capacitance of an infinitesimal plate is analogous to that of a parallel-plate capacitor, $\varepsilon_0 \frac{w dx}{d + x \sin \theta}$. The total capacitance of the system is obtained by determining the equivalent capacitance of these infinitesimal elements. Since the infinitesimal capacitors are connected in parallel, the equivalent capacitance is simply the sum (integral) of the individual capacitances.

$$
\begin{aligned}
C_{eq} &= \int_0^l \frac{\varepsilon_0 w}{d + x \sin \theta} dx \\
&= \left[\frac{\varepsilon_0 w}{\sin \theta} \ln |d + x \sin \theta| \right]_0^l \\
&= \frac{\varepsilon_0 w}{\sin \theta} \ln \left(1 + \frac{l \sin \theta}{d} \right).
\end{aligned}
$$

19. Half-Filled Capacitor**

Define the x-axis to be positive rightwards. At a distance x from the left end, the height of the dielectric is $h - \frac{xh}{l}$ while that of vacuum is $\frac{xh}{l}$. Now, consider the section between x-coordinates x and $x + dx$. It is composed of two capacitors of capacitances $\varepsilon_0 \frac{w dx}{\frac{hx}{l}} = \varepsilon_0 \frac{w l dx}{hx}$ and $\varepsilon_1 \frac{w dx}{h - \frac{hx}{l}}$ in series (the capacitances are again analogous to that of a parallel-plate capacitor because $h \ll l$ and $h \ll w$ such that the electric field is approximately vertical between the two plates). The equivalent capacitance of this section is then

$$
dC = \frac{1}{\frac{hx}{\varepsilon_0 w l dx} + \frac{h - \frac{hx}{l}}{\varepsilon_1 w dx}} = \frac{w \varepsilon_0 \varepsilon_1 dx}{(\varepsilon_1 - \varepsilon_0)\frac{hx}{l} + h \varepsilon_0}.
$$

The total capacitance of the system is then obtained by integrating the above over all sections.

$$
\begin{aligned}
C &= \int_0^l \frac{w \varepsilon_0 \varepsilon_1 dx}{(\varepsilon_1 - \varepsilon_0)\frac{hx}{l} + h \varepsilon_0} \\
&= \frac{w l \varepsilon_0 \varepsilon_1}{(\varepsilon_1 - \varepsilon_0)h} \ln \frac{\varepsilon_1}{\varepsilon_0}
\end{aligned}
$$

as the sections are connected in parallel, since the potential differences across them are identical due to the equipotential nature of the plates.

20. Rising Water*

We have shown that the force on a dielectric that partially fills a parallel-plate capacitor is

$$F = \frac{(\kappa - 1)\varepsilon_0 V^2 a}{2d}$$

and tends to pull the dielectric further into the capacitor. In this case, the fluid level rises until the weight of the additional fluid above the level in the beaker balances F.

$$\frac{(\kappa - 1)\varepsilon_0 V^2 a}{2d} = \rho g a^2 h$$

$$h = \frac{(\kappa - 1)\varepsilon_0 V^2}{2\rho g a d}.$$

21. Parallel Plates with Dielectrics*

Let the upper and lower plates be those with charge densities σ and $-\sigma$ respectively. The electric displacements, electric fields and polarizations within each slab should be uniform as the plates are effectively infinite. Due to this infinite nature, these quantities should only be perpendicular to the plates. Drawing a Gaussian pillbox with one base inside the upper plate and another base within the first dielectric (the bases are parallel to the plates), Gauss' law for electric displacement yields

$$-D_1 A = \sigma A$$

$$D_1 = -\sigma$$

where the negative sign indicates that D_1 points in the downwards direction (from the upper plate to the lower plate). Applying a similar procedure to a Gaussian pillbox that straddles the lower plate and lies within the second dielectric, the electric displacement in the second dielectric is

$$D_2 = -\sigma.$$

Correspondingly, the electric fields within the two dielectrics are

$$E_1 = \frac{D_1}{\varepsilon_1} = -\frac{\sigma}{\varepsilon_1},$$

$$E_2 = \frac{D_2}{\varepsilon_2} = -\frac{\sigma}{\varepsilon_2}.$$

The polarizations are

$$P_1 = D_1 - \varepsilon_0 E_1 = -\frac{(\varepsilon_1 - \varepsilon_0)\sigma}{\varepsilon_1},$$

$$P_2 = D_2 - \varepsilon_0 E_2 = -\frac{(\varepsilon_2 - \varepsilon_0)\sigma}{\varepsilon_2}.$$

There are no volume bound charges anywhere as the polarizations are uniform within each dielectric. The bound surface charge density at the upper plate is

$$P_1 = -\frac{(\varepsilon_1 - \varepsilon_0)\sigma}{\varepsilon_1},$$

while the bound surface charge density at the lower plate is

$$-P_2 = \frac{(\varepsilon_2 - \varepsilon_0)\sigma}{\varepsilon_2}.$$

Since there are no surface free charges at the interface between the two dielectrics, the discontinuity in electric field at the interface must purely be engendered by the bound surface charges there. The bound surface charge density at the interface is thus

$$\varepsilon_0(E_1 - E_2) = \frac{\varepsilon_0(\varepsilon_1 - \varepsilon_2)\sigma}{\varepsilon_1\varepsilon_2}.$$

Another way to derive this is to see it as the superposition of the bound charges that P_1 and P_2 would have caused at the interface.

$$-P_1 + P_2 = \frac{\varepsilon_0(\varepsilon_1 - \varepsilon_2)\sigma}{\varepsilon_1\varepsilon_2}.$$

22. Dielectric with Cavity*

Since this problem is effectively equivalent to a dielectric sphere, surrounded by vacuum, under a uniform external field $\boldsymbol{E}_0$ with the roles of the dielectric and vacuum interchanged, we propose that the desired electric field is equal to answer for the electric field of the dielectric sphere, after replacing κ by $\frac{1}{\kappa}$. Let us check if this suggested solution is coherent with the free charge distribution in this problem.

Since there are no free volume charges everywhere, there are no bound volume charges everywhere. Hence, the surface integral of the electric field must be zero for all Gaussian surfaces lying entirely in the cavity or in the dielectric. Our proposed solution easily satisfies this requirement as there were also no free volume charges everywhere in the case of a dielectric sphere (swapping κ for $\frac{1}{\kappa}$ does not change this as it is equivalent to choosing a dielectric sphere with dielectric constant $\frac{1}{\kappa}$ instead of one with κ). Similarly, the line integral of our proposed electric field is also guaranteed to be path-independent.

Now that we are sure that our proposed solution is valid inside the cavity and inside the dielectric, we have to ensure that it fulfils the boundary conditions. The only condition that can conceivably fail (the continuity of potential and the parallel component of electric field are natural since they were satisfied by the dielectric sphere) is

$$\varepsilon_2 E_{\perp 2} = \varepsilon_1 E_{\perp 1} + \sigma_f.$$

If we let $E_{\perp 2}$ denote the field outside of the cavity or the dielectric sphere and $E_{\perp 1}$ denote the field inside the sphere or dielectric sphere, we require

$$\kappa \varepsilon_0 E_{\perp 2} = \varepsilon_0 E_{\perp 1}$$

for our current problem (we have noted that $\sigma_f = 0$). However, we also know that for the dielectric sphere,

$$\varepsilon_0 E_{\perp 2} = \kappa \varepsilon_0 E_{\perp 1},$$

so the substitution of $\frac{1}{\kappa}$ for the κ's in the electric field of the dielectric sphere is appropriate. It is paramount to observe that this substitution is valid only because σ_f was zero — otherwise, no such analogy can satisfy the boundary condition. Replacing κ in the electric field of the dielectric sphere with $\frac{1}{\kappa}$, the electric field due to the current set-up is

$$\boldsymbol{E}_{in} = \frac{3\kappa}{2\kappa + 1} \boldsymbol{E}_0$$

within the cavity and

$$\boldsymbol{E}_{out}(\boldsymbol{r}) = \left(\frac{2(1 - \kappa)R^3}{(1 + 2\kappa)r^3} + 1 \right) E_0 \cos\theta \hat{\boldsymbol{r}} + \left(\frac{(1 - \kappa)R^3}{(1 + 2\kappa)r^3} - 1 \right) E_0 \sin\theta \hat{\boldsymbol{\theta}}$$

outside the cavity, with respect to spherical coordinates about the center of the cavity.

23. Dipole in Dielectric Sphere**

Firstly, consider the effects of embedding a single point charge q at the center of the spherical dielectric. Due to the spherical symmetry of this system, applying Gauss' law for electric displacement to a spherical Gaussian surface would yield

$$D = \frac{q}{4\pi r^2}$$

at a radial distance r from the center of the sphere. This result is valid for both regions outside and inside the sphere. The electric fields outside and inside the sphere are thus

$$E_{r\geq R} = \frac{D}{\varepsilon_0} = \frac{q}{4\pi\varepsilon_0 r^2},$$

$$E_{r<R} = \frac{D}{\kappa\varepsilon_0} = \frac{q}{4\pi\kappa\varepsilon_0 r^2}.$$

We see that the dielectric effectively reduces the magnitude of the point charge $\frac{1}{\kappa}$ for regions within the sphere while leaving the electric field outside the sphere unchanged. Since an ideal dipole can be seen as the superposition of two opposite point charges with a separation that tends to zero, the field outside of the sphere in the original problem must simply be that due to $\boldsymbol{p}$ in vacuum!

$$\boldsymbol{E}_{r\geq R}(\boldsymbol{r}) = \frac{1}{4\pi\varepsilon_0 r^3}[3(\boldsymbol{p}\cdot\hat{\boldsymbol{r}})\hat{\boldsymbol{r}} - \boldsymbol{p}].$$

Meanwhile, the field inside the sphere is akin to that produced by an ideal dipole with a scaled dipole moment $\frac{\boldsymbol{p}}{\kappa}$.

$$\boldsymbol{E}_{r<R}(\boldsymbol{r}) = \frac{1}{4\pi\kappa\varepsilon_0 r^3}[3(\boldsymbol{p}\cdot\hat{\boldsymbol{r}})\hat{\boldsymbol{r}} - \boldsymbol{p}].$$

24. Spherical Conductor in Dielectric**

Let us first check if a potential of the form $V(r) = \frac{A}{r}$ (outside the conductor) is coherent with the charge distribution outside the conductor. Because there are no free volume charges anywhere, no net volume charges exist in the region outside the conductor. This potential naturally fulfils this condition as it is akin to that produced by a point charge located at the origin and no charge elsewhere.

A more rigorous proof is to use Gauss' law. Applying Gauss' law to an infinitesimal cuboid, we have

$$\nabla \cdot \boldsymbol{E} = \frac{\rho}{\varepsilon_0}$$

where ρ is the volume charge density. Since $\boldsymbol{E} = -\nabla V$,

$$\nabla \cdot \nabla V = \nabla^2 V = -\frac{\rho}{\varepsilon_0},$$

where ∇^2 is known as the Laplacian operator. For a scalar field that only depends on the radial distance from the origin r,

$$\nabla^2 V = \frac{1}{r^2}\frac{\partial}{\partial r}\left(r^2\frac{\partial V}{\partial r}\right) = \frac{1}{r^2}\frac{\partial}{\partial r}(-A) = 0$$

everywhere outside the conductor — proving that $\rho = 0$ outside the conductor. Next, we have to check if $\frac{A}{r}$ satisfies the various boundary conditions.

Firstly, it trivially causes the surface of the spherical conductor to be equipotential. Secondly, the electric field associated with $V(r) = \frac{A}{r}$ is

$$\boldsymbol{E}(r) = -\nabla V = \frac{A}{r^2}\hat{\boldsymbol{r}},$$

which is directed solely along the radial direction. This also trivially satisfies the condition that the normal component of electric displacement is continuous over the vacuum-dielectric interface (because there is no normal component in the first place since $\boldsymbol{D}$ is in the direction of the strictly radial $\boldsymbol{E}$). Furthermore, the parallel components of electric field and potentials on the two sides of the interface are also naturally continuous.

Therefore, we see that the potential $V(r) = \frac{A}{r}$ meets the various conditions outside of the conductor! Now, we simply have to tweak A such that it is consistent with the total free charge on the conductor and assert that $V(r)$ is the correct potential since the potential is guaranteed to be unique by the conditions in the problem.

Our proposed potential implies that the electric displacement at the surface of the conductor is

$$\varepsilon_0 \boldsymbol{E} = \frac{\varepsilon_0 A}{R^2}\hat{\boldsymbol{r}}$$

along the conductor-vacuum interface and

$$\varepsilon \boldsymbol{E} = \frac{\varepsilon A}{R^2}\hat{\boldsymbol{r}}$$

along the conductor-dielectric interface. Since the surface integral of the electric displacement over the surface of the spherical conductor must reflect the total free charge on the conductor by Gauss' law,

$$\frac{\varepsilon_0 A}{R^2} \cdot 2\pi R^2 + \frac{\varepsilon A}{R^2} \cdot 2\pi R^2 = Q$$

$$\implies A = \frac{Q}{2\pi(\varepsilon_0 + \varepsilon)},$$

where we have noted that half of the conductor surface is covered by the dielectric while the other half is surrounded by vacuum. Thus, the potential is

$$V(r) = \frac{Q}{2\pi(\varepsilon_0 + \varepsilon)r}$$

for regions outside the conductor and

$$V = \frac{Q}{2\pi(\varepsilon_0 + \varepsilon)R}$$

within the conductor (since it is equipotential).

25. Charge Above Infinite Dielectric Plane**

There are no free charges on the surface of the dielectric or within the dielectric. Therefore, there are no bound volume charges and hence no net volume charges within the dielectric by Eq. (6.36). Only bound surface charges can exist on the interface between the dielectric and vacuum.

Define the origin such that coordinates of q are $(0, 0, d)$. Now, since the potential at infinity should be zero and we know the free charge distribution in all space (which is just q), the electric field is unique in all space by the (modified) first uniqueness theorem. Before proceeding to guess a solution, it is beneficial to write down the boundary conditions we have to satisfy. If we let $\boldsymbol{E}_{top}(x, y, z)$ and $\boldsymbol{E}_{bot}(x, y, z)$ denote the electric fields above and below the dielectric-vacuum interface. By Eq. (6.37),

$$\varepsilon_0 \boldsymbol{E}_{top}(x, y, 0) = \kappa \varepsilon_0 \boldsymbol{E}_{bot}(x, y, 0)$$

$$\boldsymbol{E}_{top}(x, y, 0) = \kappa \boldsymbol{E}_{bot}(x, y, 0).$$

Furthermore, the potentials must be continuous at the interface.

$$V_{top}(x, y, 0) = V_{bot}(x, y, 0).$$

Now, we can construct image charge configurations for $\boldsymbol{E}_{top}$ and $\boldsymbol{E}_{bot}$. As always, we must be cautious in not placing image charges in a volume of interest as that would change the charge distribution in that volume. For $\boldsymbol{E}_{top}$ in the volume $z \geq 0$, we can imagine that the entire $z < 0$ region is equivalent to an image charge q_b being placed at $(0, 0, -d)$. q_b is a variable to be determined but it represents the total charge plastered on the dielectric-vacuum interface (this is a consequence of Gauss' law applied to an infinite pillbox that contains the entire interface but excludes q). The field along the interface due to this configuration is

$$\boldsymbol{E}_{top}(x, y, 0) = \frac{(q_b - q)d}{4\pi\varepsilon_0(x^2 + y^2 + d^2)^{\frac{3}{2}}}.$$

For $\boldsymbol{E}_{bot}$ in the volume $z < 0$, we can imagine that the entire region $z \geq 0$ region is equivalent to an image charge $q + q_b$ placed at $(0, 0, d)$. Note that we choose a charge of magnitude $q + q_b$ as it is ordained by Gauss' law applied to a surface that bounds $z \geq 0$ (as such a surface contains the charge q and all charges on the interface which amounts to q_b). The field along the interface due to this is

$$\boldsymbol{E}_{bot}(x, y, 0) = \frac{-(q_b + q)d}{4\pi\varepsilon_0(x^2 + y^2 + d^2)^{\frac{3}{2}}}.$$

For $\boldsymbol{E}_{top}(x, y, 0) = \kappa \boldsymbol{E}_{bot}(x, y, 0)$, we must have

$$q_b = \frac{1 - \kappa}{1 + \kappa} q.$$

Furthermore, it is easy to see that the potentials due to the top and bottom image configurations are continuous along $z = 0$. Therefore, the force experienced by q is

$$F_z = \frac{q_b q}{4\pi\varepsilon_0(2d)^2} = \frac{(1 - \kappa)q^2}{16\pi\varepsilon_0(1 + \kappa)d^2}.$$

Actually, this is one of the rare problems that can be directly solved without any guessing. Since the bound charge distribution purely consists of surface charges along the interface, the normal component of electric field above or below $(x, y, 0)$ is solely due to the superposition of that produced by q at $(x, y, 0)$ and that produced by the surface charge $\sigma_b(x, y, 0)$ (the surface charges at other locations do not contribute to a normal electric field here). The contribution due to the latter is simply $\frac{\sigma_b}{2\varepsilon_0}$ (outwards) by Gauss' law and symmetry. Therefore, the net normal component of electric field directly below the interface, at coordinates $(x, y, 0^-)$, is

$$E_n(x, y, 0^-) = -\frac{\sigma_b}{2\varepsilon_0} - \frac{qd}{4\pi\varepsilon_0(x^2 + y^2 + d^2)^{\frac{3}{2}}}.$$

Next, the normal component of polarization P_n in the dielectric at $z = 0$ is

$$P_n(x, y, 0^-) = (\kappa - 1)\varepsilon_0 E_n(x, y, 0^-).$$

However, we also know that $P_n(x, y, 0^-)$ is precisely $\sigma_b(x, y, 0)$ as $\sigma_b = \boldsymbol{P} \cdot \hat{n}$!
Thus,

$$E_n(x, y, 0^-) = \frac{\sigma_b}{(\kappa - 1)\varepsilon_0}.$$

Substituting this into the first equation,

$$\frac{\sigma_b}{(\kappa - 1)\varepsilon_0} = -\frac{\sigma_b}{2\varepsilon_0} - \frac{qd}{4\pi\varepsilon_0(x^2 + y^2 + d^2)^{\frac{3}{2}}}$$

$$\sigma_b = -\frac{(\kappa - 1)qd}{2\pi\varepsilon_0(1 + \kappa)(x^2 + y^2 + d^2)^{\frac{3}{2}}}.$$

Observe that this is simply the induced surface charge on an infinite conducting plane, due to a charge q placed a distance d above it, scaled by a factor $\frac{\kappa-1}{1+\kappa}$. Consequently, the force on q in this situation must be the force on q due to an infinite conducting plane (which was previously derived), multiplied by $\frac{\kappa-1}{1+\kappa}$.

$$F_z = \frac{(1 - \kappa)q^2}{16\pi\varepsilon_0(1 + \kappa)d^2}.$$

Chapter 7

Magnetism

In this chapter, we will study another form of charge-dependent interactions between moving charges and the concept of a magnetic field.

7.1 Lorentz Force Law and the Definition of Magnetic Field

Empirically, it is observed that two current-carrying wires exert forces on each other. In search for an explanation, one might be tempted to say that this may be due to each wire carrying a net charge and that the force is purely electrostatic in nature. However, stationary charges placed next to a current-carrying wire or two experience no such force. Only when the charges start moving do they experience a force. Hence, there must be another interaction that is non-electrostatic and velocity-dependent in some manner.

The Lorentz force law states that the total charge-dependent force on a charge that has a certain instantaneous position and instantaneous velocity v is

$$F = qE + qv \times B, \qquad (7.1)$$

where E and B are the electric field and magnetic field at the instantaneous position of the charge, respectively. The first term is the familiar Coulomb force. The second term is the magnetic force on which the magnetic field is defined. That is, B at a point (with a known E) is such that the expression for the force on a test charge q placed at that point is given by Eq. (7.1). Again, the magnetic field is a vector field whose utility lies in the fact that identical charges with the same velocities respond to identical magnetic fields in the same manner, regardless of the exact source of the magnetic fields.

Zero Work Done by Magnetic Force

A corollary of the above expression for the magnetic force is that a magnetic force does no work on a charged particle as it always acts in the direction perpendicular to a charge's instantaneous velocity and thus instantaneous displacement.

$$W = \int \mathbf{F} \cdot d\mathbf{s} = \int q(\mathbf{v} \times \mathbf{B}) \cdot \mathbf{v} \, dt = 0. \tag{7.2}$$

This implies that if a charge is influenced solely by a magnetic force, its speed remains constant though its direction of velocity may vary.

Force on a Current-Carrying Wire

Stemming from Eq. (7.1), the magnetic force on a thin wire carrying current I can be calculated as

$$\mathbf{F} = \int \mathbf{v} \times \mathbf{B} \, dq = \int I \, d\mathbf{s} \times \mathbf{B}, \tag{7.3}$$

where $d\mathbf{s}$ is an infinitesimal current element along the wire since the current of a thin wire is the rate of charge transport across a particular point on the wire.

$$I d\mathbf{s} = \frac{dq}{dt} d\mathbf{s} = dq \frac{d\mathbf{s}}{dt} = \mathbf{v} dq.$$

The integration in Eq. (7.3) is performed over the entire current-carrying wire. For a straight wire of length L which carries a current I at an angle θ with respect to a uniform magnetic field $\mathbf{B}$, the magnitude of the magnetic force on the wire is

$$F = BIL \sin \theta.$$

The direction of this force can be determined by evaluating the cross product. Actually, this result is valid for any arbitrarily shaped wire carrying current between two terminals in a uniform magnetic field, as we shall prove in the following problem. In such cases, L becomes the straight line distance between the two terminals and the direction of the magnetic force on the wire is the same as that of a straight wire delivering current from the starting point to the ending point.

Problem: A finite wire of an arbitrary shape, with terminals A and B, lies in a region containing a uniform external magnetic field, $\mathbf{B}$. The wire carries a current I from A to B. If the vector pointing from A to B is $\mathbf{l}$, find the magnetic force on the wire. Hence, what can you say about the force on a current loop in a uniform magnetic field?

The magnetic force on the wire is given by

$$\boldsymbol{F} = I \int d\boldsymbol{s} \times \boldsymbol{B}$$

$$= I \left(\int d\boldsymbol{s} \right) \times \boldsymbol{B}$$

$$= I\boldsymbol{l} \times \boldsymbol{B},$$

where the second equality is possible because $\boldsymbol{B}$ is uniform in space. This is exactly the force on a straight wire, which has length l, that carries current I from the first point to the second. Since $\boldsymbol{F} = \boldsymbol{0}$ when $l = \boldsymbol{0}$, an interesting corollary is that there is no net magnetic force on a current loop in a uniform magnetic field.

In fact, we can do much better than a single wire. Suppose an arbitrary arrangement of thin wires carries a total current I from A to B. The wires can merge and split freely. We claim that the total force $\boldsymbol{F}$ on these wires is still given by the previous expression. To see why, suppose that we track an infinitesimal current dI along an arbitrary path P from A to B. The contribution to $\boldsymbol{F}$ due to this current is $d\boldsymbol{F} = dI\boldsymbol{l} \times \boldsymbol{B}$. Summing over all currents,

$$\boldsymbol{F} = \int dI\boldsymbol{l} \times \boldsymbol{B} = \left(\int dI \right) \boldsymbol{l} \times \boldsymbol{B} = I\boldsymbol{l} \times \boldsymbol{B}.$$

7.2   Magnetic Field

Now that we have understood how a charge responds to a magnetic field, let us analyze how a magnetic field arises. Surprisingly, moving charges are not only the victims of a magnetic field but are also the sources as well.

Unfortunately, there are no empirical laws for the magnetic field of an isolated point charge so we cannot really determine the magnetic field of a system of moving charges by piling the contributions of point charges. However, there are laws for systems involving steady currents which are adequate forms of consolation as our everyday experiences with moving charges often involve currents in wires.

A steady current refers to a continuous flow of charge that is not varying with time — this requires no accumulation or loss of charge at any point in space. Examples would include currents that are cyclic and currents that travel to infinity (which is not really realistic). Finite, non-cyclic wires, on the other hand, do not carry a steady current as there is charge retrieved from one end and deposited on the other end. Moving on, the significance of a steady current is to generate a steady magnetic field that is not varying with respect to time, just like how stationary charges produce steady electric fields. A situation involving steady magnetic fields is often known as magnetostatics.

The Biot-Savart law quantifies the magnetic field at a point P due to a steady thin current I as

$$B = \frac{\mu_0}{4\pi} \int \frac{I d\boldsymbol{s} \times \hat{\boldsymbol{r}}}{r^2} \tag{7.4}$$

where $d\boldsymbol{s}$ is an infinitesimal displacement along the current-carrying element and $\boldsymbol{r}$ is the vector pointing from that infinitesimal element to the point of interest, P. The integration is performed over the entire current-carrying entity. The constant μ_0 is known as the permeability of free space which has a numerical value of $1.257 \times 10^{-6} \, \mathrm{m\,kg\,s^{-2}A^{-2}}$. More generally, the current may not be transmitted along a thin strip.

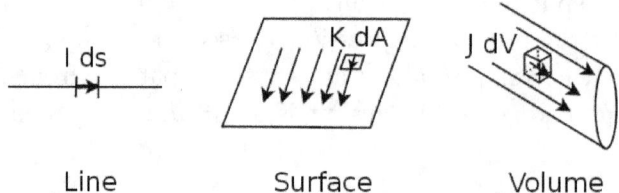

Line   Surface   Volume

Figure 7.1:   Line, surface and volume currents

Depicted in Fig. 7.1 are three forms of infinitesimal current elements which have different spatial dimensions. The simplest is a current line element that flows over an infinitesimal cross section — this is the case that we have considered hitherto. Next, current could also flow on a thin sheet. Such currents are known as surface currents and can be parameterized by the surface current density $\boldsymbol{K}$ whose magnitude is the current per unit perpendicular length $\frac{dI}{dl}$ (dl is the infinitesimal length perpendicular to the current flow). $\boldsymbol{K}$ is a vector that points in the direction of the current. In such situations, we can then substitute $I d\boldsymbol{s} = \boldsymbol{K} dA$ into the Biot-Savart law where dA is the area of an infinitesimal surface element on the current sheet. Finally, we can have a volume current which flows across a finite cross sectional area — a realistic wire is a vivid example. Then, the volume current density $\boldsymbol{J}$, which is the current per unit cross sectional area in the direction of current flow, can be used to characterize the current. An infinitesimal current-carrying element of volume dV then corresponds to $I d\boldsymbol{s} = \boldsymbol{J} dV$.

In light of the above expressions, the Biot-Savart law can be expressed in the following forms for line, surface and volume currents respectively.

$$B = \frac{\mu_0 I}{4\pi} \int \frac{ds \times \hat{r}}{r^2} = \frac{\mu_0}{4\pi} \iint \frac{K dA \times \hat{r}}{r^2} = \frac{\mu_0}{4\pi} \iiint \frac{J dV \times \hat{r}}{r^2}, \quad (7.5)$$

where we have included double and triple integrals to explicitly indicate that we are integrating over area and volume distributions. Incidentally, though a single moving point charge q traveling at a velocity v is the ultimate antithesis of a steady current as the charge in all space is evidently not constant, the non-relativistic approximation[1] of the magnetic field produced at a point P by it is

$$B = \frac{\mu_0 q}{4\pi} \frac{v \times \hat{r}}{r^2} \quad (7.6)$$

where r is the vector pointing from the point charge to point P. Note that this is a completely different equation from the Biot-Savart law (which cannot be applied here) that is derived from special relativity. One can obtain the above expression from the Biot-Savart law via the sleight-of-hand $Ids = qv$, but this is an incorrect application of the Biot-Savart law.

To obtain a better understanding of the meaning of the terms in the Biot-Savart law, consider the following examples.

Problem: A thin conducting ring of radius R carries an anti-clockwise current I. Determine the magnetic field along a perpendicular axis passing through the center of the ring.

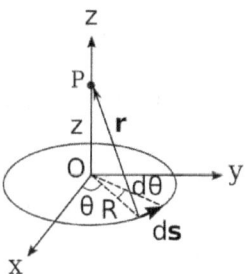

Figure 7.2: Current-carrying ring

Due to the circular symmetry of the set-up in Fig. 7.2, there can only be a magnetic field along the axis of the ring which we define as the z-direction, but we shall pretend that we do not know this for now to better illustrate the

---

[1] This is the regime where $v \ll c$, with c being the speed of light in vacuum.

terms in the Biot-Savart law. Consider a point P that is a distance z above the center of the plane of the ring. The contribution to the magnetic field at P due to an infinitesimal segment $Rd\theta$ of the ring at angular coordinate θ is

$$d\boldsymbol{B} = \frac{\mu_0 I}{4\pi} \frac{d\boldsymbol{s} \times \hat{\boldsymbol{r}}}{r^2}.$$

The infinitesimal length element $d\boldsymbol{s}$ in this case is

$$d\boldsymbol{s} = Rd\theta \begin{pmatrix} -\sin\theta \\ \cos\theta \\ 0 \end{pmatrix},$$

while the vector pointing from this infinitesimal segment to point P is

$$\hat{\boldsymbol{r}} = \begin{pmatrix} -\frac{R}{\sqrt{R^2+z^2}}\cos\theta \\ -\frac{R}{\sqrt{R^2+z^2}}\sin\theta \\ \frac{z}{\sqrt{R^2+z^2}} \end{pmatrix}.$$

Furthermore, the distance between this infinitesimal segment and P is $r = \sqrt{R^2 + z^2}$. The contribution to the magnetic field at P by this infinitesimal current element is then

$$d\boldsymbol{B} = \frac{\mu_0 I}{4\pi(R^2 + z^2)} \cdot Rd\theta \begin{pmatrix} -\sin\theta \\ \cos\theta \\ 0 \end{pmatrix} \times \begin{pmatrix} -\frac{R}{\sqrt{R^2+z^2}}\cos\theta \\ -\frac{R}{\sqrt{R^2+z^2}}\sin\theta \\ \frac{z}{\sqrt{R^2+z^2}} \end{pmatrix}$$

$$= \frac{\mu_0 I R d\theta}{4\pi(R^2 + z^2)^{\frac{3}{2}}} \begin{pmatrix} z\cos\theta \\ z\sin\theta \\ R \end{pmatrix}.$$

Technically, we cannot say that this is the magnetic field due to this infinitesimal segment, as the Biot-Savart law only applies to steady currents. We can only say that the total magnetic field at point P due to the entire ring is obtained from integrating the above from $\theta = 0$ to $\theta = 2\pi$. The $\sin\theta$ and $\cos\theta$ terms vanish after this integration — only the z-component of the magnetic field survives as expected.

$$B_z(z) = \int_0^{2\pi} \frac{\mu_0 I R^2 d\theta}{4\pi(R^2 + z^2)^{\frac{3}{2}}} = \frac{\mu_0 I R^2}{2(R^2 + z^2)^{\frac{3}{2}}}.$$

Problem: Determine the magnetic field everywhere due to a thin, long wire that carries a current I.

Orient the wire such that the current I flows along the x-axis, towards the positive direction. The magnetic fields at all points with the same y and z-coordinates but different x-coordinates are identical due to the infinite nature of the wire such that if we are interested in the magnetic field at a certain point, we can consider a corresponding point at $x = 0$. Furthermore, we can always orient the y-axis such that this new point of interest lies on the y-axis. Therefore, if we determine the magnetic field at point P with coordinates $(0, \rho, 0)$, we will have solved for the magnetic field everywhere.

Now, consider an infinitesimal segment of wire $d\boldsymbol{s} = (dx, 0, 0)$ at x-coordinate x. The separation vector between this segment and P is $\boldsymbol{r} = (-x, \rho, 0)$ such that $\hat{\boldsymbol{r}} = (-\frac{x}{x^2+\rho^2}, \frac{\rho}{x^2+\rho^2}, 0)$ and $r^2 = x^2 + \rho^2$. Integrating the contributions from all such segments with x-coordinates ranging from $-\infty$ to $+\infty$, the Biot-Savart law yields the magnetic field at P as

$$\boldsymbol{B} = \frac{\mu_0 I}{4\pi} \int_{-\infty}^{+\infty} \frac{1}{x^2 + \rho^2} \begin{pmatrix} dx \\ 0 \\ 0 \end{pmatrix} \times \begin{pmatrix} -\frac{x}{\sqrt{x^2+\rho^2}} \\ \frac{\rho}{\sqrt{x^2+\rho^2}} \\ 0 \end{pmatrix}$$

$$= \frac{\mu_0 I}{4\pi} \int_{\infty}^{+\infty} \frac{\rho \, dx}{(x^2 + \rho^2)^{\frac{3}{2}}} \begin{pmatrix} 0 \\ 0 \\ 1 \end{pmatrix}.$$

To evaluate the integral for the z-component, use the substitutions $x = \rho \tan \theta$ and $dx = \rho \sec^2 \theta \, d\theta$.

$$B_z = \frac{\mu_0 I}{4\pi} \int_{-\frac{\pi}{2}}^{\frac{\pi}{2}} \frac{\rho}{\rho^3 \sec^3 \theta} \cdot \rho \sec^2 \theta \, d\theta$$

$$= \frac{\mu_0 I}{4\pi\rho} \int_{-\frac{\pi}{2}}^{\frac{\pi}{2}} \cos \theta \, d\theta$$

$$= \frac{\mu_0 I}{2\pi\rho}.$$

Applying this result for general P, the magnetic field strength is purely dependent on the radial distance ρ from the wire.

$$B(\rho) = \frac{\mu_0 I}{2\pi\rho}.$$

Furthermore, the magnetic field is solely azimuthal, as seen from the fact that it is perpendicular to the plane containing the wire and a line joining a point on the wire to P. Writing the above in vector form,

$$\boldsymbol{B}(\rho) = \frac{\mu_0 I}{2\pi\rho} \hat{\phi},$$

where $\hat{\phi}$ is the azimuthal unit vector whose positive direction is given by applying the right-hand-grip rule with your thumb pointing in the direction of the current I.

Currents Formed by Rotations

In some cases, currents are formed by rotating charge distributions about an axis with a certain angular velocity $\boldsymbol{\omega}$. Then, there are elegant expressions for the surface and volume current densities $\boldsymbol{K}$ and $\boldsymbol{J}$. The magnitude of the current carried by an infinitesimal surface element with area dA and surface charge density σ can be written as

$$dI = \frac{dq}{dt} = \frac{dq}{dA} \cdot \frac{ds}{dt} \cdot dl = \sigma \cdot \frac{ds}{dt} \cdot dl = \sigma v dl,$$

where ds and dl are the infinitesimal lengths of the element, parallel and perpendicular to its velocity. It can then be seen that the magnitude of the surface current density is

$$K = \frac{dI}{dl} = \sigma v.$$

Furthermore, one can easily see that the direction of $\boldsymbol{K}$ should be parallel to the velocity of this element $\boldsymbol{v}$ as well. Therefore,

$$\boldsymbol{K} = \sigma \boldsymbol{v}. \tag{7.7}$$

One can use a similar analysis to conclude that the volume current density of an infinitesimal volume element with volume charge density ρ and velocity $\boldsymbol{v}$ is

$$\boldsymbol{J} = \rho \boldsymbol{v}. \tag{7.8}$$

These expressions are completely general. Now, if we define our origin to be at a point along the axis of rotation, the velocity of an infinitesimal element at position vector $\boldsymbol{s}$ is

$$\boldsymbol{v} = \boldsymbol{\omega} \times \boldsymbol{s}.$$

The respective current densities are then

$$\boldsymbol{K} = \sigma \boldsymbol{\omega} \times \boldsymbol{s}, \tag{7.9}$$

$$\boldsymbol{J} = \rho \boldsymbol{\omega} \times \boldsymbol{s}. \tag{7.10}$$

One can then substitute these expressions for the current densities in the Biot-Savart law.

Problem: In Fig. 7.3, an insulating disk of radius R possesses a uniform surface charge density σ and is rotating in the anti-clockwise direction with an angular velocity ω about its central axis. Find the magnetic field of a point that is at a distance h above the disk, along the axis of the disk.

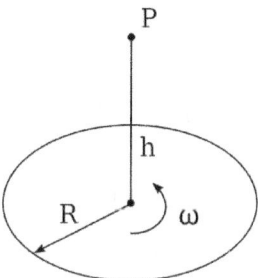

Figure 7.3: Rotating disk

Referring to Fig. 7.4, consider the plane of the disk in polar coordinates. We will use s and ϕ to denote the radial distance and azimuthal angle.

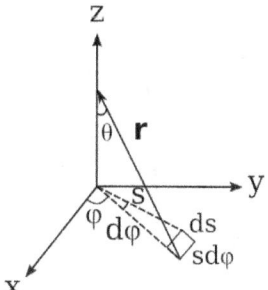

Figure 7.4: Vectors

Consider an infinitesimal area element $dA = s\,d\phi\,ds$ at (s, ϕ). It has position vector $\boldsymbol{s} = (s\cos\phi, s\sin\phi, 0)$ and thus surface current density

$$\boldsymbol{K} = \sigma \begin{pmatrix} 0 \\ 0 \\ \omega \end{pmatrix} \times \begin{pmatrix} s\cos\phi \\ s\sin\phi \\ 0 \end{pmatrix} = \sigma \begin{pmatrix} -\omega s\sin\phi \\ \omega s\cos\phi \\ 0 \end{pmatrix}.$$

The separation vector $\boldsymbol{r}$ pointing from this infinitesimal element to the point of interest P is

$$\boldsymbol{r} = r \begin{pmatrix} -\sin\theta\cos\phi \\ -\sin\theta\sin\phi \\ \cos\theta \end{pmatrix},$$

where we will defer the substitutions of the expressions for r and θ till later. Applying the Biot-Savart law,

$$\mathbf{B} = \frac{\mu_0\sigma}{4\pi} \int_0^R \int_0^{2\pi} \frac{sd\phi ds}{r^2} \begin{pmatrix} -ws\sin\phi \\ ws\cos\phi \\ 0 \end{pmatrix} \times \begin{pmatrix} -\sin\theta\cos\phi \\ -\sin\theta\sin\phi \\ \cos\theta \end{pmatrix}$$

$$= \frac{\mu_0\sigma}{4\pi} \int_0^R \int_0^{2\pi} \frac{ws^2 d\phi ds}{r^2} \begin{pmatrix} \cos\theta\cos\phi \\ \cos\theta\sin\phi \\ \sin\theta \end{pmatrix}.$$

The x- and y-components disappear as the integrals of $\sin\phi$ and $\cos\phi$ over an entire period yield zero. Then, we only need to evaluate the z-component.

$$B_z = \frac{\mu_0\sigma w}{2} \int_0^R \frac{s^2\sin\theta}{r^2} ds.$$

Substituting $s = h\tan\theta$, $ds = h\sec^2\theta d\theta$ and $r = h\sec\theta$,

$$B_z = \int_0^{\tan^{-1}\frac{R}{h}} \frac{\mu_0\sigma w h}{2} \frac{\sin^3\theta}{\cos^2\theta} d\theta$$

$$= \int_0^{\tan^{-1}\frac{R}{h}} \frac{\mu_0\sigma w h}{2} \frac{\sin\theta(1-\cos^2\theta)}{\cos^2\theta} d\theta$$

$$= \int_0^{\tan^{-1}\frac{R}{h}} \frac{\mu_0\sigma w h}{2} \frac{\sin\theta}{\cos^2\theta} d\theta - \int_0^{\tan^{-1}\frac{R}{h}} \frac{\mu_0\sigma w h}{2} \sin\theta d\theta.$$

Using the substitution $u = \cos\theta$, $du = -\sin\theta d\theta$ for the first integral,

$$B_z = \int_1^{\frac{h}{\sqrt{h^2+R^2}}} -\frac{\mu_0\sigma w h}{2} \frac{1}{u^2} du - \int_0^{\tan^{-1}\frac{R}{h}} \frac{\mu_0\sigma w h}{2} \sin\theta d\theta$$

$$= \left[\frac{\mu_0\sigma w h}{2} \frac{1}{u} \right]_1^{\frac{h}{\sqrt{h^2+R^2}}} + \left[\frac{\mu_0\sigma w h}{2} \cos\theta \right]_0^{\tan^{-1}\frac{R}{h}}$$

$$= \frac{\mu_0\sigma w h}{2} \left(\frac{\sqrt{h^2+R^2}}{h} - 1 \right) + \frac{\mu_0\sigma w h}{2} \left(\frac{h}{\sqrt{h^2+R^2}} - 1 \right)$$

$$= \frac{\mu_0\sigma w}{2} \left(\frac{2h^2+R^2}{\sqrt{h^2+R^2}} - 2h \right).$$

7.2.1   *Magnetic Field Lines*

With Biot-Savart's law, the magnetic field in all space can be calculated. However, how can we visualize this magnetic field? To illustrate the direction of the magnetic field, magnetic field lines, which are continuous curves which point in the direction of the magnetic field at each point, can be drawn. In magnetism, we usually consider current-carrying entities as our elementary building blocks, as opposed to point charges, because most laws are only valid for steady currents. Thus, we shall study the magnetic field lines due to a steady state current-carrying wire. As the wire is axial-symmetric, the magnetic field generated by it should also obey such a property. The magnetic field lines due to current-carrying wires are depicted in Figs. 7.5 and 7.6 (they are concentric circles in a transverse cross section, as seen from the result derived previously for a long wire).

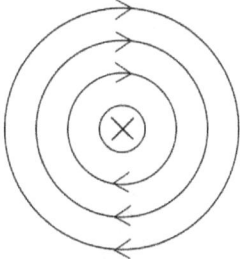

Figure 7.5:   Current into the page

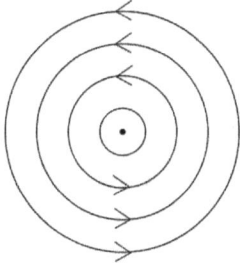

Figure 7.6:   Current out of the page

The cross and dot represent current going into and coming out from the page respectively. One can imagine firing an arrow into the page — the arrow head, which corresponds to the cross, will point into the page while the tail, which corresponds to the dot, will point out of the page.

It can be observed that magnetic field lines always exist in loops. To remember the direction of the magnetic field line loops, you can grasp your right hand to form a spiral and point your thumb in the direction of the current. Then, your fingers will curl in the direction of the magnetic field. This is known as the right-hand-grip rule.

Magnetic field lines do not originate from or terminate anywhere. In fact, one of Maxwell's equations states that the closed surface integral of a magnetic field over an arbitrary closed surface S is zero.[2] This is intuitive as magnetic field lines, which exist in loops, that enter a volume must also exit from it.

$$\oiint_S \boldsymbol{B} \cdot d\boldsymbol{A} = 0. \tag{7.11}$$

Electric field lines, on the other hand, face no such constraint and instead, emanate from a positive charge and return to a negative charge. We have established that electric field lines cannot have any loops from the line integral of an electric field. Another perspective to this is that the existence of a loop requires electric field lines to start and end at the same point and hence implies that the charge there must be simultaneously positive and negative — an absurd criterion!

7.2.2 *Magnetic Energy Density*

Similar to the electric field, a magnetic field can be associated with a volume energy density $u_B = \frac{B^2}{2\mu_0}$ such that an infinitesimal volume element dV is associated with energy

$$dU = \frac{\boldsymbol{B} \cdot \boldsymbol{B}}{2\mu_0} dV = \frac{B^2}{2\mu_0} dV.$$

Integrating this over all space, the energy "stored" in a magnetic field can be computed as

$$U = \int_{allspace} \frac{B^2}{2\mu_0} dV. \tag{7.12}$$

Again, this expression can be derived from considering the work done by an external agent in introducing currents but the proof requires tools beyond the scope of this book. However, we will understand why work is required

---

[2]You can actually prove this directly for magnetostatic fields given by Biot-Savart's law (though this requires some vector calculus) like how Gauss' law can be proven for electrostatic fields from Coulomb's law. However, we should treat the nullity of the closed surface integral of the magnetic field (alas, this equation does not have a name) and Gauss' law as fundamental laws rather than corollaries as we will soon discover that magnetic and electric fields can literally be induced by another mechanism (i.e. the sources of magnetic and electric fields need not only be currents and charges). Even in these non-magnetostatic and non-electrostatic regimes, the nullity of the closed surface integral of the magnetic field and Gauss' law still hold — the magnetic and electric fields now comprise both the static and induced portions.

to create currents and deduce another expression for the energy stored in a magnetic field in the next chapter.

7.3   Ampere's Law

In a system that is constituted by steady currents, an equivalent but somewhat more convenient form of the Biot-Savart law exists. Ampere's law states that the closed loop integral of the magnetic field along an arbitrary cyclic contour is proportional to the amount of current I_{enc} cutting perpendicularly to a surface[3] S bounded by the loop (i.e. surface integral of the current density over S), with the constant of proportionality being μ_0. For planar loops, we usually choose S to be the plane of the loop so we shall refer to I_{enc} as the current enclosed or encased in the case of planar loops henceforth (but keep in mind that only the perpendicular component matters). Recall that a closed loop integral refers to the sum of values of the dot product of a vector field with infinitesimal length elements — which are vectors — computed along a cyclic path.

$$\oint \boldsymbol{B} \cdot d\boldsymbol{s} = \mu_0 I_{enc}. \tag{7.13}$$

The loop chosen to evaluate the above integral is often known as the Amperian loop. Now, there seemingly lies an ambiguity regarding the positive direction of I_{enc}. To rectify this muddy point, we use the right-hand-grip rule — curl your palm in the direction of the integration along the Amperian loop. Your straightened thumb will point in the positive direction of current.

Generally, the integral on the left-hand side is non-trivial and onerous to calculate. However, in systems with certain forms of symmetry, we can choose a convenient Amperian loop such that the integral can be evaluated virtually effortlessly — enabling an elegant method to calculate magnetic fields. Hopefully, the following common illustrations of this technique will be enlightening.

Infinitely Long Wire

Ampere's law can be applied to an infinitely long current-carrying wire, as it can be regarded as a steady state current due to the fact that no charge

---

[3]You might think that there is an ambiguity in the surface S here but I_{enc} is actually independent of the exact surface S chosen for steady currents (as long as it spans the contour) — refer to the next chapter for more details.

accumulates anywhere in space. A finite wire, on the other hand, engenders accumulation of charges at its ends which renders Ampere's law inapplicable.

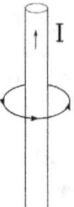

Figure 7.7:   Long wire

The first step to solving for a magnetic field is obviously to determine its direction. In this case, the magnetic field due to an infinite wire can only be in the azimuthal direction (i.e. in circles perpendicular to the wire) due to its axial symmetry. Let us prove this rigorously. Firstly, as the magnetic field produced by a current is always perpendicular to the current (due to the expression given by the Biot-Savart law), there must not be an axial component of magnetic field. Next, suppose that in the plane containing a certain cross section of a wire, the magnetic field had a radial component. For purposes of illustration, let's say that this is radially outwards. Now, we can imagine flipping the entire wire such that current now runs downwards — the radial magnetic field must still be outwards due to this flip. This implies that if we superpose this flipped wire and the original wire, there will be a radially outwards magnetic field, even though there is no current — hence invalidating our claim!

Therefore, the magnetic field due to an infinitely long wire can only be azimuthal. To solve for this, we can draw a circular Amperian loop as shown in Fig. 7.7. The magnitude of the magnetic field is uniform throughout the loop due to axial symmetry and its direction is always along the loop. Thus,

$$\oint \boldsymbol{B} \cdot d\boldsymbol{s} = B \cdot 2\pi r = \mu_0 I.$$

In writing this equation, the direction of integration along the loop was anti-clockwise — implying that the upwards direction is taken to be positive for I_{enc} by the right-hand-grip rule. Rearranging,

$$B = \frac{\mu_0 I}{2\pi r},$$

which agrees with what we have previously derived from the Biot-Savart law.

Infinite Current Sheet

Now, consider a current sheet with a thickness t that propagates indefinitely in the directions perpendicular to its thickness. It carries a current density J in a single direction along the plane. To determine the magnetic field due to this set-up, we first deliberate its direction as always. Evidently, there cannot be a component parallel to the direction of the current density as the magnetic field involves the cross product of the current density (Biot-Savart's law). Furthermore, there cannot be a component of magnetic field normal to the infinite plane due to the infinite nature of the sheet. Consider the cross section of the sheet perpendicular to the current density.

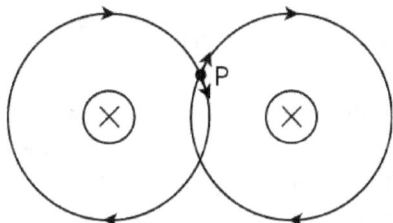

Figure 7.8:   Magnetic field at point along line of symmetry of cross section

Referring to Fig. 7.8, suppose we wish to determine the magnetic field at a point P along a vertical line of symmetry. Then, the contribution to the vertical magnetic field by the left current cancels that of the right element (we mean the entire lines of currents that run into the page) — this nullification holds for all pairs symmetric about the vertical axis. Furthermore, this argument works everywhere as the plane is infinite. Therefore, there must not be a vertical component of magnetic field.

Actually, there is a more elegant argument, similar to our previous argument for the long wire, as follows. Suppose that the above sheet of current going into page produces a magnetic field emanating from the sheet. Flipping this sheet would produce a current sheet that comes out of the page and still produces a magnetic field emanating from the plane. Superimposing these set-ups would produce a system with zero current but a non-zero normal magnetic field — an evident contradiction.

Now that we have determined that the direction of magnetic field due to the sheet must be perpendicular to both the normal and the current density, we can draw a rectangular Amperian loop of width w and arbitrary length as shown in Fig. 7.9. The component of magnetic field along lines 2 and 4 is zero as there is no vertical component of magnetic field. For lines 1 and 3, the magnetic field throughout each line should be constant due to the

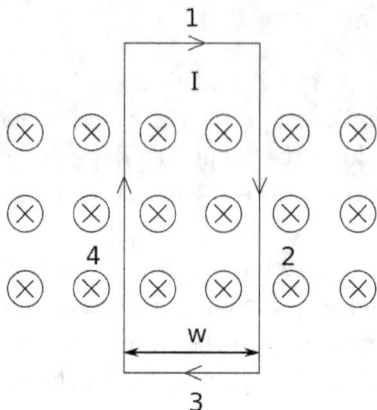

Figure 7.9:   Cross section of infinite sheet

infinite nature of the system. Furthermore, they must be purely horizontal. Thus, if we draw the loop such that lines 1 and 3 are equidistant from the set-up, the magnetic field strength should be identical along both lines with the only exception being that the magnetic fields must point in different horizontal directions (based on the right-hand-grip rule for the magnetic field of a current). Thus, the closed loop integral of magnetic field along this Amperian loop is

$$\oint \boldsymbol{B} \cdot d\boldsymbol{s} = \int_1 \boldsymbol{B} \cdot d\boldsymbol{s} + \int_2 \boldsymbol{B} \cdot d\boldsymbol{s} + \int_3 \boldsymbol{B} \cdot d\boldsymbol{s} + \int_4 \boldsymbol{B} \cdot d\boldsymbol{s}$$

$$= B \cdot w + 0 + B \cdot w + 0$$

$$= 2Bw.$$

The enclosed current depends on whether the Amperian loop protrudes out of the sheet. If it does,

$$I_{enc} = Jwt.$$

Thus, by applying Ampere's law,

$$2Bw = \mu_0 Jwt$$

$$B = \frac{\mu_0 Jt}{2}$$

for regions outside the sheet. The magnetic field above the plane is right-wards for positive J while that below the plane is leftwards. Interestingly, the magnetic field strength due to an infinite current sheet is independent of the distance from the point of concern to the sheet for regions outside the sheet. If the loop is within the sheet, let the length of the loop be $2l$. Then, the enclosed current is

$$I_{enc} = 2Jwl,$$

$$B = \mu_0 Jl.$$

The magnetic field strength scales linearly with the distance from the center of the sheet, in regions within the sheet, as the Amperian loop encloses more current with increasing length. Again, the magnetic field is rightwards above the middle and leftwards below the middle.

Infinitely Long Solenoid

A solenoid is a coil that takes the shape of a spring. For the sake of our purposes, we assume that the coil is densely wound such that the gaps between rings are small. Then, we can approximate a solenoid as a collection of rings, without taking into account the effects of the segments that connect them.

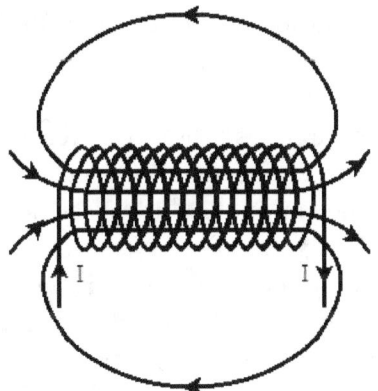

Figure 7.10: Solenoid

The magnetic field lines due to a solenoid are depicted in Fig. 7.10. One end of a solenoid is similar to a North pole (emits magnetic field lines) while the other is similar to a South pole of a magnet (receives magnetic field lines). To easily remember the "poles", apply the right-hand-grip rule to the current — your thumb will point towards the North pole.

To understand why the magnetic field is tenuous outside the solenoid and concentrated within it, consider the side view of the solenoid in Fig. 7.11. The currents at the top come out while those at the bottom go into the page. Therefore, their magnetic fields counteract in regions beyond the coil but are reinforced within the coil (use the right-hand-grip rule). In fact,

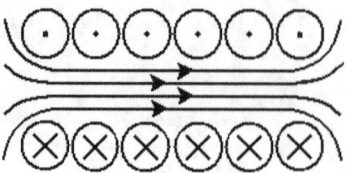

Figure 7.11:   Side view

for an infinitely long solenoid, the magnetic field outside is zero while the magnetic field everywhere inside the solenoid is uniform and horizontal.

To prove these, we first determine the direction of the magnetic field produced by a long solenoid. Firstly, there should not be a radial component of magnetic field — if not, we can flip the solenoid (such that the current runs in the reverse direction) and superpose it with the original solenoid to obtain a set-up with no current but non-zero magnetic field. Secondly, there should not be an azimuthal component of magnetic field. Due to the solenoid's axial symmetry, we can draw an Amperian loop that is concentric with the solenoid axis. The azimuthal component of magnetic field (if it exists) should be uniform throughout this loop due to symmetry so the closed loop integral of magnetic field over this Amperian loop is simply the azimuthal component of magnetic field multiplied by the perimeter of the loop. However, this Amperian loop does not contain any net current that crosses it (regardless of whether the loop is inside or outside[4] the solenoid) which implies that the azimuthal component of magnetic field is zero by Ampere's law. In conclusion, the magnetic field due to a long solenoid can only be axial.

For physical reasons, the magnetic field due to the solenoid should vanish at points which are infinitely far away from the solenoid axis. In light of this, we can prove that the magnetic field outside the solenoid is zero everywhere. Draw the upper Amperian loop depicted in Fig. 7.12, which extends radially to infinity. As the radial magnetic field is zero and the magnetic field disappears at infinity, the only contribution to the closed loop integral of the magnetic field over this Amperian loop is the bottom horizontal segment — along which the axial magnetic field must be uniform due to the infinite nature of the solenoid. Since no current crosses through this loop, the axial

---

[4]Actually, when the loop is outside the solenoid windings, there should be some current flowing through it due to the segments that connect different layers of the solenoid. However, the current cutting normally to the plane of the loop is negligible when the turns are densely wound such that the gradient of these segments are very small (they are essentially horizontal and parallel to the loop).

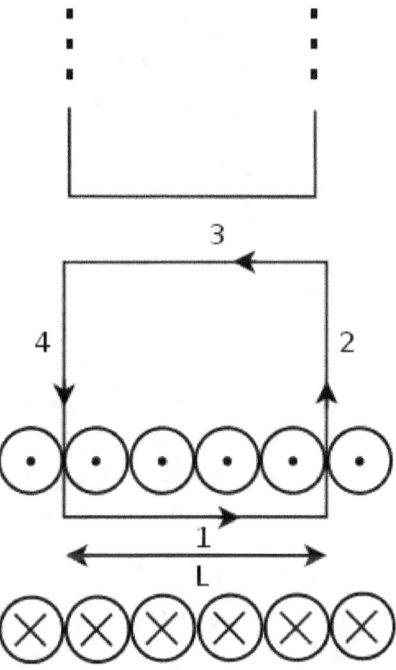

Figure 7.12: Amperian loop

magnetic field along the bottom segment must be zero — a fact that holds for all points outside the solenoid.

To evaluate the magnetic field inside a solenoid, we can draw an Amperian loop of length L that protrudes out of the solenoid (we can be conservative and choose segment 3 at infinity to ensure that the magnetic field is zero along it) as shown in Fig. 7.12. The components of magnetic field along lines 2 and 4 are zero as we have shown that the magnetic field can only be horizontal. Furthermore, the magnetic field along segment 3 is also zero as we have shown that no magnetic field lies outside the solenoid. Since the magnetic field is uniform along line 1 and is always parallel to the path of integration, the closed loop integral anti-clockwise along this loop evaluates to

$$\oint \boldsymbol{B} \cdot d\boldsymbol{s} = \int_1 \boldsymbol{B} \cdot d\boldsymbol{s} + \int_2 \boldsymbol{B} \cdot d\boldsymbol{s} + \int_3 \boldsymbol{B} \cdot d\boldsymbol{s} + \int_4 \boldsymbol{B} \cdot d\boldsymbol{s}$$
$$= B \cdot L + 0 + 0 + 0$$
$$= BL.$$

Let η be the number of turns per unit length of the solenoid. The current enclosed is then $I_{enc} = \eta L I$. Applying Ampere's law,

$$BL = \mu_0 \eta L I$$

$$B = \mu_0 \eta I.$$

7.3.1   *Magnetostatic Boundary Conditions*

Similar to how a surface charge engenders a discontinuity in the normal component of the electrostatic field, a surface current results in a discontinuity in the magnetic fields on the two sides of the surface. Firstly, we can draw a pillbox, of infinitesimal length and a small base area, that straddles a point on the surface and exploit the fact that the closed surface integral of the magnetic field is zero, to conclude that the normal component of the magnetic field must be continuous across the surface.

$$B_{\perp 2} = B_{\perp 1}$$

where the subscripts 1 and 2 refer to the two sides of the surface. Now, suppose that the surface current density at the point of concern on the surface is $\boldsymbol{K}$ such that the current flowing across an infinitesimal length segment dl, that is along the surface and perpendicular to $\boldsymbol{K}$, at that point is $K dl$. Drawing an Amperian loop, with two essentially zero-length edges normal to the surface and one edge of length dl (parallel to the length segment in the previous statement) on each side of the surface, around this point and applying Ampere's law,

$$B_{\|2} dl - B_{\|1} dl = K dl$$

$$B_{\|2} - B_{\|1} = K,$$

where $B_{\|2}$ and $B_{\|1}$ are the magnetic fields along the two edges of length dl. Their positive directions are aligned and are defined to be along the direction of the closed loop integral along the edge of length dl on side 2 (anti-clockwise relative to $\boldsymbol{K}$). We can rewrite the above in vector notation as

$$\boldsymbol{B}_{\|2} - \boldsymbol{B}_{\|1} = \boldsymbol{K} \times \hat{\boldsymbol{n}},$$

where $\hat{\boldsymbol{n}}$ is the normal unit vector, pointing towards side 2. $\boldsymbol{B}_{\|2}$ and $\boldsymbol{B}_{\|1}$ are now the magnetic field vectors on sides 2 and 1, tangential to the surface. Combining this boundary condition with the fact that the normal component of magnetic field is continuous, the net magnetic fields, $\boldsymbol{B}_1$ and $\boldsymbol{B}_2$, on sides 1 and 2 are related by

$$\boldsymbol{B}_2 - \boldsymbol{B}_1 = \boldsymbol{K} \times \hat{\boldsymbol{n}}.$$

7.4   Motion in Magnetic Fields

Now that we understand how moving charges respond to and generate magnetic fields, certain special forms of motion shall be analyzed.

7.4.1   *Charge in Uniform Magnetic Field*

A particle with charge q and mass m traveling perpendicularly to a uniform magnetic field $\boldsymbol{B}$ will exhibit uniform circular motion, as the magnetic force on it is always perpendicular to its velocity, and also because its speed remains constant, since the magnetic force does no work on the charge.

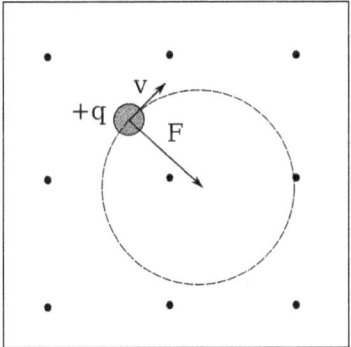

Figure 7.13:   Charge moving in a magnetic field

Referring to Fig. 7.13, since the magnetic force provides the centripetal force, the radius of orbit can be computed. Letting the constant speed of the charge in the plane perpendicular to $\boldsymbol{B}$ be $\boldsymbol{v}$,

$$F = qvB = \frac{mv^2}{r}$$

$$r = \frac{mv}{qB}. \tag{7.14}$$

The angular velocity is

$$\omega = \frac{v}{r} = \frac{qB}{m} \tag{7.15}$$

which is known as the cyclotron (angular) frequency. If the charge has a component of velocity parallel to the magnetic field, that component of velocity will be constant as the direction of the magnetic force on the charge must be normal to the magnetic field. Therefore, the particle will still orbit circularly in the plane perpendicular to the magnetic field, while traveling at a constant velocity along the magnetic field — the trajectory of the particle is then helicoidal.

To be completely rigorous about our claims, we can solve the particle's equation of motion. Align the z-axis with the external magnetic field and let the instantaneous coordinates of the particle be (x, y, z) with initial coordinates (x_0, y_0, z_0) and initial velocity (u_x, u_y, u_z). The net force on the charge is

$$\boldsymbol{F} = q \begin{pmatrix} \dot{x} \\ \dot{y} \\ \dot{z} \end{pmatrix} \times \begin{pmatrix} 0 \\ 0 \\ B \end{pmatrix} = \begin{pmatrix} q\dot{y}B \\ -q\dot{x}B \\ 0 \end{pmatrix}.$$

By Newton's second law,

$$\ddot{x} = \frac{qB}{m}\dot{y},$$

$$\ddot{y} = -\frac{qB}{m}\dot{x},$$

$$\ddot{z} = 0.$$

The z-component can be solved easily.

$$\dot{z} = u_z$$

$$z = z_0 + u_z t.$$

For the other two components, we can devise the following formulation, which is especially insightful when the equation of motion involves the cross product of certain quantities. Define a new complex variable $\eta = x + iy$ where x and y are the instantaneous x and y-coordinates of the charge. Since x and y are real, the coordinates of the charge is simply that of η in the Argand plane. The slick part of this approach is that we can now combine the two coupled equations of motion into an equation involving a single variable. Multiplying the y-component of the equation of motion by i and adding it to the x-component,

$$\ddot{x} + i\ddot{y} = \frac{qB}{m}(\dot{y} - i\dot{x}) = -\frac{iqB}{m}(\dot{x} + i\dot{y})$$

$$\ddot{\eta} = -\frac{iqB}{m}\dot{\eta}.$$

We can immediately guess[5] the general solution to this differential equation as

$$\dot{\eta} = \dot{\eta}_0 e^{-\frac{iqB}{m}t},$$

where $\dot{\eta}_0 = u_x + iu_y$ is $\dot{\eta}$ at $t = 0$. The above can be integrated once again to yield

$$\eta = \eta_0 - \frac{im}{qB}\dot{\eta}_0 + \frac{im}{qB}\dot{\eta}_0 e^{-\frac{iqB}{m}t}$$

$$= \eta_0 - \frac{m}{qB}(-u_y + iu_x) + \frac{m}{qB}(-u_y + iu_x)e^{-\frac{iqB}{m}t}.$$

Expressing the time-independent portion in Cartesian form and the time-dependent portion in polar form via Euler's formula,

$$\eta = x_0 + \frac{mu_y}{qB} + i\left(y_0 - \frac{mu_x}{qB}\right) + \frac{m\sqrt{u_x^2 + u_y^2}}{qB}e^{-i\left(\frac{qB}{m}t+\phi\right)}$$

where $\phi = \tan^{-1}\frac{u_x}{u_y}$. To interpret this solution, observe that $\eta = x + iy$ is the addition of a constant term $x_0 + \frac{mu_y}{qB} + i(y_0 - \frac{mu_x}{qB})$ that represents a constant vector $(x_0 + \frac{mu_y}{qB}, y_0 - \frac{mu_x}{qB})$ in the Argand plane and another vector associated with $\frac{m\sqrt{u_x^2+u_y^2}}{qB}e^{-i(\frac{qB}{m}t+\phi)}$ that represents a vector of constant length $\frac{m\sqrt{u_x^2+u_y^2}}{qB}$ rotating at a constant clockwise angular velocity $\frac{qB}{m}$, beginning at angle ϕ below the positive real axis. Therefore, the above solution describes a circular motion of radius $\frac{m\sqrt{u_x^2+u_y^2}}{qB}$ in the xy-plane about the center $(x_0 + \frac{mu_y}{qB}, y_0 - \frac{mu_x}{qB})$. The angular frequency of this motion can also be deduced from the rotating vector in the Argand plane as $\frac{qB}{m}$ clockwise. Finally, one can retrieve the instantaneous x and y-coordinates of the particle by taking the real and imaginary components of the above.

$$x = \text{Re}(\eta) = x_0 + \frac{mu_y}{qB} + \frac{m\sqrt{u_x^2 + u_y^2}}{qB}\cos\left(\frac{qB}{m}t + \phi\right),$$

$$y = \text{Im}(\eta) = y_0 - \frac{mu_x}{qB} - \frac{m\sqrt{u_x^2 + u_y^2}}{qB}\sin\left(\frac{qB}{m}t + \phi\right),$$

---

[5] One may be tempted to separate variables and integrate but one will be delving into the unfamiliar realm of integrating over a complex variable! Instead, it is simpler to directly guess the solution and assert that we have found the general solution. The number of independent solutions to a differential equation involving η is equal to that involving x or y (if they can be combined into a single equation in η) as the former differential equation can be seen as the composition of the equations in x and y through its real and imaginary parts.

where $\phi = \tan^{-1}\frac{u_x}{u_y}$. In fact, one can also manipulate the above set of parametric equations (by isolating the terms of each equation that involve t, squaring and adding them together) to directly conclude that the trajectory of the particle is a circle in the xy-plane.

7.4.2   *Charge in Uniform and Perpendicular Electric and Magnetic Fields*

In this section, we will analyze the motion of a particle with charge q and mass m in a mutually perpendicular and uniform pair of magnetic field $\boldsymbol{B} = (0,0,B)$ and electric field $\boldsymbol{E} = (0,E,0)$. By the Lorentz force law, the equations of motion of the charge are

$$\ddot{x} = \frac{qB}{m}\dot{y},$$

$$\ddot{y} = \frac{qE}{m} - \frac{qB}{m}\dot{x},$$

$$\ddot{z} = 0.$$

Evidently, the motion of the particle in the z-direction is not particularly interesting as it travels at a constant velocity. Hence, we will simply consider the motion of the particle in the x and y-directions. To solve this set of coupled differential equations, we can introduce a new complex variable $\eta = x + iy$. Multiplying the y-component in the equation of motion by i and adding it to the x-component,

$$\ddot{\eta} = -\frac{iqB}{m}\dot{\eta} + \frac{iqE}{m}$$

$$\implies \ddot{\eta} + \frac{iqB}{m}\dot{\eta} = \frac{iqE}{m}.$$

The particular solution to this differential equation is evidently $\dot{\eta} = \frac{E}{B}$ while the homogeneous solution is $\dot{\eta} = Ae^{-\frac{iqB}{m}t}$ where A is a certain complex constant that can be solved from the initial conditions. The general solution to $\dot{\eta}$ is

$$\dot{\eta} = Ae^{-\frac{iqB}{m}t} + \frac{E}{B}.$$

Enforcing the initial condition that $\dot{\eta}$ at $t = 0$ is $\dot{\eta}_0 = u_x + iu_y$,

$$\dot{\eta} = \left(\dot{\eta}_0 - \frac{E}{B}\right)e^{-\frac{iqB}{m}t} + \frac{E}{B}.$$

Integrating the above with respect to time,

$$\eta = \eta_0 + \frac{im}{qB}\left(\dot{\eta}_0 - \frac{E}{B}\right)\left(e^{-\frac{iqB}{m}t} - 1\right) + \frac{E}{B}t,$$

where $\eta_0 = x_0 + iy_0$. Substituting the expression for $\dot{\eta}_0$,

$$\eta = x_0 + iy_0 + \frac{m}{qB}\left(-u_y + iu_x - i\frac{E}{B}\right)e^{-\frac{iqB}{m}t} + \frac{E}{B}t - \frac{m}{qB}\left(-u_y + iu_x - i\frac{E}{B}\right).$$

Applying Euler's formula to simplify $-u_y + i(u_x - \frac{E}{B}) = \sqrt{(u_x - \frac{E}{B})^2 + u_y^2}\,e^{-i\phi}$

with $\phi = \tan^{-1}\frac{u_x - \frac{E}{B}}{u_y}$ in the time-dependent term and grouping the real and complex terms,

$$\eta = x_0 + \frac{mu_y}{qB} + \frac{E}{B}t + i\left(y_0 - \frac{m\left(u_x - \frac{E}{B}\right)}{qB}\right) + \frac{m\sqrt{\left(u_x - \frac{E}{B}\right)^2 + u_y^2}}{qB}e^{-i\left(\frac{qB}{m}t + \phi\right)}.$$

Let us try to interpret this trajectory physically by first considering the simplest case where the particle is initially stationary, such that $u_x = u_y = 0$ and $\phi = -\frac{\pi}{2}$. Then,

$$\eta = x_0 + \frac{E}{B}t + i\left(y_0 + \frac{mE}{qB^2}\right) + \frac{mE}{qB^2}e^{-i\left(\frac{qB}{m}t - \frac{\pi}{2}\right)}.$$

The particle appears to undergo uniform circular motion with a radius $r = \frac{mE}{qB^2}$ and angular velocity $\omega = \frac{qB}{m}$ about a moving center that was initially at $(x_0, y_0 + \frac{mE}{qB^2})$ and is traveling at a constant velocity in the positive x-direction, $v = \frac{E}{B}$, which is also coincidentally equal to $r\omega$.

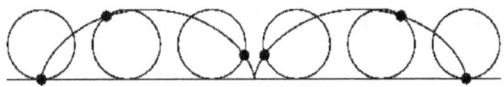

Figure 7.14: Cycloid

Hence, the charge's trajectory is identical to that of a particle attached to a rigid circle with a radius r that is rolling without slipping in the xy-plane on a flat ground parallel to the x-axis, such that the center of the circle is maintained at a constant y-coordinate $y_0 + \frac{mE}{qB^2}$. Note that the particle is initially located vertically below the instantaneous center of rotation (i.e. the bottom point of the left-most circle in Fig. 7.14) as $\phi = -\frac{\pi}{2}$ represents the negative imaginary direction in the Argand plane. This trajectory is known as a cycloid and these equations of motion arise when an initially stationary charge is placed in uniform electric and magnetic fields that are mutually perpendicular. If the charge has a z-component of velocity, its trajectory will be a cycloidal motion in the xy-plane, superimposed with a constant z-velocity.

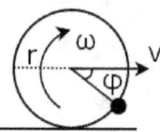

Figure 7.15:   Initial position of particle on translating and rotating circle

In the general case where the particle starts with non-zero x and y-velocities, the trajectory of the particle is tantamount to that of one attached to a circle with radius

$$r = \frac{m\sqrt{\left(u_x - \frac{E}{B}\right)^2 + u_y^2}}{qB} \tag{7.16}$$

that is rotating at a constant clockwise angular frequency

$$\omega = \frac{qB}{m}, \tag{7.17}$$

with its center translating at a constant positive x-velocity

$$v = \frac{E}{B}. \tag{7.18}$$

Its initial angular position is a clockwise angle $\phi = \tan^{-1}\frac{u_x - \frac{E}{B}}{u_y}$ with respect to the horizontal, as shown in Fig. 7.15. The constant y-coordinate of the center of the circle is $y_0 - \frac{m(u_x - \frac{E}{B})}{qB}$.

7.4.3   *Current Loop in Uniform Magnetic Field*

Though we have shown that a current loop experiences no net magnetic force due to a uniform magnetic field, it will generally still experience a torque. Let us first consider a simple rectangular current loop whose area vector $\boldsymbol{A}$ makes an angle θ with respect to a uniform magnetic field $\boldsymbol{B}$. Two particular parallel segments of the loop are perpendicular to $\boldsymbol{B}$.

We have oriented $\boldsymbol{B}$ in the y-direction and segments 2 and 4 in the x-direction in Fig. 7.16. Segments 2 and 4 have lengths a while segments 1 and 3 have lengths b. Evidently, the magnetic forces along lines 1 and 3 produce no net torque. Thus, we can simply consider the side view of the rectangular loop in Fig. 7.17 where segments 2 and 4 extend into the page.

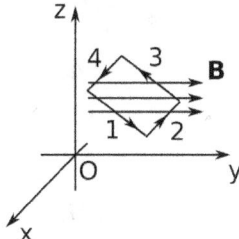

Figure 7.16: Loop in magnetic field

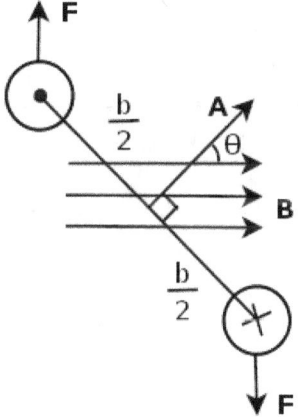

Figure 7.17: Side view

The magnetic forces along lines 2 and 4 are of common magnitude

$$F = BIa$$

as the currents along those lines are always perpendicular to the magnetic field. Then, the total torque on the loop about its center can be calculated as

$$\tau = \frac{Fb}{2} \cos\left(\frac{\pi}{2} - \theta\right) \times 2 = BIab\sin\theta.$$

This is the torque about any point in space as the torque on an object is the same with respect to all pivots if the forces producing the torque sum to zero vectorially — a condition that is satisfied by the zero net force on a current loop due to the uniform magnetic field. Next, we notice that if we define an area vector $\boldsymbol{A}$ that points perpendicularly from the plane of the loop as

$$\boldsymbol{A} = ab\hat{n}$$

and the magnetic dipole moment as

$$\mu = NIA, \tag{7.19}$$

where N refers to the number of loops ($N = 1$ in this case), the torque on the loop can be rewritten as

$$\tau = \mu \times B \tag{7.20}$$

for a rectangular coil with N loops. The direction of the normal vector $\hat{n}$ is given by the right-hand-grip rule. If you grip your right hand along the current loop, your straightened thumb will point in its direction.

Now, we proceed with a more general formulation — we now claim that the torque, about any pivot, on an arbitrarily shaped planar coil with N densely wound turns and carrying current I is given by Eq. (7.20) when it is placed in a uniform magnetic field B. We will provide the proof for this below but you can in fact extend Eq. (7.20) to non-planar loops (see Problem 16) with a modified but similar definition for the magnetic dipole moment μ.

Proof: To derive the above result, suppose that we put a single turn of a coil carrying anti-clockwise current I in the xy-plane in a region of uniform magnetic field B, as shown in Fig. 7.18.

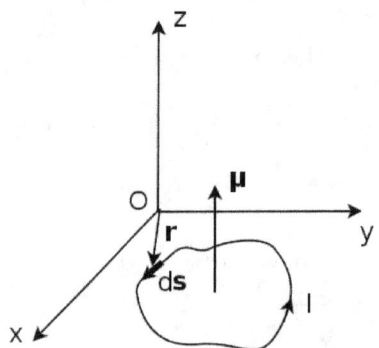

Figure 7.18:   Loop in xy-plane

We can always arrange for B to not have an x-component. Then, $B = (0, B_y, B_z)$. Now, consider an infinitesimal segment of the coil $ds = (dx, dy, 0)$ at position vector $r = (x, y, 0)$. The magnetic force on this segment is

$$dF = Ids \times B.$$

The torque on this segment about the origin is

$$d\tau = Ir \times (ds \times B) = I(r \cdot B)ds - I(r \cdot ds)B$$

by the BAC-CAB rule. Now, observe that

$$\mathbf{r} \cdot d\mathbf{s} = r dr,$$

where r is the radial distance from the origin. Then, the above can be simplified into

$$d\boldsymbol{\tau} = I y B_y d\mathbf{s} - I r \mathbf{B} dr.$$

The total torque about the origin is then obtained from integrating the above over the entire loop. The second term vanishes after integration as the initial and final r's are the same for a loop. The contribution to the total torque only stems from the first term.

$$\boldsymbol{\tau} = \oint I y B_y \begin{pmatrix} dx \\ dy \\ 0 \end{pmatrix}.$$

Considering the different components,

$$\tau_x = \oint I y B_y dx = -I A B_y$$

where A is the positive area of the coil drawn in Fig. 7.18. The integral yields the negative area as the current I points in the negative x-direction at larger values of y and points in the positive x-direction at smaller values of y. Next,

$$\tau_y = \oint I y B_y dy = 0$$

as the initial and final y values are the same for a loop. There is also no z-component of torque as there is no z-component of $d\mathbf{s}$. Then, the net torque on the loop is

$$\boldsymbol{\tau} = \begin{pmatrix} -I A B_y \\ 0 \\ 0 \end{pmatrix},$$

which is consistent with

$$\boldsymbol{\mu} \times \mathbf{B} = \begin{pmatrix} 0 \\ 0 \\ I A \end{pmatrix} \times \begin{pmatrix} 0 \\ B_y \\ B_z \end{pmatrix} = \begin{pmatrix} -I A B_y \\ 0 \\ 0 \end{pmatrix}.$$

For a coil with N turns, we can simply substitute $N I$ for I — hence proving Eq. (7.20).

7.5   Magnetic Fields in Matter

Similar to how electric dipoles arise when matter is placed in an electric field, magnetic dipoles are induced in matter when an external magnetic field is imposed. The three main mechanisms through which this occurs are diamagnetism, paramagnetism and ferromagnetism — we will discuss the former two here.

Diamagnetism can be understood from the elementary model of electrons orbiting a nucleus in an atom. The rapid revolution of the electron appears like a current when averaged over time and hence constitutes a magnetic dipole. Normally, the orbits are randomly oriented such that there is no net orbital magnetic dipole moment. However, when a magnetic field is turned on, an induced emf will be generated by Faraday's law (introduced in the next chapter), in an attempt to oppose the external magnetic field. This induced electric field then accelerates all electrons (regardless of their original directions of revolution) in a manner such that a net magnetic dipole moment is generated **antiparallel** to the imposed magnetic field (to reduce the change in magnetic flux through an imaginary loop). Such a phenomenon is universal and affects all atoms.

Paramagnetism, on the other hand, occurs because an external magnetic field exerts a torque on magnetic dipoles that tends to align their magnetic dipole moments with the field. It turns out that an electron not only orbits about its nucleus but also possesses an intrinsic angular momentum known as spin. This is a purely quantum mechanical effect that has no classical explanation, but you can picture an electron as a rotating ball of charge if it helps. This spin endows an electron with a magnetic dipole moment and causes it to behave like a magnetic dipole. Without an external magnetic field, the orientation of the spin magnetic dipole moment is arbitrary and thus results in zero net dipole moment over time. However, when an external magnetic field is present, a lone electron tends to be oriented such that its magnetic dipole moment is parallel to the external field, which is the lowest energy configuration (it is not completely aligned due to thermal fluctuations, so there is a compromise here). Therefore, a net dipole moment parallel to the external magnetic field is induced. Now, it happens that most electrons are stuck together as a pair with opposite spins which are mutually nullified — implying that this effect does not ascribe them a net dipole moment. As a result, paramagnetism predominantly occurs in molecules with an odd number of electrons (such that there is an unpaired electron).

7.5.1 *Bound Currents*

Now that we have understood the mechanisms of creating magnetic dipoles in matter, we can describe the effects of the dipoles quantitatively. Similar to the electric polarization, we can define a quantity known as the magnetization M that describes the magnetic dipole per unit volume.

Consider an infinitesimal cuboid of edge lengths dx, dy and dz with a uniform magnetization M in the positive z-direction, depicted in Fig. 7.19.

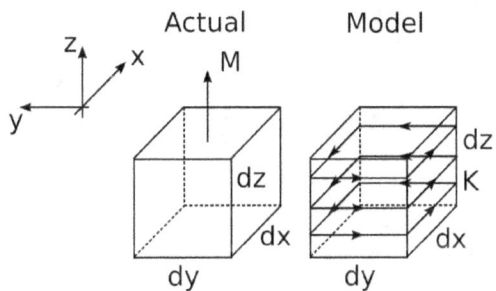

Figure 7.19:   Actual cuboid and equivalent current distribution

As we shall show in Section 8.10.3 that the distant magnetic field (at length scales much further away than the size of the dipole) due to a magnetic dipole is only dependent on the magnetic dipole moment of the dipole and independent of the internal structure of the dipole (which could be current loops, multiple current loops, non-planar loops, built-in dipoles, etc), the magnetic field due to this cuboid, outside itself, is akin to a magnetic dipole constituted by current loops with surface current density $K = M$ (on the right). To show this equivalence, firstly note that the magnetic dipole moment of the actual cuboid is

$$\mu = M dx dy dz \hat{k}.$$

To compute the magnetic dipole moment of our model, observe that each layer at a constant height constitutes a current loop of area $dxdy$. Therefore, the total magnetic dipole moment of our model is

$$\mu = (K dz) dx dy$$

where Kdz is the total surface current running along a face. Therefore, we require

$$K = M$$

for the magnetic dipole moments to be identical. Similar to our agenda for electric fields in matter, we proceed with the analysis of the equivalent surface current distributions of cuboids with wedges attached. This is because any volume can be seen as an array of cuboids in the interior and a collection of cuboids with wedges attached on the surface (as the surface is generally inclined). To this end, consider the wedge in Fig. 7.20, still with a magnetization M in the positive z-direction.

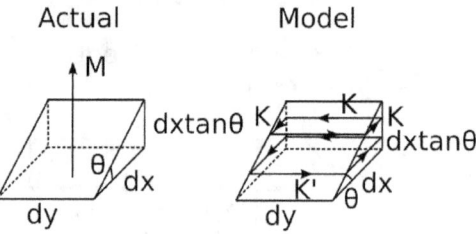

Figure 7.20:   Actual wedge and equivalent current distribution

Its total magnetic dipole moment is

$$\mu = \frac{1}{2}M dx^2 dy \tan\theta \hat{k}.$$

Now, we wish to propose an equivalent surface charge distribution that generates this magnetic dipole moment. For the sake of simplicity, we hope that our model can be seen as a collection of current loops (when sliced at different heights along the wedge). Then, we need the total current flowing along each face to be identical. If the uniform surface current density on the "hypotenuse" face is K' while the uniform surface current densities on the other three faces are K, we require

$$K' = K \sin\theta,$$

since the length of the hypotenuse, perpendicular to the flow of current, is larger than that of the other faces by a factor of $\frac{1}{\sin\theta}$. With this, we can proceed with calculating the magnetic dipole moment of our model. Since the surface current density is uniform over each face, the total magnetic dipole moment is akin to that produced by the entire current $I = K dx \tan\theta$ running in the loop at half the height of the wedge (since this represents the average).

$$\mu = K dx \tan\theta \cdot \frac{dx dy}{2}.$$

To match the actual and proposed magnetic dipole moments,

$$K = M,$$

$$K' = M \sin \theta.$$

Notice that the surface current density on each face of the wedge or cuboid can be nicely summarized by

$$\boldsymbol{K} = \boldsymbol{M} \times \hat{\boldsymbol{n}},$$

where $\hat{\boldsymbol{n}}$ is the normal unit vector pointing outwards of the face. By the principle of superposition, this expression for $\boldsymbol{K}$ is valid for any element constructed from attaching wedges to a cuboid. What's more, we can extend this result to a magnetization $\boldsymbol{M}$ in a general direction, as we can always divide it into its components and apply the principle of superposition.

Armed with this knowledge, we can analyze a piece of magnetic material with significant volume and a certain magnetization $\boldsymbol{M}$ that is a function of position. Since this magnetic substance can be divided into an array of cuboids in the interior and an array of cuboids with wedges attached along the surface, the magnetic field (due to magnetization) outside this material is equivalent to that produced by a bound surface current density

$$\boldsymbol{K}_b = \boldsymbol{M} \times \hat{\boldsymbol{n}}$$

along the exterior surface and a certain current configuration in the interior. We have included a subscript b to emphasize that the origin of this bound surface current density is the magnetization of the material (for which the bound current is merely a model).

To determine the interior currents, consider two juxtaposed cuboids with edge lengths dx, dy and dz at (x, y, z) and $(x, y + dy, z)$. The cuboids are tiny enough such that the magnetizations $\boldsymbol{M}(x, y, z)$ and $\boldsymbol{M}(x, y + dy, z)$ are uniform over each of their volumes.

Referring to the left diagram of Fig. 7.21, due to the discrepancy in the z-components of magnetization, there is a surface current of total current

$$I_{x1} = [M_z(x, y + dy, z) - M_z(x, y, z)]dz = \frac{\partial M_z}{\partial y} dy dz$$

flowing along the overlapping surface, in the positive x-direction. Similarly, we can show that for two cuboids stacked on top of each other (depicted in the right diagram of Fig. 7.21), the discrepancy in the y-components of magnetization generates another current

$$I_{x2} = -\frac{\partial M_y}{\partial z} dy dz$$

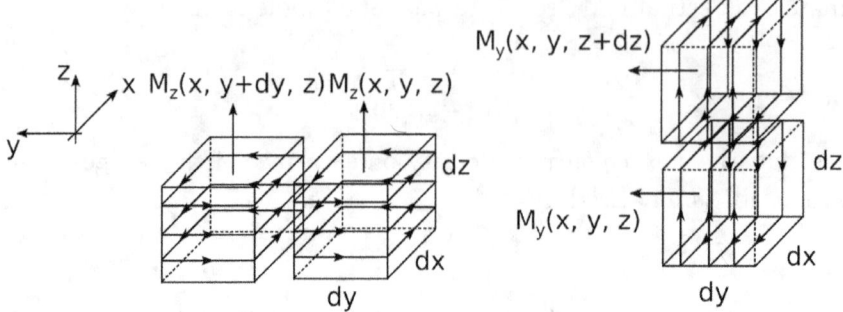

Figure 7.21:   Juxtaposed cuboids

in the positive x-direction. The net current in the x-direction that is associated with the cuboid at (x, y, z), due to magnetization, is thus

$$I_x = \left(\frac{\partial M_z}{\partial y} - \frac{\partial M_y}{\partial z} \right) dy dz,$$

which is equivalent to the cuboid possessing a bound volume current density

$$J_x = \frac{\partial M_z}{\partial y} - \frac{\partial M_y}{\partial z}$$

in the positive x-direction. Repeating the above procedure for the other components of current, the equivalent bound volume current density of the cuboid satisfies

$$J_y = \frac{\partial M_x}{\partial z} - \frac{\partial M_z}{\partial x},$$

$$J_z = \frac{\partial M_y}{\partial x} - \frac{\partial M_x}{\partial y}.$$

More compactly, the bound volume current density is

$$\boldsymbol{J_b} = \nabla \times \boldsymbol{M}.$$

To summarize what we have derived so far, the magnetic field outside a magnetic material is equivalent to that produced by a bound surface current density

$$\boldsymbol{K_b} = \boldsymbol{M} \times \hat{\boldsymbol{n}} \tag{7.21}$$

on the exterior surface and a bound volume current density

$$\boldsymbol{J_b} = \nabla \times \boldsymbol{M} \tag{7.22}$$

in the interior of the material. Now, what can we say about the magnetic field (due to magnetization) inside the magnetic material? It is important to remember that our proposed model of a magnetized cuboid as one with bound surface currents only produces the correct magnetic field at distances much larger than the size of the cuboid, from the cuboid. Therefore, our model definitely produces the wrong microscopic field within the substance. However, we shall show that the macroscopic magnetic field, which is the magnetic field averaged over a small volume containing many molecules, can also be taken to be that due to bound current distribution afore!

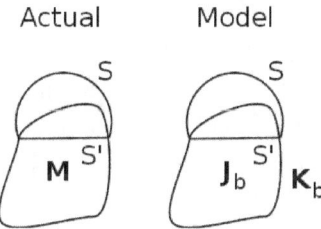

Figure 7.22:   Surface S on actual magnetized matter and bound current model

Consider a small volume of magnetic material around a relevant point in the substance in Fig. 7.22 (the rest of it is not depicted) that contains a multitude of molecules. The magnetic field (due to magnetization) within this region has two causes — due to the magnetized material outside of this region and due to the material within this region. Since the bound current model is already accurate for the material outside of this region, we simply have to show that the bound current model for the material within the region produces the correct macroscopic field to prove our claim.

Draw a Gaussian surface S that protrudes out of the substance in this region in both the actual system (with a certain magnetization M) and our model of bound currents (K_b and J_b). The portion S' of the surface within the demarcated volume is perpendicular to the vertical. If we let the magnetic fields of the original system and our model be B and B' respectively, we have

$$\oiint_S B \cdot dA = 0,$$

$$\oiint_S B' \cdot dA = 0$$

over the Gaussian surface S.

$$\implies \oiint_S B \cdot dA = \oiint_S B' \cdot dA.$$

Since we already know that $B = B'$ outside of this region, the surface integrals of B and B' over the portion S' within the highlighted region must be identical!

$$\iint_{S'} \boldsymbol{B} \cdot d\boldsymbol{A} = \iint_{S'} \boldsymbol{B}' \cdot d\boldsymbol{A}.$$

If we define the z-direction to be along the vertical, notice that the above is stating that the averages of B_z and B_z' are equal over the surface S'. Therefore, if we choose other surfaces similar to S' at different heights along the demarcated region, we can guarantee that the averages of B_z and B_z' over the entire region are identical! Repeating this process with surfaces S' oriented in the other directions, we can subsequently prove that the average magnetic fields of the original system and our proposed model, over this isolated region, are equal! Therefore, the macroscopic field within a magnetic material is also given by the bound currents $\boldsymbol{K}_b$ and $\boldsymbol{J}_b$.

Ampere's Law in Magnetic Materials

Now that we have understood the magnetic field produced by magnetized matter, we are ready to study the net magnetic field within a magnetic material which can be due to both magnetization and free currents channeled to the material by an external entity. In magnetic materials, the nullity of the closed surface integral of the net magnetic field still holds.

$$\oiint_S \boldsymbol{B} \cdot d\boldsymbol{A} = 0. \tag{7.23}$$

However, it is often more edifying to express Ampere's law (within the material) purely in terms of free currents, just like what we did for dielectrics. Applying Ampere's law to a contour C encased in the interior of the magnetic material,

$$\oint_C \boldsymbol{B} \cdot d\boldsymbol{s} = \mu_0 \iint_S \boldsymbol{J} \cdot d\boldsymbol{A}$$

where S is a surface that spans C. The volume current density comprises both free and bound currents.

$$\boldsymbol{J} = \boldsymbol{J}_f + \boldsymbol{J}_b.$$

Since $\boldsymbol{J}_b = \nabla \times \boldsymbol{M}$,

$$\oint_C \boldsymbol{B} \cdot d\boldsymbol{s} = \mu_0 \iint_S \boldsymbol{J}_f \cdot d\boldsymbol{A} + \mu_0 \iint_S (\nabla \times \boldsymbol{M}) \cdot d\boldsymbol{A}.$$

It happens that Stokes' theorem in vector calculus asserts that

$$\iint_S (\nabla \times \boldsymbol{F}) \cdot d\boldsymbol{A} = \oint_C \boldsymbol{F} \cdot d\boldsymbol{s}$$

for any vector field $\boldsymbol{F}$, where C is a contour that bounds surface S. Applied to the current situation, this implies that

$$\iint_S (\nabla \times \boldsymbol{M}) \cdot d\boldsymbol{A} = \oint_C \boldsymbol{M} \cdot d\boldsymbol{s}$$

$$\oint_C \left(\frac{\boldsymbol{B}}{\mu_0} - \boldsymbol{M} \right) \cdot d\boldsymbol{s} = \iint_S \boldsymbol{J}_f \cdot d\boldsymbol{A}.$$

This suggests that we define a new quantity

$$\boldsymbol{H} = \frac{\boldsymbol{B}}{\mu_0} - \boldsymbol{M} \qquad (7.24)$$

which shall be referred to as the H-field (some call this quantity the magnetic field but it is not as fundamental as $\boldsymbol{B}$ and is rather confusing). Then, we retrieve an equation analogous to the original form of Ampere's law.

$$\oint_C \boldsymbol{H} \cdot d\boldsymbol{s} = I_f \qquad (7.25)$$

where I_f is the free current passing through a surface bounding C. It is interesting to note that the above equation is much more useful than its electric counterpart $\oiint_S \boldsymbol{D} \cdot d\boldsymbol{A} = q_f$ in practice. This is because I_f is much easier to measure empirically than q_f. In electrostatic experiments, we usually connect a battery of a known emf to a dielectric which tells us nothing about the distribution of free charge. At most, we can only determine the line integral of $\boldsymbol{E}$ through the dielectric but this is still not directly related to $\boldsymbol{D}$. On the other hand, in magnetostatic experiments, (free) current is transferred to a magnetic material via external wires and this current can be conveniently measured via an ammeter!

Boundary Conditions

The following boundary conditions must be fulfilled at a surface carrying a certain free surface current. Firstly, imposing the nullity of the surface integral of the magnetic field over an infinitesimal pillbox with negligible length around a point on the interface (its small bases are parallel to the surface around that point), the normal components of magnetic fields on the two sides of that point must be continuous.

$$B_{\perp 2} = B_{\perp 2}. \qquad (7.26)$$

Next, suppose that a free surface current density $\boldsymbol{K}_f$ runs at a point of interest on the interface. Draw a loop that is perpendicular to $\boldsymbol{K}_f$ around that point, with the edges perpendicular to the surface being negligible. If we let the non-negligible length of this loop along the surface be dl, Ampere's law for the H-field yields

$$H_{\|2}dl - H_{\|1}dl = K_f dl$$

$$H_{\|2} = H_{\|1} + K_f$$

where $H_{\|1}$ and $H_{\|2}$ are positive in the positive direction defined along the segment on side 2 when the right-hand-grip rule is applied to K_f. In vector form,

$$\boldsymbol{H}_{\|2} = \boldsymbol{H}_{\|1} + \boldsymbol{K}_f \times \hat{\boldsymbol{n}} \qquad (7.27)$$

where $\hat{\boldsymbol{n}}$ is the normal unit vector pointing from side 1 to side 2.

7.5.2 *Magnetic Susceptibility*

In a preponderance of materials, the magnetization is directly proportional to the net magnetic field. You might think that we will adopt a definition, parallel to the electric susceptibility, for the constant of proportionality such as

$$\boldsymbol{M} = \frac{\chi_b \boldsymbol{B}}{\mu_0} \quad \text{(incorrect)}$$

where $\boldsymbol{M}$ is proportional to the B-field. However, convention ordains us to define the magnetic susceptibility χ_b as

$$\boldsymbol{M} = \chi_b \boldsymbol{H}. \qquad (7.28)$$

With this definition, the relationship $\boldsymbol{H} = \frac{\boldsymbol{B}}{\mu_0} - \boldsymbol{M}$ yields

$$(1 + \chi_b)\boldsymbol{H} = \frac{\boldsymbol{B}}{\mu_0}$$

$$\boldsymbol{B} = \mu_0(1 + \chi_b)\boldsymbol{H}. \qquad (7.29)$$

The material-dependent constant of proportionality $\mu_0(1 + \chi_b)$ is known as the magnetic permeability μ of the substance.

$$\mu = \mu_0(1 + \chi_b), \qquad (7.30)$$

$$\boldsymbol{B} = \mu \boldsymbol{H}. \qquad (7.31)$$

With this definition, we can rewrite Ampere's law for the H-field in a rather simple form, if the Amperian loop lies entirely in a region of homogeneous material (else the magnetic permeability is a function of location and cannot be extracted from the integral).

$$\oint_C \frac{\boldsymbol{B}}{\mu} \cdot d\boldsymbol{s} = I_f$$

$$\oint \boldsymbol{B} \cdot d\boldsymbol{s} = \mu I_f,$$

so it is as if we can replace μ_0 and I in our original Ampere's law with μ and I_f for a region of homogeneous material!

Boundary Conditions at Interface between Two Magnetic Media

At the interface between two magnetic media with permeabilities μ_1 and μ_2, the magnetic fields on the two sides must satisfy

$$B_{\perp 1} = B_{\perp 2} \tag{7.32}$$

and

$$\mu_2 \boldsymbol{B}_{\parallel 2} = \mu_1 \boldsymbol{B}_{\parallel 1} + \boldsymbol{K}_f \times \hat{\boldsymbol{n}}, \tag{7.33}$$

where $\hat{\boldsymbol{n}}$ is the normal unit vector pointing from medium 1 to medium 2.

7.5.3 *Analogy between Magnetic and Electric Fields*

An enlightening analogy exists between the electric field of a configuration that is completely absent of free charges and the magnetic field of a set-up devoid of free currents. The former obeys the equations

$$\oiint_S \boldsymbol{D} \cdot d\boldsymbol{A} = 0,$$

$$\oint_C \boldsymbol{E} \cdot d\boldsymbol{s} = 0,$$

for any surface S and contour C, with

$$\varepsilon_0 \boldsymbol{E} = \boldsymbol{D} - \boldsymbol{P}.$$

Meanwhile, the latter obeys the equations

$$\oiint_S \boldsymbol{B} \cdot d\boldsymbol{A} = 0,$$

$$\oint_C \boldsymbol{H} \cdot d\boldsymbol{s} = 0,$$

for any surface S and contour C, with

$$\mu_0 \boldsymbol{H} = \boldsymbol{B} - \mu_0 \boldsymbol{M}.$$

Exploiting the fact that the substitutions $\boldsymbol{D} \to \boldsymbol{B}$, $\boldsymbol{E} \to \boldsymbol{H}$, $\boldsymbol{P} \to \mu_0 \boldsymbol{M}$ and $\varepsilon_0 \to \mu_0$ transform the electrostatic problem into the magnetostatic one, the answers to many problems involving permeable magnetic materials can be directly stated from this parallelism. However, be wary that there must indeed be a correlation between $\boldsymbol{P}$ and $\mu_0 \boldsymbol{M}$ for this analogy to work (e.g. they are directly proportional to each other in all space).

For linear media,

$$\boldsymbol{E} = \frac{\boldsymbol{D}}{\varepsilon},$$

$$\boldsymbol{H} = \frac{\boldsymbol{B}}{\mu},$$

so we make the substitution $\varepsilon \to \mu$ instead of $\boldsymbol{P} \to \mu_0 \boldsymbol{M}$ and $\varepsilon_0 \to \mu_0$. In such cases, we then have to ensure that there is a valid parallelism between ε and μ.

Problem: Determine the magnetic field in all space due to a sphere of radius R with a uniform magnetization $\boldsymbol{M}$.

The electric field of a sphere of radius R and uniform polarization $\boldsymbol{P}$ is

$$\boldsymbol{E}(\boldsymbol{r}) = \frac{2R^3 P \cos\theta}{3\varepsilon_0 r^3}\hat{\boldsymbol{r}} + \frac{R^3 P \sin\theta}{3\varepsilon_0 r^3}\hat{\boldsymbol{\theta}}$$

outside the sphere (with respect to spherical coordinates about the center) and

$$\boldsymbol{E} = -\frac{P}{3\varepsilon_0}\hat{\boldsymbol{k}}$$

inside the sphere, where the positive z-direction has been defined to be along $\boldsymbol{P}$. Making the appropriate substitutions, the H-field due to the uniformly magnetized sphere is

$$\boldsymbol{H} = \frac{2R^3 M \cos\theta}{r^3}\hat{\boldsymbol{r}} + \frac{R^3 M \sin\theta}{3r^3}\hat{\boldsymbol{\theta}}$$

outside of the sphere and

$$\boldsymbol{H} = -\frac{M}{3}\hat{\boldsymbol{k}}$$

inside the sphere. Therefore, the magnetic fields are

$$\boldsymbol{B} = \mu_0 \boldsymbol{H} = \frac{2\mu_0 R^3 M \cos\theta}{r^3} \hat{\boldsymbol{r}} + \frac{\mu_0 R^3 M \sin\theta}{3r^3} \hat{\boldsymbol{\theta}}$$

outside the sphere and

$$\boldsymbol{B} = \mu_0 \boldsymbol{H} + \mu_0 \boldsymbol{M} = \frac{2\mu_0 M}{3} \hat{\boldsymbol{k}}$$

within the sphere.

Problems

Lorentz Force and Miscellaneous

1. *An Apparent Paradox**

Determine the magnetic force per unit length between two parallel, thin and long current-carrying wires that are separated by a distance r and held stationary. The currents in the wires are I_1 and I_2 respectively, possibly traveling in different directions. Now, wire 1 is still held fixed but wire 2 is gently released. The current in wire 1 is maintained at I_1 by an external battery. We know from the previous result that wire 2 will begin to move towards or away from wire 1. Since the kinetic energy of wire 2 increases, have we violated the fact that a magnetic force cannot produce work? If not, suggest possible forms of energy that this kinetic energy originated from, in the case where wire 2 is not connected to any external entity and the case where the current in wire 2 is also maintained at I_2 by an external battery. Do not worry about how these energies are actually converted to the kinetic energy of wire 2.

2. *Force Between Moving Charges**

A charge q moves along the positive y-axis while another charge $-q$ moves along the positive x-axis on fixed rails. Both start at the same time from the origin, and move with constant speed $0 < v \ll c$. What is the magnetic force between the charges? Remember to indicate the direction too. Can you spot something wrong? Now, what if $v = 0$? Argue, physically and mathematically, why your expressions are invalid.

3. *Current on Cube**

Referring to the left figure, a current I flowing along the edges of one face of a cube produces a magnetic field B at the center of the cube. What is the magnetic field at the center of the same cube in the right figure? The cubes are isolated systems.

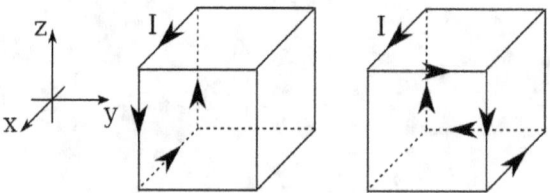

4. Lorentz Force on a Short Wire**

An infinitely long and thin wire carrying current I_1 in the negative y-direction lies along the y-axis ($x = 0$, $z = 0$). Another short and thin wire carrying current I_2 in the positive x-direction lies parallel to the x-axis. Its two ends are at $(0, 0, h)$ and $(l, 0, h)$. Find the force on the short wire due to the long wire.

5. Torque on Arbitrary Wires**

A uniform external magnetic field B permeates all space in the positive z-direction. Consider two points A and B in the xy-plane. An arbitrary connection of thin wires that lie entirely in the xy-plane is used to transfer I amount of total current from A to B. No charge is accumulated anywhere, except for the terminals A and B, possibly. If the linear distance between A and B, is l, determine the total torque experienced by the wires connecting A to B, about terminal A. Note that the wires may merge or split freely.

Biot-Savart's Law

6. Helmholtz Coil**

A pair of two identical coils of radius R are placed symmetrically along a common axis and are separated by a distance d. The common axis coincides with the z-axis, with the origin located at the center of the coils. The coils carry identical currents I in the same direction and each has N turns in total. Find the magnetic field strength along the axis of the coils $B(z)$ as a function of z. Determine the distance d such that $\frac{\partial^2 B}{\partial z^2} = 0$ at the center of the two coils. This set-up is known as the Helmholtz coil and is useful in generating a relatively uniform magnetic field between the coils (possibly to cancel Earth's local field).

7. Bent Wire**

An infinite wire is bent as shown in the figure below. Find the magnetic field at point P.

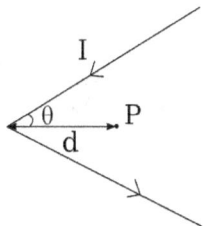

8. Finite Solenoid***

A thin, vertical solenoid with η turns per unit length, length l and radius R carries a current I anti-clockwise with respect to the positive z-axis which is along its symmetrical axis. Determine the magnetic field everywhere along the z-axis. To enforce the steady current condition, assume that thin wires — whose contributions to the magnetic field can be neglected — are used to transfer current to and from infinity.

Now, consider a vertical solenoid with η turns per unit length, inner radius r_0 and outer radius r_1 which carries a uniform current I anti-clockwise. Determine the magnetic field everywhere along the z-axis.

9. Rotating Sphere***

An insulating sphere with a uniform volume charge density ρ and radius R rotates about an axis through its center with a constant angular frequency ω. Find the magnetic field at a point along the axis of rotation.

Ampere's Law

10. Thick Infinite Wire*

A long cylinder, with its axis oriented in the z-direction, carries an axial current. The current density, although symmetric about the cylindrical axis, is not uniform but varies with radial distance r from the axis according to

$$J(r) = \begin{cases} \frac{2I_0}{\pi a^2}\left(1 - \frac{r^2}{a^2}\right) & \text{for } r < a \\ 0 & \text{for } r \geq a \end{cases}$$

where a is the radius of the cylinder and I_0 is a constant with units of amperes. Determine the magnetic field due to the cylinder everywhere.

11. Wire with Cavity*

A small long cylinder of radius $\frac{R}{2}$ is carved out of a long cylinder of radius R as shown on the next page (a cross section is depicted). The "wire" then carries a uniform current I, coming out of the page. Find the magnetic field at point P.

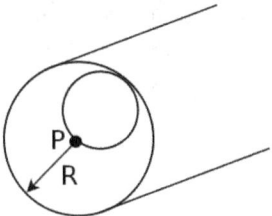

12. Opposite Currents*

The figure below shows the cross section of a current distribution that extends indefinitely in the z-direction. It is given by two overlapping circles of radius b with a separation $2a$ between their centers. The shaded regions of the left and right circles carry a uniform current density J into and out of the page respectively. No current flows across the intersection of the two circles. Determine the magnetic field at a point in the overlapping region, in this cross section.

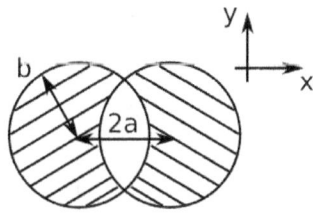

13. Toroid**

A toroid is a donut with a uniform cross section. For example, a toroid with a circular cross section can be formed by bending a solenoid along a circular ring and connecting both ends together. Now, suppose that we coil a wire into a toroid with an arbitrary cross section and $N \gg 1$ densely wound turns. If the resultant toroid carries a current I, determine the magnetic field due to this set-up everywhere.

14. Magnetic Flux and Field Lines**

Consider a long solenoid with η turns per unit length and radius R that carries a steady current I. We categorize the solenoid into two halves about its center — one half contains the North pole while the other half contains the South. What is the net magnetic flux that leaves the solenoid through the lateral surface of the North half? Next, a field line propagates at a radial distance r from the solenoid axis, in the direction of the solenoid axis and towards the North pole, at the central cross section of the solenoid. For what values of r will the field line exit from the lateral surface of the solenoid? If it does not, what is its radial distance from the solenoid axis as it leaves the North end of the solenoid?

Motion of Charges in Magnetic Fields

15. *Bouncing Particle* *

A particle of mass m and charge q is currently located on the interior of a square room of side length l as shown in the figure below (top view). A uniform magnetic field B is directed into the page in the figure. Suppose that the particle is now propelled at an initial velocity $v_0 = \frac{qBl}{8m}$ perpendicular to a wall of the room, as depicted in the figure. Determine the time it requires to return to its initial position, assuming that its collisions with the walls are perfectly elastic.

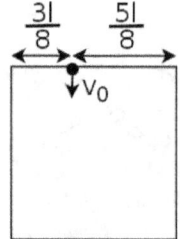

16. *Magnetic Dipole Moment* **

Determine the instantaneous torque experienced by the loop below when it is placed in a uniform magnetic field B, whose direction is depicted in the figure. The loop is composed of two semicircles of radius r that subtend a right angle. B bisects the angle between the semicircles.

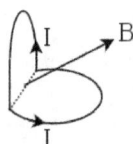

In light of the previous set-up, propose a definition for the magnetic dipole moment μ of a non-planar loop C with N turns, carrying a current

I, such that the torque that it experiences in a uniform magnetic field B is

$$\tau = \mu \times B.$$

Hint: Consider the vector area of the loop C, defined as $A = \iint_S dA$, where the integral is performed over a surface S that spans C and where dA is the area vector of an infinitesimal element on the surface S. Notice that we did not specify which surface S to consider (there are infinitely many possibilities). Provide a physical argument for why we could make such an "ambiguous" statement.

17. Homing Charge**

A particle with a charge magnitude q and mass m is moving in the xy-plane. It is launched from $(-a, 0)$ with speed v. The region above $y = f(x)$ is filled with a uniform and constant magnetic B pointing in the negative z-direction. It is observed that regardless of the direction of its initial velocity, as long as the charge enters the magnetic field region in the region $x \leq 0$, it will pass through $(a, 0)$ via a path symmetric about the y-axis. (International Physics Olympiad)

(a) What is the sign of the charge?
(b) With what speed does the charge pass through point $(a, 0)$?
(c) Find the function $f(x)$.

18. Rolling Sphere**

A solenoid of N turns carrying a constant current I is wound around the equatorial plane of a uniform sphere of radius R and mass m. The sphere is then placed on a rough, inclined plane with an angle of inclination θ, as shown in the figure below. A uniform magnetic field B points upwards everywhere. The magnetic dipole moment of the coil is initially directed perpendicularly outwards from the plane and the sphere is initially stationary. If the sphere rolls without slipping subsequently, determine the clockwise angular velocity of the sphere $\dot{\phi}$, as a function of the clockwise angle ϕ that the sphere has rotated.

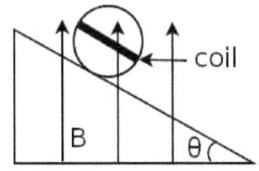

19. *Charge and Wire***

A fixed infinite wire carries current I along the positive z-axis. Supposing that a charged particle of mass m and charge q is initially launched at a radial velocity $v_0 > 0$ (positive outwards), at a perpendicular distance r_0 from the wire, determine the maximum and minimum radial distances that the charged particle can attain from the wire.

20. *Two Identical Charges***

Two identical particles of mass m and charge q are placed along the x-axis, in a region of uniform magnetic field B in the positive z-direction, with arbitrary initial velocities. The charges still lie in the xy-plane in their resulting motion.

By denoting r_1 and r_2 as the position vectors of the two charges, write down the equations of motion of the charges. Now, express these in terms of the position vector of the center of mass r_{CM} and the separation vector $r = r_1 - r_2$. What is the motion of the center of mass of the two charges?

Now, supposing that we wish to ensure that the distance between the two charges is a constant d, show that there is a minimum d for which this is possible. The angular velocity of the separation vector r in the lab frame, that corresponds to the minimum d, undertakes a certain constant value ω. After determining this ω, show that the original position of the center of mass and the instantaneous positions of the two particles are always collinear if the initial velocity of the center of mass (which is non-zero) does not have a y-component.

21. *Two Opposite Charges****

Two particles of the same mass m and charges q and $-q$ are placed along the x-axis, in a region of uniform magnetic field B in the positive z-direction. Let the position vectors of the charges q and $-q$ be r_1 and r_2 respectively and define the separation vector as $r = r_1 - r_2$. Only under certain special initial conditions can the two charges remain in the xy-plane while their separation vector remains at a constant magnitude d and rotates at a constant angular velocity ω. Determine the initial velocity of the center of mass of the two particles that results in such a motion. Given ω, find d and show that there exists a minimum magnitude of the angular velocity ω for such a motion to be possible.

22. *Magnetic Lens* ****

A thin, vertical solenoid carrying anti-clockwise current I with η turns per unit length, length l and radius R can be used to focus off-axis charges. Define the origin at the bottom of the solenoid and the z-axis to be positive upwards (towards the top of the solenoid).

(a) Determine the longitudinal and radial magnetic field at a small perpendicular distance $r \ll R$ from the cylindrical axis as a function of z-coordinate z. **Hint:** we have calculated one of these fields in a previous problem.

(b) Now, suppose that particles with charge q and mass m are placed at radial distances $r \ll R$ at the bottom of the solenoid. They are given a large velocity v_0 in the positive z-direction such that their radial coordinates are approximately constant throughout the solenoid. If $l \gg R$, approximately how large should v_0 be as compared to the other parameters for this to occur? Now, determine the instantaneous velocities of these particles at the instance they exit from the top of the solenoid.

(c) Determine the time t after this instance, at which they coincide with the z-axis, assuming that the magnetic field outside the solenoid is zero. Would this set-up function as a good lens to focus charges (of the same magnitude) with different initial radial distances?

Solutions

1. An Apparent Paradox*

Apply Ampere's law to a circular loop, perpendicular to and centered about wire 1, that passes through wire 2.

$$B \cdot 2\pi r = \mu_0 I_1.$$

The magnetic field due to wire 1 at wire 2 is then

$$B = \frac{\mu_0 I_1}{2\pi r}.$$

The force per unit length on wire 2 due to wire 1 is

$$f = BI_2 = \frac{\mu_0 I_1 I_2}{2\pi r}.$$

If the two currents are in the same direction, the wires tend to attract each other. Otherwise, they tend to repel.

   For the second part of the problem, observe that the expression for the force per unit length on wire 2 is only valid when it is stationary. Once it begins moving, the charges in wire 2 acquire a component of velocity transverse to the wire, in addition to the existing component along it. The magnetic force on these charges due to wire 1 is perpendicular to their net velocities which are obtained from the vector sum of the transverse and longitudinal components. Therefore, a longitudinal force will be generated on the charges in wire 2, parallel and anti-parallel to the current I_2 in wire 2 for negative and positive charges respectively, on top of the transverse force discussed in the first part of the problem. In other words, wire 2 gains velocity due to the transverse component of the magnetic force but the net magnetic force on wire 2 is not purely transverse and definitely still performs zero work. This is akin to how we can claim that a component of the normal force on a block sliding down a frictionless, inclined plane "accelerates" the block in the horizontal direction (parallel to the flat ground), though the normal force itself really does no work on the block.

   But still, the kinetic energy of wire 2 must come from somewhere! Since the longitudinal component of magnetic force tends to reduce the current in wire 2, if wire 2 is not connected to any external entity, the current in wire 2 will decrease — suggesting a decrease[6] in the energy stored in

---

[6] Actually, our question is ill-posed at the moment as the energies stored in the wires are actually infinite. One should use the more realistic picture of a wire having a non-negligible cross section for concrete calculations.

the magnetic field (empirically, it always takes positive external work to increase a current[7]). Therefore, the decrease in potential energy stored in the magnetic field balances the increase in kinetic energy of wire 2.

Otherwise, if wire 2 is connected to a battery, the battery needs to produce an additional longitudinal force on the charges in wire 2 to maintain its current at I_2. A possible source of the kinetic energy is then the work done by the battery attached to wire 2. However, note that the energy stored in the magnetic field still varies in this case (as the distance between the wires changes) so it remains a factor here.

2. Force Between Moving Charges*

Label the charges q and $-q$ as 1 and 2 respectively. At time t, charges 1 and 2 are at $(0, vt)$ and $(vt, 0)$. The magnetic field at the location of charge 2 due to charge 1, in the non-relativistic regime, is

$$\boldsymbol{B}_1 = \frac{\mu_0 q \boldsymbol{v}_1 \times \boldsymbol{r}}{4\pi r^3} = -\frac{\mu_0 q v(vt)}{4\pi(\sqrt{2}vt)^3}\hat{\boldsymbol{k}} = -\frac{\mu_0 q}{8\sqrt{2}\pi v t^2}\hat{\boldsymbol{k}}.$$

Keep in mind that we are not applying the Biot-Savart law in saying this, as a moving point charge surely does not constitute a steady current. The magnetic force acting on charge 2 is thus

$$\boldsymbol{F}_{21} = -q\boldsymbol{v}_2 \times \boldsymbol{B}_1 = -q(v\hat{\boldsymbol{i}}) \times \left(-\frac{\mu_0 q}{8\sqrt{2}\pi v t^2}\hat{\boldsymbol{k}}\right) = -\frac{\mu_0 q^2}{8\sqrt{2}\pi t^2}\hat{\boldsymbol{j}}.$$

However, if we repeat this procedure, we will find that the magnetic force on charge 1 is

$$\boldsymbol{F}_{12} = -\frac{\mu_0 q^2}{8\sqrt{2}\pi t^2}\hat{\boldsymbol{i}},$$

so Newton's third law is seemingly violated! This thought experiment led to the modern idea of treating the electromagnetic field as an entity of its own such that the charges interact with the field rather than each other. By associating a momentum with the electromagnetic field, the conservation of momentum (in the system comprising both charges and the field) can then be retained.

---

[7]However, if we wish to nit-pick further, a smaller current, in this case, does not necessarily imply a decrease in energy stored in the magnetic field since the location of the current in wire 2 varies. We should just accept that the kinetic energy of wire 2 comes from the decrease in energy stored in the magnetic field as that is the only possible source here.

Even though $\boldsymbol{F}_{12}$ and $\boldsymbol{F}_{21}$ are both independent of v, they are incorrect when $v = 0$. This is because stationary charges do not produce a magnetic field and do not experience a magnetic force even in the presence of an external magnetic field. Mathematically, we cannot cancel the v's in the numerator and denominator of the magnetic fields (e.g. the third equality in the computation of $\boldsymbol{B}_1$) when $v = 0$. Instead, the more rigorous statement is that $v \to 0$ and $r \to 0$ but when the numerator and denominator of a function both tend to zero, the value of the function in such a limit depends on how v and r scale to zero and is generally indeterminate. However, we usually require the magnetic field to be zero for physical reasons.

3. Current on Cube*

We can see the relevant current distribution as the superposition of three current loops on the left, front and bottom faces of the cube, as shown in Fig. 7.23. Each current loop produces a magnetic field of magnitude B at the center, parallel to the area vector associated with the current loop (whose direction is determined by the right-hand-rule). Therefore, the relevant magnetic field is

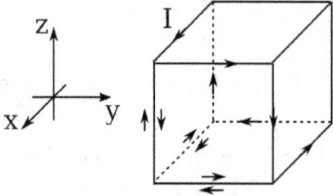

Figure 7.23:   Superposition of currents

$$\boldsymbol{B}' = -B\hat{\boldsymbol{i}} + B\hat{\boldsymbol{j}} + B\hat{\boldsymbol{k}},$$

which is of magnitude $\sqrt{3}B$ and directed from the center towards the top-right vertex on the back face in Fig. 7.23.

4. Lorentz Force on a Short Wire**

Applying Ampere's law to the infinitely long wire, the magnetic field at coordinate (x, y, z) due to the infinite long wire is

$$B(x, y, z) = \frac{\mu_0 I_1}{2\pi\sqrt{x^2 + z^2}} \begin{pmatrix} \frac{z}{\sqrt{x^2+z^2}} \\ 0 \\ \frac{x}{\sqrt{x^2+z^2}} \end{pmatrix}.$$

The wire of finite length spans from $(0, 0, h)$ to $(l, 0, h)$. Therefore, the force on an infinitesimal segment of the finite wire between $(x, 0, h)$ and $(x + dx, 0, h)$ is

$$d\mathbf{F} = I_2 d\mathbf{s} \times \mathbf{B}$$

$$= I_2 \begin{pmatrix} dx \\ 0 \\ 0 \end{pmatrix} \times \frac{\mu_0 I_1}{2\pi\sqrt{x^2 + h^2}} \begin{pmatrix} \frac{h}{\sqrt{x^2+h^2}} \\ 0 \\ \frac{x}{\sqrt{x^2+h^2}} \end{pmatrix}$$

$$= -\frac{\mu_0 I_1 I_2 x}{2\pi(x^2 + h^2)} dx \begin{pmatrix} 0 \\ 1 \\ 0 \end{pmatrix}.$$

Integrating along the entire wire,

$$F_y = \int_0^l -\frac{\mu_0 I_1 I_2 x}{2\pi(x^2 + h^2)} dx$$

$$= \left[-\frac{\mu_0 I_1 I_2}{4\pi} \ln(x^2 + h^2) \right]_0^l$$

$$= -\frac{\mu_0 I_1 I_2}{4\pi} \ln \frac{l^2 + h^2}{h^2}.$$

5. Torque on Arbitrary Wires**

Suppose that we follow an infinitesimal current dI along a certain path P from terminal A towards terminal B, which lies entirely in the xy-plane. Consider an infinitesimal segment $d\mathbf{r}$ along P which lies at a position vector $\mathbf{r}$ relative to A. Since the torque on this infinitesimal current segment about A is $dI \mathbf{r} \times (d\mathbf{r} \times \mathbf{B})$, the total torque on the current dI along path P, with respect to A, is

$$dI \int_P \mathbf{r} \times (d\mathbf{r} \times \mathbf{B}) = dI \int_P [d\mathbf{r}(\mathbf{r} \cdot \mathbf{B}) - \mathbf{B}(\mathbf{r} \cdot d\mathbf{r})]$$

$$= -\mathbf{B} dI \int_P \mathbf{r} \cdot d\mathbf{r}$$

$$= -\mathbf{B} dI \int_0^l r dr$$

$$= -\frac{dI l^2 \mathbf{B}}{2},$$

where we have noted that $r \cdot B = 0$ in the second equality and $r \cdot dr = \frac{1}{2}d(r \cdot r) = \frac{1}{2}d(r^2) = rdr$ in the third equality. Summing the contributions of all currents from A, the total torque on the wires about A is

$$\tau = -\frac{\int l^2 B dI}{2} = -\frac{Il^2 B}{2}$$

which is identical to the torque, about A, on a straight wire delivering current I from A to B. An interesting property is that the torque about A is always opposite in direction to the external field B.

6. Helmholtz Coil**

The magnetic field strength at a distance z away from the axis of a circular ring of current I was previously derived to be

$$B = \frac{\mu_0 I R^2}{2(R^2 + z^2)^{\frac{3}{2}}}.$$

For a coil with N turns, the corresponding magnetic field strength is then

$$B = \frac{\mu_0 N I R^2}{2(R^2 + z^2)^{\frac{3}{2}}}.$$

The Helmholtz set-up consists of two coils with currents running in the same direction. Therefore, the magnetic field strength in the region within the coils is reinforced. The magnetic field strength at a perpendicular distance z from the center of one coil is

$$B(z) = \frac{\mu_0 N I R^2}{2(R^2 + z^2)^{\frac{3}{2}}} + \frac{\mu_0 N I R^2}{2\left(R^2 + (d-z)^2\right)^{\frac{3}{2}}},$$

where we assume that the coils carry anti-clockwise currents I relative to the positive z-direction.

$$\frac{dB}{dz} = -\frac{3\mu_0 N I R^2 z}{2(R^2 + z^2)^{\frac{5}{2}}} + \frac{3\mu_0 N I R^2(d-z)}{2\left(R^2 + (d-z)^2\right)^{\frac{5}{2}}},$$

$$\frac{d^2 B}{dz^2} = -\frac{3\mu_0 N I R^2}{2(R^2 + z^2)^{\frac{5}{2}}} + \frac{15\mu_0 N I R^2 z^2}{2(R^2 + z^2)^{\frac{7}{2}}} - \frac{3\mu_0 N I R^2}{2\left(R^2 + (d-z)^2\right)^{\frac{5}{2}}}$$

$$+ \frac{15\mu_0 N I R^2(d-z)^2}{2\left(R^2 + (d-z)^2\right)^{\frac{7}{2}}}.$$

For $\frac{d^2 B}{dz^2}\big|_{z=\frac{d}{2}} = 0$,

$$-\left(R^2 + \frac{d^2}{4}\right) + \frac{5d^2}{4} - \left(R^2 + \frac{d^2}{4}\right) + \frac{5d^2}{4} = 0$$

$$d = R.$$

7. Bent Wire**

Firstly, consider the following auxiliary problem.

Auxiliary Problem: A thin, conducting wire carries a uniform current I. Determine the magnetic field at a point P that is situated at a perpendicular, vertical distance h from the wire, while assuming that the Biot-Savart law holds.[8] The ends of the wire subtend anti-clockwise angles θ_1 and θ_0 with respect to the vertical.

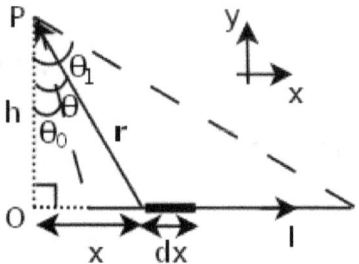

Figure 7.24: Thin wire

Define the x and y-axes to be the horizontal and vertical axes, positive rightwards and upwards, in Fig. 7.24. Consider an infinitesimal segment of wire between x-coordinates x and $x + dx$. The infinitesimal vector of this segment is $d\mathbf{s} = (dx, 0, 0)$ while the separation vector between this segment and point P is $\mathbf{r} = (-\frac{x}{\sqrt{x^2+h^2}}, \frac{h}{\sqrt{x^2+h^2}}, 0)$. Applying the Biot-Savart law, the magnetic field at P due to the wire is

$$\mathbf{B} = \frac{\mu_0 I}{4\pi} \int_{x_0}^{x_1} \frac{1}{x^2 + h^2} \begin{pmatrix} dx \\ 0 \\ 0 \end{pmatrix} \times \begin{pmatrix} -\frac{x}{\sqrt{x^2+h^2}} \\ \frac{h}{\sqrt{x^2+h^2}} \\ 0 \end{pmatrix} = \frac{\mu_0 I}{4\pi} \int_{x_0}^{x_1} \frac{h\,dx}{(x^2 + h^2)^{\frac{3}{2}}} \begin{pmatrix} 0 \\ 0 \\ 1 \end{pmatrix}$$

where x_0 and x_1 are the x-coordinates of the ends of the rod. To evaluate the integral for the z-component, use the substitutions $x = h\tan\theta$ and

---

[8]The current in this set-up is definitely not steady but we will be applying our result to a steady current configuration that this set-up is a part of. We are thus finding the contribution due to this finite wire, in a certain sense.

$dx = h\sec^2\theta d\theta$.

$$B_z = \frac{\mu_0 I}{4\pi}\int_{\theta_0}^{\theta_1}\frac{h}{h^3\sec^3\theta}\cdot h\sec^2\theta d\theta$$

$$= \frac{\mu_0 I}{4\pi h}\int_{\theta_0}^{\theta_1}\cos\theta d\theta$$

$$= \frac{\mu_0 I(\sin\theta_1 - \sin\theta_0)}{4\pi h}.$$

The magnetic field at P points out of the page (recall that $\hat{i}\times\hat{j} = \hat{k}$ for a conventional coordinate system).

Returning to our original context, break the bent wire into two at the kink. The above result can then be applied to the two resultant wires with $\theta_1 = \frac{\pi}{2} - \theta$ and $\theta_0 = -\frac{\pi}{2}$ for the top wire and $\theta_1 = \frac{\pi}{2}$ and $\theta_0 = \theta - \frac{\pi}{2}$ for the bottom wire. The perpendicular distance between P and the wires, h, is $d\sin\theta$ in this case. Observing that the contributions of the two half-infinite wires are reinforced at P, the magnetic field at P is then

$$B_z = \frac{\mu_0 I(\cos\theta + 1)}{2\pi d\sin\theta} = \frac{\mu_0 I\cdot 2\cos^2\frac{\theta}{2}}{4\pi d\sin\frac{\theta}{2}\cos\frac{\theta}{2}} = \frac{\mu_0 I}{2\pi d}\cot\frac{\theta}{2}$$

pointing out of the page.

8. Finite Solenoid**

Define the origin at the bottom end of the cylinder (smaller z-coordinate) and consider the solenoid in cylindrical coordinates. Slice the solenoid into rings of thickness dh. Suppose that we wish to calculate the magnetic field at a certain point P with coordinates $(0, 0, z)$. We have previously derived in an example problem that a circular ring of radius R that carries an anti-clockwise current I produces a vertical magnetic field

$$B_z = \frac{\mu_0 I R^2}{2(z^2 + R^2)^{\frac{3}{2}}}$$

at a point that is at a perpendicular height z from the ring, along the symmetrical axis. Applying this result to this problem, the contribution to the magnetic field at the origin due to an infinitesimal ring between z-coordinates h and $h + dh$ is

$$dB_z = \frac{\mu_0 \eta I dh R^2}{2((h-z)^2 + R^2)^{\frac{3}{2}}},$$

as the particular infinitesimal ring carries current $\eta I dh$ and because point P is a distance $z - h$ away from the ring at z-coordinate h. Then, the total magnetic field at P is obtained by integrating the above over all rings from $h = 0$ to $h = l$.

$$B_z = \int_0^l \frac{\mu_0 \eta I R^2}{2((h-z)^2 + R^2)^{\frac{3}{2}}} dh.$$

Making the substitutions $h - z = R \tan \theta$ and $dh = R \sec^2 \theta d\theta$,

$$B_z = \int_{\tan^{-1} \frac{-z}{R}}^{\tan^{-1} \frac{l-z}{R}} \frac{\mu_0 \eta I}{2} \cos \theta d\theta = \left[\frac{\mu_0 \eta I}{2} \sin \theta \right]_{\tan^{-1} \frac{-z}{R}}^{\tan^{-1} \frac{l-z}{R}}$$

$$= \frac{\mu_0 \eta I}{2} \left(\frac{l-z}{\sqrt{(l-z)^2 + R^2}} + \frac{z}{\sqrt{z^2 + R^2}} \right).$$

To solve the next problem, slice the solenoid into thin cylindrical shells with radial distances between r and $r+dr$. Observe that the current per unit cross sectional area of the solenoid is $J = \frac{\eta I}{r_1 - r_0}$. Therefore, the current per unit length of this cylindrical shell is $J dr$. With the same origin at the bottom of the solenoid, the contribution to the magnetic field at a point P along the symmetrical axis by this cylindrical shell is

$$dB_z = \frac{\mu_0 J dr}{2} \left(\frac{l-z}{\sqrt{(l-z)^2 + r^2}} + \frac{z}{\sqrt{z^2 + r^2}} \right),$$

where we have replaced the previous current per unit length ηI with $J dr$, and R with the variable r. The total magnetic field at P is then

$$B_z = \int_{r_0}^{r_1} \frac{\mu_0 J}{2} \left(\frac{l-z}{\sqrt{(l-z)^2 + r^2}} + \frac{z}{\sqrt{z^2 + r^2}} \right) dr.$$

To evaluate the integral $\int \frac{a}{\sqrt{a^2 + x^2}} dx$, make the substitutions $x = a \tan \theta$ and $dx = a \sec^2 \theta$. Then,

$$\int_{x_0}^{x_1} \frac{a}{\sqrt{a^2 + x^2}} dx = \int_{\tan^{-1} \frac{x_0}{a}}^{\tan^{-1} \frac{x_1}{a}} a \sec \theta d\theta$$

$$= [a \ln | \sec \theta + \tan \theta |]_{\tan^{-1} \frac{x_0}{a}}^{\tan^{-1} \frac{x_1}{a}} = a \ln \left| \frac{\sqrt{x_1^2 + a^2} + x_1}{\sqrt{x_0^2 + a^2} + x_0} \right|.$$

Therefore,

$$B_z = \frac{\mu_0 \eta I(l-z)}{2(r_1 - r_0)} \ln \left| \frac{\sqrt{r_1^2 + (l-z)^2} + r_1}{\sqrt{r_0^2 + (l-z)^2} + r_0} \right| + \frac{\mu_0 \eta I z}{2(r_1 - r_0)} \ln \left| \frac{\sqrt{r_1^2 + z^2} + r_1}{\sqrt{r_0^2 + z^2} + r_0} \right|.$$

9. Rotating Sphere***

Define the origin at the center of the sphere and adopt spherical coordinates. We will use s to denote the radial distance of an infinitesimal element from the origin. The position vector of an infinitesimal element at coordinates (s, ϕ, θ) is $\boldsymbol{s} = (s \sin \theta \cos \phi, s \sin \theta \sin \phi, s \cos \theta)$. Its velocity is

$$\boldsymbol{v} = \begin{pmatrix} 0 \\ 0 \\ \omega \end{pmatrix} \times \begin{pmatrix} s \sin \theta \cos \phi \\ s \sin \theta \sin \phi \\ s \cos \theta \end{pmatrix} = \begin{pmatrix} -\omega s \sin \theta \sin \phi \\ \omega s \sin \theta \cos \phi \\ 0 \end{pmatrix}.$$

Suppose that we wish to compute the magnetic field at a point P with a z-coordinate z along the z-axis. Then, the separation vector is

$$\boldsymbol{r} = \begin{pmatrix} 0 \\ 0 \\ z \end{pmatrix} - \begin{pmatrix} s \sin \theta \cos \phi \\ s \sin \theta \sin \phi \\ s \cos \theta \end{pmatrix} = \begin{pmatrix} -s \sin \theta \cos \phi \\ -s \sin \theta \sin \phi \\ z - s \cos \theta \end{pmatrix}.$$

Applying the Biot-Savart law and substituting $\rho \boldsymbol{v}$ for the current density $\boldsymbol{J}$ of an infinitesimal volume element $dV = s^2 \sin \theta d\phi d\theta ds$ in spherical coordinates,

$$\boldsymbol{B} = \frac{\mu_0}{4\pi} \int_0^R \int_0^\pi \int_0^{2\pi} \frac{\rho \omega s \cdot s^2 \sin \theta d\phi d\theta ds}{(z^2 + s^2 - 2zs \cos \theta)^{\frac{3}{2}}} \begin{pmatrix} -\sin \theta \sin \phi \\ \sin \theta \cos \phi \\ 0 \end{pmatrix}$$

$$\times \begin{pmatrix} -s \sin \theta \cos \phi \\ -s \sin \theta \sin \phi \\ z - s \cos \theta \end{pmatrix}.$$

After integration, only the z-component remains due to the symmetry of the sphere.

$$\begin{aligned} B_z &= \frac{\mu_0 \rho \omega}{4\pi} \int_0^R \int_0^\pi \int_0^{2\pi} \frac{s^4 \sin^3 \theta d\phi d\theta ds}{(z^2 + s^2 - 2zs \cos \theta)^{\frac{3}{2}}} \\ &= \frac{\mu_0 \rho \omega}{2} \int_0^R \int_0^\pi \frac{s^4 \sin^3 \theta d\theta ds}{(z^2 + s^2 - 2zs \cos \theta)^{\frac{3}{2}}}. \end{aligned}$$

This double integral is non-trivial to evaluate. We first evaluate the inner integral via multiple integrations-by-parts.

$$\int_0^\pi \frac{s^4 \sin^3 \theta d\theta}{(z^2 + s^2 - 2zs \cos \theta)^{\frac{3}{2}}} = \left[-\frac{s^3 \sin^2 \theta}{z\sqrt{z^2 + s^2 - 2zs \cos \theta}} \right]_0^\pi$$

$$+ \int_0^\pi \frac{2s^3 \sin \theta \cos \theta}{z\sqrt{z^2 + s^2 - 2zs \cos \theta}} d\theta$$

$$= 0 + \left[\frac{2s^2 \cos \theta}{z^2} \sqrt{z^2 + s^2 - 2zs \cos \theta} \right]_0^\pi$$

$$+ \int_0^\pi \frac{2s^2 \sin \theta}{z^2} \sqrt{z^2 + s^2 - 2zs \cos \theta} d\theta$$

$$= -\frac{2s^2}{z^2} \left(\sqrt{z^2 + s^2 - 2zs} + \sqrt{z^2 + s^2 + 2zs} \right)$$

$$+ \int_0^\pi \frac{2s^2 \sin \theta}{z^2} \sqrt{z^2 + s^2 - 2zs \cos \theta} d\theta.$$

To evaluate the final integral, make the substitutions $u = z^2 + s^2 - 2zs \cos \theta$ and $du = 2zs \sin \theta d\theta$.

$$\int_0^\pi \frac{2s^2 \sin \theta}{z^2} \sqrt{z^2 + s^2 - 2zs \cos \theta} d\theta$$

$$= \int_{z^2+s^2-2zs}^{z^2+s^2+2zs} \frac{s}{z^3} \sqrt{u} du$$

$$= \frac{2s}{3z^3} \left[(z^2 + s^2 + 2zs)^{\frac{3}{2}} - (z^2 + s^2 - 2zs)^{\frac{3}{2}} \right].$$

Note that we cannot evaluate the square root yet as we do not know the relative magnitude of z and s. Actually, another way to evaluate the integral is to directly adopt the same substitutions from the start such that

$$\int_0^\pi \frac{s^4 \sin^3 \theta d\theta}{(z^2 + s^2 - 2zs \cos \theta)^{\frac{3}{2}}}$$

$$= \int_0^\pi \frac{s^4 (1 - \cos^2 \theta) \sin \theta d\theta}{(z^2 + s^2 - 2zs \cos \theta)^{\frac{3}{2}}}$$

$$= \int \frac{s^4 \left[1 - \left(\frac{z^2 + s^2 - u}{2zs} \right)^2 \right] \frac{du}{2zs}}{u^{\frac{3}{2}}}$$

$$= \frac{s}{8z^3}\left[-(z^2-s^2)^2\int u^{-\frac{3}{2}}\,du + 2(z^2+s^2)\int u^{-\frac{1}{2}}\,du - \int u^{\frac{1}{2}}\,du\right]$$

$$= \frac{s}{8z^3}\left[2(z^2-s^2)^2 u^{-\frac{1}{2}} + 4(z^2+s^2)u^{\frac{1}{2}} - \frac{2}{3}u^{\frac{3}{2}}\right]_{z^2+s^2-2zs}^{z^2+s^2+2zs}$$

and so on. However, it just happens that the resulting terms require less simplification if we integrate-by-parts first. In any case, the magnetic field at P is

$$B_z = \frac{\mu_0\rho\omega}{2}\int_0^R \left(-\frac{2s^2}{z^2}\left(\sqrt{z^2+s^2-2zs}+\sqrt{z^2+s^2+2zs}\right)\right.$$

$$\left. + \frac{2s}{3z^3}[(z^2+s^2+2zs)^{\frac{3}{2}} - (z^2+s^2-2zs)^{\frac{3}{2}}]\right)ds.$$

To actually compute this integral, we have to consider two regimes of z: $z \geq R$ (point P is outside the sphere) and $z < R$. In the case of the former, $\sqrt{(z-s)^2} = z - s$ for the entire regime of integration

$$B_z = \frac{\mu_0\rho\omega}{2}\int_0^R \left(-\frac{4s^2}{z} + \frac{2s}{3z^3}\left[(z+s)^3 - (z-s)^3\right]\right)ds$$

$$= -\frac{2\mu_0\rho\omega R^3}{3z} + \int_0^R \frac{2\mu_0\rho\omega s^2}{3z^3}\left(3z^2+s^2\right)ds$$

$$= \frac{2\mu_0\rho\omega R^5}{15z^3}$$

for $z \geq R$ (i.e. outside the sphere). If $z < R$, we have to split the integrals in s into two parts — from 0 to z and z to R. $\sqrt{z^2+s^2-2zs} = z - s$ for the first region and $\sqrt{z^2+s^2-2zs} = s - z$ for the second region. Therefore,

$$B_z = \frac{\mu_0\rho\omega}{2}\int_0^z -\frac{4s^2}{z}\,ds + \frac{\mu_0\rho\omega}{2}\int_z^R -\frac{4s^3}{z^2}\,ds$$

$$+ \frac{\mu_0\rho\omega}{2}\int_0^z \frac{2s}{3z^3}[(z+s)^3 - (z-s)^3]ds$$

$$+ \frac{\mu_0\rho\omega}{2}\int_z^R \frac{2s}{3z^3}\left[(z+s)^3 - (s-z)^3\right]ds$$

$$= -\frac{\mu_0\rho\omega z^2}{5} + \frac{\mu_0\rho\omega R^2}{3}.$$

10. Thick Infinite Wire*

The magnetic field can only be azimuthal for the same reasons as those argued for an infinite thin wire. Draw a circular Amperian loop of radius r perpendicular to the cylindrical axis, with its center along the axis. If $r < a$, the current that this loop encloses is

$$I_{enc} = \int_0^r \frac{2I_0}{\pi a^2}\left(1 - \frac{r^2}{a^2}\right)\cdot 2\pi r\, dr$$

$$= \frac{2I_0 r^2}{a^2} - \frac{I_0 r^4}{a^4}$$

$$= \frac{I_0 r^2}{a^2}\left(2 - \frac{r^2}{a^2}\right).$$

For $r \geq a$, the enclosed current is obtained from substituting $r = a$ in the preceding expression.

$$I_{enc} = \int_0^a \frac{2I_0}{\pi a^2}\left(1 - \frac{r^2}{a^2}\right)\cdot 2\pi r\, dr = I_0.$$

Applying Ampere's law to this loop (regardless of whether $r < a$ or $r \geq a$),

$$B\cdot 2\pi r = \mu_0 I_{enc}$$

$$\implies B = \frac{\mu_0 I_{enc}}{2\pi r} = \begin{cases} \frac{\mu_0 I_0 r}{2\pi a^2}\left(2 - \frac{r^2}{a^2}\right) & \text{for } r < a \\ \frac{\mu_0 I_0}{2\pi r} & \text{for } r \geq a \end{cases}.$$

11. Wire with Cavity*

The current density of the wire is

$$J = \frac{I}{\pi R^2 - \pi\frac{R^2}{4}} = \frac{4I}{3\pi R^2}.$$

Now, imagine filling up the hole with a wire that carries the same current density J. Applying Ampere's law to a circle perpendicular to the axis of a solid wire,

$$B(r)\cdot 2\pi r = \mu_0 J \pi r^2$$

$$B(r) = \frac{\mu_0 J r}{2}$$

for $r \leq R$, anti-clockwise in the azimuthal direction (referring to the figure in the problem) as the current density J is coming out of the page. Since $B = 0$

when $r = 0$, the magnetic field at P due to a solid wire must be zero. Now, this is the magnetic field at P due to both the original and imaginary wires. To determine the magnetic field due to only the original wire, we subtract the contribution by the imaginary wire which is $B = \frac{\mu_0 JR}{4}$ anti-clockwise along the circular boundary of the cavity (substitute $r = R$ into the above and replace R with $\frac{R}{2}$). Therefore, the magnetic field at P due to the wire with a cavity is

$$B' = 0 - \frac{\mu_0 JR}{4} = -\frac{\mu_0 I}{3\pi R},$$

where the negative sign indicates that this field is clockwise along the circular boundary of the cavity.

12. Opposite Currents*

The current distribution in the problem can be seen as the superposition of two currents with a circular cross section of radius b and current densities $-J$ and J in the positive z-direction. Consider the latter current. Applying Ampere's law to a circular loop of radius $r < b$ (whose center is cut by the cylindrical axis of the current) in the xy-plane, we have

$$B \cdot 2\pi r = \mu_0 J \pi r^2$$

$$B = \frac{\mu_0 Jr}{2}.$$

The magnetic field is azimuthal everywhere. Expressing it in vector form,

$$\boldsymbol{B} = \frac{\mu_0 J}{2} \hat{k} \times \boldsymbol{r}$$

where $\boldsymbol{r}$ is the vector pointing from the center of the circle, in the cross section, to the point of interest for $r < b$. The magnetic field of the other current can be obtained from substituting $-J$ for J and adopting a similar definition for $\boldsymbol{r}$. If we denote the left and right currents as 1 and 2 respectively, the net magnetic field at a point within the overlapping region is

$$\boldsymbol{B} = \frac{\mu_0 J}{2} \hat{k} \times (\boldsymbol{r}_2 - \boldsymbol{r}_1) = \frac{\mu_0 J}{2} \hat{k} \times (-2a\hat{i}) = -\mu_0 aJ\hat{j}.$$

13. Toroid**

Firstly, we can deduce the direction of the magnetic field produced by the toroid — we shall simply state the procedure for this and the reader should fill the details in. Define the z-axis as the axis that the toroid is rotationally symmetric about. Suppose that we wish to compute the magnetic field at a certain point P. Orient the positive x-axis such that point P lies in the xz-plane. One can show that the net magnetic field at P due to an infinitesimal current segment and its counterpart corresponding to its reflection about xz-plane only has a y-component (azimuthal with respect to the z-axis). Dividing the toroid into two halves along the xz-plane and applying this argument to all of such pairs, we conclude that the net magnetic field at P must only be azimuthal. Since the toroid is circularly symmetric, the magnetic field lines must take the form of circles around the z-axis.

Now, consider the circle (whose radius is denoted as r) obtained from rotating point P about the z-axis for a complete revolution. Due to circular symmetry, the magnetic field strengths at all points along this loop (the direction of the magnetic field is azimuthal) must be identical. Therefore, if we let this common value be B and use this loop as an Amperian loop to apply Ampere's law to,

$$B \cdot 2\pi r = \mu_0 I_{enc}$$

where I_{enc} is the current enclosed by this circle. If the circle lies outside the toroid, it encloses no net current — implying that the magnetic field is zero everywhere outside the toroid. On the other hand, if the circle lies within the toroid, it encloses a total current NI. Thus,

$$B \cdot 2\pi r = \mu_0 NI$$

$$B(r) = \frac{\mu_0 NI}{2\pi r},$$

where we reiterate that r is the perpendicular distance between the point of concern and the z-axis and that this result is only valid if the point of concern is enclosed within the toroid.

14. Magnetic Flux and Field Lines**

By Ampere's law, the magnetic field at the center of a long solenoid is

$$B = \mu_0 \eta I$$

and is uniform within the central cross section. The total magnetic flux cutting across the central cross section is thus

$$\Phi_{B,center} = B \cdot \pi R^2 = \mu_0 \pi \eta I R^2.$$

Now, we simply have to subtract the magnetic flux exiting the North end of the solenoid from the above to compute the magnetic flux leaving from the lateral surface since the net magnetic flux emanating from a closed surface, which is the North half of the solenoid in this case, must be zero. To this end, we have to determine the component of magnetic field along the solenoid axis at the North end (name this the z-component). This can be easily accomplished by exploiting the principle of superposition. Observe that if we break a long solenoid into two at its center, we obtain two semi-infinite solenoids whose z-components of magnetic field are reinforced at the center. Therefore, twice the z-component of magnetic field, B_z, at any point on the end of a solenoid is $\mu_0 \eta I$, which implies that

$$B_z = \frac{\mu_0 \eta I}{2}$$

and is uniform over the cross section at the end of the solenoid. Therefore, the total magnetic flux leaving the North end is

$$\Phi_{B,end} = B_z \cdot \pi R^2 = \frac{\mu_0 \pi \eta I R^2}{2}.$$

The net magnetic flux leaving the North part of the solenoid from its lateral surface is then

$$\Phi_{B,center} - \Phi_{B,end} = \frac{\mu_0 \pi \eta I R^2}{2}.$$

We can adopt a similar approach to the second part. Suppose that a magnetic field line that begins at a radial distance r at the central cross section departs from the North end of the solenoid at a radial distance r'. Since the solenoid is symmetrical about its axis, we can rotate this field line for a complete revolution to generate other axial-symmetric field lines. This set of lines, coupled with the two circles of radii r and r' at the cross sections of the solenoid at the center and the North end, forms a closed surface. Requiring the net magnetic flux crossing this closed surface to be zero while observing that the magnetic flux is only non-zero at the two circles (since we purposely constructed part of our surface to be parallel to the field lines), the magnetic fluxes through the two circles must be equal in absolute magnitude.

$$B_z \cdot \pi r'^2 = B \cdot \pi r^2$$

$$r' = \sqrt{2}r.$$

The field line will only leave via the lateral surface if $\sqrt{2}r > R$.

15. Bouncing Particle*

As analyzed before, the particle will undergo circular motion with radius

$$r = \frac{mv_0}{qB} = \frac{l}{8}$$

in the uniform magnetic field. When the particle undergoes a perfectly elastic, head-on collision with the wall, its velocity is reversed and the center of rotation shifts by $2r = \frac{l}{4}$. Furthermore, observe that near the edges of the square, the particle still rotates with its center of rotation being a vertex of the square — causing its subsequent collision with another edge of the square to also be head-on. Therefore, the particle (if q is positive) follows the trajectory in Fig. 7.25.

Figure 7.25: Clockwise movement of positive charge

If the particle is negatively charged, it would take the reverse path — hence requiring the same amount of time to complete a cycle. In total, the particle would have rotated $\frac{\frac{4l}{l}}{\frac{8}{4}} = 8$ complete rounds. The period of a single rotation is 2π divided by the cyclotron angular frequency $\frac{qB}{m}$.

$$T = \frac{2\pi}{\frac{qB}{m}} = \frac{2\pi m}{qB}.$$

Therefore, the total time taken by the particle to return to its original position is

$$8T = \frac{16\pi m}{qB}.$$

16. Magnetic Dipole Moment**

We can deem the original loop given in the problem as the composition of two cyclic semi-circular loops in Fig. 7.26. The fictitious currents cancel such that the current distribution is the same as the set-up in the problem.

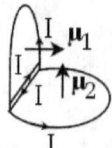

Figure 7.26:　Two current loops

Suppose that we orient our positive x and y-axes rightwards and upwards such that the two loops have magnetic dipole moments

$$\boldsymbol{\mu}_1 = I\pi R^2 \begin{pmatrix} 1 \\ 0 \\ 0 \end{pmatrix},$$

$$\boldsymbol{\mu}_2 = I\pi R^2 \begin{pmatrix} 0 \\ 1 \\ 0 \end{pmatrix},$$

while the external magnetic field is

$$\boldsymbol{B} = B \begin{pmatrix} \frac{\sqrt{2}}{2} \\ \frac{\sqrt{2}}{2} \\ 0 \end{pmatrix}.$$

The total torque experienced by the original loop is the sum of the torques experienced by these semi-circular loops

$$\boldsymbol{\tau} = (\boldsymbol{\mu}_1 + \boldsymbol{\mu}_2) \times \boldsymbol{B}$$

$$= I\pi R^2 \begin{pmatrix} 1 \\ 1 \\ 0 \end{pmatrix} \times \begin{pmatrix} \frac{\sqrt{2}}{2} \\ \frac{\sqrt{2}}{2} \\ 0 \end{pmatrix}$$

$$= \boldsymbol{0}.$$

For a general loop C, we can leverage on the previous idea to divide any non-planar loop into planar loops by replacing regions that are devoid of current with two fictitious line currents that travel in opposite directions. In fact, why not further divide the planar loops into smaller and smaller planar loops such that we obtain infinitesimal planar loops which carry current I (all in the same direction — either clockwise or anti-clockwise)? This process is depicted in Fig. 7.27.

Within the loop, the adjacent sides of the infinitesimal current loops nullify each other such that the net effect of the entire group of infinitesimal

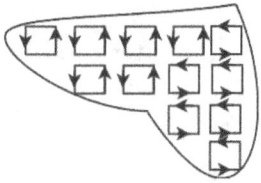

Figure 7.27: Dividing a non-planar loop into infinitesimal loops

loops is to deliver current I along the boundary line (the original loop C). Each infinitesimal current loop with area vector $d\mathbf{A}$ experiences a torque $I d\mathbf{A} \times \mathbf{B}$ such that one turn of the coil experiences a torque $I \iint_S d\mathbf{A} \times \mathbf{B}$ where S is the surface constructed from patching the infinitesimal planar loops together (it is a surface that spans C). Therefore, the torque experienced by a coil with N turns is

$$\boldsymbol{\tau} = \left(NI \iint_S d\mathbf{A} \right) \times \mathbf{B} = \boldsymbol{\mu} \times \mathbf{B}$$

where

$$\boldsymbol{\mu} = NI \iint_S d\mathbf{A}$$

is the magnetic dipole moment of the coil. Now, we claim that $\iint_S d\mathbf{A}$ is only dependent on the loop C and is independent of the surface that we choose, as long as it spans C. To this end, we simply need to prove that the surface integral of $d\mathbf{A}$ is zero for any closed surface, where the positive direction is taken to be outwards. This is because, if we have $\oiint d\mathbf{A} = \mathbf{0}$ and two surfaces S_1 and S_2 that both span C, gluing these surfaces together forms a closed surface — implying that

$$\iint_{S_1} d\mathbf{A} - \iint_{S_2} d\mathbf{A} = 0$$

$$\implies \iint_{S_1} d\mathbf{A} = \iint_{S_2} d\mathbf{A},$$

which shows that all surfaces which span the same loop C result in the same surface integral. In writing the first equation, the negative sign (rather than positive) stems from the fact that we take outwards to be the positive direction in writing $\oiint d\mathbf{A}$. However, for a current loop C, we already have a predetermined positive direction for the infinitesimal area vector of a surface S that spans it (given by the right-hand-grip rule applied to the proposed positive direction of current) such that $d\mathbf{A}$ for one of the surfaces is actually inwards so we need to include a minus sign to reverse it.

With this clarification, let us prove that $\oiint d\boldsymbol{A} = \boldsymbol{0}$ — we shall first do this with a physical argument and afterwards, a mathematical one. Imagine that we impose a uniform electric field $\boldsymbol{E}$ in the volume enclosed by a closed surface S. Gauss' law states that

$$\oiint_S \boldsymbol{E} \cdot d\boldsymbol{A} = \frac{q_{enc}}{\varepsilon_0}$$

where q_{enc} is the charge enclosed within S. However, observe that by applying Gauss' law to a small cube within every point inside S, the net charge at each point within a region of uniform electric field must be zero since no net flux cuts across the surface of the small cubes! Another way to see this is that the field lines are straight within S such that they cannot start at a positive charge or end at a negative charge within S. Therefore, $q_{enc} = 0$.

$$\oiint_S \boldsymbol{E} \cdot d\boldsymbol{A} = \boldsymbol{E} \cdot \oiint_S d\boldsymbol{A} = 0.$$

Since $\boldsymbol{E}$ is arbitrary (it just has to be uniform),

$$\oiint_S d\boldsymbol{A} = \boldsymbol{0}$$

for any closed surface S. This jump is most obvious if we choose three electric fields that are along the x, y and z-directions of our Cartesian coordinate system such that we can conclude that the x, y and z-components of

$$\oiint_S d\boldsymbol{A}$$

are zero. Now, this nullity is rather intuitive as it asserts that an ideal gas with a uniform pressure p does not exert a net force on the container that contains it (the net force exerted is $\oiint pd\boldsymbol{A}$). Well, we certainly do not see containers of gas moving around on their own.

A rigorous mathematical proof requires elementary vector calculus (which we assume that the reader has prior knowledge of, since we have already presented a physical proof for other readers). The divergence theorem states that for any vector field $\boldsymbol{v}$,

$$\iiint_V (\nabla \cdot \boldsymbol{v}) \, d\tau = \oiint_S \boldsymbol{v} \cdot d\boldsymbol{A}$$

where the left-hand side is the volume integral of $\nabla \cdot \boldsymbol{v}$, known as the divergence of $\boldsymbol{v}$ which is $\frac{\partial v_x}{\partial x} + \frac{\partial v_y}{\partial y} + \frac{\partial v_z}{\partial z}$ in Cartesian coordinates, over a volume V while the right-hand side is the surface integral of $\boldsymbol{v}$ over the closed surface

S that bounds V. In particular, if we choose the vector field $\boldsymbol{v}$ to be uniform everywhere, its divergence vanishes while the right-hand side becomes

$$\oiint_S \boldsymbol{v} \cdot d\boldsymbol{A} = \boldsymbol{v} \cdot \oiint_S d\boldsymbol{A}$$

such that

$$\boldsymbol{v} \cdot \oiint_S d\boldsymbol{A} = 0$$

for any uniform vector field $\boldsymbol{v}$, from which we arrive at the conclusion $\oiint_S d\boldsymbol{A} = \boldsymbol{0}$.

17. Homing Charge**

(a) Within the magnetic field, the charge moves along the arc of a circle. For the charge to reach the positive x-axis when it starts with an initial velocity with a positive y-component, its trajectory must be clockwise. Using the right-hand-rule for the magnetic force, the charge must be negative.

(b) Since the only force that can act on the charge is the magnetic force that does no work, its speed must remain at v at all times.

(c) Since the path of the particle is symmetric about the y-axis, the boundaries of the magnetic field region must also follow suit. That is, $f(x)$ must be an even function. This is because we can play a movie of the charge's motion backwards and the condition in the problem would still be fulfilled (with the two terminals reversed). Now, referring to Fig. 7.28, suppose the particle enters the magnetic field region at $(-x, y)$ and leaves it at (x, y) where $x \geq 0$. Note that the particle's velocities at the junctures of entering and exiting the magnetic field must be locally tangential to the circular arc.

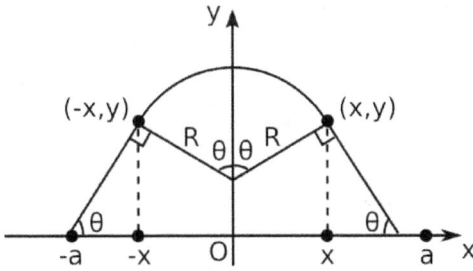

Figure 7.28: Trajectory of particle

From Fig. 7.28, we have

$$\tan\theta = \frac{x}{\sqrt{R^2 - x^2}} = \frac{y}{|x - a|}.$$

Even though the diagram depicts the case where $x < a$, the above equation captures the case where $x \geq a$ too. Solving,

$$y = \frac{x|x - a|}{\sqrt{R^2 - x^2}}.$$

Thus, for $x \geq 0$,

$$f(x) = f(-x) = y$$

so we can write for general x

$$f(x) = \frac{|x|||x| - a|}{\sqrt{R^2 - x^2}}.$$

18. Rolling Sphere**

There are three types of forces on the sphere — its weight, friction and the magnetic force on the coil. We have shown that the last interaction produces no net force on a current loop. Applying Newton's second law along the direction parallel to the plane, while taking the positive direction as downwards,

$$mg\sin\theta - f = ma$$

where f is the friction force on the sphere and a is the acceleration of the center (of mass) of the sphere. The torques on the sphere about its center are due to friction and the magnetic force on the coil. When the sphere has rotated a clockwise angle ϕ, the magnetic dipole moment $\boldsymbol{\mu}$ has similarly rotated ϕ clockwise. Then, the magnetic torque on the sphere is $-\mu B\sin(\phi + \theta)$ clockwise where $\mu = NI\pi R^2$. Combining this with the mechanical torque,

$$fR - \mu B\sin(\phi + \theta) = \frac{2}{5}mR^2\ddot{\phi}.$$

Imposing the non-slip condition $a = R\ddot{\phi}$ and solving the above equations simultaneously,

$$\ddot{\phi} = \frac{5g}{7R}\sin\theta - \frac{5\mu B}{7mR^2}\sin(\phi + \theta).$$

Substituting $\ddot{\phi} = \dot{\phi}\frac{d\dot{\phi}}{d\phi}$,

$$\int_0^{\dot{\phi}} \dot{\phi}d\dot{\phi} = \int_0^{\phi} \left(\frac{5g}{7R}\sin\theta - \frac{5\mu B}{7mR^2}\sin(\phi + \theta) \right) d\phi$$

$$\frac{\dot{\phi}^2}{2} = \frac{5g}{7R}\phi\sin\theta + \frac{5\mu B}{7mR^2}\cos(\phi + \theta) - \frac{5\mu B}{7mR^2}\cos\theta$$

$$\dot{\phi} = \sqrt{\frac{10g}{7R}\phi\sin\theta + \frac{10\mu B}{7mR^2}\cos(\phi + \theta) - \frac{10\mu B}{7mR^2}\cos\theta}$$

$$= \sqrt{\frac{10g}{7R}\phi\sin\theta + \frac{10\pi NIB}{7m}\cos(\phi + \theta) - \frac{10\pi NIB}{7m}\cos\theta}.$$

19. Charge and Wire**

Consider the set-up in cylindrical coordinates with the axis aligned with the wire. Let the radial, azimuthal and longitudinal unit vectors be $\hat{r}$, $\hat{\phi}$ and $\hat{k}$ respectively. Define the origin to be at the point on the wire such that the z-coordinate of the particle is zero. The magnetic field at a radial distance r is

$$\boldsymbol{B} = \frac{\mu_0 I}{2\pi r}\hat{\phi}$$

by Ampere's law. The force on the particle is then

$$\boldsymbol{F} = q\boldsymbol{v} \times \boldsymbol{B}$$

$$= q\left(\dot{r}\hat{r} + r\dot{\phi}\hat{\phi} + \dot{z}\hat{k} \right) \times \frac{\mu_0 I}{2\pi r}\hat{\phi}$$

$$= \frac{q\dot{r}\mu_0 I}{2\pi r}\hat{k} - \frac{q\dot{z}\mu_0 I}{2\pi r}\hat{r}.$$

A crucial observation is that this force does not lead to a torque in the longitudinal direction. Therefore, this component of angular momentum is conserved and the particle does not acquire an azimuthal velocity. When the particle is at the greatest or shortest distance away from the wire, its radial velocity must be zero — implying that its longitudinal speed is v_0 as the kinetic energy of the particle must be conserved (the magnetic force does no work). Writing the equation of motion of the particle in the z-direction,

$$F_z = m\ddot{z} = \frac{q\dot{r}\mu_0 I}{2\pi r}.$$

Applying the impulse-momentum theorem in the z-direction,

$$m\Delta\dot{z} = \int F_z dt = \int_{r_0}^r \frac{q\mu_0 I}{2\pi r}dr = \frac{q\mu_0 I}{2\pi}\ln\frac{r}{r_0},$$

where $\Delta \dot{z}$ is the change in the z-component of the particle's velocity when it is currently at a radial distance r. When the particle attains the greatest distance r_{max} away from the wire, $\Delta \dot{z} = v_0$ since its initial longitudinal velocity is zero (this direction is correct as the force is evidently initially positive in the z-direction as $\dot{r} \geq 0$). Then, substituting $r = r_{max}$ at this juncture,

$$r_{max} = r_0 e^{\frac{2\pi m v_0}{q \mu_0 I}}.$$

After the particle reaches r_{max}, it will acquire a negative radial velocity and hence experience a force in the $-\hat{k}$ direction. Its final z-velocity at the point of closest approach with the wire is $-v_0$. Substituting $\Delta \dot{z} = -v_0$ into the impulse-momentum equation,

$$r_{min} = r_0 e^{-\frac{2\pi m v_0}{q \mu_0 I}}.$$

20. Two Identical Charges**

The equations of motion of the two particles are

$$m\ddot{\boldsymbol{r}}_1 = \frac{q^2}{4\pi\varepsilon_0 r^3}\boldsymbol{r} + q\dot{\boldsymbol{r}}_1 \times \boldsymbol{B},$$

$$m\ddot{\boldsymbol{r}}_2 = -\frac{q^2}{4\pi\varepsilon_0 r^3}\boldsymbol{r} + q\dot{\boldsymbol{r}}_2 \times \boldsymbol{B}.$$

Adding the two equations together and using the fact that $\boldsymbol{r}_{CM} = \frac{\boldsymbol{r}_1 + \boldsymbol{r}_2}{2}$,

$$2m\ddot{\boldsymbol{r}}_{CM} = 2q\dot{\boldsymbol{r}}_{CM} \times \boldsymbol{B}$$

$$\ddot{\boldsymbol{r}}_{CM} = \frac{q}{m}\dot{\boldsymbol{r}}_{CM} \times \boldsymbol{B}.$$

This is analogous to the equation of motion of a positive charge q in a uniform magnetic field! Since the motion of the two charges is confined to the xy-plane, there must not be a component of the velocity of the center of mass parallel to the magnetic field. Then, the center of mass simply undergoes circular motion with the cyclotron frequency $\omega_{CM} = \frac{qB}{m}$! The radius of rotation is indeterminate with arbitrary initial velocities.

Next, subtracting the equation of motion of the second particle from that of the first yields

$$m\ddot{\boldsymbol{r}} = \frac{q^2}{2\pi\varepsilon_0 r^3}\boldsymbol{r} + q\dot{\boldsymbol{r}} \times \boldsymbol{B}.$$

Now, if the distance between the two particles is a constant $r = d$, the separation vector $\boldsymbol{r}$ can only rotate. Suppose that it rotates at an initial angular velocity ω (this must be in the z-direction). Then, from the kinematics of rotations,

$$\dot{\boldsymbol{r}} = \boldsymbol{\omega} \times \boldsymbol{r} = \begin{pmatrix} 0 \\ 0 \\ \omega \end{pmatrix} \times \begin{pmatrix} r_x \\ r_y \\ 0 \end{pmatrix} = \begin{pmatrix} -\omega r_y \\ \omega r_x \\ 0 \end{pmatrix}$$

$$\dot{\boldsymbol{r}} \times \boldsymbol{B} = \begin{pmatrix} -\omega r_y \\ \omega r_x \\ 0 \end{pmatrix} \times \begin{pmatrix} 0 \\ 0 \\ B \end{pmatrix} = \begin{pmatrix} \omega B r_x \\ \omega B r_y \\ 0 \end{pmatrix} = \omega B \boldsymbol{r}.$$

The second time derivative of $\boldsymbol{r}$ is

$$\ddot{\boldsymbol{r}} = \frac{d(\boldsymbol{\omega} \times \boldsymbol{r})}{dt} = \frac{d\boldsymbol{\omega}}{dt} \times \boldsymbol{r} + \boldsymbol{\omega} \times \dot{\boldsymbol{r}} = \frac{d\boldsymbol{\omega}}{dt} \times \boldsymbol{r} - \omega^2 \boldsymbol{r}.$$

The angular velocity vector cannot possibly change in direction as that would imply that the motion of the charges at a later instance no longer lies in the xy-plane. Then, the time derivative of the angular velocity vector can only be due to its change in magnitude.

$$\frac{d\boldsymbol{\omega}}{dt} = \frac{d\omega}{dt}\hat{\boldsymbol{k}}.$$

Substituting the expressions for $\ddot{\boldsymbol{r}}$ and $\boldsymbol{\omega} \times \boldsymbol{r}$ into the previous differential equation in $\boldsymbol{r}$,

$$m\left(\frac{d\omega}{dt}\hat{\boldsymbol{k}} \times \boldsymbol{r} - \omega^2 \boldsymbol{r}\right) = \frac{q^2}{2\pi\varepsilon_0 r^3}\boldsymbol{r} + q\omega B \boldsymbol{r}.$$

Observe that the first term is perpendicular to $\boldsymbol{r}$ while all other terms are parallel to $\boldsymbol{r}$. As $\boldsymbol{r}$ cannot be the null vector, $\frac{d\omega}{dt}$ must be zero! That is, the angular velocity of the separation vector is a constant. Then,

$$\frac{q^2}{2\pi\varepsilon_0 r^3}\boldsymbol{r} + q\omega B \boldsymbol{r} + m\omega^2 \boldsymbol{r} = \boldsymbol{0}.$$

Simplifying and since $r = d$,

$$\omega^2 + \frac{qB}{m}\omega + \frac{q^2}{2\pi m\varepsilon_0 d^3} = 0$$

$$\omega = \frac{-\frac{qB}{m} \pm \sqrt{\frac{q^2 B^2}{m^2} - \frac{2q^2}{\pi m\varepsilon_0 d^3}}}{2}.$$

For ω to be real, the discriminant must be larger than or equal to zero. This implies that

$$\frac{q^2 B^2}{m^2} \geq \frac{2q^2}{\pi m \varepsilon_0 d^3}$$

$$d \geq \sqrt[3]{\frac{2m}{\pi \varepsilon_0 B^2}}.$$

When d is minimum, the discriminant is zero and the angular velocity of the separation vector is $\omega = -\frac{qB}{2m}$. The negative sign indicates that the separation vector rotates clockwise. Consider the initial set-up in the xy-plane in Fig. 7.29.

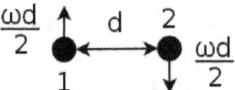

Figure 7.29:   Initial y-velocities (x-velocities not shown)

In the context of the problem, there must be no y-component of velocity of the center of mass which implies that the first and second particles travel at initial y-velocities $\frac{\omega d}{2}$ and $-\frac{\omega d}{2}$ upwards (so that the separation vector instantaneously rotates at angular velocity ω clockwise). Now suppose that the center of mass has an initial velocity rightwards (the other situation is similar), the center of mass then undergoes circular motion at clockwise angular velocity $\omega_{CM} = \frac{qB}{m}$ about a center O, located below the initial position of the center of mass, A in Fig. 7.30. At the same time, the separation vector undergoes a rotation at angular velocity $\omega = \frac{qB}{2m}$ clockwise.

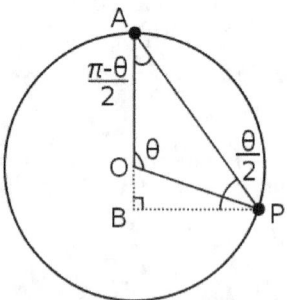

Figure 7.30:   Trajectory of center of mass

Now, consider a point P along the trajectory of the center of mass, after the center of mass has rotated a clockwise angle θ. At this juncture, the separation vector would have rotated a clockwise angle $\frac{\theta}{2}$. Therefore, in order to show that the original position of mass and the instantaneous positions of the two particles are collinear, we simply have to show that the line AP subtends an angle $\frac{\theta}{2}$ with respect to the horizontal via some geometry. As $\triangle AOP$ is isosceles,

$$\angle OAP = \angle OPA = \frac{\pi - \theta}{2},$$

$$\angle APB = \pi - \frac{\pi}{2} - \angle OAP = \frac{\theta}{2},$$

hence proving the claim that these three points are always collinear.

21. Two Opposite Charges***

The equations of motion of the two charges are

$$m\ddot{\boldsymbol{r}}_1 = -\frac{q^2}{4\pi\varepsilon_0 r^3}\boldsymbol{r} + q\dot{\boldsymbol{r}}_1 \times \boldsymbol{B},$$

$$m\ddot{\boldsymbol{r}}_2 = \frac{q^2}{4\pi\varepsilon_0 r^3}\boldsymbol{r} - q\dot{\boldsymbol{r}}_2 \times \boldsymbol{B}.$$

The instantaneous position vector of the center of mass is $\frac{\boldsymbol{r}_1 + \boldsymbol{r}_2}{2}$. Adding the two equations of motion,

$$2m\ddot{\boldsymbol{r}}_{CM} = q\dot{\boldsymbol{r}} \times \boldsymbol{B}.$$

Subtracting the second equation from the first,

$$m\ddot{\boldsymbol{r}} = -\frac{q^2}{2\pi\varepsilon_0 r^3}\boldsymbol{r} + 2q\dot{\boldsymbol{r}}_{CM} \times \boldsymbol{B}.$$

To decouple the two differential equations above, integrate the former with respect to time

$$2m(\dot{\boldsymbol{r}}_{CM} - \boldsymbol{v}_{CM}^0) = q(\boldsymbol{r} - \boldsymbol{r}_0) \times \boldsymbol{B},$$

where $\boldsymbol{v}_{CM}^0$ and $\boldsymbol{r}_0$ are the initial velocity of the center of mass and the initial separation vector respectively. Substituting the expression for $\dot{\boldsymbol{r}}_{CM}$ obtained from this equation into the second differential equation,

$$\ddot{\boldsymbol{r}} = -\frac{q^2}{2\pi m\varepsilon_0 r^3}\boldsymbol{r} + \frac{q^2}{m^2}[(\boldsymbol{r} - \boldsymbol{r}_0) \times \boldsymbol{B}] \times \boldsymbol{B} + \frac{2q\boldsymbol{v}_{CM}^0 \times \boldsymbol{B}}{m}.$$

Applying the BAC-CAB rule to $(\boldsymbol{r} \times \boldsymbol{B}) \times \boldsymbol{B} = -\boldsymbol{B} \times (\boldsymbol{r} \times \boldsymbol{B})$,

$$(\boldsymbol{r} \times \boldsymbol{B}) \times \boldsymbol{B} = \boldsymbol{B}(\boldsymbol{B} \cdot \boldsymbol{r}) - \boldsymbol{r}(\boldsymbol{B} \cdot \boldsymbol{B}) = -B^2\boldsymbol{r}$$

since r is perpendicular to B. A similar statement holds for $(r_0 \times B) \times B$. Therefore, the above can be rewritten as

$$\ddot{r} = \left(-\frac{q^2}{2\pi m \varepsilon_0 r^3} - \frac{q^2 B^2}{m^2} \right) r + c,$$

where c is the constant vector given by

$$c = \frac{q^2 B^2}{m^2} r_0 + \frac{2q v_{CM}^0 \times B}{m}.$$

Now, we impose the condition that the separation vector can only rotate at a constant angular velocity. Then, $\ddot{r}$ must only correspond to the centripetal acceleration.

$$\ddot{r} = -\omega^2 r.$$

Substituting this expression for $\ddot{r}$,

$$\left(\omega^2 - \frac{q^2}{2\pi \varepsilon_0 m r^3} - \frac{q^2 B^2}{m^2} \right) r = -c.$$

Note that the left-hand side involves a variable vector r (though its magnitude is constant) while the right-hand side involves a constant vector c. For this equation to be valid at all instances, c must be the null vector. This implies that

$$\frac{q^2 B^2}{m^2} r_0 + \frac{2q v_{CM}^0 \times B}{m} = 0.$$

Let the constant magnitude of r be d. Furthermore, without any loss of generality, suppose that r_0 is in the x-direction. Substituting $r_0 = (\pm d, 0, 0)$, $v_{CM}^0 = (v_{CMx}^0, v_{CMy}^0, 0)$ and $B = (0, 0, B)$ yields

$$\frac{q^2 B^2}{m^2} \begin{pmatrix} \pm d \\ 0 \\ 0 \end{pmatrix} = -\frac{2q}{m} \begin{pmatrix} v_{CMx}^0 \\ v_{CMy}^0 \\ 0 \end{pmatrix} \times \begin{pmatrix} 0 \\ 0 \\ B \end{pmatrix} = -\frac{2q}{m} \begin{pmatrix} B v_{CMy}^0 \\ -B v_{CMx}^0 \\ 0 \end{pmatrix},$$

$$v_{CM}^0 = \begin{pmatrix} 0 \\ \mp \frac{qB}{2m} d \\ 0 \end{pmatrix}.$$

Moving on, since c is a null vector, we must have

$$\omega^2 - \frac{q^2}{2\pi \varepsilon_0 m d^3} - \frac{q^2 B^2}{m^2} = 0.$$

Solving for d,

$$d = \sqrt[3]{\frac{2\pi\varepsilon_0 m}{q^2}\left(\omega^2 - \frac{q^2 B^2}{m^2}\right)}.$$

For d to be positive,

$$|\omega| \geq \frac{qB}{m}.$$

22. Magnetic Lens****

Let the point P at which the magnetic field is to be determined be at $(r, 0, z)$ in Cartesian coordinates (i.e. we choose the x and y-axes such that its y-coordinate is zero). We will determine the magnetic field at point P by integrating over the solenoid in cylindrical coordinates. Consider an infinitesimal surface element on the solenoid of sides $Rd\phi$ and dh at radial distance R, azimuthal angle ϕ and z-coordinate h. The separation vector between this element and P is

$$\mathbf{r'} = \begin{pmatrix} r \\ 0 \\ z \end{pmatrix} - \begin{pmatrix} R\cos\phi \\ R\sin\phi \\ h \end{pmatrix} = \begin{pmatrix} r - R\cos\phi \\ -R\sin\phi \\ z - h \end{pmatrix}.$$

Note that we use $\mathbf{r'}$ to denote the separation vector, instead of $\mathbf{r}$ whose symbol (the scalar) has been used to denote the radial coordinate of P. The surface current density along the surface of the solenoid is $\mathbf{K} = \eta I(-\sin\phi, \cos\phi, 0)$ in the azimuthal direction. Applying Biot-Savart's law and integrating over the entire solenoid, the magnetic field at P is

$$\mathbf{B} = \frac{\mu_0}{4\pi} \iint \frac{\mathbf{K}\,dA \times \hat{\mathbf{r'}}}{r'^2}$$

$$= \frac{\mu_0\eta I}{4\pi} \int_0^l \int_0^{2\pi} \frac{Rd\phi dh}{[R^2 - 2Rr\cos\phi + r^2 + (h-z)^2]^{\frac{3}{2}}}$$

$$\times \begin{pmatrix} -\sin\phi \\ \cos\phi \\ 0 \end{pmatrix} \begin{pmatrix} r - R\cos\phi \\ -R\sin\phi \\ z - h \end{pmatrix}$$

$$= -\frac{\mu_0\eta I}{4\pi} \int_0^l \int_0^{2\pi} \frac{Rd\phi dh}{[R^2 - 2Rr\cos\phi + r^2 + (h-z)^2]^{\frac{3}{2}}} \begin{pmatrix} (h-z)\cos\phi \\ (h-z)\sin\phi \\ r\cos\phi - R \end{pmatrix}.$$

We are only interested in the longitudinal and radial components which are B_z and B_x, respectively.

$$B_z = -\frac{\mu_0 \eta I}{4\pi} \int_0^l \int_0^{2\pi} \frac{R(r \cos\phi - R)}{[R^2 - 2Rr \cos\phi + (h-z)^2]^{\frac{3}{2}}} d\phi dh$$

$$= -\frac{\mu_0 \eta I}{4\pi} \int_0^l \int_0^{2\pi} \frac{R(r \cos\phi - R)}{[R^2 + (h-z)^2]^{\frac{3}{2}} \left(1 - \frac{2Rr \cos\phi}{R^2 + (h-z)^2}\right)^{\frac{3}{2}}} d\phi dh$$

$$\approx -\frac{\mu_0 \eta I}{4\pi} \int_0^l \int_0^{2\pi} \frac{R(r \cos\phi - R)}{[R^2 + (h-z)^2]^{\frac{3}{2}}} \left(1 + \frac{3Rr \cos\phi}{R^2 + (h-z)^2}\right) d\phi dh,$$

where we have used the binomial expansion and discarded second order and higher terms in $\frac{r}{R}$. The integral of $\cos\phi$ over an entire period yields zero. Therefore, the above integral reduces to (after discarding another second order term in $\frac{r}{R}$)

$$B_z = \frac{\mu_0 \eta I}{4\pi} \int_0^l \int_0^{2\pi} \frac{R^2}{[R^2 + (h-z)^2]^{\frac{3}{2}}} d\phi dh$$

$$= \frac{\mu_0 \eta I}{2} \left[\frac{l - z}{\sqrt{z^2 + (l-z)^2}} + \frac{z}{\sqrt{R^2 + z^2}}\right],$$

which is just the magnetic field along the axis of the solenoid that we have previously computed in Problem 8. To determine the radial component of the magnetic field at point P, we need to evaluate B_x (we purposely defined P to be along the x-direction).

$$B_x = -\frac{\mu_0 \eta I}{4\pi} \int_0^{2\pi} \int_0^l \frac{R(h-z) \cos\phi \, dh d\phi}{[R^2 - 2Rr \cos\phi + (h-z)^2]^{\frac{3}{2}}}.$$

It is easier to integrate over h first. Making the substitutions $u = R^2 - 2Rr \cos\phi + (h-z)^2$ and $du = 2(h-z)dh$, the radial magnetic field (we shall use B_r to denote this now) is

$$B_r = -\frac{\mu_0 \eta I}{4\pi} \int_0^{2\pi} \int_{R^2 - 2Rr \cos\phi + z^2}^{R^2 - 2Rr \cos\phi + (l-z)^2} \frac{R \cos\phi}{2u^{\frac{3}{2}}} du d\phi$$

$$= \int_0^{2\pi} \frac{\mu_0 \eta I R \cos\phi}{4\pi}$$

$$\times \left[\frac{1}{\sqrt{R^2 - 2Rr\cos\phi + (l-z)^2}} - \frac{1}{\sqrt{R^2 - 2Rr\cos\phi + z^2}} \right] d\phi$$

$$= \int_0^{2\pi} \frac{\mu_0 \eta I R \cos\phi}{4\pi}$$

$$\times \left[\frac{1}{\sqrt{R^2 + (l-z)^2}\left(1 - \frac{2Rr\cos\phi}{R^2+(l-z)^2}\right)^{\frac{1}{2}}} - \frac{1}{\sqrt{R^2 + z^2}\left(1 - \frac{2Rr\cos\phi}{R^2+z^2}\right)^{\frac{1}{2}}} \right] d\phi$$

$$\approx \int_0^{2\pi} \frac{\mu_0 \eta I R \cos\phi}{4\pi} \left[\frac{1 + \frac{Rr\cos\phi}{R^2+(l-z)^2}}{\sqrt{R^2 + (l-z)^2}} - \frac{1 + \frac{Rr\cos\phi}{R^2+z^2}}{\sqrt{R^2 + z^2}} \right] d\phi$$

$$= \frac{\mu_0 \eta I R^2 r}{4} \left[\frac{1}{[R^2 + (l-z)^2]^{\frac{3}{2}}} - \frac{1}{(R^2 + z^2)^{\frac{3}{2}}} \right],$$

as the integral of $\cos\phi$ and $\cos^2\phi$ over a single period yield 0 and π respectively. To estimate the magnitude of v_0 required for the charges' radial coordinate to remain approximately constant throughout the solenoid, observe that the azimuthal and radial forces on the charges at a radial coordinate r are approximately

$$F_\phi \approx q v_0 B_r,$$

$$F_r \approx q v_\phi B_z,$$

where we have assumed that the radial velocity is negligible. The azimuthal acceleration is thus

$$a_\phi = \frac{q v_0 B_r}{m}.$$

Since the time required for the particle to exit the solenoid is of the order of $\frac{l}{v_0}$, the azimuthal velocity is of the order

$$v_\phi \approx \frac{q B_r l}{m}.$$

The radial acceleration is then approximately

$$a_r = \frac{q v_\phi B_z}{m} \approx \frac{q^2 B_r B_z l}{m^2}.$$

The radial distance covered by the particle, during its time is the solenoid, is then of order

$$\left| \Delta r \right| \approx \left| \frac{q^2 B_r B_z l}{m^2} \cdot \frac{l^2}{v_0^2} \right| = \left| \frac{q^2 B_r B_z l^3}{m^2 v_0^2} \right|.$$

This must be much smaller than r for the radial distance to be approximately constant. Furthermore, we know from the expressions for the components of the magnetic field that

$$\left| B_r \right| \approx \frac{\mu_0 \eta I R^2 r}{l^3},$$

$$B_z \approx \mu_0 \eta I$$

for $l \gg R$. Then, the required condition is that

$$\frac{q^2 l^3}{m^2 v_0^2} \cdot \frac{\mu_0^2 \eta^2 I^2 R^2 r}{l^3} \ll r$$

$$v_0 \gg \frac{q \mu_0 \eta I R}{m}.$$

Now to analyze the focusing of charges, we must first determine the impact of the magnetic field on the velocities of the charges. We first prove that the azimuthal velocity of a charge is still zero after it has left the solenoid. The z-component of torque is $qr\dot{z}B_r$. By the angular-impulse-momentum theorem, the z-component of angular momentum at the top end of the solenoid is

$$L_z = \int qr\dot{z}B_r dt$$

$$= \int_0^l qr B_r dz$$

$$= \int_0^l \frac{\mu_0 \eta I q R^2 r^2}{4} \left[\frac{1}{[R^2 + (l-z)^2]^{\frac{3}{2}}} - \frac{1}{(R^2 + z^2)^{\frac{3}{2}}} \right] dz,$$

where we have used the fact that the initial z-component of angular momentum is zero. At this point, we claim that this integral is zero as $\int_0^l \frac{1}{[R^2 + (l-z)^2]^{\frac{3}{2}}} dz = \int_0^l \frac{1}{(R^2 + z^2)^{\frac{3}{2}}} dz$. This is due to the fact that

$$\int_a^b f(x)dx = \int_a^b f(a+b-x)dx$$

for any function $f(x)$, as one can easily show via a substitution. Then,

$$\Delta L_z = 0$$

which means that the final azimuthal velocities of the particles are still zero! However, note that the azimuthal velocities are not zero at intermediate points within the solenoid. To determine $\dot{\phi}$ within the solenoid, we integrate the above with more general limits,

$$L_z = \int_0^z \frac{\mu_0 \eta I q R^2 r^2}{4} \left[\frac{1}{[R^2 + (l-z)^2]^{\frac{3}{2}}} - \frac{1}{(R^2 + z^2)^{\frac{3}{2}}} \right] dz$$

$$= \frac{\mu_0 \eta I q r^2}{4} \left(\frac{l}{\sqrt{l^2 + R^2}} - \frac{l-z}{\sqrt{(l-z)^2 + R^2}} + \frac{z}{\sqrt{R^2 + z^2}} \right),$$

where the integrations can be performed via a substitution. Since $L_z = mr\dot{\phi}$ where $\dot{\phi}$ is the azimuthal angular velocity,

$$\dot{\phi} = \frac{\mu_0 \eta I q r}{4m} \left(\frac{l}{\sqrt{l^2 + R^2}} - \frac{l-z}{\sqrt{(l-z)^2 + R^2}} + \frac{z}{\sqrt{R^2 + z^2}} \right).$$

Moving on, there is a change in the radial velocity of the particles (small but non-negligible in the context of the particles' motion after leaving the solenoid, which can possibly last indefinitely). To compute this, we apply the impulse-momentum theorem to the radial direction.

$$\Delta p_r = \int q r \dot{\phi} B_z dt = \int_0^l q r \dot{\phi} B_z \frac{dt}{dz} \cdot dz \approx \int_0^l \frac{q r \dot{\phi} B_z}{v_0} dz.$$

Substituting the expressions for B_z and $\dot{\phi}$ and integrating from $z = 0$ to $z = l$ would yield

$$\Delta p_r = -\frac{\mu_0^2 \eta^2 I^2 q^2 R^2 r^2 l}{2 m v_0 \sqrt{R^2 + l^2}} \left(\sqrt{R^2 + l^2} - R \right).$$

Again the integrations in this step can be vastly simplified by finding pairs of the form $\int_a^b f(x) dx - \int_a^b f(a + b - x) dx = 0$. The above expression implies that the final radial velocities of the charges after leaving the solenoid are

$$v_r = -\frac{\mu_0^2 \eta^2 I^2 q^2 R^2 r^2 l}{2 m^2 v_0 \sqrt{R^2 + l^2}} \left(\sqrt{R^2 + l^2} - R \right).$$

This radial velocity is directed radially inwards. The time t required for these particles to coincide with the z-axis is then

$$t = \frac{r}{|v_r|} = \frac{2 m^2 v_0 \sqrt{R^2 + l^2}}{\mu_0^2 \eta^2 I^2 q^2 R^2 r l \left(\sqrt{R^2 + l^2} - R \right)}.$$

This set-up would not function as a good lens as the distance above the top end of the solenoid at which charges with initial radial coordinate r coincide with the z-axis is roughly $v_0 t$ which is inversely proportional to r. Therefore, charges with different initial radial coordinates are focused to different points along the z-axis.

Chapter 8

Currents and EMI

In the previous chapter, we used steady currents as our fundamental building blocks to generate magnetic fields. In this chapter, we will analyze how currents actually arise in media and how they are related to electromagnetic induction engendered by time-varying electric and magnetic fields. We will be considering a more general system beyond stationary charges (electrostatics) and steady currents (magnetostatics).

8.1 Voltage

The voltage between points A and B along a certain path L is defined as

$$V_{AB} = - \int_L \boldsymbol{E} \cdot d\boldsymbol{s} \tag{8.1}$$

where $\boldsymbol{E}(\boldsymbol{r}, t)$ is the electric field permeating the relevant space. Now, the above expression looks inordinately similar to the potential difference between two points in an electrostatic field. However, they are really two different concepts as the electric field $\boldsymbol{E}$ in the above expression may not be due to an electrostatic system (we are considering a more general case). Then, the line integral of the electric field may actually depend on the path of integration taken and an electric potential function cannot be assigned to the system. As a result, we define the voltage in such a way that it may be dependent on the path of integration, in order to accommodate such a possibility.

Finally, some authors interpret the above as the work done per unit charge by an external force against the electric field $\boldsymbol{E}$ in bringing a test charge from points A to B along path L, without a change in kinetic energy, but such a definition is slightly misleading in this case where $\boldsymbol{E}(\boldsymbol{r}, t)$ is a function of time.[1] This is because the act of moving a test charge implies that the electric field should be evaluated progressively in time but the line integral above must be computed at a particular time t where the electric field is "paused"!

8.2   Current

An electrical current refers to the flow of charges, transported by charge carriers which are usually electrons in a circuit and possibly ions in certain cases (e.g. inside an electrochemical cell). By convention, a positive current is defined in the direction of positive charge flow, contrary to the direction of the flow of electrons.

There are a few conditions that must be satisfied in order for a current to flow in a branch of a circuit. Firstly, the circuit components in branches which transport currents must possess mobile charge carriers that are capable of carrying charge. This in turn affects a quantity known as the conductivity of a material which describes its ability to carry a current. Metals are generally good conductors as they possess a multitude of free electrons in their electron clouds and thus high conductivities while insulating materials, such as plastic, have low conductivity and "obstruct" current flow as their electrons are tightly bound to the nuclei.

Next, there must be a net external force on charges in any non-ideal conductor to sustain their motion as the charges are constantly redirected by their collisions with each other and the atoms of the conductor, which lead to the random redirection of velocities. Usually, this force is not mechanical as we do not have access to charges inside the wire. Instead, the force is electromagnetic and the force due to an electric field usually dominates the magnetic force due to the relative magnitudes of the electric permittivity and magnetic permeability, coupled with the slow average velocity of charge carriers (as we shall see). Therefore, an electric field must usually be present inside a non-ideal conductor — implying that there needs to be a voltage across any two points of a non-ideal conductor.

Lastly, there must be a cyclic conducting path in a circuit for a non-zero, steady current to flow. A steady current refers to a current that is unvarying with respect to time and requires no charge to accumulate at any point in the circuit. To understand the physical reason behind this cyclic requirement, consider Fig. 8.1.

---

[1]You may recall that we adopted a similar interpretation of the electric potential but that was possible because electrostatic fields do not change over time.

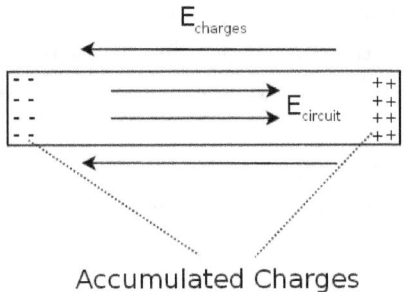

Accumulated Charges

Figure 8.1:   Accumulated charges at ends.

Assume that there is already an electric field that is driving charges from one end of the non-cyclic circuit to the other. A transient flow of charges may occur but in a short amount of time, charges, with appropriate signs, will accumulate at the ends of the circuit to produce an electric field that cancels out the initial electric field. Compounded with the fact that the electric field, which drives charges in a conductor, is usually small, there will no longer be any net current flowing in this non-cyclic circuit almost instantaneously. When there is a closed conducting path in the circuit, charges are no longer accumulated at any point in the circuit and a steady current can flow when the rates of charge entering and leaving each section of the circuit are the same.

The above situation is analogous to pumping air into a soccer ball. As more fluid is pumped into a container, the pressure of the fluid inside the container is increased until it eventually attains the same pressure of the pump, after which no more net transfer of fluid will occur. However, if another hole is made on the container, a dynamic equilibrium (with a constant influx of fluid) can be attained when the rate of fluid escaping the hole is equal to that entering the container.

Interestingly, the above mechanism is actually responsible for the smooth flow of current in a closed conducting loop. Despite its cyclic nature, a loop would not be able to carry a steady current if the charges were to be "lost" at the bends. So how do the charges know that they must turn at specific locations? Well, it is definitely not due to a messenger "informing" them of a predetermined route at an earlier instance, as a bend can be made at any juncture (even after the charges have started moving). The answer to this question is the existence of charges on the surface of the wire. Though the wire may be neutral as a whole, charges can be redistributed on various surfaces such that the wire is not entirely neutral, but positively charged in certain regions and negatively charged in others as long as the total charge is zero. When a bend is made and a current initially flows, some charge is piled up at the bends in this transitory period — creating their own electric fields which provide the necessary centripetal force for future charges to make the turn!

As a physical quantity, the current intensity, which is often denoted as just the current, is the rate of flow of charge across a surface.

$$I = \frac{dQ}{dt}. \tag{8.2}$$

Now, this is a scalar quantity with no mention of the direction of the velocities of charges whatsoever. In light of the three-dimensional motion of charges, the current density should be considered. The current density, J, is defined as the electric current per unit perpendicular area at a given point in space and is a vector quantity whose direction points in the direction of positive charge flow. The total current through a surface S is then the surface integral of current density J over S.

$$I = \iint_S J \cdot d\boldsymbol{A}. \tag{8.3}$$

Lastly, there is an important, intuitive property regarding the above current density integral in the context of steady currents. Referring to Fig. 8.2, consider two surfaces S_1 and S_2 which are bounded by the same contour C; we claim that the current across each surface will be identical in the case of steady currents.

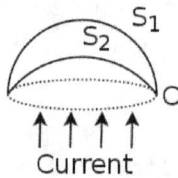

Figure 8.2:   Surfaces S_1 and S_2 bounded by the same contour C

Observe that gluing the two surfaces forms a closed surface. Since a steady current dictates that charge cannot accumulate anywhere, the current entering surface S_1 must be equal to the current leaving surface S_2 for the amount of charge confined in the combined closed surface to remain constant. Therefore, we conclude that the integral in Eq. (8.3) is only dependent on the contour that bounds the surface and not the exact surface itself, in the case of steady currents.

8.2.1   *Microscopic View*

Let us analyze the constitution of a current density microscopically. Consider a region of charges with charge q traveling at a velocity $\boldsymbol{u}$ with a number density n. What is the total current crossing an arbitrary rectangular surface with area vector $\boldsymbol{A}$?

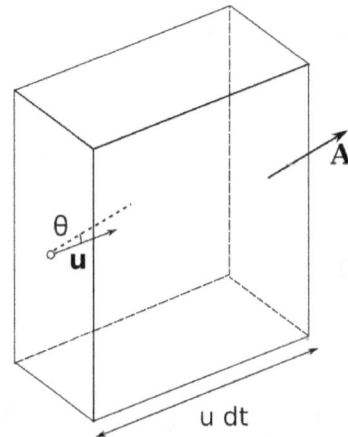

Figure 8.3:   Moving charges that pass through a given area

In time dt, the volume of charge that would have cut through the surface is given by the parallelepiped in Fig. 8.3. Its volume is

$$dV = uA \cos \theta dt = \boldsymbol{u} \cdot \boldsymbol{A} dt.$$

The net amount of charge passing through this surface in the time interval is then the volume of this parallelepiped multiplied by the number density n of the charges and the common charge q carried by each particle.

$$dQ = nq\boldsymbol{u} \cdot \boldsymbol{A} dt.$$

The current that is transported across the rectangular surface is then

$$I = \frac{dQ}{dt} = nq\boldsymbol{u} \cdot \boldsymbol{A}.$$

In general, a region consists of charged particles with varying $\boldsymbol{u}$ and n (but usually with the same charge such as $-e$ in the case of electrons). Let the ith class of such charges have velocity $\boldsymbol{u_i}$ and number density n_i and let there be k classes of charges in total. Then,

$$I = \boldsymbol{A} q \cdot \sum_{i=1}^{k} n_i \boldsymbol{u_i}.$$

For an infinitesimal A (i.e. dA), the above implies that the current density at a point is

$$J = q \cdot \sum_{i=1}^{k} n_i u_i.$$

Finally, we define the average velocity of the charges, which is also known as the drift velocity, and the total number density of the charges to be

$$v_d = \frac{\sum_{i=1}^{k} n_i u_i}{n},$$

$$n = \sum_{i=1}^{k} n_i,$$

respectively. We can then rewrite the current density as

$$J = nqv_d. \tag{8.4}$$

Hence, the current density at a point on a surface is directly proportional to the number density and the drift velocity of charges in the infinitesimal section surrounding the surface when the particles possess equal charge.

8.2.2   *Drude's Model of Conduction*

In this section, we will develop a simple model for the drift velocity of charged particles. We assume that the force on the charges (if any) is due to the most common cause — the electric field. In the absence of an electric field in a conductor, electrons travel randomly in all directions, resulting in no net average velocity and thus no net charge transfer.

As we shall show, a constant electric field[2] is required to drive a constant current through any non-ideal conductor, under certain assumptions. Now this should set off some bells ringing in your mind. The above statement implies that an electric field causes the charges to travel at a constant average velocity. This is contrary to the common expectation that the average velocity of the charges should be increasing as they are accelerated! Actually, the presence of an electric field is mandated by the frequent collisions of charges with surrounding atoms, ions and each other, which cause them to lose the "memory" of their previous velocity distributions. In the ideal case,

---

[2]You may now recall that the electric field should be zero inside a conductor at electrostatic equilibrium. However, in this case, the conductor may not have attained static equilibrium yet, either because it is in the process of redistributing charges or because it achieves a state of dynamic equilibrium.

the resultant velocity of a charge after a collision is along a random direction. Then, these charges are accelerated by the electric field for a certain period of time before undergoing collisions again. This process repeats itself indefinitely so we can associate a mean velocity averaged over all particles at all times, which is the drift velocity.

Let $\boldsymbol{u}_i^{aft}$ denote the velocity of the ith charged particle of mass m and charge q immediately after its last collision in a certain section of the circuit. Let t_i denote the time elapsed since its last collision and N be the total number of particles. If the charges are placed in a region of uniform electric field $\boldsymbol{E}$, the velocity of the ith particle at this instant, $\boldsymbol{u}_i^{now}$, is given by the impulse-momentum theorem as

$$m\boldsymbol{u}_i^{now} = m\boldsymbol{u}_i^{aft} + q\boldsymbol{E}t_i.$$

Thus the average velocity of the particles in a certain section at this instant is

$$\bar{\boldsymbol{u}}^{now} = \bar{\boldsymbol{u}}^{aft} + \frac{q\boldsymbol{E}}{Nm}\sum_{i=1}^{N} t_i.$$

$\frac{1}{N}\sum_{i=1}^{N} t_i$ is the average time elapsed between the current instance and a particle's last collision under the influence of $\boldsymbol{E}$, but for small values of $\boldsymbol{E}$ (which is usually the case), it can also be deemed as that when $\boldsymbol{E}$ is absent. The latter is equal to the average time between consecutive collisions of a particle[3] which is known as the mean free time, τ. Note that τ is a constant that depends on the properties of the medium or material that contains the charges.

Moving on, the average velocities of the particles immediately after their last collision, $\bar{\boldsymbol{u}}^{aft}$, should be zero as there are abundant particles which travel in random directions after their last collision. Thus,

$$v_d = \bar{\boldsymbol{u}}^{now} = \frac{q\tau}{m}\boldsymbol{E}. \tag{8.5}$$

We conclude that the drift velocity of the charges and thus, current density should be proportional to the local electric field from the above analysis.

---

[3] A rigorous proof of this requires some probabilistic analysis (this can be obtained from the fact that the probability of a collision at time t follows the exponential distribution, similar to our estimation of the mean free time of gas molecules in kinetic theory). For the exact value of the mean free time, we would need to know the distribution of speeds among the electrons. However, a rough idea of why this claim should be valid can be obtained if we imagine observing a particle to collide at this instant. We do not know exactly when its previous collision was and only know its probabilistic distribution. Then, the average time elapsed between the current instant and the particle's last collision should be equal to the average time between consecutive collisions as the particle collides at this instant.

Substituting this expression for the drift velocity into the current density in Eq. (8.4), the above model suggests that the current density is directly proportional to the net electric field.

$$J = \frac{nq^2\tau}{m}E. \tag{8.6}$$

Now, a typical order of magnitude of the drift velocity of electrons in a conductor can be millimeters per second. In response to this, some may think of the following apparent paradox. If the electrons are traveling so slowly, how is that a light bulb lights up almost instantaneously when a battery is connected via long wires far away? In fact, there is even a more glaring "paradox" in the context of AC circuits — if the electrons are traveling so slowly, they are confined to an extremely small region due to their tiny amplitudes of oscillations! How is it possible for a current to flow along the circuit? Well, this confusion is best clarified by the following analogy involving an open, hollow cylinder packed with metal balls.

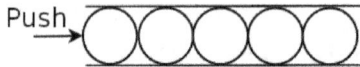

Figure 8.4:   Cylinder of balls

The cylinder in Fig. 8.4 represents the wire while the balls represent the stationary (on average) electrons which are already present in the wire before any external battery is connected to the set-up. Now, in this mechanical analog, give the left ball a slight push. The right ball moves and falls off almost instantaneously though the left ball has hardly moved. Therefore, it is not the speed of the left ball that determines how fast the right ball responds to the push — it is the speed of force propagation in this case. Similarly, in the case of a wire which is suddenly connected to an external battery, it is not the speed of electrons in one region that determines when the electrons in another region respond and start moving — it is the speed of the propagation of the electric field inside the wire which can be on the order of one-tenth of the speed of light! Therefore, the crux of this resolution is that the electrons were already in the wire in the first place (so as to respond to an incoming electric field and propagate it) and that it is the speed of propagation of the electric field that determines when current begins to flow in a region of the wire.

To resolve the paradox regarding how a current can flow in an AC circuit, observe that since we have established the fact that "electromagnetic news" spreads rapidly, the electrons in a circuit are approximately always moving in the same direction (clockwise or anti-clockwise along the circuit), even though they are confined to a certain region. Therefore, even though each electron only moves a tiny distance, the entire group of electrons moving for a small distance looks as if a current is flowing on the large scale!

8.2.3 *Ohm's Law*

The model in the previous section was in fact developed to explain the following experimental rule which holds for most materials. It is empirically observed that the current density at a point J in a medium is proportional to the force per unit charge f on charged particles at that point.

$$J = \sigma f. \tag{8.7}$$

The constant of proportionality σ is known as the conductivity of the material. The magnitude of σ is large for a conductor, which implies that f need not be very large. This force f is usually electromagnetic in nature, with the magnetic force often being negligible. The Lorentz force law then gives

$$J = \sigma E. \tag{8.8}$$

The above empirical law is known as Ohm's law and states that current density at each point in a general material is proportional to the electric field at that point. This equation, in a certain sense, is the microscopic version of the common form of Ohm's law as it describes the current density at any single point in some material, instead of the total current across a surface. The magnitude of E need not be very large in conductors as σ is large.

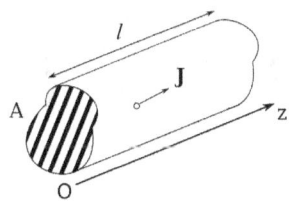

Figure 8.5: Wire with constant cross section

Now, consider a uniform wire of length l and a constant cross sectional area A of an arbitrary shape, with one of its ends located at the origin and with its axis pointing in the z-direction in Fig. 8.5. The wire carries a steady current. We will first consider the case where this current is driven by a strictly electrostatic field. Ultimately, our objective is to relate the current flowing through the cross section to the potential difference across the ends.

Firstly, what constitutes a wire? A conducting wire must be enveloped by an insulator so that an isolated conducting path can be defined. Thus, current will not flow out of the surface of the wire. This gives us a boundary condition for all points on the surface on the wire, less the two terminals. At these points, there must not be any normal component of current and hence electric field. The partial derivatives of the electric potential with respect to the normal direction at these points are then zero.

Next, the two terminals correspond to equipotential surfaces as the ends of the wire are usually connected to a perfect conductor. Let the reference potential of the end at the origin be V_0 and the potential of the other end be 0. Combining this boundary condition regarding the potentials at the ends with the previous requirement, the boundary conditions of the volume enclosed by the wire are completely specified.

Moving on, let us analyze what a steady current implies for the charge distribution within the wire. Consider an infinitesimal box element inside the wire. The steady current conditions mandates that the surface integral of the current density over this box is zero. Since the current density at a point is proportional to the electric field at that point, the previous statement implies that the surface integral of the electric field (i.e. electric flux) over the box is zero[4]! Therefore, the charge contained within the infinitesimal box must be zero by Gauss' law. Since this holds for all infinitesimal boxes, there is no net charge anywhere within the wire.

It turns out that specifying the charge distribution inside the wire and the boundary conditions afore is sufficient to guarantee a unique potential function for the wire.[5] Now, let us guess a solution to the potential function which satisfies these boundary conditions. An obvious guess would be

$$V(x, y, z) = V_0 - \frac{V_0}{l} z.$$

One can easily check that this solution satisfies the boundary conditions. Since this must be the correct solution, the electric field in the wire is directed

---

[4]This conclusion requires the premise that the conductivity is uniform over the box. If this does not hold (e.g. at the interface of two different materials), the surface integral of the electric field is generally not zero and some charge is stored at the interface. We cannot conclude from the nullity of the surface integral of the current density that the surface integral of the electric field is also zero as the constant of proportionality changes over different surfaces.

[5]This can be seen as a combination of the first and third uniqueness theorems (see solution to Problem 11 in Chapter 6).

solely along the z-axis and is uniform.

$$E = -\nabla V = \frac{V_0}{l}\hat{k}.$$

It follows directly that

$$J = \frac{\sigma V_0}{l}\hat{k}.$$

Thus, it can be concluded that the current density and electric field are identical at all points in the wire. Then, the current I through the wire and the decrease in potential V across the wire can be evaluated trivially in terms of the "microscopic quantities" — current density and electric field.

$$I = JA,$$

$$V = El,$$

where we have dropped the vector notations. Using the relation $J = \sigma E$, we obtain the most common form of Ohm's law for a uniform wire.

$$V = \frac{l}{\sigma A}I.$$

$\frac{1}{\sigma}$ is known as the resistivity of the material, ρ. It can be seen that the potential drop across the wire is proportional to the current flowing through it. This constant of proportionality is known as the resistance of the wire and is defined as

$$R = \frac{\rho l}{A}. \tag{8.9}$$

Thus, the relationship between the decrease in potential across the wire and the steady current through it is given by

$$V = IR \tag{8.10}$$

where R is the resistance of the wire. This equation is valid for a general resistive circuit component with a constant cross sectional area as we could have easily forgone the assumption of a wire. However, note that the expression for R is invalid for a component with a non-uniform cross sectional area as the current density may vary at different points, both in magnitude and direction. Despite this, the potential across two surfaces is, in general, also proportional to the current flowing through them, in the form of Eq. (8.10), as the current density is proportional to the electric field. This pervasive relationship is also commonly known as Ohm's law. However, note that the resistance has to be derived via more fundamental methods, as we shall see later.

As an aside, the equivalent resistances of k resistors connected in series and parallel are respectively

$$R_{eq} = \sum_{i=1}^{k} R_i, \tag{8.11}$$

$$\frac{1}{R_{eq}} = \sum_{i=1}^{k} \frac{1}{R_i}. \tag{8.12}$$

These can be easily proven by applying Ohm's law and exploiting the facts that the currents through circuit components are the same in a series connection and that the potential differences across parallel branches are the same in a parallel connection.

Now, what about the case where the force on the charge carriers in a resistor with a uniform cross section is no longer purely electrostatic? If the cross section is small enough, the force per unit charge $\boldsymbol{f}$ is usually uniform throughout the wire. Then, the current density is also uniform over the wire by Eq. (8.7). Therefore,

$$I = JA = \sigma A f.$$

Since the line integral of $\boldsymbol{f}$ from one end of the resistor to the other is simply[6] $\int \boldsymbol{f} \cdot d\boldsymbol{s} = fl = \frac{l}{\sigma A} I$ where l is the length of the resistor,

$$\int \boldsymbol{f} \cdot d\boldsymbol{s} = IR \tag{8.13}$$

with $R = \frac{\rho l}{A}$. Most notably, if $\boldsymbol{f}$ is purely electrical and consists of electrostatic and non-electrostatic fields, the decrease in voltage (note that V is no longer the decrease in potential) is

$$V = \int \boldsymbol{E} \cdot d\boldsymbol{s} = IR \tag{8.14}$$

where $\boldsymbol{E}$ represents the total electric field. However, a caveat here is that the notion of an equivalent resistance for parallel connections no longer exists because the voltage across parallel branches need not be identical (since the line integral of a non-electrostatic field is generally path-dependent).

Finally, the instantaneous power P dissipated in a resistor R carrying a steady current I at any given time is the rate of work done by $\boldsymbol{f}$ on the charges traveling through the resistor at that time. This is because, since the circuit does not gain any energy overall, the rate of work done

---

[6]Note that the line integral is path-independent as $\boldsymbol{f}$ is uniform in the relevant region.

on these charges by f must be equal to the rate of heat generated by the resistor! Consider a cross section of the wire with an infinitesimal thickness ds. Suppose that a total charge dq crosses this cross section in time dt, traversing a displacement ds. The work done on this set of charges is

$$dW = dq f \cdot ds,$$

such that the power delivered to these charges is

$$dP = \frac{dq}{dt} f \cdot ds = I f \cdot ds,$$

which is also the contribution to the instantaneous power dissipated in the resistor. Summing the above over different cross sections, the total instantaneous power dissipated in the resistor is thus

$$P = \int I f \cdot ds = I \cdot \int f \cdot ds. \tag{8.15}$$

Equation (8.15) is sometimes incorrectly stated as the work done per unit charge in bringing a charge through the resistor (for which $\int f \cdot ds$ is mistakenly[7] used) multiplied by the rate of charge flowing through it (I). Such a definition suggests that we are evaluating the work done on a single group of charges as it moves through the resistor but we can see from the above process that we are actually summing the powers delivered to various groups of charges at different cross sections within the resistor at a particular time. After all, the term "instantaneous" in instantaneous power means that we do not have the privilege of waiting for a particular group of charges to pass through the entire resistor. Moving on, by Eq. (8.13),

$$P = I^2 R. \tag{8.16}$$

This can be expressed in three equivalent forms via Ohm's law in case where f is electrical in nature.

$$P = VI = \frac{V^2}{R} = I^2 R \tag{8.17}$$

where V is the voltage across the ends of a resistor. Moving on, let us return to an example about determining the resistance of a more general configuration.

Problem: Two thin, conducting spherical shells are concentric and have radii a and b with $a < b$. The gap between them is filled by a material of

---

[7]See Section 8.1.

conductivity σ. Then, the two shells are maintained at a constant potential difference with respect to each other such that a steady current flows between them. Determine the "resistance" between the two shells.

Let the charge stored in the inner shell be Q. This charge should be distributed evenly over the surface of the inner shell due to the isotropic nature of the set-up. Next, a crucial observation is that due to the steady current condition, the net electric flux cutting through the surfaces of an infinitesimal box element within the mediating material (similar to the case of the wire with a constant cross sectional area earlier) is zero. Therefore, there must be no net charge stored anywhere within the mediating material. Then, the electric field at a radius r $(a \leq r \leq b)$ from the common center is given by Gauss' law to be

$$E = \frac{Q}{4\pi\varepsilon_0 r^2}.$$

The current density is thus

$$J(r) = \frac{\sigma Q}{4\pi\varepsilon_0 r^2}.$$

Therefore, the current through each spherical shell, which should be identical for all spherical shells due to the steady current condition, is

$$I = J(r) \cdot 4\pi r^2 = \frac{\sigma Q}{\varepsilon_0}.$$

The potential drop from the inner shell to the outer shell is

$$V = \int_a^b E \, dr = \int_a^b \frac{Q}{4\pi\varepsilon_0 r^2} dr = \frac{Q}{4\pi\varepsilon_0 a} - \frac{Q}{4\pi\varepsilon b}.$$

Note the absence of a negative sign in front of the line integral as we are now determining the decrease in potential. Observe that V is related to I by

$$V = I \cdot \frac{1}{4\pi\sigma} \left(\frac{1}{a} - \frac{1}{b} \right).$$

The resistance of this set-up, which is evidently only dependent on the physical properties of the shells and mediating material and independent of the potential difference and current, is then

$$R = \frac{1}{4\pi\sigma} \left(\frac{1}{a} - \frac{1}{b} \right).$$

Alternatively, consider a spherical shell of conducting material with radius r and thickness dr. Since the electric field is normal at each point on the surface of this spherical shell, we can see the entire configuration as spherical shells with radii ranging from a to b, connected in series. When we claim that resistors are connected in "series", the qualification in the previous sentence must be checked for as we need to ensure that the surfaces of individual resistors (to be joined) are firstly equipotential, as the notion of resistance is based on the fact that the two terminals of the resistor in question are, foremost, equipotential. Having clarified this, because the flow of current is radially outwards and essentially one-dimensional,[8] we can apply $R = \frac{\rho l}{A}$ with the "area" of a spherical shell, with radius r being $4\pi r^2$ and its thickness being dr. The resistance of a spherical shell is hence

$$dR = \frac{\rho l}{A} = \frac{dr}{\sigma 4\pi r^2}.$$

The total "resistance" between the two shells is then $\int dR$ by the rule for computing the equivalent resistance of resistors in series.

$$R = \int_a^b \frac{dr}{\sigma 4\pi r^2} = \frac{1}{4\pi\sigma}\left(\frac{1}{a} - \frac{1}{b}\right).$$

This is similar to what we have discussed in the chapter on heat conduction.

8.3 Electromotive Force

A force is required to drive charges around a closed conducting path and to sustain a constant average velocity due to the frequent collisions of charges with the molecules and ions of a conductor. However, the force on the charges cannot be purely electrostatic throughout the entire loop. This is because an electrostatic field will be directed from a point of higher potential to a point of lower potential. There needs to be a non-electrostatic force to bring a positive charge from a lower potential to a higher potential and vice-versa for a negative charge. This force can either be due to a non-conservative electric field or a non-electrical force altogether. Therefore, a component is needed to maintain a voltage while circulating the charges. Such a component is known as an electromotive force source or emf source for short.

The electromotive force (emf) in a loop is defined as the closed loop integral of the total force per unit charge along the entire loop at a certain

---

[8]More rigorously, the equation of concern is $\boldsymbol{J} = \sigma\boldsymbol{E} = -\sigma\nabla V$ where V is the potential. For a strictly one-dimensional flow in the case of a uniform wire along the z-direction, $J = -\sigma\frac{dV}{dz}$. In this case, the set-up is isotropic such that $\nabla V = \frac{dV}{dr}\hat{r}$ and $J = -\sigma\frac{dV}{dr}$. It is thus evident that these two equations have analogous solutions if the boundary conditions are analogous (which is indeed true as we are imposing the potentials at the extreme z and r values in the two set-ups).

instance. Letting the instantaneous force per unit charge due to an emf source be $\boldsymbol{f}$ along an infinitesimal segment $d\boldsymbol{s}$ of the loop of concern, the emf in a loop is

$$\varepsilon = \oint \boldsymbol{f} \cdot d\boldsymbol{s}. \tag{8.18}$$

Note that the integral is evaluated over the relevant loop at a specific instance where it is paused — the motion of the loop is ignored. For a loop with multiple emf sources, ε on the left-hand side should be replaced with the sum of the individual emfs of the sources but we shall only consider a single emf source for now. In light of the previous discussion, the force on a charge along a closed path comprises both electrostatic and non-electrostatic forces, which include non-conservative electric fields and other forms of interactions. Then, $\boldsymbol{f}$ can be divided into the electrostatic and non-electrostatic parts.

$$\varepsilon = \oint \boldsymbol{E}_{elec} \cdot d\boldsymbol{s} + \oint \boldsymbol{f}_{non} \cdot d\boldsymbol{s}$$

where $\boldsymbol{E}_{elec}$ is the electrostatic field. Since the closed loop integral of an electrostatic field is zero,

$$\varepsilon = \oint \boldsymbol{f}_{non} \cdot d\boldsymbol{s}. \tag{8.19}$$

Usually if the emf source has distinct terminals, the non-electrostatic forces, such as chemical forces, are confined within the region of the source while the electrostatic forces exist everywhere. Therefore, the integral of the non-electrostatic force can be performed solely inside the source.

$$\varepsilon = \int_{A}^{B} \boldsymbol{f}_{non} \cdot d\boldsymbol{s} \tag{8.20}$$

where A and B refer to the terminals of the source, if they exist. Now, in an ideal emf source, the net force per unit charge in its interior is zero so that it is just able to bring charges across itself without any waste of additional resources. Since the total force inside the source comprises both electrostatic and non-electrostatic forces,

$$\boldsymbol{f}_{source} \;=\; \boldsymbol{E}_{elec} + \boldsymbol{f}_{non} = 0$$

$$\Longrightarrow \boldsymbol{f}_{non} = -\boldsymbol{E}_{elec},$$

where $\boldsymbol{E}_{elec}$ is the electrostatic electric field inside the source. The emf in the direction from terminal A to B of the source can then be written as

$$\varepsilon = -\int_{A}^{B} \boldsymbol{E}_{elec} \cdot d\boldsymbol{s} = V_B - V_A. \tag{8.21}$$

Thus, we see that a bounded emf source produces a potential difference (not voltage) across its ends in accordance with its emf. This relationship, coupled with Ohm's law, enables us to compute the current flowing in a circuit consisting of resistors and emf sources via the fact that the sum of potential differences V_p along a loop is zero.

$$\sum V_p = 0. \tag{8.22}$$

The potential difference across an emf source with terminals is simply its emf while the potential drop across the ends of a resistor carrying current I is IR in the direction of the current. Finally, the instantaneous power delivered, P, by an emf source with terminals, transporting a steady current I, can be derived in a similar manner as Eq. (8.15) to be

$$P = \varepsilon I. \tag{8.23}$$

Now, what about the case where certain emf sources are unbounded? A good example would be a conducting circuit moving through a magnetic field. Evidently, it is no longer the case that the net force on the charges is zero throughout the loop (recall that a net force is required to sustain a current in most cases). Then, we cannot generally determine the potential difference across each circuit component. We must return to the more fundamental definition of the emf in Eq. (8.18). For multiple emf sources in a loop,

$$\sum \varepsilon = \oint \boldsymbol{f} \cdot d\boldsymbol{s}. \tag{8.24}$$

The left-hand side includes all emfs (both with terminals and without terminals). For the right-hand side, the line integral of the force per unit charge through a resistor carrying current I is IR in the direction of the current while the force per unit charge through an ideal emf source with terminals is zero. The current through resistors can then be determined by specifying all ε and R.

In the case where $\boldsymbol{f}$ is purely electrical in nature,

$$\sum \varepsilon = \oint \boldsymbol{E} \cdot d\boldsymbol{s}$$

where $\boldsymbol{E}$ is the total electric field (both electrostatic and non-electrostatic). Shifting the line integral to the left and using the definition of voltage being the negative line integral of the electric field,

$$\sum \varepsilon + \sum V = 0 \tag{8.25}$$

where V is the voltage across each component and the sum is evaluated over the entire loop. Note that the voltage across an ideal emf source with terminals, which produces a non-electrostatic field, is zero due to the lack of a net force on charges within itself and hence the absence of a net electric field. The voltage drop across a resistor carrying current I is IR in the direction of the current, by Eq. (8.10), given that the non-electrostatic field is uniform over the resistor.

8.4 Motional Emf

A conductor moving in a magnetic field can be an electromotive source — such emfs are known as motional emfs. Consider a metal rod moving with a velocity $\boldsymbol{v}$ in a magnetic field, $\boldsymbol{B}$, that is not time-varying. However, the magnetic field may vary across different points in space.

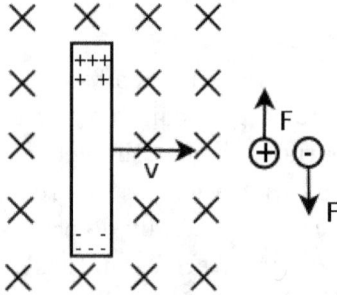

Figure 8.6:   Rough illustration of the deflected charges

Charges of different signs experience magnetic forces in opposite directions and hence diverge. In the particular case of Fig. 8.6, the positive charges will be deflected upwards while the negative charges will be deflected downwards. Assuming that the system eventually attains an equilibrium state,[9] this process will ensue until the electric field caused by the redistribution of charges produces a conservative Coulomb force that equalizes the magnetic force. In general, the velocity of charges in a conductor arises from their motion along the conductor and the motion of the conductor itself. In this situation, there should no longer be any movement of charges in the direction of the rod at static equilibrium. Therefore, the magnetic force on the charges will solely be due to the horizontal velocity $\boldsymbol{v}$ imposed by the rod. The emf due to the two ends of the conductor is consequently

$$\varepsilon = \int_A^B \boldsymbol{f}_{non} \cdot d\boldsymbol{s} = \int_A^B \boldsymbol{v} \times \boldsymbol{B} \cdot d\boldsymbol{s}$$

---

[9]There can only be a static equilibrium state and not a dynamic one as there is no cyclic path for the charges to circulate to produce steady currents.

as the non-electrostatic force in this case is the magnetic force. A and B represent the terminals of the rod. Given the exact function for $\boldsymbol{B}$, we can evaluate the emf. Note that the integral is performed over the required path inside the rod at a particular instance in time — that is, a snapshot of the system is taken. Moving on, we can say much more about this set-up. For example, an interesting question to ask would be whether this motional emf depends on the path of integration between the two terminals. Applying the condition for static equilibrium, we balance the forces on the charges in the rod.

$$q(\boldsymbol{E} + \boldsymbol{v} \times \boldsymbol{B}) = 0$$

$$\boldsymbol{E} = -\boldsymbol{v} \times \boldsymbol{B} \tag{8.26}$$

where $\boldsymbol{E}$ is the electrostatic field stemming from the redistribution of charges. This equation is valid for all points on the conductor and for conductors of arbitrary shapes, given that a static equilibrium state exists. We can verify that the potential difference between the ends of the conductor is equal to the emf.

$$\Delta V = - \int_A^B \boldsymbol{E} \cdot d\boldsymbol{s} = \int_A^B \boldsymbol{v} \times \boldsymbol{B} \cdot d\boldsymbol{s} = \varepsilon.$$

Since $\boldsymbol{E}$ is conservative, the potential difference is independent of the path taken such that the motional emf follows suit! The moral of the story is that we have not analyzed any new laws of physics — the above result is merely derived from the Lorentz force law and the condition for static equilibrium. Besides the path independence of the emf, another defining feature of a motional emf source is that it can possess terminals, as its carrier is a conductor which is not necessarily cyclic.

Problem: A conducting disk of radius R rotates about a vertical axis through its center (defined as the positive z-axis) with an angular velocity ω while placed in a uniform magnetic field $\boldsymbol{B}$ in the positive z-direction. Find the emf between the center and the edge of the disk. Assuming that a static equilibrium state exists (relative to the disk), find the potential difference between these two points and show that the line integral of the electric field between these two points under such an assumption is indeed path-independent. For the second part, neglect the centripetal force required for the charges to rotate as their masses are negligible.

   Set the origin at the center of the disk and define $\boldsymbol{r}$ as the position vector of the current infinitesimal displacement $d\boldsymbol{r}$ along the path in the conductor that we are integrating over. The velocity of the conducting element at this position is

$$v = \omega \times r$$

where ω is the angular velocity of the disk (pointing along the axis). Therefore,

$$\varepsilon = \int_0^R v \times B \cdot dr$$

$$= \int_0^R (\omega \times r) \times B \cdot dr$$

$$= -\int_0^R B \times (\omega \times r) \cdot dr$$

$$= \int_0^R (\omega \cdot B) r \cdot dr - \int_0^R (r \cdot B) \omega \cdot dr$$

by the BAC-CAB rule. Since the plane of the conductor is perpendicular to its axis, $r \cdot B = 0$ — only the first term survives. Next, observe that

$$r \cdot dr = \frac{1}{2} d(r \cdot r) = \frac{1}{2} d(r^2) = r dr.$$

Since ω is parallel to B, $\omega \cdot B = \omega B$. Then,

$$\varepsilon = \int_0^R \omega B r dr = \frac{\omega B R^2}{2}$$

which is independent of the path of integration taken inside the disk. Assuming that a static situation exists, the electric field at an arbitrary point within the conductor is

$$E = -v \times B.$$

Therefore, the potential difference between the center and the rim is

$$\Delta V = -\int_0^R E \cdot dr$$

$$= \int_0^R v \times B \cdot dr$$

$$= \frac{\omega B R^2}{2},$$

which is path-independent as the line integral of $v \times B$ is path-independent.

Now, let us not get carried away by the assumption that a static equilibrium always exists. Consider the set-up in Fig. 8.7, which blatantly violates such a claim.

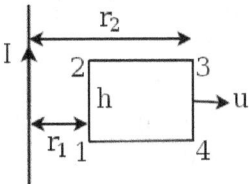

Figure 8.7: Loop in magnetic field

A rectangular conducting loop of height h travels perpendicularly to the magnetic field produced by an infinite wire at a constant horizontal speed u. We know that the magnetic field at a radial distance r due to a long wire carrying a steady current I is $\frac{\mu_0 I}{2\pi r}$. The only locations at which part of the magnetic force is directed along the loop are the vertical edges — meaning that only the vertical sides contribute to the emf. The contribution to the clockwise emf in the loop along the left vertical edge is

$$\varepsilon_l = \int_1^2 \boldsymbol{v} \times \boldsymbol{B} \cdot d\boldsymbol{s} = \int_1^2 uB(r_1)ds = \frac{\mu_0 I u h}{2\pi r_1}.$$

It is paramount to understand that $\boldsymbol{v}$ refers to the net velocity of charges and possibly has a vertical component (which constitutes a current). However, the magnetic force due to this vertical component is in the horizontal direction and thus does not contribute to the line integral — enabling us to write the second equality. Similarly, the contribution to the clockwise emf by the right vertical edge is

$$\varepsilon_r = \int_3^4 \boldsymbol{v} \times \boldsymbol{B} \cdot d\boldsymbol{s} = -\int_3^4 uB(r_2)ds = -\frac{\mu_0 I u h}{2\pi r_2}$$

where the negative sign stems from the fact that we are integrating along the opposite direction now (from 3 to 4 instead of from 4 to 3) such that the component of $\boldsymbol{v} \times \boldsymbol{B}$ along the loop now opposes the direction of integration. The total clockwise emf in this loop is then

$$\varepsilon = \varepsilon_l + \varepsilon_r = \frac{\mu_0 I u h}{2\pi} \left(\frac{1}{r_1} - \frac{1}{r_2} \right)$$

which is non-zero. Now, we shall prove by contradiction that a static situation does not exist. If the electrostatic field produced by the charges indeed balanced the magnetic force at all locations,

$$\boldsymbol{E} = -\boldsymbol{v} \times \boldsymbol{B}.$$

Then, the clockwise closed loop integral of $\boldsymbol{E}$ is simply $-\varepsilon$ which is non-zero. However, we know that the closed loop integral of an electrostatic field must be zero — leading to a contradiction! In such a situation, a dynamic equilibrium exists. You might think that this emf ε circulates infinite current along the conducting loop but we have neglected another effect of current loops. A changing current in a cyclic conductor induces a time-varying magnetic field which then generates an induced emf through itself (this is another form of emf that shall be explored later)! This phenomenon is known as self-inductance. In fact, the correct condition in such cases of perfectly conducting loops is that no net emf is generated within the loop. What occurs in this set-up is that a finite amount of current circulates within the loop such that the total magnetic flux through the loop is constant! We shall delve further into this in the section on perfect conductors.

Besides the precursors to the later sections, a point to be made here is that the obvious drawback of such a dynamic equilibrium is that we can no longer determine the electrostatic field at each point in the loop via Eq. (8.26). If we let $\boldsymbol{E}$ denote the net electric field (comprising both electrostatic and non-electrostatic components) however, Eq. (8.26) is still valid in perfect conductors.

Flux Rule for Motional Emf in a Loop

Speaking of loops, there is an elegant rule for the motional emf produced in a conducting loop moving in a time-independent magnetic field. Recall that the magnetic flux Φ_B through a surface S is the surface integral of the magnetic field over S.

$$\Phi_B = \iint_S \boldsymbol{B} \cdot d\boldsymbol{A}. \tag{8.27}$$

Now, there is an important property of the magnetic flux. Because the magnetic flux over a closed surface is guaranteed to be zero, the magnetic flux over an open surface is only dependent on the contour C bounding the surface and is the same over all surfaces that span the same contour C (a similar property was established for the current flux in Section 8.2).

We claim that the motional emf induced in a conducting loop moving through a magnetic field is simply the negative rate of change of magnetic flux through the loop.

$$\varepsilon = -\frac{d\Phi_B}{dt}.$$

Now, how do we specify the direction of ε and the area vector in evaluating the magnetic flux? The answer is the right-hand-grip rule as always. Choose a particular direction (clockwise or anti-clockwise) to evaluate the motional emf for. Then, curl your fingers in this direction and straighten your thumb — your thumb will point in the positive direction of the area vectors to use for the magnetic flux integral. Another more intuitive way of finding the direction of ε is Lenz's law, but we shall postpone its discussion till later as it is usually associated with Faraday's law. Moving on, for a coil with N densely wound turns,

$$\varepsilon = -N\frac{d\Phi_B}{dt}. \tag{8.28}$$

The quantity $N\Phi_B$ is known as the magnetic flux linkage through a coil. Therefore, the motional emf generated in a coil is the negative rate of change of magnetic flux linkage.

Proof: Let u be the instantaneous velocity of a segment on the conducting loop of interest at time t. Consider the loop at times t and $t + dt$, indicated by contours C and C'. Define the magnetic fluxes at these two instances to be Φ_C and $\Phi_{C'}$ respectively (recall that the fluxes are only dependent on the boundaries). In the time interval dt, the loop sweeps an infinitesimal strip that is shaded in Fig. 8.8.

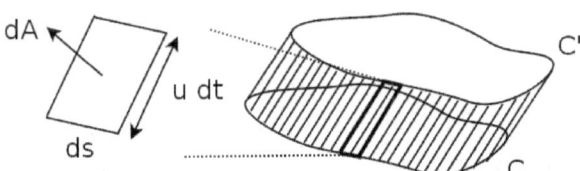

Figure 8.8: Flux at two instances

The magnetic flux through this surface, denoted as $d\Phi_{is}$, can be computed via

$$d\Phi_{is} = (d\Phi_{is} + \Phi_{C'}) - \Phi_{C'}.$$

When evaluating the magnetic fluxes in this section, we will choose the anti-clockwise direction as our reference direction and apply the right-hand-grip rule to determine the positive direction of the infinitesimal area vectors. Observe that $d\Phi_{is} + \Phi_{C'}$ is simply the magnetic flux through a surface bounded by contour C. Replacing this expression with Φ_C,

$$d\Phi_{is} = \Phi_C - \Phi_{C'} = -d\Phi,$$

where $d\Phi$ is the change in magnetic flux through the physical conducting loop. Then, we can evaluate $d\Phi_{is}$ to find $d\Phi$. The former is the magnetic flux through the infinitesimal strip

$$d\Phi_{is} = \iint_{strip} \boldsymbol{B} \cdot d\boldsymbol{A}.$$

The infinitesimal area $d\boldsymbol{A}$ swept by an infinitesimal segment $d\boldsymbol{s}$ on the loop can be expressed as

$$d\boldsymbol{A} = -\boldsymbol{u} \times d\boldsymbol{s} dt$$

where $\boldsymbol{u}$ is the instantaneous velocity of the particular loop segment. We include a negative sign to ensure that the infinitesimal area vectors point outwards, in accordance with the anti-clockwise direction. We can then integrate over $d\boldsymbol{s}$ (i.e. the entire loop) to determine the surface integral. Then,

$$\frac{d\Phi}{dt} = -\frac{d\Phi_{is}}{dt} = \oint \boldsymbol{B} \cdot (\boldsymbol{u} \times d\boldsymbol{s}).$$

Now, let the net velocity of the charges in a particular segment $d\boldsymbol{s}$ of the loop be $\boldsymbol{v}$ and the velocity of the charges along the loop at that segment be $\boldsymbol{w}$. Then,

$$\boldsymbol{v} = \boldsymbol{u} + \boldsymbol{w},$$

as the net velocity comes from both the velocity of the conductor and the velocity of the charges flowing along the conductor. Since $\boldsymbol{w}$ is parallel to $d\boldsymbol{s}$ such that $\boldsymbol{w} \times d\boldsymbol{s} = \boldsymbol{0}$,

$$\frac{d\Phi}{dt} = \oint \boldsymbol{B} \cdot (\boldsymbol{v} \times d\boldsymbol{s}).$$

We can rewrite the above via the scalar product rule $\boldsymbol{a} \cdot (\boldsymbol{b} \times \boldsymbol{c}) = \boldsymbol{c} \cdot (\boldsymbol{a} \times \boldsymbol{b})$ as

$$\frac{d\Phi}{dt} = -\oint (\boldsymbol{v} \times \boldsymbol{B}) \cdot d\boldsymbol{s}.$$

The expression in brackets is simply the magnetic force per unit charge on the charges in the loop. The closed loop integral of that gives the motional emf. Thus,

$$\frac{d\Phi}{dt} = -\varepsilon$$

$$\varepsilon = -\frac{d\Phi}{dt}.$$

The motional emf for a coil with N turns follows accordingly. A paramount point to understand is that we have not introduced any new law at all. Equation (8.28) is only a neat way of expressing the line integral of the magnetic force per unit charge along a loop, which gives the motional emf in such set-ups. Now, the potency of this expression arises from the fact that we have not assumed anything about the motion of the loop or the geometry of the loop. Equation (8.28) is valid when the conducting loop translates, rotates or even expands (as we did not impose the rigid body condition on u). It is also valid regardless of whether the loop is planar or non-planar.

Emf due to a Rotating Planar Loop

A classic application of the flux rule entails a rotating planar coil of area A in a region uniform magnetic field B.

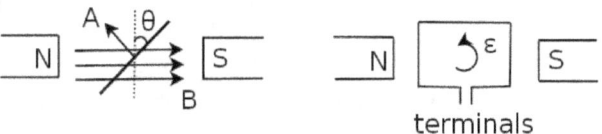

Figure 8.9:   Front and top view of rotating loop

Referring to the left diagram of Fig. 8.9, when the loop is tilted at an angle θ with respect to the vertical, the magnetic flux crossing the loop is

$$\Phi_B = -BA\cos\theta.$$

Therefore, the anti-clockwise motional emf (this is anti-clockwise because we chose the area vector to point to the left in the left figure) is

$$\varepsilon = -N\frac{d\Phi_B}{dt} = -NBA\sin\theta\dot\theta$$

if the coil has N turns and where $\dot\theta$ is its instantaneous angular velocity. At this point, we notice an interesting application of this set-up. If the loop is not completely closed and is instead connected to an external circuit through wires that form a small gap in the loop, the rotating loop can function as an alternating-current generator, if maintained at a constant angular velocity!

Now, let us assume that this coil is a perfect conductor and connect it to an external resistor R via long, ideal wires. The clockwise current that flows in this circuit is

$$I = \frac{\varepsilon}{R} = \frac{NBA\sin\theta\dot\theta}{R}.$$

The power dissipated in the resistor is

$$P = I^2 R = \frac{N^2 B^2 A^2 \sin^2\theta\dot\theta^2}{R}.$$

Now, where does this power come from? Fret not, we have not violated the conservation of energy. Recall that a current loop forms a magnetic dipole which experiences a torque (but no net force) due to an external magnetic field. The torque experienced by the loop is given by

$$\boldsymbol\tau = \boldsymbol\mu \times \boldsymbol B$$

where $\boldsymbol\mu = NI\boldsymbol A$ is the magnetic dipole moment and $\boldsymbol A$ is the area vector of a single turn, whose direction is determined by applying the right-hand-grip rule to the current loop. Since θ is the angle between $\boldsymbol A$ and $\boldsymbol B$, the magnitude of the magnetic torque on the rotating loop is

$$\tau = -\frac{N^2 B^2 A^2 \sin^2\theta\dot\theta}{R}$$

whose direction opposes the current rotation of the loop. Since the instantaneous angular velocity of the loop is $\dot\theta$, the power delivered by this torque is

$$P = \tau\dot\theta = -\frac{N^2 B^2 A^2 \sin^2\theta\dot\theta^2}{R},$$

which is commensurate with the instantaneous power dissipated in the resistor. Therefore, if no external mechanical force is exerted to sustain the motion of the loop, its rotation slows down while the resistor heats up, in accordance with the conservation of energy. On the other hand, this implies that if we deliver $\frac{N^2 B^2 A^2 \sin^2\theta\dot\theta^2}{R}$ amount of external mechanical power via an external torque to maintain the angular velocity of the loop at a constant $\dot\theta$, we are converting mechanical work to electrical energy (followed by less useful heat).

Finally, you may notice a more fundamental issue here. How is it that the magnetic force seemingly does work through the torque produced? The answer is that we have divided the magnetic force into two components and calculated their individual works. The net velocity of the charges in the coil has two causes — due to the rotation of the loop and due to the component

along the loop (which constitutes current). The magnetic force is perpendicular to this net velocity and does no work. However, the rotational velocity of
the loop leads to a component of magnetic force along the loop — generating
the emf and producing a power $\frac{N^2 B^2 A^2 \sin^2 \theta \dot\theta^2}{R}$ that is dissipated in the resistor. Meanwhile, the latter part of the velocity leads to a force perpendicular
to the segments and hence a torque which delivers the power $-\frac{N^2 B^2 A^2 \sin^2 \theta \dot\theta^2}{R}$
calculated above. These components each perform work individually but the
sum of their works must obviously add to zero as they are quintessentially
manifestations of the magnetic force.

8.5 Induced Emf

We have deduced that a moving conductor in a magnetic field that is not
time-varying at each point in space can be a potential emf source. For example, motional emf will be produced when a coil is moved closer towards a
stationary magnet. However, we can also view this situation from the perspective of the coil. In the coil's frame, it is stationary while the magnet
moves towards it — the notion of a motional emf is no longer a suitable
explanation for the emf generated in the coil.[10] However, observe that this
situation is also slightly different from the coil's frame in another way — the
magnetic field is no longer steady at each point in space as it "translates"
through space, with the magnet. This hints at the fact that the induced emf
has something to do with the time-varying magnetic field.

Proceeding with this investigation, it is natural to ask if an emf is induced
in a conducting loop that is either moving or stationary in a time-varying

---

[10] Actually, special relativity dictates that a magnetic field in one frame transforms into
an electric field in another. In the case of a conducting rod moving in a uniform but time-
independent magnetic field in the lab frame, an external, uniform electric field exists in
the frame of the conductor. This uniform electric field (which is conservative) then leads
to the redistribution of charges (seen previously) and produces the emf. In fact, for non-
relativistic speeds of the conductor in the lab frame, the external electric field at a point
in its frame is $E_{ext} = v \times B$, where v and B are the velocity of the conductor and the
magnetic field at the same point in the lab frame — a quantity that makes sense as we
know that electric field produced by the redistribution of charges within the conductor
is $E = -v \times B$ such that the net electric field within the conductor is $E + E_{ext} = 0$
in order for electrostatic equilibrium to be attained. This is the correct explanation for
the motional emf as seen from the rod's frame. However, in the case of a conducting loop
moving in a non-uniform but time-independent magnetic field in the lab frame, a solely
conservative electric field in the loop's frame is not a plausible explanation as it would
imply zero emf in the loop when a non-zero emf is observed. This second situation is of
concern here as we seek to delve further into the non-conservative electric field produced.

magnetic field. Faraday observed through his experiments that the two situations above indeed produced currents which deflected his galvanometers. This led to his ingenious hypothesis that a changing magnetic field induces a non-conservative electric field in space. Faraday's law states that the emf induced in an imaginary contour C in space is given by

$$\varepsilon = -\frac{d\Phi_B}{dt},$$

where Φ_B refers to the magnetic flux through a surface, S, that spans the contour C. There may be myriad such surfaces but the magnetic flux through them will be the same in light of the discussion previously. Such emfs engendered by a time-changing magnetic field are known as induced emfs.

Now, the nature of the induced emf is a non-conservative electric field $\boldsymbol{E_{non}}$. That is, a changing magnetic field produces a non-conservative electric field. Therefore, there are really two forms of electric fields — one produced by electrostatic charges and one produced by a changing magnetic field. In light of this phenomenon, the emf induced can be expressed as

$$\varepsilon = \oint \boldsymbol{E_{non}} \cdot d\boldsymbol{s} = \oint \boldsymbol{E} \cdot d\boldsymbol{s}$$

where $\boldsymbol{E}$ is the total electric field along the contour C, since the closed loop integral of a conservative electric field is zero. A pivotal difference between motional and induced emfs is the fact that induced emfs are produced independently of the presence of a conductor. The contour C need not be contained within a conductor, contrary to the case of applying the flux rule to motional emfs. This is because the non-conservative electric field is simply generated in space due to a changing magnetic field and is not associated with the movement of charges in a conductor (which leads to a magnetic force).

In conclusion, for a stationary coil with N densely wound turns, Faraday's law states that the emf induced in the coil is proportional to the rate of change of magnetic flux linkage.

$$\varepsilon = -N\frac{d\Phi_B}{dt}. \tag{8.29}$$

However, note that the line integral of the electric field is unquestionably not reliant on N.

$$\oint \boldsymbol{E} \cdot d\boldsymbol{s} = -\frac{d\Phi_B}{dt} \tag{8.30}$$

as the emf increases with the number of turns only because the turns stack up along the non-conservative electric field. The non-conservative electric field is most definitely not amplified by more turns as it must be independent of the coil.

Similarly, there is now an ambiguity regarding the direction of the electromotive force. More specifically, the direction of the infinitesimal area vector $d\mathbf{A}$ in calculating the magnetic flux is vague for a given reference direction for the emf. Technically, its direction can again be specified using the right-hand-grip rule. That said, a much more intuitive alternative approach can be utilized. It is observed that nature detests the change in magnetic flux. Thus, an emf will tend to be induced in a direction such that it generates a current (if a conductor were to be placed there) and thus a magnetic field that opposes the initial change in flux. This is known as Lenz's law. Therefore, one can determine the direction of an induced emf by leveraging on this empirical fact.

Now, there is a haunting similarity between Faraday's law and the flux rule for motional emfs (we shall reserve the term "Faraday's law" for induced emfs) which are fundamentally different. The former is a new physical law which ascribes a non-conservative electric field to every point in space with a time-varying magnetic field while the latter is simply a convenient way of expressing the motional emf generated in a conducting loop moving in a magnetic field. However, since the emfs produced by these disparate effects have similar forms, we can combine Eqs. (8.28) and (8.29) into a "universal flux rule". The total emf generated in a conducting loop with N turns moving through a possibly time-varying magnetic field can be computed as follows. Let the contour delineating the conducting loop be C and consider the time interval between t and $t + dt$ where the conducting loop moves from $C(t)$ to $C(t + dt)$. The emf induced in the conducting loop at time t is the sum of the induced and motional emfs

$$\varepsilon = \varepsilon_{ind} + \varepsilon_{mot}.$$

By Faraday's law, the induced emf along the location of the loop at t is

$$\varepsilon_{ind} = -N \lim_{\Delta t_1 \to 0} \frac{\Phi_{B,C(t)}(t + \Delta t_1) - \Phi_{B,C(t)}(t)}{\Delta t_1}$$

where $\Phi_{B,C(t)}(t)$ and $\Phi_{B,C(t)}(t + \Delta t_1)$ refer to the magnetic fluxes through $C(t)$ (the location of the loop at time t) at times t and $t + \Delta t_1$. Do not brood too much over the sign of Δt_1 as it could be positive or negative, as long as $\Phi_{B,C(t)}(t + \Delta t_1)$ lies in an interval where $\Phi_{B,C(t)}$ is differentiable. Furthermore, do not worry too much about the physical meaning of $\Phi_{B,C(t)}(t + \Delta t_1)$. Even though the physical loop may not be at $C(t + \Delta t_1)$ at time t, we are pre-empting its movement in a certain sense, to compute the induced emf through it at time t. Next, the flux rule for motional emf states that

$$\varepsilon_{mot} = -N \lim_{\Delta t_2 \to 0} \frac{\Phi_{B,C(t+\Delta t_2)}(t) - \Phi_{B,C(t)}(t)}{\Delta t_2}.$$

Note that both fluxes are computed at time t, as you can tell from the fundamental expression for the motional emf ($\varepsilon = \oint (\boldsymbol{v} \times \boldsymbol{B}) \cdot d\boldsymbol{s}$) that it only relies on the magnetic field at the current instance where we pause the loop to evaluate the line integral (it does not know that the magnetic field may change at the next instance). Correspondingly, the total emf in the conducting loop at time t is

$$\varepsilon = -N \lim_{\Delta t_1 \to 0} \frac{\Phi_{B,C(t)}(t + \Delta t_1) - \Phi_{B,C(t)}(t)}{\Delta t_1}$$

$$-N \lim_{\Delta t_2 \to 0} \frac{\Phi_{B,C(t+\Delta t_2)}(t) - \Phi_{B,C(t)}(t)}{\Delta t_2}.$$

In particular, we can set $\Delta t_1 = -\Delta t_2 = -dt$ such that

$$\varepsilon = -N \lim_{dt \to 0} \frac{\Phi_{B,C(t)}(t) - \Phi_{B,C(t)}(t - dt) + \Phi_{B,C(t+dt)}(t) - \Phi_{B,C(t)}(t)}{dt}$$

$$\varepsilon = -N \lim_{dt \to 0} \frac{\Phi_{B,C(t+dt)}(t) - \Phi_{B,C(t)}(t - dt)}{dt} = -N \frac{d\Phi_{B,cond}}{dt},$$

where $\frac{d\Phi_{B,cond}}{dt}$ is the rate of change of magnetic flux cutting across the physical conducting loop. This reveals a very important point in applying the universal flux rule — the loop that we choose to compute the magnetic flux over must be "attached" to the physical conductor such that they travel together. This makes sense as it signifies that an isolated, current-carrying loop that is moving cannot generate an emf through itself. Ultimately, we conclude that the total emf generated in a conducting loop is the rate of change of magnetic flux linkage through it, even in regions with time-varying magnetic fields.

Analogy between Non-Conservative Electric Field and Magnetostatic Field

It turns out empirically that Gauss' law is valid (where the electric field now comprises both conservative and non-conservative fields) even in the presence of a non-conservative electric field. This implies that the surface integral of a non-conservative electric field is always zero over any arbitrary surface.

$$\oiint E_{non} \cdot dA = 0.$$

Now, there is an interesting parallel between the magnetic field due to a steady current and the induced electric field. The former obeys

$$\oint_C B \cdot ds = \mu_0 \oiint_S J \cdot dA$$

$$\oiint B \cdot dA = 0.$$

The first equation is Ampere's law, after replacing the current I on the right hand side with the surface integral of the current density (where S is a surface that spans contour C). On the other hand, the induced electric field obeys

$$\oint_C E_{non} \cdot ds = -\oiint_S \frac{\partial B}{\partial t} \cdot dA$$

$$\oiint E_{non} \cdot dA = 0.$$

It turns out that the closed loop and surface integrals of a field are sufficient in uniquely defining a vector field, provided that it vanishes at infinity. Therefore, since the above laws are similar, E_{non} and B have analogous solutions, with the rate of change of magnetic field $\frac{dB}{dt}$ playing the role of a "current density" (more technically, $-\mu_0 J$). Then, we can transfer all of our tools from magnetostatics in determining the induced electric field.

Problem: A long solenoid of radius R has its axis aligned with the z-axis. A time-varying current runs through it such that a varying magnetic field, given by the expression $B = \alpha t$ in the positive z-direction (out of the page in Fig. 8.10), is generated within itself. Determine the non-conservative electric field everywhere.

The induced electric field should be azimuthal as the rate of change of magnetic field is parallel to the solenoid axis (the induced field should be similar to the magnetic field generated by a thick current carrying wire). As always, there are two regions to consider — the interior and the exterior of the solenoid. We begin with the latter. Draw a circular loop of radius $r \geq R$ outside of the solenoid, with its axis aligned with the axis of the solenoid. By Faraday's law, the closed loop integral of the non-conservative electric

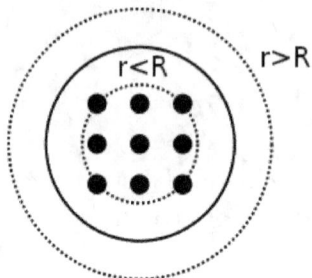

Figure 8.10:   Changing magnetic flux in long solenoid

field in the anti-clockwise direction along the loop is

$$E_{non} \cdot 2\pi r = -\frac{d\Phi_B}{dt} = -\frac{dB}{dt} \cdot \pi R^2 = -\alpha\pi R^2.$$

Therefore,

$$E_{non} = \frac{-\alpha R^2}{2r}$$

in the azimuthal direction. The negative sign indicates that it should be clockwise for positive α and anti-clockwise for negative α. This direction can also be swiftly determined from Lenz's law. Without the loss of generality, assuming that α is positive, the magnetic flux coming out of the page through the loop increases with time. Therefore, a clockwise emf tends to be induced to produce a clockwise current to oppose this increase in magnetic flux. The argument for negative α is also similar.

Next, for the interior region, we draw a similar circular loop of radius $r \leq R$ and apply Faraday's law. However, the loop this time does not enclose the entire cross section of the solenoid.

$$E_{non} \cdot 2\pi r = -\frac{dB}{dt} \cdot \pi r^2 = -\alpha\pi r^2$$
$$E_{non} = -\frac{\alpha r}{2}$$

in the anti-clockwise azimuthal direction.

8.6   Self-Inductance

Armed with the idea of induced electric fields, we can refine our analysis of current-carrying loops. The current through a coil, I, generates a magnetic flux, Φ_B, through its own turns. Thus, a change in the loop's current I will lead to a change in the magnetic flux linkage through itself — which in turn, engenders an induced emf to oppose this change in flux linkage. The emf induced is known as a back emf as it tends to oppose the change in the current I by Lenz's law. Since the magnetic flux, Φ_B, through the system is proportional to I, the total magnetic flux linkage $N\Phi_B$ through the coil is proportional to NI where N is the number of turns of the coil. Therefore, if we define the self-inductance L of the system as

$$L = \frac{N\Phi_B}{I}, \tag{8.31}$$

the back emf in the system can be expressed as

$$\varepsilon = -N\frac{d\Phi_B}{dt} = -N\frac{d\left(\frac{LI}{N}\right)}{dt} = -L\frac{dI}{dt} \tag{8.32}$$

assuming that L remains constant. The positive direction of ε is defined to be aligned with I. The self-inductance of a system is purely a geometric quantity that depends only on intrinsic factors such as the number of turns and the shape of the loop. It does not depend on the amount of current that the loop carries, as the linear relationship between Φ_B and I leads to the cancellation of I in Eq. (8.31).

Energy Stored in an Inductor

Since a varying current generates a back emf ε in a loop, work must be done against the back emf in increasing the current through it. Concretely, we need to deliver $-I\varepsilon$ amount of power when the system is carrying a current I to overcome the back emf ε which delivers $I\varepsilon$ amount of power. Thus, we can ascribe a potential energy stored in an inductor, from increasing its current from an initial state 0 to a final state I, as

$$U = \int -I\varepsilon\, dt = \int_0^I LI\, dI = \frac{1}{2}LI^2. \tag{8.33}$$

The reason behind the necessity of external work in producing a current should be apparent by now. Even though the initial and final configurations may lead to purely magnetic fields (which do no work), the process of changing a current induces a non-electrostatic field which can do work! This argument holds even when there are multiple coils and non-steady currents.

The above deposited energy is associated with the magnetic field of the coil and must be consistent with integrating the magnetic energy density $u_B = \frac{B^2}{2\mu_0}$ over all space. Such a component which stores energy in its magnetic field is known as an inductor.

Finally, as energy should only be a function of state, the energy stored in an inductor is independent of how the system evolved to carry current I and solely depends on the current configuration (pun intended) of the system.

Problem: A long solenoid of radius R has η turns per unit length. Determine its self-inductance per unit length l.

By Ampere's law, if we draw an Amperian loop that encloses all turns of the solenoid,

$$B = \eta\mu_0 I.$$

The magnetic flux through the solenoid is then

$$\Phi_B = B \cdot \pi R^2 = \eta\mu_0 I \pi R^2.$$

Thus, the self-inductance of the solenoid per unit length is

$$l = \frac{\eta\Phi_B}{I} = \eta^2\mu_0\pi R^2,$$

which, as expected, is independent of I and is only affected by the physical parameters of the solenoid.

8.7   Mutual Inductance

Now, consider a set-up comprising two current-carrying loop systems which generate magnetic fields. Let us call them systems 1 and 2. The current in system 2, I_2, produces magnetic field lines that cut through the loops in system 1. Thus, if there is a change in I_2, there will be a change in the magnetic flux of system 1 and thus an emf induced in system 1. Furthermore, the magnetic flux linkage through the N_1 loops in system 1 due to system 2, $N_1\Phi_{B12}$, is proportional to I_2. Thus, if we define the mutual inductance of system 1 due to a change in the current of system 2, M_{12}, as the positive ratio of the magnetic flux linkage in the N_1 loops of system 1 to I_2,

$$M_{12} = \left| \frac{N_1\Phi_{B12}}{I_2} \right|$$

$$\implies N_1\Phi_{B12} = \pm M_{12}I_2,$$

we can express the induced emf in system 1 due to a change in current in system 2 as

$$\varepsilon_1 = -N_1\frac{d\Phi_{B12}}{dt} = \mp M_{12}\frac{dI_2}{dt},$$

where ε_1 is positive in the direction of the proposed positive direction of the current I_1 flowing through system 1 (it doesn't matter if $I_1 = 0$ currently as we can still assign a positive direction for it). The ambiguity in sign stems from the fact that M_{12} is coerced to be a positive value and thus cannot reflect the sign of magnetic flux.

Next, the emf in system 2 due to a change in the current of system 1 is similarly

$$\varepsilon_2 = -N_2 \frac{d\Phi_{B21}}{dt} = \mp M_{21} \frac{dI_1}{dt},$$

where ε_2 is defined to be positive in the positive direction ascribed to I_2. The signs of $\mp$ in front of ε_1 and ε_2 must be identical (this is most obvious if we express the magnetic field of a magnetic dipole in terms of its magnetic dipole moment). Given the proposed positive directions of I_1 and I_2 (which determine the positive directions of the infinitesimal area vectors on the respective surfaces spanning these loops by the right-hand-grip rule for the computation of magnetic flux), we can determine the appropriate sign to use in ε_1 and ε_2.

The mutual inductance between two systems is, once again, a purely geometric quantity which depends on factors such as the physical parameters of the systems and their relative orientations.

Reciprocity Theorem

You might think that we need two calculations to determine M_{12} and M_{21}. However, the reciprocity theorem saves us a lot of effort as it states that these two mutual inductances are in fact equal.

$$M_{12} = M_{21} = M.$$

Thus, we will drop the subscripts and just refer to the mutual inductance between two systems as M, henceforth.

Proof: We can exploit the fact that the energy stored in a system of two inductors should only be a function of state (i.e. the final currents). Let the final currents in systems 1 and 2 be I_1 and I_2 respectively. We will determine the amounts of work required to bring an initial configuration with zero current for both systems to the final configuration through two different processes. In the first process, we begin by increasing the current in system 1 from 0 to I_1 while maintaining the current in system 2 at zero. Since the external power required to maintain a current I in the presence of an opposing emf ε is $-I\varepsilon$, it takes zero work to maintain a current at zero. The total amount of work required in this step is thus $\frac{1}{2}L_1 I_1^2$ where L_1 is the self-inductance of the first system. Next, we bring the current in system 2 from 0 to I_2 while maintaining I_1 current in system 1. Now, in addition to overcoming the back emf in system 2 (which requires $\frac{1}{2}L_2 I_2^2$ amount of energy), notice that changing the current in system 2 also induces an emf

$\varepsilon = \mp M_{12}\frac{dI_2}{dt}$ in system 1. Then, the total energy needed in maintaining I_1 current in system 1 is $\int -\varepsilon I_1 dt = \int_0^{I_2} \pm M_{12} I_1 dI_2 = \pm M_{12} I_1 I_2$. Thus, the overall external energy delivered to these two systems is

$$U = \frac{1}{2} L_1 I_1^2 + \frac{1}{2} L_2 I_2^2 \pm M_{12} I_1 I_2.$$

We can then reverse the order of incrementing the currents (i.e. raise the current in system 2 before raising that in system 1). The overall external work delivered this way would be

$$U = \frac{1}{2} L_1 I_1^2 + \frac{1}{2} L_2 I_2^2 \pm M_{21} I_1 I_2.$$

Since energy is only a function of state, these two energies must be equal — implying that

$$M_{12} = M_{21} = M,$$

where M is the unequivocal mutual inductance of the two systems. The total potential energy stored in these two systems is thus

$$U = \frac{1}{2} L_1 I_1^2 + \frac{1}{2} L_2 I_2^2 \pm M I_1 I_2. \tag{8.34}$$

Relationship between Mutual and Self-Inductances

Since $M_{12} = M_{21} = M$, observe that

$$M^2 = M_{12} M_{21} = \frac{N_1 \Phi_{B21}}{I_1} \cdot \frac{N_2 \Phi_{B12}}{I_2}.$$

If the magnetic fluxes through both systems, due to the magnetic field of system 1, are identical such that $\Phi_{B21} = \Phi_{B11}$ where Φ_{B11} is the magnetic flux through the first system due to its own magnetic field and ditto for the magnetic field of system 2 (i.e. $\Phi_{B12} = \Phi_{B22}$), the above can be rewritten as

$$M^2 = \frac{N_1 \Phi_{B11}}{I_1} \cdot \frac{N_2 \Phi_{B22}}{I_2} = L_1 L_2$$

$$M = \sqrt{L_1 L_2}.$$

Under such conditions, the two systems are known to be perfectly coupled. Usually, some of the flux produced by one system is not captured by other. Thus, the mutual inductance is usually written as

$$M = k\sqrt{L_1 L_2}, \tag{8.35}$$

where $0 \leq k \leq 1$ is known as the coupling constant which describes how well the fluxes produced by the coils are linked to each other. The fact that the mutual inductance must be smaller than the geometric mean of the self-inductances can be deduced from the following physical arguments. The energy of the two systems can be written as

$$U = \frac{1}{2}L_1 I_1^2 + \frac{1}{2}L_2 I_2^2 \pm M I_1 I_2 = \frac{1}{2}L_1\left(I_1 \pm \frac{M}{L_1}I_2\right)^2 + \frac{1}{2}\left(L_2 - \frac{M^2}{L_1}\right)I_2^2,$$

and is valid for any values of I_1 and I_2. Most notably, we can choose the particular value $I_2 = \mp\frac{L_1 I_1}{M}$ such that the potential energy is

$$U = \frac{1}{2}\left(L_2 - \frac{M^2}{L_1}\right)\left(\frac{L_1 I_1}{M}\right)^2.$$

Now, it is highly unlikely that the total potential energy of two carrying-current systems is negative as that would imply that it can act as a source of energy (it can deliver power and produce more current at the same time). More formally, the potential energy of two inductors is "stored" in the magnetic field in all space. Since the magnetic energy density must be non-negative, we expect $U \geq 0$ such that

$$M^2 \leq L_1 L_2$$

$$M \leq \sqrt{L_1 L_2}.$$

Problem: Two circular loops of radius r_1 and r_2, with $r_1 \gg r_2$, are centered at two points along the z-axis and are perpendicular to the z-axis. If the two loops are separated by a perpendicular distance z, determine their mutual inductance.

The small loop produces a relatively complicated magnetic field around it, causing the magnetic flux through the large loop to be more tedious to calculate. On the other hand, the magnetic field due to the large loop through the small loop is approximately uniform due to its small size. Therefore, we can calculate the mutual inductance of the small loop due to a change in current in the large loop as we have the prerogative of choosing the loop whose current we want to vary in calculating the mutual inductance, by virtue of the reciprocity theorem.

Assume that a current I_1 flows through the large loop of radius r_1. Now, the magnetic field due to a ring of radius R, carrying a current I, at a height z along the central axis of the ring was computed in the previous chapter (Section 7.2) as

$$B_z = \frac{\mu_0 I R^2}{2(R^2 + z^2)^{\frac{3}{2}}}.$$

Therefore, the magnetic field at the center of the small loop (which is along the z-axis) due to the current I_1 in the large loop is

$$B_z = \frac{\mu_0 I_1 r_1^2}{2(r_1^2 + z^2)^{\frac{3}{2}}}.$$

The magnetic flux through the small loop is simply the above multiplied by its area as the magnetic field throughout the entire region encased by it can be assumed to be that at the center.

$$\Phi_{B21} = B_z \cdot \pi r_2^2 = \frac{\mu_0 I_1 \pi r_1^2 r_2^2}{2(r_1^2 + z^2)^{\frac{3}{2}}},$$

$$M = \frac{\Phi_{B21}}{I_1} = \frac{\mu_0 \pi r_1^2 r_2^2}{2(r_1^2 + z^2)^{\frac{3}{2}}}.$$

8.8   Ampere–Maxwell Law

As highlighted in the chapter on magnetostatics, Ampere's law is only valid for steady currents and cannot be applied to non-steady situations. Consider the following instructive set-up illustrated in Fig. 8.11. A capacitor is connected to a battery and a current I flows through the circuit. We would like to determine the magnetic field along a circular loop around the wire at a point in space.

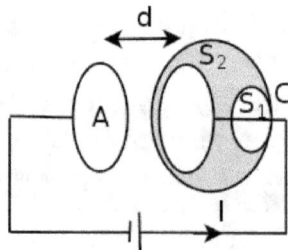

Figure 8.11:   Circuit with a capacitor (contour C is the boundary of the white circle)

Suppose we draw the above circular Amperian loop C, what is the enclosed current? If we choose the simplest surface — which is the planar surface demarcated by the loop, S_1 — to compute the enclosed current, $I_{enc} = I$. Applying Ampere's law along C and integrating the current density over this surface then yields

$$\oint_C \boldsymbol{B} \cdot d\boldsymbol{s} = \mu_0 I.$$

Now, we wish to obtain a more general law that holds for all situations (beyond magnetostatics). Ideally, we hope to find a law that yields the same result for any surface S that spans the loop C. Evidently, Ampere's law is not the right choice here. If the semi-ovoid surface S_2 shown above is chosen instead, $I_{enc} = 0$ as no current flows between the capacitor plates — leading to the contradictory conclusion that $\oint \boldsymbol{B} \cdot d\boldsymbol{s} = 0$! This discrepancy arises due to charges piling up somewhere, namely the capacitor plates, which causes the current to be non-steady. Ampere's law has only worked so far because of the steady current condition which ensures that the surface integral of the current density is independent of the surface chosen.

To rectify this error, Maxwell devised a correction to Ampere's law which then came to be known as the Ampere–Maxwell Law which is also valid for non-steady systems. Its integral form states that

$$\oint_C \boldsymbol{B} \cdot d\boldsymbol{s} = \mu_0 \left(I_{enc} + \varepsilon_0 \iint_S \frac{\partial \boldsymbol{E}}{\partial t} \cdot d\boldsymbol{\Lambda} \right),$$

where the surface integral on the right-hand side is performed over a surface S that spans the contour C on the left-hand side and $I_{enc} = \iint_S \boldsymbol{J} \cdot d\boldsymbol{A}$ is the current flowing across the same surface S. The second term in the brackets is known as the displacement current.

$$I_d = \varepsilon_0 \iint_S \frac{\partial \boldsymbol{E}}{\partial t} \cdot d\boldsymbol{A}.$$

It is the rate of change of electric flux through the surface S, multiplied by ε_0. Maxwell's correction leads to an appealing symmetry. A changing electric field also generates a magnetic field!

Let us see how this correction resolves the discrepancy in the situation above. For the simple surface S_1, $I_d = 0$ as there should be no electric field (and hence change in electric field) outside of the capacitors in the ideal case. If we choose the contour C to be anti-clockwise relative to the leftwards direction,

$$\oint_C \boldsymbol{B} \cdot d\boldsymbol{s} = \mu_0 I.$$

To find the corresponding equation for the semi-ovoid surface S_2, we first find the electric field between the plates which are assumed to be close together.

The standard result is

$$E = \frac{Q}{A\varepsilon_0}$$

where Q is the instantaneous charge on the plates and A is the surface area of a plate. Then,

$$I_d = \varepsilon_0 \iint_{S_2} \frac{\partial E}{\partial t} \cdot dA = \varepsilon_0 \iint_{S_2} \frac{I}{A\varepsilon_0} \hat{k} \cdot dA,$$

where I is the current in the wire connecting the plates. Note that the positive directions of the infinitesimal area elements dA are such that their horizontal components point leftwards (apply the right-hand-grip rule to the contour C which we define to be anti-clockwise relative to the left again). In writing the second equality, we have asserted that the rate of change of electric field only stems from the part of S_2 that is sandwiched between the two plates, as the electric field beyond the two plates is virtually zero. The unit vector $\hat{k}$ is along the perpendicular direction pointing from the positively charged plate to the negatively charged plate (i.e. leftwards). The integral $\int k \cdot dA$ over the portion of S_2 in between the plates yields the projection of this portion onto the capacitor plates, which is just the area of a plate A. Therefore,

$$I_d = \varepsilon_0 \iint_{S_2} \frac{I}{A\varepsilon_0} \hat{k} \cdot dA = I.$$

On the other hand, $I_{enc} = 0$ for surface S_2. Thus, we arrive at the same result as previously derived.

$$\oint_C B \cdot ds = \mu_0 (I_{enc} + I_d) = \mu_0 I.$$

In conclusion, similar to how there are two forms of electric fields, there are now two forms of magnetic fields — namely, one type is produced by currents while the other is produced by time-varying electric fields. It turns out that even with this modification, the closed surface integral of the total magnetic field (of both types) is empirically determined to always be zero.

$$\oiint B \cdot dA = 0.$$

Summary of Maxwell's Equations

We have officially assembled all the pieces of Maxwell's equations — the fundamental laws which describe all of electromagnetism — and are no longer restricted to laws which are only valid in the realm of electrostatics or magnetostatics, such as Coulomb's law and Ampere's law. Maxwell's equations in integral form are summarized below. For electric fields,

$$\oiint_S \boldsymbol{E} \cdot d\boldsymbol{A} = \frac{q_{enc}}{\varepsilon_0}, \qquad \text{(Gauss)}$$

$$\oint_C \boldsymbol{E} \cdot d\boldsymbol{s} = -\iint_S \frac{\partial \boldsymbol{B}}{\partial t} \cdot d\boldsymbol{A}. \qquad \text{(Faraday)}$$

For magnetic fields,

$$\oiint_S \boldsymbol{B} \cdot d\boldsymbol{A} = 0, \qquad \text{(no name)}$$

$$\oint_C \boldsymbol{B} \cdot d\boldsymbol{s} = \mu_0 \left(\iint_S \boldsymbol{J} \cdot d\boldsymbol{A} + \varepsilon_0 \iint_S \frac{\partial \boldsymbol{E}}{\partial t} \cdot d\boldsymbol{A} \right), \qquad \text{(Ampere–Maxwell)}$$

where $\boldsymbol{E}$ and $\boldsymbol{B}$ encapsulate both forms of electric and magnetic fields. Furthermore, we have the Lorentz force law and Ohm's law (which is not a rigorous law and only a rule of thumb) which describe the response of charges to the fields governed by Maxwell's equations.

$$\boldsymbol{F} = q\left(\boldsymbol{E} + \boldsymbol{v} \times \boldsymbol{B}\right). \qquad \text{(Lorentz Force)}$$

$$\boldsymbol{J} = \sigma \boldsymbol{f} = \sigma\left(\boldsymbol{E} + \boldsymbol{v} \times \boldsymbol{B}\right). \qquad \text{(Ohm)}$$

8.9 Perfect Conductors and Superconductors

Let us now apply our knowledge of electrodynamics to perfect conductors and superconductors. A perfect conductor refers to one whose conductivity is infinite. For finite volume current densities,[11] Ohm's law implies that inside the conductor,

$$\boldsymbol{J} = \sigma(\boldsymbol{E} + \boldsymbol{v} \times \boldsymbol{B})$$

$$\implies \boldsymbol{E} + \boldsymbol{v} \times \boldsymbol{B} = 0$$

as $\lim_{\sigma \to \infty} \frac{|\boldsymbol{J}|}{\sigma} = 0$ for finite $|\boldsymbol{J}|$. Note that both $\boldsymbol{E}$ and $\boldsymbol{B}$ refer to the electric and magnetic fields **inside** the conductor. Now, we can perform a closed loop integral along a contour that lies entirely within the conductor (the loop has

---

[11] Note that it is very difficult for infinite current to flow in a perfect conductor, however counter-intuitive it may seem, due to the self-inductance of the conductor.

to be inside the "meat" of the conductor but the surfaces that it bounds do not).

$$\oint \boldsymbol{E} \cdot d\boldsymbol{s} + \oint (\boldsymbol{v} \times \boldsymbol{B}) \cdot d\boldsymbol{s} = 0.$$

Observe that the first and second terms correspond to the induced and motional emfs along the loop. Applying the universal flux rule, we obtain

$$\frac{d\Phi_B}{dt} = 0$$

where Φ_B is the magnetic flux through the loop.

$$\implies \Phi_B = c.$$

That is, the magnetic flux through a loop in a conductor is constant. Be wary that in applying the universal flux rule, the loop is "attached" to the conductor such that it follows the conductor if the conductor is moving. Before we revisit a familiar set-up below, it should be noted that if a perfect conductor is stationary (such that the motional emf is zero), the above implies that the magnetic field must not vary at any point within the conductor (else there will be an induced emf).

Problem: In Fig. 8.12, an infinitely long wire carries a constant current I_1 upwards. A conducting rectangular loop of self-inductance L has side lengths h and a which are in the axial and radial directions with respect to the long wire. Initially, the loop is stationary, with its left edge at a distance l from the wire, and does not carry any current. Subsequently, you bring the loop to infinity, without increasing its kinetic energy. What is the final current in the loop I_2? What is the total work delivered to the system comprising the wire and the loop in this process, and what external agent(s) is(are) responsible for this work?

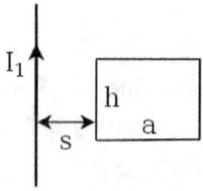

Figure 8.12:   Rectangular loop and current-carrying wire

The magnetic field as a function of radial distance r from the wire is

$$B(r) = \frac{\mu_0 I_1}{2\pi r}$$

by Ampere's law. Now, denote s as the instantaneous radial distance between the left edge of the loop and the wire. The magnetic flux through the loop, solely due to the magnetic field of the wire, as a function of s is

$$\Phi_B^{wire}(s) = \int_s^{s+a} \frac{\mu_0 I_1 h}{2\pi r} dr = \frac{\mu_0 I_1 h}{2\pi} \ln\left(1 + \frac{a}{s}\right).$$

Originally, the conducting loop had zero current when $s = l$. Therefore, the total magnetic flux through the loop, which must be constant, is $\Phi_B^{wire}(l)$. When the wire is subsequently moved to infinity, the magnetic flux through the loop due to the wire is zero — implying that the total magnetic flux through the loop must be maintained entirely by its final current I_2.

$$LI_2 = \frac{\mu_0 I_1 h}{2\pi} \ln\left(1 + \frac{a}{l}\right)$$

$$\implies I_2 = \frac{\mu_0 I_1 h}{2\pi L} \ln\left(1 + \frac{a}{l}\right)$$

where I_2 is positive clockwise (take note of the direction to ensure that the magnetic flux is correct). The total work done in this process is the increase in mechanical energy of the system comprising the wire and the loop. Since the wire and the loop can be considered as isolated systems once the loop is at infinity, the increase in energy is simply that associated with the loop's magnetic field.

$$W = \Delta U = \frac{1}{2} L I_2^2 = \frac{\mu_0^2 I_1^2 h^2}{8\pi^2 L} \left[\ln\left(1 + \frac{a}{l}\right)\right]^2.$$

The external agents responsible for this work are you and the external source used to maintain the current in the long wire at I_1 (as it too experiences an induced emf when the loop moves away from it).

Meissner effect

Superconductors have an additional property, known as the Meissner effect, that is distinct from perfect conductors. Superconductors expel magnetic fields from their interior such that $\mathbf{B} = \mathbf{0}$ inside them. Furthermore, this superconducting property only occurs when the superconductor is cooled below a certain critical temperature, T_c; so in a certain sense, there is an "on-off" switch for a superconductor.

In a manner analogous to how the absence of a net electric field inside a stationary perfect conductor implies zero net volume charge within it, the lack of magnetic fields within a superconductor implies that there are no volume currents within it. This can be easily proven by drawing small Amperian loops within the superconductor — the closed loop integral of the magnetic field yields zero, insinuating that the net current passing through the loop is zero by the Ampere–Maxwell law, assuming that there are no time-varying electric fields in the superconductor (this assertion should hold when a steady state has been reached). Since this argument applies to small loops in all directions, the volume current at a point within a superconductor must be zero. Therefore, only surface currents can exist on the boundaries of a superconductor.

Moving on, we can deduce certain properties about the magnetic field directly outside the surface of a superconductor. Drawing a small pillbox which straddles the surface at a certain point and exploiting the fact that the closed surface integral of a magnetic field is zero, we can conclude that the normal component of magnetic field at the surface of a superconductor is zero, i.e.

$$B_\perp = 0,$$

since the normal component of magnetic field within the conductor is zero. This is an extremely important boundary condition, as we shall soon see. Meanwhile, similar to the boundary condition in magnetostatics, the magnetic field vector directly above (as in outside) a point on the surface of the superconductor, tangential to the surface, is related to the surface current density $\boldsymbol{K}$ at that point by

$$\boldsymbol{B}_\| = \boldsymbol{K} \times \hat{\boldsymbol{n}},$$

where $\hat{\boldsymbol{n}}$ is the normal unit vector that points outwards from the superconductor at that point. Note that this is valid even when there is a changing electric field immediately outside the superconductor, as the displacement current through our small Amperian loop is negligible.

Problem: A long, thin cylindrical shell of radius R is made of superconducting material. Initially, it is above its critical temperature T_c and placed in a region of uniform external magnetic field B_0 in the positive z-direction, parallel to the cylindrical axis. If it is subsequently cooled below T_c, determine the current distribution on the shell. Finally, while the shell is still maintained in its superconducting state, the external field is removed. What is the current distribution on the shell now?

When the superconducting state is first "switched on", the superconductor must expel the uniform magnetic field B_0 within itself. This is achieved by inducing a $-B_0$ magnetic field in itself via the surface currents that flow on the shell. It is easy to spot a current distribution that produces such a uniform axial field — the current in an infinite solenoid. This suggests that we introduce a uniform azimuthal surface current K_{outer} throughout the outer surface of the cylindrical shell. By Ampere's law, this outer surface current generates a magnetic field

$$B_{outer} = \mu_0 K_{outer}$$

within the cylindrical shell. For $B_{outer} = -B_0$, we must have

$$K_{outer} = -\frac{B_0}{\mu_0}.$$

Now, this cannot be the only surface current as the magnetic flux through a cross section of the shell must be maintained. To correct this, we can introduce an azimuthal surface current on the inner surface of the shell.

$$K_{inner} = \frac{B_0}{\mu_0}.$$

With these currents, the conditions imposed by the superconductor are satisfied. However, how do we know that this must be the correct answer? The answer, as always, is a uniqueness theorem. Observe that our boundary conditions for the volumes inside and outside the shell are that the normal component of magnetic field must be zero along the surfaces of the shell. Furthermore, the net magnetic field is subject to the following conditions. Firstly, because there are no currents anywhere in the regions that are not within the superconductor, we must have

$$\oint_C \boldsymbol{B} \cdot d\boldsymbol{s} = 0$$

for any closed contour C. Furthermore, the closed surface integral of the magnetic field is always zero for any arbitrary surface S.

$$\oiint_S \boldsymbol{B} \cdot d\boldsymbol{A} = 0.$$

Now, compare these equations with those of the electric field in a region devoid of net charges.

$$\oint_C \boldsymbol{E} \cdot d\boldsymbol{s} = 0,$$

$$\oiint_S \boldsymbol{E} \cdot d\boldsymbol{A} = 0,$$

where the first equation is Faraday's law (with no changing magnetic field everywhere) and the second equation is Gauss' law with no net charge everywhere. In a volume V subjected to such conditions, we know that if we define the normal component of electric field throughout the surface bounding this volume, the third uniqueness theorem guarantees the uniqueness of the electric field. Observing that these are the exact conditions on the magnetic field in the current situation, the magnetic field that satisfies such conditions must be unique! Since we can determine $\boldsymbol{K}$ from $\boldsymbol{B}_\parallel = \boldsymbol{K} \times \hat{\boldsymbol{n}}$, we conclude that our proposed solution must be correct since the magnetic field is ensured to be valid.

When the external magnetic field is removed, the magnetic field in the shell must still be zero while the magnetic flux through a cross section must be constant. Therefore, the outer surface current is dispersed while the inner surface current is retained such that

$$K_{outer} = 0,$$

$$K_{inner} = \frac{B_0}{\mu_0}.$$

8.10   Force on Inductors

Having analyzed the energies stored by self-inductors and mutual inductors, it is imperative that we devise a way to compute forces via the principle of virtual work. In doing so, it is important to note that we are free to connect external entities — such as batteries or current sources — to our set-up or vary the resistance of our inductors, as long as the instantaneous currents flowing in them are consistent.

How can we be sure that these tweaks do not change the forces experienced by the inductors? Well, the forces on inductors are purely magnetic in nature and thus only depend on the instantaneous currents (and not the future currents) carried by the inductors. External entities such as batteries and current sources only affect the future currents carried by inductors (the current may change as we will be deforming our inductors) and hence do not change the instantaneous forces experienced by them. Ultimately, we only care about the instantaneous current distribution and not how the currents got there.

8.10.1 *Pressure on Self-Inductor*

As an object cannot exert a net force on itself, a self-inductor can only produce a magnetic pressure on itself. If the symmetry of the inductor assures that this magnetic pressure is uniform, the magnetic pressure on an inductor with self-inductance L and current I can be computed via the principle of virtual work. Naturally, we want this process to be as simple as possible (by performing the optimal tweaks to the system). Intuitively, we would want our inductor to be perfectly conducting to eliminate the energy dissipated as heat. Furthermore, our first intuition may hint at connecting a current source, a circuit component which maintains the current in the inductor at a constant value (by producing an appropriate emf), to the inductor so that we do not need to worry about changes in current. Let's try that.

Connecting a Current Source to a Perfect Conductor

Firstly, introduce an external pressure p (possibly varying with location) on the self-inductor to counteract the magnetic pressure such that the internal tension in the inductor is zero everywhere (else, we have to include the virtual work performed by this internal tension later). Next, connect a current source to maintain the current in the inductor, which we presume to be perfectly conducting, at I. Now, suppose that the inductor expands in a certain manner such that its self-inductance changes by δL.

To start off, there is virtual work performed by the external pressure δW_{ext}. Next, there is a (net) virtual work performed by the inductor which is negative of the change in energy stored in the magnetic field of the inductor.

$$\delta W_{ind} = -\delta U_{ind} = -\frac{1}{2}\delta L I^2$$

as $U_{ind} = \frac{1}{2}LI^2$ for an inductor and I is constant. Finally, there is a virtual work δW_{cc} performed by the current source. When the inductor expands, its magnetic flux linkage due to its own magnetic field changes — inducing a back emf ε that the current source needs to counteract to maintain the current in the inductor at I. For the sake of illustration, suppose that the expansion of the inductor occurs in time dt (the expansion is virtual and should really take no time). The total energy delivered by the current source is $\delta W_{cc} = -\varepsilon I dt$ as it must produce a counter-emf $-\varepsilon$ (so that the net emf is zero as a perfect conductor does not require a non-zero emf to carry a current). Moreover, we know from the universal flux rule that

$$\varepsilon = -\frac{d\Phi_B^{linkage}}{dt}$$

where $\Phi_B^{linkage}$ is the magnetic flux linkage through the inductor. Hence,

$$\delta W_{cc} = \delta \Phi_B^{linkage} I.$$

Now, we know from the definition of self-inductance that

$$L = \frac{\Phi_B^{linkage}}{I}$$

such that, as I remains constant,

$$\delta \Phi_B^{linkage} = \delta L I$$

$$\implies \delta W_{cc} = \delta L I^2.$$

By the principle of virtual work, the sum of all forms of virtual work must be equal to zero as the inductor was originally in static equilibrium.

$$\delta W_{ext} + \delta W_{ind} + \delta W_{cc} = 0$$

$$\implies \delta W_{ext} = -\frac{1}{2}\delta L I^2.$$

By choosing a suitable expansion, one can in principle compute W_{ext} in terms of the external pressure p and solve for p through the above equation if it is uniform. Then, one can take the negative of p to determine the magnetic pressure everywhere. Another way to see this is that the virtual work performed by the magnetic pressure must be negative of the virtual work performed by the external pressure, which opposes the magnetic pressure everywhere.

$$\delta W_{magp} = -\delta W_{ext} = \frac{1}{2}\delta L I^2.$$

Note that even though a real magnetic force produces no work, it can produce virtual work as we are considering a virtual displacement that is not necessarily parallel to the instantaneous velocity of the charge that the magnetic force acts on. At this juncture, one might raise the following question: why is δW_{magp} different from $\delta W_{ind} = -\frac{1}{2}\delta L I^2$ that we have computed previously? That is, couldn't we have computed δW_{magp} directly from the virtual work performed by the inductor which is also the negative change in energy stored in the inductor, $\delta W_{ind} = -\delta U_{ind}$? The answer to the latter query is yes and no.

It is paramount to understand that $\delta W_{ind} = -\delta U_{ind}$ is the net work done by the inductor and comes from various factors, of which the virtual work due to magnetic pressure is merely one. δW_{ind} depends on the actual configuration of the system (e.g. the external connections) while δW_{magp} does not (which enables us to consider various set-ups). In this case, the presence of the current source changes the net virtual work performed by the inductor as it delivers energy to the inductor. Therefore, if we want to compute δW_{magp} from the net virtual work done by the inductor, we must consider a separate set-up where the only virtual work performed by the inductor is due to precisely magnetic pressure — this requires there to be no external connections to the inductor. It seems that we have dug a hole for ourselves with our first intuition as we not only have to include the work done by the current source, δW_{magp} is no longer related to δW_{ind}. In light of this, let us now consider an isolated, perfectly conducting inductor.

Isolated Perfect Conductor

Consider a separate set-up where the inductor is now isolated (i.e. no external entity is connected) and perfectly conducting (to ensure that no energy is dissipated as heat) such that the only forms of virtual work are due to the external pressure exerted to balance the magnetic pressure and due to the inductor itself. The virtual work performed by the magnetic pressure is the negative of the former factor such that the principle of virtual work ensures that $\delta W_{magp} = \delta W'_{ind}$, where $\delta W'_{ind}$ is the virtual work performed by the inductor in this set-up due to a virtual change in its self-inductance δL.

$$\delta W'_{ind} = -\delta U'_{ind} = -\frac{1}{2}\delta L I^2 - LI\delta I$$

where the current I in the inductor can also vary now (due to the lack of a connected current source). We can relate δL and δI by using the fact that the magnetic flux linkage through a perfect conductor must be constant.

$$\Phi_B^{linkage} = LI = c$$

$$\delta L I + L\delta I = 0 \implies L\delta I = -\delta LI.$$

Therefore,

$$\delta W_{magp} = \delta W'_{ind} = \frac{1}{2}\delta L I^2,$$

and we arrive at the same answer. We emphasize that we should unequivocally obtain the same answer, regardless of the tweaks we impose, as long as we compute the virtual work correctly.

Constant Emf Source Connected to Resistive Inductor

As another (ludicrously complicated) example, suppose that we connect a constant emf source ε_0 to the inductor, which we now presume has a certain finite resistance, that generates the current I in it. If we vary the self-inductance of the inductor by δL in a time interval dt while assuming that its resistance remains approximately constant, the virtual work performed by the inductor is now

$$W''_{ind} = -\frac{1}{2}\delta LI^2 - (\varepsilon_0 + \varepsilon)Idt$$

where the first term is its change in potential energy and the second term is the heat dissipated due to its resistance (its potential difference $\varepsilon_0 + \varepsilon$ multiplied by the constant current I, where ε is the induced emf due to the expansion of the inductor). Note that even though there is an abrupt increase in the net emf through the inductor, the current through the inductor can no longer instantaneously change due to its self-inductance, coupled with its resistance. Now, there is also virtual work performed by the emf source

$$W_{emf} = \varepsilon_0 Idt$$

such that by the principle of virtual work,

$$W_{ext} + W''_{ind} + W_{emf} = 0$$

$$W_{ext} = \frac{1}{2}\delta LI^2 + \varepsilon Idt$$

where $\varepsilon dt = -\delta\Phi_B^{linkage} = -\delta LI$ again.

$$W_{ext} = -\frac{1}{2}\delta LI^2$$

$$\implies W_{magp} = -W_{ext} = \frac{1}{2}\delta LI^2,$$

and we yet again retrieve the same answer. All in all, the moral of these three examples is that it is usually the easiest to apply the principle of virtual work by assuming that our set-up is an isolated, perfect conductor! Let us finally apply this result to a concrete example.

Problem: Determine the magnetic pressure on a long solenoid with η turns per unit length that is carrying a current I.

Firstly, let us grasp some intuition about the origin of this magnetic pressure. We know that a solenoidal current produces an axial magnetic field within it. This axial magnetic field exerts a radially outwards force on the turns of a solenoid — generating the magnetic pressure. However, note that we cannot simply take the magnetic field inside the solenoid multiplied by the current per unit length ηI of the solenoid to compute the magnetic pressure on the solenoid as the magnetic field is discontinuous across the surface of the solenoid (we must multiply ηI by the magnetic field **at** the solenoid surface). This can be achieved by imposing the boundary conditions on the magnetic field,[12] similar to how we computed the force felt by a charged surface, but we shall solve this using the principle of virtual work instead.

Due to the axial symmetry and the infinite nature of the solenoid, the magnetic pressure p should be uniform throughout its surface. Let the radius of the solenoid be R. We have previously computed the self-inductance per unit of a solenoid as

$$l = \eta^2 \mu_0 \pi R^2,$$

which implies that if we expand the solenoid radially by δR,

$$\delta l = 2\eta^2 \mu_0 \pi R \delta R.$$

Applying the previous result, the work done per unit length of the solenoid by the magnetic pressure, due to this virtual expansion is

$$w_{magp} = \frac{1}{2}\delta l I^2 = \eta^2 \mu_0 I^2 \pi R \delta R.$$

Another expression for w_{magp} is

$$w_{magp} = p \cdot 2\pi R \delta R$$

as the work done by magnetic pressure on a cross section of the solenoid of thickness ds in pushing it radially outwards by distance δR is $p \cdot 2\pi R ds \cdot \delta R$ such that the work per unit length is $2\pi p R \delta R$. Equating the two expressions for w_{magp},

$$p = \frac{1}{2}\eta^2 \mu_0 I^2.$$

---

[12]In fact, you will similarly find out that the magnetic field at the solenoid windings is the average of the fields on its two sides.

8.10.2   Force Between Mutual Inductors

In a similar vein, there is an elegant way of computing the force between two mutual inductors with mutual inductance M that carry currents I_1 and I_2 respectively via the principle of virtual work.

Having learnt our lesson from the previous section, we propose that the two inductors are perfect conductors in computing our virtual works. Suppose that we wish to compute the force on inductor 2 due to inductor 1. Exert an external force $\boldsymbol{F}_{ext}$ on inductor 2 to ensure that it is in static equilibrium. Subsequently, consider a virtual displacement $\delta\boldsymbol{r}$ of inductor 2. The virtual works done are that due to the external force

$$\delta W_{ext} = \boldsymbol{F}_{ext} \cdot \delta\boldsymbol{r}$$

and that due to the system of inductors

$$\delta W_{sys} = -\delta U_{sys}.$$

We know that for a system of two mutual inductors, their total potential energy is

$$U_{sys} = \frac{1}{2}L_1 I_1^2 + \frac{1}{2}L_2 I_2^2 \pm MI_1 I_2,$$

where L_1 and L_2 are the respective self-inductances of inductors 1 and 2. Since the self-inductances remain constant as the shapes of the inductors are unchanged,

$$\delta U_{sys} = L_1 I_1 \delta I_1 + L_2 I_2 \delta I_2 \pm \delta MI_1 I_2 \pm M\delta I_1 I_2 \pm MI_1 \delta I_2.$$

Now, we can relate some of these terms by imposing the condition that the magnetic flux linkages through the inductors must be invariant. The magnetic flux linkages through inductors 1 and 2 are respectively

$$L_1 I_1 \pm MI_2 = c_1 \implies L_1 \delta I_1 \pm \delta MI_2 \pm M\delta I_2 = 0,$$

$$L_2 I_2 \pm MI_1 = c_2 \implies L_2 \delta I_2 \pm \delta MI_1 \pm M\delta I_1 = 0,$$

where c_1 and c_2 are constants. Multiplying the first equation by I_1 and adding it to the second equation multiplied by I_2,

$$L_1 I_1 \delta I_1 \pm \delta MI_1 I_2 \pm MI_1 \delta I_2 + L_2 I_2 \delta I_2 \pm \delta MI_1 I_2 \pm M\delta I_1 I_2 = 0$$

$$\implies \delta U_{sys} \pm \delta MI_1 I_2 = 0$$

$$\delta U_{sys} = \mp \delta MI_1 I_2.$$

Therefore, from the principle of virtual work,

$$\delta W_{ext} + \delta W_{sys} = 0$$

$$\boldsymbol{F}_{ext} \cdot \delta \boldsymbol{r} = \delta U_{sys} = \mp \delta M I_1 I_2.$$

Since the force $\boldsymbol{F}_{21}$ exerted on inductor 2 by inductor 1 must be negative of the external force required to balance it,

$$\boldsymbol{F}_{21} \cdot \delta \boldsymbol{r} = \pm \delta M I_1 I_2.$$

The sign on the right-hand side depends on the linkage of fluxes between the inductors. Given the positive directions of I_1 and I_2, if a positive current I_1 flowing in inductor 1 generates a magnetic flux through inductor 2 that reinforces the magnetic flux produced by the positive current I_2 in the inductor through itself, we choose the positive sign. Otherwise, we choose the negative sign.

By choosing $\delta \boldsymbol{r}$ wisely and computing the change in mutual inductance subsequently, one can determine the components of $\boldsymbol{F}_{21}$ and thus solve for it!

We can now combine the tricks on our table to solve the following problem — a feat that usually requires considerable vector calculus and seems intractable otherwise.

8.10.3 *Distant Magnetic Field of Magnetic Dipole*

Our objective in this section is to derive the distant magnetic field of a stationary magnetic dipole (i.e. a current loop) with magnetic dipole moment $\boldsymbol{\mu}$. Define the positive z-axis to be aligned with $\boldsymbol{\mu}$ and the origin at the infinitesimal magnetic dipole as shown in Fig. 8.13.

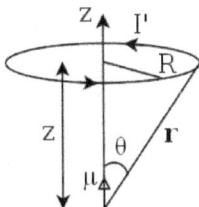

Figure 8.13:   Magnetic dipole

The magnetic dipole moment is depicted by the white arrow. For distances much larger than the length dimension of the magnetic dipole, this set-up is axially symmetric about the z-axis. Therefore, by applying Ampere's law to a circle centered at and perpendicular to the z-axis, we can conclude that the azimuthal component of magnetic field is zero everywhere, as all of such circles enclose zero current.

To proceed further, we start by first assuming that we are working with a planar magnetic dipole. Recall that for a planar loop with N turns and current I, its magnetic dipole moment is defined as

$$\boldsymbol{\mu} = NI\boldsymbol{A}$$

where $\boldsymbol{A}$ is its area vector whose positive direction can be determined by applying the right-hand-grip rule along its current I. Now, imagine that we place a circular conducting loop perpendicular to the z-axis, centered at a z-coordinate z, and run an anti-clockwise current I' through it. We have previously computed the mutual inductance between a large circular ring of radius r_1 and a parallel small circular ring of radius $r_2 \ll r_1$ that are separated by a perpendicular distance z as

$$M = \frac{\mu_0 \pi r_1^2 r_2^2}{2(r_1^2 + z^2)^{\frac{3}{2}}}.$$

One can repeat the same procedure to show that if we replace the small ring with a parallel planar coil, of an arbitrary shape, small area A and N turns, the mutual inductance is

$$M = \frac{N\mu_0 A r_1^2}{2(r_1^2 + z^2)^{\frac{3}{2}}}.$$

Therefore, the mutual inductance between our imaginary ring and the planar magnetic dipole in this case is

$$M = \frac{N\mu_0 A R^2}{2(R^2 + z^2)^{\frac{3}{2}}}.$$

We known from the previous section that the z-component of the magnetic force experienced by the circular ring due to the magnetic dipole is

$$F_z \delta z = \delta M I I'$$

where δM is the virtual change in mutual inductance if we increase the z-coordinate of the ring by δz. We have chosen the positive sign as the magnetic fluxes of the two loops reinforce each other. For example, the anti-clockwise current in the magnetic dipole generates a magnetic field at the imaginary ring with an upwards component — bolstering the upwards magnetic flux produced by its own anti-clockwise current I'. We can compute δM as

$$\delta M = \frac{\partial M}{\partial z}\delta z = -\frac{3N\mu_0 AR^2 z}{2(R^2+z^2)^{\frac{5}{2}}}\delta z$$

$$\implies F_z = -\frac{3N\mu_0 II' AR^2 z}{2(R^2+z^2)^{\frac{5}{2}}}.$$

This force only comes from the component of magnetic field on the imaginary ring, perpendicularly outwards from the z-axis, B_ρ at the ring (this must be uniform throughout the ring due to axial symmetry). Therefore,

$$F_z = -B_\rho I' \cdot 2\pi R$$

$$B_\rho = \frac{3N\mu_0 IARz}{4\pi(R^2+z^2)^{\frac{5}{2}}} = \frac{3\mu_0\mu\sin\theta\cos\theta}{4\pi r^3},$$

where r is the distance between a point on the rim and the origin and θ is the angle defined in the figure above. To determine the axial magnetic field (in the z-direction) B_z, we can use the following trick. Label the magnetic dipole as inductor 1 and the imaginary ring as inductor 2. From the definition of mutual inductance,

$$M = \frac{\Phi_{B21}}{I}$$

where Φ_{B21} is the magnetic flux through inductor 2 due to the magnetic field of inductor 1. Now, suppose that we slightly increase the radius of the ring by δR while maintaining the currents in the two inductors (we do not care about how we do this) such that the mutual inductance changes by δM.

$$\delta M = \frac{\partial M}{\partial R}\delta R = \left(\frac{N\mu_0 AR}{(R^2+z^2)^{\frac{3}{2}}} - \frac{3N\mu_0 AR^3}{2(R^2+z^2)^{\frac{5}{2}}}\right)\delta R.$$

We know that

$$\delta M = \frac{\delta\Phi_{B21}}{I}$$

where $\delta\Phi_{B21}$ is the change in magnetic flux through the ring, due to the magnetic field of the dipole, caused by the expansion of the ring's radius by δR. However, $\delta\Phi_{B21}$ is simply the z-component of magnetic field at the rim of the ring, B_z, multiplied by the circular shell of area $2\pi R\delta R$. Therefore,

$$\frac{B_z \cdot 2\pi R\delta R}{I} = \frac{\partial M}{\partial R}\delta R = \left(\frac{N\mu_0 AR}{(R^2+z^2)^{\frac{3}{2}}} - \frac{3N\mu_0 AR^3}{2(R^2+z^2)^{\frac{5}{2}}}\right)\delta R$$

$$B_z = \frac{N\mu_0 I A}{2\pi(R^2 + z^2)^{\frac{3}{2}}} - \frac{3N\mu_0 I A R^2}{4\pi(R^2 + z^2)^{\frac{5}{2}}}$$

$$B_z = \frac{\mu_0 \mu}{2\pi r^3} - \frac{3\mu_0 \mu \sin^2 \theta}{4\pi r^3}.$$

Therefore, if we are interested in the magnetic field at a point with position vector r relative to the origin, we can consider the plane containing r and the z-axis and define the polar coordinate system as shown in the figure above. Then, the magnetic field can be resolved into $\hat{r}$ and $\hat{\theta}$ components as

$$B_r = B_z \cos\theta + B_\rho \sin\theta = \frac{\mu_0 \mu}{2\pi r^3} \cos\theta,$$

$$B_\theta = B_\rho \cos\theta - B_z \sin\theta = \frac{\mu_0 \mu}{4\pi r^3} \sin\theta.$$

In vector form, the magnetic field at the point of interest is

$$\boldsymbol{B} = \frac{\mu_0 \mu}{4\pi r^3}(2\cos\theta \hat{r} + \sin\theta \hat{\theta}), \tag{8.36}$$

$$\boldsymbol{B} = \frac{\mu_0}{4\pi r^3}[3(\boldsymbol{\mu} \cdot \hat{r})\hat{r} - \boldsymbol{\mu}]. \tag{8.37}$$

For a magnetic dipole that takes the form of a non-planar loop, we can divide it into many planar loops and apply the principle of superposition (see Problem 16 in Chapter 7) to retain the above expression for the magnetic field, except that

$$\boldsymbol{\mu} = N I A = N I \iint_S d\boldsymbol{A}$$

now. $\boldsymbol{A}$ is now replaced by the vector area of the dipole loop C — it is the integral of $d\boldsymbol{A}$ (infinitesimal area vectors) over a surface S that spans the dipole loop C.

Gilbert's Model of Magnetic Monopoles

Astute readers may observe that the previous expression for the distant magnetic field of a magnetic dipole is analogous to the distant electric field of an electric dipole after the substitutions $\boldsymbol{E} \to \boldsymbol{B}$, $\boldsymbol{p} \to \boldsymbol{\mu}$ and $\frac{1}{\varepsilon_0} \to \mu_0$. Does this mean that we can model a magnetic dipole, as two "magnetic charges" q_m and $-q_m$ separated by a distance d such that $q_m d = \mu$, as in the case of an electric dipole?

Well, not really. The existence of "magnetic charges" or magnetic monopoles is forbidden by one of Maxwell's equations: $\oiint \boldsymbol{B} \cdot d\boldsymbol{A} = 0$ over any closed surface. In nature, no magnetic monopoles have hitherto been discovered. Therefore, even though the above model, known as Gilbert's model, is rather accurate for large distances away from the magnetic dipole, it is fundamentally wrong. This difference is amplified when we approach the magnetic dipole on scales comparable with its length dimensions such that the answers we obtain from Gilbert's model are completely wrong. For example, based on Gilbert's analogy between magnetic and electric dipoles, we would expect the force on a magnetic dipole $\boldsymbol{\mu}$ due to an external magnetic field $\boldsymbol{B}$ to be

$$F = (\boldsymbol{\mu} \cdot \nabla)\boldsymbol{B}.$$

However, the correct force (obtained from simplifying $I \oint d\boldsymbol{s} \times \boldsymbol{B}$, where the integral is performed over the magnetic dipole loop, via some vector calculus) is actually

$$\boldsymbol{F} = \nabla(\boldsymbol{\mu} \cdot \boldsymbol{B}). \tag{8.38}$$

This discrepancy evidently stems from the fact that we need to consider the true local structure of the magnetic dipole, which is a current loop (known as Ampere's model), in computing the force it experiences. That said, we can rewrite the second expression as

$$\boldsymbol{F} = \nabla(\boldsymbol{\mu} \cdot \boldsymbol{B}) = (\boldsymbol{\mu} \cdot \nabla)\boldsymbol{B} + \boldsymbol{\mu} \times (\nabla \times \boldsymbol{B}),$$

so the discrepancy vanishes when the curl of the magnetic field $\nabla \times \boldsymbol{B} = \boldsymbol{0}$ or when $\boldsymbol{\mu}$ is parallel to $\nabla \times \boldsymbol{B}$. The former condition is more commonly adopted as we want the force to be correct for any orientation of the dipole. It turns out that the differential form of the Ampere–Maxwell law is

$$\nabla \times \boldsymbol{B} = \mu_0 \boldsymbol{J} + \mu_0 \varepsilon_0 \frac{\partial \boldsymbol{E}}{\partial t},$$

so we require the right-hand side to disappear for Gilbert's model to be correct in determining forces. To this end, we usually require there to be no external current (the magnetic dipole is so small that it produces no current density) and changing electric field at the location of the dipole.

Having said all of this, Gilbert's model is usually proscribed as it is fundamentally invalid. However, if one insists on applying it, the following are some rules of thumb that should be obeyed. Firstly, the magnetic field due to Gilbert's model is only valid at distances much larger than the length dimensions of the dipole. Secondly, if we want to compute the magnetic force on a dipole with Gilbert's model, we must ensure that there are no volume currents or changing electric fields at its location. Finally, even though this has not been discussed before, it is best to not talk about the potential energy of a magnetic dipole (whether using Gilbert's or Ampere's model) as there are many subtleties involved. Refer to Problem 19 in this chapter for an example.

Problems

Currents

1. *Interface**

Consider a thin, long cylindrical shell branching along its axis. Half of the shell is filled with a material of resistivity ρ_1 while the other half is filled with one of resistivity ρ_2 (the cross section of the shell is uniform). If a steady and uniform current density J flows from the former material to the latter, parallel to the axis of the cylindrical shell, determine the surface charge density of the charges trapped at the interface.

2. *Cylindrical Current***

Consider two thin, concentric cylindrical shells, made of conducting material, that have length l and radii a and b, with $a < b$. The gap between the two shells is filled with a material with conductivity $\sigma(r) = \frac{k}{r}$ where r is the perpendicular distance from the axis. The inner and outer shells are maintained at a constant potential difference V_0 with the inner shell having the higher potential. Determine the resistance of this set-up when a steady current flows between the shells. Determine the electric field between the two shells and hence the charge density ρ everywhere within the mediating material.

3. *Four Point Probe***

A homogeneous matter of unknown resistivity ρ fills the region $x \geq 0$. Physicists then work in the other half-space to determine ρ. They draw a pale square with corners A, B, C, D at coordinates $(0, -\frac{a}{2}, \frac{a}{2})$, $(0, \frac{a}{2}, \frac{a}{2})$, $(0, \frac{a}{2}, -\frac{a}{2})$, $(0, -\frac{a}{2}, -\frac{a}{2})$. Subsequently, they inject a steady current I into point A and withdraw the same amount from point B via tiny electrodes. If the (positive) potential difference between points C and D is measured as V, determine ρ. **Hint:** apply the principle of superposition.

4. *Leaking Charge***

If a volume V is filled with a conducting material of uniform conductivity σ, determine the total electric charge $q(t)$ enclosed by the volume V, given that the total initial charge inside V is q_0.

Emfs

5. *Flipping Switch* *

Two bulbs are connected to opposite sides of a circular loop of wire, as shown in the figure below. A changing magnetic field (confined within the loop) induces an emf in the loop that causes the two bulbs to light at first. All wires lie in the same plane. When the switch is closed, what happens to the bulbs? Subsequently, the wires containing the closed switch remain connected at points A and B and are lifted up (out of the page) and moved gradually towards the other side, until the configuration in the right figure is finally attained (where all wires again lie in the same plane). Describe the responses of the bulbs throughout this process.

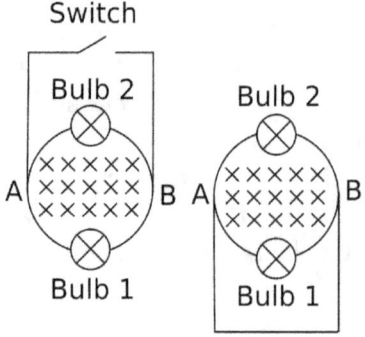

6. *Voltmeters* *

There are two long solenoids that are encircled by loops with two resistors R_1 and R_2, as shown in the figure below. The right solenoid produces a magnetic flux $\Phi_B^r = \alpha t$ out of the page through itself where t is the time elapsed from a reference point while the left solenoid produces a magnetic flux $\Phi_B^l = \beta t$ out of the page through itself. The voltmeters are ideal and possess infinite resistance. If each voltmeter measures the voltage across its ends via a path through itself, determine the readings of the voltmeters.

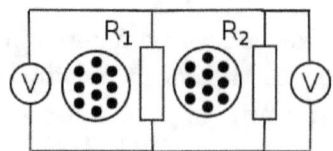

7. *Compressing Square Loop* *

Imagine a square loop of edge length 3.00m and resistance 10.0Ω in this page. It is placed in a uniform external 0.100T magnetic field that is directed perpendicularly into this page. The loop is then compressed (the lengths of the edges remain unchanged but the angles between them may vary) into a rhombus in the same plane, with a separation 3.00m between two opposite vertices. Note that the edges can possibly protrude out of this page en route but they must still remain in this page in the final configuration. If this process takes 0.100s, what is the average current generated in the loop? What is its direction? Remember to account for the self-inductance of the loop.

8. *Voltage due to Solenoid* *

A long solenoid of radius R produces a magnetic field $B(t) = B_0 \sin \omega t$ out of the page (defined as the positive z-axis). Now, consider two points A and B in the plane of the cross section of the solenoid. If points A and B are outside the solenoid and their position vectors from the center of the solenoid in the current plane subtend an angle θ, determine the voltage from points A to B along any line that does not cut through the solenoid. B is located anti-clockwise of A, where the anti-clockwise direction is determined by applying the right-hand-grip rule to the positive z-direction.

9. *Moving Loop* *

A rectangular loop of dimensions l and h moves with a constant velocity u away from a long wire that carries a steady current I_1 in the plane of the loop. The total resistance of the loop is R. Derive an expression for the current I_2 in the loop at the instant the closer side of the loop is a distance r from the wire. We have done a similar problem before but use the flux rule this time. Notice that the resistor heats up. What is providing this energy or rather, doing work on the system? The magnetic force seems to be doing work! Resolve this apparent paradox.

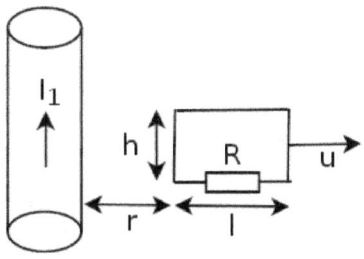

10. *Loop Exiting Magnetic Field**

A square loop is exiting a constant and uniform magnetic field B with a constant velocity v perpendicular to the magnetic field. Find the emf induced in the loop when the left end of the loop is at a distance x from the right boundary of the magnetic field. Find the force required to maintain the velocity of the loop. The loop has resistance R and negligible self-inductance.

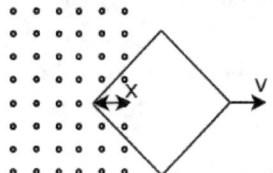

11. *Rotating Wheel**

A wheel with six spokes is placed in a perpendicular magnetic field $B = 0.5$T as shown in the figure. The field is directed into the plane of the paper and permeates the entire wheel. The external wires are connected to the wheel's center and a point on its rim. When the switch S is closed, there is an initial current of 6A through the battery and the wheel begins to rotate. The resistance of the spokes and the rim may be neglected. You may find the result of Problem 5 (Chapter 7) to be helpful in the following questions.

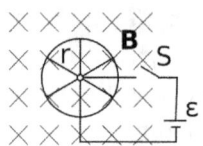

(a) What is the direction of rotation of the wheel? Explain.
(b) The radius of the wheel is $r = 0.2$m. Calculate the initial torque on the wheel about its center.
(c) Describe qualitatively the angular velocity of the wheel as a function of time. Let the emf of the battery be $\varepsilon = 1$V.

12. *Coil in Magnetic Field***

A loop of width w and self-inductance L is exiting a constant and uniform magnetic field B, that is perpendicular to the plane of the loop. An external force is exerted on the loop to sustain its velocity at a constant v. Determine the current in the loop as a function of time t if there was no initial current. Determine the external force required to maintain the loop at a constant velocity as a function of time. Lastly, verify that the power delivered by the external force to the coil is consistent with the rate of increase in energy in all other entities, except that which exerts the external force.

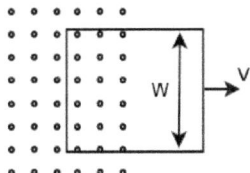

13. Rotating Ring**

A ring with mass m, radius r and resistance R is rotating about a diameter in a region of constant and uniform external magnetic field B. Initially, the magnetic field passes through the ring perpendicularly, resulting in maximum flux through the ring, and the ring rotates with initial angular speed w_0. Neglect the self-inductance of the ring.

(a) Find the relation between the total angle ϕ that the ring rotates before it stops, and the other given variables.
(b) Find the number of complete rounds that the ring manages to rotate when it initially spins at 8 rotations per second, $m = 1\text{kg}$, $R = 1\Omega$, $r = 30\text{cm}$, and $B = 1\text{T}$.

14. Railgun**

A bar of mass m and resistance R slides without friction in a horizontal plane that is perpendicular to a constant and uniform magnetic field B, moving on parallel rails as shown. The perfectly conductive rails are separated by a distance d and a battery maintains a constant ε between these two rails. Assuming that the bar starts from rest at $t = 0$, find the speed of the bar as a function of t. Neglect the self-inductance of this set-up.

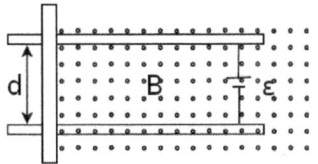

15. *Inclined Loop* **

A uniform external magnetic field B is directed vertically downwards, passing through the slope of an inclined plane which supports a wire loop as shown in the figure. A rod, of mass m, length l and resistance R_3 is sliding down without friction, starting from rest. The wire loop behaves like a "rail" for the rod. Neglect all self- and mutual inductances in this set-up.

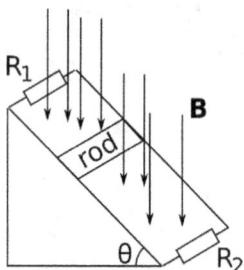

(a) When the rod has an instantaneous velocity v (positive down the slope), find the magnitude and direction of the induced current flowing in each wire segment.

(b) Find the magnitude and direction of the magnetic force acting on the rod.

(c) Find the velocity and acceleration of the rod as functions of time.

(d) Find the instantaneous "power" of the magnetic force[13] on the rod. You can leave the expression in terms of v.

(e) Show that the rate of change of the total mechanical energy of the rod equals to negative of the dissipated power.

16. *Magnetic Wave* **

A rectangular loop of width h and length l is currently traveling in the xy-plane (the length is aligned with the x-direction). The resistance of the entire loop is R. A magnetic field $B = B_0 \cos(kx - \omega t)$, where k and ω are constants, in the positive z-direction (out of the page) permeates all space. Given that the only possible force on the loop is the magnetic force, determine the condition for the loop to be able to travel at a constant velocity in the positive x-direction at all possible velocities. When the previous condition is

---

[13] We really mean the power delivered by the component of magnetic force caused by the current in the rod.

not met, there is a terminal velocity of the loop when the loop is given an initial velocity in the x-direction. Determine this terminal velocity. Neglect the self-inductance of the loop.

17. *Pulling a Solenoid***

An outer solenoid of cross sectional area A_1, η_1 turns per unit length and anti-clockwise current I_1 (relative to the positive z-axis defined along its solenoid axis) encloses an inner, coaxial solenoid of cross sectional area A_2, η_2 turns per unit length and anti-clockwise current I_2. The currents in both perfectly conducting solenoids are maintained at their respective values by current sources. If the inner solenoid is now pulled out of the outer solenoid at a constant velocity v in the positive z-direction, determine the emfs produced by the current sources. You may assume the individual magnetic fields of the solenoids to be uniform inside them and zero outside of them. Determine the ratio of these emfs and explain why it makes sense.

18. *Ring on Ring***

On a smooth, insulating and neutral large ring of radius R, there is a small ring of mass m which carries charge q. The large ring is placed in a uniform magnetic field of strength $B(t)$ and perpendicular to the plane of the ring (xy-plane), $B(t) = B_0 + \alpha t$ in the positive z-direction. Find the force of the small ring acting on the big ring thereafter and describe the motion of the small ring. (Singapore Physics Olympiad)

19. *Work Done on Magnetic Dipole***

A coil carrying a constant current I with N densely wound rounds of area A is placed in a region of uniform magnetic field $\mathbf{B}$. Define θ to be the angle between the magnetic dipole moment and the magnetic field. Determine the external mechanical work required to rotate the magnetic dipole from $\theta = \frac{\pi}{2}$ to a general angle θ along a single axis of rotation, without any increase in kinetic energy. Now, there is an apparent paradox as the dipole does not gain any energy, and yet we are supplying (possibly negative) external work to it. Furthermore, we know that the magnetic force cannot perform any work on the coil (to counteract the mechanical work). Where does this external mechanical work then go? Note that we have analyzed a similar set-up before.

Inductance

20. Rectangular Toroid*

Find the self inductance of a N-turn toroid which has a rectangular cross section with inner radius r_1, outer radius r_2 and height h.

21. Solenoid in Solenoid**

A finite solenoid of length l, radius r and N turns is placed in a long solenoid with turns per unit length η and radius R, $r \ll R$. The axes of the solenoids coincide. Find the mutual inductance of this system.

22. Inductance of Tetrahedron Sides**

If the self-inductance of an equilateral triangle loop of side length l is L, determine the self-inductance of the loop shown in the figure below. It is akin to two sides of a tetrahedron formed by four equilateral triangles of length l.

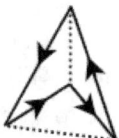

23. A Neat Mutual Inductance**

If the self-inductance of an equilateral triangle loop of side length l is L, determine the mutual inductance of the two loops that are placed side by side as shown in the figure below. One loop is an equilateral triangle of length l while the other loop is a rhombus formed by joining two of such equilateral triangles together (and removing the common side).

24. Transformer**

A primitive transformer is made by overlapping two solenoids of the same length and radius (that are not connected) and inserting an iron core within them to ensure that the magnetic fluxes through the turns of the solenoids are identical. The primary circuit is connected to one solenoid with N_1 turns while the secondary circuit is connected to the other solenoid with N_2 turns. To specify the direction of coupling of the mutual inductors, define I_1 to be the clockwise current in the primary circuit and I_2 to be the anti-clockwise current in the secondary circuit. The magnetic flux through the secondary solenoid due to a positive current I_1 in the primary solenoid reinforces the magnetic flux through the secondary solenoid due to its own positive current I_2 — the converse statement holds as well. Show that the ratio of the emfs produced by the two solenoids in the primary and secondary circuits is $\frac{\varepsilon_2}{\varepsilon_1} = \frac{N_2}{N_1}$.

Now, you may think that this set-up violates the conservation of energy as we can ramp up the emf in the secondary circuit by increasing N_2. To ease your worries, consider the situation where the primary solenoid is connected to an AC source with a clockwise emf $\varepsilon = \varepsilon_0 \cos \omega t$ (i.e. no resistance in the primary circuit) while the secondary solenoid is connected to a resistor R. Find the steady state currents in the two circuits. Show that the conservation of energy holds in the steady state situation by computing the various rates of changes of energy. You may have to define some quantities of your own.

25. *Magnetic Field of Point Charge***

By applying Ampere–Maxwell law, show that the magnetic field (in the non-relativistic approximation $v \ll c$) due to a point charge q traveling at a constant velocity v is

$$B = \frac{\mu_0 q v \times \hat{r}}{r^2}$$

where r is the instantaneous vector joining the point charge to the location at which the magnetic field is of concern. You may assume that the electric field due to the moving charge is still given by Coulomb's law in the non-relativistic regime.

Perfect Conductors and Superconductors

26. *Bringing Two Loops**

Two identical square loops that are made of perfectly conducting material initially carry currents I_1 and I_2 when they are infinitely far apart. They are then brought closer such that the second loop now carries current I_2'. Determine the current I_1' in the first loop at this juncture.

27. Exiting Conducting Loop**

A conducting rectangular loop of self-inductance L and mass m is initially stationary at the edge of a region of uniform magnetic field B (pointing out of the page) as shown in the figure below. The length of the loop along the vertical is l. The loop initially carries no current. If you now displace the loop towards the right, describe the motion that the loop will undergo. You do not need to solve the equation of motion of the loop.

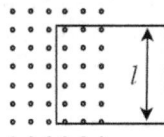

28. Levitating Magnet**

Defining the z-axis to be positive upwards, the region $z \leq 0$ is covered with an infinite superconducting material. A small magnet, which can be modeled as a small magnetic dipole with a magnetic dipole moment $\boldsymbol{\mu}$ that is pointing in the positive z-direction, is currently levitating at a z-coordinate h. If the mass of the magnet is M, determine h.

29. Magnetic Compression***

Consider a finite solenoid, with η turns per unit length, that is carrying a current I. The length of its cylindrical axis squared is much larger than its cross sectional area A. Argue qualitatively why this finite solenoid should experience a compressive force in the axial direction, in addition to the magnetic pressure discussed in Section 8.10.1. Determine the magnitude of this compressive force via the principle of virtual work. Finally, determine the force between two of such solenoids with their ends placed near each other and their axes aligned. The two solenoids carry currents I_1 and I_2, which are not necessarily in the same direction.

Solutions

1. Interface*

We know that the electric field and current density at a point inside a material are related by

$$E = \rho J$$

where ρ is the resistivity. Thus, the electric fields inside the two materials are $E_1 = \rho_1 J$ and $E_2 = \rho_2 J$ respectively. Finally, we know from the chapter on electrostatics that the discontinuity in the normal component of electric field ΔE is related to the surface charge density of the surface charges σ_{surf} on a sheet (which, in this case, is the interface) by

$$\Delta E = \frac{\sigma_{surf}}{\varepsilon_0}.$$

Thus,

$$\sigma_{surf} = \varepsilon_0 \Delta E = \varepsilon_0 J(\rho_2 - \rho_1).$$

2. Cylindrical Current**

As the conductivity is now non-uniform, there will be charges located in the mediating material (we cannot conclude that the surface integral of the electric field over an infinitesimal cube is zero from the fact that the surface integral of the current density over the cube is zero due to the steady current condition). However, we can still exploit the steady current condition and the axial symmetry of the set-up. Let the total current flowing between the shells be I — this must be the current flowing out of every cylindrical shell of radius r, $a < r < b$. Due to axial symmetry, the current density is

$$J = \frac{I}{2\pi r l}$$

at a radial distance r, radially outwards. The electric field is thus

$$E(r) = \frac{J}{\sigma} = \frac{I}{2\pi k l}$$

radially outwards. The potential drop from the inner shell to the outer shell is then

$$V = \int_a^b E dr = \int_a^b \frac{I}{2\pi k l} dr = \frac{I(b-a)}{2\pi k l}.$$

Evidently, the resistance between the shells is

$$\frac{b-a}{2\pi kl}.$$

From above, when the potential difference is $V = V_0$, the current flowing between the shells is

$$I = \frac{2\pi klV_0}{b-a}.$$

The current density and electric field at a radius r are thus

$$J = \frac{I}{2\pi rl} = \frac{kV_0}{(b-a)r},$$

$$E = \frac{J}{\sigma} = \frac{V_0}{(b-a)}$$

radially outwards. Now, consider a small box element in cylindrical coordinates at a radial distance r, with side lengths dr, dz and $rd\theta$ where θ is the azimuthal angle. The total electric flux through this box only stems from the faces perpendicular to the radial direction.

$$\Phi_E = E(r+dr) \cdot (r+dr)d\theta dz - E(r) \cdot rd\theta dz.$$

Since $E(r) = \frac{V_0}{(b-a)}$ is a constant,

$$\Phi_E = \frac{V_0}{(b-a)} dr d\theta dz.$$

Let the volume density of the infinitesimal box element be ρ. Then, the total charge stored in this box is

$$Q = \rho rd\theta dzdr.$$

By Gauss' law,

$$\Phi_E = \frac{Q}{\varepsilon_0}$$

which implies that

$$\rho = \frac{\varepsilon_0 E}{r} = \frac{\varepsilon_0 V_0}{(b-a)r}.$$

3. Four Point Probe**

We can apply the principle of superposition by considering the two following set-ups. Firstly, inject current I into A and withdraw it from infinity ($x \to +\infty$). The current I will be dispersed uniformly in a hemispherical shape such that the current density, as a function of radial distance r from A, is

$$J(r) = \frac{I}{2\pi r^2}$$

radially outwards from A in this first set-up. By Ohm's law, the electric field strength $E(r)$ is

$$E(r) = \rho J(r) = \frac{\rho I}{2\pi r^2}.$$

The potential difference between C and D in this set-up is

$$V_{DC} = V_C - V_D = -\int_{\sqrt{2}a}^{a} \frac{\rho I}{2\pi r^2} dr = \frac{\rho I}{2\pi}\left(\frac{1}{a} - \frac{1}{\sqrt{2}a}\right)$$

where we have exploited the path independence of the line integral of an electrostatic field to integrate along a strictly radial path. We have also used the fact that C and D are radial distances a and $\sqrt{2}a$ away from A.

Now, consider a separate set-up where current I is injected at infinity ($x \to +\infty$) and withdrawn from B. The current density, as a function of radial distance r from B, is similarly

$$J'(r) = -\frac{I}{2\pi r^2}$$

where the negative sign indicates that the $\boldsymbol{J'}$ is radially inwards with respect to B.

$$E'(r) = -\frac{\rho I}{2\pi r^2}.$$

The potential difference between C and D in this set-up is

$$V'_{DC} = V'_C - V'_D = -\int_{a}^{\sqrt{2}a} -\frac{\rho I}{2\pi r^2} dr = \frac{\rho I}{2\pi}\left(\frac{1}{a} - \frac{1}{\sqrt{2}a}\right)$$

where C and D are now radial distances $\sqrt{2}a$ and a away from B. Finally, we can superpose these set-ups to obtain a combined set-up where current I is injected into A and withdrawn from B (i.e. our original set-up). The currents injected at and withdrawn from infinity in the two set-ups nullify each other. The potential difference between C and D in the original set-up is hence

$$V_{DC} + V'_{DC} = 2 \times \frac{\rho I}{2\pi} \left(\frac{1}{a} - \frac{1}{\sqrt{2}a} \right) = V$$

$$\rho = \frac{\sqrt{2}\pi a V}{(\sqrt{2} - 1)I}.$$

4. Leaking Charge**

The rate of charge flowing out of an infinitesimal surface element dA on the surface bounding the volume V is $\sigma E_n dA$ by Ohm's law, where E_n is the normal component of electric field at that surface element. Therefore,

$$\dot{q} = - \oiint_S \sigma E_n dA = -\sigma \oiint_S E_n dA$$

where S is the surface bounding volume V. Observe that the last integral is simply the total electric flux cutting across S! We know from Gauss' law that this must be $\frac{q}{\varepsilon_0}$. Thus,

$$\dot{q} = -\frac{\sigma}{\varepsilon_0} q.$$

Separating variables and integrating,

$$\int_{q_0}^{q} \frac{1}{q} dq = - \int_{0}^{t} \frac{\sigma}{\varepsilon_0} dt$$

$$\implies q = q_0 e^{-\frac{\sigma}{\varepsilon_0} t}.$$

This result is extremely surprising in that we have not assumed anything about the shape of volume V!

5. Flipping Switch*

Denote the segments connecting each of bulbs 1 and 2 between A and B as segments 1 and 2. Denote the external wire containing the switch between A and B as segment 3. When the switch is closed, there is no emf induced in the loop formed by segments 2 and 3, as there is no rate of change of magnetic flux. Therefore, by Eq. (8.25), the voltage across bulb 2 is zero (as there is no emf in the loop and the voltage across an ideal wire is zero) and it is extinguished. However, applying the same argument to the loop formed by segments 1 and 3, in which the emf is non-zero, would lead to the conclusion that a non-zero voltage exists across bulb 1 — causing it to be illuminated.

When segment 3 is in the midst of being flipped, applying the previous argument would yield non-zero voltage across both bulbs — implying that they are both lit up (note that the gradual movement of segment 3 implies that any motional emf is negligible). However, once segment 3 reaches the final state, there would be zero voltage across bulb 1 and non-zero voltage across bulb 2. Therefore, bulb 1 is extinguished while bulb 2 is illuminated.

6. Voltmeters*

Applying Faraday's law to the middle loop, the anti-clockwise emf through this loop is

$$\varepsilon = -\frac{d\Phi^r_B}{dt} = -\alpha.$$

By Eq. (8.25), the clockwise current through the resistors is

$$I = \frac{\alpha}{R_1 + R_2}.$$

Therefore, the voltage drops across the resistors are

$$V_1 = IR_1 = \frac{\alpha R_1}{R_1 + R_2}$$

from the bottom to top end and

$$V_2 = IR_2 = \frac{\alpha R_2}{R_1 + R_2}$$

from the top to bottom end. Now, consider the right-most mesh. Since there is no changing magnetic flux through this loop, the total induced emf is zero — implying that the sum of voltages along this loop is zero. If the voltage across the voltmeter is V_{volt2} when going from the bottom terminal to the top terminal,

$$V_{volt2} - V_2 = 0$$

when applying Eq. (8.25) to an anti-clockwise loop.

$$V_{volt2} = V_2 = \frac{\alpha R_2}{R_1 + R_2}.$$

Next, the anti-clockwise emf induced in the left-most mesh is $-\beta$. Hence, if we let the voltage across the left voltmeter be V_{volt1} from the top to bottom terminal, applying Eq. (8.25) in the anti-clockwise direction to the left-most loop yields

$$-\beta - V_1 + V_{volt1} = 0$$

$$V_{volt1} = \beta + \frac{\alpha R_1}{R_1 + R_2}.$$

7. Compressing Square Loop*

Let the instantaneous self-inductance and current of the loop be L and I respectively. By the universal flux rule, the total clockwise emf through the loop is

$$\varepsilon = -\frac{d\Phi_B}{dt} - \frac{d(LI)}{dt}.$$

The first term is the motional emf caused by the loop being compressed in the presence of the external magnetic field while the second term is the emf (partly motional due to the loop being compressed under its own field and partly induced due to its varying current) caused by the loop itself. Denoting the loop's resistance as $R = 10.0\Omega$, the clockwise current I through the loop is

$$I = \frac{\varepsilon}{R}.$$

The average anti-clockwise current is thus

$$\langle I \rangle = -\frac{1}{R}\left\langle \frac{d\Phi_B}{dt} \right\rangle - \frac{1}{R}\left\langle \frac{d(LI)}{dt} \right\rangle$$
$$= -\frac{1}{R} \cdot \frac{\Delta\Phi_B}{\Delta t} - \frac{1}{R}\frac{\Delta(LI)}{\Delta t}$$
$$= -\frac{B}{R} \cdot \frac{\Delta A}{\Delta t}$$
$$= 0.121\text{A (3sf)},$$

where ΔA is the change in the loop's area and $B = 0.100\text{T}$ is the external magnetic field. We have used the fact that the initial and final currents through the loop are both zero in writing the third equality and substituted $B = 0.100\text{T}$, $R = 10.0\Omega$, $\Delta A = 9(\frac{\sqrt{3}}{2} - 1)\text{m}^2$ (since the area of the rhombus is $3^2 \cdot \sin 60° = \frac{9\sqrt{3}}{2}\text{m}^2$) and $\Delta t = 0.100\text{s}$ in the last step. The positive value indicates that the current is clockwise.

8. Voltage due to Solenoid*

In the cross section of the solenoid that contains A and B, draw two straight lines OA and OB from the center of the solenoid O and consider an arbitrary path joining A and B, AB, that does not cross the solenoid. From Faraday's law, we know that the total emf ε induced along the loop produced by these three lines is simply the rate of change of magnetic flux through the circular sector of angle θ. Since the area of the sector is $\frac{\theta}{2}R^2$, the anti-clockwise emf induced is

$$\varepsilon = -\frac{dB}{dt} \cdot \frac{\theta}{2} R^2$$

$$\varepsilon = -\frac{\theta B_0 \omega R^2}{2} \cos \omega t.$$

Now, we know from Eq. (8.25) that

$$\varepsilon + V_{OA} + V_{AB} + V_{BO} = 0.$$

We know from the axial symmetry of the solenoid that the induced non-conservative electric field should only be directed in the azimuthal direction. Therefore, $V_{AB} = V_{BO} = 0$ as the paths are purely radial. Then,

$$V_{AB} = -\varepsilon = \frac{\theta B_0 \omega R^2}{2} \cos \omega t.$$

9. Moving Loop*

Ampere's law gives the magnetic field at a distance x from the wire as

$$B = \frac{\mu_0 I_1}{2\pi x}.$$

Thus, the total magnetic flux through the loop is

$$\Phi_B = \iint B \cdot dA$$

$$= h \int_r^{r+l} \frac{\mu_0 I_1}{2\pi x} dx$$

$$= \frac{\mu_0 I_1 h}{2\pi} \ln \frac{r+l}{r}.$$

The emf is then

$$\varepsilon = -\frac{d\Phi_B}{dt}$$

$$= -\frac{\mu_0 I_1 h}{2\pi} \cdot \frac{r}{r+l} \cdot -\frac{l}{r^2} \cdot \frac{dr}{dt}$$

$$= \frac{\mu_0 I_1 l h u}{2\pi r(r+l)}$$

as $\frac{dr}{dt} = u$. Lastly, by Eq. (8.25),

$$I_2 = \frac{\varepsilon}{R} = \frac{\mu_0 I_1 h l u}{2\pi r(r+l)R}.$$

The answer to the second part is that you, or whatever entity is pulling the loop, are providing the work by exerting a force in maintaining the loop's velocity! Consider a charge in the left end of the loop. The top and bottom parts of the loop have no net effect on the loop as the magnetic force on them cancels out.

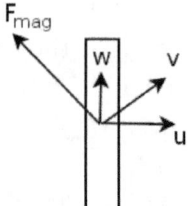

Figure 8.14:   Magnetic force on left edge

A charge travels at a velocity $\boldsymbol{w}$ along the wire but actually travels with a velocity $\boldsymbol{v} = \boldsymbol{u} + \boldsymbol{w}$ as the wire is also moving. Thus, the magnetic force is perpendicular to $\boldsymbol{v}$ and is in the direction as shown in Fig. 8.14. At equilibrium, the vertical component of the magnetic force provides the emf to drive the current. Thus, you must supply a force that is equal to the horizontal component of the magnetic force on the charge. The horizontal component of the magnetic force on the left part of the loop is simply given by

$$F_{xleft} = -B \cdot I_2 \cdot h = -\frac{\mu_0 I_1}{2\pi r} \cdot I_2 \cdot h,$$

as I_2 is defined to be the current along the loop. Similarly, for the right part of the loop (take note of the direction),

$$F_{xright} = \frac{\mu_0 I_1}{2\pi(r+l)} \cdot I_2 \cdot h.$$

The external force needed to maintain the loop at a constant velocity is

$$F_{ext} = -(F_{xleft} + F_{xright}) = \frac{\mu_0 I_1 I_2 hl}{2\pi r(r+l)}.$$

The power delivered by the force is

$$P = F_{ext} \cdot u = \frac{\mu_0 I_1 I_2 hlu}{2\pi r(r+l)} = \varepsilon I_2$$

which is the power dissipated in the resistor.

10. Loop Exiting Magnetic Field*

The magnetic flux as a function of x is

$$\Phi_B = B \cdot A = Bx^2.$$

Thus,

$$\varepsilon = -\frac{d\Phi_B}{dt} = -2Bx\frac{dx}{dt} = 2Bxv$$

as $v = -\frac{dx}{dt}$. As shown in the chapter on magnetism, the magnetic force on a current-carrying wire of an arbitrary shape in a uniform magnetic field is simply given by

$$F_{mag} = BIh,$$

where h is the linear distance between the terminals of the wire. Its direction is akin to that of the force on a straight wire carrying current I from the start point to the end point. In this case, the terminals of the wire are the two intersections of the loop with the boundary of the magnetic field. The distance between them is then $h = 2x$. Thus,

$$|F_{ext}| = |F_{mag}| = B \cdot \frac{\varepsilon}{R} \cdot 2x = \frac{4B^2 x^2 v}{R}.$$

The direction of the external force is along the velocity of the loop as the emf generated tends to reduce the velocity of the exiting loop.

11. Rotating Wheel*

(a) Applying the result of Chapter 7, Problem 5, the torque about the center of the wheel is opposite in direction to the external magnetic field so the wheel rotates clockwise. A more intuitive explanation is that closing the switch generates a current in the wheel and thus a magnetic flux. The wheel will subsequently try to reduce this increase in magnetic flux by decreasing the current via a motional emf, across the center and the rim, that opposes the emf of the battery. The wheel must hence rotate clockwise to produce such a motional emf.

(b) Applying the result of Chapter 7, Problem 5,

$$\tau = -\frac{BIr^2}{2} = -\frac{0.5 \cdot 6 \cdot 0.2^2}{2} = -0.06 \text{Nm},$$

where the negative sign indicates that the torque about the center is clockwise.

(c) The clockwise angular velocity of the wheel increases at a decreasing rate until it reaches a steady value $\omega = \frac{2\varepsilon}{Br^2} = \frac{2 \cdot 1}{0.5 \cdot 0.2^2} = 100 \text{s}^{-1}$. Firstly, the clockwise angular velocity of the wheel increases due to the clockwise torque about its center. However, as the angular velocity of the wheel increases, the motional emf produced across the center and the rim increases and opposes the emf of the battery to a greater extent — causing the current flowing in the wheel to decrease and thus the torque to decrease. Therefore, the wheel's clockwise angular velocity increases at a decreasing rate. The wheel's angular velocity will cease to change once the motional emf across its center and rim, $\frac{\omega Br^2}{2}$ (identical to that previously computed for a rotating disk), balances the emf ε of the battery such that no current flows in the wheel. Therefore, the terminal angular velocity of the wheel is $\omega = \frac{2\varepsilon}{Br^2} = \frac{2 \cdot 1}{0.5 \cdot 0.2^2} = 100 \text{s}^{-1}$.

12. Coil in Magnetic Field**

The induced emf in the coil is

$$\varepsilon = -\frac{d\Phi_B}{dt} - L\frac{dI}{dt} = Bwv - L\frac{dI}{dt}$$

where Φ_B is the magnetic flux due to the external field (coming out of the page). We take all quantities to be positive anti-clockwise. Furthermore, by Eq. (8.25),

$$\varepsilon = IR$$

where I is the anti-clockwise current.

$$\frac{dI}{dt} = \frac{Bwv}{L} - \frac{R}{L}I.$$

Separating variables and integrating,

$$\int_0^I \frac{1}{I - \frac{Bwv}{R}} dI = -\int_0^t \frac{R}{L} dt$$

$$\ln \left| \frac{I - \frac{Bwv}{R}}{\frac{Bwv}{R}} \right| = -\frac{R}{L}t.$$

Observe that since $I = 0$ at $t = 0$, $I \le \frac{Bwv}{R}$ for all subsequent times. Then,

$$\left| \frac{I - \frac{Bwv}{R}}{\frac{Bwv}{R}} \right| = \frac{\frac{Bwv}{R} - I}{\frac{Bwv}{R}}.$$

$$I = \frac{Bwv}{R}(1 - e^{-\frac{R}{L}t}).$$

The magnetic force pulling the loop back into the magnetic field is

$$F_{mag} = -BIw = -\frac{B^2w^2v}{R}(1 - e^{-\frac{R}{L}t}).$$

The external force required to oppose this magnetic force is equal in magnitude and opposite in direction. Therefore, the power delivered by this external force is

$$P = -F_{mag}v = \frac{B^2w^2v^2}{R}(1 - e^{-\frac{R}{L}t}).$$

Now, let us compare this with the rate of increase in energy of the other entities. Energy is mainly dispersed as heat to the external surroundings and stored in the magnetic field of the inductor. Thus, the rate of increase in energy of the other entities is

$$\frac{dE}{dt} = I^2R + \frac{d\left(\frac{1}{2}LI^2\right)}{dt},$$

$$= I^2R + LI\frac{dI}{dt},$$

$$= \frac{B^2w^2v^2}{R}e^{-\frac{2R}{L}t} + \frac{B^2w^2v^2}{R} - \frac{2B^2w^2v^2}{R}e^{-\frac{R}{L}t},$$

$$+ L\frac{Bwv}{R}(1 - e^{-\frac{R}{L}t})\frac{Bwv}{L}e^{-\frac{R}{L}t}$$

$$= \frac{B^2w^2v^2}{R} - \frac{B^2w^2v^2}{R}e^{-\frac{R}{L}t}$$

which is consistent with the power delivered by the external force.

13. Rotating Ring**

(a) Define θ as the angle between the magnetic field vector and the area vector of the ring. There are two possible choices for the area vector, so choose the one that produces the maximum positive flux initially. The magnetic flux through the ring when it has rotated an angle θ is

$$\Phi_B = B\pi r^2 \cos\theta.$$

The emf through the ring is thus

$$\varepsilon = -\frac{d\Phi_B}{dt} = B\pi r^2 \dot{\theta}\sin\theta.$$

The current through the ring is then

$$I = \frac{\varepsilon}{R} = \frac{B\pi r^2}{R}\dot{\theta}\sin\theta.$$

As such, the magnetic dipole moment of the ring is

$$\boldsymbol{\mu} = I \cdot \pi r^2 \hat{\boldsymbol{n}} = \frac{B\pi^2 r^4}{R}\dot{\theta}\sin\theta\hat{\boldsymbol{n}},$$

where $\hat{\boldsymbol{n}}$ is the unit vector in the direction of the area vector that we have defined previously. The torque on the ring is

$$\boldsymbol{\tau} = \boldsymbol{\mu} \times \boldsymbol{B}$$

or

$$\tau = -\mu B\sin\theta = -\frac{B^2\pi^2 r^4}{R}\dot{\theta}\sin^2\theta,$$

where the negative sign reflects the fact that the torque tends to reduce the angular velocity of the ring in order to oppose the rate of change of magnetic flux. Applying $\tau = M\ddot{\theta}$ about the axis of rotation of the ring, where $M = \frac{1}{2}mr^2$ is the moment of inertia of a ring with respect to an axis passing through a diameter,

$$\frac{1}{2}mr^2\ddot{\theta} = -\frac{B^2\pi^2 r^4}{R}\dot{\theta}\sin^2\theta.$$

Another way to obtain this equation is to directly equate the rate of change of kinetic energy of the loop with the negative of the power dissipated due to resistive heating.

$$\frac{d\left(\frac{1}{2}M\dot{\theta}^2\right)}{dt} = M\dot{\theta}\ddot{\theta} = -I^2 R$$

$$\frac{1}{2}mr^2\dot{\theta}\ddot{\theta} = -\frac{B^2\pi^2 r^4}{R}\dot{\theta}^2 \sin^2\theta$$

$$\implies \frac{1}{2}mr^2\ddot{\theta} = -\frac{B^2\pi^2 r^4}{R}\dot{\theta}\sin^2\theta.$$

Rearranging,

$$\int_{\omega_0}^{0} d\dot{\theta} = \int_{0}^{\phi} -\frac{B^2\pi^2 r^2}{mR}(1 - \cos 2\theta)d\theta$$

$$\omega_0 = \frac{B^2\pi^2 r^2}{mR}\left(\phi - \frac{\sin 2\phi}{2}\right)$$

$$2\phi - \sin 2\phi = \frac{2mR\omega_0}{B^2\pi^2 r^2}.$$

(b) Substituting $\omega_0 = 2\pi \cdot 8 = 16\pi$ and the other parameters given in the problem,

$$\frac{2mR\omega_0}{B^2\pi^2 r^2} = \frac{3200}{9\pi} = 113.2 \text{ (3sf)}.$$

Defining $x = 2\phi$, we require

$$x - \sin x = \frac{2mR\omega_0}{B^2\pi^2 r^2}.$$

Let $f(x) = x - \sin x$, then $f'(x) = 1 - \cos x$ which shows that $f(x)$ is an increasing function, as $\cos x \leq 1$. This implies that there will only be a unique solution to our equation (if a solution exists). Now, observe that

$$36\pi < 113.1 < \frac{3200}{9\pi}$$

$$40\pi > 125 > \frac{3200}{9\pi}.$$

Since $f(x)$ is continuous, the desired solution x to our equation satisfies

$$36\pi < x < 40\pi$$

$$\implies 18\pi < \phi < 20\pi$$

so the ring only rotates 9 complete rounds.

14. Railgun**

The bar experiences a magnetic force leftwards of magnitude

$$F_{mag} = BId.$$

The current through the bar is given by

$$I = \frac{\varepsilon + \varepsilon_{ind}}{R}$$

where ε_{ind} refers to the motional emf induced.

$$\varepsilon_{ind} = -\frac{d\Phi_B}{dt}.$$

The magnetic flux through the loop when the bar is at a distance x from the battery is

$$\Phi_B = Bxd.$$

Thus,

$$\varepsilon_{ind} = -Bd\frac{dx}{dt} = -Bdv,$$

$$F_{mag} = \frac{Bd}{R}(\varepsilon - Bdv),$$

$$m\dot{v} = \frac{B^2d^2}{R}\left(\frac{\varepsilon}{Bd} - v\right),$$

$$\int_0^v \frac{1}{\frac{\varepsilon}{Bd} - v}dv = \int_0^t \frac{B^2d^2}{mR}dt$$

$$\left[-\ln\left|\frac{\varepsilon}{Bd} - v\right|\right]_0^v = \frac{B^2d^2}{mR}t.$$

From the expression for $\dot{v}$, it is evident that v is always smaller or equal to $\frac{\varepsilon}{Bd}$ at all later times if $v = 0$ at $t = 0$. Then, $\left|\frac{\frac{\varepsilon}{Bd} - v}{\frac{\varepsilon}{Bd}}\right| = \frac{\frac{\varepsilon}{Bd} - v}{\frac{\varepsilon}{Bd}}$.

$$\frac{\frac{\varepsilon}{Bd} - v}{\frac{\varepsilon}{Bd}} = e^{-\frac{B^2d^2}{mR}t}$$

$$v = \frac{\varepsilon}{Bd}\left(1 - e^{-\frac{B^2d^2}{mR}t}\right).$$

15. Inclined Loop**

(a) Let v denote the instantaneous velocity of the rod. The motional emf across the rod is

$$\varepsilon = \int (\boldsymbol{v} \times \boldsymbol{B}) \cdot d\boldsymbol{s} = vBl \cos \theta$$

where the line integral runs from the left end to right end of the rod, so that the rod acts as a battery of emf ε, with the positive terminal at its right end. Now, let I_1 and I_2 denote the currents flowing through R_1 and R_2 from their right to left ends. Then, the current in the rod is $I_1 + I_2$ from its left to right end for no charge to accumulate anywhere in the circuit. Applying Eq. (8.25) to the two loops containing one of R_1 or R_2 and the rail,

$$I_1 R_1 + (I_1 + I_2)R_3 = \varepsilon,$$
$$(I_1 + I_2)R_3 + I_2 R_2 = \varepsilon.$$

Solving,

$$I_1 = \frac{\varepsilon R_2}{R_1 R_2 + R_1 R_3 + R_2 R_3} = \frac{R_2}{R_1 R_2 + R_1 R_3 + R_2 R_3} vBl \cos \theta,$$

$$I_2 = \frac{\varepsilon R_1}{R_1 R_2 + R_1 R_3 + R_2 R_3} = \frac{R_1}{R_1 R_2 + R_1 R_3 + R_2 R_3} vBl \cos \theta,$$

$$I_1 + I_2 = \frac{\varepsilon (R_1 + R_2)}{R_1 R_2 + R_1 R_3 + R_2 R_3} = \frac{R_1 + R_2}{R_1 R_2 + R_1 R_3 + R_2 R_3} vBl \cos \theta,$$

$$= kvBl \cos \theta,$$

where we have let

$$k = \frac{R_1 + R_2}{R_1 R_2 + R_1 R_3 + R_2 R_3}.$$

(b) Since the current in the rod is perpendicular to the external magnetic field, the magnetic force on the rod is

$$F = -B(I_1 + I_2)l = -kvB^2 l^2 \cos \theta$$

in the horizontal direction (parallel to the flat ground from the side view of the inclined plane). The negative sign indicates that F is leftwards in the figure.

(c) The component of F along the slope is

$$F_\parallel = F\cos\theta = -kvB^2l^2\cos^2\theta.$$

By Newton's second law, the equation of motion of the rod is

$$m\dot{v} = mg - kvB^2l^2\cos^2\theta.$$

Separating variables,

$$\int_0^v \frac{1}{v - \frac{mg}{kB^2l^2\cos^2\theta}}dv = \int_0^t -\frac{kB^2l^2\cos^2\theta}{m}dt$$

$$\ln\left|\frac{kB^2l^2\cos^2\theta v}{mg} - 1\right| = -\frac{kB^2l^2\cos^2\theta}{m}t.$$

Due to the facts that v is initially zero, $\dot{v} > 0$ only when $v < \frac{mg}{kB^2l^2\cos^2\theta}$ and $\dot{v} \le 0$ otherwise, $\frac{kB^2l^2\cos^2\theta v}{mg} \le 1$ so we must take the negative value of the argument in removing the absolute value brackets. Rearranging,

$$v = \frac{mg}{kB^2l^2\cos^2\theta}\left(1 - e^{-\frac{kB^2l^2\cos^2\theta}{m}t}\right)$$

$$a = \dot{v} = ge^{-\frac{kB^2l^2\cos^2\theta}{m}t}.$$

(d)

$$P_{mag} = \mathbf{F}\cdot\mathbf{v} = -Fv\cos\theta = -kv^2B^2l^2\cos^2\theta.$$

(e) By the work-energy theorem, the rate of change of mechanical energy of the rail is the total power delivered by non-conservative forces which is simply that due to F in this case. Thus, we just have to show that $-P_{mag} = P$, where P is the rate of heat dissipated in the resistors.

$$P = I_1^2 R_1 + I_2^2 R_2 + (I_1 + I_2)^2 R_3$$

$$= \frac{\varepsilon^2}{(R_1 R_2 + R_1 R_3 + R_2 R_3)^2}(R_2^2 R_1 + R_1^2 R_2 + R_1^2 R_3 + R_2^2 R_3 + 2R_1 R_2 R_3)$$

$$= \frac{\varepsilon^2(R_1 + R_2)}{R_1 R_2 + R_1 R_3 + R_2 R_3}$$

$$= kv^2 B^2 l^2 \cos^2\theta$$

$$= -P_{mag}.$$

16. Magnetic Wave**

Let the left and right ends of the loop be at x-coordinates x and $x+l$. The total magnetic flux through this loop at the current instance is

$$\Phi_b = h \int_x^{x+l} B_0 \cos(kx - \omega t)\,dx$$

$$= \left[\frac{hB_0}{k}\sin(kx - \omega t)\right]_x^{x+l}$$

$$= \frac{hB_0}{k}\left[\sin\left[k(x+l) - \omega t\right] - \sin(kx - \omega t)\right].$$

The rate of change of magnetic flux through this loop is

$$\frac{d\Phi_B}{dt} = hB_0 v\left[\cos[k(x+l) - \omega t] - \cos(kx - \omega t)\right]$$

$$-\frac{hB_0\omega}{k}\left[\cos[k(x+l) - \omega t] - \cos(kx - \omega t)\right]$$

$$= hB_0\left(v - \frac{\omega}{k}\right)\left[\cos\left[k(x+l) - \omega t\right] - \cos(kx - \omega t)\right]$$

where $v = \dot{x}$ is the loop's instantaneous x-velocity (remember that x changes with time too as the loop moves). We can immediately conclude now that the condition for the loop to travel at a constant velocity is for the above expression to be zero such that no emf, and thus no current, is induced (and finally no force). However, we shall hold this off for now to see why the loop tends to a terminal velocity for the second part of the question. The anti-clockwise emf induced in the loop is simply negative of the above expression. The anti-clockwise current induced is

$$I = \frac{hB_0}{R}\left(\frac{\omega}{k} - v\right)\left[\cos[k(x+l) - \omega t] - \cos(kx - \omega t)\right].$$

The net force on the loop is due to magnetic force on the two vertical segments along the y-direction. The end at the larger x-coordinate experiences a force $B_0 \cos\left[k(x+l) - \omega t\right] Ih$ while the other end experiences a force $-B_0 \cos(kx - \omega t)Ih$. Therefore, the force in the positive x-direction is

$$F = \frac{h^2 B_0^2}{R}\left(\frac{\omega}{k} - v\right)\left[\cos[k(x+l) - \omega t] - \cos(kx - \omega t)\right]^2.$$

The first condition in the question occurs when

$$\cos\left[k(x+l) - \omega t\right] = \cos(kx - \omega t)$$

$$l = n\frac{2\pi}{k}$$

for some integer n. The second condition occurs when the terms in the round brackets yield zero. The terminal velocity is

$$v_0 = \frac{\omega}{k}.$$

To see why the loop tends to this x-velocity, observe that when $v > v_0$, the force is negative and when $v < v_0$, the force is positive.

17. Pulling a Solenoid**

The solenoids produce uniform individual magnetic fields $B_1 = \mu_0 \eta_1 I_1$ and $B_2 = \mu_0 \eta_2 I_2$ within themselves. Let us first determine the emf in the outer solenoid generated by pulling out the inner solenoid. By the universal flux rule,

$$\varepsilon_1 = -\frac{d\Phi_{B1}^{linkage}}{dt}$$

where $\Phi_{B1}^{linkage}$ is the magnetic flux linkage through the outer solenoid. As the inner solenoid is withdrawn from the outer solenoid (i.e. more of it protrudes outside), the magnetic field that parts of the inner solenoid previously occupied in the outer solenoid decreases from $B_1 + B_2$ to B_1. In time dt, the volume in the outer solenoid for which this happens is $A_2 v dt$. Therefore, the change in magnetic flux linkage through the outer solenoid is

$$d\Phi_{B1}^{linkage} = -\eta_1 B_2 A_2 v dt$$

where we have noted that there are $\eta_1 v dt$ turns of the outer solenoid that "experience" this change in magnetic field. Correspondingly,

$$\varepsilon_1 = \eta_1 B_2 A_2 v = \eta_1 \eta_2 \mu_0 I_2 A_2 v.$$

The current source must supply an emf negative of this $-\eta_1 \eta_2 \mu_0 I_2 A_2 v$. Similarly, applying the universal flux rule to the inner solenoid, the emf generated in the inner solenoid due to its motion is

$$\varepsilon_2 = -\frac{d\Phi_{B2}^{linkage}}{dt}.$$

During a time interval dt, the magnetic field within $\eta_2 v dt$ turns of the inner solenoid (those which have just left the outer solenoid) decreases from $B_1 + B_2$ to B_2. Therefore, the change in magnetic flux linkage through the inner solenoid is

$$d\Phi_{B2}^{linkage} = -\eta_2 B_1 A_2 v dt$$

$$\varepsilon_2 = \eta_2 B_1 A_2 v = \eta_1 \eta_2 \mu_0 I_1 A_2 v$$

such that the current source connected to the inner solenoid must deliver an emf $-\eta_1 \eta_2 \mu_0 I_1 A_2 v$. The ratio of the emfs delivered by the current sources to the solenoids is also equal to the ratio of ε_1 and ε_2.

$$\frac{\varepsilon_1}{\varepsilon_2} = \frac{I_2}{I_1}.$$

This makes sense for the following reason. The change in the magnetic flux linkage of each solenoid is solely due to the magnetic field of the other solenoid as its self-inductance and current do not vary. Therefore, if the mutual inductance between the solenoids is $M(t)$,

$$\varepsilon_1 = -\frac{d(MI_2)}{dt} = -\frac{dM}{dt} I_2,$$

$$\varepsilon_2 = -\frac{d(MI_1)}{dt} = -\frac{dM}{dt} I_1,$$

as I_1 and I_2 are constant.

$$\implies \frac{\varepsilon_1}{\varepsilon_2} = \frac{I_2}{I_1}.$$

This ratio is hence an innate consequence of the reciprocity of mutual inductance.

18. Ring on Ring**

The key observation is ironically the lack of interaction between the small ring and the large ring since the large ring is not charged. Thus, the force that the small ring exerts on the large ring is solely the normal force.

The changing magnetic field induces a non-conservative electric field which accelerates the charge in the tangential direction. Then, a normal force is needed to constrain the small ring such that it only moves along the large ring (circular motion). If we take anti-clockwise to be positive along the large ring (xy-plane), the negative change in magnetic flux through the ring is

$$-\frac{d\Phi_B}{dt} = -\frac{d((B_0 + \alpha t)\pi R^2)}{dt} = -\alpha \pi R^2.$$

Since

$$\varepsilon = \oint \boldsymbol{E} \cdot d\boldsymbol{s} = E \cdot 2\pi R$$

due to the axial symmetry of the system, applying Faraday's law: $\oint \boldsymbol{E} \cdot d\boldsymbol{s} = -\frac{d\Phi_B}{dt}$ yields the non-conservative electric field as

$$E = -\frac{\alpha R}{2}.$$

Note that $\boldsymbol{E}$ is purely tangential and is taken to be positive anti-clockwise. Considering the tangential forces on the small ring,

$$ma_\theta = qE = -q\frac{\alpha R}{2}$$

$$\implies v_\theta = -\frac{q\alpha Rt}{2m},$$

where v_t is the anti-clockwise tangential velocity as a function of time (note that the radial distance of the ring cannot change — causing the $2m\dot{r}\dot{\theta}$ term in the equation of motion in polar coordinates to vanish). When the small ring has a tangential velocity v_θ, the magnetic force it experiences is

$$\boldsymbol{F}_{mag} = q\boldsymbol{v}_\theta \times \boldsymbol{B} = -\frac{q^2\alpha RtB}{2m}\hat{\boldsymbol{r}}$$

where $\hat{\boldsymbol{r}}$ is the unit vector directed radially outwards. Lastly, the combination of the normal force on the small ring by the large ring and the magnetic force provides the centripetal force required for the small ring to exhibit circular motion. Thus,

$$N - \frac{q^2\alpha RtB}{2m} = -\frac{mv_\theta^2}{R}.$$

Shifting the second term on the left over and substituting $B = B_0 + \alpha t$,

$$N = \frac{q^2\alpha Rt}{4m}(2B_0 + \alpha t).$$

19. Work Done on Magnetic Dipole**

The torque on a magnetic dipole with a dipole moment μ in an external magnetic field B is

$$\tau = -\mu B \sin\theta$$

where $\mu = NIA$. The negative sign comes from the fact that the torque tends to reduce θ. Then, the external torque required to nullify this magnetic torque is

$$\tau_{ext} = -\tau = \mu B \sin\theta.$$

The work done by the external torque in rotating the dipole from $\theta = \frac{\pi}{2}$ to θ is then

$$W = \int_{\frac{\pi}{2}}^{\theta} \mu B \sin\theta = -\mu B \cos\theta.$$

This energy supplied by the mechanical work is used up in maintaining the current in the coil at a constant I. To see why an entity is required to maintain the current from a microscopic perspective, notice that when the coil rotates such that θ increases, the charges in the coil acquire a component velocity perpendicular to the coil, in additional to a velocity along the coil. This additional component of velocity leads to a Lorentz force that tends to changing the velocities of the charges along the coil — hence changing the current in the coil. Therefore, an external entity or circuit component is required to sustain the current. We can be quantitative about the power consumption of such a component by determining the induced emf. The magnetic flux through a single turn is

$$\Phi_B = BA \cos\theta.$$

By Faraday's law, the induced emf is

$$\varepsilon_{ind} = -N\frac{d\Phi_B}{dt} = NBA \sin\theta\dot{\theta}.$$

The emf that the circuit component needs to supply is then negative of this.

$$\varepsilon_{ext} = -NBA \sin\theta\dot{\theta}.$$

The power delivered by this component is then the potential difference across its ends (which is equal to its emf) multiplied by the current I that it sustains.

$$P_{ext} = -NIAB \sin\theta\dot{\theta} = -\mu B \sin\theta\dot{\theta}.$$

We compare this with the rate of work done by the external torque.

$$P_{torque} = \tau_{ext}\dot{\theta} = \mu B \sin\theta\dot{\theta}.$$

The two powers are equal in magnitude and opposite in sign. Therefore, when the mechanical work is positive, the work performed by the circuit component in maintaining the current in the coil is negative and perfectly negates the mechanical work. Vice versa for negative mechanical work.

20. Rectangular Toroid*

Draw a circular Amperian loop inside the toroid of radius r from the center of the toroid. Applying Ampere's law, the azimuthal magnetic field (actually, the magnetic field only has an azimuthal component as we have discussed in the previous chapter) as a function of radial distance r (for $r_1 \leq r \leq r_2$) is

$$B \cdot 2\pi r = \mu_0 NI$$

$$B = \frac{\mu_0 NI}{2\pi r}.$$

The magnetic flux through a turn of the toroid is

$$\Phi_B = \iint \boldsymbol{B} \cdot d\boldsymbol{A} = h \int_{r_1}^{r_2} \frac{\mu_0 NI}{2\pi r} dr = \frac{\mu_0 NIh}{2\pi} \ln\frac{r_2}{r_1}.$$

Thus, the self-inductance is given by

$$L = \frac{N\Phi_B}{I} = \frac{\mu_0 N^2 h}{2\pi} \ln\frac{r_2}{r_1}.$$

21. Solenoid in Solenoid**

Since it is difficult to calculate the magnetic field due to the short solenoid, we shall calculate the mutual inductance of the system due to a change in current of the long solenoid. Let the current in the long solenoid be I. The essentially constant magnetic field through the small solenoid due to the long solenoid is given by Ampere's law as

$$B = \mu_0 \eta I$$

and is parallel to the axes of both solenoids. Thus, the magnetic flux through the small solenoid is

$$\Phi_B = B \cdot A = \mu_0 \eta I \pi r^2.$$

The mutual inductance of the system is then given by

$$M = \frac{N\Phi_B}{I} = \mu_0 N\eta\pi r^2.$$

Note that N refers to the number of turns of the short solenoid while η refers to the turns per unit length of the long solenoid.

22. Inductance of Tetrahedron Sides**

Suppose that we run an anti-clockwise current in the loop given in the problem. We can see it as the composition of two anti-clockwise triangular current loops shown in Fig. 8.15. In this case, the anti-clockwise direction for each face is defined such that applying the right-hand-grip rule, along the edges of the face in that direction, produces an area vector normally outwards from the tetrahedron.

Figure 8.15: Two anti-clockwise current loops

Let the desired self-inductance be L'. By definition of the self-inductance, $L'I$ must be the total flux through the loop in the problem. This can be written as

$$L'I = 2\Phi_{self} + 2\Phi_{neigh},$$

where Φ_{self} is the magnetic flux through a triangular loop due to an anti-clockwise current I through itself and Φ_{neigh} is the magnetic flux through one triangular loop due to the anti-clockwise current flowing in the other triangular loop. To determine a relationship between Φ_{self} and Φ_{neigh}, observe that we can piece together two more anti-clockwise triangular current loops (on the "missing" sides) to form a tetrahedron with zero net current flowing through its edges. Since the surface of the tetrahedron is closed, the net magnetic flux emanating from it must be zero. Since the magnetic flux through each face of the complete tetrahedron is $\Phi_{self} + 3\Phi_{neigh}$, we must have

$$4\left(\Phi_{self} + 3\Phi_{neigh}\right) = 0$$

$$\implies \Phi_{neigh} = -\frac{1}{3}\Phi_{self}$$

$$L'I = \frac{4}{3}\Phi_{self}.$$

By definition of the self-inductance of a triangular loop, $\Phi_{self} = LI$. Thus,

$$L' = \frac{4}{3}L.$$

23. A Neat Mutual Inductance**

Run an anti-clockwise current I through the equilateral triangle loop and divide the rhombus into upper and lower equilateral triangles. Let the magnetic flux through the lower and upper portions due to the current I in the equilateral triangle loop be Φ_{near} and Φ_{far} respectively. We need to determine $\left|\frac{\Phi_{near}+\Phi_{far}}{I}\right|$ to compute the mutual inductance of this system.

To this end, let us adopt a new perspective to this problem. In light of the fact that the equilateral triangle and rhombus appear to be part of a large equilateral triangle of length $2l$, let us consider what happens to the self-inductance of an equilateral triangle loop if we scale its side lengths by a factor of two. By dimensional analysis, the magnetic flux through an equilateral triangle current loop due to itself should be proportional to its length l (magnetic flux has the dimensions of magnetic field, which has an inverse-length dimension, multiplied by area which has a squared length dimension). Therefore, the self-inductance of an equilateral triangle of side length $2l$ should be $2L$.

Next, an equilateral triangle loop of side length $2l$ and anti-clockwise current I can be deemed as the combination of four equilateral triangle loops of side length l that each carry an anti-clockwise current I. Let Φ_{self} denote the magnetic flux through an equilateral triangle loop due to its own anti-clockwise current I. From this perspective, the magnetic flux through the equilateral loop of side length $2l$ should be $4\Phi_{self} + 6\Phi_{near} + 6\Phi_{far}$ ($\Phi_{self} + \Phi_{near} + 2\Phi_{far}$ through each of the three triangles near the vertices and $\Phi_{self} + 3\Phi_{near}$ through the middle triangle). Since the self-inductance of an equilateral triangle loop of side length $2l$ has been asserted to be $2L$, this quantity must be $2LI$. Thus,

$$2LI = 4\Phi_{self} + 6\Phi_{near} + 6\Phi_{far}$$
$$-2LI = 6\Phi_{near} + 6\Phi_{far}$$

as $\Phi_{self} = LI$.

$$\Phi_{near} + \Phi_{far} = -\frac{1}{3}LI.$$

The desired mutual inductance is hence

$$M = \left| \frac{\Phi_{near} + \Phi_{far}}{I} \right| = \frac{1}{3}L.$$

24. Transformer**

Define L_1, L_2 and M as the respective self-inductances of the primary and secondary solenoids and the mutual inductance of the solenoids. The clockwise emf in the primary circuit due to the primary solenoid is

$$\varepsilon_1 = -L_1 \frac{dI_1}{dt} - M \frac{dI_2}{dt}.$$

The anti-clockwise emf in the secondary circuit due to the secondary solenoid is

$$\varepsilon_2 = -L_2 \frac{dI_2}{dt} - M \frac{dI_1}{dt}.$$

In this case, the solenoids are perfectly coupled such that

$$M = \sqrt{L_1 L_2}.$$

Thus,

$$\frac{\varepsilon_2}{\varepsilon_1} = \sqrt{\frac{L_2}{L_1}}.$$

Now, we assert that the self-inductance of a solenoid is proportional to its squared number of turns. This can be seen directly from the previous result in Section 8.6 regarding the self-inductance of a solenoid or from the following scaling arguments. When you scale the number of turns of a solenoid by a factor of k, the magnetic field within itself increases by a factor of k accordingly while the number of turns that the magnetic field passes through also increases by a factor of k — causing the magnetic flux linkage of the solenoid and hence its self-inductance to increase by a factor of k^2. This discussion implies

$$L_1 \propto N_1^2,$$

$$L_2 \propto N_2^2,$$

with all other parameters held constant. Thus,

$$\frac{\varepsilon_2}{\varepsilon_1} = \frac{N_2}{N_1}.$$

In the given set-up, ε_1 must be negative of the emf produced by the AC source as the net emf must be zero along the perfectly conducting primary circuit.

$$\varepsilon_1 = -\varepsilon_0 \cos \omega t$$

$$\implies \varepsilon_2 = -\frac{N_2}{N_1}\varepsilon_0 \cos \omega t.$$

Applying Eq. (8.25) to the secondary circuit, we have

$$\varepsilon_2 - I_2 R = 0$$

$$I_2 = -\frac{N_2 \varepsilon_0}{N_1 R} \cos \omega t$$

$$\implies \frac{dI_2}{dt} = \frac{N_2 \varepsilon_0 \omega}{N_1 R} \sin \omega t.$$

We know from the previous part that

$$\varepsilon_1 = -L_1 \frac{dI_1}{dt} - M \frac{dI_2}{dt}.$$

Therefore,

$$L_1 \frac{dI_1}{dt} + M \frac{dI_2}{dt} = \varepsilon_0 \cos \omega t$$

$$\frac{dI_1}{dt} = \frac{\varepsilon_0}{L_1} \cos \omega t - \frac{M}{L_1} \cdot \frac{N_2 \varepsilon_0 \omega}{N_1 R} \sin \omega t.$$

Since $\frac{M}{L_1} = \sqrt{\frac{L_2}{L_1}} = \frac{N_2}{N_1}$, the above is equivalent to

$$\frac{dI_1}{dt} = \frac{\varepsilon_0}{L_1} \cos \omega t - \frac{N_2^2 \varepsilon_0 \omega}{N_1^2 R} \sin \omega t$$

$$\implies I_1 = \frac{\varepsilon_0}{L_1 \omega} \sin \omega t + \frac{N_2^2 \varepsilon_0}{N_1^2 R} \cos \omega t,$$

where we do not include a constant of integration as we are looking at the particular solution to the current (which should not depend on initial conditions). We will compute the various rates of changes of energy but leave out the substitutions of the expressions for I_1 and I_2 which are unnecessary. The rate of power delivered by the AC source is

$$P_{AC} = \varepsilon_0 \cos \omega t \cdot I_1 = \varepsilon_0 I_1 \cos \omega t.$$

The rate of heat dissipated in the resistor R is

$$P_R = I_2^2 R.$$

The total potential energy stored in the pair of mutual inductors is

$$U = \frac{1}{2}L_1 I_1^2 + \frac{1}{2}L_2 I_2^2 + M I_1 I_2,$$

$$\begin{aligned}
\frac{dU}{dt} &= L_1 I_1 \frac{dI_1}{dt} + L_2 I_2 \frac{dI_2}{dt} + M \frac{dI_1}{dt} I_2 + M I_1 \frac{dI_2}{dt} \\
&= I_1 \left(L_1 \frac{dI_1}{dt} + M \frac{dI_2}{dt} \right) + I_2 \left(L_2 \frac{dI_2}{dt} + M \frac{dI_1}{dt} \right) \\
&= -\varepsilon_1 I_1 - I_2 \varepsilon_2 \\
&= \varepsilon_0 I_1 \cos \omega t - I_2^2 R.
\end{aligned}$$

The conservation of energy evidently holds as

$$P_{AC} = \frac{dU}{dt} + P_R.$$

25. Magnetic Field of Point Charge**

Align the positive x-axis with the charge $q's$ velocity and suppose that we wish to determine the magnetic field at an instantaneous distance r from charge q, whose position vector from the charge subtends an angle θ with the positive x-axis as shown in Fig. 8.16.

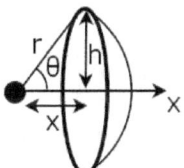

Figure 8.16: Spherical cap of radius r and half-angle θ about q

Due to the axial symmetry of this set-up and because the magnetic field should solely be azimuthal about the x-axis, we can consider the magnetic field of all points that are axially symmetric to the current one under consideration. The magnetic field strengths at these points should undertake a common value B and the direction of the magnetic fields should be azimuthal. This set of points form the bolded circle above, over which the line integral of the magnetic field is then $B \cdot 2\pi h$. Next, we want to find a convenient surface that spans this circle to apply the Ampere–Maxwell law to. An intuitive surface to choose is a spherical cap of radius r and half-angle θ, centered about charge q, as depicted in the diagram above. The real current crossing this surface is zero but the displacement current crossing it is

$$I_d = \varepsilon_0 \frac{d\Phi_E}{dt},$$

where Φ_E is the electric flux cutting across the spherical cap. We have shown in the chapter on electrostatics that a spherical cap of half-angle θ captures $\sin^2 \frac{\theta}{2}$ of the total electric flux emitted by a charge located at its center. Therefore,

$$\Phi_E = \frac{q}{\varepsilon_0} \sin^2 \frac{\theta}{2} = \frac{q}{2\varepsilon_0}(1 - \cos\theta) = \frac{q}{2\varepsilon_0}\left(1 - \frac{x}{\sqrt{h^2 + x^2}}\right),$$

where x and h are the distances defined in Fig. 8.16 above.

$$\frac{d\Phi_E}{dt} = -\frac{q}{2\varepsilon_0}\left(\frac{1}{\sqrt{h^2 + x^2}} - \frac{x^2}{(h^2 + x^2)^{\frac{3}{2}}}\right) \cdot \frac{dx}{dt} = \frac{qh^2 v}{2\varepsilon_0(h^2 + x^2)^{\frac{3}{2}}},$$

where we have used the fact that $\frac{dx}{dt} = -v$.

$$\implies I_d = \frac{qh^2 v}{2(h^2 + x^2)^{\frac{3}{2}}}.$$

By the Ampere–Maxwell law,

$$B \cdot 2\pi h = \mu_0 I_d = \frac{\mu_0 q h^2 v}{2(h^2 + x^2)^{\frac{3}{2}}}$$

$$\implies B = \frac{\mu_0 q h v}{4\pi(h^2 + x^2)^{\frac{3}{2}}} = \frac{\mu_0 q v \sin\theta}{4\pi r^2}.$$

Since the magnetic field is azimuthal and θ is the angle subtended by the position vector $\boldsymbol{r}$ of the point of concern from the charge and the velocity $\boldsymbol{v}$ of the charge, the above expression in vector form is

$$B = \frac{\mu_0 q \boldsymbol{v} \times \hat{\boldsymbol{r}}}{4\pi r^2}.$$

26. Bringing Two Loops*

Let the common self-inductance of the loops be L. Initially, the magnetic fluxes through the loops are

$$\Phi_{B1} = LI_1,$$

$$\Phi_{B2} = LI_2,$$

as their mutual inductance is zero when they are infinitely far apart. When the loops are brought closer together, they will possess a non-negligible mutual inductance M. However, the magnetic fluxes through the loops must be unchanged as they are perfectly conducting. Therefore,

$$LI_1 = LI_1' \pm MI_2',$$

$$LI_2 = LI_2' \pm MI_1'.$$

Multiplying the former equation by I_1' and subtracting the latter equation, multiplied by I_2', from it,

$$LI_1I_1' - LI_2I_2' = LI_1'^2 - LI_2'^2$$

$$I_1'^2 - I_1I_1' + I_2'(I_2 - I_2') = 0$$

$$I_1' = \frac{I_1 \pm \sqrt{I_1^2 + 4I_2'(I_2' - I_2)}}{2}.$$

We choose the positive root as $I_1' = I_1$ when $I_2' = I_2$.

$$I_1' = \frac{I_1 + \sqrt{I_1^2 + 4I_2'(I_2' - I_2)}}{2}.$$

27. Exiting Conducting Loop**

The magnetic flux through the loop is constant. When the loop is displaced towards the right by a distance x, the magnetic flux through the loop (positive out of the page) decreases by Blx. Therefore, for the magnetic flux in the loop to be maintained, the anti-clockwise current I in the loop must satisfy

$$LI = Blx$$

$$I = \frac{Bl}{L}x.$$

Since the magnetic field is uniform in the region, the magnetic force on the loop is BI multiplied by the distance between the two end points of the loop's intersection with the magnetic field. Therefore,

$$F = -BIl,$$

where the negative sign indicates that the force is directed leftwards. Applying Newton's second law,

$$\ddot{x} = -\frac{BIl}{m} = -\frac{B^2l^2}{mL}x,$$

which indicates a simple harmonic motion with angular frequency

$$\omega = \frac{Bl}{\sqrt{mL}}.$$

28. Levitating Magnet**

This set-up should remind you of a point charge placed above an infinite conducting plane and it can, in fact, be similarly solved via the method of mirror images. Since we are only interested in the magnetic field in the region $z \geq 0$, what are the boundary conditions that must be satisfied? Well, the normal component of magnetic field must be zero along $z = 0$ (technically, directly above the superconducting plane) by the properties of a supercon-ductor. The third uniqueness theorem then guarantees the unique solution to the magnetic field in the region $z \geq 0$ as long as we can find a magnetic field that satisfies this boundary condition. To make the normal component of magnetic field vanish along $z = 0$, we can introduce an imaginary mag-netic dipole moment $-\boldsymbol{\mu}$ at a z-coordinate $-h$, directly below the original magnetic dipole. The magnetic field in the region $z \geq 0$ is hence the superpo-sition of the fields of the two magnetic dipoles. The force felt by the physical magnetic dipole is

$$\boldsymbol{F} = \nabla(\boldsymbol{\mu} \cdot \boldsymbol{B}) = (\boldsymbol{\mu} \cdot \nabla)\boldsymbol{B}$$

where $\boldsymbol{B}$ is the magnetic field due to the image dipole. The second equality comes from the fact that there are no currents or changing electric fields at the location of the physical dipole (see Section 8.10.3). In terms of polar coordinates centered about the image dipole,

$$\boldsymbol{B} = -\frac{\mu_0 \mu}{4\pi r^3} \left(2\cos\theta\hat{\boldsymbol{r}} + \sin\theta\hat{\boldsymbol{\theta}}\right),$$

$$B_r = -\frac{\mu_0 \mu}{2\pi r^3} \cos\theta,$$

$$B_\theta = -\frac{\mu_0 \mu}{4\pi r^3} \sin\theta,$$

by Eq. (8.36), where θ is measured from the positive z-axis and $\boldsymbol{r}$ is the position vector joining the image dipole to the point of concern. Since the gradient in spherical coordinates is

$$\nabla = \frac{\partial}{\partial r}\hat{\boldsymbol{r}} + \frac{\partial}{r\partial\theta}\hat{\boldsymbol{\theta}} + \frac{\partial}{r\sin\theta\partial\phi}\hat{\boldsymbol{\phi}}$$

where ϕ is the azimuthal angle,

$$\boldsymbol{F} = \boldsymbol{\mu} \cdot \nabla B_r + \boldsymbol{\mu} \cdot \nabla B_\theta$$

$$= \boldsymbol{\mu} \cdot \left(\frac{3\mu_0\mu}{2\pi r^4} \cos\theta \hat{\boldsymbol{r}} + \frac{\mu_0\mu}{2\pi r^4} \sin\theta \hat{\boldsymbol{\theta}} \right)$$

$$+ \boldsymbol{\mu} \cdot \left(\frac{3\mu_0\mu}{4\pi r^4} \sin\theta \hat{\boldsymbol{r}} - \frac{\mu_0\mu}{4\pi r^4} \cos\theta \hat{\boldsymbol{\theta}} \right).$$

When $\theta = 0$ and $r = 2h$ (corresponding to the location of the physical dipole),

$$\boldsymbol{F} = \frac{3\mu_0\mu^2}{32\pi h^4}\hat{\boldsymbol{k}}.$$

For force balance,

$$\frac{3\mu_0\mu^2}{32\pi h^4} = Mg$$

$$h = \sqrt[4]{\frac{3\mu_0\mu^2}{32\pi Mg}}.$$

29. Magnetic Compression***

The finite solenoid experiences an axial compression due to the fringe fields at its ends which have radial components. To determine this compressive force, cut the solenoid into two parts that are not necessarily identical. The axial forces on the upper portion are the magnetic compressive force and the mechanical compressive force T exerted on it by the lower portion to balance the former. Similar to the section on finding the forces on an inductor, we can presume the solenoid to be perfectly conducting to simplify the process of applying the principle of virtual work. Suppose that the upper portion experiences a virtual displacement δl away from the bottom portion (which thus undergoes a virtual extension δl). The virtual work performed by the mechanical force due to the lower portion is $T\delta l$ while the virtual work performed by the magnetic force due to the lower portion is $-\frac{B^2}{2\mu_0}A\delta l$ since the increase in potential energy stored in its magnetic field is $\frac{B^2}{2\mu_0}dV = \frac{B^2}{2\mu_0}A\delta l$ (note that the magnitude of B does not change as there cannot be a change in magnetic flux[14] through a turn of a perfect conductor). By the

---

[14]Therefore, what really happens during the virtual extension is that the current in the solenoid must increase to compensate for the decrease in the number of turns per unit length.

principle of virtual work, the sum of all forms of virtual work on the upper portion must be zero.

$$T\delta l - \frac{B^2}{2\mu_0} A\delta l = 0$$

$$T = \frac{1}{2}\mu_0 \eta^2 I^2 A,$$

where $B = \mu_0 \eta I$ by Ampere's law. This compressive force is uniform throughout the entire solenoid as we have not assumed anything about the division of the solenoid into upper and lower portions. The magnetic compressive force also has magnitude T and tends to push any two pieces of the solenoid together (corresponding to the definition of a compressive force). For the last part of the problem, observe that the force between the two solenoids scales with the product of their currents. Furthermore, negating the direction of one current reverses the direction of the force between them. Deeming the two solenoids as a complete solenoid of twice their individual lengths, the force between them is $\frac{1}{2}\mu_0 \eta^2 I^2 A$ and is attractive in nature if the two solenoids carry the same current I in the same direction by the previous argument. Therefore, the force between the solenoids carrying currents I_1 and I_2 is

$$F = \frac{1}{2}\mu_0 \eta^2 I_1 I_2 A$$

and is attractive if the directions of the currents are identical, and repulsive otherwise. Actually, there is another elegant way of solving this problem — we shall only show the solution for the first part; the second part should follow accordingly. We know from Problem 14 in Chapter 7, that if we divide the solenoid into two portions, the upper portion (where the North pole lies) has a total magnetic flux of $\Phi = \frac{1}{2}\mu_0 \eta I A$ (half the magnetic flux cutting through its central cross section) leaking from its lateral surface (this leaking virtually happens entirely at the North pole if the squared length dimension of the solenoid is much larger than A). Meanwhile, let B_r denote the radial magnetic field as a function of axial coordinate along the upper portion of the solenoid. If we let the radius of the solenoid be r, the magnetic force on the upper portion is

$$F = \int \eta B_r \cdot 2\pi r I dl = \eta I \int 2\pi r B_r dl,$$

where dl is an infinitesimal length along the axial direction. However, notice that the right-most integral is simply the magnetic flux leaking through the lateral surface of the upper portion! Therefore,

$$F = \eta I \cdot \Phi = \frac{1}{2}\mu_0 \eta^2 I^2 A.$$

Chapter 9

DC Circuits

In this chapter, we will be analyzing the movement of charges in the form of direct electric currents. Only Direct Current (DC) circuits, in which currents are constantly unidirectional, will be considered in this chapter. Furthermore, only circuits involving emf sources and resistors will be discussed in this chapter — other circuit elements such as capacitors and inductors will be saved for the next chapter. An important assumption in the next two chapters would be that all sources are presumed to be independent such that their individual responses do not affect each other's.

Useful methods in solving circuitry problems will be discussed. It is recommended for the reader to attempt the problems at the end of the chapter as they illustrate how different problem-solving methods can be applied and function as good practice. Besides, they are really fun!

9.1 Kirchhoff's Laws

Kirchhoff's laws are a set of rules that govern the macroscopic view of charge flow in terms of current and voltage instead of current density and electric field.

9.1.1 *Kirchhoff's Loop Rule*

Kirchhoff's loop rule requires the sum of potential differences (due to a conservative electric field) along a closed loop to be equal to zero.

$$\sum V = 0. \tag{9.1}$$

It is basically restating the fact that the closed loop integral of a conservative electric field is zero.

$$\oint \boldsymbol{E} \cdot d\boldsymbol{s} = 0.$$

In evaluating the summation, note that the potential difference across the terminals of an ideal voltage source is simply equal, in both magnitude and direction, to its emf produced as shown in Chapter 8. Furthermore, there is a potential drop of $V = IR$ across a resistor carrying current I, in the direction of the current.

9.1.2   *Kirchhoff's Junction Rule*

For steady currents, Kirchhoff's junction rule states that the sum of currents flowing into a junction is equal to the sum of currents flowing out of that node. This is essentially the conservation of charge with an additional constraint that there can be no charge accumulation, analogous to the continuity equation. Mathematically,

$$\sum I = 0 \tag{9.2}$$

at every junction where a consistent sign convention is adopted (e.g. current flowing out of the junction is positive).

9.1.3   *Sign Conventions*

In applying Kirchhoff's loop rule, there are certain sign conventions that most adopt. They aid in ensuring clarity and preventing confusion. Firstly, one must choose a loop around the circuit to apply Kirchhoff's loop rule. This Kirchhoff loop must have a certain direction — either clockwise or counter-clockwise. Next, one must also propose a current across each segment of the circuit. When adding the potential differences along a loop, the potential difference across a battery is positive if the loop cuts a battery from the negative to the positive terminal. When the loop runs into a resistor, R, the voltage across the resistor is negative if the proposed current direction is the same as that of the loop and is positive otherwise.

9.1.4   *Definitions*

- A **branch** connects the two terminals of a circuit component.
- A **node** is the intersection of two or more branches.
- A **mesh** is a simplest planar loop in a circuit that does not contain any other smaller loops.

9.1.5   *Circuit Elements*

Figure 9.1 summarizes common components in circuits (capacitors, inductors and AC sources will be reserved for the next chapter). Note that the end of the battery that is represented by the longer line is the positive terminal.

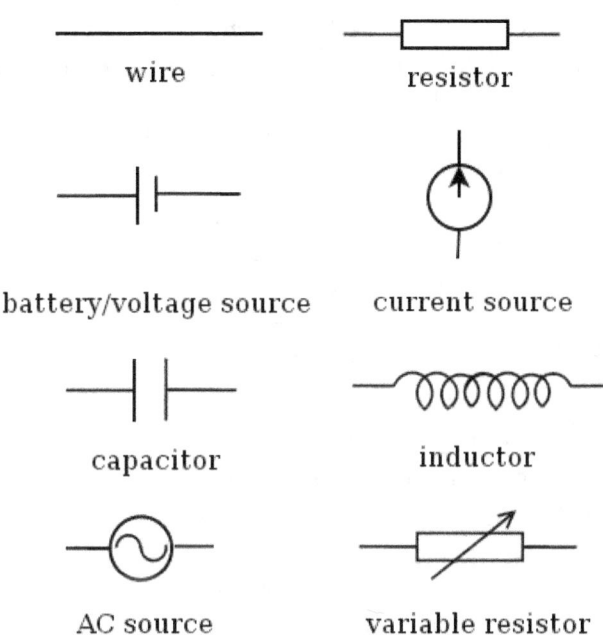

Figure 9.1: Circuit elements

In this chapter, we will assume that our circuit elements are ideal. There is no voltage across two ends of a wire as its resistance is negligible. The resistors obey Ohm's law perfectly. Batteries have no internal resistance and produce a constant emf. Perhaps the component that requires further elaboration would be the current source. A current source can be thought of as a hidden emf source that maintains the current through its branch at a certain value. It is merely a construct used to complicate problems and is infeasible in practice. An ideal current source has infinite internal resistance so that it is perpetually able to produce the same current regardless of external connections.

9.1.6 *Mesh Analysis*

Let us apply Kirchhoff's laws to some examples via mesh analysis. A mesh refers to a loop in the circuit that does not contain other smaller loops while a supermesh is defined as a loop in the circuit that contains multiple smaller meshes. The method of mesh analysis can be summarized with the following steps:

(1) Evaluate whether to use mesh analysis or node analysis. Mesh analysis is usually preferred when there are fewer voltage sources and more current sources.

(2) Define current variables in each branch of the circuit. One can either define the current in certain branches and compute the currents in other branches via Kirchhoff's junction rule or define the current in each branch via current loops in each of the meshes, which are known as **mesh currents**. If the method of ascribing mesh currents is utilized for a circuit with current sources, the sum of mesh currents must be equal to the current enforced at each current source — producing new equations that must be obeyed. These two methods are depicted in the fourth example below. The total number of current variables should be equal to the number of meshes minus the number of non-redundant current sources.

(3) Draw appropriate loops, which are known as Kirchhoff loops, in the circuit. Kirchhoff loops are usually drawn in each mesh when current sources are absent. When a circuit contains current sources, some Kirchhoff loops should be drawn in supermeshes to deliberately avoid current sources whose potential differences are unknown. The total number of linearly independent loops drawn should correspond to the number of current variables.

(4) Apply Kirchhoff's loop rule to each of the loops to generate a set of simultaneous equations which can be used to solve for the current variables.

Before we begin with formal applications of Kirchhoff's laws, let us go through the following simple example to highlight a special property regarding parallel connections.

Problem: In Fig. 9.2, two resistors are connected in parallel to a current source that delivers a current I. Find the current through each resistor.

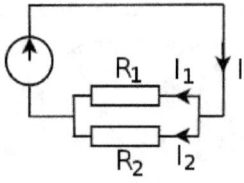

Figure 9.2:   Current divider principle

The potential differences across parallel branches must be identical as an electric potential is uniquely defined for each point in the system and as there is zero electric field inside an ideal wire — resulting in no change in potential when moving along wires. Therefore,

$$I_1 R_1 = I_2 R_2.$$

Furthermore, we know from Kirchhoff's junction rule that

$$I = I_1 + I_2$$

for no charge accumulation at the junction where the currents split. Then,

$$I_1 = \frac{R_2}{R_1 + R_2}I,$$

$$I_2 = \frac{R_1}{R_1 + R_2}I.$$

We shall term this set of equations the current divider principle. Next, we shall begin with mesh analysis formally by deriving the voltage divider principle in the following simple circuit.

Problem: In Fig. 9.3, two resistors are connected in series to a voltage source. Find the voltage across each resistor.

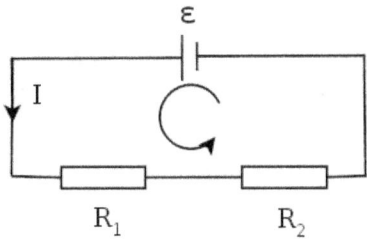

Figure 9.3: Voltage divider principle

We first define a current variable I and propose its direction to be in the anti-clockwise direction. Drawing an anti-clockwise loop in the mesh and applying Kirchhoff's law, we obtain

$$\varepsilon - IR_1 - IR_2 = 0$$

$$I = \frac{\varepsilon}{R_1 + R_2}.$$

The voltages[1] across the resistors are

$$V_1 = IR_1 = \frac{R_1}{R_1 + R_2}\varepsilon,$$

$$V_2 = IR_2 = \frac{R_2}{R_1 + R_2}\varepsilon,$$

---

[1]We shall use voltage and potential difference interchangeably here despite their slightly different meanings. The voltage is defined as the line integral of the electric field (both conservative and non-conservative) along a path. The voltage across two points is equal to the potential difference if there is no non-conservative field — a condition that is satisfied by all set-ups in this chapter.

we see that the voltages across the resistors are allocated according to the ratio of their resistances.

Problem: Find the current flowing through branch 23 in Fig. 9.4.

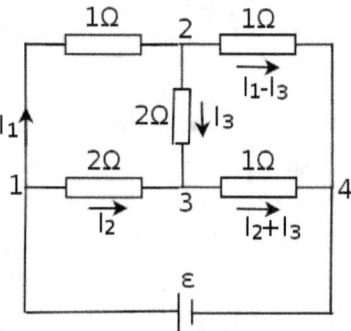

Figure 9.4:   Circuit

We first assign variables along with proposed directions to the currents through each branch. If the surmised direction for a current turns out to be incorrect, the current will have a negative value which indicates that current is actually flowing in the direction opposite to that proposed. We only need three current variables in this case, as the currents in other branches can be determined by Kirchhoff's junction rule. Then, three linearly independent loops[2] are required to be drawn to form three linearly independent simultaneous equations to solve for this system. Applying Kirchhoff's loop rule to loop 1231,

$$-I_1 - 2I_3 + 2I_2 = 0.$$

For loop 2432,

$$I_3 - I_1 + I_2 + I_3 + 2I_3 = 0.$$

For loop 4134,

$$\varepsilon - 2I_2 - I_2 - I_3 = 0.$$

Solving,

$$I_3 = \frac{\varepsilon}{19}.$$

---

[2]Three distinct loops do not guarantee three linearly independent equations. For example, choosing loops 1231, 12431 and 2342 will result in one redundant equation that is a linear combination of the other two.

We could have chosen any other loops, such as one in the supermesh 23412, and we would have obtained the same answer as long as three linearly independent loops were chosen. This means that any linear combination of any subset of the loops (multiplication by a constant, subtraction and addition) must not give another loop that we have already chosen. Three linearly independent loops are required to provide three linearly independent simultaneous equations to solve for the three current variables. Choosing more than three loops will generate equations that are redundant, or in linear algebra terms, linearly dependent.

Usually, a systematic method in choosing the Kirchhoff loops would be to draw a loop in each of the meshes in the circuit if current sources are absent — similar to what we have done above. When a circuit contains current sources, we require one less equation and current variable for every additional non-redundant current source in the circuit as each current source provides information about the current flowing in a particular branch. However, it is paramount to tactfully avoid current sources in a Kirchhoff loop as the potential difference across a current source is usually unknown. Furthermore, as the current in the branch containing a current source is already known, that branch can be avoided entirely without any harm as there are no more variables to solve for in that particular branch. Thus, we may need to consider loops in supermeshes when the circuit contains current sources which we deliberately want to avoid.

On another note, a common way in assigning currents is to draw a current loop within each mesh (usually with a coherent clockwise or anti-clockwise direction). This is illustrated in Fig. 9.5.

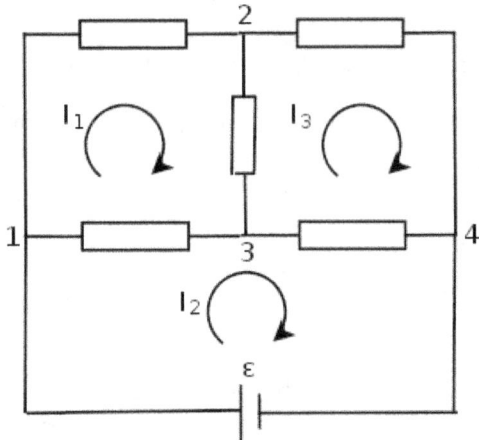

Figure 9.5: Current loops

To obtain the current in a branch at the border of two meshes, we just add the contributions from each of the current loops while taking note of their directions. For example, the current through branch 23 is $I_1 - I_3$ downwards (as I_1 is going downwards in branch 23 while I_2 is going upwards) and the current through branch 13 is $I_2 - I_1$ rightwards. The advantage of this allocation would be that Kirchhoff's junction rule will be automatically satisfied. At each junction, current that enters must also leave as we have drawn them in loops beforehand.

Let us analyze a circuit containing a current source. In such situations, a Kirchhoff loop in a supermesh should be drawn to avoid current sources.

Problem: Find the currents in branches 12 and 24 in Fig. 9.6.

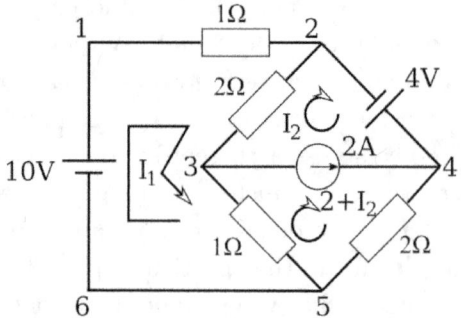

Figure 9.6:   Circuit with current source

We define the currents in the circuit via the current loops as shown above. Note that the current in the bottom triangular mesh is $2 + I_2$ in the clockwise direction so that the current in branch 34 is 2A, rightwards, as enforced by the current source.

Applying Kirchhoff's loop rule to mesh 123561,

$$10 - I_1 - 2(I_1 - I_2) - (I_1 - I_2 - 2) = 0.$$

Applying Kirchhoff's loop rule to supermesh 24532,

$$4 - 2(2 + I_2) - (2 + I_2 - I_1) - 2(I_2 - I_1) = 0.$$

Solving,

$$I_1 = \frac{54}{11}\text{A},$$

$$I_2 = \frac{28}{11}\text{A}.$$

Observe that we did not and did not need to consider meshes 2342 and 3453 as the potential difference across the current source is unknown and because the current in branch 34 was already predetermined. Finally, a systematic way of choosing meshes and supermeshes in such context is to remove branches containing current sources and then choose Kirchhoff loops in all resultant meshes. We did this above — by removing branch 34, there are two meshes 123561 and 24532.

9.1.7 *Nodal Analysis*

Mesh analysis solves for the branch currents in a circuit in order to describe the response in each branch. Another approach to circuitry problems would be to determine the voltage across every two nodes in a circuit. Usually one node in the entire circuit is chosen to be the reference node, whose potential is zero, and is denoted by a ground symbol. The potentials of all other nodes in the circuit are then defined relative to the potential of the reference node; these relative potentials are known as **nodal voltages**. The method of nodal analysis can be summarized with the following steps:

(1) Evaluate whether to use mesh analysis or node analysis. Nodal analysis is usually preferred when there are more voltage sources and fewer current sources.

(2) Identify an appropriate reference node in the circuit. As a rule of thumb, the node that is connected to the most voltage sources or circuit components is usually denoted as the reference node. Afterwards, define nodal voltage variables or compute the nodal voltages directly at each node of the circuit. The nodal voltage of the node at the positive terminal of a battery is more than that of the node at the negative terminal by the emf of the battery. The total number of nodal voltage variables should be equal to the total number of nodes minus one (due to the reference node) minus the number of voltage sources.

(3) For each node whose voltage is unknown, apply Kirchhoff's junction rule to ensure that the net current flowing into or out of the node is zero. The current through a branch emanating from the node can be determined by the difference in node voltages divided by the resistance between them or directly from the magnitude of current enforced by the current source in that branch. When there exist nodes that are connected via a voltage source to another node that is not the reference node, Kirchhoff's junction rule cannot directly be applied to such nodes as the current through the voltage source is unknown. Hence, the **method of supernodes**, which will be elaborated later, should be used instead. The

total number of nodes or supernodes, to which Kirchhoff's junction rule is applied, should correspond to the number of nodal voltage variables.

(4) A set of simultaneous equations, that can be used to solve for the voltage variables, will be obtained.

Problem: Determine the current through the 6Ω resistor in Fig. 9.7.

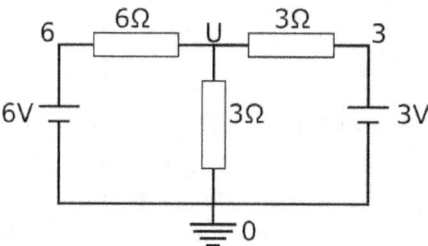

Figure 9.7:   Circuit with two emfs

We denote the reference node to be that attached to the ground symbol. Then the nodal voltages of the two nodes connected to the reference node via voltage sources are 6V and 3V respectively. Let the nodal voltage of the node connected to the three resistors be U. Applying Kirchhoff's junction rule to that particular node,

$$\frac{U-6}{6} + \frac{U-0}{3} + \frac{U-3}{3} = 0.$$

The three terms on the left-hand side correspond to the currents flowing out of that particular node to the 6V, 0V and 3V nodes respectively. The net current flowing out of any node is required to be zero by Kirchhoff's junction rule. Solving,

$$U = \frac{12}{5}\text{V}.$$

Thus, the current through the 6Ω resistor is

$$I = \frac{6 - \frac{12}{5}}{6} = \frac{3}{5}\text{A}$$

rightwards.

Let us now analyze a circuit in which the method of supernodes is necessary. When neither of the two terminals of a voltage source is connected to the reference node, the nodal voltages of both nodes need to be determined but Kirchhoff's junction rule cannot be applied to each individual node directly as the current through the voltage source is unbeknownst to us. However, these nodes can be "compressed" to a single supernode as the total net current entering these nodes should also be zero by Kirchhoff's junction rule. Though only one equation is obtained from a single supernode, there is no harm here as the nodal voltages of the nodes included in the supernode are related by the emfs of the voltage sources. Hence, there will still be enough equations to solve for all the nodal voltages.

Problem: Determine the current through the 4Ω resistor and the 12V battery in Fig. 9.8.

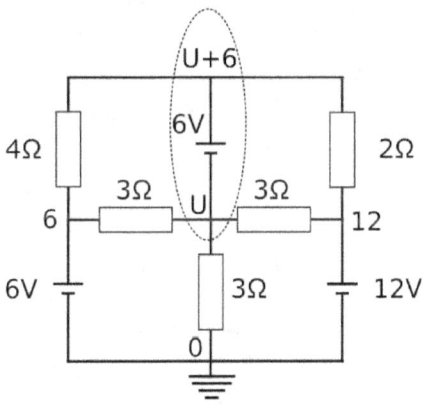

Figure 9.8: Circuit with supernode

Again, we define a convenient reference node that is connected to the most voltage sources. Then, each nodal voltage can either be computed directly or assigned a variable as labeled above. Now, observe the Kirchhoff's junction rule cannot be directly applied to the nodes of nodal voltages U and $U+6$, as the current through the battery connecting them is unknown.

However, it can be seen that there should also be no net current flowing into or out from the region demarcated by the dotted lines that enclose the two nodes. Hence, the two nodes can be treated as a "supernode". The total current flowing out of this supernode is

$$\frac{U+6-6}{4} + \frac{U+6-12}{2} + \frac{U-6}{3} + \frac{U-0}{3} + \frac{U-12}{3} = 0$$

$$U = \frac{36}{7}\text{V}.$$

The current through the 4Ω resistor is

$$I_1 = \frac{\frac{36}{7}+6-6}{4} = \frac{9}{7}\text{A}$$

downwards. The current through the 12V battery can be computed from the sum of currents entering the node at its positive terminal.

$$I_2 = \frac{\frac{36}{7} - 12}{3} + \frac{\frac{36}{7} + 6 - 12}{2} = -\frac{19}{7}\text{A}$$

which implies that the current flows from the negative terminal to the positive terminal.

Technically, we are done here — all circuits can be solved via the two simple yet meaningful Kirchhoff's laws. Besides the fact that it would be extremely boring if everything can be reduced to trivial applications of Kirchhoff's laws, complex circuits often generate complicated systems of equations that are tedious to solve. Solely applying Kirchhoff's laws to such circuits is merely a brute force method. Thus, there are several sleights-of-hand that can be applied to simplify intricate circuits before applying Kirchhoff's laws to the simplified circuits. These will be discussed in the next few sections.

9.2   The Principle of Superposition

The principle of superposition for electrical circuits states that the linear response (current or voltage through or across ideal ohmic resistors, capacitors and inductors) in any branch of a system that has two or more independent sources equals the sum of the responses induced by each independent source acting alone, with all other sources replaced by their internal resistances. That is, we turn off the emf components of all other sources while retaining their resistances when considering the effects of a single source. As ideal voltage sources have zero internal resistance, they can be short-circuited (connected by a wire) when they are replaced. Ideal current sources on the other hand must be open-circuited (disconnected) as they have infinite internal resistance. The principle of superposition is a powerful tool that allows us to divide a problem pertaining to multiple independent sources into smaller sub-circuits involving single sources.

To prove the principle of superposition, we simply have to verify that the solution obtained from piecing together the contributions of various sources is indeed valid and subsequently assert the uniqueness of the solution to Kirchhoff's laws (loosely speaking, we have equal numbers of variables as independent equations). Firstly, we can decouple all internal resistances from the sources, by connecting a resistor in series to a voltage source or parallel to a current source, such that they become either ideal voltage or current sources. Consider the ith sub-circuit associated with the ith source (ideal voltage or current source) that is obtained from replacing all other ideal sources with their internal resistances. Since Kirchhoff's loop and junction rules are satisfied in all sub-circuits, they are automatically satisfied in the superposition of the sub-circuits. Furthermore, notice that the potential difference across the ith source, if it is a voltage source, is zero[3] in all sub-circuits except the ith one (which yields exactly the required voltage) since the ith voltage source is replaced by an ideal wire in the other sub-circuits. Similarly, the current across the ith source, if it is a current source, is zero in all sub-circuits except the ith one (which yields exactly the required current) since the ith current source is open-circuited in the other sub-circuits. Finally, because Ohm's law is satisfied for each Ohmic resistor in all sub-circuits and because Ohm's law describes a linear relationship between voltage and current, the superposition of voltages and currents also naturally satisfies Ohm's law for all resistors. The same can be said for capacitors and inductors as the equation governing their voltages are also linear in variables which are linearly related to current, as we shall see in the next chapter. Since all conditions imposed by Kirchhoff's laws have been satisfied, the superposition is valid and thus the unique solution to the original circuit.

Problem: Find the current through branch XY in Fig. 9.9.

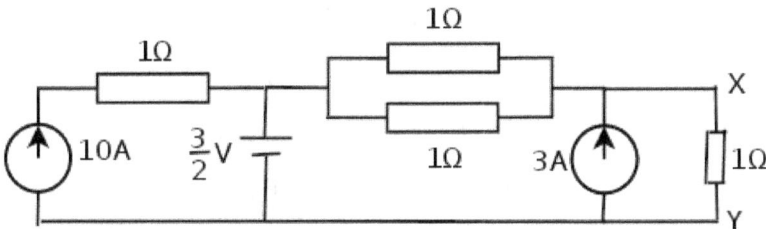

Figure 9.9: Complete circuit

The above circuit is the superposition of the three circuits in Figs. 9.10–9.12.

The current through XY in the first sub-circuit is zero as the current would prefer to flow through the branch without any resistors.

$$I_1 = 0.$$

---

[3]This point is murky in the case where infinite current flows through a wire. A situation like this occurs when there is a closed loop consisting solely of a voltage source but we shall ignore such ill-defined cases where the principle of superposition cannot be applied meaningfully.

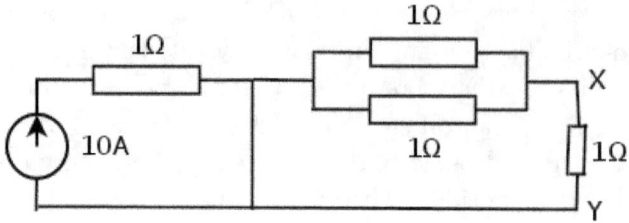

Figure 9.10:   Sub-circuit 1

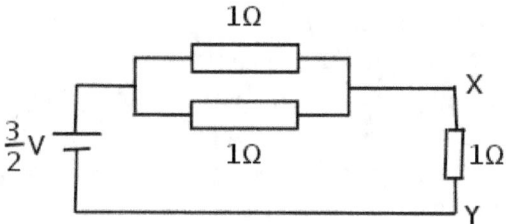

Figure 9.11:   Sub-circuit 2

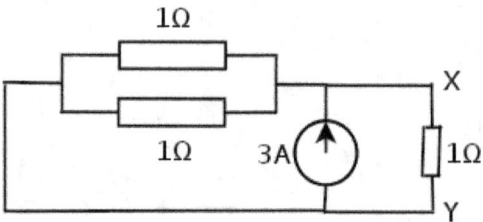

Figure 9.12:   Sub-circuit 3

We define the current through XY to be positive if it flows from X to Y. The current through XY due to the second circuit is

$$I_2 = \frac{\frac{3}{2}}{\frac{1}{2} + 1} = 1\text{A}.$$

The current through XY in the last circuit can be computed via the current divider principle.

$$I_3 = \frac{\frac{1}{2}}{1 + \frac{1}{2}} \cdot 3 = 1\text{A}.$$

Thus, the total current through XY is

$$I = I_1 + I_2 + I_3 = 2\text{A}.$$

Note that the principle of superposition is only valid for linear responses in a circuit. The power dissipated in a resistor in a circuit is not the linear sum of the powers dissipated in that resistor in different sub-circuits. Furthermore, the principle of superposition cannot be applied to circuit components whose I-V characteristics are non-linear. For example, the resistance of a realistic resistor increases with temperature as the atoms in a conductor vibrate more vigorously and thus collide with the electrons more frequently — obstructing current flow. Lastly, the response in a component must be symmetrical in both possible directions of connection for the principle of superposition to hold true. This is satisfied in most cases, except for diodes which ideally restrict the flow of current to a single direction.

The principle of superposition can also be applied to determine the equivalent resistance of symmetric networks. It is instructive to consider the infinite grid of identical resistors R in Fig. 9.13. Each node is connected to four neighboring nodes by four resistors. We wish to calculate the equivalent resistance between two adjacent nodes of this grid, such as that between X and Y.

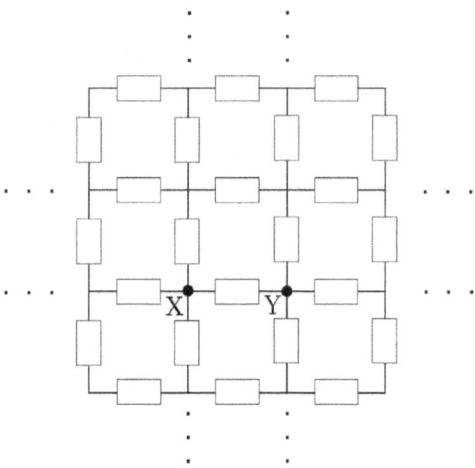

Figure 9.13:   Infinite grid of resistors

Here is our line of attack. We can imagine connecting two current sources to the infinite grid as shown in Fig. 9.14.

Both current sources are connected to the infinite grid at infinity. However, one is injecting 1A of current into node X while the other is withdrawing 1A current from node Y. Consider the boundary at infinity, there is 1A of current entering it due to the current source on the right and another 1A of current being withdrawn from the boundary due to the current source on the

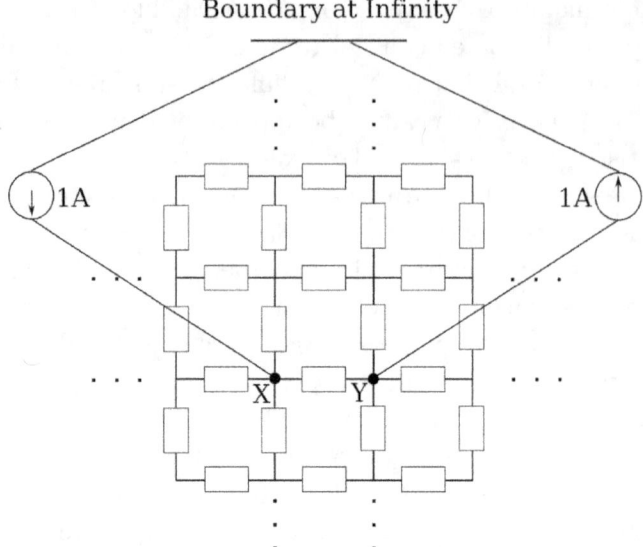

Figure 9.14:   Infinite grid of resistors

left. Hence, no net current leaks out of the infinite grid into the two external wires by Kirchhoff's junction rule. Effectively, 1A of current is injected into node X and is then circulated in the infinite grid (while circulating, no current leaves via the external wires — they all go back to node Y) before being withdrawn from node Y. If the voltage across nodes X and Y, V_{XY}, can be determined, the equivalent resistance, R_{eq} can be computed as

$$R_{eq} = \frac{V_{XY}}{1},$$

as the voltage V_{XY} is required to be applied between nodes X and Y to inject 1A current into the network through node X, circulate it through the network and withdraw it from node Y.

To determine V_{XY}, we can consider the superposition of two sub-circuits, each with one current source open-circuited as the internal resistance of a current source is infinite. First, we can imagine injecting 1A current into node X and withdrawing all of it at infinity (the other external wire is now disconnected). The 1A of injected current will flow symmetrically in the immediate surroundings of node X as shown in Fig. 9.15.

Next, in a completely new set-up, withdraw 1A current from node Y while injecting all of it at infinity. Similarly, the current will distribute itself evenly in the immediate surroundings of node Y as shown in Fig. 9.16.

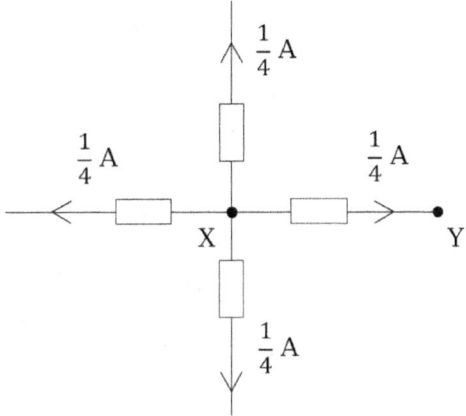

Figure 9.15: Injection of 1A into X and withdrawal from infinity

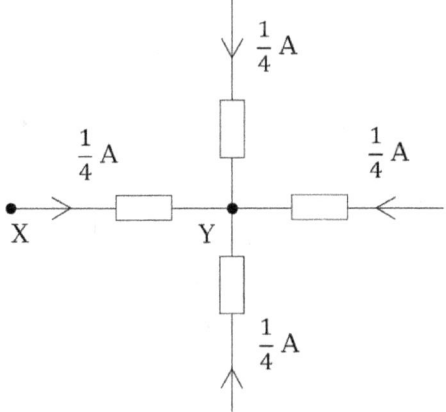

Figure 9.16: Injection of 1A into infinity and withdrawal from Y

Lastly, if we superpose the two set-ups together, we obtain the equivalent system that inputs 1A of current into node X and withdraws it from node Y, as depicted in Fig. 9.14. There is no net current flowing out from or into the boundary at infinity to and from external connections. Since a total of $\frac{1}{4} + \frac{1}{4} = \frac{1}{2}$A of current flows from nodes X to Y through the resistor directly connecting them, the voltage across nodes X and Y is

$$V_{XY} = IR = \frac{R}{2}.$$

Since the voltage across nodes X and Y is V_{AB} when 1A circulates in the infinite grid through them, the equivalent resistance of the grid with respect to points X and Y is

$$R_{eq} = \frac{V_{XY}}{1} = \frac{R}{2}.$$

Note that this analysis is only limited to the equivalent resistance of the grid with respect to adjacent points. If we wish to calculate the equivalent resistance between node X and the node directly above Y for instance, a completely different approach is required. Here's why. Let's say we inject 1A of current into node X and withdraw it from infinity again, it is true that $\frac{1}{4}$A of current will flow directly from node X to node Y but we cannot conclude that $\frac{1}{12}$A of current flows from node Y to the node above Y. This is due to the limited symmetry of the grid. The node above Y is symmetrical to the node below Y but not to the node on the right of Y in this sub-system. Thus, it is almost impossible to invoke symmetrical arguments in this case.

In general, in approaching circuits with some form of symmetry, appropriate methods of injecting and withdrawing current in a sub-system should be devised such that the currents flow symmetrically in that sub-system. In finite circuits, the injected currents cannot be entirely withdrawn from a single node as that would often lead to an asymmetrical distribution of current. A common way of constructing the two sub-systems to be used for the superposition is illustrated below.

Problem: A regular polyhedron (e.g. tetrahedron, dodecahedron) with N vertices is made of wires which form its edges. If the resistance of an edge is R, determine the equivalent resistance of the polyhedron across the two terminals of an edge. Next, determine the equivalent resistance across the same two terminals if the edge directly connecting them is removed. Each vertex of a polyhedron is directly connected to $n \geq 2$ neighbouring vertices.

Let the two adjacent vertices that we will compute the equivalent resistance with respect to, be X and Y. Inject 1A of current into X and withdraw $\frac{1}{N-1}$A current from all other vertices. Due to the symmetrical nature of the polyhedron and because each vertex is connected to n neighbouring vertices, the current that flows directly from X to Y through the edge connecting them in this set-up is $\frac{1}{n}$A. Next, consider a new set-up where 1A current is withdrawn from Y and $\frac{1}{N-1}$A current is injected into all other vertices. The current flowing between the edge connecting X and Y is also $\frac{1}{n}$A in this set-up.

Superposing the two set-ups afore, $1 + \frac{1}{N-1} = \frac{N}{N-1}$A current is effectively injected into X, circulated within the polyhedron and entirely withdrawn from Y by an appropriate external battery with some emf ε. No current enters or leaks from any other vertices via external connections as the $\frac{1}{N-1}$A current withdrawn in the first set-up negates the $\frac{1}{N-1}$A current injected in the second set-up. In this superposed set-up, $\frac{2}{n}$A current flows from X to Y via the edge directly connecting them — implying that the voltage across them and hence the emf ε of the external battery is $\frac{2R}{n}$. Since $\frac{N}{N-1}$A current circulates within the polyhedron when a battery with emf $\varepsilon = \frac{2R}{n}$ is connected to the two terminals, the equivalent resistance of the polyhedron between X and Y is

$$R_{eq} = \frac{\frac{2R}{n}}{\frac{N}{N-1}} = \frac{2(N-1)}{nN} R.$$

Next, we can deem the second scenario, where the edge between X and Y is removed, as the first scenario connected in parallel with a $-R$ resistor between X and Y. This is because the parallel connection of two resistors with resistances R and $-R$ (which are the two direct edges between X and Y in this case) yields a diverging equivalent resistance $\frac{R \cdot -R}{R-R} \to \infty$ which effectively open-circuits the immediate connection between A and B. Therefore, the new equivalent resistance is

$$R'_{eq} = \frac{R_{eq} \cdot -R}{R_{eq} - R} = \frac{2N - 2}{nN + 2 - 2N} R.$$

The veracity of introducing a hypothetical component with a negative resistance is obvious from a mathematical perspective as we are basically solving a set system of linear equations in current variables (mesh analysis) obtained from Kirchhoff's laws — there is completely no regard for whether the coefficients (the resistances) in front of the current variables are positive or negative as long as we obey the rules set by Kirchhoff's laws (the voltage drop across a resistor R carrying current I is IR in the direction of the current where R, in this context, is merely a coefficient).

Finally, it is important to highlight the key takeaway of this problem. For finite networks, the principle of superposition can often be applied by injecting 1A of current into a single node and withdrawing equal proportions of this current from the rest of the nodes. The same goes for withdrawing 1A of current.

9.3 Equipotential Points

Nodes of the same potential in a circuit are essentially the same point. The circuit's response does not vary if these equipotential points are combined into a single point while removing the connections, such as resistors and wires, between them. This is because the response in a branch can be uniquely defined by the potential difference across its ends. Thus we can tweak the circuit as much as we want as long as the potential differences across components remain the same. This allows us to tidy up and transform messy and obfuscated circuits into more tractable and lucid diagrams. Perhaps, the following examples will elucidate this point.

Problem: Find the equivalent resistance between terminals 1 and 6 in Fig. 9.17. All resistors are identical and possess a resistance R.

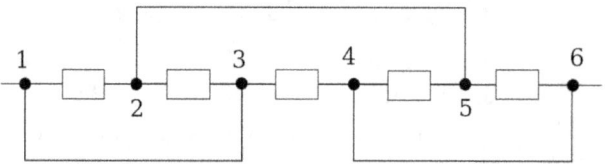

Figure 9.17:   Circuit

Nodes 1 and 3 are at the same potential as they are connected by a wire. Similarly, nodes 2, 5 and 4, 6 are equipotential pairs. We shall compress these pairs mentioned afore into combined nodes A, B and C respectively.

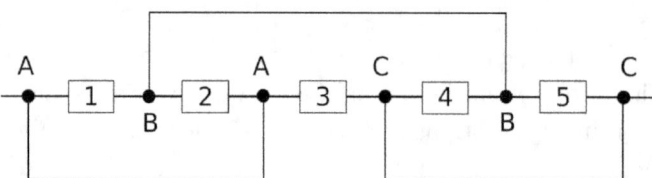

Figure 9.18:   Labelled circuit

To determine the equivalent circuit, we label nodes 1–6 to their corresponding equivalent nodes as shown in Fig. 9.18. Whenever there is a resistor between a node from 1–6 to another, we add the corresponding resistor between the equivalent nodes in an appropriate manner (while taking note of series and parallel configurations). Eventually, the equivalent circuit is obtained as Fig. 9.19.

Thus, the equivalent resistance is

$$R_{eq} = \frac{1}{\frac{1}{R} + \frac{1}{\frac{R}{2} + \frac{R}{2}}} = \frac{R}{2}.$$

Sometimes, equipotential points are not as easy to spot as they are not directly connected by wires. Instead, symmetry should be abused to find such points.

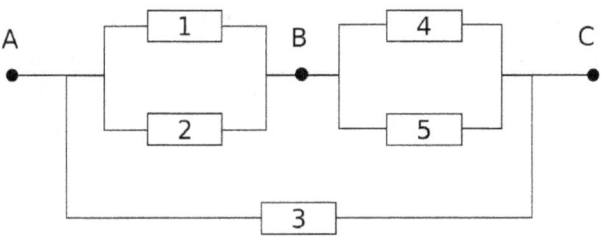

Figure 9.19: Equivalent circuit

Mirror Symmetry

Consider a plane of resistors, with respect to two terminals, which is symmetrical about an axis — that is perpendicular to the line joining the two terminals — which divides it into identical halves. The half of the circuit, including one terminal of interest, on one side of the axis is essentially the reflection of the other half about the line of symmetry. Hence, this form of symmetry is known as mirror symmetry. In such situations, the points along the symmetrical axis must be equipotential points. In fact, their potentials must actually be the average of the potentials at the two terminals of the circuit when connected to an external voltage source. Note that though there may be connections along the line of symmetry, they can simply be removed since points along the axis are guaranteed to be equipotential — preventing current from flowing through such links.

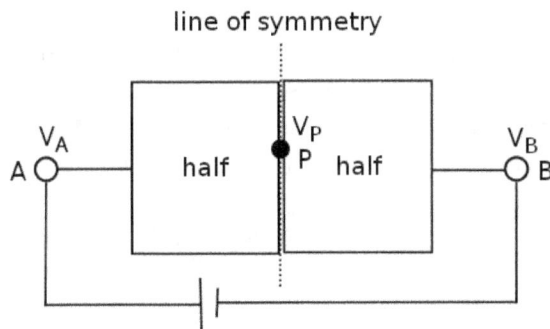

Figure 9.20: Mirror symmetry

An elegant proof is as follows. Suppose that the connected voltage source in Fig. 9.20 causes the potentials of terminals A and B and an arbitrary point P along the line of symmetry to be V_A and V_B and V_P respectively. Now, we can conjure a separate set-up with the emf reversed. Then, terminals A and B will have potentials V_B and V_A while point P still has potential

V_P due to symmetry. Then, superposing these two set-ups would produce potential $V_A + V_B$ for both terminals and potential $2V_P$ for point P. Since, the external emfs in opposite directions nullify each other, there should be no current flowing in the superposed circuit and hence no potential difference across any two points. Then,

$$2V_P = V_A + V_B$$

$$V_P = \frac{V_A + V_B}{2}.$$

Since this argument holds for all points along the line of symmetry, they must be equipotential.

Problem: Find the equivalent resistance of the arrangement in Fig. 9.21 with respect to points A and B.

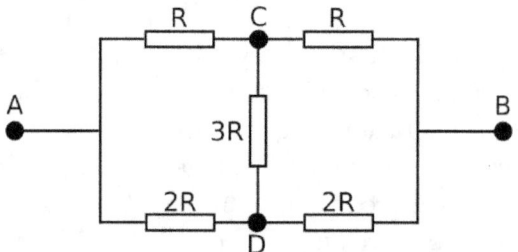

Figure 9.21:   Circuit with mirror symmetry

In the circuit, an axis that passes through CD divides the circuit into two identical and symmetrical halves. Then, points C and D must form an equipotential pair — enabling the removal of the resistor joining them. The equivalent resistance of the circuit with respect to terminals A and B is then

$$R_{eq} = \frac{1}{\frac{1}{4R} + \frac{1}{2R}} = \frac{4}{3}R.$$

Actually, if a network of resistors, with respect to two terminals, has a line of symmetry such that one half of the network is the mirror-image of the other, coupled with a constant scaling of resistance (e.g. twice of all resistances of the other side), all points along this line of symmetry must be equipotential, regardless of the connections between them. This equipotential property is evidently true when there are no connections between points along the line. Then, one can wait for this condition to be established before forming the connections (with wires or resistors) between points along the symmetrical axis. Since the points were equipotential before the connections, they should remain equipotential after the connections as no current tends to flow across the connections and because Kirchhoff's laws make no mention of the order of connections. This can be taken as an intuitive physical argument of the above claim. Mathematically, one can show that the equipotential solutions work when there are no connections between points along the line of symmetry. Furthermore, when such connections exist, one can show that the equipotential solutions afore, in combination with zero current across the connections, yield a valid solution to Kirchhoff's laws (which only have a unique solution).

Therefore, if the top two resistors in the circuit above were $1R$ and $2R$ while the bottom two were $2R$ and $4R$, nodes C and D would still form an equipotential pair.

Path Symmetry

When determining the equivalent resistance of a network of resistors with respect to two terminals, one may identify symmetrical paths from a starting terminal to an ending terminal and correspondingly, abuse such symmetry. If you were a charge at the starting node, some paths to a terminal node look indistinguishable from your perspective. You are equally likely to take any of the paths, analogous to how identical currents flow through corresponding segments along those paths. Therefore, corresponding points along those paths must be equipotential points. Below is an instructive example that exploits this fact.

Problem: Consider the cube formed by identical resistors of resistance R as its edges in Fig. 9.22. Find the equivalent resistances of the cube between nodes 1 and 2, 1 and 4 and 1 and 8.

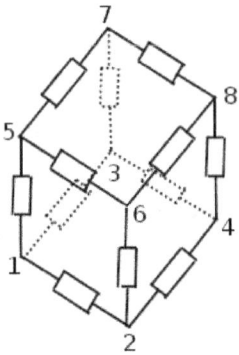

Figure 9.22:   Cube of resistors

When considering the equivalent resistance between nodes 1 and 2, nodes 3, 5 and 4, 6 are equipotential pairs by symmetry as they are indistinguishable from each other on a path from nodes 1 to 2. Thus, we can combine those pairs to obtain the equivalent circuit in Fig. 9.23.

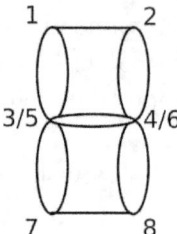

Figure 9.23:   Equivalent circuit 1

Each connection, indicated by a line, represents a resistor. To construct the above diagram, simply identify the connections between corresponding points (including compressed ones). For example, there are two resistors in parallel between node 1 and nodes 3/5 in branches 13 and 15. Similarly, there are two resistors in parallel between nodes 3/5 and nodes 4/6 in branches 34 and 56. The above diagram transforms to Fig. 9.24.

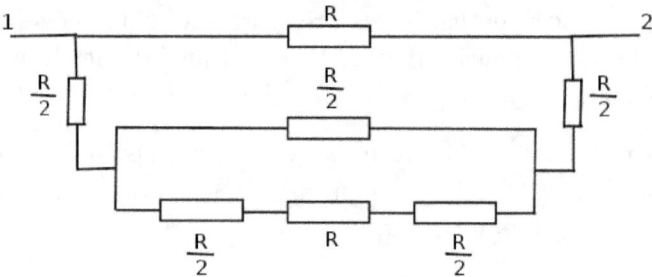

Figure 9.24:   Equivalent circuit 2

The equivalent resistance of the parallel connection comprising the four resistors at the bottom of Fig. 9.24 is

$$R' = \frac{1}{\frac{2}{R} + \frac{1}{\frac{R}{2} + \frac{R}{2} + R}} = \frac{2}{5}R.$$

Thus,

$$R_{eq} = \frac{1}{\frac{1}{R} + \frac{1}{\frac{2}{5}R + \frac{R}{2} \times 2}} = \frac{7}{12}R.$$

This agrees with our previous formula for the equivalent resistance of a polyhedron network with N vertices, each with n neighbours. For a cube, substituting $N = 8$ and $n = 3$ into $\frac{2(N-1)}{nN}R = \frac{2(8-1)}{24}R = \frac{7}{12}R$ verifies the result above. To solve for the equivalent resistance between nodes 1 and 4, we observe that nodes 2, 3 and 6, 7 are equipotential pairs by symmetry. Thus, the simplified diagram is depicted in Fig. 9.25.

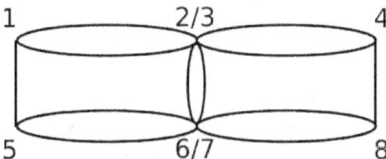

Figure 9.25:   Equivalent circuit between nodes 1 and 4

Lastly, we observe that nodes 2/3 and 6/7 in the above diagram must also be equipotential points due to mirror symmetry. Thus, we can remove the resistors between them since no current will flow across them anyway. We then obtain the corresponding circuit in Fig. 9.26.

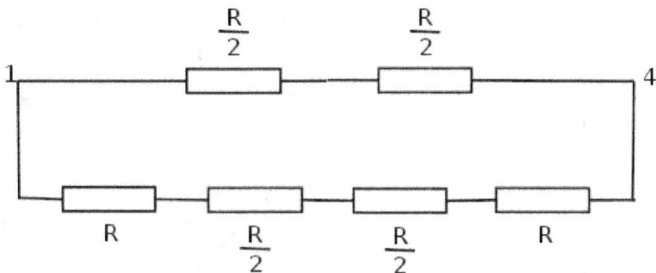

Figure 9.26:   Second equivalent circuit between nodes 1 and 4

$$R_{eq} = \frac{1}{\frac{1}{\frac{R}{2}+\frac{R}{2}} + \frac{1}{R+R+\frac{R}{2}\times 2}} = \frac{3}{4}R.$$

Finally, when considering the equivalent resistance of the cube between nodes 1 and 8, we observe that nodes 2,3,5 and nodes 4,6,7 are equipotential triplets. Thus, the simplified circuit is shown in Fig. 9.27.

The equivalent resistance is then

$$R_{eq} = \frac{R}{3} + \frac{R}{6} + \frac{R}{3} = \frac{5}{6}R.$$

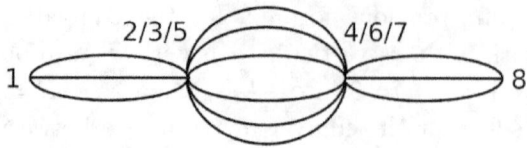

Figure 9.27:   Equivalent circuit between nodes 1 and 8

Dividing Nodes

Besides combining equipotential nodes, we can also perform the reverse process of dividing a single node into two or more nodes that will be equipotential after the split. This is valid as we can always conjoin these resultant nodes back into the original node due to their equipotential property. A common way of identifying such divisions is to split the nodes along a line of symmetry while maintaining the mirror symmetry, as new nodes along the line of symmetry are guaranteed to be equipotential.

Problem: Determine the equivalent resistance between nodes 1 and 16 in the 3 × 3 grid of resistors depicted on the left of Fig. 9.28. Each edge in the network has resistance R.

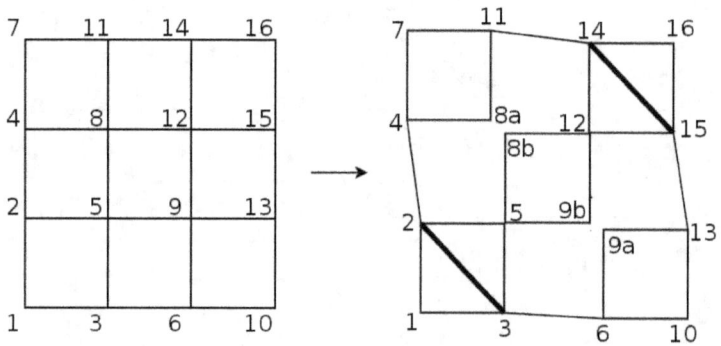

Figure 9.28:   Combining pairs 2,3 and 14,15 while splitting 8 and 9

It is tempting at the first glance to combine nodes along the diagonals of the grid but be wary that the grid has limited symmetry. We can say that nodes 4 and 6 are equipotential due to path symmetry but we cannot say that nodes 4, 5 and 6 are equipotential, as node 5 is not indistinguishable from nodes 4 and 6 along a path from node 1 to 16 due to the two resistors connecting nodes 2 and 3 to node 5. Despite this, we can combine nodes 2,3 and nodes 14, 15 (depicted by the thick lines) as they are equipotential by path symmetry (we do not need to combine nodes 4, 6 and 11, 13 in our analysis).

Next, observe that we can break node 8 along the diagonal direction into two nodes 8a and 8b (see right diagram of Fig. 9.28) as the resultant network will still be symmetrical about the diagonal connecting nodes 7 and 10 (relative to terminals 1 and 16) such that the two nodes 8a and 8b are ensured to be equipotential. A similar statement holds for the division of node 9 into nodes 9a and 9b. Observe that through this procedure, we obtain three parallel branches (which grow apart at nodes 2/3 and 14/15) with resistances $3R$, $2R$ and $3R$, connected in series with two $\frac{R}{2}$ resistors (one between nodes 1 and 2/3 and another between nodes 16 and 14/15). Therefore, the equivalent resistance between nodes 1 and 16 is

$$\left(1 + \frac{1}{\frac{1}{3} + \frac{1}{3} + \frac{1}{2}}\right) R = \frac{13}{7}R.$$

9.4 Thevenin's Theorem

For any network that comprises purely independent emf sources, current sources and resistors between two terminals, Thevenin's theorem states it can be transformed into an equivalent network that consists of a single Thevenin voltage source ε_{eq} connected in series with an internal Thevenin resistance R_{eq} with respect to the two terminals. The network that Thevenin's theorem is applied to is usually a sub-circuit extracted from a larger circuit — the choice of terminals is up to our own discretion.

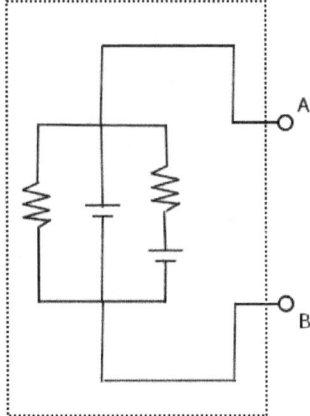

Figure 9.29: Arbitrary circuit in a "black box"

For example, the network in Fig. 9.29 can be transformed into the equivalent circuit in Fig. 9.30 with respect to terminals A and B.

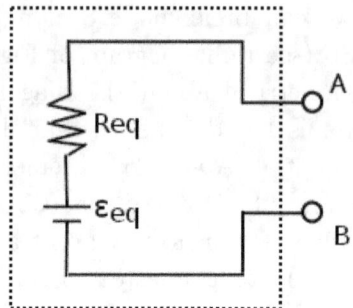

Figure 9.30:   Equivalent Thevenin circuit

We can imagine surrounding these networks in a black box depicted by the dotted lines in the figures. From the outside, we will not be able to determine any difference in the effects produced by the two black boxes. That is, for any voltage imposed between terminals A and B, the currents flowing through the terminals in the two circuits will be identical. Vice-versa, for any current driven into terminals A and B, the voltages across A and B in the two circuits will be identical.

The existence of such an equivalent system shall be proven soon, but let us first ponder how we could determine the equivalent emf and resistance. The Thevenin equivalent emf, ε_{eq}, can be obtained by computing the voltage between points A and B if the external connection between A and B is open-circuited (removed). This voltage is known as the open-circuit voltage V_{oc} and corresponds to the voltage measured by an ideal voltmeter of an infinite resistance connected externally to points A and B. There will be negligible voltage across R_{eq} and thus all of the voltage is consumed and measured by the voltmeter. Finally, note that the polarity of the Thevenin equivalent emf $\varepsilon_{eq} = V_{oc}$ is oriented such that its positive terminal points in a direction of higher voltage in the original system (when the terminals are open-circuited).

Next, to calculate R_{eq}, two different approaches can be utilized. Firstly, we can imagine connecting an external ideal wire between points A and B and measuring the short-circuit current in that wire, I_{sc}. Then, we can compute R_{eq} by dividing ε_{eq} by I_{sc}.

$$R_{eq} = \frac{\varepsilon_{eq}}{I_{sc}}. \tag{9.3}$$

However, there is a much more efficient alternative. The equivalent resistance, R_{eq}, is equal to the resistance between terminals A and B with all ideal voltage sources short-circuited and all ideal current sources open-circuited inside the black box. Physically, this is equivalent to replacing the Thevenin equivalent emf with an ideal wire by muting all sources. This method is the more common and advised approach.

Proof: Now, we shall show the existence of such an equivalent circuit and at the same time, justify the second method of computing R_{eq}. Referring to Fig. 9.31, the primary circuit, which we are interested in finding an equivalent Thevenin circuit for, is connected to a secondary circuit. Both circuits in general can have ideal emf sources (ideal voltage and current sources) which are denoted by ε's and resistors which are denoted by R's. Notice that we can add two voltage sources V_{oc} in opposing directions to a branch connecting the primary and secondary circuits and the response of the system will be unaffected, as the voltage sources nullify each other.

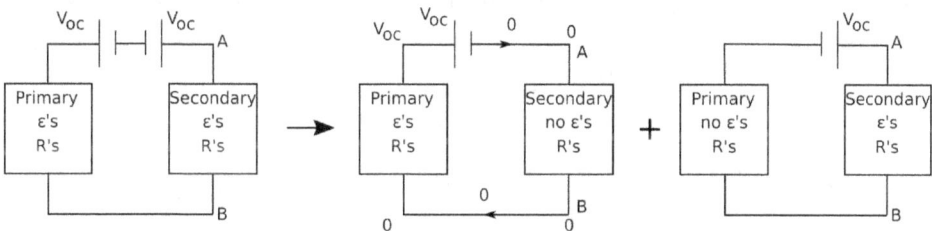

Figure 9.31: Superposition of circuits

Now, we can decompose the original circuit into the left and right sub-circuits by the principle of superposition. The left sub-circuit includes the emf sources of the primary circuit and the voltage source V_{oc} that opposes the direction of the potential difference across the primary circuit when it is open-circuited between terminals A and B. All other emf sources are replaced by their internal resistances. Observe that the potentials labeled at the vertices of the left sub-circuit, coupled with zero current flowing between the primary and secondary circuits and zero potential in the entire secondary circuit, is a valid solution to Kirchhoff's laws, since the potential difference across the primary circuit is V_{oc} when there is no current flowing into terminals A and B (by definition of the open-circuit voltage), and because the secondary circuit only consists of resistors now. Thus, there is no voltage across and current through the resistors of the secondary circuit in the left sub-circuit. By the principle of superposition, the response in the secondary circuit is then that in the right sub-circuit.

The right sub-circuit includes the leftover sources — the other voltage source V_{oc} and the emf sources of the secondary circuit. Observe that it is basically the original circuit, with the secondary circuit unchanged, except that the emf sources of the primary circuit have been replaced by their internal resistances and a voltage source V_{oc} has been added in series to its remaining resistors. As such, the Thevenin equivalent emf ε_{eq} is evidently V_{oc}, while R_{eq} can be computed by computing the resistance of the primary circuit between terminals A and B when all sources in it have been replaced by their internal resistances. As we have not assumed much about the constituents of the secondary circuit, the above argument actually works even if the secondary circuit has other linear components such as ideal capacitors and inductors.

Proceeding with its actual applications, Thevenin's theorem is a potent strategy in simplifying complex circuits, especially when Kirchhoff's laws produce a plethora of simultaneous equations.

Problem: Find the current across branch AB in Fig. 9.32.

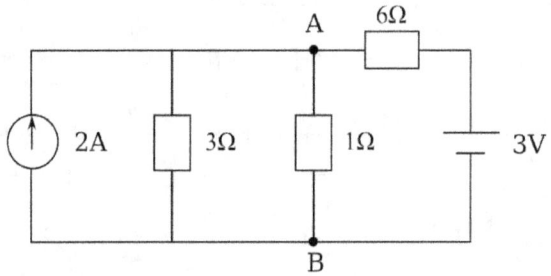

Figure 9.32:   Circuit

Generally when applying Thevenin's theorem, the branch of concern, which in this case is branch AB, is excluded in the transformations as the information within the branch will be lost. We first transform the loop on the left in Fig. 9.33.

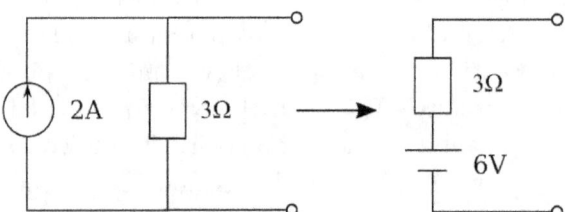

Figure 9.33:   Transformation of branches left of AB

The equivalent voltage of the above component is the voltage across the two terminals when open-circuited. This is simply the voltage across the resistor when it carries 2A current.

$$V_{eq} = I \cdot R = 2 \cdot 3 = 6\text{V}.$$

To compute the equivalent resistance with respect to the terminals, the current source is disconnected. Then,

$$R_{eq} = 3\Omega.$$

Then, we concatenate this equivalent component back into the original circuit and consider the network with points A and B as its terminals in Fig. 9.34.

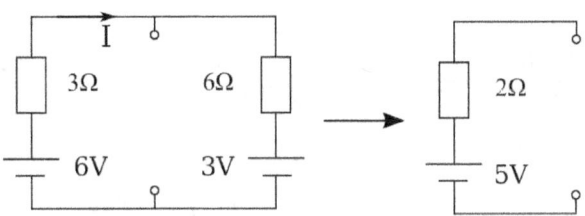

Figure 9.34: Thevenin circuit with respect to terminals A and B

The current that flows through the loop is

$$I = \frac{6-3}{3+6} = \frac{1}{3}\text{A}$$

clockwise. Thus, the voltage between the two terminals is

$$V_{eq} = 6 - 3 \times \frac{1}{3} = 5\text{V}.$$

Short-circuiting the batteries, the equivalent Thevenin resistance is equal to that of the 3Ω and 6Ω resistors connected in parallel.

$$R_{eq} = \frac{1}{\frac{1}{3} + \frac{1}{6}} = 2\Omega.$$

Finally, we splice this circuit with branch AB to obtain Fig. 9.35.

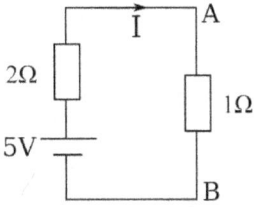

Figure 9.35: Equivalent circuit

Thus, the current flowing through the 1Ω resistor in branch AB is

$$I = \frac{5}{2+1} = \frac{5}{3} A.$$

Problem: Find current I in Fig. 9.36.

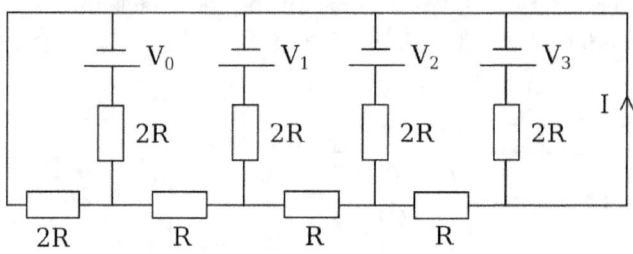

Figure 9.36:   Circuit

We apply Thevenin's theorem to one loop at a time, beginning from the left.

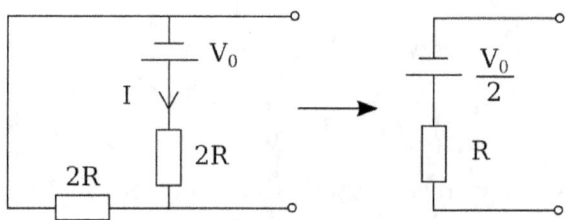

Figure 9.37:   First transformation

$$I = \frac{V_0}{4R},$$

$$V_{eq} = V_0 - I \cdot 2R = \frac{V_0}{2},$$

$$R_{eq} = \frac{1}{\frac{1}{2R} + \frac{1}{2R}} = R.$$

Substituting this equivalent component into the original circuit and applying Thevenin's theorem again in Fig. 9.38,

$$I = \frac{V_1 - \frac{V_0}{2}}{2R + R + R} = \frac{V_1}{4R} - \frac{V_0}{8R},$$

$$V_{eq} = V_1 - I \cdot 2R = \frac{V_0}{4} + \frac{V_1}{2},$$

$$R_{eq} = R.$$

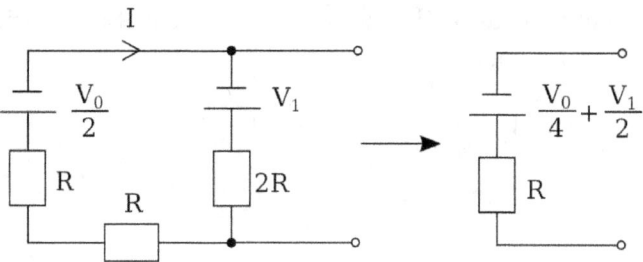

Figure 9.38:   Second transformation

Repeating this process again in Fig. 9.39,

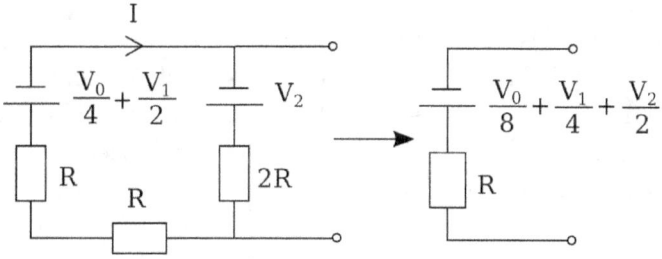

Figure 9.39:   Third transformation

$$I = \frac{V_2 - \frac{V_0}{4} - \frac{V_1}{2}}{2R + R + R} = \frac{V_2}{4R} - \frac{V_0}{16R} - \frac{V_1}{8R},$$

$$V_{eq} = V_2 - I \cdot 2R = \frac{V_0}{8} + \frac{V_1}{4} + \frac{V_2}{2},$$

$$R_{eq} = R.$$

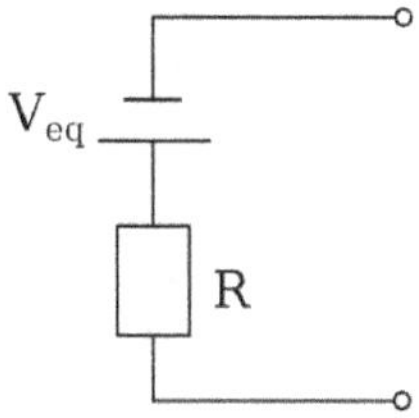

Figure 9.40:   Result of n transformations

Referring to Fig. 9.40, we observe that in the general case, the equivalent Thevenin circuit for the first n voltage sources on the left possesses

$$V_{eq} = \frac{V_0}{2^n} + \frac{V_1}{2^{n-1}} + \cdots + \frac{V_{n-1}}{2}$$

and an equivalent resistance R. The equivalent circuit for the original system occurs when $n = 4$. Thus,

$$V_{eq} = \frac{V_0}{16} + \frac{V_1}{8} + \frac{V_2}{4} + \frac{V_3}{2},$$

$$I = \frac{1}{R}\left(\frac{V_0}{16} + \frac{V_1}{8} + \frac{V_2}{4} + \frac{V_3}{2}\right).$$

This is actually a digital to analog converter! In practice, V_i is either some constant V or 0. Then, this circuit is able to process the output from many digital sources (which produce either 0 or V) into an approximately analog signal (which has a range of output from 0 to V) through different binary combinations!

Ultimately, there is a compromise between the number of times you have to apply Thevenin's theorem and the complexity of the simultaneous equations obtained from Kirchhoff's laws. However, applying Thevenin's theorem multiple times is generally expeditious as the simultaneous equations that need to be solved are drastically simplified.

9.4.1   *Source Transformations*

Referring to Fig. 9.41, a voltage source ε connected in series with a resistor R between two terminals A and B can be transformed into a current source I connected in parallel to a resistor R', across the same two terminals. The reverse transformation holds as well.

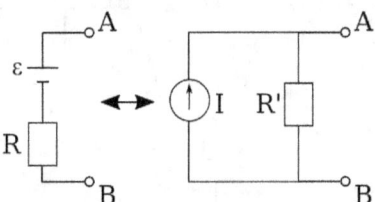

Figure 9.41:   Source transformation

The existence of such an equivalence and the relationship between the above variables is given by Thevenin's theorem. Applying Thevenin's theorem to the right circuit,

$$R' = R, \tag{9.4}$$

$$I = \frac{\varepsilon}{R'} = \frac{\varepsilon}{R}. \tag{9.5}$$

These equations are known as the source transformations. Due to the above interconversion, an equivalent of Thevenin's theorem, known as Norton's theorem, can be stated as follows. For any network that comprises purely independent emf sources, current sources and resistors between two terminals, Norton's theorem states that it can be transformed into an equivalent network that consists of a single Norton current source I_{eq} connected in parallel to an internal Norton resistance R_{eq} with respect to the two terminals. From the source transformation rules, R_{eq} is the Thevenin equivalent resistance which can be computed as previously discussed. Furthermore, if we denote ε_{eq} as the Thevenin equivalent emf,

$$I_{eq} = \frac{\varepsilon_{eq}}{R_{eq}} = I_{sc}$$

by Eq. (9.3). The Norton equivalent current is thus the short-circuit current between the two terminals.

Similar to how voltage sources connected in series can be reduced into an equivalent voltage source trivially (just add the internal resistances, because they are connected in series, and the emfs), Norton's theorem paves a way to reduce current sources connected in parallel into an equivalent current source.

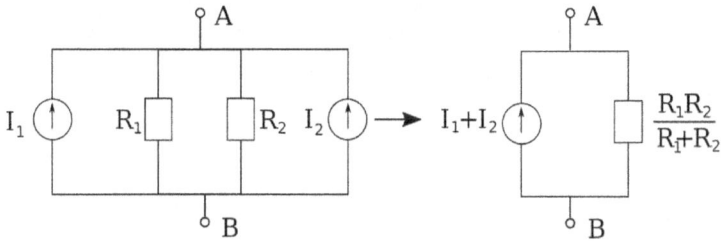

Figure 9.42:   Current sources in parallel

Referring to Fig. 9.42, consider two current sources I_1 and I_2, with internal resistances R_1 and R_2 connected in parallel. The Norton equivalent resistance between terminals A and B is evidently

$$R_{eq} = \frac{1}{\frac{1}{R_1} + \frac{1}{R_2}} = \frac{R_1 R_2}{R_1 + R_2}.$$

Suppose that we connected an ideal wire between terminals A and B. The currents from the two sources would both flow through the ideal wire, with zero current going through the internal resistances. Therefore, the short-circuit current is $I_{sc} = I_1 + I_2$ and the equivalent Norton current is

$$I_{eq} = I_1 + I_2.$$

Notice that even if the branches containing the current sources included an additional resistor connected in series, the above results would not change. Therefore, any resistor connected in series to a current source can be ignored (this is why the internal resistance of a current source must be connected in parallel). Finally, we can repeat this algorithm $(k-1)$ times when there are k current sources connected in parallel between terminals A and B, with the jth current source I_j possessing internal resistance R_j, to obtain the following equivalent Norton resistance and current.

$$R_{eq} = \frac{1}{\sum_{j=1}^{k} \frac{1}{R_j}},$$

$$I_{eq} = \sum_{j=1}^{k} I_j.$$

In light of the above transformations, we have the following idea: if we want to add components in parallel, we convert everything into current sources to reduce them into a single current source. Conversely, if we want to add components in series, we convert everything into voltage sources. Therefore, we can simplify the circuit in Fig. 9.43 in the following manner, where the last two are the equivalent Thevenin and Norton circuits with respect to the two terminals.

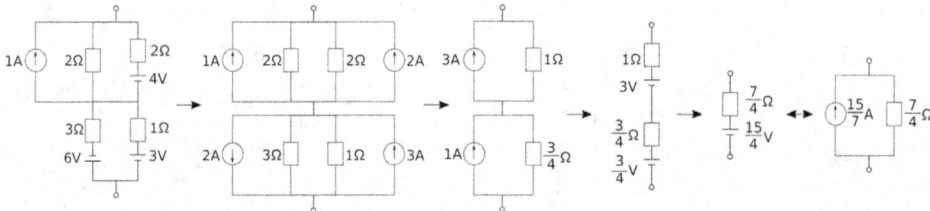

Figure 9.43:   Simplification of complex circuit

9.5   Y-Δ Transformations

The Y-Δ transformations are a set of mathematical rules and simplifications to convert between a circuit consisting of resistors arranged in a "Y-shape" (Fig. 9.44) and another that is arranged in a "Δ-shape" (Fig. 9.45).

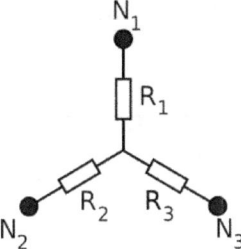

Figure 9.44: Y-circuit

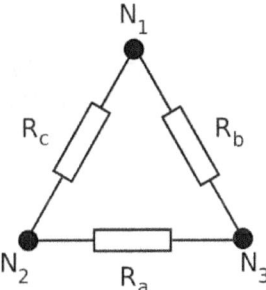

Figure 9.45: Δ-circuit

By convention, the alphabetical subscripts of the resistors in the Δ-circuit correspond to the nodes opposite to the sides containing the resistors. For example, the opposite of R_a is node 1 while the opposite of R_b is node 2. The transformations from the Δ to Y circuit are

$$R_1 = \frac{R_b R_c}{R_a + R_b + R_c},$$

$$R_2 = \frac{R_a R_c}{R_a + R_b + R_c},$$

$$R_3 = \frac{R_a R_b}{R_a + R_b + R_c}.$$

An easy way to remember the resistance of a resistor directly adjacent to a particular node in the Y-circuit is to take the product of the resistances of the resistors adjacent to that node in the Δ circuit and divide it by the sum of all the resistances. The inverse transformations from the Y to Δ circuit are

$$R_a = \frac{R_1 R_2 + R_2 R_3 + R_3 R_1}{R_1},$$

$$R_b = \frac{R_1 R_2 + R_2 R_3 + R_3 R_1}{R_2},$$

$$R_c = \frac{R_1 R_2 + R_2 R_3 + R_3 R_1}{R_3}.$$

Another easy way to remember the resistance of a particular resistor in the Δ-circuit is to take the sum of all possible combinations of the product of pairs of resistances in the Y-circuit and divide it by the resistance in the Y-circuit that corresponds to the node opposite to that particular resistor.

Proof: The existence of these equivalent transformations and the equivalent resistances can be proven by the principle of superposition. These two circuits are said to be equivalent if the voltages between pairs of nodes (V_{12}, V_{23}, V_{31}) are the same in the two circuits for any currents (I_1, I_2, I_3) entering the corresponding nodes (N_1, N_2, N_3) and vice-versa (identical currents, given fixed voltages).

The resistances of the two circuits can be tuned to satisfy the forward condition by considering the superposition of three different set-ups with currents

$$\left(\frac{I_1 - I_2}{3}, \frac{I_2 - I_1}{3}, 0 \right),$$

$$\left(0, \frac{I_2 - I_3}{3}, \frac{I_3 - I_2}{3} \right),$$

$$\left(\frac{I_1 - I_3}{3}, 0, \frac{I_3 - I_1}{3} \right),$$

which in combination gives

$$\left(\frac{2I_1 - I_2 - I_3}{3}, \frac{2I_2 - I_1 - I_3}{3}, \frac{2I_3 - I_1 - I_2}{3} \right).$$

Furthermore, $I_1 + I_2 + I_3 = 0$ as required by Kirchhoff's junction rule which implies that the superposition of those circuits give an equivalent circuit with currents

$$(I_1, I_2, I_3)$$

entering nodes (N_1, N_2, N_3) which is the general set-up of concern. Thus, if we are able to show that the two circuits satisfy the first condition (identical voltages given incoming currents) in the three sub-problems, we will also be able to prove that the two circuits satisfy the first condition for any general currents flowing into the nodes.

Let us consider the first sub-problem with currents $(\frac{I_1-I_2}{3}, \frac{I_2-I_1}{3}, 0)$ flowing into the nodes. This is equivalent to connecting the ends of a battery of a certain emf to nodes N_1 and N_2. The voltage between N_1 and N_3 in the Δ-circuit can be determined as

$$V_{13} = I_{13} \cdot R_b$$

where I_{13} is the current flowing from node 1 to node 3. I_{13} can be calculated from the current divider principle as

$$I_{13} = \frac{R_c}{R_a + R_b + R_c} \cdot \frac{I_1 - I_2}{3},$$

$$V_{13} = \frac{R_b R_c}{R_a + R_b + R_c} \cdot \frac{I_1 - I_2}{3}.$$

Next, the voltage between N_1 and N_3 in the Y-circuit in this sub-problem is

$$V_{13} = R_1 \cdot \frac{I_1 - I_2}{3}.$$

In order for the two V_{13} in the two circuits to be the same,

$$R_1 = \frac{R_b R_c}{R_a + R_b + R_c}.$$

A similar process can be applied to ensure that V_{23} is the same in both circuits. The criterion for this is

$$R_2 = \frac{R_a R_c}{R_a + R_b + R_c}.$$

We do not need to find another condition for the two V_{12}'s in the two circuits to be equal, as the equivalence of the two voltages above already guarantees so ($V_{12} = V_{13} - V_{23}$). Lastly, this entire procedure can be used to determine the appropriate resistances for the two circuits in the other sub-problems. Then, six equations for three variables, which are thankfully coherent, are obtained. The solutions are

$$R_1 = \frac{R_b R_c}{R_a + R_b + R_c}, \tag{9.6}$$

$$R_2 = \frac{R_a R_c}{R_a + R_b + R_c}, \tag{9.7}$$

$$R_3 = \frac{R_a R_b}{R_a + R_b + R_c}. \tag{9.8}$$

This shows that the resistances in the circuits can be tuned to satisfy the first condition. Moving on, we then need to prove that the currents (I_1, I_2, I_3)

entering the nodes (N_1, N_2, N_3) are equal in both circuits for any voltages between pairs of nodes (V_{12}, V_{23}, V_{13}). The appropriate resistances that fulfil this requirement can be determined by considering the superposition of the following set-ups with voltages:

$$(V_{12}, 0, 0),$$

$$(0, V_{23}, 0),$$

$$(0, 0, V_{13}).$$

Let us consider the first sub-problem. The current I_2 in the Δ circuit is simply

$$I_2 = \frac{V_{12}}{R_c}.$$

The current I_2 in the Y-circuit can also be computed as

$$I_2 = \frac{V_{12}}{\frac{R_1 R_2 + R_2 R_3 + R_3 R_1}{R_3}}.$$

Equating these, we obtain

$$R_c = \frac{R_1 R_2 + R_2 R_3 + R_3 R_1}{R_3}.$$

Similarly, if we impose the requirement that the two I_3's must be equal in the two circuits under the conditions of this sub-problem, it can be concluded that

$$R_b = \frac{R_1 R_2 + R_2 R_3 + R_3 R_1}{R_2}.$$

Again, the condition for I_1 to be equal in both circuits is automatically satisfied as a consequence of Kirchhoff's junction rule. Then, a similar process can be applied to the rest of the sub-problems to obtain a total of six equations which can be reduced to the following three unique equations.

$$R_a = \frac{R_1 R_2 + R_2 R_3 + R_3 R_1}{R_1}, \tag{9.9}$$

$$R_b = \frac{R_1 R_2 + R_2 R_3 + R_3 R_1}{R_2}, \tag{9.10}$$

$$R_c = \frac{R_1 R_2 + R_2 R_3 + R_3 R_1}{R_3}. \tag{9.11}$$

Finally, it can be shown that the set of Eqs. (9.6)–(9.8) is entirely coherent with the set of Eqs. (9.9)–(9.11) after some algebraic manipulation. Therefore, the Δ and Y-circuits are equivalent if either set of equations is satisfied.

The two sets of equations can each be used as a transformation rule between the circuits. Given a particular direction of transformation, the more convenient set of equations is usually preferred. Equations (9.6)–(9.8) are usually used to transform the Δ-circuit to the Y-circuit while Eqs. (9.9)–(9.11) usually function as the inverse transformations.

Application

The Y-Δ transformations are often used to simplify circuits with nodes that are interlinked by resistors. They act as a slightly more efficient substitute for Kirchhoff's laws, though the calculation of the equivalent resistances can sometimes be tedious. Most of the time, the Y-Δ transformations should be used when the direct application of Kirchhoff's laws is the only other feasible method and when the other sleights-of-hand discussed earlier are inapplicable. However, note that a conversion from a Y-circuit to a Δ-circuit eliminates the node at the center of the "Y". Thus, information that pertains to that eliminated node is harder to be retrieved from the equivalent Δ-circuit.

Problem: Determine currents I_1 and I_2 in Fig. 9.46.

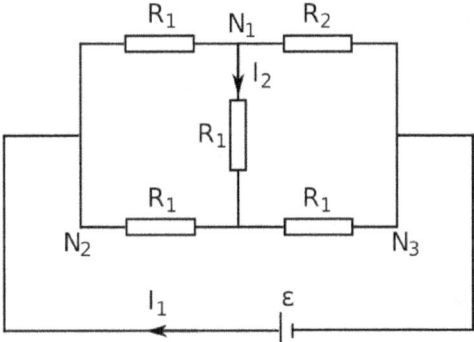

Figure 9.46: Initial circuit (Y-circuit)

The Y-circuit demarcated by the nodes N_1, N_2 and N_3 can be transformed to a Δ configuration with resistances

$$R_a = R_b = R_c = 3R_1.$$

Therefore, the equivalent circuit in Fig. 9.47 can be obtained.

The equivalent resistance of the entire circuit can be computed to be

$$R_{eq} = \frac{3R_1^2 + 5R_1R_2}{5R_1 + 3R_2}.$$

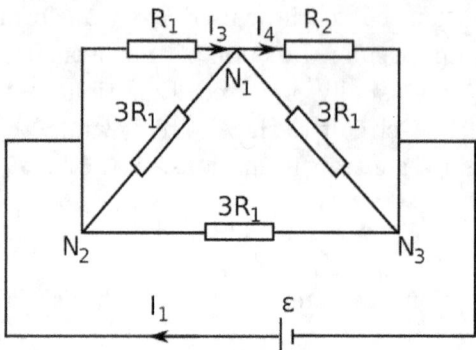

Figure 9.47:   Equivalent circuit (Δ-circuit)

Thus,

$$I_1 = \frac{5R_1 + 3R_2}{3R_1^2 + 5R_1R_2}\varepsilon.$$

However, I_2 cannot be computed directly from the equivalent Δ-circuit and must instead be determined by subtracting I_4 from I_3. I_3 and I_4 can eventually be calculated as the following expressions from the rules regarding series and parallel connections of resistors.

$$I_3 = \frac{3R_1 + R_2}{3R_1^2 + 5R_1R_2}\varepsilon,$$

$$I_4 = \frac{4R_1}{3R_1^2 + 5R_1R_2}\varepsilon.$$

Hence,

$$I_2 = I_3 - I_4 = \frac{R_2 - R_1}{3R_1^2 + 5R_1R_2}\varepsilon.$$

Reduction of Circuits

The utility of the Y-Δ transformations is not only restricted to the interconversion between the two types of circuits. In fact, the Y-Δ transformations imply that any network of resistors can be converted into an equivalent Y or Δ-circuit with respect to three terminals, analogous to how an arbitrary network of resistors can be reduced to a single equivalent resistor with respect to two terminals. To show this, let the three terminals of concern be A, B and C. If there is another node D that is connected to all of A, B and C via paths of resistors, observe that we can see A, B, C and D as a Y-circuit with D at the center and transform it into a Δ-circuit to eliminate node D.

Repeating this for all other such nodes, we will only be left with nodes that are connected to a pair of terminals (in A, B and C) or a lone terminal. Nodes in the former classification can be reduced to equivalent resistors between the corresponding pairs of terminals to form the corresponding sides of the equivalent Δ-circuit while nodes in the latter classification are meaningless in the context of determining the network's response when terminals A, B and C are connected to external entities as they are not linked to at least one pair of terminals. Hence, any network can be reduced to an equivalent Δ and thus Y-circuit, with respect to three terminals.

In fact, the equivalent Y-circuit can be easily constructed if we know the equivalent resistance between the three possible pairs of terminals. Adopting the notation in Fig. 9.44, if the equivalent resistances between (N_1, N_2), (N_1, N_3) and (N_2, N_3) are R_{12}, R_{13} and R_{23}, we have the following set of linear equations

$$R_1 + R_2 = R_{12},$$

$$R_1 + R_3 = R_{13},$$

$$R_2 + R_3 = R_{23},$$

whose solutions are

$$R_1 = \frac{R_{12} + R_{13} - R_{23}}{2}, \tag{9.12}$$

$$R_2 = \frac{R_{12} - R_{13} + R_{23}}{2}, \tag{9.13}$$

$$R_3 = \frac{-R_{12} + R_{13} + R_{23}}{2}. \tag{9.14}$$

This equivalence can be applied in tandem with the previous techniques to solve harder variations of problems such as the following.

Problem: Determine the equivalent resistance between points A and C in the infinite triangular grid of resistors depicted in Fig. 9.48. Each edge of a triangle has resistance R and the edge between points A and B has been removed.

We can deem the absent connection between points A and B as two resistors R and $-R$ connected in parallel across these points. Then, we can determine the equivalent Y-circuit of the imaginary resistor R in this branch and the rest of the grid (i.e. a complete grid) with respect to the three terminals A, B and C. The resistance between any pair of these terminals is simply that between two adjacent nodes in a complete infinite triangular grid. This can be computed via the principle of superposition.

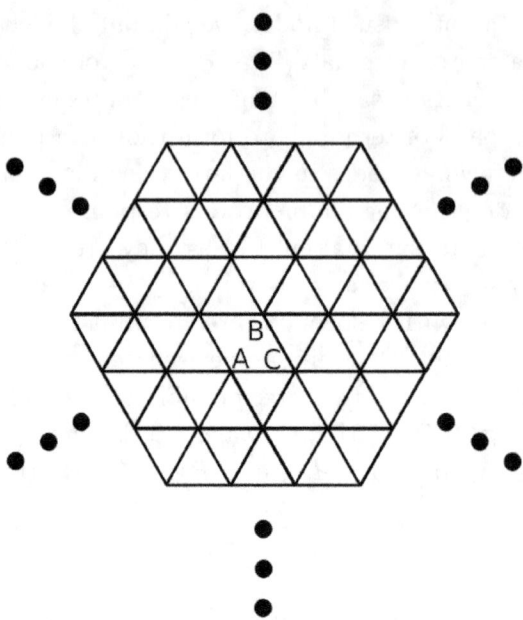

Figure 9.48:   Infinite circuit with edge AB removed

Suppose that we inject 1A current into a certain node 1 in the complete grid and withdraw it entirely at infinity — $\frac{1}{6}$A current will flow from node 1 to a neighboring node (name this node 2) via the resistor directly connecting them as a result of symmetry. In a similar vein, consider a new set-up where we withdraw 1A current from node 2 and inject 1A current at infinity. $\frac{1}{6}$A current again flows from node 1 to node 2 via the branch directly connecting them. Superposing these set-ups, 1A current is injected into node 1, circulated within the infinite grid and finally withdrawn from node 2. No current enters or leaves from infinity. In this process, the current directly flowing across the resistor R connecting nodes 1 and 2 is $\frac{1}{6} + \frac{1}{6} = \frac{1}{3}$A which implies that the voltage between these nodes is $\frac{R}{3}$. Since $\frac{R}{3}$ voltage is required to circulate 1A current through the grid via nodes 1 and 2, the equivalent resistance of the infinite grid with respect to these nodes is $\frac{R}{3}$.

Returning to the original problem, since the equivalent resistance between any pair out of the terminals A, B and C is $\frac{R}{3}$ for a complete infinite grid, the equivalent circuit is shown in Fig. 9.49.

The Y-circuit is the equivalent of the complete grid with respect to A, B and C ($\frac{R}{6}$ resistors because the equivalent resistance between any two terminals is $\frac{R}{3}$). Remember that we have to include the $-R$ resistor between A and B as they are actually disconnected. The equivalent resistance between

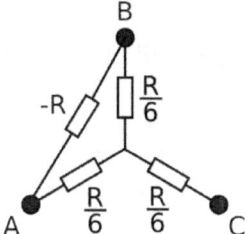

Figure 9.49:   Equivalent circuit between terminals A, B and C

A and C can be computed from the rules of series and parallel connections.

$$R_{AC} = \frac{R}{6} + \frac{-\frac{5R}{6} \cdot \frac{R}{6}}{-\frac{4R}{6}} = \frac{3}{8}R.$$

9.6   Infinite Networks

Often, it is extremely important to astutely abuse symmetrical properties when tackling infinitely large and repeating circuits. Usually, this involves defining the quantity of interest as a variable and constructing a equation in that variable by utilizing the fact that breaking off or adding one sub-unit to the infinite network does not change the resultant quantity, since the network extends forever.

Problem: Find the equivalent resistance of the infinite resistor ladder in Fig. 9.50 across nodes A and B.

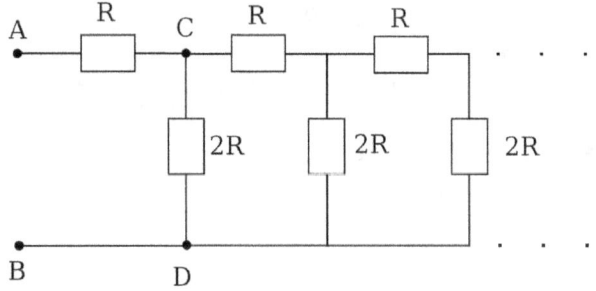

Figure 9.50:   Infinite ladder of resistors

We observe that if we break off the right side of the circuit along line CD, we obtain the exact same network.[4] Furthermore, this excised

---

[4]Equivalently, we could have added another two resistors on the right.

component was previously connected in parallel to the $2R$ resistor between nodes C and D. Thus, if we let the equivalent resistance of the original circuit with respect to A and B be R_{eq}. The original circuit can be transformed into Fig. 9.51.

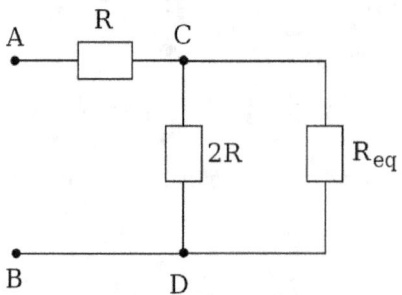

Figure 9.51:   Modified circuit

The equivalent resistance of this circuit with respect to terminals A and B should also be R_{eq}. Thus,

$$R_{eq} = \frac{2RR_{eq}}{2R + R_{eq}} + R.$$

Simplifying,

$$R_{eq}^2 - RR_{eq} - 2R^2 = 0$$

$$(R_{eq} - 2R)(R_{eq} + R) = 0$$

$$\implies R_{eq} = 2R,$$

where the physically incorrect solution, $R_{eq} = -R$, has been rejected. For those who find mathematics more appealing, what we are actually doing is as follows. The equivalent resistance with respect to terminals A and B is

$$\frac{R_{eq}}{R} = 1 + \cfrac{1}{\frac{1}{2} + \cfrac{1}{1 + \cfrac{1}{\frac{1}{2} + \frac{1}{\ddots}}}}.$$

Observe that we can rewrite the above as

$$\frac{R_{eq}}{R} = 1 + \cfrac{1}{\frac{1}{2} + \frac{1}{\frac{R_{eq}}{R}}} = 1 + \frac{2R_{eq}}{R_{eq} + 2R}.$$

since the infinite fraction extends forever.

$$\implies R_{eq}^2 - RR_{eq} - 2R^2 = 0$$

$$R_{eq} = 2R.$$

We choose the positive solution as the infinite fraction is evidently positive. Ultimately, in both of the above cases, we define a variable for the attribute we wish to solve for and then generate an equation in this variable by exploiting the infinite nature of the question.

Problems

Kirchhoff's Laws

1. *Connecting a Resistor**

Two resistors R_1 and R_2 are connected in series with a constant voltage source. We do not know the exact values of R_1 and R_2 but only their ratio $r = \frac{R_1}{R_2}$. If we subsequently connect a certain resistor in parallel to R_2, the current flowing through R_1 changes by ΔI_1. What is the current flowing through the new resistor?

2. *Circuit 1**

Find the current through the 2Ω resistor in the circuit below.

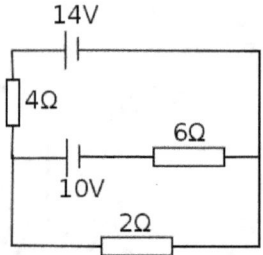

3. *Circuit 2**

Determine currents I_1 and I_2 in the circuit below.

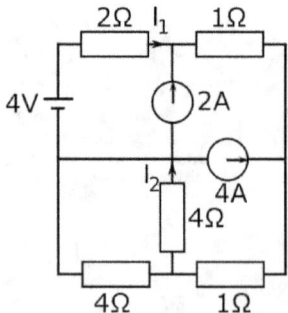

4. *Bridge**

Determine the current flowing through the battery with emf ε in the circuit below.

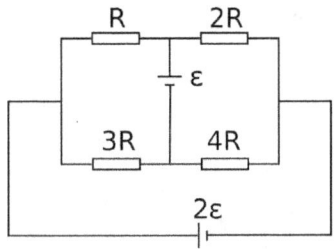

5. *Zero Current**

Show that no current flows in every branch of the circuit below. All batteries are identical and have emf ε.

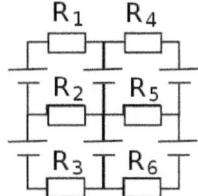

6. *Circuit 3***

Determine current I in the circuit below via nodal analysis.

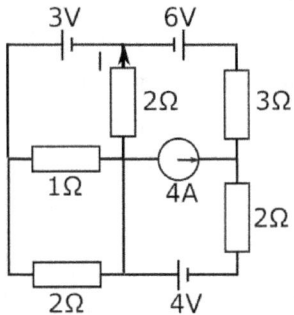

Special Techniques

7. *Triangle Circuit**

Exploiting the principle of superposition, determine the currents through all resistors on the circuit below.

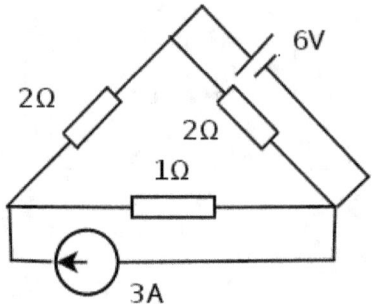

8. *Infinite Grid Revisited**

Find the equivalent resistance of the infinite grid of resistors R between adjacent nodes A and B if the resistor R between A and B were replaced by a resistor R' instead.

9. *Finite Grid**

Find the equivalent resistances between nodes 1, 9 and nodes 1, 5. All resistors are identical and possess resistance R.

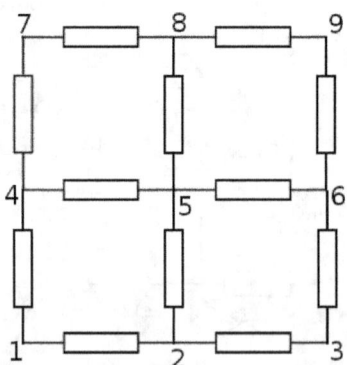

10. *Tetrahedron* *

A wire of resistivity ρ and cross sectional area A is bent into a tetrahedron as shown below. If all sides of the tetrahedron are of length l, find the equivalent resistance between points A and M which is the midpoint of BC.

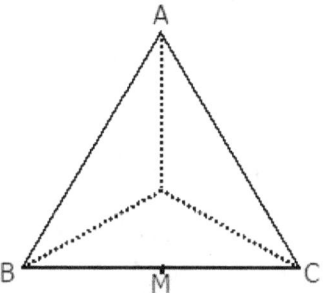

11. *20 Resistors* *

Twenty identical resistors R are connected as shown in the figure below. Calculate the equivalent resistance between points (a) A and B (b) A and C (c) A and D.

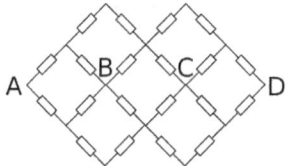

12. *Scaling the Ladder* *

Determine the equivalent resistance of the infinite ladder below between terminals A and B. The resistors in each section of the ladder (except the left-most one) have k times the resistance of those in the preceding section. Verify your answer in the limit $k \to \infty$.

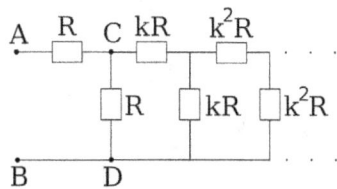

13. *Unknown Circuit**

An unknown circuit, consisting of independent emf sources and resistors, is placed into a black box with two terminals A and B. When an ideal ammeter is connected between A and B, its reading is I. When a resistor R is connected instead, the current through that resistor is i. What would be the reading V of an ideal voltmeter connected to A and B?

14. *Hexagon***

Each line in the figure below represents a 1Ω resistor.

(a) Determine the equivalent resistance between A and C.
(b) Determine the equivalent resistance between B and C.
(c) If a voltage source of 10V is connected between A and C, what is the potential difference between D and E?
(d) If a voltage source of 10V is connected between B and C, what is the potential difference between B and D?

15. *Infinite Hexagonal Tiles***

Find the equivalent resistance between points A and B in an infinite grid consisting of identical resistors R arranged in hexagonal tiles.

16. *2018 Nodes***

There are 2018 nodes with a resistor R connected between each pair of nodes. Find the equivalent resistance of the entire network between any two nodes.

17. *Cube of Resistors Revisited***

Find the equivalent resistance between any two points in a cube of resistors by utilizing the principle of superposition.

18. *Fractal Resistance***

A piece of wire with cross sectional area A and resistivity ρ is bent into the equilateral triangle fractal below. Each successive triangle has half the side length of its predecessor triangle and the pattern repeats indefinitely. Find the equivalent resistance between points A and B.

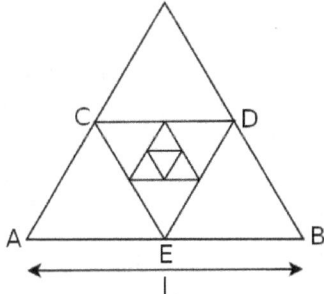

19. *Double Cube***

Determine the equivalent resistance between vertices A and B in the figure below. All resistors have resistance R. Hint: make use of vertex C.

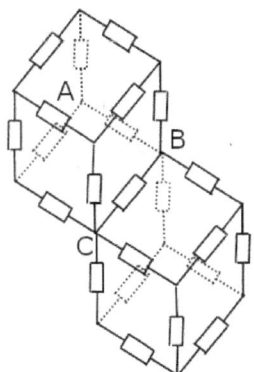

20. *Wheatstone Bridge***

By applying Thevenin's theorem, determine the current flowing through branch AB in the circuit below.

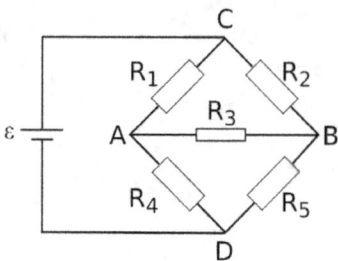

21. *Equivalent Battery***

N batteries are connected in parallel, with their terminals oriented in the same direction. If the ith battery has an emf ε_i and internal resistance r_i and if the entire set-up can be reduced to a single equivalent battery with emf ε_{eq} and internal resistance r_{eq} with respect to two terminals at the ends of a parallel branch, determine ε_{eq} and r_{eq}.

22. *Maximum Power Transfer***

Determine the resistor r that should be connected across the two terminals in the figure below such that the power through it is maximal across all possible values of r.

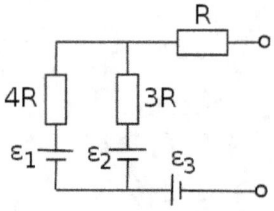

23. *Another Infinite Ladder**

Find the current I in the circuit below.

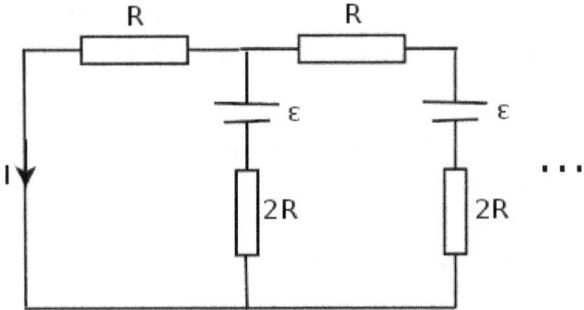

24. *Thick Infinite Ladder***

Determine the equivalent resistance between two adjacent nodes in the middle row of the infinite network of identical resistors R in the figure below. The circuit only extends to infinity in the horizontal direction.

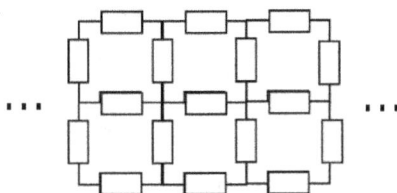

25. *N-gon with Spokes***

The N edges of a regular N-gon are constructed by N resistors R. Furthermore, all N vertices of the N-gon are connected to its geometric center O via spokes of resistance $2R$. Determine the equivalent resistance of this set-up between a vertex and O.

Solutions

1. Connecting a Resistor*

The trick in this problem is to apply Ohm's law to the differences in voltages and currents. Since the current through resistor 1 changes by ΔI when the new resistor is added, the change in voltage across resistor 1 is

$$\Delta V_1 = \Delta I_1 R_1.$$

The change in voltage across resistor 2 must be negative of this as the voltage source produces a constant emf.

$$\Delta V_2 = -\Delta I_1 R_1.$$

This implies that the change in current flowing through R_2 is

$$\Delta I_2 = -\Delta I_1 \frac{R_1}{R_2} = -r\Delta I_1.$$

The current flowing through the new resistor is thus

$$\Delta I_1 - \Delta I_2 = (1 + r)\Delta I_1.$$

2. Circuit 1*

We define the currents I_1 and I_2 using the clockwise loops in Fig. 9.52. Applying Kirchhoff's loop rule to the two clockwise loops within the meshes,

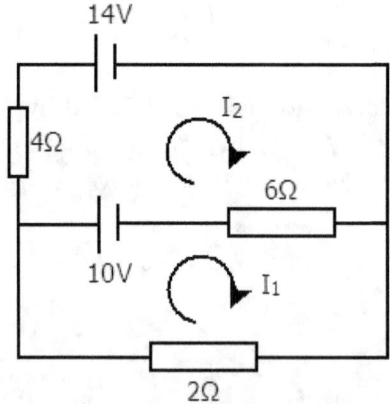

Figure 9.52:   Circuit

$$10 - 4I_2 - 14 - 6(I_2 - I_1) = 0$$

$$10I_2 - 6I_1 = -4$$

$$-10 + 6(I_1 - I_2) + 2I_1 = 0$$

$$8I_1 - 6I_2 = 10.$$

Solving,

$$I_1 = -\frac{31}{11}A$$

which means that the direction of the current through the 2Ω resistor is towards the right.

3. Circuit 2*

We shall use mesh analysis to solve this problem, since there are relatively many current sources. Labeling the currents in each branch in Fig. 9.53,

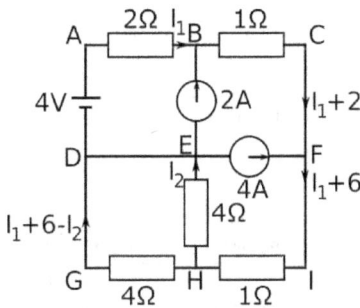

Figure 9.53: Labeled circuit

Applying Kirchhoff's loop rule to supermesh DABCFIHED,

$$4 - 2I_1 - (I_1 + 2) - (I_1 + 6) - 4I_2 = 0.$$

Considering mesh HEDGH,

$$-4I_2 + 4(I_1 + 6 - I_2) = 0.$$

Solving,

$$I_1 = -\frac{8}{3}A,$$

$$I_2 = \frac{5}{3}A.$$

4. Bridge*

Since there are many batteries relative to the number of meshes, we shall use nodal analysis here. Define the potentials of the positive and negative terminals of the 2ε battery as 2ε and 0 respectively. Let the potentials of the positive and negative terminals of the ε battery be $V + \varepsilon$ and V. Then, treat the nodes at the two ends of the ε battery as a supernode. For there to be no net current flowing out of the supernode,

$$\frac{V + \varepsilon - 2\varepsilon}{R} + \frac{V + \varepsilon - 0}{2R} + \frac{V - 2\varepsilon}{3R} + \frac{V - 0}{4R} = 0$$

$$V = \frac{14}{25}\varepsilon.$$

The current through the ε battery can then be determined by subtracting the current through the $4R$ resistor from that through the $3R$ resistor.

$$I_{bat} = \frac{2\varepsilon - V}{3R} - \frac{V - 0}{4R} = \frac{17\varepsilon}{50R}$$

from the negative to the positive terminal.

5. Zero Current*

Let the potential of the middle node be zero. Then, label the potentials of the other nodes according to Fig. 9.54.

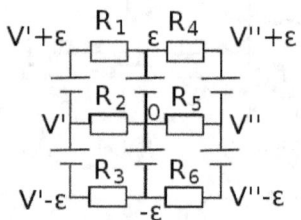

Figure 9.54:   Circuit with labeled potentials

Consider the left three nodes as a supernode. There must be zero net current emanating from the supernode. Therefore,

$$\frac{V' + \varepsilon - \varepsilon}{R_1} + \frac{V'}{R_2} + \frac{V' - \varepsilon + \varepsilon}{R_3} = 0$$

$$\implies V' = 0.$$

A similar analysis for the three right nodes would show that the potentials of all nodes on the first, second and third rows are ε, 0 and $-\varepsilon$. Then, no current flows across all resistors. By Kirchhoff's junction rule, the current should be zero across all batteries as well.

6. Circuit 3**

Again, we first choose a reference node and label all other nodal voltages according to Fig. 9.55.

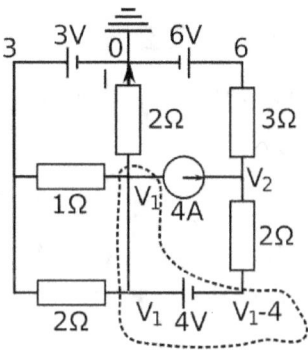

Figure 9.55: Labeled circuit

Applying Kirchhoff's junction rule to the supernode demarcated by the dashed lines in the figure above,

$$\frac{V_1 - 3}{1} + \frac{V_1 - 0}{2} + 4 + \frac{V_1 - 3}{2} + \frac{V_1 - 4 - V_2}{2} = 0.$$

Next, considering the node with nodal voltage V_2,

$$\frac{V_2 - 6}{3} - 4 + \frac{V_2 - V_1 + 4}{2} = 0.$$

Solving,

$$V_1 = \frac{49}{22} V,$$

$$V_2 = \frac{135}{22} V.$$

Hence,

$$I = \frac{V_1 - 0}{2} = \frac{49}{44} A.$$

7. Triangle Circuit*

The circuit is composed of the two sub-circuits in Figs. 9.56 and 9.57.

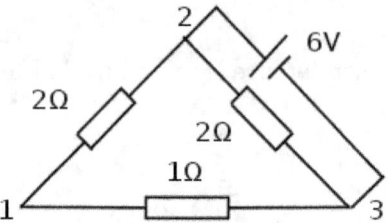

Figure 9.56:   Sub-circuit 1

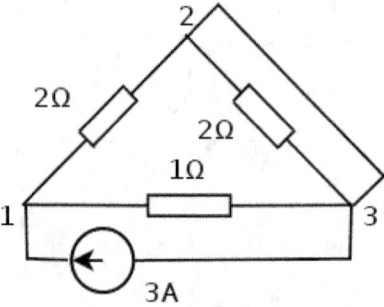

Figure 9.57:   Sub-circuit 2

In the first circuit, the currents through the resistors are

$$I_{12,1} = -\frac{6}{2+1} = -2\text{A},$$

$$I_{23,1} = \frac{6}{2} = 3\text{A},$$

$$I_{13,1} = \frac{6}{2+1} = 2\text{A}.$$

For the second circuit, the currents through the resistors can be determined via the current divider principle (while ignoring the resistor in branch 23, as current would prefer to flow in the ideal wire).

$$I_{12,2} = \frac{1}{1+2} \cdot 3 = 1\text{A},$$

$$I_{23,2} = 0,$$

$$I_{13,2} = \frac{2}{1+2} \cdot 3 = 2\text{A}.$$

The superposition of the above two circuits gives the currents in the branches as

$$I_{12} = I_{12,1} + I_{12,2} = -1\text{A},$$

$$I_{23} = I_{23,1} + I_{23,2} = 3\text{A},$$

$$I_{13} = I_{13,1} + I_{13,2} = 4\text{A}.$$

8. Infinite Grid Revisited*

We have calculated that the equivalent resistance of an infinite grid with identical resistors R between two adjacent nodes is

$$R_{eq} = \frac{R}{2}.$$

Let the equivalent resistance of the entire infinite grid, excluding the resistor between A and B, be R'_{eq}. R'_{eq} can be obtained from R_{eq} by connecting a resistor $-R$ across A and B in a new branch (this $-R$ in parallel with R effectively cuts off the direct connection between A and B).

$$R'_{eq} = \frac{\frac{R}{2} \cdot -R}{\frac{R}{2} - R} = R.$$

The equivalent resistance of the network in question, R''_{eq}, is that of R'_{eq} connected in parallel to R'. Thus,

$$R''_{eq} = \frac{RR'}{R + R'}.$$

9. Finite Grid*

For the equivalent resistance between nodes 1 and 9, observe that nodes 3, 5 and 7 are equipotential points by mirror symmetry. Furthermore, nodes 2, 4 and nodes 6, 8 are equipotential pairs due to path symmetry when traveling from nodes 1 to 9. Thus, the equivalent circuit is shown in Fig. 9.58.

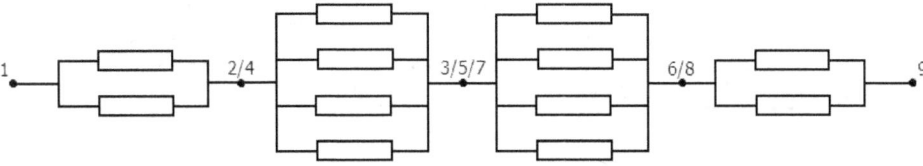

Figure 9.58: Equivalent circuit

The equivalent resistance is then

$$R_{eq} = \frac{R}{2} \times 2 + \frac{R}{4} \times 2 = \frac{3R}{2}.$$

In the second case, nodes 2,4 and nodes 3,7 as well as nodes 6,8 are equipotential pairs while node 9 is useless as it is connected to an equipotential pair. The equivalent circuit is illustrated in Fig. 9.59.

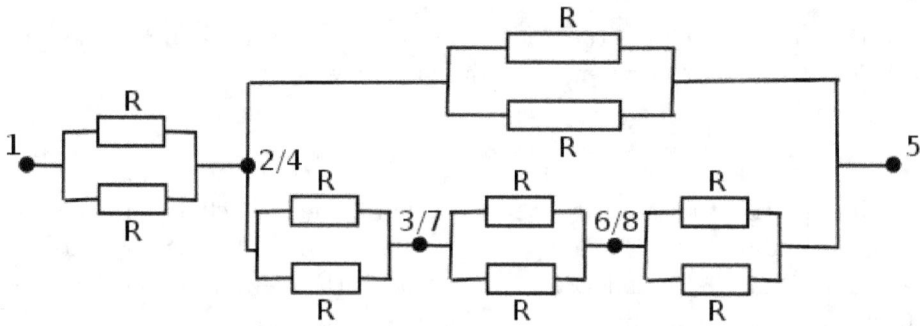

Figure 9.59:   Equivalent circuit

$$R_{eq} = \frac{R}{2} + \frac{\frac{1}{2} \cdot \frac{3}{2}}{\frac{1}{2} + \frac{3}{2}} R = \frac{7}{8} R.$$

10. Tetrahedron*

Let the resistance of a wire of length l be R. Furthermore, we observe that nodes B and C must be equipotential points. Thus, the circuit can be reduced to Fig. 9.60.

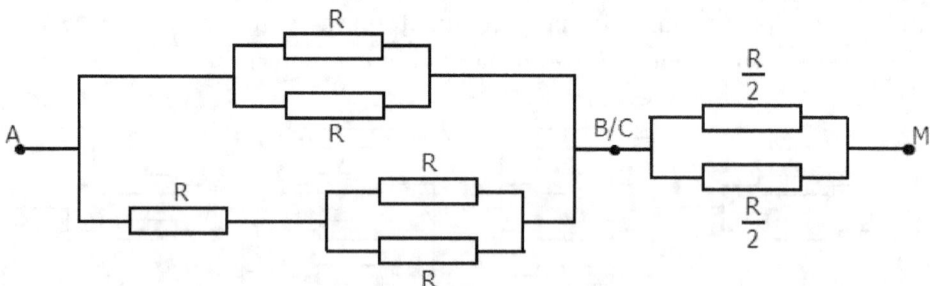

Figure 9.60:   Equivalent circuit

The equivalent resistance is then

$$R_{eq} = \frac{1}{\frac{2}{R} + \frac{2}{3R}} + \frac{R}{4} = \frac{5}{8}R$$

where

$$R = \frac{\rho l}{A},$$

$$R_{eq} = \frac{5\rho l}{8A}.$$

11. 20 Resistors*

The network exhibits mirror symmetry about the horizontal line joining A to D with respect to any two terminals that lie on that line (for which all three cases in the question satisfy). Therefore, the circuit can be reduced to Fig. 9.61 via combining equipotential points.

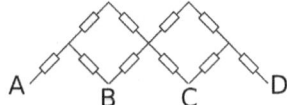

Figure 9.61: Each resistor represents $\frac{R}{2}$

(a) The equivalent resistance between A and B is thus

$$R_{AB} = \frac{R}{2} + \frac{\frac{R}{2} \cdot \frac{3R}{2}}{\frac{R}{2} + \frac{3R}{2}} = \frac{7}{8}R.$$

(b) The equivalent resistance between A and C is

$$R_{AC} = \frac{R}{2} + \frac{R \cdot R}{R + R} + \frac{\frac{R}{2} \cdot \frac{3R}{2}}{\frac{R}{2} + \frac{3R}{2}} = \frac{11}{8}R.$$

(c) Between points A and D, we can find the equivalent resistance directly from Fig. 9.61 to be

$$R_{AD} = \frac{R}{2} + \frac{R \cdot R}{R + R} + \frac{R \cdot R}{R + R} + \frac{R}{2} = 2R,$$

or combine all nodes that lie along the same vertical line in the original circuit by path symmetry to obtain

$$R_{AD} = \frac{R}{2} + \frac{R}{4} + \frac{R}{4} + \frac{R}{4} + \frac{R}{4} + \frac{R}{2} = 2R.$$

12. Scaling the Ladder*

Let R_{eq} denote the equivalent resistance between terminals A and B. Then, the equivalent resistance of the circuit on the right of branch CD (excluding the resistor R in branch CD) with respect to terminals C and D is kR_{eq} by scaling arguments. Thus,

$$R_{eq} = R + \frac{kRR_{eq}}{R + kR_{eq}}$$

$$RR_{eq} + kR_{eq}^2 = R^2 + 2kRR_{eq}$$

$$kR_{eq}^2 - (2k - 1)RR_{eq} - R^2 = 0$$

$$R_{eq} = \frac{2k - 1 + \sqrt{4k^2 + 1}}{2k}R,$$

where we have rejected the negative solution. In the limit $k \to \infty$,

$$R_{eq} = \left(1 - \frac{1}{2k} + \sqrt{1 + \frac{1}{4k^2}}\right) R \to 2R,$$

which is correct since the circuit in such a limit comprises two resistors R in series.

13. Unknown Circuit*

By Thevenin's theorem, the unknown circuit can be reduced into a single equivalent Thevenin voltage source ε_{eq} connected in series with an equivalent Thevenin resistor R_{eq}. From the first clue,

$$\varepsilon_{eq} = IR_{eq}.$$

From the second clue,

$$\varepsilon_{eq} = i(R_{eq} + R).$$

Eliminating R_{eq},

$$\varepsilon_{eq} = \frac{iIR}{I - i}.$$

The ideal voltmeter with infinite resistance measures ε_{eq} by the voltage divider principle. Thus,

$$V = \varepsilon_{eq} = \frac{iIR}{I - i}.$$

14. Hexagon**

The network exhibits mirror symmetry about line ABC with respect to both terminals A,C and B,C. Therefore, D/F and E/G are equipotential pairs when determining the equivalent resistances between A,C and B,C. Thus, the circuit can be redrawn as Fig. 9.62.

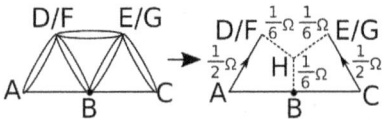

Figure 9.62: Solid, arrowed and dashed lines represent 1Ω, $\frac{1}{2}\Omega$ and $\frac{1}{6}\Omega$

We can further perform a Y-Δ transformation to obtain the circuit on the right, with a new vertex H.

(a) With respect to terminals A and C, the circuit exhibits mirror symmetry about a vertical line passing through B. Therefore, the middle $\frac{1}{6}\Omega$ can be removed as it is connected between a pair of equipotential points. The equivalent resistance between A and C is then

$$R_{AC} = \frac{\frac{4}{3} \cdot 2}{\frac{4}{3} + 2} = \frac{4}{5}\Omega.$$

(b) The equivalent resistance of the resistors between nodes A, B, D/F and H with respect to A and H is

$$r = \frac{\frac{5}{3} \cdot \frac{1}{6}}{\frac{5}{3} + \frac{1}{6}} = \frac{5}{33}\Omega.$$

Thus,

$$R_{BC} = \frac{1 \cdot \left(\frac{5}{33} + \frac{1}{6} + \frac{1}{2}\right)}{1 + \left(\frac{5}{33} + \frac{1}{6} + \frac{1}{2}\right)} = \frac{9}{20}\Omega.$$

(c) Looking at the right circuit after removing the middle $\frac{1}{6}\Omega$, the voltage divider principle yields

$$V_{AC} = \frac{\frac{1}{6} + \frac{1}{6}}{\frac{1}{2} + \frac{1}{6} + \frac{1}{6} + \frac{1}{2}} \times 10 = 2.5\text{V}.$$

(d) By the voltage divider principle, the voltage between B and H is

$$V_{BH} = \frac{r}{r + \frac{1}{6} + \frac{1}{2}} \times 10 = \frac{50}{27}V.$$

By the voltage divider principle again, the voltage between B and D is

$$V_{BD} = \frac{\frac{3}{2}}{\frac{3}{2} + \frac{1}{6}} \times V_{BH} = \frac{5}{3}V = 1.7V \text{ (2sf)}.$$

15. Infinite Hexagonal Tiles**

We inject 1A of current into node A and withdraw it from infinity. Then, the current flowing through branch AC in this set-up is

$$I_{AC} = \frac{1}{3}A$$

by symmetry. The current in branch CB is then

$$I_{CB} = \frac{1}{6}A.$$

Similarly, we consider a new set-up where we withdraw 1A of current from node B and inject it at infinity. The currents flowing through branches AC and CB in this set-up are

$$I'_{AC} = \frac{1}{6}A,$$

$$I'_{CB} = \frac{1}{3}A.$$

We can then superpose these two set-ups to obtain a combined set-up where we inject 1A of current into node A and withdraw it from node B. The voltage between nodes A and B in this case is

$$V = (I_{AC} + I'_{AC})R + (I_{CB} + I'_{CB})R = R.$$

Thus, the equivalent resistance of this circuit with respect to nodes A and B is

$$R_{eq} = \frac{V}{1} = R,$$

since 1A of current flows in the network of resistors when a potential difference V is applied across nodes A and B.

16. 2018 Nodes**

Equipotential Points: Let us label the nodes from 1 to 2018. If we wish to find the equivalent resistance between nodes 1 and 2018, nodes 2 to 2017 are equipotential points by path symmetry. There is one resistor connecting each of nodes 1 and 2018 to each of the nodes from 2 to 2017. Furthermore, there is one resistor directly connecting node 1 to node 2018. Hence the equivalent resistance between nodes 1 and node 2018 is

$$R_{eq} = \frac{\frac{R}{1008} \cdot R}{R + \frac{R}{1008}} = \frac{R}{1009}.$$

Superposition: We first inject 1A of current into node 1 and withdraw $\frac{1}{2017}$A from each of the other 2017 nodes. By symmetry, the current that flows to the immediate neighbors of node 1 is

$$I = \frac{1}{2017}\text{A}.$$

Next we consider another set-up where we withdraw 1A of current from node 2018 and inject $\frac{1}{2017}$A into each of the other 2017 nodes. By symmetry, the current that flows from the node 1 to this node is also

$$I = \frac{1}{2017}\text{A}.$$

Thus, if we superpose these two circuits, $\frac{2018}{2017}$A of current goes into node 1 and $\frac{2018}{2017}$A of current flows out of node 2018. The current that directly flows in the resistor between them is

$$I' = \frac{2}{2017}\text{A}.$$

Consequently, the voltage across the two nodes is

$$V = I'R = \frac{2R}{2017}.$$

This is equal to the emf required of an external battery whose ends are connected to the two nodes and drives a current of $\frac{2018}{2017}$A in the circuit. Thus, the equivalent resistance of the network is

$$R_{eq} = \frac{V}{\frac{2018}{2017}} = \frac{R}{1009}.$$

17. Cube of Resistors Revisited**

Referring to Fig. 9.63, inject 1A of current into node 1 and withdraw $\frac{1}{7}$A of current from each of the other nodes.

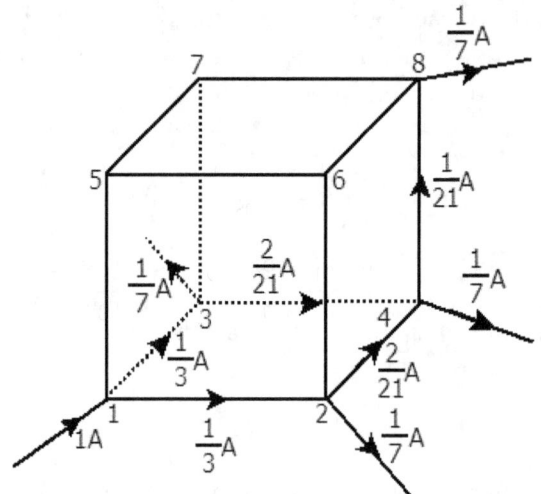

Figure 9.63:   Injection of 1A of current

By symmetry, the current that flows from node 1 to 2 is

$$I_{12} = \frac{1}{3}\text{A}.$$

Then, $\frac{1}{7}$A of current is removed from node 2 and half of the remainder goes to node 4. Thus,

$$I_{24} = \frac{\frac{1}{3} - \frac{1}{7}}{2} = \frac{2}{21}\text{A}.$$

Next, node 4 receives $\frac{2}{21}$A of current from both nodes 2 and 3. $\frac{1}{7}$A of current is then withdrawn from node 4 while the remainder flows to node 8. Thus,

$$I_{48} = \frac{2}{21} \times 2 - \frac{1}{7} = \frac{1}{21}\text{A}.$$

Finally, we can choose another point to withdraw 1A of current from and inject $\frac{1}{7}$A into all other 7 nodes in a new set-up. Whatever node is chosen, the distribution of current will be similar to that above. To calculate the equivalent resistance between nodes 1 and 2, we select node 2.

After superposing these two set-ups, we get a combined set-up that involves injecting $\frac{8}{7}$A into node 1 and removing it from node 2. The current in the resistor between nodes 1 and 2 is

$$I = 2 \times \frac{1}{3} = \frac{2}{3}\text{A}.$$

Thus, the voltage between them is

$$V = \frac{2}{3}R$$

which implies that the equivalent resistance is

$$R_{eq} = \frac{\frac{2}{3}}{\frac{8}{7}} = \frac{7}{12}R.$$

In a similar vein, if we had chosen node 4 as the point where we withdraw current, the voltage difference between nodes 1 and 4, after the superposition of the two separate set-ups, will be

$$V = \left(\frac{1}{3} + \frac{2}{21}\right) \cdot 2 \cdot R = \frac{6}{7}R.$$

The equivalent resistance of the cube with respect to nodes 1 and 4 is then

$$R_{eq} = \frac{V}{\frac{8}{7}} = \frac{3}{4}R.$$

Lastly, if we had chosen node 8 as the node to withdraw current from in the second set-up, the voltage between nodes 1 and 8 will be

$$V = \left(\frac{1}{3} + \frac{2}{21} + \frac{1}{21}\right) \cdot 2 \cdot R = \frac{20}{21}R$$

after the superposition. The equivalent resistance of the cube between nodes 1 and 8 is then

$$R_{eq} = \frac{V}{\frac{8}{7}} = \frac{5}{6}R.$$

18. Fractal Resistance**

If we let the resistance of the whole fractal with respect to A and B be R_{eq}, a fractal with side length $\frac{l}{2}$ will possess an equivalent resistance $\frac{R_{eq}}{2}$ by scaling arguments as

$$R \propto \frac{L}{A}.$$

The resistance of the circuit is proportional to the length dimension of the wires L. Therefore, scaling the length of the sides by a factor of half (while maintaining the cross sectional area) halves the equivalent resistance. Thus, we can transform the original circuit into the equivalent circuit in Fig. 9.64, with the center fractal replaced with $\frac{R_{eq}}{2}$.

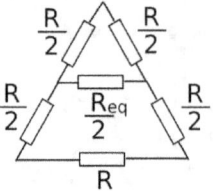

Figure 9.64:   Equivalent circuit

where

$$R = \frac{\rho l}{A}.$$

Now you may find the above transformation to be dubious, as there should be current flowing from points C to E. However, there is seemingly no such current in the above circuit. The trick is that we have divided node E into two equipotential nodes, and one of the nodes was transformed into the $\frac{R_{eq}}{2}$ resistor. The reason why the division of that particular node is possible is because the two nodes formed by disconnecting the top two branches from the bottom two will be at the same potential of the original combined node (exactly the average of the potentials at A and B) due to mirror symmetry. Thus, the current that originally traversed from C to E now flows through the $\frac{R_{eq}}{2}$ resistor. Calculating the equivalent resistance of the triangle of resistors at the top,

$$R' = \frac{1}{\frac{2}{R_{eq}} + \frac{1}{R}} = \frac{RR_{eq}}{R_{eq} + 2R}.$$

Then,

$$R_{eq} = \frac{\left(R + \frac{RR_{eq}}{R_{eq} + 2R}\right)R}{2R + \frac{RR_{eq}}{R_{eq} + 2R}}.$$

Simplifying,

$$3R_{eq}^2 + 2RR_{eq} - 2R^2 = 0$$

$$R_{eq} = \frac{\sqrt{7} - 1}{3}R = \frac{(\sqrt{7} - 1)\rho l}{3A},$$

where the negative solution has been rejected.

19. Double Cube**

We have derived the equivalent resistances of a cube with respect to two adjacent vertices and with respect to two opposite vertices of the same face as $\frac{7}{12}R$ and $\frac{3}{4}R$ respectively. Furthermore, the equivalent resistance of a cube with respect to two terminals, with the resistor in the edge directly connecting the two terminals removed, can be computed as

$$\frac{\frac{7}{12}R \cdot -R}{\frac{7}{12}R - R} = \frac{7}{5}R$$

by connecting a resistor $-R$ in parallel across the two terminals to disconnect the edge directly connecting the terminals. With this information, we can construct the equivalent circuit in Fig. 9.65 between terminals A, B and C.

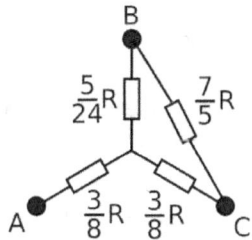

Figure 9.65: Equivalent circuit

The Y-circuit represents the equivalent circuit for the complete upper cube with respect to A, B and C (by Eqs. (9.12)–(9.14)) while the $\frac{7}{5}R$ resistor is the equivalent resistance of the lower cube, with edge BC removed, that is connected in parallel with the direct resistor in edge BC. Therefore, the equivalent resistance between A and B is

$$R_{eq} = \frac{3}{8}R + \frac{\frac{71}{40} \cdot \frac{5}{24}}{\frac{71}{40} + \frac{5}{24}}R = \frac{1069}{1904}R.$$

20. Wheatstone Bridge**

We apply Thevenin's theorem with terminals at nodes A and B. We first determine the potential difference between nodes A and B when R_3 is disconnected. Let the potentials of the positive and negative terminals of the battery be ε and 0. Then, the potentials of nodes A and B are, by the voltage divider principle,

$$V_A = \frac{R_4}{R_1 + R_4} \varepsilon,$$

$$V_B = \frac{R_5}{R_2 + R_5} \varepsilon.$$

The Thevenin emf is the voltage across nodes A and B which is

$$\varepsilon_{th} = \left(\frac{R_4}{R_1 + R_4} - \frac{R_5}{R_2 + R_5} \right) \varepsilon.$$

Now, we determine the Thevenin resistance between the terminals by short-circuiting the battery.

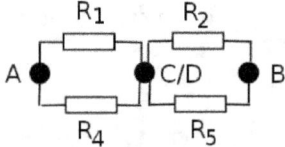

Figure 9.66:   Equivalent resistance between A and B

Observe that the wire formed by short-circuiting the battery causes nodes C and D to become equipotential points. Then, the equivalent network between A and B is shown in Fig. 9.66. The Thevenin resistance is consequently

$$R_{th} = \frac{R_1 R_4}{R_1 + R_4} + \frac{R_2 R_5}{R_2 + R_5}.$$

Finally, we splice this Thevenin-equivalent circuit back with the resistor R_3. The current flowing through R_3 from nodes A to B is then

$$I = \frac{\varepsilon_{th}}{R_{th} + R_3}$$

$$= \frac{R_2 R_4 - R_1 R_5}{R_1 R_2 R_4 + R_1 R_2 R_5 + R_1 R_4 R_5 + R_2 R_4 R_5 + (R_1 + R_4)(R_2 + R_5) R_3} \varepsilon.$$

21. Equivalent Battery**

Apply Thevenin's theorem across the two terminals mentioned in the problem. To determine r_{eq}, short-circuit all voltage sources — the set-up is consequently left with the N resistors connected in parallel, implying that

$$r_{eq} = \frac{1}{\sum_{i=1}^{N} \frac{1}{r_i}}.$$

It is not convenient to directly determine the open-circuit voltage to compute ε_{eq} in this case. Instead, it is expeditious to connect an ideal external wire to the two terminals and find the short-circuit current I_{sc} first. Observing that the ends of each parallel branch are equipotential, the current flowing through the ith parallel branch is thus $\frac{\varepsilon_i}{r_i}$. Following from this, the total short-circuit current is

$$I_{sc} = \sum_{i=1}^{N} \frac{\varepsilon_i}{r_i}.$$

The equivalent emf is then

$$\varepsilon_{eq} = I_{sc} \cdot r_{eq} = \frac{\sum_{i=1}^{N} \frac{\varepsilon_i}{r_i}}{\frac{1}{\sum_{i=1}^{N} \frac{1}{r_i}}}.$$

22. Maximum Power Transfer**

Let us first consider a separate problem. If a battery with a constant emf ε is connected in series with a fixed resistor R_1 and a variable resistor R_2, what value should R_2 undertake to maximize the power dissipated in R_2? The current through the circuit is $\frac{\varepsilon}{R_1+R_2}$ which implies that the power through R_2 is

$$P = \frac{\varepsilon^2 R_2}{(R_1 + R_2)^2}.$$

$$\frac{dP}{dR_2} = \frac{\varepsilon^2}{(R_1 + R_2)^2} - \frac{2\varepsilon^2 R_2}{(R_1 + R_2)^3} = \frac{\varepsilon^2(R_1 - R_2)}{(R_1 + R_2)^3}.$$

Therefore, P is maximum when $R_2 = R_1$ (you can easily check that this is indeed a maximum point by checking adjacent values of $\frac{dP}{dR_2}$). Now, returning to the original problem, we know that the circuit can be reduced to an equivalent emf ε_{eq}, connected in series with an equivalent resistor r_{eq}, across the two terminals by Thevenin's theorem. r_{eq} in this case is

$$r_{eq} = R + \frac{4R \cdot 3R}{4R + 3R} = \frac{19}{7} R.$$

By the previous result, the resistor r that should be connected to the two terminals to maximize the power dissipated by itself should be

$$r = r_{eq} = \frac{19}{7} R.$$

In summary, the key takeaway of this problem is that whenever we want to find the resistor r to be connected to two terminals to maximize the power through it, we simply have to determine the Thevenin-equivalent resistance of the rest of the circuit with respect to those two terminals.

23. Another Infinite Ladder**

Instead of conventionally defining an equivalent resistance, let the Thevenin-equivalent of the original circuit with terminals at the nodes of the branch carrying the current I have a Thevenin-equivalent resistance R_{eq} and emf ε_{eq}. Then, we perform the same sleight of hand as before and replace the right portion of the circuit that we are applying Thevenin's theorem on with the Thevenin-equivalent circuit to obtain Fig. 9.67.

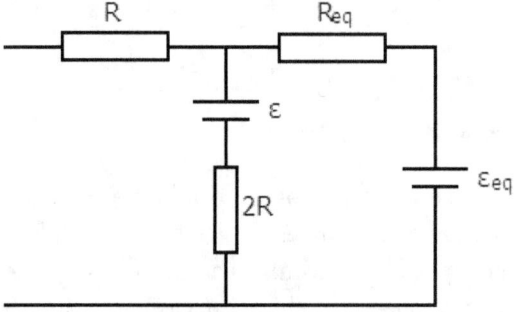

Figure 9.67:   Modified circuit whose Thevenin's equivalent should be identical

Now we apply Thevenin's theorem to the circuit above with respect to the two terminals on the left. We should still obtain ε_{eq} and R_{eq} as the Thevenin-equivalent emf and resistance respectively. The equivalent resistance is

$$R_{eq} = R + \frac{2RR_{eq}}{2R + R_{eq}}$$

$$(R_{eq} - 2R)(R_{eq} + R) = 0$$

$$R_{eq} = 2R.$$

Once again, we have rejected the infeasible negative solution. The clockwise current through the loop is

$$I = \frac{\varepsilon - \varepsilon_{eq}}{2R + R_{eq}} = \frac{\varepsilon - \varepsilon_{eq}}{4R}.$$

Thus, the open-circuit voltage between the two terminals is

$$\varepsilon_{eq} = \varepsilon - I \cdot 2R$$

$$\varepsilon_{eq} = \varepsilon.$$

We splice this equivalent circuit with the branch carrying I. Then applying Ohm's law,

$$I = \frac{\mathcal{E}_{eq}}{R_{eq}} = \frac{\mathcal{E}}{2R}.$$

24. Thick Infinite Ladder***

Number the nodes according to Fig. 9.68.

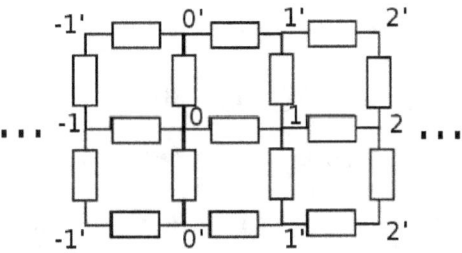

Figure 9.68: Labelled nodes

Suppose that we wish to determine the equivalent resistance of the network between nodes 0 and 1. Observe that primed nodes with the same number are equipotential pairs by path symmetry. Then, the above circuit can be reduced to the infinite ladder in Fig. 9.69.

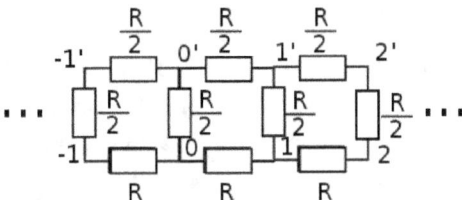

Figure 9.69: Equivalent ladder

Now, observe that the above circuit comprises two infinite ladders — on the left of 00′ and on the right of 11′. We need to determine the equivalent resistance R_{eq} of these ladders. Observe that part of this ladder can be broken off to form an identical infinite ladder with resistance R_{eq}. Therefore, the infinite ladder in Fig. 9.70 is equivalent to the circuit on the right — yielding a quadratic equation in R_{eq}.

The equivalent resistance of the set-up on the right with respect to nodes 1 and 1′ should still be R_{eq}.

$$R_{eq} = \frac{1}{\frac{2}{R} + \frac{1}{R_{eq} + \frac{3R}{2}}}.$$

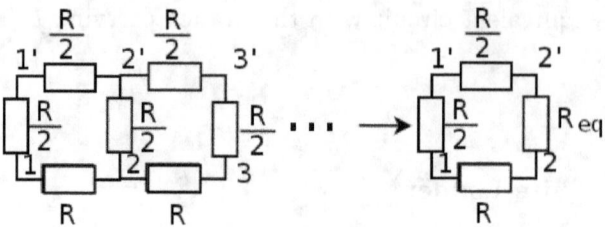

Figure 9.70:   Infinite ladder on the right of $11'$

Simplifying,

$$4R_{eq}^2 + 6RR_{eq} - 3R^2 = 0$$

$$R_{eq} = \frac{-3 + \sqrt{21}}{4}R$$

where the unphysical negative solution has been rejected. Now, the equivalent infinite ladder comprised two of such ladders with resistance R_{eq} in series with a resistor of resistance $\frac{R}{2}$, connected in parallel to a resistor of resistance R. The equivalent resistance with respect to nodes 0 and 1 is then

$$R_{tot} = \frac{\left(2R_{eq} + \frac{R}{2}\right) \cdot R}{2R_{eq} + \frac{R}{2} + R} = \frac{21 - 2\sqrt{21}}{21}R.$$

25. N-gon with Spokes***

Label the outer vertices from 1 to N and denote the center as O. Suppose that we wish to determine the equivalent resistance between nodes 1 and O. Then, observe that nodes i and $N + 2 - i$ form an equipotential pair, whose combined node shall be denoted as $i/(N + 2 - i)$, for all $1 \le i \le \lfloor \frac{N-1}{2} \rfloor$. Now, we have to consider two different cases — namely, when N is odd and when N is even. If N is even, there will be a lone node $\frac{N}{2}$. Thus, the circuit becomes Fig. 9.71 after combining the equipotential pairs together.

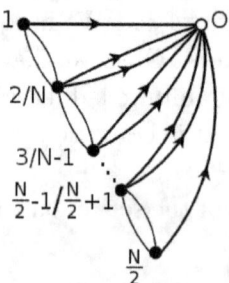

Figure 9.71: Equivalent circuit (thin lines represent $1R$ while thick lines with arrows represent $2R$)

Its equivalent resistance is

$$R_{eq} = R \cdot \cfrac{1}{\frac{1}{2} + \cfrac{1}{\frac{1}{2} + \cfrac{1}{1 + \cfrac{1}{\ddots + \cfrac{1}{\frac{1}{2} + \cfrac{1}{1 + \cfrac{1}{\frac{1}{2} + 2}}}}}}},$$

where the $\frac{1}{2}$'s appear $\frac{N}{2}$ times in the denominator (excluding the top-most layer due to the connection between nodes 1 and O which destroys the pattern). To simplify the continued fraction, let us work out a few cases from the bottom.

$$\cfrac{1}{\frac{1}{2} + 2} = \frac{2}{5},$$

$$\cfrac{1}{1 + \cfrac{1}{\frac{1}{2} + 2}} = \frac{5}{7},$$

$$\cfrac{1}{\frac{1}{2} + \cfrac{1}{1 + \cfrac{1}{\frac{1}{2} + 2}}} = \frac{14}{17},$$

$$\cfrac{1}{1 + \cfrac{1}{\frac{1}{2} + \cfrac{1}{1 + \cfrac{1}{\frac{1}{2} + 2}}}} = \frac{17}{31},$$

$$\cfrac{1}{\frac{1}{2} + \cfrac{1}{1 + \cfrac{1}{\frac{1}{2} + \cfrac{1}{1 + \cfrac{1}{\frac{1}{2} + 2}}}}} = \frac{62}{65}.$$

Now, we can begin to observe a pattern. It seems like for the fractions which terminate with an addition of half at the top of the denominator, the numerator is seemingly always smaller than the denominator by 3. Therefore, define $\frac{a_n}{b_n}$ to be the simplified expression for the continued fraction with n $\frac{1}{2}$'s that ends with an addition of $\frac{1}{2}$ in the highest layer (we do not define this for the $\frac{1}{2}$ in the top-most layer of the continued fraction which spoils the pattern). Subsequently, notice that

$$\frac{a_{n+1}}{b_{n+1}} = \cfrac{1}{\frac{1}{2} + \cfrac{1}{1 + \frac{a_n}{b_n}}} = \frac{2(a_n + b_n)}{a_n + 3b_n}.$$

If $a_n = b_n - k$ for some constant k,

$$\frac{a_{n+1}}{b_{n+1}} = \frac{4b_n - 2k}{4b_n - k}$$

$$\implies a_{n+1} = b_{n+1} - k,$$

which shows that the numerator is always a constant smaller than the denominator throughout the sequence! Furthermore, this result is independent of the bottom-most number in the continued fraction (which is 2 in this case) as we have not made any assumptions about it. From the above, we also obtain

$$b_{n+1} = 4b_n - k,$$

from which we can solve for b_n in terms of n explicitly. In the current case, $k = 3$ and the base case is $b_1 = 5$ (from $\frac{1}{\frac{1}{2}+2} = \frac{2}{5}$). The above recurrence relation can be rewritten as

$$b_{n+1} = 4b_n - 3$$

$$b_{n+1} - 1 = 4(b_n - 1).$$

Thus, if we define $c_n = b_n - 1$, the sequence c_n is a geometric progression with base case $c_1 = 4$. Therefore,

$$c_n = 4^{n-1}c_1 = 4^n,$$

$$b_n = 4^n + 1.$$

In this case, we are interested in

$$\frac{a_{\frac{N}{2}}}{b_{\frac{N}{2}}} = \frac{b_{\frac{N}{2}} - 3}{b_{\frac{N}{2}}} = \frac{4^{\frac{N}{2}} - 2}{4^{\frac{N}{2}} + 1},$$

as

$$R_{eq} = R \cdot \frac{1}{\frac{1}{2} + \frac{a_{\frac{N}{2}}}{b_{\frac{N}{2}}}} = \frac{2 \cdot 4^{\frac{N}{2}} + 2}{3 \cdot 4^{\frac{N}{2}} - 3} R.$$

When N is odd, all other vertices besides vertex 1 are matched in pairs. The circuit becomes Fig. 9.72 after combining the equipotential pairs.

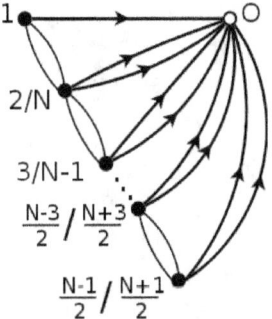

Figure 9.72: Equivalent circuit (thin lines represent $1R$ while thick lines with arrows represent $2R$)

$$R_{eq} = R \cdot \cfrac{1}{\frac{1}{2} + \cfrac{1}{\frac{1}{2} + \cfrac{1}{1 + \cfrac{1}{\ddots + \cfrac{1}{\frac{1}{2} + \cfrac{1}{1 + \cfrac{1}{\frac{1}{2}+1}}}}}}}$$

where there are $\frac{N-1}{2}$ $\frac{1}{2}$'s in the pattern (excluding the top-most one). This continued fraction can be simplified through the exact same procedure. Adopting the same definition of $\frac{a_n}{b_n}$, we have

$$\frac{a_1}{b_1} = \frac{1}{\frac{1}{2} + 1} = \frac{2}{3}$$

$$\implies a_n = b_n - 1$$

by the previous argument $(k = 1)$. The recurrence relation is thus

$$b_{n+1} = 4b_n - 1$$

$$b_{n+1} - \frac{1}{3} = 4\left(b_n - \frac{1}{3}\right)$$

with base case $b_1 = 3$. Therefore,

$$b_n - \frac{1}{3} = 4^{n-1}\left(3 - \frac{1}{3}\right)$$

$$b_n = 4^n \cdot \frac{2}{3} + \frac{1}{3}$$

$$\implies \frac{a_{\frac{N-1}{2}}}{b_{\frac{N-1}{2}}} = \frac{b_{\frac{N-1}{2}} - 1}{b_{\frac{N-1}{2}}} = \frac{2 \cdot 4^{\frac{N-1}{2}} - 2}{2 \cdot 4^{\frac{N-1}{2}} + 1},$$

$$R_{eq} = R \cdot \frac{1}{\frac{1}{2} + \frac{a_{\frac{N-1}{2}}}{b_{\frac{N-1}{2}}}} = \frac{4^{\frac{N+1}{2}} + 2}{3 \left(2 \cdot 4^{\frac{N-1}{2}} - 1 \right)} R.$$

We can combine the results of the odd and even cases to write for general N,

$$R_{eq} = \frac{2^{N+1} + 2}{3 \cdot 2^N - 3} R.$$

Chapter 10

RLC and AC Circuits

The role of capacitors and inductors will be discussed in this chapter. In addition to DC circuits, Alternating Current circuits (AC circuits), in which currents perpetually vary in direction, will also be analyzed. In this process, we will observe an enlightening analogy between AC circuits and DC circuits!

10.1 Roles of Capacitors and Inductors

Capacitors

A capacitor usually consists of two conductors and possesses an ability to store charge, given a potential difference between the conductors. Conversely, it also generates a potential difference across its ends via stored charges. Note that in the process of charging, charges are transferred from one plate to another through an external wire connecting the two plates — insignificant charge flows directly across the plates. The fundamental relation between the stored charge Q and the potential difference ΔV is governed by an intrinsic and geometric property known as the capacitance of the capacitor.

$$C = \left| \frac{Q}{\Delta V} \right|. \tag{10.1}$$

The potential energy stored in a capacitor is

$$U = \frac{1}{2} C \Delta V^2 = \frac{Q^2}{2C} = \frac{1}{2} Q \Delta V. \tag{10.2}$$

The function of a capacitor in a circuit is to oppose an incoming or outgoing current. Referring to Fig. 10.1, suppose a current I_1 is incident on the left plate of a capacitor at the current instance. At the next instance, some additional positive charge would have been deposited on the left plate

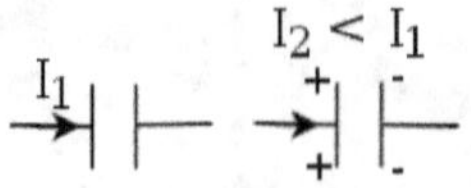

$$I_2 < I_1$$

Figure 10.1:   Current at two instances

while some positive charge would have departed from the right plate —
leaving net additional negative charges behind. The capacitor then generates
an increased potential difference, as compared to before, which opposes the
incoming current (by Kirchhoff's loop rule). Therefore, the current at the
next instance, I_2, is smaller than I_1.

Inductor

In the context of circuits, an inductor usually[1] refers to a component with a
certain self-inductance L. If the current flowing through the inductor is $I(t)$,
the magnetic field produced by the inductor varies with time and hence,
generates a non-conservative electric field. This non-conservative electric field
then generates an emf in the inductor that opposes the change in magnetic
flux linkage in accordance with Faraday's law. Recall that the induced emf
in an inductor with self-inductance L is given by

$$\varepsilon = -L\frac{dI}{dt}, \tag{10.3}$$

where the negative sign indicates that the induced emf is opposite in direction to the change in current through the inductor. In an ideal emf source, the
potential difference (due to the conservative electric field within the source)
is equal to the emf generated. The induced emf thus produces a potential difference which hinders the change in current — this is the main responsibility
of an inductor in a circuit.

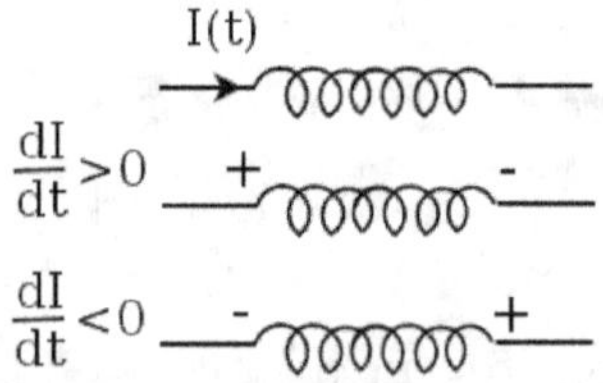

Figure 10.2:   Polarity of an inductor

---

[1]When there are multiple inductors, the mutual inductance between inductors is sometimes taken into account. This is explored in a later section.

Suppose a current $I(t)$ flows rightwards through the inductor depicted in Fig. 10.2. If $I(t)$ is increasing at the current instance ($\frac{dI}{dt} > 0$), the left end of the inductor will be at a higher potential than the right — in an attempt to reduce the increase in current. Similarly, if $\frac{dI}{dt} < 0$, the right end of the inductor will be at a higher potential.

Finally, the potential energy stored inside an inductor when current I flows through it is

$$U = \frac{1}{2}LI^2. \tag{10.4}$$

10.1.1 *Series and Parallel Configurations*

Series Connections

The equivalent capacitance and self-inductance for n capacitors and inductors connected in series are

$$\frac{1}{C_{eq}} = \sum_{i=1}^{n} \frac{1}{C_i}, \tag{10.5}$$

$$L_{eq} = \sum_{i=1}^{n} L_i. \tag{10.6}$$

Parallel Connections

For parallel connections,

$$C_{eq} = \sum_{i=1}^{n} C_i, \tag{10.7}$$

$$\frac{1}{L_{eq}} = \sum_{i=1}^{n} \frac{1}{L_i}. \tag{10.8}$$

These can be proven by utilizing the facts that the current through and the voltage across circuit components connected in series and parallel are the same respectively. In the case of capacitors in series, one can use the fact that the total charges in the segments connecting adjacent capacitor plates are conserved as they are electrically isolated.

Analogy with Resistors

Because of the analogous formulae for equivalent capacitance and self-inductance, we can devise a way to transform a problem involving the determination of equivalent capacitances and self-inductances into problems of finding equivalent resistances. This analogy is evident in the case of self-inductance as we can simply substitute $R \to L$. That is, we change every self-inductor L into a resistor R that corresponds to L and compute the equivalent resistance R_{eq} as a function of the various resistances. The equivalent self-inductance can then be retrieved by substituting the corresponding self-inductances back into the resistances and $L_{eq} \to R_{eq}$. Similarly, in the case of capacitance, we can make the substitution $R \to \frac{1}{C}$. That is, we change every capacitor C into a resistor R that corresponds to $\frac{1}{C}$ as $\frac{1}{C}$ obeys the rules of adding resistors in series and parallel.

In fact, such a correspondence becomes lucid when we reach the section on impedance in AC circuits. Basically, at steady state, a capacitor C and inductor L respond like resistors with "complex resistance" which are proportional to $\frac{1}{C}$ and L respectively. These "complex resistances" follow the normal rules of adding resistors in series and parallel. Thus, we can naturally make the substitutions above.

Problem: Determine the equivalent self-inductance and capacitance of an infinite square grid of self-inductors L and an infinite square grid of capacitors C between two adjacent points on the grid.

Transforming this into an equivalent resistance problem, we know that the equivalent resistance of an infinite square grid of resistors R between two adjacent points is $\frac{R}{2}$.

$$R_{eq} = \frac{R}{2}.$$

Drawing the analogy $R_{eq} \to L_{eq}$ and $R \to L$, the equivalent self-inductance of the first network is

$$L_{eq} = \frac{L}{2}.$$

Adopting the substitution $R_{eq} \to \frac{1}{C_{eq}}$ and $R \to \frac{1}{C}$, the equivalent capacitance of the second network obeys

$$\frac{1}{C_{eq}} = \frac{1}{2C}$$

$$\implies C_{eq} = 2C.$$

10.1.2   *Sign Conventions*

Kirchhoff's loop and junction rules are applicable to circuits with capacitors and inductors as well. Though inductors inherently produce non-conservative electric fields within themselves, Kirchhoff's loop rule only speaks about the potential difference due to the conservative electric field and is thus still valid. Recall that in applying Kirchhoff's loop rule, we first choose a Kirchhoff loop with a certain direction and propose currents in each branch of the circuit. The sign convention for the potential difference across a capacitor is identical to that for batteries. When the Kirchhoff loop runs from a plate with charge $-Q$ to the other plate with charge Q, there is an increase in potential of $\frac{Q}{C}$ (note that Q could be negative). In evaluating the potential difference across an inductor, we have to also take note of the proposed direction of current in addition to the direction of the Kirchhoff loop. If the Kirchhoff loop runs in the same direction as the current I across the inductor, the potential difference across the inductor in the direction of the Kirchhoff loop is $-L\frac{dI}{dt}$. Otherwise, if the Kirchhoff loop opposes the proposed current I, the potential difference is $L\frac{dI}{dt}$.

In practice, it is easier to assign positive and negative signs to the ends of an inductor in a fashion similar to the terminals of a battery. The end, at which the proposed current first crosses the inductor, is denoted as the positive end. Then, the potential difference across the inductor is akin to that of a battery with an emf $L\frac{dI}{dt}$ — the negative and positive terminals take care of the sign of this emf.

10.1.3 *Short-term and Long-term Effects*

The qualitative effects of an inductor and capacitor in the short and long run in a DC circuit can be analyzed in light of their roles in a circuit. Immediately after a swift change, an inductor will respond by ensuring that the current through itself is the same as before, by producing a potential difference to resist the change. Thus, inductors become ideal current sources in the short run. A capacitor, on the other hand, produces the same voltage as before, as charges have not been transferred in the form of currents during the short time interval, most of the time. In the rare case where there is a direct path comprising only batteries and capacitors that is newly established, there must be a discontinuity in the stored charges of the capacitor in order for Kirchhoff's loop rule to be satisfied — the capacitors are then no longer ideal voltage sources. Physically, infinite current flows through the path during a short time interval — leading to a non-negligible deposition of charges on the capacitors.

In a DC circuit, ideal inductors essentially become ideal wires after a long time. This is because the system will eventually reach a steady state such that the current through the circuit remains constant. Then, $\frac{dI}{dt} = 0$ which causes the potential difference across the inductor to be zero.

Capacitors, on the other hand, can eventually be reduced to open-circuits. When the system has equilibrated, the charge on the capacitor remains constant. Thus, the current flowing from its plates must be zero — signifying that capacitors can just be disconnected in the long run. In a certain sense, charging a capacitor is analogous to pumping a ball via an air pump. It becomes progressively harder to pump air into the ball as the pressure in the ball increases (analogous to potential) due to the increase in air (analogous to charge). Eventually, no additional air can be pumped (and analogously no current) when the pressure in the ball is equal to the pressure of the pump (analogous to the external emf).

Often, we will be tasked to determine the charge stored in a capacitor in the long run. To do so, we can first disconnect the capacitors and solve for the resultant currents in the circuit. Then, Kirchhoff's loops, that cut through the capacitors, can be drawn to generate simultaneous equations regarding the potential differences across the capacitors and correspondingly, the charges stored in the capacitors. The last point to take note of would be the conservation of charge in adjacent, connected capacitor plates as there cannot be any charge flow directly across the plates of an individual capacitor. The following example will illustrate this process.

Problem: The capacitors in Fig. 10.3 were initially neutral. Then, the circuit is allowed to reach steady state. After a long time, what is the charge stored in the 10mF capacitor? (Chinese Physics Olympiad)

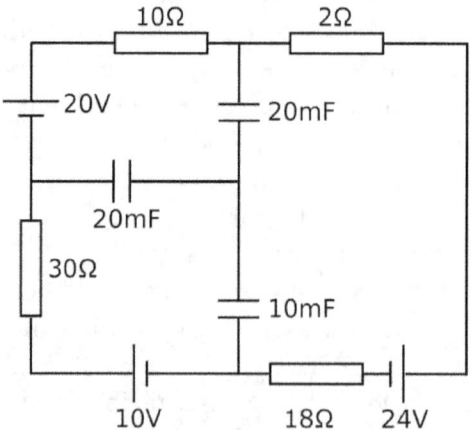

Figure 10.3:   Circuit with capacitors

After the system has reached steady state, a current flows in only the outer loop as the capacitors are essentially disconnected.

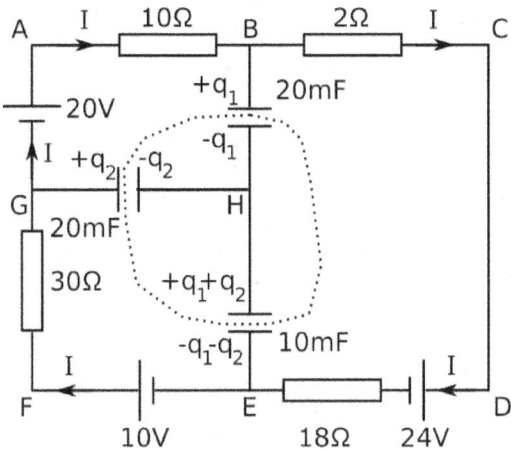

Figure 10.4: Circuit with capacitors with labeled charges

Referring to Fig. 10.4, we can solve for the current I that flows through the outer loop by applying Kirchhoff's loop rule to cycle GABCDEFG,

$$20 - 10I - 2I - 24 - 18I + 10 - 30I = 0$$

$$I = \frac{1}{10}A.$$

Next, notice that the region enclosed by the dotted surface is electrically isolated from the rest of the circuit. Therefore, the quantity of charge encased is conserved, resulting in the charge distribution on the capacitors as labeled in Fig. 10.4 (i.e. neutral collectively since it was initially so). Lastly, we can draw two more Kirchhoff loops that cross the capacitors, to solve for q_1 and q_2. Using loop GABHG,

$$20 - 10I - \frac{q_1}{20 \times 10^{-3}} + \frac{q_2}{20 \times 10^{-3}} = 0.$$

For loop EFGHE,

$$10 - 30I - \frac{q_2}{20 \times 10^{-3}} - \frac{q_1}{10 \times 10^{-3}} - \frac{q_2}{10 \times 10^{-3}} = 0.$$

Solving, we obtain

$$q_1 = \frac{32}{125}C,$$

$$q_2 = -\frac{31}{250}C.$$

The quantity of charge stored in the $10mF$ capacitor is

$$|q_1 + q_2| = \frac{1}{250}C.$$

Problem: The switch in Fig. 10.5 is closed for a long time and a steady state has been reached. (Estonian-Finnish Olympiad)

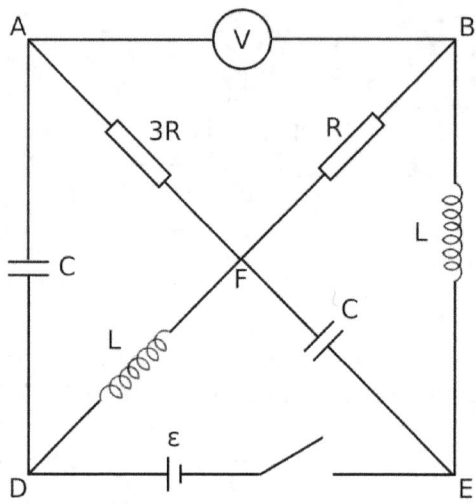

Figure 10.5:   Circuit

(1) Find the reading of the voltmeter.
(2) The switch is opened; find the reading of the voltmeter immediately after opening the switch.
(3) Find the total amount of heat dissipated in each resistor after opening the switch and after a new equilibrium state has been reached.

(1) When steady state is reached originally, the inductors are identical to wires and the capacitors are essentially disconnected. There is only a current that flows through loop EDFBE. Thus, the voltmeter measures the voltage across the resistor in branch FB which is simply ε. Note that the $3R$ resistor does not reduce the voltage across the voltmeter by the voltage divider principle as the ideal voltmeter has infinite resistance.

(2) Immediately after the switch is opened, the inductors maintain the currents through themselves which are of magnitude

$$I = \frac{\varepsilon}{R}.$$

The circuit can now be divided into two sub-circuits. A current I flows in the anti-clockwise direction in loop DFAD while a current of the same magnitude flows clockwise in loop FBEF. In the right triangular loop FBEF, the voltage between F and B is

$$V_B - V_F = -I \cdot R = -\varepsilon.$$

Considering the triangular loop DFAD, the voltage difference between F and A is

$$V_A - V_F = -I \cdot 3R = -3\varepsilon.$$

Thus, the voltmeter measures

$$\Delta V = |V_B - V_A| = 2\varepsilon.$$

(3) Remember that the system essentially consists of two separate circuits, DFAD and FBEF. Therefore, the total heat dissipated in the $3R$ resistor is simply the difference in energies stored in the left capacitor and inductor while that dissipated in the R resistor is simply the difference in energies stored in the right capacitor and inductor.

We need to compute the energies stored by the inductors and capacitors before or immediately after the switch is opened (initial stored energies). Note that the potential differences across the capacitors and current through the inductors are the same immediately before and after opening the switch. Thus, the amounts of energy stored at these two instances are identical.

The potential across the capacitor on the left is zero before the switch is opened. This is evident if we draw a Kirchhoff loop AFDA; the current through AF is zero while the voltage across the inductor is also zero in the long run when the switch is closed — causing the voltage across the capacitor to be zero too, by Kirchhoff's loop rule. Meanwhile, the inductor in loop DFAD originally carried a current

$$I = \frac{\varepsilon}{R}.$$

Therefore, the total initial potential energy stored in the left capacitor and inductor (using $U = \frac{1}{2}CV^2$ and $U = \frac{1}{2}LI^2$) is

$$U_l = \frac{L\varepsilon^2}{2R^2}.$$

Now, let us compute the final total potential energy of these components. The final current through loop DFAD is zero in order for the capacitor to reach a steady state. Since the potential differences across the resistor and inductor are zero, there will also be no potential difference across and thus charge stored in the capacitor. The final total potential energy is then zero.

The heat dissipated in the $3R$ resistor is the change in potential energy which is

$$Q_l = U_l - 0 = \frac{L\varepsilon^2}{2R^2}.$$

Moving on, we shall determine the initial potential energy in the loop FBEF. Before the switch is opened, a steady current $\frac{\varepsilon}{R}$ flows in loop DFBED. Applying Kirchhoff's loop rule to DFED, the potential difference across the right capacitor must be ε (there is no potential difference across the left inductor as the current is steady). This, combined with the fact that the right inductor carries current $\frac{\varepsilon}{R}$, implies that the initial total potential energy is

$$U_r = \frac{1}{2}C\varepsilon^2 + \frac{1}{2}\frac{L\varepsilon^2}{R^2}.$$

The final potential energy of the system is also zero as there must eventually be zero current through the capacitor — implying that no current exists in loop FBEF eventually. Next, the charge in the capacitor in branch EF must also eventually be dispersed so that no current flows in the loop FBEF by Kirchhoff's loop rule (as the potential differences across the right resistor and inductor are both zero). Since the final total potential energy is zero, the total energy dissipated in resistor R is

$$Q_r = \frac{1}{2}C\varepsilon^2 + \frac{1}{2}\frac{L\varepsilon^2}{R^2}.$$

Finally, let us consider a problem where a capacitor is not an ideal constant voltage source in the short run.

Problem: Initially, switch S in Fig. 10.6 is closed at terminal 1. If switch S is now turned towards terminal 2, determine the total energy lost by the components in the circuit. The emfs of the ideal batteries are ε_1 and ε_2, respectively, while the capacitance of the capacitor is C.

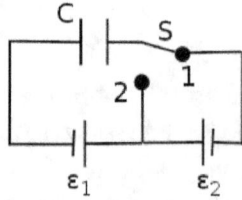

Figure 10.6:   Circuit with capacitor and batteries

Initially, the total charges stored on the right and left plates of the capacitor are, respectively

$$\pm Q = \pm(\varepsilon_1 - \varepsilon_2)C,$$

and the total initial potential energy stored by the capacitor is

$$U = \frac{1}{2}C(\varepsilon_1 - \varepsilon_2)^2.$$

Now, you might think that the total energy lost by the circuit when switch S is changed from terminals 1 to 2 is simply the change in the capacitor's potential energy. However, we have to remember to account for the work done by the battery ε_1 as well. When switch S is turned, the voltage across the capacitor is initially less than the emf ε_1. Therefore, the emf source ε_1 drives infinite current in the resultant loop and the charges crash into the capacitor plates — resulting in a loss in kinetic energy. Therefore, to compute the energy loss by the entire system, we have to return to the fundamental work-energy theorem. Noting that the above process occurs until the charges on the right and left plates of the capacitor become

$$\pm Q' = \pm \varepsilon_1 C,$$

the battery ε_1 delivers a total amount of charge $\varepsilon_2 C$ from the left plate to the right plate of the capacitor, across a potential difference ε_1 — implying that it does work:

$$W_{bat1} = qV = (\varepsilon_2 C) \cdot \varepsilon_1 = \varepsilon_1 \varepsilon_2 C.$$

If there were no energy loss, the final potential energy of the capacitor would have been $U + W_{bat1}$. However, the actual final potential energy of the capacitor is

$$U' = \frac{1}{2}C\varepsilon_1^2.$$

This indicates that the energy loss is

$$U + W_{bat1} - U' = \frac{1}{2}C\varepsilon_2^2.$$

10.1.4 *Effects at All Times*

In this section, we shall explicitly determine the characteristics of certain circuits involving resistors, inductors and capacitors at all times. This form of analysis, which entails drawing Kirchhoff loops and solving the resultant differential equations, can be applied to all circuits in general.

RC Circuits

Consider an emf source, resistor and capacitor connected in series in Fig. 10.7. The switch was open for time $t < 0$ and is closed at time $t = 0$. Defining a clockwise current and drawing a clockwise Kirchhoff loop, Kirchhoff's loop rule requires

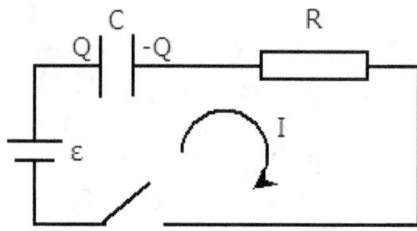

Figure 10.7:   RC series circuit

$$\varepsilon - IR - \frac{Q}{C} = 0,$$

where we have defined Q to be the charge on the left plate of the capacitor. Furthermore, for the capacitor,

$$I = \frac{dQ}{dt},$$

as the proposed current is flowing into the left plate. If this was otherwise (proposed current is flowing out of the left plate), $I = -\frac{dQ}{dt}$. Always take note of the sign!

$$\varepsilon - \frac{Q}{C} = R\frac{dQ}{dt}$$

$$\frac{1}{RC}dt = \frac{1}{\varepsilon C - Q}dQ$$

$$\frac{1}{RC}t = [-\ln|\varepsilon C - Q|]_{Q_0}^{Q} = -\ln\left|\frac{\varepsilon C - Q}{\varepsilon C - Q_0}\right|.$$

The solution to this is independent of whether $\varepsilon C > Q_0$ or $\varepsilon C \leq Q_0$, where Q_0 is the initial charge on the left plate of the capacitor. This is because, if $\varepsilon C > Q_0$, $\varepsilon C > Q$ at all instances afterwards (deduced from $\frac{dQ}{dt} = \frac{\varepsilon}{R} - \frac{Q}{RC}$ such that Q only decreases until εC). On the other hand, if $\varepsilon C \leq Q_0$, $\varepsilon C \leq Q$ at all following instances. Therefore, $\frac{\varepsilon C - Q}{\varepsilon C - Q_0}$ is definitely positive.

$$\varepsilon C - Q = (\varepsilon C - Q_0)e^{-\frac{1}{RC}t},$$

$$Q = \varepsilon C(1 - e^{-\frac{1}{RC}t}) + Q_0 e^{-\frac{1}{RC}t},$$

$$I = \frac{dQ}{dt} = \frac{\varepsilon}{R}e^{-\frac{1}{RC}t} - \frac{Q_0}{RC}e^{-\frac{1}{RC}t}.$$

If $Q_0 > \varepsilon C$, the capacitor discharges until its final charge reaches εC. Otherwise if $Q_0 < \varepsilon C$, the capacitor charges until it attains εC amount of charge. When the charge on the capacitor is eventually εC (though this takes an infinite amount of time), there will no longer be any current in the circuit. Lastly, if $Q_0 = \varepsilon C$, no current flows in the circuit at all instances. The quantity RC is known as the time constant of the RC circuit and appears in the denominator of the decay exponent.

RL Circuits

Consider an emf source, resistor and inductor connected in series in Fig. 10.8. The switch is open for time $t < 0$ and closed at time $t = 0$. Applied in a clockwise fashion, Kirchhoff's loop rule requires

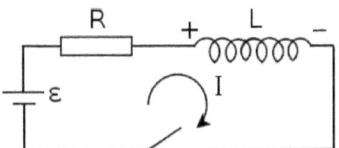

Figure 10.8: RL series circuit

$$\varepsilon - IR - L\frac{dI}{dt} = 0.$$

Note that we have proposed the current to run clockwise. Thus, the left hand side of the inductor is also proposed to be the "positive terminal" with a voltage $L\frac{dI}{dt}$ across its two ends (refer to the section on sign conventions).

$$\int_0^I \frac{1}{\frac{\varepsilon}{R} - I}dI = \int_0^t \frac{R}{L}dt$$

$$\left[-\ln\left|\frac{\varepsilon}{R} - I\right|\right]_0^I = \frac{R}{L}t.$$

Note that the current I will be smaller than $\frac{\varepsilon}{R}$ at all instances as the inductor obstructs the current from reaching its maximum value of $\frac{\varepsilon}{R}$ (this is most obvious if you write $\frac{dI}{dt} = \frac{\varepsilon}{L} - \frac{R}{L}I$ which shows that I stops increasing once

$I = \frac{\varepsilon}{R}$). Thus $\left|\frac{\varepsilon}{R} - I\right| = \frac{\varepsilon}{R} - I$.

$$\frac{\varepsilon}{R} - I = \frac{\varepsilon}{R}e^{-\frac{R}{L}t}$$

$$I = \frac{\varepsilon}{R}(1 - e^{-\frac{R}{L}t}).$$

RLC Circuit

Lastly, consider an emf source, resistor, inductor and capacitor connected in series in Fig. 10.9. For time $t < 0$, the switch is open. At time $t = 0$, the switch is closed. Defining the charge on the left capacitor plate as Q and applying Kirchhoff's loop rule in the clockwise direction (the current I is also clockwise),

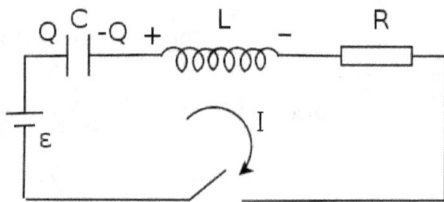

Figure 10.9:   RLC series circuit

$$IR + L\frac{dI}{dt} + \frac{Q}{C} = \varepsilon.$$

Furthermore, for the capacitor,

$$I = \frac{dQ}{dt},$$

$$L\frac{d^2Q}{dt^2} + R\frac{dQ}{dt} + \frac{Q}{C} = \varepsilon.$$

Observe that this second order linear differential equation is similar to that of a damped oscillation.

$$m\ddot{x} + b\dot{x} + kx = 0.$$

The inverse of the capacitance is analogous to the elastic constant k of the restoring force in a damped oscillation, the resistance R is analogous to the coefficient of the damping force b and the inductance is analogous to the mass m which provides the inertia that resists change.

Thus, the solution to Q in this differential equation is analogous to that of the displacement x in the case of damped oscillations, except with an additional εC constant to account for the constant ε on the right-hand side of the equation (particular solution). Again, we have to consider three cases — namely, underdamping, critical damping and overdamping. Referring to the chapter on oscillations,

When $R < 2\sqrt{\frac{L}{C}}$, it is a case of underdamping. The solution for Q is

$$Q = e^{-\frac{R}{2L}t} c_0 \cos\left(\sqrt{\frac{1}{L^2C^2} - \frac{R^2}{4L^2}}t + \phi\right) + \varepsilon C.$$

When $R = 2\sqrt{\frac{L}{C}}$, it is a case of critical damping. The solution for Q is

$$Q = e^{-\frac{R}{2L}t}(c_1 + c_2 t) + \varepsilon C.$$

When $R > 2\sqrt{\frac{L}{C}}$, it is a case of overdamping. The solution for Q is

$$Q = c_1 e^{-\frac{R}{2L} + \sqrt{\frac{R^2}{4L^2} - \frac{1}{L^2C^2}}} + c_2 e^{-\frac{R}{2L} - \sqrt{\frac{R^2}{4L^2} - \frac{1}{L^2C^2}}} + \varepsilon C.$$

The constants c_0, c_1, c_2 and ϕ are determined by initial conditions such as the initial charge on the capacitor and the initial current in the circuit. Finally, the expression for the current $I = \frac{dQ}{dt}$ can be obtained by differentiating the appropriate expression for Q above with respect to time.

Capacitor in Parallel

When a capacitor is connected in parallel to another component, it is usually expeditious to adopt nodal analysis rather than mesh analysis and express the capacitor relationship $Q = C\Delta V$ as $I = \pm C\frac{dV}{dt}$, where V is the potential difference across the capacitor and the choice of sign depends on the direction of the current I. This is because we often cannot relate the charge Q stored in the capacitor to other variables in the circuit directly — but working with the potential difference V enables us to do so. Furthermore, the equation $I = \pm C\frac{dV}{dt}$ nicely parallels the inductor equation $V = \pm L\frac{dI}{dt}$!

Problem: Find the current $I(t)$ through the inductor and the charge stored by the capacitor $Q(t)$ if $I(0) = -1$A and $Q(0) = 0$ in Fig. 10.10.

Let the potential difference across the capacitor be $V(t)$, positive if the left plate has the higher potential. Imposing Kirchhoff's junction rule to the node on the left of the resistor,

$$I_g - \frac{V}{R} - I_C - I = 0$$

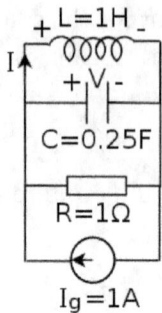

Figure 10.10:   RLC parallel circuit

where $I_C = C\frac{dV}{dt} = \frac{1}{4}\frac{dV}{dt}$ is the rightwards current through the capacitor. Differentiating the above with respect to t,

$$\frac{1}{4}\frac{d^2V}{dt^2} + \frac{dV}{dt} + \frac{dI}{dt} = 0.$$

Since $L\frac{dI}{dt} = V$, $\frac{dI}{dt} = V$ when $L = 1\text{H}$. Thus,

$$\frac{1}{4}\frac{d^2V}{dt^2} + \frac{dV}{dt} + V = 0.$$

The characteristic equation associated with this linear differential equation is

$$\frac{1}{4}\alpha^2 + \alpha + 1 = 0$$

$$\implies (\alpha + 2)^2 = 0,$$

which only has one unique root $\alpha = -2$. Therefore, the general solution to V is

$$V = (A + Bt)e^{-2t}$$

for some constants A and B determined by initial conditions. Since $Q(0) = 0$, $V(0) = 0$.

$$\implies A = 0.$$

$$V = Bte^{-2t}.$$

As $I(0) = -1\text{A}$ and $V(0) = 0$ (such that the initial current through the resistor is zero), the initial rightwards current through the capacitor must be $I_C(0) = I_g - I(0) = 2A$. This implies

$$\frac{dV}{dt}(0) = 4I_C(0) = 8$$

$$\implies B = 8.$$

Thus,

$$V = 8te^{-2t},$$

$$Q = CV = 2te^{-2t},$$

$$I = I_g - \frac{V}{R} - \frac{1}{4}\frac{dV}{dt} = 1 - 8te^{-2t} - 2e^{-2t} + 4te^{-2t} = 1 - 2e^{-2t} - 4te^{-2t}.$$

10.1.5 *Mutual Inductance*

Recall from the previous chapter that two inductors can be coupled with each other such that the change in current through one inductor generates an induced emf in the circuit containing the other inductor. To be exact about the mechanism, the change in current through one inductor leads to a change in magnetic field due to that inductor which in turn, results in a change in the magnetic flux linkage in the other coupled inductor — inducing an emf in its circuit. Recall that the mutual inductance of two inductors is denoted as M and that the magnitude of emf induced in a second inductor due to the change in the current I_1 in a first inductor is given by

$$|\varepsilon_2| = \left| M \frac{dI_1}{dt} \right|. \tag{10.9}$$

Furthermore, the mutual inductance is related to the self-inductances of the two inductors, L_1 and L_2 by

$$M = k\sqrt{L_1 L_2}, \tag{10.10}$$

where k is known as the coupling constant. It is equal to one in the case of ideal coupling, and between zero and one in realistic situations.

As there are two possible orientations of the coupling between inductors, the dot notation (depicted by black circles) is used to denote the direction of the mutually induced emfs. If the proposed current is flowing into a dot of an inductor, the reference polarity of the mutual induced emf at the end of the other inductor, that is also marked by a dot, is positive. Otherwise if the proposed current is flowing out from a dot of an inductor, the reference polarity of the mutual inductance at the corresponding end of the other inductor is negative. This will be illustrated in the following examples.

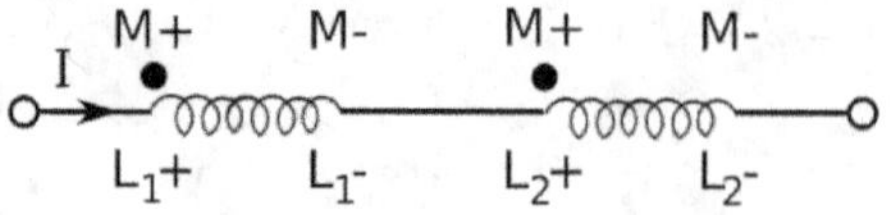

$$\text{M+} \qquad \text{M-} \qquad \text{M+} \qquad \text{M-}$$

$$L_1\text{+} \qquad L_1\text{-} \qquad L_2\text{+} \qquad L_2\text{-}$$

Figure 10.11:   Cumulatively coupled inductors in series

We wish to determine the equivalent inductance of the set-up in Fig. 10.11. As the current is flowing into the dot of the left inductor, the reference polarity of the mutual inductance at the left end of the right inductor is positive. A similar logic allows us to conclude that the reference polarity at the left end of the left inductor is positive. If the current through this particular branch is I, the voltage between its ends (through a segment of a Kirchhoff loop that runs from the left to right) is given by

$$\Delta V = -L_1 \frac{dI}{dt} - M \frac{dI}{dt} - L_2 \frac{dI}{dt} - M \frac{dI}{dt}$$
$$= -(L_1 + L_2 + 2M) \frac{dI}{dt}.$$

The equivalent inductance is then

$$L_{eq} = L_1 + L_2 + 2M.$$

In this case, the magnetic fields of the two inductors aid each other in opposing the change in current; this configuration is sometimes described as cumulatively coupled inductors. Next, we can determine the equivalent inductance of a similar series configuration of two inductors which are oriented such that their magnetic fields oppose each other. This is illustrated in Fig. 10.12.

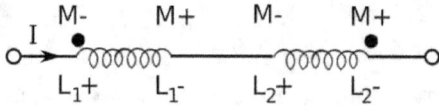

$$\text{M-} \qquad \text{M+} \qquad \text{M-} \qquad \text{M+}$$

$$L_1\text{+} \qquad L_1\text{-} \qquad L_2\text{+} \qquad L_2\text{-}$$

Figure 10.12:   Differentially coupled inductors in series

Again, the current I flows into the dot of the left inductor, which now causes the right end of the right inductor to have a positive reference polarity. Furthermore, the current I now flows out of the dot of the right inductor which causes the left end of the left inductor to have a negative reference polarity. The voltage across this branch is now

$$\Delta V = -L_1 \frac{dI}{dt} + M \frac{dI}{dt} - L_2 \frac{dI}{dt} + M \frac{dI}{dt}$$

$$= -(L_1 + L_2 - 2M) \frac{dI}{dt}$$

$$\implies L_{eq} = L_1 + L_2 - 2M.$$

In this scenario where the magnetic fields of the two inductors oppose each other, the inductors are known to be differentially coupled. Moving on, we shall now analyze the two possible orientations of two coupled inductors in parallel. These scenarios truly reflect the essence of mutual inductance as the currents across each individual inductor are now different. Take note that it is the change in current across one inductor that induces an emf in the branch of the other coupled inductor.

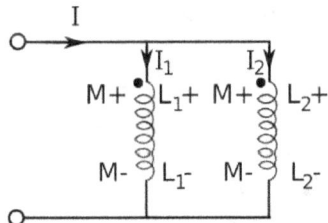

Figure 10.13: "Aiding" inductors in parallel

Consider two coupled inductors that are connected in parallel as depicted in Fig. 10.13. A current I that originates from a terminal splits into two smaller currents, I_1 and I_2, across the branches containing the coupled inductors of self-inductances L_1 and L_2 respectively. Kirchhoff's junction rule requires

$$I = I_1 + I_2.$$

The voltage V across the two parallel branches must be the same. Analyzing the branch on the left, the voltage across this branch is due both to the self-induced emf due to the change in I_1 and also the mutually-induced emf due to the change in I_2.

$$V = -L_1 \frac{dI_1}{dt} - M \frac{dI_2}{dt}.$$

Similarly for the branch on the right,

$$V = -M \frac{dI_1}{dt} - L_2 \frac{dI_2}{dt}.$$

Solving for $\frac{dI_1}{dt}$ and $\frac{dI_2}{dt}$,

$$\frac{dI_1}{dt} = -\frac{V(L_2 - M)}{L_1 L_2 - M^2},$$

$$\frac{dI_2}{dt} = -\frac{V(L_1 - M)}{L_1 L_2 - M^2}.$$

We wish to find the equivalent inductance of these coupled inductors, across the two terminals in the diagram, which satisfies the following relationship.

$$V = -L_{eq}\frac{dI}{dt}.$$

Since $I = I_1 + I_2$,

$$L_{eq} = -\frac{V}{\frac{dI_1}{dt} + \frac{dI_2}{dt}}$$

$$= \frac{L_1 L_2 - M^2}{L_1 + L_2 - 2M}.$$

Once again, there is another possible configuration (depicted in Fig. 10.14) in which one of the inductors is reversed — causing the individual magnetic field produced by one inductor to oppose the magnetic flux linkage in the other inductor due to the other inductor's own current.

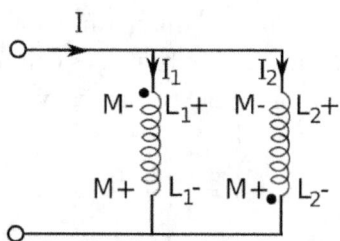

Figure 10.14:   "Opposing" inductors in parallel

Similarly, the voltages across the two branches are the same.

$$V = -L_1\frac{dI_1}{dt} + M\frac{dI_2}{dt},$$

$$V = M\frac{dI_1}{dt} - L_2\frac{dI_2}{dt}.$$

Solving,

$$\frac{dI_1}{dt} = -\frac{V(L_2 + M)}{L_1 L_2 - M^2},$$

$$\frac{dI_2}{dt} = -\frac{V(L_1 + M)}{L_1 L_2 - M^2}.$$

Then, the equivalent inductance is given by

$$L_{eq} = -\frac{V}{\frac{dI_1}{dt} + \frac{dI_2}{dt}}$$

$$= \frac{L_1 L_2 - M^2}{L_1 L_2 + 2M}.$$

When more than two coupled inductors are given, the effect of coupling and the mutual inductance between each pair of inductors must be accounted for. The procedure is still similar to the above process, albeit much more tedious.

10.2 AC Circuits

The DC circuits in the previous sections will eventually stabilize such that their properties, such as currents and voltages, eventually reach constant values. However, if the circuit is connected to an AC source which produces an oscillating emf, the linear properties of the system will eventually reach a steady state with the same angular frequency of oscillation as the AC source, though there may be a phase difference.

Solving an AC circuit problem similarly involves solving the equations obtained from Kirchhoff's laws. However, the germane equations are now non-homogeneous linear differential equations instead of homogeneous ones. The general solution of such a system comprises a particular solution and the homogeneous solution. However, we will only consider the particular solution, as that is the determining factor of the system's steady state response. The solution to the homogeneous part is merely a transient response that will usually undergo exponential decay (as seen from the previous sections) until it is eventually negligible — after which the system will exhibit a response governed by only the particular solution.

There are two methods in procuring the particular solution to the non-homogeneous second order differential equations that we will encounter. The first approach entails guessing a sinusoidal function of the driving angular frequency and solving for the amplitude and phase difference. The second approach leverages the linearity of the equations and modifies the differential equations to include complex variables. Subsequently, the actual properties of the system are computed by taking the real component of their corresponding complex counterparts. This second method then hints at an elegant method — that extends to general, intricate circuits — of introducing the notion of complex admittances and impedances.

10.2.1   *Real Variables*

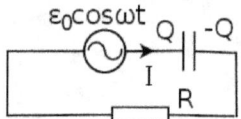

Figure 10.15:   RC circuit

Consider the RC circuit in Fig. 10.15 with an alternating current source that produces an emf $\varepsilon = \varepsilon_0 \cos \omega t$. The exact polarity of the emf depends on the origin of time but it doesn't really matter since the set-up is oscillatory. We shall just define ε to be positive clockwise henceforth, by default. Next, we define the left plate of the capacitor to possess a positive charge $Q(t)$. Applying Kirchhoff's loop rule in the clockwise direction,

$$\varepsilon_0 \cos \omega t - \frac{Q}{C} - IR = 0.$$

Furthermore,

$$I = \frac{dQ}{dt},$$

$$\frac{Q}{C} + R\frac{dQ}{dt} = \varepsilon_0 \cos \omega t.$$

To solve for the particular solution of the above equation, we can try a solution of the form $Q = A \sin(\omega t + \phi)$.

$$\frac{A}{C} \sin(\omega t + \phi) + RA\omega \cos(\omega t + \phi) = \varepsilon_0 \cos \omega t.$$

To solve for A, we equate the magnitude of the left-hand side, after applying the trigonometric R-formula, with ε_0.

$$\sqrt{\frac{1}{C^2} + R^2\omega^2} \cdot A = \varepsilon_0$$

$$A = \frac{\varepsilon_0}{\sqrt{\frac{1}{C^2} + R^2\omega^2}}.$$

To solve for the phase difference ϕ, we set $\omega t = \frac{\pi}{2}$. Then,

$$\frac{A}{C}\cos\phi - RA\omega \sin\phi = 0$$

$$\tan\phi = \frac{1}{R\omega C}.$$

Thus,

$$Q = \frac{\varepsilon_0}{\sqrt{\frac{1}{C^2} + R^2\omega^2}}\sin\left(\omega t + \phi\right)$$

$$I = \frac{\varepsilon_0}{\sqrt{\frac{1}{\omega^2 C^2} + R^2}}\cos\left(\omega t + \phi\right),$$

where $\phi = \tan^{-1}\frac{1}{R\omega C}$, which is a positive value. Thus, it is said that the current $I(t)$ leads the driving voltage $\varepsilon_0 \cos\omega t$ in a capacitive circuit. Finally, notice that we did not need to substitute any initial conditions as the particular solution does not depend on the beginning state of the system. Another way to see this is that the initial conditions, such as the initial charge Q, are lost and unrecoverable as the circuit stabilizes to a standardized steady state.

10.2.2 Complex Variables

Next, consider the RL circuit in Fig. 10.16.

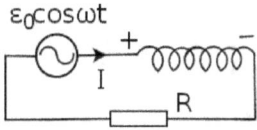

Figure 10.16: RL circuit

Propose a clockwise current I and define the positive and negative terminals of the inductor accordingly. Kirchhoff's loop rule in the clockwise direction requires

$$IR + L\frac{dI}{dt} = \varepsilon_0 \cos\omega t.$$

A slick way to solve this differential equation is to consider a complex driving voltage $\varepsilon_0 e^{i\omega t}$ and a complex current $\tilde{I}$ such that

$$\tilde{I}R + L\frac{d\tilde{I}}{dt} = \varepsilon_0 e^{i\omega t}.$$

We claim that if $\tilde{I}$ is the particular solution to the above equation, its real component $\mathrm{Re}(\tilde{I})$ is the particular solution to the previous equation.

$$I = \mathrm{Re}(\tilde{I}).$$

This is valid because of the linearity of the differential equation and the addition of complex numbers. Consider two complex number z_1 and z_2. Then,

$$\mathrm{Re}(z_1) + \mathrm{Re}(z_2) = \mathrm{Re}(z_1 + z_2).$$

Thus,

$$\mathrm{Re}\left(\tilde{I}R + L\frac{d\tilde{I}}{dt}\right) = \mathrm{Re}(\tilde{I})R + L\frac{d\mathrm{Re}(\tilde{I})}{dt}.$$

We can bring Re into the differentiation as the order of the real operator and differentiation does not matter. Then,

$$\mathrm{Re}(\tilde{I})R + L\frac{d\mathrm{Re}(\tilde{I})}{dt} = \mathrm{Re}(\varepsilon_0 e^{i\omega t}) = \varepsilon_0 \cos \omega t,$$

which is of the same form as the original equation — implying $\mathrm{Re}(\tilde{I})$ is a valid particular solution to the purely real differential equation. Since a linear differential equation only has one particular solution, this must be the unique solution. To appreciate why the linearity of the differential equation is necessary to exploit this method, consider two complex numbers of the form $z_1 = x_1 + iy_1$, $z_2 = x_2 + iy_2$ where x_1, x_2, y_1 and y_2 are real. Then,

$$\mathrm{Re}(z_1) \cdot \mathrm{Re}(z_2) = x_1 x_2.$$

However,

$$\mathrm{Re}(z_1 \cdot z_2) = \mathrm{Re}(x_1 x_2 - y_1 y_2 + i(x_1 y_2 + x_2 y_1)) = x_1 x_2 - y_1 y_2$$

$$\implies \mathrm{Re}(z_1 \cdot z_2) \neq \mathrm{Re}(z_1) \cdot \mathrm{Re}(z_2).$$

This means that if we have a product of two variables in our differential equation, we cannot substitute complex variables for them and hope to retrieve the physical solution by taking the real components of their complex solutions.

Moving on, we can guess a solution[2] for the complex current of the form $\tilde{I} = I_0 e^{i\omega t}$. I_0 may be a complex number, but it is time-independent.

---

[2]The whole point of replacing $\varepsilon_0 \cos \omega t$ with $\varepsilon_0 e^{i\omega t}$ is to facilitate such an exponential guess. Technically, we could have considered any other differential equation $\tilde{I}R + L\frac{d\tilde{I}}{dt} = \varepsilon_0 \cos \omega t + ik$ where k is real and $\mathrm{Re}(\tilde{I})$ will be the particular solution to our desired equation — a drawback of this general form is that the solution for $\tilde{I}$ is difficult to determine.

Substituting this trial solution,

$$RI_0 e^{i\omega t} + iL\omega I_0 e^{i\omega t} = \varepsilon_0 e^{i\omega t}.$$

Cancelling $e^{i\omega t}$ and solving for I_0,

$$I_0 = \frac{\varepsilon_0}{i\omega L + R}$$

$$= \frac{\varepsilon_0}{\sqrt{R^2 + \omega^2 L^2} e^{-i\phi}}$$

$$= \frac{\varepsilon_0}{\sqrt{R^2 + \omega^2 L^2}} e^{i\phi}$$

where $\tan\phi = -\frac{\omega L}{R}$. Splicing this complex amplitude with the exponential term of $\tilde{I}$,

$$\tilde{I} = \frac{\varepsilon_0}{\sqrt{R^2 + \omega^2 L^2}} e^{i(\omega t + \phi)}.$$

The actual current is the real component of this

$$I = \mathrm{Re}(\tilde{I}) = \frac{\varepsilon_0}{\sqrt{R^2 + \omega^2 L^2}} \cos(\omega t + \phi),$$

where $\phi = \tan^{-1}(-\frac{\omega L}{R})$ which, in this case, is a negative value. We say that the current lags behind the driving voltage in an inductive circuit.

10.2.3 *Method of Complex Admittance and Impedance*

The methods above usually suffice for most simple circuits and in fact, can also be applied in finding the solution for driven mechanical oscillations. However, when they are applied to complex circuits, the equations may turn out to be extremely messy. In light of this limitation, the idea of introducing a complex "resistance" for each component generates elegant solutions and rectifies such a cumbersome bottleneck, as we shall see.

Let us first formulate the general AC circuit problem with a single sinusoidal AC source of driving angular frequency ω and emf $\varepsilon = \varepsilon_0 \cos\omega t$. Kirchhoff's loop and junction rules require

$$\sum V = 0,$$

$$\sum I = 0,$$

for every loop and junction respectively. Now, the voltage V across an arbitrary circuit component (resistor, inductor or capacitor) is always linear

---

However, we have the liberty to choose k and hence pick $k = \varepsilon_0 \sin\omega t$ to expedite the process of solving for $\tilde{I}$.

with respect to the current I through it, its derivative or integral. Therefore, Kirchhoff's loop rule generates a set of linear non-homogeneous equations in general. Since this set involves a "driving" sinusoidal emf, the particular solution to V and I of each circuit component is sinusoidal with an angular frequency equal to the driving frequency ω. Hence, the voltage V across and current I through an arbitrary component is of the form

$$I = I_0 \cos(\omega t + \phi_i),$$

$$V = V_0 \cos(\omega t + \phi_v),$$

where the amplitudes and phase offsets are unknown. Now, we define complex variables $\tilde{V}$ and $\tilde{I}$ in replacement of V and I in each branch. In particular, we also replace the emf of the AC source, $\varepsilon = \varepsilon_0 \cos \omega t$, with $\tilde{\varepsilon} = \varepsilon_0 e^{i\omega t}$. Then, consider the differential equations obtained by substituting the complex variables for the real ones in the equations generated by Kirchhoff's laws above. That is,

$$\sum \tilde{V} = 0,$$

$$\sum \tilde{I} = 0.$$

Now, observe that since $\sum V = 0$ is linear in I, its derivative or integral, $\sum \tilde{V}$ must also be linear in $\tilde{I}$, its derivative or integral, as the latter is just obtained from substituting $\tilde{I}$ for I. Due to this linear property, if $\tilde{I}$ is a solution to $\sum \tilde{V} = 0$ and $\sum \tilde{I} = 0$, $\mathrm{Re}(\tilde{I})$ is a solution to $\sum V = 0$ and $\sum I = 0$. Furthermore, as we have chosen $\tilde{\varepsilon} = \varepsilon_0 e^{i\omega t}$, the solutions to the complex variables are also exponential with angular frequency ω. Further matching the real component of these exponential variables with the sinusoidal solutions of the physical variables (e.g. $I = I_0 \cos(\omega t + \phi_i)$) yields

$$\tilde{I} = I_0 e^{i(\omega t + \phi_i)},$$

$$\tilde{V} = V_0 e^{i(\omega t + \phi_v)},$$

for each component. We haven't done anything fancy up till now. However, the crucial component of this method lies in the fact that though V is linear in I, its derivative or integral for a circuit component, the drop in $\tilde{V}$ is always proportional (except for the AC source) to $\tilde{I}$ by a possibly complex number Z, which is known as the impedance of the component!

$$\tilde{V} = \tilde{I} Z.$$

We shall prove this claim soon enough but let us first examine its ramifications to understand why such a proportionality is so useful. Ultimately, we seek for solutions to

$$\sum \tilde{V} = 0,$$

$$\sum \tilde{I} = 0,$$

after which we can take the real components to obtain the physical solutions. Sheerly by our choice, $\tilde{V} = \tilde{\varepsilon} = \varepsilon_0 e^{i\omega t}$ across the AC source. Furthermore, if the previous claim is true, the complex voltage drop across a circuit component (other than the AC source) is always proportional to the current through it and is given by $\tilde{V} = \tilde{I}Z$.

Now, compare this with a DC circuit problem involving solely emf sources and resistors. Kirchhoff's laws require

$$\sum V = 0,$$

$$\sum I = 0.$$

Furthermore, V across an emf source is simply its emf while the voltage drop across a resistor is $V = IR$. It can be seen that the AC problem is exactly identical to a DC network of "resistors" with "complex resistance" Z, when expressed in terms of complex variables! Then, all our machinery in DC circuits can be migrated to AC circuits!

For example, we can determine the equivalent impedance of components connected in series and parallel in the exact same manner as the case of real resistors. For series connections, with n elements,

$$Z_{eq} = \sum_{i=1}^{n} Z_i.$$

Similarly for parallel connections,

$$\frac{1}{Z_{eq}} = \sum_{i=1}^{n} \frac{1}{Z_i}.$$

Now, we shall prove the paramount proposition that the complex voltage drop $\tilde{V}$ across a circuit component (resistor, inductor or capacitor) is proportional to the complex current $\tilde{I}$ through it in a circuit with a single AC source of angular frequency ω.

$$\tilde{V} = \tilde{I}Z. \tag{10.11}$$

In this process, we shall also determine the impedance Z for the various components. The trivial case occurs in the case of resistors where Ohm's law holds for the real voltage and current.

$$V = IR.$$

Replacing the real properties with the complex counterparts,

$$\tilde{V} = \tilde{I}R.$$

Therefore, the impedance of a resistor is simply its resistance. For inductors, the voltage drop is

$$V = L\frac{dI}{dt}$$

which implies that in complex variables,

$$\tilde{V} = L\frac{d\tilde{I}}{dt}.$$

Now, we can exploit the exponential form of $\tilde{I}$. Substituting $\tilde{I} = I_0 e^{i(\omega t + \phi_i)}$,

$$\tilde{V} = i\omega L I_0 e^{i(\omega t + \phi_i)} = i\omega L\tilde{I}$$

which is coherent with Eq. (10.11). Therefore, the complex impedance of an inductor with inductance L is $i\omega L$. Finally, in the case of a capacitor with capacitance C, the voltage drop between the Q and $-Q$ plate is

$$V = \frac{Q}{C} = \frac{\int I dt}{C}.$$

In terms of complex variables,

$$\tilde{V} = \frac{\int \tilde{I} dt}{C}$$

$$= \frac{\int I_0 e^{i(\omega t + \phi_i)} dt}{C}$$

$$= \frac{\frac{I_0}{i\omega} e^{i(\omega t + \phi_i)} + c}{C}.$$

Now, we claim that the constant of integration c is zero. The first suggestion of this is the fact that we are looking at the particular solution of the AC circuit, which should not involve any initial conditions (which determine c). Mathematically, substituting this expression for $\tilde{V}$ for a capacitor into the

$$\sum \tilde{V} = 0$$

equations generated by Kirchhoff's laws would yield a series of terms that vary with time and a constant term associated with c. In order for this equation to be satisfied at all times, c must be zero to eliminate the constant term. Physically, one can also understand that $c = 0$, because the average charge on a capacitor in the long run should be zero as the response is oscillatory, regardless of the initial charge on the capacitor, as the charge should eventually even out (e.g. more charge will lead to a larger current outflow which decreases the amount of charge stored). Therefore,

$$\tilde{V} = \frac{1}{i\omega C}I_0 e^{i(\omega t + \phi_i)} = \frac{\tilde{I}}{i\omega C}.$$

The impedance of a capacitor is hence $\frac{1}{i\omega C}$. The inverse of the impedance is known as the admittance $Y = \frac{1}{Z}$ and the impedances and admittances of various circuit elements are summarized below.

Table 10.1: Admittances and impedances

	Admittance, Y	Impedance, Z
Resistor, R	$\frac{1}{R}$	R
Inductor, L	$\frac{1}{i\omega L}$	$i\omega L$
Capacitor, C	$i\omega C$	$\frac{1}{i\omega C}$

Finally, the phase difference $\Delta\phi = \phi_v - \phi_i$ between the voltage across a component and the current flowing through it can be directly calculated from the impedance. Since $\tilde{I} = I_0 e^{i(\omega t + \phi_i)}$, $\tilde{V} = V_0 e^{i(\omega t + \phi_v)}$ and $\tilde{V} = \tilde{I}Z$,

$$\frac{\tilde{V}}{\tilde{I}} = \frac{V_0}{I_0}e^{i\Delta\phi} = Z = |Z|e^{i\tan^{-1}\frac{\mathrm{Im}(Z)}{\mathrm{Re}(Z)}}.$$

Comparing the exponents,

$$\tan\Delta\phi = \frac{\mathrm{Im}(Z)}{\mathrm{Re}(Z)}. \tag{10.12}$$

Let us apply this technique of complex impedances to the RLC circuit in Fig. 10.17.

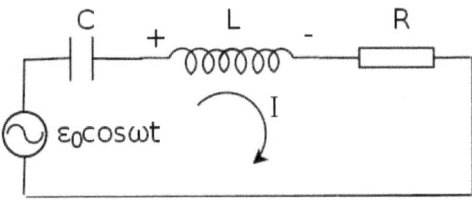

Figure 10.17: RLC circuit

The impedance of this circuit is

$$Z_{eq} = i\omega L + \frac{1}{i\omega C} + R.$$

Therefore,

$$\tilde{I} = \frac{\varepsilon_0}{i\left(\omega L - \frac{1}{\omega C}\right) + R}e^{i\omega t}$$

$$= \frac{\varepsilon_0}{\sqrt{R^2 + \left(\omega L - \frac{1}{\omega C}\right)^2}e^{i\phi}}e^{i\omega t}$$

$$= \frac{\varepsilon_0}{\sqrt{R^2 + \left(\omega L - \frac{1}{\omega C}\right)^2}}e^{i(\omega t - \phi)}$$

where

$$\tan\phi = \frac{\omega L}{R} - \frac{1}{R\omega C}.$$

Thus,

$$I = \mathrm{Re}(\tilde{I}) = \frac{\varepsilon_0}{\sqrt{R^2 + (\omega L - \frac{1}{\omega C})^2}}\cos(\omega t - \phi).$$

There is an interesting geometric relationship between the complex voltages — across the resistor, inductor, capacitor and emf source — and the complex current in the complex plane. Specifically,

$$\tilde{V}_R = \tilde{I} \cdot R,$$

$$\tilde{V}_L = i\omega L \cdot \tilde{I} = \tilde{I}\omega L e^{i\frac{\pi}{2}},$$

$$\tilde{V}_C = \frac{1}{i\omega C} \cdot \tilde{I} = \frac{1}{\omega C}\tilde{I}e^{-i\frac{\pi}{2}},$$

$$\tilde{V}_\varepsilon = \tilde{I} \cdot Z_{eq} = ||Z_{eq}||\tilde{I}e^{i\phi}.$$

We see that the complex voltage across the inductor leads the complex current by a phase angle of $\frac{\pi}{2}$ and the complex voltage across the capacitors lags behind the complex current by a phase angle $\frac{\pi}{2}$. Note that the complex current still lags a phase difference ϕ with respect to the complex voltage of the emf source. If we draw these complex voltages as vectors on a single Argand diagram, along with the complex current,[3] we obtain Fig. 10.18 at the time when the complex current is purely real.

---

[3]The complex current is only used as a reference direction. It has different units from the voltages and should really not be drawn in the diagram.

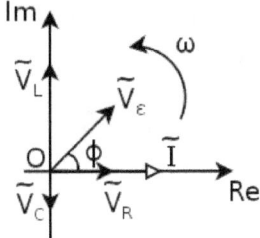

Figure 10.18: Complex vectors

These vectors all rotate at an angular frequency ω anti-clockwise (due to the $e^{i\omega t}$ term in $\tilde{I}$) and thus maintain a fixed shape with respect to each other. Furthermore,

$$\tilde{V}_\varepsilon = \tilde{V}_L + \tilde{V}_R + \tilde{V}_C$$

$$\mathrm{Re}(\tilde{V}_\varepsilon) = \mathrm{Re}(\tilde{V}_L) + \mathrm{Re}(\tilde{V}_R) + \mathrm{Re}(\tilde{V}_C)$$

at the current instance, in accordance with Kirchhoff's laws. Since these vectors all rotate at the same rate, if the above relations are true for a particular instance, it is true for all moments. Equivalently, if Kirchhoff's loop rule is satisfied by the complex voltages at a certain instance in time, it is perpetually fulfilled.

Resonance

Observing the expression for the previous complex current, we see that the circuit responds with the greatest amplitude when the driving angular frequency is

$$\omega_r = \frac{1}{\sqrt{LC}},$$

as the denominator of the amplitude, which is the only variable in ω, is minimized. This is the resonant driving frequency of the LC circuit. It is easy to see why this should be the condition for resonance from the vantage point of impedances. Given an AC source with this angular frequency, the impedances of the inductor and the capacitor effectively cancel out, reducing the circuit to a simple circuit with just a resistor. The maximum amplitude is then

$$I_{max} = \frac{\varepsilon_0}{R},$$

and there is no phase difference between the current through the circuit and the emf of the AC source.

$$I = \frac{\varepsilon_0}{R} \cos \omega t$$

as the set-up effectively consists of a single resistor.

Problem: Determine the current through the resistor in Fig. 10.19 as a function of time. Given fixed R and L, for what value of C is the amplitude of this current the largest? For this particular C, determine the power dissipated in the resistor as a function of time.

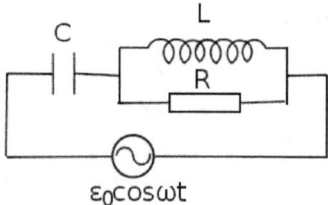

Figure 10.19:   C and LR circuit

The impedance of the capacitor is

$$Z_C = -\frac{i}{\omega C}$$

while the equivalent impedance of the inductor and resistor is

$$Z_{RL} = \frac{i\omega RL}{R + i\omega L} = \frac{(R - i\omega L)i\omega RL}{R^2 + \omega^2 L^2} = \frac{\omega^2 RL^2}{R^2 + \omega^2 L^2} + \frac{i\omega R^2 L}{R^2 + \omega^2 L^2}.$$

By the voltage divider principle, the complex voltage across the resistor is

$$\tilde{V} = \frac{Z_{RL}}{Z_{RL} + Z_C} \varepsilon_0 e^{i\omega t}$$

$$= \frac{\frac{\omega^2 RL^2}{R^2 + \omega^2 L^2} + \frac{i\omega R^2 L}{R^2 + \omega^2 L^2}}{\frac{\omega^2 RL^2}{R^2 + \omega^2 L^2} + i\left(\frac{\omega^2 R^2 L}{R^2 + \omega^2 L^2} - \frac{1}{\omega C}\right)} \varepsilon_0 e^{i\omega t}.$$

The complex current through the resistor is thus

$$\tilde{I} = \frac{\tilde{V}}{R}$$

$$= \frac{\frac{\omega^2 L^2}{R^2 + \omega^2 L^2} + \frac{i\omega RL}{R^2 + \omega^2 L^2}}{\frac{\omega^2 RL^2}{R^2 + \omega^2 L^2} + i\left(\frac{\omega^2 R^2 L}{R^2 + \omega^2 L^2} - \frac{1}{\omega C}\right)} \varepsilon_0 e^{i\omega t}.$$

At this point, we can conclude that the maximum amplitude of current occurs when $\frac{\omega^2 R^2 L}{R^2 + \omega^2 L^2} - \frac{1}{\omega C} = 0$, as $\left|\frac{Z_1}{Z_2}\right| = \frac{|Z_1|}{|Z_2|}$ for any two complex numbers Z_1 and Z_2. The only variable in C in this case is the denominator whose magnitude is minimized when

$$C = \frac{R^2 + \omega^2 L^2}{\omega^3 R^2 L}.$$

When this condition is satisfied, the complex current is

$$\tilde{I} = \left(\frac{1}{R} + \frac{i}{\omega L}\right) \varepsilon_0 e^{i\omega t}$$

$$= \sqrt{\frac{1}{R^2} + \frac{1}{\omega^2 L^2}} \varepsilon_0 e^{i(\omega t + \phi)}$$

where $\phi = \tan^{-1} \frac{R}{\omega L}$. The real current flowing through the resistor is thus

$$I = \operatorname{Re}(\tilde{I}) = \sqrt{\frac{1}{R^2} + \frac{1}{\omega^2 L^2}} \varepsilon_0 \cos(\omega t + \phi).$$

The power dissipated is

$$P = I^2 R = \left(\frac{1}{R} + \frac{R}{\omega^2 L^2}\right) \varepsilon_0^2 \cos^2(\omega t + \phi).$$

Note that $\operatorname{Re}(\tilde{I}^2 R)$, $\operatorname{Re}(\tilde{I}\tilde{V})$ and $\operatorname{Re}(\frac{\tilde{V}^2}{R})$ are all invalid expressions for the power dissipated as these expressions are no longer linear in $\tilde{I}$ and $\tilde{V}$. The real component of the complex variables must be taken before applying $P = VI = I^2 R = \frac{V^2}{R}$.

10.2.4   *Root-Mean-Square Values*

For an AC circuit, it is convenient to define the root-mean-square (rms) values of certain properties of a circuit as it is a measure of the "average" value. This may be useful in certain cases, such as in determining whether a component will melt due to overheating by calculating the average power. For a sinusoidal function of the form

$$A = A_0 \cos(\omega t + \phi),$$

the mean-square value is defined as the square of A, averaged over a single period. The root-mean-square is then the square root of the mean-square value.

$$A_{rms} = \sqrt{\langle A^2 \rangle} = \sqrt{\langle A_0^2 \cos^2(\omega t + \phi) \rangle}$$

$$A_{rms} = \frac{1}{\sqrt{2}} A_0 \tag{10.13}$$

as the average of a squared sinusoidal function is $\frac{1}{2}$ over a period.[4] Thus, for sinusoidal currents and voltages,

$$I_{rms} = \frac{1}{\sqrt{2}} I_0,$$

$$V_{rms} = \frac{1}{\sqrt{2}} V_0.$$

To calculate the average power dissipated in a resistor in a sinusoidal AC circuit, first note that the current I through a resistor and the voltage across it will have no phase difference. Thus if we let

$$V = V_0 \cos \omega t,$$

$$I = \frac{V_0}{R} \cos \omega t,$$

$$P = VI = I^2 R,$$

as $V = IR$. This step seems trivial but we will see the significance of this soon enough. Taking the root-mean-squared value of both sides,

$$\langle P \rangle = \langle I^2 \rangle R = I_{rms}^2 R = V_{rms} I_{rms} = \frac{V_{rms}^2}{R}, \tag{10.14}$$

as both V and I are sinusoidal with no phase difference. Next, let us compute the average power delivered by an emf source. In general, the current in the emf source may have a phase difference with respect to the emf supplied by it. They then take the general form of

$$\varepsilon = \varepsilon_0 \cos \omega t,$$

$$I = I_0 \cos (\omega t - \phi).$$

---

[4]One way to do so is to observe that $\sin^2(\omega t + \phi) + \cos^2(\omega t + \phi) = 1$. Taking the time-average of both sides over a single period and noting that $\langle \cos^2(\omega t + \phi) \rangle = \langle \sin^2(\omega t + \phi) \rangle$ as the cos function is simply the sin function shifted by $\frac{\pi}{2}$ phase, we obtain $\langle \sin^2(\omega t + \phi) \rangle = \frac{1}{2}$. Alternatively, the reader should try envisioning a graphical proof. Hint: slice the graph of a squared sinusoidal function by a horizontal line $y = \frac{1}{2}$ and shift the portions above this line to fill up the "holes."

The power delivered by the emf source is

$$P = \varepsilon I$$
$$= \varepsilon_0 I_0 \cos \omega t \cos (\omega t - \phi)$$
$$= \varepsilon_0 I_0 (\cos^2 \omega t \cos \phi + \cos \omega t \sin \omega t \sin \phi).$$

The time average of $\cos^2 \omega t$ is $\frac{1}{2}$ while that of $\sin \omega t \cos \omega t = \frac{1}{2} \sin 2\omega t$ is zero. Thus,

$$\langle P \rangle = \frac{1}{2} \varepsilon_0 I_0 \cos \phi = \varepsilon_{rms} I_{rms} \cos \phi. \tag{10.15}$$

As seen from the above, the phase difference between the current and emf leads to an additional $\cos \phi$ term. As a final reminder, always remember to take the real component of the complex variables first (if they are used) before computing the power, as the instantaneous power P is no longer linear in V or I.

Problems

Short-term and Long-term Effects

1. *Infinite Capacitor Ladder**

Find the equivalent capacitance between the two left-most terminals in the following infinite ladder of capacitors.

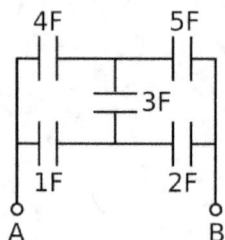

2. *Equivalent Capacitance**

Determine the equivalent capacitance of the circuit, shown in the figure, across terminals A and B. Determine the charge stored by each capacitor when a battery with an emf of $21V$ is connected between A and B, with its positive terminal pointing towards A.

3. *Circuit 1**

The switch S is initially closed towards terminal A until the system has reached a steady state. Afterwards, the switch is changed to terminal B. Find the final charges on each of the capacitors with capacitances $4F$, $6F$ and $3F$.

4. *Circuit 2**

Determine the charges stored by the capacitors if the switch is closed for a long time, given that the capacitors start from a configuration with zero stored charge. What if the switch is opened from the start instead?

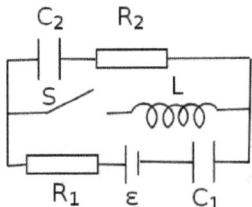

5. *Gargantuan Circuit***

The system below has reached a steady state after a long time. Find the final charge on capacitor A. All batteries, resistors and capacitors have an emf, resistance and capacitance of ε, R and C respectively.

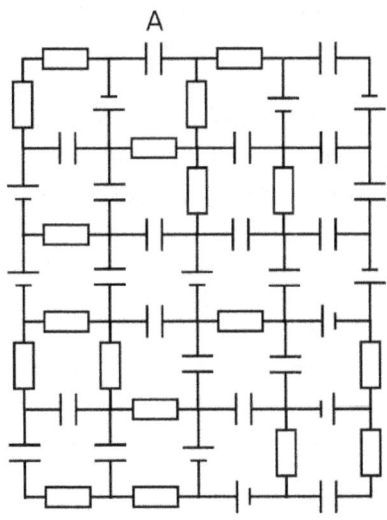

Figure 10.20: Gargantuan circuit

6. *εRC Cube***

Four ideal batteries of emfs $\varepsilon_1 = 4V$, $\varepsilon_2 = 8V$, $\varepsilon_3 = 12V$ and $\varepsilon_4 = 16V$, four capacitors with identical capacitances $C_1 = C_2 = C_3 = C_4 = 1F$, and four identical resistors are connected in the form of a cube as shown in the figure. Compute the total energy U stored by the capacitors after a steady state has been attained. Now, suppose points H and B are connected by an ideal wire. Find the charge stored by capacitor C_2 in the new steady state configuration. (International Physics Olympiad)

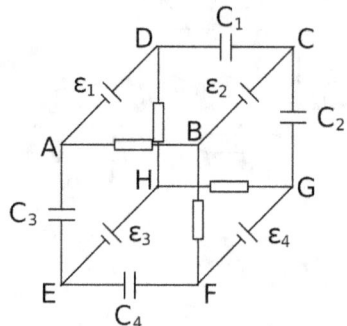

7. Inserting a Plate**

The capacitors on the right all have the same surface area, A. The separations between the two plates of the capacitors with capacitance C are d. Now, a new capacitor plate, of total charge Q_0 and surface area A, is inserted at a distance x from the left plate of the $\frac{C}{2}$ capacitor. After the system has equilibrated, what is the final charge on the left plate of the capacitor (labeled as B on the diagram) that had an original capacitance $\frac{C}{2}$? (Chinese Physics Olympiad)

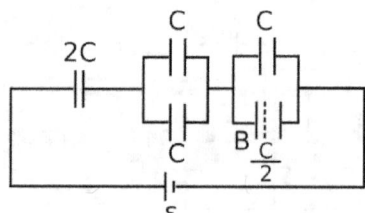

Effects at all Times

8. RC Circuit*

Determine the potential difference $V(t)$ across the capacitor as a function of time t for $t \geq 0$ if the capacitor does not store any charge at $t = 0$.

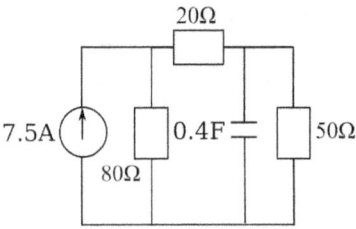

9. *RL Circuit 1**

Determine the potential difference $V_3(t)$ across resistor R_3 and the current $I_1(t)$ through resistor R_1 for $t \geq 0$ if no current flows through the inductor at $t = 0$.

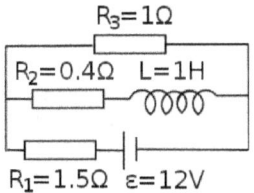

10. *RL Circuit 2**

Before $t = 0$, the circuit is in steady state with the switch S open. At $t = 0$, the switch S is closed. Determine the current I_L through and voltage V_L across the inductor at $t = 0^+$. Next, find $I_L(t)$ and $V_L(t)$ for $t \geq 0$.

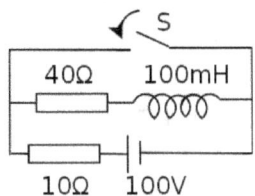

11. *R and LC Circuit***

For $t < 0$, the switch in the set-up on the right is open and the capacitor stores no charge. At $t = 0$, the switch is closed. Determine the current through the inductor as a function of time. The relevant emf, resistance, capacitance and inductance are ε, R, C and L respectively, with $L > 4R^2C$.

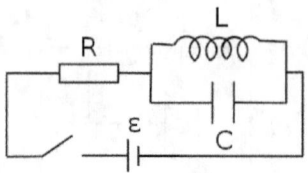

12. *C and RL Circuit* **

Determine the current through the inductor in the below figure for $t \geq 0$ if it is 1A (from the left to right end) at $t = 0$. Furthermore, the potential difference across the capacitor at $t = 0$ is 2V, with the left plate having the higher potential.

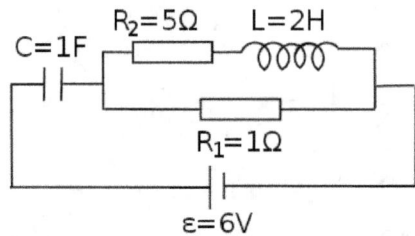

13. *Parallel RLC Circuit* **

For $t < 0$, the switch in the set-up below is open for a long time. At $t = 0$, the switch is closed. Determine the charges stored on the two capacitors as functions of time. Note that you will have to consider three regimes.

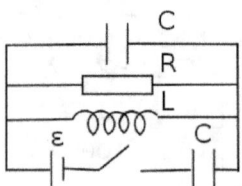

14. *Contracting Capacitor* ***

Two capacitors are arranged as shown in the circuit on the next page. The bottom capacitor has capacitance C_1 while the top capacitor has initial plate separation d_0 and area A (the gap is filled by vacuum). The capacitors are initially held fixed and each store an equal amount of charge Q_0 such that there is no net charge in the portion containing the left plates of the capacitors. Determine Q_0. Now, suppose that the top capacitor is released such that it is free to move — with the mass of each plate being m. The massless wires are coiled into two heaps such that the wires are slack. Determine time as a function of the charges on the capacitors (it is difficult to invert this relationship). Warning: heavy math ahead.

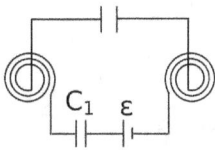

AC Circuits

15. *Current in Parallel RLC Circuit**

Consider a circuit where a resistor, inductor and capacitor of resistance R, inductance L and capacitance C are connected in parallel to an AC source with emf $\varepsilon = \varepsilon_0 \cos \omega t$. Suppose that we forgot the impedance of a capacitor but know that the impedance of the inductor is $i\omega L$. Determine the current through the AC source as a function of time by determining the rate of energy stored or lost by each component.

16. *Bridge**

Determine the current through the AC source as a function of time in the long run. The capacitor has a capacitance $C = \frac{1}{2\omega^2 L}$.

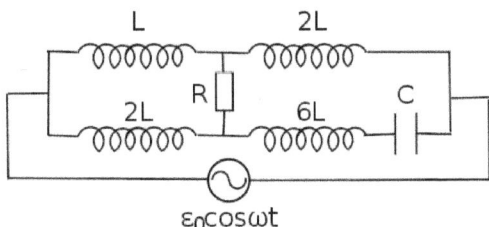

17. *Transformer Circuit***

Consider the circuit on the next page. The resistors have resistances R while the left and right inductors have self-inductances L_1 and L_2. The mutual inductance between the inductors is M and their polarities are indicated by the dot convention. Finally, the capacitor has capacitance $C = \frac{1}{\omega^2 L_2}$. By applying Kirchhoff's laws and substituting complex exponential trial solutions, determine the currents through each loop in the long run. From the perspective of impedances, what is the effect of the capacitor in this set-up? Determine the phase difference between the currents.

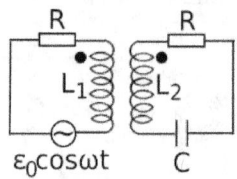

18. *Mutual Inductors***

A resistor R and two parallel inductors L_1 and L_2 are connected as shown in the circuit below. The two inductors have a mutual inductance M and are constructively coupled. Determine the current through inductor L_1 as a function of time by deriving the effective impedance of **each** inductor.

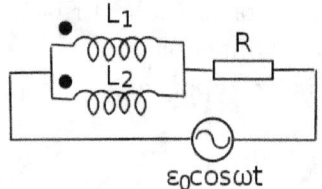

Solutions

1. Infinite Capacitor Ladder*

Similar to the question on an infinite resistor ladder, if we let the equivalent capacitance of the circuit be C_{eq}, we can replace the right part of the original circuit, only leaving a single branch containing a single set of C and $2C$ capacitors, with a capacitor C_{eq} in parallel with the remaining $2C$ capacitor — resulting in Fig. 10.21. Then, we can form an equation in C_{eq} as the equivalent capacitance of this modified circuit should also be C_{eq}.

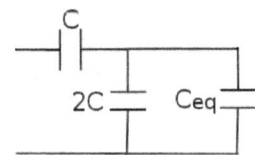

Figure 10.21: Modified circuit

$$C_{eq} = \frac{C(2C + C_{eq})}{C + (2C + C_{eq})} = \frac{2C^2 + CC_{eq}}{3C + C_{eq}}$$

$$\implies C_{eq}^2 + 2C_{eq}C - 2C^2 = 0.$$

$$C_{eq} = (\sqrt{3} - 1)C$$

as we reject the negative solution which is physically incorrect.

2. Equivalent Capacitance*

One can obtain a direct solution to the problem by imposing an external voltage V across terminals A and B and computing the sum of the charges stored by the 1F and 2F capacitors, divided by V, to deduce the equivalent capacitance. In doing so, one would have to use the conservation of charge in a manner akin to the example problems in the section on the long-term behaviour of capacitors. However, there is a slicker method which exploits the analogy between resistance and the reciprocal of capacitance.

Recall that we can transform a capacitor problem into a resistor problem by changing each capacitor C into a resistor $R = \frac{1}{C}$. From this, we can construct a Y-Δ transformation for capacitors. The Y to Δ transformations for resistors are

$$R_a = \frac{R_1 R_2 + R_1 R_3 + R_2 R_3}{R_1}$$

and its cyclic permutations. Using the analogy $R \to \frac{1}{C}$,

$$\frac{1}{C_a} = \frac{\frac{1}{C_1 C_2} + \frac{1}{C_1 C_3} + \frac{1}{C_2 C_3}}{\frac{1}{C_1}} = \frac{C_1 + C_2 + C_3}{C_2 C_3}$$

$$\implies C_a = \frac{C_2 C_3}{C_1 + C_2 + C_3}.$$

The above equation and its cyclic permutations form the Y to Δ transformations for capacitors. Applying this to the 1F, 2F and 3F capacitors, we obtain Fig. 10.22.

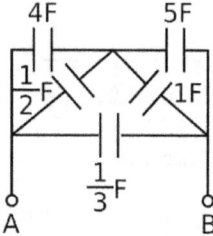

Figure 10.22:   Circuit after Y-Δ transformation

The equivalent capacitance of all other capacitors besides the $\frac{1}{3}$F one is

$$C = \frac{\left(4 + \frac{1}{2}\right) \cdot (5 + 1)}{4 + \frac{1}{2} + 5 + 1} = \frac{18}{7}\text{F},$$

which implies that the equivalent capacitance of the circuit across A and B is

$$C_{eq} = C + \frac{1}{3} = \frac{61}{21}\text{F}.$$

Before we compute the charge stored by each capacitor, we first develop an important tool — the charge divider principle. Suppose that we have two capacitors C_1 and C_2 connected in parallel to two external terminals and we have a total charge q stored between them. What are the charges q_1 and q_2 stored on each capacitor? Well, the potential across the capacitors must be identical so

$$\frac{q_1}{C_1} = \frac{q_2}{C_2}.$$

Solving this with $q_1 + q_2 = q$,

$$q_1 = \frac{C_1}{C_1 + C_2} q,$$

$$q_2 = \frac{C_2}{C_1 + C_2} q.$$

There is another way to obtain these without any calculations. Since $V = \frac{q}{C}$ is analogous to $V = IR$, with $R \to \frac{1}{C}$ and $I \to q$, we can make the above substitutions in the current divider principle for resistors.

$$I_1 = \frac{R_2}{R_1 + R_2} I,$$

$$I_2 = \frac{R_1}{R_1 + R_2} I,$$

$$q_1 = \frac{\frac{1}{C_2}}{\frac{1}{C_1} + \frac{1}{C_2}} q = \frac{C_1}{C_1 + C_2} q,$$

$$q_2 = \frac{C_2}{C_1 + C_2} q.$$

Actually, this equivalence even accounts for the fact that both currents through components in series and charges stored by capacitors in series must be identical! Armed with the charge divider principle, we can compute the charges stored by each capacitor in the original circuit. Firstly, the total charge deposited through terminal A is

$$q = C_{eq} \cdot 21 = 61C$$

By the charge divider principle, the total charge stored by the 4F and $\frac{1}{2}$F capacitors, which is identical to the total charge on the 5F and 1F capacitors, in the equivalent circuit is

$$q' = \frac{C}{C_{eq}} q = \frac{\frac{18}{7}}{\frac{61}{21}} \cdot 61 = 54C.$$

Applying the current divider principle again, the charges stored by the 4F, $\frac{1}{2}$F, 5F and 1F capacitors in the equivalent circuit are

$$q_4 = \frac{4}{4 + \frac{1}{2}} q' = 48C,$$

$$q_{\frac{1}{2}} = \frac{\frac{1}{2}}{4 + \frac{1}{2}} q' = 6C,$$

$$q_5 = \frac{5}{5 + 1} \cdot q' = 45C,$$

$$q_{1'} = \frac{1}{5 + 1} \cdot q' = 9C,$$

where we prime the subscript in $q_{1'}$ to emphasize the fact that this is the charge stored by the 1F capacitor in the equivalent circuit and not the original circuit. Finally, the charge stored in the $\frac{1}{3}$F capacitor in the equivalent circuit is

$$q_{\frac{1}{3}} = \frac{\frac{1}{3}}{C_{eq}} q = 7\text{C}.$$

At this juncture, note that q_4 and q_5 are indeed the correct charges stored by the 4F and 5F capacitors in the original circuit as these capacitors were unchanged. The charge stored by the original 1F capacitor is the sum of the charges on the left plates of the $\frac{1}{2}$F and $\frac{1}{3}$F capacitors in the equivalent circuit.

$$q_1 = q_{\frac{1}{2}} + q_{\frac{1}{3}} = 13\text{C}.$$

Similarly,

$$q_2 = q_{\frac{1}{3}} + q_{1'} = 16\text{C},$$

$$q_3 = q_{1'} - q_{\frac{1}{2}} = 3\text{C}.$$

Note the negative sign in the last equation as the charge on the top plate of the original 3F capacitor is the sum of those on the right plate of the $\frac{1}{2}$F capacitor (which is $-q_{\frac{1}{2}}$) and the left plate of the 1F capacitor in the equivalent circuit. Another way to compute q_3 is to take $q_4 - q_5 = 3\text{C}$ by the conservation of charge.

3. Circuit 1*

Initially, the capacitor C_1 is charged to

$$Q_0 = 4\varepsilon.$$

Next, after the switch is turned to terminal B, charges will flow from the positive plate of C_1(the top plate) to the other capacitors — causing the top plates of C_2 and C_3 to also be positively charged. The key observation is that the net charge is conserved between the adjacent plates of different capacitors. This implies that the final positive charges on C_2 and C_3 are the same since the total initial charge on the two plates of C_2 and C_3 that are directly connected by a wire is zero. We define that final identical charge on C_2 and C_3 as Q_2. The final charge on C_1, Q_1, is then given by the conservation of charge.

$$Q_1 = Q_0 - Q_2.$$

Applying Kirchhoff's loop rule to a cycle through all capacitors,

$$\frac{Q_1}{4} - \frac{Q_2}{6} - \frac{Q_2}{3} = 0$$

$$Q_2 = \frac{Q_0}{3} = \frac{4\varepsilon}{3}$$

$$Q_1 = \frac{2Q_0}{3} = \frac{8\varepsilon}{3}.$$

4. Circuit 2*

In the long run, the inductors and capacitors are effectively short-circuited and open-circuited respectively. Therefore, when the switch is closed for a long time, the inductor becomes an ideal wire while the capacitors are disconnected. Then, no current flows through the circuit. Drawing a clockwise Kirchhoff loop across the emf source, ideal wire and capacitor C_1,

$$\varepsilon - \frac{Q_1}{C_1} = 0$$

where Q_1 is the charge on the right plate of the capacitor C_1. Thus,

$$Q_1 = \varepsilon C_1.$$

Now, draw a Kirchhoff loop through the ideal wire and the capacitor C_2. Since no current flows everywhere, the charge stored by capacitor C_2 must be zero by Kirchhoff's loop rule.

Next, in a separate set-up where the switch is opened for a long time, no current flows everywhere once again. Let the charges on the right plate of capacitor C_1 and the left plate of capacitor C_2 be q. Note that they must possess the same charge as the segment of the circuit between the right plates of the capacitors are electrically isolated from the rest of the circuit (this didn't occur in the previous case due to the ideal wire). Applying Kirchhoff's loop rule to a clockwise loop through the battery and the capacitors,

$$\varepsilon - \frac{q}{C_1} - \frac{q}{C_2} = 0.$$

Therefore, the charges stored by the capacitors are of quantity

$$q = \frac{\varepsilon C_1 C_2}{C_1 + C_2}.$$

5. Gargantuan Circuit**

The crux of this question is to remain composed. When the system has reached steady state, we can effectively remove all capacitors as no current will flow through them. The resultant circuit in Fig. 10.23 will be obtained.

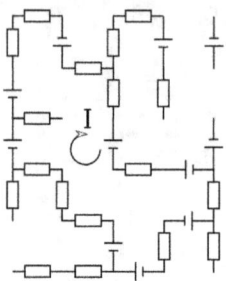

Figure 10.23:   Resultant circuit

Observe that only one loop is present in the entire circuit. Applying Kirchhoff's law to that loop in the clockwise direction, we can find the clockwise current I to be

$$I = \frac{\varepsilon}{5R}.$$

Next, we can draw the Kirchhoff loop, depicted by the white arrow in Fig. 10.24, in the original circuit.

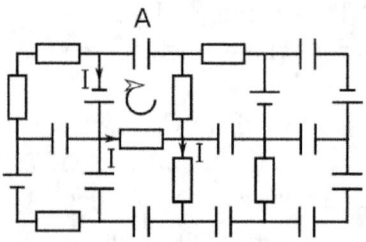

Figure 10.24:   Resultant circuit

Applying Kirchhoff's loop rule and defining the charge on the right capacitor plate to be Q,

$$\frac{Q}{C} + IR - \varepsilon = 0$$

$$Q = \frac{4}{5}\varepsilon C.$$

6. εRC Cube**

In the long run, the capacitors can be disconnected. Therefore, we obtain Fig. 10.25.

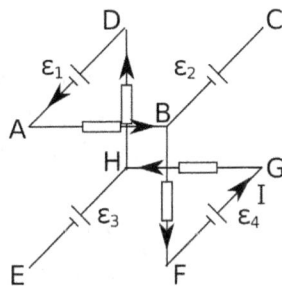

Figure 10.25: Cube after removing capacitors

If we let R denote the resistance of a resistor, the current I depicted in Fig. 10.25 is

$$I = \frac{\varepsilon_4 - \varepsilon_1}{4R} = \frac{3}{R}.$$

Therefore, if we let the potential of vertex A be zero ($V_A = 0$),

$$V_B = V_A - IR = -3\text{V},$$

$$V_C = V_B + \varepsilon_2 = 5\text{V},$$

$$V_D = V_A + \varepsilon_1 = 4\text{V},$$

$$V_F = V_B - IR = -6\text{V},$$

$$V_G = V_F + \varepsilon_4 = 10\text{V},$$

$$V_H = V_G - IR = 7\text{V},$$

$$V_E = V_H - \varepsilon_3 = -5\text{V}.$$

The total energy stored by the capacitors at steady state is

$$U = \frac{1}{2} \cdot 1 \cdot \left[(V_C - V_D)^2 + (V_G - V_C)^2 + (V_A - V_E)^2 + (V_E - V_F)^2 \right]$$

$$= \frac{1}{2} \cdot 1 \cdot (1^2 + 5^2 + 5^2 + 1^2) = 26\text{J}.$$

After an ideal wire is connected between B and H, vertices B and H become equipotential and thus can be compressed into a single point B/H. The loop in Fig. 10.25 then becomes Fig. 10.26,

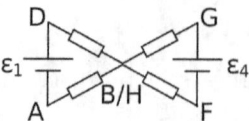

Figure 10.26:   Loop after combining nodes B and H

which can be further decomposed into two isolated circuits ADB/H and FGB/H (this is most obvious when applying the principle of superposition and considering one emf source at a time). Therefore, the potential difference between B and G is

$$V_G - V_B = \frac{\varepsilon_4}{2} = 8\text{V}.$$

The potential difference across C and G is then

$$V_G - V_C = V_G - (V_B + \varepsilon_2) = 8 - 8 = 0\text{V}.$$

Therefore, the capacitor C_2 stores no charge in the new steady state configuration, as $q_2 = C_2|V_G - V_C| = 0$.

7. Inserting a Plate**

Let the final charge on plate B be Q. Then the charge on the left surface of the inserted plate is $-Q$ (by Gauss' law) which results in the right surface containing charge $Q + Q_0$. The capacitor plate on the right of the inserted plate then has charge $-Q - Q_0$. The three plates essentially form two capacitors in series with separations x and $2d - x$ respectively. Let the charge on the left plate of the capacitor in the branch above the three plates be q'. In order for the voltages across the two branches to be the same,

$$\frac{Q}{\varepsilon_0 \frac{A}{x}} + \frac{Q + Q_0}{\varepsilon_0 \frac{A}{2d-x}} = \frac{q'}{\varepsilon_0 \frac{A}{d}}$$

$$q' = 2(Q + Q_0) - \frac{x}{d}Q_0.$$

The total amount of charge contained in the right plates of the two left capacitors with capacitance C, the left plate of the remaining capacitor of capacitance C which contains charge q' and capacitor plate B, which carries a charge Q, must be conserved. Furthermore, the voltages across the two C capacitors connected in parallel must be the same. Hence, the quantity of charge stored in each of these two capacitors must be $\frac{Q+q'}{2}$ (on their left plates). Following a similar logic, the charge stored in the left plate of the

$2C$ capacitor must be $Q + q'$. Hence, by drawing a clockwise Kirchhoff loop through the emf and the capacitors, we obtain

$$\frac{Q + q'}{2C} + \frac{Q + q'}{2C} + \frac{q'}{C} = \varepsilon$$

$$Q = \frac{\varepsilon C - 4Q_0 + 2\frac{x}{d}Q_0}{5}.$$

8. RC Circuit*

We shall present two solutions here. The brute force solution is to solve Kirchhoff's laws directly. Let the potential of the "negative" and "positive" terminals of the current source be 0 and U. Let the potential at the top capacitor plate be V. Imposing Kirchhoff's junction rule at the node above the 80Ω resistor,

$$\frac{U}{80} + \frac{U - V}{20} = 7.5$$

$$\frac{U}{8} = 15 + \frac{V}{10}.$$

Let q denote the charge stored in the top plate of the capacitor. Then,

$$V = \frac{q}{0.4} = \frac{5q}{2}.$$

Applying Kirchhoff's junction rule to the node above the top plate of the capacitor yields

$$\dot{q} = \frac{U - V}{20} - \frac{V}{50}$$

$$\implies \frac{dV}{dt} = \frac{5\dot{q}}{2} = \frac{U - V}{8} - \frac{V}{20} = \frac{U}{8} - \frac{7V}{40} = 15 - \frac{3V}{40},$$

$$\frac{dV}{dt} + \frac{3V}{40} = 15,$$

where we have used the equation $\frac{U}{8} = 15 + \frac{V}{10}$. The general solution to this differential equation is

$$V = Ae^{-\frac{3}{40}t} + \frac{15 \cdot 40}{3} = Ae^{-\frac{3}{40}t} + 200$$

for some constant A determined by initial conditions. Since the capacitor stores no charge initially, $V(0) = 0$.

$$\implies V = 200\left(1 - e^{-\frac{3}{40}t}\right).$$

The second method is to apply Thevenin's theorem across the terminals of the capacitor to convert the rest of the circuit into an equivalent Thevenin emf $\varepsilon_{eq} = \frac{80}{20+50+80} \cdot 7.5 \cdot 50 = 200\text{V}$ and Thevenin resistance $R_{eq} = \frac{(80+20)\cdot 50}{80+20+50} = \frac{100}{3}\Omega$. Then, the circuit becomes a series RC circuit with $\varepsilon_{eq} = 200\text{V}$, $R_{eq} = \frac{100}{3}\Omega$ and $C = 0.4\text{F}$. Substituting these values into the relevant solution derived in Section 10.1.4,

$$V = \varepsilon_{eq}\left(1 - e^{-\frac{1}{R_{eq}C}t}\right) = 200\left(1 - e^{-\frac{3}{40}t}\right).$$

9. RL Circuit 1*

We can apply Thevenin's theorem with respect to the ends of the inductor to convert the rest of the circuit (besides the inductor) into a Thevenin emf $\varepsilon_{eq} = \frac{1}{1+1.5} \cdot 12 = \frac{24}{5}\text{V}$ and Thevenin resistance $R_{eq} = \frac{1\cdot 1.5}{1+1.5} + 0.4 = 1\Omega$. Therefore, the original circuit becomes a series RL circuit with $\varepsilon_{eq} = \frac{24}{5}\text{V}$, $R_{eq} = 1\Omega$ and $L = 1\text{H}$. Applying the result from Section 10.1.4, the current through the inductor (from its left to right end) as a function of time is

$$I_L = \frac{\varepsilon_{eq}}{R}\left(1 - e^{-\frac{R}{L}t}\right) = \frac{24}{5}(1 - e^{-t}).$$

The potential difference across R_3 is that across R_2 plus that across the inductor.

$$V_3 = I_L R_2 + L\frac{dI_L}{dt} = \frac{48}{25}(1 - e^{-t}) + \frac{24}{5}e^{-t} = \frac{48}{25} + \frac{72}{25}e^{-t}.$$

The current through R_1 is I_L plus that through R_3.

$$I_1 = I_L + \frac{V_3}{R_3} = \frac{24}{5}(1 - e^{-t}) + \frac{48}{25} + \frac{72}{25}e^{-t} = \frac{168}{25} - \frac{48}{25}e^{-t}.$$

10. RL Circuit 2*

Before the switch is closed, the current through the inductor is

$$I_L(0^-) = \frac{100}{40+10} = 2\text{A}$$

from its left to right end, as the inductor is effectively an ideal wire in the long run. Immediately after the switch is closed, the inductor maintains the current through itself so

$$I_L(0^+) = I_L(0^-) = 2\text{A}.$$

After the switch is closed, the branch containing the 10Ω resistor and 100V battery can be removed since its ends become equipotential. The circuit is then effectively the resistor 40Ω connected in series with the inductor 100mH. Since the inductor drives 2A current through the 40Ω resistor at $t = 0^+$,

$$V_L(0^+) = 40 \times 2 = 80\text{V}$$

with the right end of the inductor having the higher potential. Let I denote the current through the inductor, from its left to right end. Applying Kirchhoff's loop rule in an anti-clockwise fashion through the resistor $R = 40\Omega$, inductor $L = 100$mH and the ideal wire would yield

$$-L\frac{dI_L}{dt} - I_L R = 0$$

$$\int_{I_0}^{I_L} \frac{1}{I_L} dI_L = \int_0^t -\frac{R}{L} dt$$

$$\ln\left|\frac{I_L}{I_0}\right| = -\frac{R}{L}t$$

$$I_L = I_0 e^{-\frac{R}{L}t}.$$

Substituting $I_0 = 2$A, $R = 40\Omega$ and $L = 100$mH,

$$I_L(t \geq 0) = 2e^{-400t},$$

$$V_L(t \geq 0) = -L\frac{dI_L}{dt} = 80e^{-400t}.$$

11. R and LC Circuit**

Let the currents flowing through the inductor and capacitor be I_1 and I_2 rightwards, respectively. Draw a clockwise Kirchhoff loop through the emf source, resistor and the inductor. This requires

$$\varepsilon - (I_1 + I_2)R - L\frac{dI_1}{dt} = 0.$$

Let the charge on the left capacitor plate be Q. Drawing a clockwise loop through the inductor and capacitor, we obtain

$$L\frac{dI_1}{dt} = \frac{Q}{C}.$$

Differentiating the above with respect to time and using $\frac{dQ}{dt} = I_2$,

$$I_2 = LC\frac{d^2 I_1}{dt^2}.$$

Substituting this expression for I_2 into the first equation,

$$RLC\frac{d^2 I_1}{dt^2} + L\frac{dI_1}{dt} + RI_1 = \varepsilon.$$

The particular solution for I_1 above is evidently $\frac{\varepsilon}{R}$. The solution to the homogeneous equation

$$RLC\frac{d^2 I_1}{dt^2} + L\frac{dI_1}{dt} + RI_1 = 0$$

can be deduced from the characteristic equation

$$RLC\alpha^2 + L\alpha + R = 0.$$

The solutions for α are

$$\alpha = \frac{-L \pm \sqrt{L^2 - 4R^2 LC}}{2RLC}.$$

Therefore, the general solution for I_1, obtained by combining the particular and general solutions, is

$$I_1(t) = \frac{\varepsilon}{R} + Ae^{\frac{-L+\sqrt{L^2-4R^2LC}}{2RLC}t} + Be^{\frac{-L-\sqrt{L^2-4R^2LC}}{2RLC}t}$$

for some constants A and B determined by initial conditions. Since the current through the inductor is zero at $t = 0$ (because it tries to instantaneously maintain the current through itself),

$$B = -A - \frac{\varepsilon}{R}.$$

Then,

$$I_1(t) = \frac{\varepsilon}{R} + Ae^{\frac{-L+|\sqrt{L^2-4R^2LC}}{2RLC}t} - \left(A + \frac{\varepsilon}{R}\right)e^{\frac{-L-\sqrt{L^2-4R^2LC}}{2RLC}t}.$$

The other initial condition is that the voltage across the inductor must be zero at time $t = 0$ because the voltage across the capacitor is zero at time $t = 0$, as it has yet to store any charge. Therefore,

$$L\frac{dI_1}{dt}\bigg|_{t=0} = 0$$

$$A \cdot \frac{-L + \sqrt{L^2 - 4R^2 LC}}{2RC} + \left(A + \frac{\varepsilon}{R}\right) \cdot \left(\frac{L + \sqrt{L^2 - 4R^2 LC}}{2RC}\right) = 0.$$

Solving,

$$A = -\frac{\varepsilon(L + \sqrt{L^2 - 4R^2LC})}{2R\sqrt{L^2 - 4R^2LC}}.$$

Substituting this expression for A into $I_1(t)$ above would yield the general solution.

12. C and RL Circuit**

Let $V(t)$ denote the potential difference across the capacitor C, positive if the left plate has a higher potential, and $I(t)$ denote the rightwards current through the inductor. Applying Kirchhoff's junction rule to the node on the right of the capacitor,

$$-C\frac{dV}{dt} + \frac{\varepsilon - V}{R_1} + I = 0$$

where $C\frac{dV}{dt}$ is the current emanating from the right capacitor plate and $\frac{\varepsilon - V}{R_1}$ is the rightwards current through R_1. Substituting the relevant parameters,

$$-\frac{dV}{dt} + 6 - V + I = 0.$$

Applying Kirchhoff's loop rule to the loop crossing the battery and the inductor,

$$6 - V - 5I = V_L$$

where $V_L = L\frac{dI}{dt} = 2\frac{dI}{dt}$ is the voltage across the inductor.

$$\implies V = 6 - 5I - 2\frac{dI}{dt}.$$

Substituting this expression for V into the previous equation,

$$5\frac{dI}{dt} + 2\frac{d^2I}{dt^2} + 6 - 6 + 5I + 2\frac{dI}{dt} + I = 0$$

$$\frac{d^2I}{dt^2} + \frac{7}{2}\frac{dI}{dt} + 3I = 0$$

whose characteristic equation has solutions $-\frac{3}{2}$ and -2. Thus, the general solution for I is

$$I = Ae^{-\frac{3}{2}t} + Be^{-2t}$$

for some constants A and B determined by initial conditions. Since $I(0) = 1$,

$$A + B = 1.$$

Furthermore, the voltage across the inductor at $t = 0$ is $V_L(0) = 6 - V(0) - 5I(0) = 6 - 2 - 5 \cdot 1 = -1$. Thus,

$$2\frac{dI}{dt}(0) = -3A - 4B = -1.$$

Solving,

$$A = 3$$
$$B = -2$$
$$\implies I = 3e^{-\frac{3}{2}t} - 2e^{-2t}.$$

13. Parallel RLC Circuit**

Define the charge on the right plate of the bottom capacitor as Q and that on the left plate on the top capacitor as Q'. Furthermore, let the current entering the right plate of the bottom capacitor be I and the currents entering the inductor, resistor and capacitor in the parallel branches be I_1, I_2 and I_3 from the left. We know from Kirchhoff's junction rule that

$$I = I_1 + I_2 + I_3.$$

Next, by definition,

$$I = \frac{dQ}{dt},$$
$$I_3 = \frac{dQ'}{dt}.$$

Furthermore, the three parallel branches must have a common voltage drop $V(t)$ from the left to right.

$$V(t) = L\frac{dI_1}{dt} = I_2 R = \frac{Q'}{C}.$$

Kirchhoff's loop rule through the battery, a parallel branch and the bottom capacitor dictates that

$$\varepsilon - \frac{Q}{C} - V(t) = 0.$$

Differentiating this with respect to time,

$$\frac{I}{C} + \frac{dV}{dt} = 0$$
$$\frac{dI}{Cdt} + \frac{d^2V}{dt^2} = 0$$
$$\frac{dI_1}{Cdt} + \frac{dI_2}{Cdt} + \frac{dI_3}{Cdt} + \frac{d^2V}{dt^2} = 0.$$

Substituting $\frac{dI_1}{dt} = \frac{V}{L}$, $\frac{dI_2}{dt} = \frac{dV}{Rdt}$ and $\frac{dI_3}{dt} = C\frac{d^2V}{dt^2}$,

$$2\frac{d^2V}{dt^2} + \frac{1}{RC}\frac{dV}{dt} + \frac{V}{LC} = 0,$$

which is analogous to the equation of motion of a damped oscillation. Before embarking on solving this differential equation, we should keep the initial conditions for $V(t)$ in mind. Now, there is a path solely comprising the battery and two capacitors — there must therefore be a discontinuity in the stored charges of the capacitors at $t = 0$ (i.e. the capacitors are not ideal batteries in the short run anymore). During an infinitesimal time interval at $t = 0$, a large amount of current travels through the capacitors and deposits charges on the plates. The inductor maintains zero current through itself while the current flowing through the resistor transfers negligible charge in this short time interval. Therefore, the charges stored by the two capacitors must be identical and their voltages must each be $\frac{\varepsilon}{2}$. Therefore,

$$V(0) = \frac{\varepsilon}{2}.$$

Next, from $\frac{I}{C} + \frac{dV}{dt} = 0$ and $V = \frac{Q'}{C}$,

$$I = -I_3.$$

Now, directly after the discontinuity in charges, the current flowing in the resistor can be computed by dividing the voltage (which is $\frac{\varepsilon}{2}$ as it is connected in parallel with the top capacitor) by its resistance.

$$I_2(0) = \frac{\varepsilon}{2R}.$$

Since $I(0) = I_1(0) + I_2(0) + I_3(0)$, $I_1(0) = 0$ as the inductor maintains the current through itself and $I = -I_3$,

$$I(0) = \frac{I_2(0)}{2} = \frac{\varepsilon}{4R}.$$

Then,

$$\frac{dV}{dt}\bigg|_{t=0} = -\frac{I(0)}{C} = -\frac{\varepsilon}{4RC}.$$

With $V(0)$ and $\frac{dV}{dt}\big|_{t=0}$ as our initial conditions, we proceed with solving the second order linear differential equation whose characteristic equation is

$$2\alpha^2 + \frac{1}{RC}\alpha + \frac{1}{LC} = 0$$

$$\implies \alpha = -\frac{1}{4RC} \pm \sqrt{\frac{1}{16R^2C^2} - \frac{1}{2LC}}.$$

If $\frac{1}{16R^2C^2} > \frac{1}{2LC}$, we let $\omega = \sqrt{\frac{1}{16R^2C^2} - \frac{1}{2LC}}$. The general solution for V is of the form

$$V = e^{-\frac{1}{4RC}t}(Ae^{\omega t} + B^{-\omega t}).$$

The initial conditions imply that

$$A + B = \frac{\varepsilon}{2},$$

$$A\left(-\frac{1}{4RC} + \omega\right) - B\left(\frac{1}{4RC} + \omega\right) = -\frac{\varepsilon}{4RC}.$$

Solving,

$$A = \frac{\varepsilon}{4} - \frac{\varepsilon}{16\omega RC},$$

$$B = \frac{\varepsilon}{4} + \frac{\varepsilon}{16\omega RC},$$

$$V = e^{-\frac{1}{4RC}t}\left[\left(\frac{\varepsilon}{4} - \frac{\varepsilon}{16\omega RC}\right)e^{\omega t} + \left(\frac{\varepsilon}{4} + \frac{\varepsilon}{16\omega RC}\right)e^{-\omega t}\right].$$

Moving on, when $\frac{1}{16R^2C^2} = \frac{1}{2LC}$, the general solution for V is

$$V = Ae^{-\frac{1}{4RC}t} + Bte^{-\frac{1}{4RC}t}.$$

The initial conditions imply

$$A = \frac{\varepsilon}{2},$$

$$-\frac{A}{4RC} + B = -\frac{\varepsilon}{4RC}$$

$$B = -\frac{\varepsilon}{8RC}.$$

Therefore,

$$V = \frac{\varepsilon}{2}e^{-\frac{1}{4RC}t} - \frac{\varepsilon}{8RC}te^{-\frac{1}{4RC}t}.$$

Finally, in the last case where $\frac{1}{16R^2C^2} < \frac{1}{2LC}$, let $i\omega = \sqrt{\frac{1}{16R^2C^2} - \frac{1}{2LC}}$. The general solution for V is of the form

$$V = e^{-\frac{1}{4RC}t}(Ae^{i\omega t} + Be^{-i\omega t}).$$

Since V must be real, A and B must be complex conjugates.

$$A = \frac{D}{2}e^{i\phi},$$

$$B = \frac{D}{2}e^{-i\phi},$$

for some real constants D and ϕ. Then,

$$V = \frac{D}{2}e^{-\frac{1}{4RC}t}\left(e^{i(\omega t+\phi)} + e^{-i(\omega t+\phi)}\right) = De^{-\frac{1}{4RC}t}\cos(\omega t + \phi).$$

The initial conditions yield

$$D\cos\phi = \frac{\varepsilon}{2},$$

$$-\frac{D}{4RC}\cos\phi - \omega D\sin\phi = -\frac{\varepsilon}{4RC}$$

$$\implies D\sin\phi = -\frac{\varepsilon}{8\omega RC}.$$

Thus,

$$D = \sqrt{\frac{1}{4} + \frac{1}{64\omega^2 R^2 C^2}}\,\varepsilon$$

where we have chosen the positive sign because the exact sign of D doesn't matter (it can be adjusted by a π-radian offset of ϕ). With this choice of D, ϕ is given by

$$\phi = \cos^{-1}\frac{\varepsilon}{2D} = \cos^{-1}\frac{1}{\sqrt{1 + \frac{1}{16\omega^2 R^2 C^2}}}.$$

Then,

$$V = \sqrt{\frac{1}{4} + \frac{1}{64\omega^2 R^2 C^2}}\,\varepsilon\cos(\omega t + \phi).$$

Now that we have computed $V(t)$ for all possible cases, the charge on the top capacitor is simply CV. The instantaneous voltage of the bottom capacitor is given by Kirchhoff's loop rule to be $\varepsilon - V$. Thus, it possesses charge $\varepsilon C - CV$.

14. Contracting Capacitor***

Let $Q(t)$ be the common charge stored in the left plate of the top capacitor and the right plate of the bottom capacitor. The charges stored must be identical as the segment connecting the two left plates of the capacitors is electrically isolated and neutral. Initially, $Q = Q_0$. By Kirchhoff's law,

$$\varepsilon - \frac{Q_0}{C_1} - \frac{Q_0}{C_2^0} = 0$$

where C_2^0 is the initial capacitance of the top capacitor.

$$C_2^0 = \varepsilon_0 \frac{A}{d_0}.$$

Then,

$$Q_0 = \frac{\varepsilon C_1 C_2^0}{C_1 + C_2^0} = \frac{\varepsilon \varepsilon_0 C_1 A}{C_1 d_0 + \varepsilon_0 A}.$$

Now, when the top capacitor is released, the plates attract each other which causes the plate separation to decrease — hence changing the capacitance of the top capacitor. Let the plate separation at time t be $d(t)$ and the capacitance of the top capacitor be $C_2(t)$. Then,

$$C_2(t) = \varepsilon_0 \frac{A}{d(t)}.$$

Furthermore, we know from Gauss' law that the electric field due to one plate at the location of the other is $\frac{Q}{2A\varepsilon_0}$. Therefore, the acceleration of each plate is $\frac{Q^2}{2mA\varepsilon_0}$ towards one another — implying that

$$\ddot{d} = -\frac{Q^2}{mA\varepsilon_0}$$

where we have multiplied by two as the second time derivative of the plate separation is compounded by the accelerations of the two plates. By Kirchhoff's loop rule,

$$\varepsilon - \frac{Q}{C_1} - \frac{Q}{C_2(t)} = 0,$$

$$\varepsilon - \frac{Q}{C_1} - \frac{Qd}{\varepsilon_0 A} = 0.$$

Dividing by Q,

$$\frac{\varepsilon}{Q} - \frac{1}{C_1} - \frac{d}{\varepsilon_0 A} = 0.$$

Differentiating the above with respect to time,

$$-\frac{\varepsilon}{Q^2}\dot{Q} = \frac{\dot{d}}{\varepsilon_0 A}.$$

The above equation implies that the initial current $\dot{Q}$ is zero as the initial velocities of the plates are zero (this will be an initial condition later). Differentiating once again,

$$\frac{2\varepsilon}{Q^3}\dot{Q}^2 - \frac{\varepsilon}{Q^2}\ddot{Q} = \frac{\ddot{d}}{\varepsilon_0 A}.$$

Substituting $\ddot{d} = -\frac{Q^2}{A\varepsilon_0}$,

$$\frac{\varepsilon}{Q^2}\ddot{Q} - \frac{2\varepsilon}{Q^3}\dot{Q}^2 = \frac{Q^2}{mA^2\varepsilon_0^2}.$$

Using the trick $\ddot{Q} = \frac{d\dot{Q}^2}{2dQ}$ and simplifying,

$$\frac{d\dot{Q}^2}{dQ} - \frac{4}{Q}\dot{Q}^2 = \frac{2Q^4}{mA^2\varepsilon_0^2\varepsilon}.$$

Multiplying the above by the integrating factor $\frac{1}{Q^4}$,

$$\frac{1}{Q^4}\frac{d\dot{Q}^2}{dQ} - \frac{4}{Q^5}\dot{Q}^2 = \frac{d\left(\frac{\dot{Q}^2}{Q^4}\right)}{dQ} = \frac{2}{mA^2\varepsilon_0^2\varepsilon}.$$

Therefore,

$$\int_0^{\frac{\dot{Q}^2}{Q^4}} d\left(\frac{\dot{Q}^2}{Q^4}\right) = \int_{Q_0}^{Q} \frac{2}{mA^2\varepsilon_0\varepsilon}\,dQ$$

$$\frac{\dot{Q}^2}{Q^4} = \frac{2(Q - Q_0)}{mA^2\varepsilon_0^2\varepsilon}$$

$$\dot{Q} = \sqrt{\frac{2(Q - Q_0)Q^4}{mA^2\varepsilon_0^2\varepsilon}},$$

where we have used the facts that $Q(0) = Q_0$ and $\dot{Q}(0) = 0$. We have chosen the positive value, as the equivalent capacitance of the system increases such that the capacitors can store more charge for a given total potential difference. Separating variable and integrating,

$$\int_{Q_0}^{Q} \frac{1}{\sqrt{Q - Q_0}Q^2}\,dQ = \int_0^t \sqrt{\frac{2}{mA^2\varepsilon_0^2\varepsilon}}\,dt.$$

The integral on the left can be evaluated via the following procedure — we shall leave out the limits of integration lest the expressions get too cluttered.

First, we use the substitution $x = Q - Q_0$ and $dx = dQ$. Then,

$$\int \frac{1}{\sqrt{Q - Q_0}Q^2} dQ = \int \frac{1}{\sqrt{x}(x + Q_0)^2} dx.$$

Next, use the substitution $y = \sqrt{x}$ such that $x = y^2$ and $dx = 2y \, dy$. Then,

$$\int \frac{1}{\sqrt{x}(x + Q_0)^2} dx = \int \frac{2}{(y^2 + Q_0)^2} dy$$

which is a standard integral that can be solved by substituting $y = \sqrt{Q_0} \tan \theta$ for some variable θ. Overall, the integral evaluates to

$$\int_{Q_0}^{Q} \frac{1}{\sqrt{Q - Q_0}Q^2} dQ = \frac{1}{\sqrt{Q_0^3}} \left(\frac{\sin\left(2 \tan^{-1} \sqrt{\frac{Q}{Q_0} - 1}\right)}{2} + \tan^{-1} \sqrt{\frac{Q}{Q_0} - 1} \right),$$

which can also be expressed as

$$\int_{Q_0}^{Q} \frac{1}{\sqrt{Q - Q_0}Q^2} dQ = \frac{1}{\sqrt{Q_0^3}} \left(\sqrt{\frac{Q_0}{Q} - \frac{Q_0^2}{Q^2}} + \tan^{-1} \sqrt{\frac{Q}{Q_0} - 1} \right).$$

Thus,

$$t = \sqrt{\frac{mA^2 \varepsilon_0^2 \varepsilon}{2Q_0^3}} \left(\sqrt{\frac{Q_0}{Q} - \frac{Q_0^2}{Q^2}} + \tan^{-1} \sqrt{\frac{Q}{Q_0} - 1} \right).$$

This expression for t in terms of Q is only valid until the plates of the top capacitor converge (it breaks down at the first assumption that the charges stored by the two capacitors are equal as the segment connecting their left plates is no longer electrically isolated). After this juncture, the top capacitor essentially becomes an ideal wire — causing it to store zero charge and the bottom capacitor to store a constant εC_1 charge.

15. Current in Parallel RLC Circuit*

The power dissipated by the resistor is

$$P = \frac{V^2}{R} = \frac{\varepsilon_0^2}{R} \cos^2 \omega t.$$

The energy stored in a capacitor with capacitance C across a potential difference V is

$$U_C = \frac{1}{2} CV^2.$$

The rate of change of the energy stored in a capacitor is thus

$$\frac{dU_C}{dt} = CV\frac{dV}{dt}.$$

Since $V = \varepsilon_0 \cos \omega t$,

$$\frac{dU_C}{dt} = -\varepsilon_0^2 \omega C \sin \omega t \cos \omega t.$$

Finally, the energy stored in an inductor carrying current I is

$$U_L = \frac{1}{2}LI^2.$$

The rate of change of energy stored is then

$$\frac{dU_L}{dt} = LI\frac{dI}{dt} = V_L I = \varepsilon_0 I \cos \omega t,$$

where $V_L = L\frac{dI}{dt} = \varepsilon_0 \cos \omega t$ is the voltage across the inductor. The current through the inductor in this case can be computed via the complex impedance method. Since the impedance of an inductor is $i\omega L$, the complex current through it is

$$\tilde{I} = \frac{\varepsilon_0 e^{i\omega t}}{i\omega L}.$$

The actual current through the inductor is the real part of this which is

$$I = \text{Re}(\tilde{I}) = \frac{\varepsilon_0}{\omega L} \sin \omega t.$$

The rate of change of energy stored in the capacitor is then

$$\frac{dU_L}{dt} = \frac{\varepsilon_0^2}{\omega L} \sin \omega t \cos \omega t.$$

The current through the AC source, I_{AC}, can be computed by equating the power delivered by the AC source ($\varepsilon_0 \cos \omega t I_{AC}$) with the rate of change of the other forms of energy. Thus,

$$I_{AC} = \frac{P + \frac{dU_C}{dt} + \frac{dU_L}{dt}}{\varepsilon_0 \cos \omega t} = \frac{\varepsilon_0}{R}\cos \omega t + \varepsilon_0\left(\frac{1}{\omega L} - \omega C\right)\sin \omega t.$$

16. Bridge*

The equivalent impedance of the $6L$ inductor and the capacitor is

$$6i\omega L - \frac{i}{\omega C} = 6i\omega L - 2i\omega L = 4i\omega L.$$

Notice that the ratio between this equivalent impedance and the impedance of the $2L$ inductor in the bottom row is $2 : 1$ — a value that is equal to that between the $2L$ inductor and the L inductor in the top row. Due to this equal ratio of impedances, the two ends of the resistor must be "equipotential points" and the resistor can effectively be removed. The total impedance of the set-up is then

$$Z_{eq} = \frac{3i\omega L \cdot 6i\omega L}{3i\omega L + 6i\omega L} = 2i\omega L.$$

The complex current through the AC source is consequently

$$\tilde{I} = \frac{\varepsilon_0}{2i\omega L} e^{i\omega t}.$$

The actual current is the real component of this.

$$I = \mathrm{Re}(\tilde{I}) = \frac{\varepsilon_0}{2\omega L} \sin \omega t.$$

17. Transformer Circuit**

Define I_1 and I_2 as the clockwise and anti-clockwise currents in the left and right loops respectively. Applying Kirchhoff's loop rule to the left loop in the clockwise direction,

$$\varepsilon_0 \cos \omega t - I_1 R - L_1 \frac{dI_1}{dt} - M \frac{dI_2}{dt} = 0.$$

Denoting the charge on the left plate of the capacitor as Q, we apply Kirchhoff's loop rule to the right loop in the anti-clockwise direction.

$$-L_2 \frac{dI_2}{dt} - M \frac{dI_1}{dt} - I_2 R - \frac{Q}{C} = 0.$$

Now, exploiting the linear nature of these equations, we consider the complex forms of the above.

$$\varepsilon_0 e^{i\omega t} - \tilde{I}_1 R - L_1 \frac{d\tilde{I}_1}{dt} - M \frac{d\tilde{I}_2}{dt} = 0$$

$$L_2 \frac{d\tilde{I}_2}{dt} + M \frac{d\tilde{I}_1}{dt} + \tilde{I}_2 R + \frac{\int \tilde{I}_2 dt}{C} = 0$$

where we have used the fact that $I_2 = \frac{dQ}{dt}$. We then guess exponential solutions for the complex currents.

$$\tilde{I}_1 = A_1 e^{i(\omega t + \phi_1)},$$

$$\tilde{I}_2 = A_2 e^{i(\omega t + \phi_2)}.$$

Substituting these expressions into the equations above,

$$\varepsilon_0 e^{i\omega t} - \tilde{I}_1 R - i\omega L_1 \tilde{I}_1 - i\omega M \tilde{I}_2 = 0$$

$$i\omega L_2 \tilde{I}_2 + i\omega M \tilde{I}_1 + \tilde{I}_2 R + \frac{\tilde{I}_2}{i\omega C} = 0,$$

where we have used the fact that the constant of integration in $\int \tilde{I}_2 dt$ must be zero, for the same reason in Section 10.2.3. Simplifying and substituting $C = \frac{1}{\omega^2 L_2}$,

$$(R + i\omega L_1)\tilde{I}_1 + i\omega M \tilde{I}_2 = \varepsilon_0 e^{i\omega t}$$

$$i\omega M \tilde{I}_1 + R\tilde{I}_2 = 0.$$

Solving these equations simultaneously,

$$\tilde{I}_1 = \frac{\varepsilon_0 R}{(R^2 + \omega^2 M^2) + i\omega R L_1} e^{i\omega t},$$

$$\tilde{I}_2 = -\frac{\varepsilon_0 i\omega M}{(R^2 + \omega^2 M^2) + i\omega R L_1} e^{i\omega t}.$$

The actual currents are the real components of the above.

$$I_1 = \text{Re}(\tilde{I}_1) = \frac{\varepsilon_0 R}{\sqrt{(R^2 + \omega^2 M^2)^2 + \omega^2 R^2 L_1^2}} \cos(\omega t - \phi),$$

$$I_2 = \text{Re}(\tilde{I}_2) = \frac{\varepsilon_0 \omega M}{\sqrt{(R^2 + \omega^2 M^2)^2 + \omega^2 R^2 L_1^2}} \sin(\phi - \omega t),$$

where $\phi = \tan^{-1} \frac{\omega R L_1}{R^2 - \omega^2 M^2}$. The role of the capacitor is to nullify the self-inductance L_2 in this case. I_2 leads I_1 by $\frac{\pi}{2}$-phase since $\cos(\omega t - \phi + \frac{\pi}{2}) = -\cos(\frac{\pi}{2} - \omega t + \phi) = -\sin(\omega t - \phi) = \sin(\phi - \omega t)$.

18. Mutual Inductors**

We first consider real variables. Let V be the common voltage across the inductors and propose currents I_1 and I_2 to flow through the respective inductors rightwards. The voltage across each inductor is caused by its self-inductance and the mutual inductance due to the change in current through the other inductor. Thus,

$$V = -L_1 \frac{dI_1}{dt} - M \frac{dI_2}{dt},$$

$$V = -L_2 \frac{dI_2}{dt} - M \frac{dI_1}{dt}.$$

Observe that these expressions are still linear in the derivatives of I_1 and I_2. Therefore, the particular solution for the complex variables $\tilde{I}_1$ and $\tilde{I}_2$ should still be exponential — implying that the method of complex impedance should still work. Now, replace V, I_1 and I_2 with their complex counterparts $\tilde{V}$, $\tilde{I}_1$ and $\tilde{I}_2$. Then,

$$\tilde{V} = -L_1 \frac{d\tilde{I}_1}{dt} - M \frac{d\tilde{I}_2}{dt},$$

$$\tilde{V} = -L_2 \frac{d\tilde{I}_2}{dt} - M \frac{d\tilde{I}_1}{dt}.$$

Rearranging and eliminating $\frac{d\tilde{I}_2}{dt}$,

$$\tilde{V} = -\frac{L_1 L_2 + M^2}{L_2 - M} \frac{d\tilde{I}_1}{dt} = -i\omega \frac{L_1 L_2 + M^2}{L_2 - M} \tilde{I}_1$$

as $\tilde{I}_1$ should be exponential with frequency ω. Therefore, the equivalent impedance of the first inductor is $Z_1 = i\omega \frac{L_1 L_2 + M^2}{L_2 - M}$. Similarly, the impedance of the second inductor is $Z_2 = i\omega \frac{L_1 L_2 + M^2}{L_1 - M}$. The equivalent impedance of these two inductors in parallel is given by $Z_{eq} = \frac{Z_1 Z_2}{Z_1 + Z_2}$.

$$Z_{eq} = i\omega \frac{\frac{(L_1 L_2 + M^2)^2}{(L_1 - M)(L_2 - M)}}{\frac{L_1 L_2 + M^2}{L_1 - M} + \frac{L_1 L_2 + M^2}{L_2 - M}} = i\omega \frac{L_1 L_2 + M^2}{L_1 + L_2 - 2M}.$$

We can check that this expression is consistent with the equivalent inductance that we have calculated in Section 10.1.5. The total impedance of the circuit is

$$R + i\omega \frac{L_1 L_2 + M^2}{L_1 + L_2 - 2M}.$$

Therefore, the complex current flowing through the AC source is

$$\tilde{I} = \frac{\varepsilon_0 e^{i\omega t}}{R + i\omega \frac{L_1 L_2 + M^2}{L_1 + L_2 - 2M}}.$$

The current through the top inductor is given by the current divider principle (refer to DC Circuits).

$$\tilde{I}_1 = \frac{Z_2}{Z_1 + Z_2} \cdot \tilde{I}$$

$$= \frac{\frac{1}{L_1 - M}}{\frac{1}{L_2 - M} + \frac{1}{L_1 - M}} \cdot \frac{\varepsilon_0 e^{i\omega t}}{R + i\omega \frac{L_1 L_2 + M^2}{L_1 + L_2 - 2M}}$$

$$= \frac{L_2 - M}{L_1 + L_2 - 2M} \cdot \frac{\varepsilon_0 e^{i\omega t}}{R + i\omega \frac{L_1 L_2 + M^2}{L_1 + L_2 - 2M}}$$

$$= \frac{(L_2 - M)\varepsilon_0}{\sqrt{R^2(L_1 + L_2 - 2M)^2 + \omega^2(L_1 L_2 + M^2)^2}} e^{i(\omega t - \phi)}$$

where $\phi = \tan^{-1} \frac{\omega(L_1 L_2 + M^2)}{R(L_1 + L_2 - 2M)}$. The actual current is the real part of the above.

$$I_1 = \text{Re}(\tilde{I}_1) = \frac{(L_2 - M)\varepsilon_0}{\sqrt{(L_1 + L_2 - 2M)^2 + \omega^2(L_1 L_2 + M^2)^2}} \cos(\omega t - \phi).$$

Chapter 11

Relativistic Kinematics

This chapter will study relativistic kinematics from the two fundamental postulates of special relativity. Special relativity is one of the more exciting and popular topics due to its profound consequences, many of which are contrary to common sense. Many apparent paradoxes will arise but one should note that special theory is a perfectly sound and coherent theory. Most of the time, these situations are not paradoxical at all and are contradictory purely because we made them to be so. Hopefully, these puzzles will be conducive to our understanding of the theory and help us to acclimatize to the strange phenomena in relativity. It may be helpful to dispel ourselves of our "common sense" in approaching this topic and accept the concepts on a clean slate — given that many effects feel extremely counter-intuitive.

There is a ubiquitous misconception that special relativity is incapable of analyzing accelerating objects or accelerating frames of reference. The former had better be false as any kinematic theory would be utterly useless if it could not describe acceleration. In fact, accelerating objects are relatively easy to handle as their motions can still be quantified in an inertial frame. Accelerating frames are much harder but can still be dealt with, in a manner similar to classical mechanics in a non-inertial frame (notice that Newton's laws are only valid in inertial frames), though it will not be elaborated in this chapter.

Finally, you will notice that most special relativity problems do not involve gravity. Well, it turns out that special relativity was not the most accurate theory for systems with gravity — general relativity is. This is to be expected as special relativity was not designed as a theory of gravitation in the first place! In fact, Einstein's special relativity was partly inspired by electromagnetism, as evidenced by the title of his famous 1905 paper: *On the Electrodynamics of Moving Bodies*. Nevertheless, the idea of objects on

Earth experiencing a uniform, constant downwards force remains a decent approximation for our purposes.

11.1   Frames of Reference

A frame of reference is an important concept in relativity and physics in general. A frame of reference sets a standardized state of motion such that physical quantities, such as displacement and velocity can be measured relative to that frame. It is pivotal in ascribing meaning to a measurement as physical measurements are relative. For example, you may observe a car to be traveling at a certain velocity towards you when you are stationary with respect to the ground. However, if you run towards the car, you will then observe that it moves at a greater velocity with respect to yourself. Evidently, there is little meaning in proclaiming that the velocity of an object is a certain value without explicitly mentioning the frame of reference in which it was measured.

Events are of particular concern in physics and we quantify them with respect to certain frames of reference. Similar to how organizing real-life meetings requires a venue and a time, events have spatial and temporal coordinates in a certain frame of reference. However, there is a distinction between a frame of reference and a coordinate system.

Formally, the frame of an observer is a set of infinite virtual or tangible points that move rigidly with the observer such that they are perpetually at rest simultaneously in the frame of the observer. There is no relative motion (i.e. their separations do not vary) between individual particles or between a particle and the observer in the frame of the observer. These particles set a standardized state of motion at every point such that a physical quantity at a point in space can be measured with respect to a particle at that same point in space. Furthermore, there exists a universal time for all the particles in the frame such that the time of an event at a point in space in a certain frame can be defined to be that recorded by a particle of that frame at that same point in space.

A coordinate system, on the other hand, is merely a construct used to quantify measurements in a frame. A frame can have infinitely many possible coordinate systems. In that sense, a coordinate system is merely a mathematical language used to describe observations in a frame. Consider a vector in a frame, assuming that a Cartesian coordinate system is chosen, there can be many different values for the $\hat{i}$, $\hat{j}$, $\hat{k}$ components of the vector due to various possible orientations of the coordinate axes. However, these all describe the same unique vector.

To define an event, a coordinate system must have spatial axes, which are usually Cartesian in special relativity, and a temporal axis. To visualize a coordinate system, we can imagine three infinite rows of meter sticks, extending from an observer who is usually defined as the origin of the coordinate frame, and a clock held by every particle that is perennially at rest in the frame of the observer. These clocks are synchronized in the frame of the observer; the possible methods of synchronization will be elaborated later.

The spatial coordinate of an event along a certain coordinate axes is then quantified by the number of meter sticks between the origin and the location of that event along the infinite row of meter sticks extending in that direction. The temporal coordinate of an event is then the reading of a clock at the spatial location of the event.

11.2 The Two Postulates

11.2.1 *The Principle of Relativity*

The first postulate in special relativity states that

- All inertial frames of reference are equivalent. That is, the laws of physics hold as well in one inertial frame as in any other inertial frame.

First and foremost, we have to understand the meaning of the term "relativity". The principle of relativity is a creed that physicists believe in — we trust that the laws of nature are symmetric and elegant on the fundamental level. The principle of relativity is an intuitive axiom that ordains all laws of physics to exist in similar forms to observers in certain frames of reference. If this were not the case, physical laws would have severely limited utility and predictive power.

The notion of relativity extends way back to the times of Galileo and Newton. Galileo identified an extremely important class of frames of reference, known as inertial frames, in which the laws of motion are observed to be the same.[1] Formally, an inertial frame is a frame of reference, whose geometry is Euclidean, in which all laws of physics appear in their simplest forms (i.e. no fictitious forces). Free particles, which are not subjected to net forces, undergo rectilinear motion at a constant velocity or remain stationary in an inertial frame. Furthermore, any frame that moves rectilinearly at a constant velocity relative to an inertial frame is also an inertial frame. In

---

[1]Note that there is a distinction here between the laws of motion (Newton's laws) and all laws of physics. Galilean relativity was proposed to only apply for mechanical laws.

his development of classical mechanics, Newton hypothesized the existence of an absolute space and that the distant stars were stationary relative to the frame of absolute space. Thus, by considering the frame of fixed stars, all other inertial frames can be defined. However in the context of special relativity, the notion of an absolute space seems superfluous and is thus dismissed. After all, why should the frame of fixed stars be so special? In either case, inertial frames are a class of infinite frames of reference that travel at a constant velocity with respect to each other. In order for the laws of physics to hold in their simplest forms in all space, inertial frames must be non-accelerating.

One of the defining consequences of the principle of relativity is the relativity of velocity. Because of the uniformity of the physical laws across all inertial frames, it is impossible for an observer to identify the exact inertial frame that he is in. A common depiction of this usually goes as follows. Given your adventurous and playful nature, you sneak into a train that is initially stationary with respect to the ground. You decide to settle down in your new "camp" and thus, carry on with your daily activities. You rinse your mouth with a cup of water when you wake up, read this book under a candle light and play billiards. However, on one night, the train departs while you are sleeping and then travels at a uniform velocity relative to the ground. Will you be able to conclude that you are on a train, that is moving with respect to the ground, the next day, solely by conducting experiments inside the train? Assume that the windows are clamped shut so that you are unable to peek outside the train.

From your perspective, nothing has changed. If you hit a billiard ball under the exact same conditions as those on the previous day, the exact same results will be observed. If you toss a basketball vertically upwards, it will still land at the same spot on the floor from which it was thrown. It is impossible for you to conclude that the train you are on is moving with respect to the ground — this is the crux of the principle of relativity. It guarantees that traveling at a constant velocity with respect to the ground leaves no impact on the world around you.

Since we are unable to distinguish between inertial frames due to the principle of relativity, we are unable to isolate a truly stationary inertial frame, if it even exists. Velocity then becomes an inherently relative concept as we are unable to say whether something is "moving", we can only conclude that something is moving with respect to something else — this is the relativity of velocity. When describing velocity, it is always paramount to mention what the velocity was measured with respect to.

Though velocity is relative, acceleration still remains absolute as the laws of physics are no longer presumed to be uniform across accelerating frames. Returning to the previous thought experiment, if the train speeds up or slows down abruptly, you will definitely be able to tell that a change in the train's velocity has occurred. The surface of the water in your cup tilts, the candle flame slants and you slam into your seat due to a fictitious force. Because of the mutability of the physical laws across accelerating frames, an observer is able to determine whether he or she is accelerating and even quantify the acceleration.

Now, you may oppose the absoluteness of acceleration by posing the following problem: if you, an observer, measure a particle to have a certain acceleration in your frame, how can you tell it is you who accelerates and not the particle or a combination of both? The answer is that you can observe a free particle (i.e. another particle). If it has an acceleration in your frame, you know that you are accelerating (as a free particle should travel at a constant velocity when you are not accelerating). Furthermore, the magnitude of your acceleration will also be reflected by that of the free particle. Afterwards, you can determine the absolute acceleration of the first particle by taking into account your own acceleration and its acceleration in your frame. For an intuitive argument, let us return to the train analogy again. This time, you observe another train to have a certain acceleration with respect to your train. However, you can tell that you are accelerating while the other train is not, as you are the one hitting your head against your seat and feeling nauseated while a person on the other train is perfectly fine. In other words, the change in the physical laws is unique to your frame and helps you to determine your acceleration.

11.2.2 *Invariance of the Speed of Light*

The second postulate in special relativity asserts that

• The speed of light in vacuum is the same in all inertial frames of reference.

This second postulate is not at all obvious and is extremely counter-intuitive. From our everyday experiences, if a train is traveling towards us while we are traveling on a car at a constant velocity directed at the train afore, our observed speed of the train is faster than its speed measured by a stationary observer on the road per se. However, in the case of light, its observed speed will be the same with respect to any observer moving at a constant velocity with respect to the ground! This seems extremely surreal but at the same time, slightly plausible, considering the fact that we are used to dealing with speeds much smaller than the speed of light.

Now, where does this bizarre claim stem from? The revolutionary Michelson–Morley experiment[2] led to the widely-accepted conclusion that light does not require a medium to propagate in. To illustrate why this conclusion leads to the second postulate, consider the following argument. An inertial observer A, who is stationary relative to the ground, observes the speed of light in vacuum to be c in his own frame. Then, another inertial observer B that is traveling at a velocity v with respect to the first observer on a train must observe the speed of light in vacuum to be c in his own frame too. If the speed of light were to depend on the inertial frame of reference (e.g. via the Galilean transformations), observer B will be able to conclude that he is on a moving train with respect to the ground, without looking outside the window, by conducting experiments with light! This violates the principle of relativity which is a sacrosanct pillar in physics. Therefore, the speed of light must be invariant across all inertial frames.

This reasoning does not apply to sound waves as they propagate in compressible media such as air or water. Sound waves travel at 343m/s in air. When we run towards sound waves, we observe the sound waves to travel at a greater velocity, as the air that is carrying the sound is now moving with respect to us. However, sound still travels at 343m/s relative to the frame of air. Therefore, even though observer B observes sound waves to travel in air at a speed that is different from 343m/s, he is unable to conclude that his train is moving relative to the ground from this relative speed of sound as the conditions of his experiments are different (the air is now moving in his frame, which is contrary to the still air that was observed in the ground frame). If the air were stationary in his frame, because it is dragged along by the train per se, the observer will still observe sound to travel at 343m/s. On the other hand, in the case of light, there is no such medium of propagation. Hence, the conditions for a light experiment are the same in two inertial frames moving relative to each other — leading to the conclusion that light must be observed to travel at the same speed c in both frames due to the principle of relativity.

Actually, any disturbance, that does not require a medium to propagate in, will possess a speed that is invariant across inertial frames. It just happens that light undertakes this role in our universe. Finally, light also has other special properties. Due to the logical consequences of these

---

[2]See Appendix A.

two postulates, the speed of light in vacuum c is also established as the theoretically maximum possible speed of information and physical objects. If this were not the case, situations that are contrary to common human experiences will arise as a corollary of the postulates. This will be elaborated in a later section.

11.2.3 *Underlying Assumptions*

Besides the principle of relativity, there are deeper, underlying assumptions about the properties of space and time in an inertial frame. Firstly, it is presumed that an inertial frame is both spatially and temporally homogeneous. That is, an experiment conducted at a certain point in space and time will produce the exact same results at that performed at another point in space and time, ceteris paribus (with all other conditions held constant). Fundamentally, this is the epistemological basis of physics which stems from inductive reasoning.

Imagine a scenario where we toss a ball into the air and observe it to fall to the ground. If we repeat this experiment multiple times on different occasions, with all other conditions held the same, and still observe that the ball falls, we might surmise that the ball will fall to the ground at all instances in time, ceteris paribus. However, there is actually no guarantee that the ball will actually do so — this is a striking and inherent flaw in inductive reasoning. Observing a certain phenomenon to occur at a certain instance, given certain conditions, does not ordain the same phenomenon to occur at the next instance, ceteris paribus. The ball could possibly accelerate into space and crash into the Moon the next time we throw it, for all we know. However, we believe in the validity of inductive reasoning — that the ball still falls to the ground when thrown at the next instance — when backed by a reasonable amount of empirical evidence. In that sense, scientists are hardly free from bias as they possess an intrinsic predilection towards elegant and general theories that describe the world around them. If the same results were not obtained from experiments performed at different times, with all other conditions held constant, physical laws would be useless as they would have to be constantly modified. A similar statement can be made about the properties of space. Therefore, the homogeneity of space and time is a fundamental assumption of physics.

Due to the presumed homogeneity of space and time in an inertial frame, spatial and temporal translations of the coordinate axes of a frame of reference or the observer in that frame do not change the observed results of an experiment.

Next, a frame of reference is isotropic in space and time. Experiments that are conducted at rotationally symmetric spatial locations will produce identical observed results, ceteris paribus. Similarly, experiments that are conducted n seconds in the future will produce the same observed results as those n seconds in the past. In other words, all spatial and temporal directions are equal — there is no preferred direction in space and time. Concomitantly, a rotation of the spatial coordinate axes of a frame of reference will not affect the observed results of an experiment. Actually, homogeneity necessarily implies isotropy but the reverse is not true.

The assumptions afore have a direct impact on our study of relativity. Generally, a coordinate system in an inertial frame may undergo a translational transformation, rotational transformation or a Lorentz boost — a process during which one changes from one inertial frame to another with a constant relative velocity, without any rotation of the coordinate axes. Due to the homogeneity and isotropy of space and time, only the last form of transformation is of primary concern in this chapter as the previous two can be performed trivially.

11.3   Consequences of the Postulates

In this section, we will "start physics anew" and deduce the consequences of the postulates on the nature of space and time. Notice that the notion of a universal time, that is invariant across all inertial frames, is not presumed as part of the theory. Therefore, it is beneficial for us to dispel ourselves of such preconceptions about time in approaching this section. As a last precursor, observe that half of our postulates talks about light. As such, light will be a fundamental part of our thought experiments as it is the only entity whose nature we are sure of, as of now. In this sense, we are about to be enlightened by light.

11.3.1   *Conventions*

Several conventions regarding the definitions of coordinate systems in inertial frames will be adopted in the following sections. Generally, there are three spatial coordinates, which are Cartesian, and one time coordinate that is of interest. Furthermore, we are often concerned about how coordinates in one inertial frame transform to those in another inertial frame. Thus, we use primed coordinates to denote the coordinates of a primed inertial frame. Usually, we will have two inertial frames, S and S', that are moving with respect to each other and have coordinate axes in x, y, z, t and x', y', z', t' respectively. The axes in S' are defined to be parallel to the corresponding axes in

S by default. Furthermore, the x and x'-axes are oriented such that S' travels at a velocity v purely along the x-direction, as observed in frame S. Moreover, the origins of the two coordinate axes are assumed to coincide (i.e. $x' = x = 0$, $y' = y = 0$, $z' = z = 0$) when $t' = t = 0$ unless otherwise stated. A pair of coordinate systems that obeys these guidelines will be known as the standard configuration. In the analysis of the fundamental effects of the postulates, we will be referring to the frames of observers instead of S and S' so that they can be better associated with the physical situation. Despite this, these observers' frames follow similar conventions to those stated above.

Lastly, as the coordinates along the y and z-directions are unchanged across different inertial frames and hence uninteresting (as we shall discover), we will primarily be concerned with the x and x'-coordinates and neglect the other spatial coordinates. Therefore, our analysis is essentially reduced to a one-dimensional problem in spatial terms but can also be easily be extended to the three-dimensional case.

11.3.2 Time

Before we analyze a concrete application of the postulates, let us first understand how the time difference between two events that occur in different positions in space can be measured in a particular frame. The occurrence of an event is defined by its position and its time with respect to an inertial frame. Note that absolute time does not exist and we really mean the time elapsed between the occurrence of a certain event and that of another event which we use as a reference when we refer to the time of an event.

We can imagine placing miniature clocks at every point in space that are stationary with respect to the given inertial frame. Then, the clocks can be synchronized. There are various methods to accomplish this. For example, we can place a light source at the middle of two points in space. The light source emits a flash which simultaneously triggers the starts of the clocks at the two points in space when they receive the signal. Alas, this method only works for synchronizing two clocks. For a more general set-up, one method would be to first start many clocks simultaneously at the same point in space. Then, one can move the clocks ever so slowly towards their respective destinations. Finally, another famous method is due to Einstein. In order to synchronize two clocks, send a light signal from clock A when it reads t_A towards clock B. When clock B receives the light signal, its reading t_B is recorded and it reflects the light signal back towards clock A which receives it at t'_A. The observers at the locations of the clocks can then meet up to exchange their findings about t_A, t_B and t'_A. If $t_B = \frac{1}{2}(t_A + t'_A)$, they can conclude that the clocks are synchronized. Otherwise, the poor engineers then have to go back to tweaking the readings of the clocks until this condition is eventually fulfilled! This process can be repeated for all pairs of clocks in an inertial frame to synchronize them.

Now that we have synchronized clocks that are operating in all space in a certain frame (it doesn't matter how this is achieved as long as the clocks are synchronized), whenever an event transpires at a position in that frame, its time of occurrence in that frame can be defined to be the recorded reading of the clock (of that frame) at that particular point in space. The time when the clocks were started can be used as a temporal reference point in this case. Accordingly, the time difference between two events in a frame is simply the difference between the times recorded by the clocks of that frame at the corresponding positions.

When there are multiple inertial frames, an array of synchronized clocks can be defined for each frame in general. These arrays may or may not be identical. In fact, we will discover that they are vastly different in the following sections due to the ramifications of the postulates. Figure 11.1 depicts two arrays of clocks synchronized with respect to two inertial frames, S and S', with conventional definitions. Note that the clocks in frame S appear like the following with respect to frame S. An observer in frame S' may or may not observe clocks in frame S to be the same as that observed by an observer in frame S. The reverse is also true.

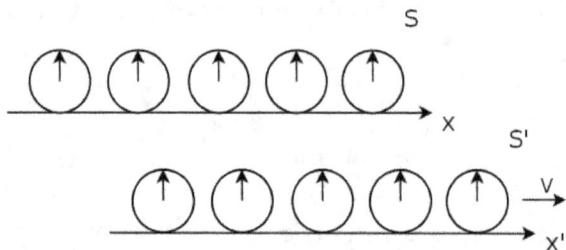

Figure 11.1:   Two arrays of clocks viewed in their own frames

Finally, a core aspect of special relativity is the grounding of time on a firmer observable foundation. Time is no longer an abstract concept that is independent of physical processes. It is necessarily measured by physical systems such as sandglasses and oscillating pendulums. Therefore, since an event requires a time of occurrence in order to be defined, the observation of the same event in different inertial frames can conversely be used to relate the times in different frames (e.g. in S and S'). This will be a key component of the following section. To this end, keep in mind that the time of an event, as recorded by a clock, must be independent of the frame that we observe the same clock[3] from (we need not only observe it from its rest frame). However, the process through which the reading on the clock undergoes in attaining the final coherent reading may differ across inertial frames.

11.3.3   *The Relativity of Simultaneity*

As a consequence of the postulates, events that are simultaneous in one inertial frame are not simultaneous in another! Then, clocks that are synchronized in one inertial frame of reference are not synchronized in another!

Consider the following situation: observer A sits on a train that is traveling at a speed v towards the right in observer B's frame. Similarly, A observes B to move towards the left at a speed v. Subsequently, A observes lightning to simultaneously strike the opposite ends of the train in his own frame. However, observer B will conclude that the lightning does not synchronously strike both ends of the train in his frame, as we shall see!

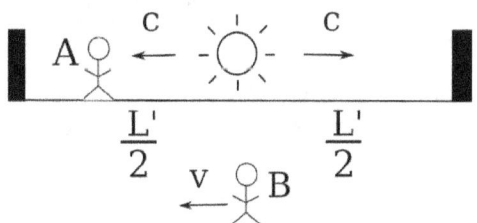

Figure 11.2:   Observer A's frame

Referring to Fig. 11.2, we can imagine placing a light source at the center of the train, that is stationary relative to the train. The light source then emits two photons towards the two ends of the train. We define the times of these emissions to be zero in both observers' frames. Then, we can define the times of occurrence of the lightning strikes in a particular frame to be those when the corresponding photons collide with the walls of the train in that frame. It does not matter if there isn't an actual light source in the set-up. What matters is that we could have placed one if we wanted to and used it as a temporal yardstick to "call lightning to strike upon a wall" when that wall receives a photon. Thus, the following analysis is valid regardless of whether an actual physical light source is used.

---

[3] Clocks of different frames can be observed to possess different readings for the same event as they are different clocks.

In the frame of A, the train is not moving and if we define the train to be of length L', the photons will strike the ends of the wall in time

$$t_A = \frac{L'}{2c}$$

simultaneously in A's frame (this is the reason why we placed the light source in the middle of the train). Now consider the frame of B in Fig. 11.3, both walls of the train move towards the right at speed v.

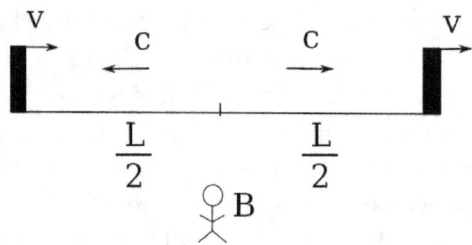

Figure 11.3:   Observer B's frame

As the speed of light remains at a constant value c in B's frame of reference, the times taken for the photons to reach the left and right ends of the train are respectively

$$t_L = \frac{L}{2(c+v)},$$

$$t_R = \frac{L}{2(c-v)},$$

where L is the length of the train in B's frame. L may or may not be equal to L' (we can't be certain right now as the postulates did not state so). We see that these two events are in fact not simultaneous with respect to B's frame of reference as long as $v \neq 0$. In fact,

$$\Delta t = t_R - t_L = \frac{L}{2(c-v)} - \frac{L}{2(c+v)} = \frac{Lv}{c^2 - v^2}.$$

It turns out that L is indeed different from L'. We shall just invoke the result from a later argument that

$$L = \frac{L'}{\gamma}$$

where

$$\gamma = \frac{1}{\sqrt{1 - \frac{v^2}{c^2}}}.$$

Thus,

$$\Delta t = \frac{L'v}{c^2} \cdot \frac{\sqrt{1 - \frac{v^2}{c^2}}}{1 - \frac{v^2}{c^2}} = \frac{\gamma L'v}{c^2}.$$

This is a solemn admonishment that events that we consider simultaneous in one inertial frame are not simultaneous in another. We should always take care in identifying the frame that is currently under consideration. It makes no sense to say that two events occur concurrently without explicitly mentioning the inertial frame of observation.

Now, let A's frame be S' and B's frame be S, under the standard configuration. Define the origins O' and O to coincide at $t' = t = 0$ at the instantaneous location of the left end of the train. Attach two clocks, synchronized in S', to the left and right ends of the train and consider two clocks, synchronized and stationary in S (i.e. these clocks do not move with the train but rest on the ground), that coincide with the instantaneous locations of the ends of the train at $t' = 0$. If the left and right clocks of S' (by clocks of a frame, we mean clocks synchronized in that frame) record $t'_L = t'_R = 0$ when lightning strikes the two ends of the train, we know from the previous scenario that the right clock of S will indicate a reading $t_R = \frac{\gamma L'v}{c^2}$ while the left clock of S will record $t_L = 0$ (as the left clocks of S and S' are synchronized under the standard configuration). From B's perspective, he would simply claim that the lightning struck the clocks of S at different junctures — resulting in the discrepancy in readings. However, A must also be able to explain the readings of the clocks synchronized in S (as the readings are physical events[4] that should be immutable across inertial frames) so it is interesting to consider his perspective. Since A observes the two lightning events to occur simultaneously, the clocks of S must have been observed by A to have differing readings in the first place! That is, A observes clocks that are synchronized in S to be asynchronous! Thus, A explains the loss of simultaneity of two concurrent events in S', as observed by B in S, as follows: since the clocks in B were asynchronous in the first place, they will naturally have a discrepancy in readings when lightning strikes them at the same instance in my frame!

---

[4]We can stop the clock once it is struck by lightning.

In retrospect, it is rather intuitive that the above argument leads to the conclusion that clocks synchronized in one frame are not synchronized when observed in another frame as our train set-up is reminiscent of a particular method of synchronizing two clocks (putting a light source at the center) of a single frame.

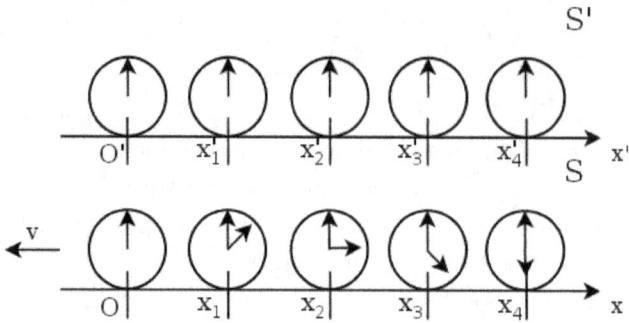

Figure 11.4:   Clocks in corresponding positions in S as viewed by observer A in S'

To illustrate what A observes at $t' = 0$, consider Fig. 11.4 which comprises an array of clocks synchronized in each of S' and S. The clocks in S' are stationary in S' while the clocks in S are moving towards the left at speed v. At $t' = 0$, the clocks in S' coincide with the corresponding clocks in S (i.e. x_1' at $t' = 0$ corresponds to a clock at x-coordinate[5] x_1 with respect to the x-axis in S). The clocks at the origins O and O' are synchronized such that $t = t' = 0$ there. When A observes the clocks of S' to read $t' = 0$, the clock of S that corresponds to the x'-coordinate x' reads $t = \frac{\gamma x' v}{c^2}$, such that the reading of the clocks of S increases towards the right as observed by A. A neat way of identifying the direction of increase is to remember that the rear clock is ahead (rear with respect to the velocity of the clocks). In summary, a moving array of clocks — synchronized in their common rest frame — is observed to possess a positive "gradient" in readings opposite to the direction of their velocities. Now, we have only established this result for $t' = 0$ (i.e. a certain juncture in S') and are unsure about other values of t'. It turns out, from the later section on time dilation, that the clocks of S tick at the same rate (as they are multiplied by the same dilation factor), as observed by A, so the "gradient" is maintained at all times t'.

---

[5]Actually, we can deduce that $x_1 = \gamma x_1'$ as x_1 is akin to the length of a train in its rest frame (L') while x_1' is the observed length of the moving train (L). Since we have used the result $L = \frac{L'}{\gamma}$, $x_1 = \gamma x_1'$ correspondingly.

As a word of caution for those who have had some exposure to time dilation, be wary that this result is not implying that the rear clock in frame S ticks at a faster rate than the front clock with respect to an observer in frame S'. They actually tick at the same rate as viewed by an observer in frame S' but the rear clock in frame S is simply a constant time ahead of the front clock, as observed by A in S', because the clocks are asynchronous as observed by A.

Simultaneous Events with Respect to Observer B

Now consider the previous situation again, except that this time, lightning strikes the ends of the train simultaneously in B's frame. How will the timings of the two lightning strikes differ in A's frame if the length of the train in A's frame is L'?

Similarly, imagine placing a light source which emits two photons in opposite directions on the train. We shall denote the time of a lightning strike at a wall in a frame to be that when a photon hits that wall. In order for the photons to strike the walls concurrently in B's frame, we know from the previous analysis that the light source must divide the train into sections of ratio $c + v : c - v$ in B's frame. This ratio must also hold in A's frame.[6] Thus, the set-up looks like Fig. 11.5 in A's frame.

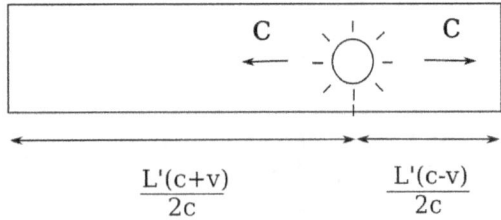

Figure 11.5: Light source in train with respect to A's frame

The time taken by the left photon to hit the left wall is longer than that required by the right photon to collide with the right wall in A's frame by

$$\Delta t = t_L - t_R = \frac{L'(c+v) - L'(c-v)}{2c^2} = \frac{L'v}{c^2}.$$

---

[6]One way to see this is that if the ratio of two relatively stationary objects is different across inertial frames, we would be able to tell if we switched between inertial frames — violating the principle of relativity.

Formally, if we define observer A's and B's frames to be S' and S respectively and their positive axes, x' and x, to be along the direction of the train's velocity in frame S, B will observe the reading of the rear clock in frame S' to lead that of the front clock by time $\frac{L'v}{c^2}$, where L' is the difference in the x'-coordinates of the two clocks in frame S'.

In order words, spatially separated events that are deemed by B to be simultaneous are events that differ by a time of $\frac{L'v}{c^2}$ in frame S'. Specifically, the spatially leading event must lag behind the spatially trailing event by $\frac{L'v}{c^2}$ in frame S' in order for them to be simultaneous in frame S.[7]

To visualize this from the perspective of B, assume that the clocks at the origins of S and S', O and O', are synchronized when they coincide such that $t = t' = 0$ at that juncture.

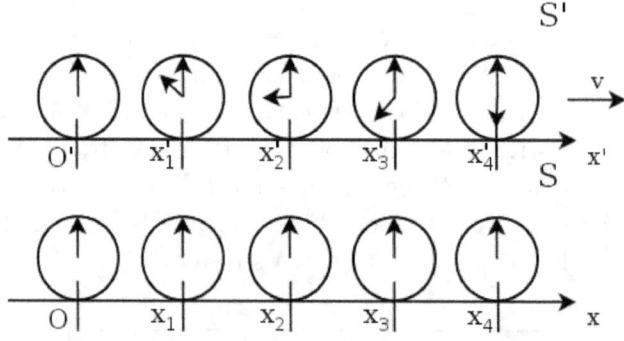

Figure 11.6:   Clocks in S' as viewed by observer B in S

At $t = 0$, if the x'-coordinate of a clock of S' is x', its reading will be $-\frac{x'v}{c^2}$ as observed by B. In Fig. 11.6, all clocks of S' — except that at O' — are displaying negative times as the reading of the clock at O' — which leads the other clocks — is zero. Finally, we comment on an aside for readers who are interested in the x-coordinate x that corresponds to x'-coordinate x' at $t = 0$. Since x' is akin to the length of a train in its own rest frame while x is the observed length of the moving train in another frame, we can deduce that $x = \frac{x'}{\gamma}$ from the equation $L = \frac{L'}{\gamma}$ that we have used earlier. It remains for the reader to check if the two time discrepancies for simultaneous events with respect to A and with respect to B are coherent.[8]

---

[7]Leading and trailing with respect to the direction of v (the velocity of frame S' with respect to frame S).

[8]Observe that the first situation becomes the second if we swap S and S' (technically, we need to reverse the direction of the velocity v but that only changes the direction of the

The Andromeda Paradox

The crux of the relativity of simultaneity is that observers in different inertial frames have different planes of simultaneity and hence observe different sets of present events. Roger Penrose proposed an argument that magnifies this effect to the extent of bizarreness.

Consider two twins that are situated at the same place on Earth, one walks in a direction towards the Andromeda galaxy while the other walks in the opposite direction. The twin that walks towards the Andromeda galaxy observes aliens traveling on spaceships en route to invade the Earth as the clock on the Andromeda galaxy is the rear clock in his frame. Thus, this twin observes events on the Andromeda galaxy to unfold much earlier than a stationary observer on the Earth. On the other hand, the other twin observes aliens convening a meeting to decide whether they should attack the meddlesome humans. This is because the Earth is now the rear clock relative to this twin. Thus, this twin will observe events at Andromeda that have already occurred in the frame of a stationary observer on Earth.

There is an apparent paradox here. How can there still be a hint of uncertainty of an alien invasion as observed by one twin while the other concludes that an imminent attack is inevitable? Before we resolve this apparent paradox, there is a clear distinction to be made between "seeing" and "observing" an event. Each twin "observes" an event on Andromeda that occurs concurrently with the present in their own inertial frame. However, he or she does not "see" that event yet as it takes time for information or photons to travel towards his or her location as the transmission of information cannot be faster than the speed of light in vacuum, c.

Well, there are usually two types of paradoxes in special relativity — those that result from fallacious reasoning and those whose consequences are so counter-intuitive that we reject them in disbelief. The situation above happens to fall into the latter category. They indeed make those observations without any contradiction. Thankfully or unfortunately, logical consistency is still maintained as it takes time for the information to reach the two twins. Suppose that the observed distance between the Earth and the Andromeda galaxy by the "prophetic" twin is L' and v is the relative velocity between him and Andromeda galaxy. The minimum time that it takes for information from the Andromeda galaxy to travel to him (assuming that information

---

"gradient" and not the magnitude). Therefore, if we switch the primed quantities in the first result $\frac{\gamma L' v}{c}$ into the unprimed quantities, we obtain $\frac{\gamma L v}{c^2} = \frac{L' v}{c^2}$ which is the second result (we have used $L = \frac{L'}{\gamma}$)

travels at the theoretically maximum speed of light) is

$$t = \frac{L'}{c} > \frac{L'v}{c^2}$$

which is greater that the $\frac{L'v}{c^2}$ "time lead into the future". Thus, the twin who "observes the future" is unable to change the fate of his planet as he is unable to "see the future" in time.

11.3.4 *Time Dilation*

The time interval between spatially coincident events in frame S', as measured by an observer in S, is larger than that as measured by an observer in S'. A direct ramification of this is that a clock that is moving with respect to an observer will be observed to run slower in that observer's frame. Consider the set-up in Fig. 11.7: observer A is in a train traveling at a speed v towards the right relative to observer B. From A's perspective, a stationary light source emits a vertical beam that is reflected normally by a mirror attached to the ceiling of the train.

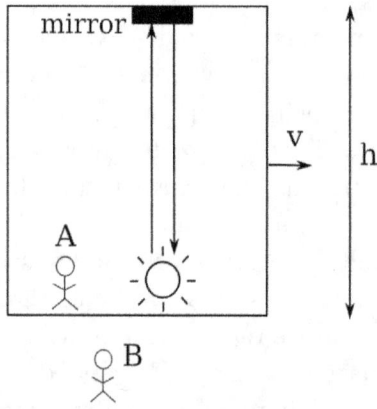

Figure 11.7:   Time dilation set-up

The time taken by the light during its roundabout trip in A's frame is simply

$$t_A = \frac{2h}{c}.$$

However, from B's perspective, the situation is shown in Fig. 11.8: the light has a component of velocity in the horizontal direction as the train is moving towards the right. However, the speed of light must still be maintained at c in B's frame.

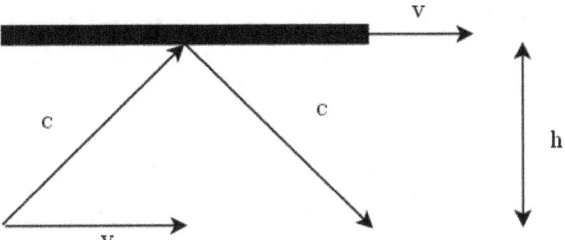

Figure 11.8: Situation in B's frame

The journey in B's frame takes

$$t_B = \frac{2h}{\sqrt{c^2 - v^2}}$$

where we have applied Pythagoras' theorem in calculating the vertical component of the velocity of light.[9] We realize that

$$t_B = \gamma t_A$$

where

$$\gamma = \frac{1}{\sqrt{1 - \frac{v^2}{c^2}}}.$$

This γ factor is ubiquitous in special relativity and thus deserves a special symbol on its own for simplicity. We see that gamma is always greater or equal to unity. Now, the above result means that B observes the time interval between two events that occur at the same spatial position in A's frame to be larger than that measured by A. Note that the only spatial position of concern is along the direction of the relative velocity between the two frames (as the result holds as long as light can traverse a straight path perpendicular to v in A's frame). In this case, it is the horizontal direction. The time dilation result still applies to the time difference between two events that are of the same horizontal position but different vertical positions in

---

[9]We have used h, which is the height of the train in A's frame, as the height of the train in B's frame without any justification here. This is because length contraction does not occur in the transverse direction. Refer to the section on length contraction for further elaboration.

A's frame (e.g. the time elapsed between the release of the beam and its incidence on the mirror).

It is paramount for the events under consideration to be at the same spatial coordinate of concern in A's frame in order for the time dilation equation above to be valid. If the two events are not at the same spatial coordinate in A's frame, there needs to be an additional correction for the loss of simultaneity of the clocks, synchronized in A's frame, as observed by B. Thus, the time dilation equation cannot be directly applied in this case.

Now, what does time dilation imply for the operation of clocks? The release and the receiving events of the light beam are analogous to the beginning and end of a clock-tick on A's "light clock." Then, B observes A's clock to tick at a slower rate than that observed by A, as the time interval between successive ticks is longer. As always, keep in mind that this statement is independent of whether a physical "light clock" is actually used. The point is that we could have used it to measure time if we wanted to.

Now, one may ask if time dilation actually happens in B's frame or is simply perceived to happen. The answer is that time dilation actually occurs in B's frame. If $\gamma = \frac{3}{2}$ and A's heart beats every 2 seconds as observed by himself (these pulsating events occur at the same location in A's frame), B will observe A's heart to beat every 3 seconds. Equivalently, it means that from B's perspective, the transition between two events that are spatially coincident in A's frame plays in slow motion. Therefore, B will actually observe A to age slower than he does as biological processes, too, slow down.

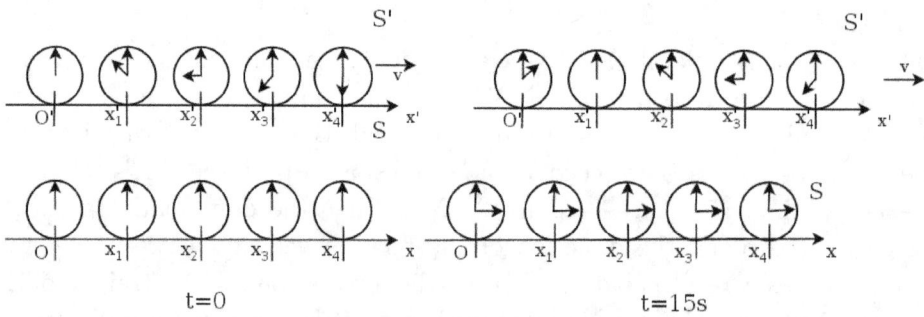

t=0   t=15s

Figure 11.9:   Clocks after 15 seconds have passed in frame S

To visualize time dilation, consider the set-up in Fig. 11.9 with $\gamma = 2$. At time $t = 0$ in frame S, the one-minute clocks in frame S' are asynchronous with respect to an observer in S due to the loss of simultaneity. Note that the four clocks on the right of O' all measure a negative reading as the rear clock — which is the clock at O' in frame S' in this case — leads the front clocks. However, the clocks at the origin are synchronized such that $t = t' = 0$ when the origins O and O' coincide. After 15 seconds have passed in frame S such that $t = 15$s, only a time interval $\Delta t' = 7.5$s has passed in frame S' as observed in S. Thus the readings on the clocks[10] in frame S' only increase by 7.5s, as tracked by an observer in S.

Apparent Paradox: Now you might argue that from A's frame, B's clock also seems to run slower. There is an apparent paradox, as we seem to be asserting that A's clock ticks at a slower rate than B's but also that B's clock ticks at a slower rate than A's. How is this possible?

The above is indeed true — as long as we define the inertial frame we are considering. In A's frame, B's clock runs slower while in B's frame, A's clock runs slower. There is no contradiction here as these are different events. One is the tick of B's clock and the other is the tick of A's. Let us consider the clock-tick of A's clock as an example. In A's frame, the start and end of the clock-tick trivially occur at the same position. Thus, we can say that $t_B = \gamma t_A$ where t_A and t_B are the times between consecutive ticks of A's clock in the frames of A and B respectively. However, the clock-ticks of A obviously do not happen at the same position in B's frame. Thus, we cannot conclude that $t_A = \gamma t_B$. We can only say so if t_A and t_B refer to the times between the ticks of B's clock in A and B's frames respectively. To illustrate this, we refer to the previous diagram. Suppose that we now consider the frame of S' such that the clocks of S are traveling towards the left at speed v. If $\gamma = 2$ and $\Delta t' = 15$s passes in S', an observer in S' will only observe the readings of the clocks of S to increase by 7.5s too (note that you need to account for the loss of simultaneity if you want to talk about the exact readings).

11.3.5 *Length Contraction*

The final piece of the puzzle concerns how moving objects are observed to be shortened longitudinally, parallel to their direction of motion. Consider the situation in Fig. 11.10. There are two twins, A and B, that are on the Earth. Twin A rapidly travels to the Moon at a speed v relative to twin B who remains on the Earth. The Moon is a distance L from the Earth as observed by twin B.

---

[10]The clocks still remain at the same x'-coordinates in S' as they are stationary in S'.

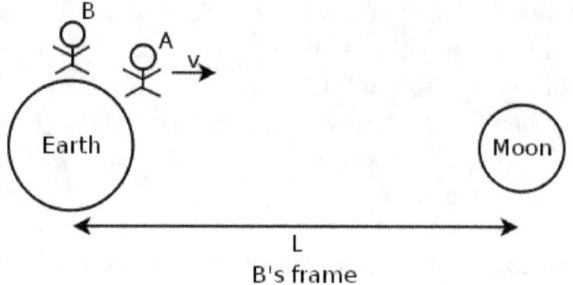

L
B's frame

Figure 11.10:   Twin A traveling to the Moon in B's frame

The time taken for twin A to reach the Moon in B's inertial frame is simply

$$t_B = \frac{L}{v}.$$

Now, notice that the starting and ending events in B's frame occur at different spatial locations. Therefore, t_B is really calculated by taking the difference in the readings of two synchronized clocks, one in B's hands that measures A's departure and one on the Moon that measures A's arrival, in B's frame. Moving on, we know from the previous section that moving clocks run slower. Thus, during this period of time, a clock held by twin A measures a reading of

$$t_A = \frac{t_B}{\gamma} = \frac{L}{\gamma v}.$$

In A's inertial frame, the situation is depicted in Fig. 11.11: the Moon is now traveling towards A.

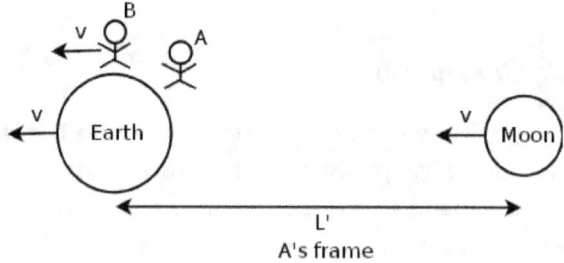

L'
A's frame

Figure 11.11:   Twin A's frame

Since A's clock records a reading of $\frac{L}{\gamma v}$ after the entire process and the Moon travels at speed v, the distance between the Earth and the Moon as observed by twin A is

$$L' = vt_A = \frac{L}{\gamma}.$$

Thus, the distance between the Earth and the Moon must have shrunk by a factor of $\frac{1}{\gamma}$ in A's frame. The observed distance between the Earth and the Moon is analogous to the observed length of an object with its ends defined to be the Earth and the Moon. Generally, if an object has a length L in its own inertial frame (rest frame), a stationary observer in another frame will observe that same object to have length $\frac{L}{\gamma}$ if the object travels at a longitudinal speed v relative to this new frame. Furthermore, length contraction is independent of the position on an object — all parts of the object are shortened by the same proportion (see Footnote 6). On another note, since the longitudinal length of an object is dependent on the inertial frame of reference, the proper length of an object is defined to be the length measured in its own rest frame (i.e. the object is stationary in that inertial frame).

Length contraction does not occur in the transverse direction. This can be proven by a simple argument that relies on the fact that physical consequences must be coherent across inertial frames, though measurements may differ. Consider a truck of proper height L traveling at a speed v into a tunnel of identical proper height L. If length contraction occurred in the transverse direction, the truck will observe the tunnel to be shortened in its inertial frame. Then in the truck's frame, the truck will crash into the tunnel. However, in the inertial frame of the tunnel, the truck is shortened and the truck passes scot-free. Evidently, there is a contradiction here. A similar argument can be used to prove that the transverse length of a moving object does not increase either. Thus, the transverse length of an object must be identical across different inertial frames.

What does Measuring Length Really Mean?

To establish a rigorous meaning for measuring length, let us return to the case of measuring the length of an object in a high school physics laboratory. We take a ruler[11] and record the readings of the ends of the object of concern via the markings on the ruler. Then, the length of the object can be obtained by taking the difference of these two readings. Now, an important qualification needs to be made here. The two readings need to be made at the same

---

[11] Recall that our coordinate system consists of meter sticks!

time. If our object were to move at a certain velocity relative to us (which is certainly possible), it makes no sense to jot down the coordinates of its ends at different times. In light of this discussion, the distance between two events, as observed in a frame S, is the difference in their spatial coordinates when they are observed simultaneously. This understanding is crucial in analyzing many situations.

For example, an intriguing question to ask is how A convinces himself that if his observed distance between the Earth and the Moon is L', the distance observed by B must be $L = \gamma L'$ from the model of clocks and meter sticks. Denote A's frame as S' and B's frame as S. Suppose that we attach two clocks synchronized in S to the Earth and Moon.

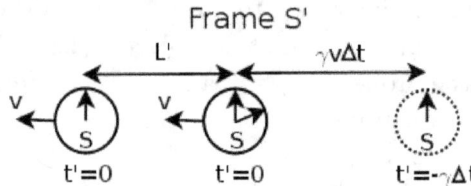

Figure 11.12:   Clocks synchronized in S, as observed in S'

Referring to Fig. 11.12, if A measures the distance between the Earth and the Moon to be L' at time $t' = 0$ and if the clock of S on the Earth reads $t = 0$ at this juncture, the rear clock (on the Moon) must have a reading $t = \Delta t = \frac{Lv}{c^2}$, as spatially separated events that are simultaneous in S' are those that are a time interval $\frac{Lv}{c^2}$ apart in S (where L is the spatial separation between the events in S). Now, A knows that B must measure the distance between these clocks at the same time t in S, such as when both clocks display $t = 0$. Therefore, A can retain the position of the left clock (which already displays $t = 0$ at $t' = 0$), while considering the position of the right clock at $t' = -\gamma \Delta t = \frac{\gamma Lv}{c^2}$ (dotted in Fig. 11.12) such that the right clock reads $t = 0$. Note that we have multiplied by a factor of γ as the ticking of the clock of S slows down by a factor of γ due to time dilation. Observer A measures the spatial separation between the left and right clocks at these specific junctures to be

$$x' = L' + \frac{\gamma L v^2}{c^2},$$

so he can reason that the distance between these events (which are now simultaneous in S and hence reflect the distance between the Earth and

Moon) will be observed by B to be

$$L = \frac{x'}{\gamma} = \frac{L'}{\gamma} + \frac{Lv^2}{c^2},$$

as A understands that meter sticks (which form the spatial coordinate axes of S') attached to his frame are shrunk by a factor of γ (because they are moving in frame S) such that the separation in x'-coordinates of two events simultaneous in S is amplified by a factor of γ as compared to the corresponding separation in x-coordinates. Solving,

$$L \left(1 - \frac{v^2}{c^2} \right) = \frac{L'}{\gamma}$$

$$\frac{L}{\gamma^2} = \frac{L'}{\gamma}$$

$$L = \gamma L'.$$

Incidentally, a pivotal concept is revealed in the analysis above. In measuring the lengths of objects by different observers (e.g. A and B), the pairs of events that are considered differ across observers, as a set of events simultaneous in one frame is asynchronous in another. Referring to the set-up that we have just dissected, even though both observers consider the clock on the Earth when it reads $t = t' = 0$, A uses the clock of S on the Moon when it reads $t = \frac{Lv}{c^2}$ (corresponding to $t' = 0$) while B uses the clock on the Moon when it reads $t = 0$ in determining their observed distances between the Earth and the Moon.

Problem: In the previous set-up, it is known that B's clock reads $\frac{L}{\gamma v}$ during the entire journey, where L is the distance between Earth and the Moon as observed by A. In B's frame, how does B reason that his journey took $\frac{L}{v}$ time in A's frame? That is, how do the relevant clocks of A's frame play out in B's frame?

B syncs his clock with A on Earth. The time of the starting event as measured by A is thus zero. The duration of B's journey in A's frame is then simply the reading of the clock on the Moon. In B's frame, the clock on the Moon is ahead of A's clock by $\frac{Lv}{c^2}$ as it is the rear clock. Furthermore, in B's frame, the time elapsed on his clock is $\frac{L}{\gamma v}$. Therefore, the time elapsed on the Moon's clock during B's journey, as observed by B, is $\frac{L}{\gamma^2 v}$ by time dilation. The final reading on the Moon's clock is

$$\frac{L}{\gamma^2 v} + \frac{Lv}{c^2} = \frac{L}{v} \left(1 - \frac{v^2}{c^2} + \frac{v^2}{c^2} \right) = \frac{L}{v}.$$

11.4   Space-Time

By now, you may have realized that our model of the arrays of clocks and meter sticks is becoming extremely complicated. It is intuitive to extend this crude model by replacing the clocks synchronized in a frame with a unified time axis, which adds a fourth dimension to our analysis. Then, we can obtain a diagram with space-time as its fabric, which is commonly known as a Minkowski diagram. Usually, it is more convenient to use ct as the time axis as opposed to t. The speed of light, c, is then similar to a conversion factor between space and time. Let us draw a space-time diagram of ct against x of an arbitrary frame S in Fig. 11.13 and superimpose part of the original model of clocks on it.

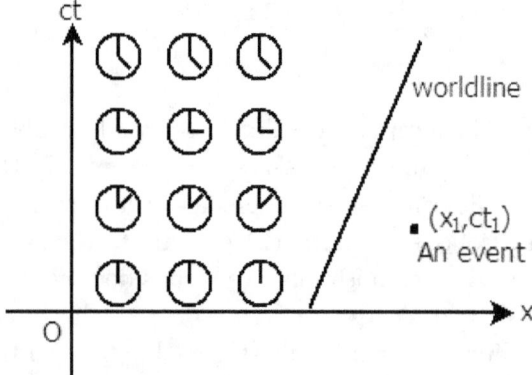

Figure 11.13:   Minkowski diagram with superimposed clocks

The array of clocks in this frame is synchronized so that they all record a reading of zero at $t = 0$ in this frame. Now imagine a vertical line, $x = k$, on the space-time diagram. This line corresponds to the space-time states of a stationary clock at coordinate k in frame S as time passes. Basically, the clock is motionless and its reading just increases at a constant rate as time elapses. The reading on the clock at a point on that line increases with the height of the vertical position of that point.

An event corresponds to a point on the space-time diagram. It has a position as indicated by its x-coordinate and a time of occurrence that is implied by its ct-coordinate. This ct-coordinate is essentially the reading on a clock (times c) that is placed at the event's spatial location when the event transpires.

Lastly, a world line of an object is the set of points $x(t)$ that corresponds to the path of the object on the space-time diagram as time elapses. For an object that is moving at a constant velocity with respect to the current inertial frame, its world line is a straight line on the space-time diagram. The world line of a photon is always a 45° line, regardless of the inertial frame of reference, due to the constancy of the speed of light. Note that the instantaneous gradient at all points on all possible world lines must have an angle[12] of inclination greater or equal to 45° as no matter or information can travel at a speed greater than the speed of light in a vacuum.

Generally, we are interested in the coordinates of certain events in an inertial frame and how they vary across different frames. When switching between inertial frames, we are essentially transforming the coordinates of a point or a set of points into those of a new inertial frame. Generally, there are two views of the transformation of coordinates. An active transformation shifts the set of points of concern into new positions on the original axes and then replaces the x and ct-axes with the new axes x' and ct'. It is obtained by transforming the coordinates of an event in frame S into that in frame S' directly.

$$x' = f_1(x, ct),$$

$$ct' = g_1(x, ct),$$

where $f_1(x, ct)$ and $g_1(x, ct)$ are the appropriate transformation functions. A passive transformation modifies the axes while leaving the set of points unchanged. Then, the coordinates of the points are read off the new axes x' and ct'. If we define $\hat{e}_x$, $\hat{e}_t$, $\hat{e}_{x'}$, $\hat{e}_{t'}$ to be unit vectors along the x, ct, x' and ct'-axes respectively (also known as the basis vectors), the transformation is obtained by performing

$$\hat{e}_{x'} = f_2(\hat{e}_x, \hat{e}_t),$$

$$\hat{e}_{t'} = g_2(\hat{e}_x, \hat{e}_t).$$

Moving on, the transformation of coordinates from one inertial frame to another must be linear as a consequence of the homogeneity of space and time in inertial frames. Suppose two events occur at coordinates x_1 and x_2 in frame S at the same time t. Then let

$$x_1' = f_1(x_1, ct),$$

$$x_2' = f_1(x_2, ct),$$

---

[12]In this paragraph, we have assumed that the length scales along the ct and x-axes are the same (i.e. measure x in units of light distances such as light seconds). This is definitely not true in general, but it is indeed a convention in drawing Minkowski diagrams.

be the x'-coordinates of the events in the new inertial frame S'. Then, the spatial separation of these events in frame S' is

$$l = f_1(x_2, ct) - f_1(x_1, ct).$$

Next, by the homogeneity of space in inertial frames, if we modify our original coordinate system by a simple translation in the x-direction such that x_1 and x_2 become $x_1 + k$ and $x_2 + k$ for some constant k, their spatial separation in S' should still be l, as it is a tangible, spatial separation between two events. Then,

$$f_1(x_2 + k, ct) - f_1(x_1 + k, ct) = f_1(x_2, ct) - f_1(x_1, ct),$$

$$f_1(x_1 + k, ct) - f_1(x_1, ct) = f_1(x_2 + k, ct) - f_1(x_2, ct).$$

Dividing the above equation by k and taking the limit as $k \to 0$,

$$\left.\frac{\partial f_1}{\partial x}\right|_{x=x_1} = \left.\frac{\partial f_1}{\partial x}\right|_{x=x_2}$$

from the first principles of calculus. Since x_1 and x_2 are arbitrary, this means that

$$\frac{\partial f_1}{\partial x} = \alpha$$

for some constant α which implies that f_1 is linear in x. A similar argument can be invoked to show that f_1 in linear in t by the homogeneity of time (by considering a translation in time). Lastly, similar arguments also be used to prove that g_1 is linear in x and t as well (by considering a temporal separation in S'). Then,

$$x' = a_1 x - a_2 ct,$$

$$ct' = a_3 ct - a_4 x,$$

for some constants a_1, a_2, a_3 and a_4, as each transformation should be linear in x and t. We shall derive these constants in the next section.

11.5   The Lorentz Transformations and Active Transformations

Instead of having to repeat the error-prone process of accounting for the relativity of simultaneity, time dilation and length contraction effects, it is much more desirable to have an integrated transformation procedure. The Lorentz transformations empower us with the ability to algebraically calculate the coordinates of an event in a new inertial frame, given its coordinates in a previous inertial frame and the relative velocity between the two frames. Formally, the spatial and temporal separation between two events in frame S as a function of those in frame S', which is moving at a velocity v with respect to the x-axis in frame S, is given by

$$\Delta x = \gamma(\Delta x' + v\Delta t'),$$

$$\Delta t = \gamma(\Delta t' + \frac{v}{c^2}\Delta x'),$$

where $\Delta x'$ and $\Delta t'$ are the spatial and temporal separations of two events in frame S' while Δx and Δt are those in frame S. These equations can be written in a more convenient and symmetric form

$$\Delta x = \gamma(\Delta x' + \beta c\Delta t'), \tag{11.1}$$

$$c\Delta t = \gamma(c\Delta t' + \beta\Delta x'), \tag{11.2}$$

where $\beta = \frac{v}{c}$. They can be expressed even more compactly with the use of matrices:

$$\begin{pmatrix} \Delta x \\ c\Delta t \end{pmatrix} = \begin{pmatrix} \gamma & \gamma\beta \\ \gamma\beta & \gamma \end{pmatrix} \begin{pmatrix} \Delta x' \\ c\Delta t' \end{pmatrix}. \tag{11.3}$$

The inverse transformations from frame S to frame S' are obtained from substituting $-v$ for v.

$$\Delta x' = \gamma(\Delta x - \beta c\Delta t), \tag{11.4}$$

$$c\Delta t' = \gamma(c\Delta t - \beta\Delta x), \tag{11.5}$$

or

$$\begin{pmatrix} \Delta x' \\ c\Delta t' \end{pmatrix} = \begin{pmatrix} \gamma & -\gamma\beta \\ -\gamma\beta & \gamma \end{pmatrix} \begin{pmatrix} \Delta x \\ c\Delta t \end{pmatrix}, \tag{11.6}$$

with the same definitions of v and γ.

The above transformations can be proven by employing the relativistic effects that we have derived before and the linearity of the transformations between inertial frames. This can be visualized better by considering space-time diagrams undergoing an active transformation. Consider two inertial frames S and S'. S' is moving at a velocity v relative to S along the positive x-axis of S. As only the separation between two events are of interest, coordinate systems in the inertial frames can be chosen such that one of the events are at the origins, O and O', in both of the inertial frames S and S'. This is due to the invariance of the separation between two events when the coordinate systems undergo a translation — a consequence of the homogeneity of inertial frames.

Following from this, two events, P and Q, occur at the origin O' and point (x', ct') on the space-time diagram in frame S'. We wish to find the coordinates of Q on the space-time diagram in frame S. Event P is again located at the origin of the space-time diagram in frame S, O.

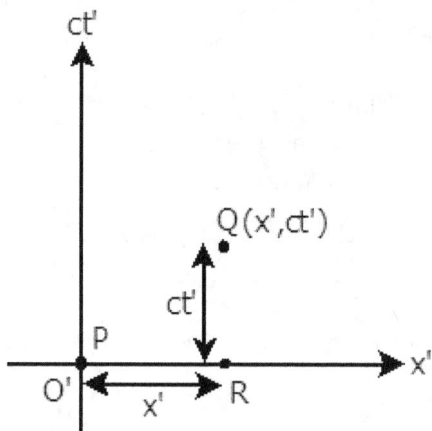

Figure 11.14:   Events P, Q and R in frame S'

Referring to Fig. 11.14, consider another point R at coordinates $(x', 0)$ in frame S'. Physically, it may represent a clock that is at coordinate x' and synchronized with the clock at the origin, O'. The world line of a clock at coordinate x' passes through the point Q in frame S'. Now, consider the same situation in frame S. In frame S, objects that are stationary in S' now travel at a velocity v. Thus, the space-time diagram for frame S is illustrated by Fig. 11.15.

The two world lines are those of the two stationary clocks at x'-coordinates 0 and x' when viewed in frame S'. We know that the slope of the world lines of the two clocks are $\frac{c\Delta t}{\Delta x} = \frac{c}{v}$. Furthermore, point Q must still lie on the world line of the clock at R in frame S as it did so in frame S'. The distance between these world lines that is measured at the same time t in frame S is $\frac{x'}{\gamma}$ due to length contraction. Furthermore, it is known from the relativity of simultaneity that two clocks that are synchronous and separated by a distance of x' in frame S' differ by a time $t = \frac{\gamma x' v}{c^2}$ in frame S. Applied to the situation at hand, the two clocks are those at P and R respectively. Moreover, the time interval between two events that are at the same x'-coordinate in frame S' is observed to be time-dilated in frame S. The two events in this case refer to Q and R (the readings of the clock at x') which differ by a

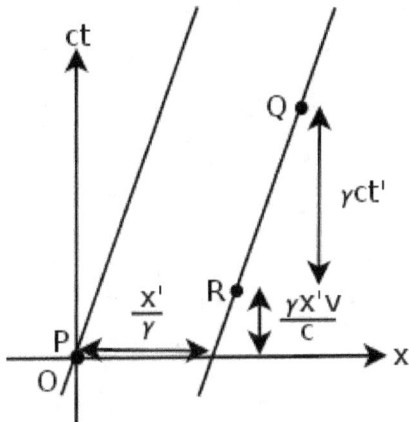

Figure 11.15: Fundamental effects and events in frame S

time interval $\gamma ct'$ in frame S. These effects are labeled appropriately in the diagram above.

Combining the information obtained from these effects, the coordinates of Q in this frame can then be easily found.

$$x = \frac{x'}{\gamma} + \frac{v}{c}\left(\frac{\gamma x' v}{c} + \gamma ct'\right)$$

$$= \gamma x'\left(\frac{1}{\gamma^2} + \frac{v^2}{c^2}\right) + \gamma vt'$$

$$= \gamma x'\left(1 - \frac{v^2}{c^2} + \frac{v^2}{c^2}\right) + \gamma vt'$$

$$= \gamma(x' + vt'),$$

$$ct = \gamma\left(ct' + \frac{v}{c}x'\right).$$

The inverse transformations from frame S to frame S' can then be obtained by substituting $-v$ for v in the equations above, as frame S travels at $-v$ relative to the positive x'-axis of frame S', the primed coordinates for the unprimed ones and vice-versa.

$$x' = \gamma(x - vt),$$

$$ct' = \gamma\left(ct - \frac{v}{c}x\right).$$

Note that the spatial separations of two events along directions that are perpendicular to the x-direction are unchanged across inertial frames.

$$\Delta y' = \Delta y,$$

$$\Delta z' = \Delta z,$$

as there is no loss of simultaneity (use a similar set-up involving light in a train) nor length contraction in the transverse direction.

Finally, let us obtain some form of closure by showing how the Lorentz transformations can be intuitively understood by our previous model of clocks and meter sticks.

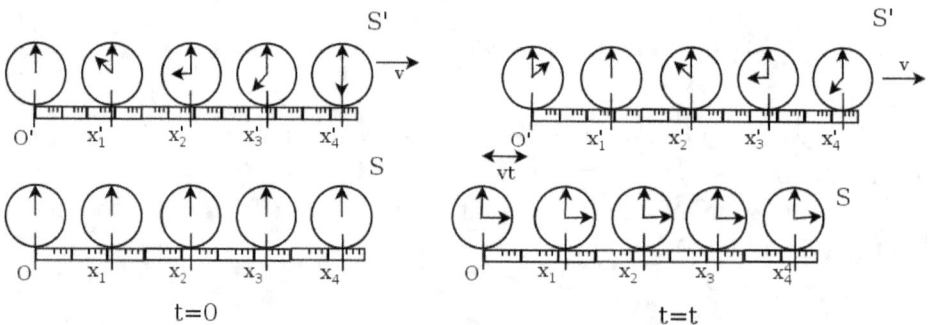

Figure 11.16:   Clocks and meter sticks of S' (top) and S (bottom) as observed in S

At time $t = 0$ in frame S, the origins of the two coordinate systems of S' and S are aligned in Fig. 11.16. At this juncture, a clock of S that is at x-coordinate x coincides (in terms of location) with a clock of S' that is at x'-coordinate $x' = \gamma x$ (x' is scaled by a factor of γ as the meter sticks of S' are shrunk by a factor of γ). However, a clock synchronized in frame S', that is located at x'-coordinate x', is observed in frame S' to possess a reading $-\frac{x'v}{c^2}$ due to the relativity of simultaneity.

At time t in frame S, origin O' of S' would have traveled towards the right, relative to O, by a distance vt. Therefore, a clock of S that is at x-coordinate x now coincides (in terms of location) with a clock of S' that is at x'-coordinate $x' = \gamma(x - vt)$. We immediately obtain the first transformation rule

$$x' = \gamma(x - vt).$$

Next, since time t has passed in frame S, the readings of the clocks of S' would have increased by $\Delta t' = \frac{t}{\gamma}$ (reduced by a factor of $\frac{1}{\gamma}$ as the clocks of S' tick slower due to time dilation). Therefore, a clock of S at x-coordinate

x and time t corresponds to a clock of S' at x-coordinate x' which displays a time

$$t' = -\frac{x'v}{c^2} + \frac{t}{\gamma}.$$

Substituting the expression for x',

$$t' = -\frac{\gamma x v}{c^2} + \frac{\gamma v^2}{c^2} t + \frac{t}{\gamma} = -\frac{\gamma x v}{c^2} + \gamma t \left(\frac{v^2}{c^2} + \frac{1}{\gamma^2} \right).$$

Plugging in $\frac{1}{\gamma^2} = 1 - \frac{v^2}{c^2}$, we retrieve the second transformation rule.

$$t' = \gamma \left(t - \frac{v}{c^2} x \right).$$

Problem: Derive the length contraction result from the Lorentz transformations.

Let the longitudinal proper length of an object be L' in its rest frame S'. Define the origin in S' such that the left and right ends of the object are located at $x' = 0$ and $x' = L'$ respectively (at all times t'). Now, suppose that we are interested in determining the length of this object in a frame S that travels at a velocity $-v$ in the x'-direction relative to S'. To do so, we need to determine the spatial separation of the two ends of the object simultaneously in S. Presuming that we want to do this when $t = 0$ in S, the left end is at $x = 0$ (since the origins of S and S' coincide at $t = t' = 0$). Now, we just need to determine the x-coordinate of the right end at $t = 0$. Notice that the (x', ct') coordinates of the right end in S' are generally (L', ct'). Applying the Lorentz transformations, the time of this event in frame S is

$$t = \gamma \left(t' + \frac{v}{c^2} L' \right)$$

so we must choose to observe the right end at $t' = -\frac{L'v}{c^2}$ in S', as it corresponds to $t = 0$ in S. Applying the Lorentz transformations to $(L', -\frac{L'v}{c})$ in S', the x-coordinate of the right end at $t = 0$ in S is then

$$x = \gamma(L' + vt') = \gamma \left(L' - \frac{L'v^2}{c^2} \right) = \frac{L'}{\gamma}.$$

The spatial separation of the two ends of the object at $t = 0$ (which is the observed length in S) is

$$L = x - 0 = \frac{L'}{\gamma}.$$

11.6   Passive Transformations

Remember that there is an alternate perspective to the transformation of coordinates. Instead of modifying the position of the points, the axes can be changed. In this section, we will consider the passive transformation from frame S to frame S'. It is then natural to determine how the new axes, x' and ct' of frame S' will look like when superimposed on the space-time diagram of frame S with x and ct as its axes.

Even without any calculations, it can be concluded that these x' and ct'-axes must be straight lines on the space-time diagram in frame S as the Lorentz transformations are linear. Thus, straight lines must be mapped onto straight lines. Assume that the origins of the two coordinate systems in the two inertial frames coincide. The equation of the x'-axis can be determined by using the fact that $t' = 0$ along it. Then, by the Lorentz transformations,

$$\gamma(ct - \beta x) = ct' = 0$$

$$\implies ct = \beta x.$$

Similarly, along the t'-axis, $x' = 0$.

$$\gamma(x - \beta ct) = x' = 0$$

$$\implies x = \beta ct.$$

These are the equations of the lines on the Minkowski diagram of frame S that delineate the x' and ct'-axes respectively. They are plotted on the space-time diagram in frame S in Fig. 11.17.

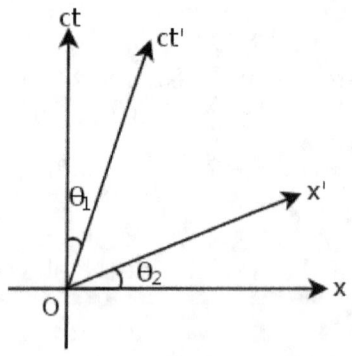

Figure 11.17:   Superimposed x' and ct'-axes on Minkowski diagram in frame S

By considering the line equations above, it can easily be proven that the two angles labeled in the figure obey the relation

$$\tan \theta_1 = \tan \theta_2 = \beta. \qquad (11.7)$$

Observe that the superimposed x' and ct'-axes are not mutually perpendicular. Orthogonal vectors in one inertial frame need not be perpendicular in another. Lines on the space-time diagram that are parallel to the superimposed x' and ct'-axes are sets of events that are simultaneous and occur at the same x'-coordinate in frame S' respectively.

The relativity of simultaneity can be easily visualized with the supcrimposed x' and ct'-axes. Consider Figs. 11.18 and 11.19.

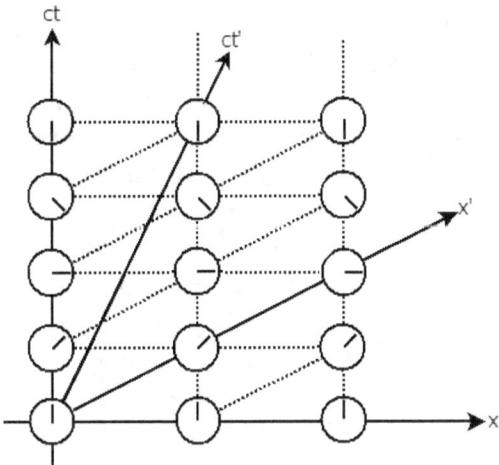

Figure 11.18:   Clocks synchronized in frame S

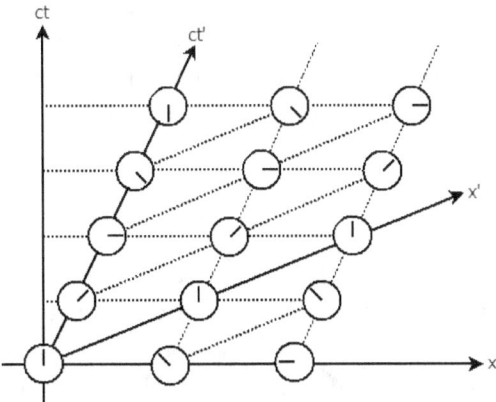

Figure 11.19:   Clocks synchronized in frame S'

In both diagrams, the horizontal and slanted lines represent the lines of simultaneity in frames S and S' in the Minkowski diagram of frame S respectively. The array of clocks on the left is synchronized in frame S. It can be seen that the events that are simultaneous in frame S, represented by horizontal lines, are not simultaneous in frame S' whose lines of simultaneity are represented by the slanted lines (parallel to the overlapping x'-axis). An observer in S' that observes certain clocks of S simultaneously will conclude that the readings on the clocks are asynchronous (consider the clocks along any slanted dotted line as an example). Similarly, the array of clocks on the right is synchronized in frame S'. Conversely, an observer in S that observes certain clocks of S' simultaneously will conclude that the readings on the clocks are asynchronous. It is natural for an observer in either frame to conclude that the clocks in the other's frame are asynchronous — a fact that is evident from the asymmetrical lines of simultaneity.

To complete our analysis of the superimposed axes, we need to find how the magnitude of one unit along the x' and ct'-axes in frame S' is reflected on the x' and ct'-axes that are superimposed on the space-time diagram of frame S. There is a need to do so as the coordinates are now measured with respect to the x and ct-axes (i.e. there is no guarantee that they will be the same). Consider the point $(1, 0)$ in frame S'; the first and second coordinates correspond to the x' and ct'-coordinates respectively. Thus by the Lorentz transformations, the coordinates of this point in frame S is $(\gamma, \gamma\beta)$. This signifies that one unit along the x'-axis in frame S' corresponds to

$$\sqrt{\gamma^2 + \gamma^2\beta^2} = \sqrt{\frac{1 + \beta^2}{1 - \beta^2}} \tag{11.8}$$

units of length as measured by the x and ct-axes in the space-time diagram in frame S. In other words, a point with coordinates $(x', 0)$ in frame S' will be a length $\sqrt{\frac{1+\beta^2}{1-\beta^2}}x'$ along the superimposed x'-axis in frame S as measured by the x- and ct-axes. Similarly, a point $(0, 1)$ in frame S' will transform to a point $(\gamma\beta, \gamma)$. Thus, one unit along the ct'-axis in frame S' also corresponds to

$$\sqrt{\gamma^2\beta^2 + \gamma^2} = \sqrt{\frac{1 + \beta^2}{1 - \beta^2}} \tag{11.9}$$

units of length as measured by the x and ct-axes in the space-time diagram in frame S. Following from this, the x' and ct'-coordinates of an event can be deduced by drawing lines that pass through that event and are parallel to the superimposed ct' and x'-axes in the Minkowski diagram of frame S respectively. Afterwards, one can identify the points of intersection of these two lines with the superimposed x' and ct'-axes and divide the distances between these points of intersection and the origin by the scaling factor $= \sqrt{\frac{1+\beta^2}{1-\beta^2}}$ to obtain the x' and ct'-coordinates of the event in frame S'.

The Twins' Paradox

Problem: Consider two twins, A and B, who are initially on Earth. In Fig. 11.20, twin B remains on Earth while twin A travels on a rocket to a distant star at a constant speed v relative to B's frame and back to Earth at a constant speed v, in the opposite direction, rapidly. When the two twins eventually compare the readings on their clocks, which twin is younger?

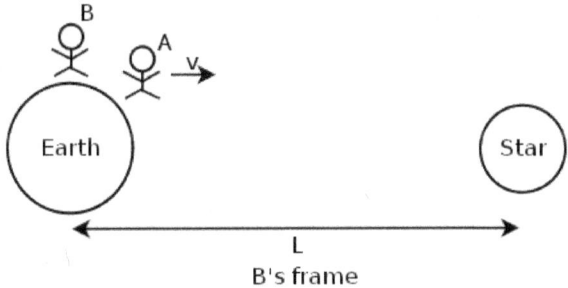

Figure 11.20: A's outbound journey in the frame of B

Let the frame of twin A be S' and that of twin B be S. There is an apparent paradox here. In frame S, twin B sees twin A's clock running slower by a factor of $\frac{1}{\gamma}$. Thus, B will conclude that A is younger. However in frame S', twin A also sees twin B's clock running slow. Thus, A will seemingly also conclude that B is younger. There appears to be a paradox here as the readings on both clocks must be the same, as observed by the twins, when they are compared at the same location. However, the correct answer is, in fact, that twin A is younger!

This is because the symmetry in this system is broken when twin A reverses the direction of his velocity, as he must experience an acceleration then. In other words, twin A is actually stationary in two different inertial frames during his outbound and inbound journeys. Thus, the above reasoning in frame S' is invalid as there are really two different inertial frames of A. However, this reasoning only explains why the latter analysis in A's frame is wrong but does not show how to correct that reasoning. There are many ways of resolving this. One argument will be presented here. Let us first consider the situation in frame S. Let the distance between the distant star and Earth in frame S be L. Then, the time of the entire journey by A in S is

$$t = \frac{2L}{v}.$$

The elapsed time of the entire journey in A's two inertial frames is then

$$t' = \frac{2L}{\gamma v},$$

as B observes A's clock to slow down by a factor of $\frac{1}{\gamma}$. Therefore, B will conclude that A is younger than him by a factor of $\frac{1}{\gamma}$. Next, we would like to consider the situation from A's perspective. This is better visualized by drawing a space-time diagram in frame S and superimposing the axes of A's inertial frames, as there are in fact two inertial frames of A.

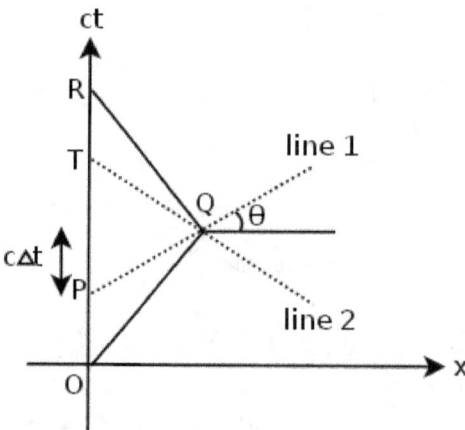

Figure 11.21:   Minkowski diagram in B's frame (S)

The line OQR in Fig. 11.21 represents the world line of twin A in frame S while the line OR is the world line of twin B whose stationary clock just ticks with time. The two inertial frames of A consist of one before the kink at point Q and one after the kink. Let the axes of the two inertial frames be x', ct' and x", ct" respectively. We know that the lines of simultaneity in those frames are represented by the lines parallel to the x'- and x"-axes. Consider the point Q at which twin A turns, causing him to switch from the first inertial frame to the second. This causes the line of simultaneity through point Q to instantaneously change from line 1 to line 2, which are parallel to the x' and x"-axes respectively, as indicated on the diagram.

Since points on the line OR correspond to the readings of B's clock, this physically means that twin A observes twin B to spontaneously age by a certain amount as he turns. Here lies the crux of this resolution. From A's perspective, B instantaneously ages by the time interval between points T and P in frame S, which is equal to $2\Delta t$ as labeled on the diagram. Δt can be easily found by utilizing the facts that the spatial separation between P and Q is L and that $\tan\theta = \beta$.

$$c\Delta t = L \cdot \beta = \frac{Lv}{c}.$$

Thus,

$$2\Delta t = \frac{2Lv}{c^2}.$$

This is the amount that twin B instantaneously ages as observed by A in his own frame when he turns around. Actually, this particular value can also be explained from the relativity of simultaneity. In A's outbound journey, the star's clock was the "rear clock" and thus led the Earth's clock by $\frac{Lv}{c^2}$. However, when A turned around, the Earth's clock became the "rear clock" — leading to a total discontinuity of $\frac{2Lv}{c^2}$ time. With this, the entire process from A's perspective can be outlined as follows — we start with the reading of A's clock. The distance between the Earth and the distant star is $\frac{L}{\gamma}$ in both of A's frames due to length contraction. Furthermore, twin A observes the distant star and Earth to approach him at a speed v in his first and second inertial frames respectively. Thus, the total time elapsed on A's clock is

$$t' = \frac{2L}{\gamma v}.$$

To investigate how the reading of B's clock changes from the perspective of A, we divide the entire process into three parts — namely before the turn, during the turn and after the turn. The situation of the first part in A's first frame appears in Fig. 11.22 — A observes both the Earth and the distant star to travel at speed v.

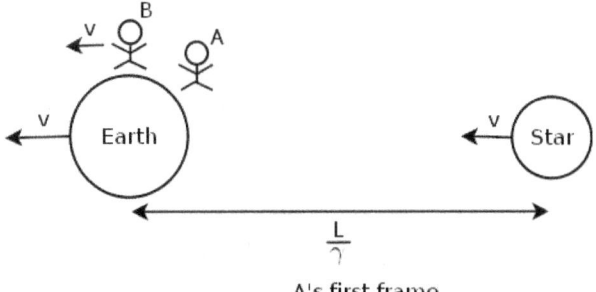

A's first frame

Figure 11.22: First half of the journey in A's first inertial frame

This takes $\frac{L}{\gamma v}$ for the distant star to reach A in A's frame due to length contraction. As twin A observes B's clock to run slow by a factor of $\frac{1}{\gamma}$, the time elapsed as measured by twin B's clock during the first part of the process is

$$t_1 = \frac{L}{\gamma^2 v}$$

which corresponds to the distance between points O and P (divided by c). Next, when twin A turns around, he observes the reading of B's clock to immediately increase by

$$t_2 = \frac{2Lv}{c^2}$$

due to the switching of inertial frames. Lastly, during twin A's return journey, the reading of B's clock increases by

$$t_3 = \frac{L}{\gamma^2 v},$$

by an argument similar to that for t_1. Thus, the total time elapsed by twin B's clock from the perspective of twin A is

$$t = t_1 + t_2 + t_3 = \frac{2Lv}{c^2} + \frac{2L}{\gamma^2 v}$$
$$= \frac{2L}{v}\left(\frac{1}{\gamma^2} + \frac{v^2}{c^2}\right) = \frac{2L}{v},$$

which is consistent with the result obtained by considering the set-up in B's frame. Hence, twin A also concludes that he is younger than twin B by a factor of $\frac{1}{\gamma}$.

11.7   The Invariant Interval

The interval Δs between two events in frame S is defined as

$$(\Delta s)^2 = (c\Delta t)^2 - (\Delta x)^2 - (\Delta y)^2 - (\Delta z)^2 \qquad (11.10)$$

where the Δ's represent the spatial and temporal separations between the two events. The interval Δs has a unique property — consider the right-hand side of the expression in a different inertial frame S' that is moving at a velocity v with respect to frame S. By applying the Lorentz transformations,

$$(c\Delta t')^2 - (\Delta x')^2 - (\Delta y')^2 - (\Delta z')^2$$

$$= \gamma^2 \left(c\Delta t - \frac{v}{c}\Delta x \right)^2 - \gamma^2 \left(\Delta x - \frac{v}{c}c\Delta t \right)^2 - (\Delta y)^2 - (\Delta z)^2$$

$$= \gamma^2 (c\Delta t)^2 \left(1 - \frac{v^2}{c^2} \right) - \gamma^2 (\Delta x)^2 \left(1 - \frac{v^2}{c^2} \right) - (\Delta y)^2 + (\Delta z)^2$$

$$= (c\Delta t)^2 - (\Delta x)^2 - (\Delta y)^2 - (\Delta z)^2$$

$$= (\Delta s)^2,$$

where we have used the fact that $\Delta y = \Delta y'$ and $\Delta z = \Delta z'$ as the y and z separations do not vary when switching between frames that travel with a relative velocity solely along the x-axis. It can be seen that the quantity $(\Delta s)^2$ is invariant under the Lorentz transformations. The invariance of this quantity across various inertial frames is similar to the invariance of the squared distance between two points in an Euclidean space under rotations.

$$(\Delta r)^2 = (\Delta x)^2 + (\Delta y)^2 + (\Delta z)^2.$$

Therefore, the interval can be treated as the "squared distance" in Minkowski space. In fact, the Lorentz transformations are hyperbolic rotations of Minkowski space which makes the analogy even more apt. Now, three specific cases of the value of $(\Delta s)^2$ between two events will be considered. In doing so, we align the x-axis of our coordinate system with the line joining the positions of the two events such that

$$(\Delta s)^2 = (c\Delta t)^2 - (\Delta x)^2$$

for the sake of convenience.

Case 1: $(\Delta s)^2 < 0$

Firstly, there is no need to worry that a squared term yields a negative value as Δs lacks physical meaning in itself and can be imaginary. In this situation, $(\Delta x)^2 > c^2(\Delta t)^2$ and these events are said to be space-like separated. This means that there exists a frame S' such that these two events occur at the same time t'. The Lorentz transformations give

$$\Delta t' = \gamma \left(\Delta t - \frac{v}{c^2}\Delta x \right).$$

Therefore, there exists a velocity with magnitude less than the speed of light, specifically $v = \frac{c^2 \Delta t}{\Delta x} < c$, that leads to $\Delta t' = 0$. However, these two events do not occur at the same x'-coordinate with respect to any inertial frame S'. This can be shown easily by contradiction. If $\Delta x' = 0$,

$$(\Delta s)^2 = c^2(\Delta t')^2 - (\Delta x')^2 = c^2(\Delta t')^2$$

where $(\Delta s)^2 < 0$, leading to a contradiction as the right-hand side consists of physical quantities that must be real.

Lastly, it makes no sense to say whether event A occurs before or after event B when these two events are space-like separated. This is because there always exist inertial frames where A precedes B and where B precedes A. By the Lorentz transformations again,

$$\Delta t' = \gamma \left(\Delta t - \frac{v}{c^2} \Delta x \right).$$

If $\Delta t > 0$, $\Delta t' < 0$ when

$$c > v > \frac{c^2 \Delta t}{\Delta x}.$$

A similar argument can be made for the case where $\Delta t < 0$ to show that it is possible for $\Delta t' > 0$. This proves that event A may precede event B in one inertial frame and that the reverse may be true in another. Concomitantly, these two events must not have a causality relationship (i.e. event A induces event B or vice-versa). If this were not the case, there will be a violation of causality as the relative order of A and B in time varies across different inertial frames.

Case 2: $(\Delta s)^2 = 0$

In this situation, $(\Delta x)^2 = c^2(\Delta t)^2$ and these events are said to be light-like separated. These events then correspond to the points of a photon's path on a space-time diagram. It is impossible to find an inertial frame S' in which the two events are simultaneous or occurs at the same x-coordinate as two points on the world line of an undisturbed photon cannot exist at different locations at the same time or at the same location at different times.

Case 3: $(\Delta s)^2 > 0$

In this scenario, $(\Delta x)^2 < c^2(\Delta t)^2$ and we say that these two events are time-like separated. Employing similar arguments as before, it can be proven that there exists an inertial frame S' in which the two events occur at the same x'-coordinate while an inertial frame S' in which the two events occur at the same time does not exist. If events A and B are time-like separated events and event A precedes B in a certain inertial frame, event A precedes B in all inertial frames. Therefore, it is possible for there to be a causality relationship between events A and B.

Proper Time and Proper Distance

Proper Time

The proper time, τ, of a point along a world line refers to the time measured by the clock which is perpetually at rest relative to the observer in the world line. Let us denote that observer as O. One can imagine observer O holding a clock that is constantly stationary to him or a fictitious time axis protruding from observer O. Then, O describes events with his own "local time". The proper time interval between two events on the world line is then simply the time interval as observed by observer O in the world line.[13] Evidently, these two events must be time-like separated in order for a proper world line to be defined. Furthermore, these two events must have the same spatial coordinates in the frame of observer O, as he is stationary. Now let observer O's world line be $r(t)$ with respect to a fixed inertial frame S. Consider an infinitesimal interval

$$(ds)^2 = c^2(dt)^2 - (dx)^2 - (dy)^2 - (dz)^2,$$

where dx, dy, dz and dt are the infinitesimal spatial and temporal separations between two neighboring events along the world line of O with respect to the frame S. Since the interval is invariant, the infinitesimal proper time $d\tau$ between these two events is given by

$$c^2(d\tau)^2 = c^2(d\tau)^2 - (dx')^2 - (dy')^2 - (dz')^2 = (ds)^2$$
$$= c^2(dt)^2 - (dx)^2 - (dy)^2 - (dz)^2$$

as the components of the spatial separation between any two events along the world line of O, as observed by O, are zero since O is stationary in his own frame $(dx' = dy' = dz' = 0)$. Then,

$$d\tau = \sqrt{1 - \frac{\left(\frac{dx}{dt}\right)^2 + \left(\frac{dy}{dt}\right)^2 + \left(\frac{dz}{dt}\right)^2}{c^2}}\, dt$$

$$= \sqrt{1 - \frac{v^2}{c^2}}\, dt$$

$$= \frac{dt}{\gamma},$$

---

[13]Note that there may not be a single inertial frame associated with observer O as he might be accelerating.

where v is the speed of observer O as observed in frame S. The proper time interval between two time-like separated events on O's world line is then obtained from integrating the expression above.

$$\int_{\tau_1}^{\tau_2} d\tau = \int_{t_1}^{t_2} \frac{dt}{\gamma}.$$

If O is traveling at a constant speed, the integral can be performed trivially.

$$\Delta\tau = \frac{\Delta t}{\gamma},$$

which is simply the time dilation equation. It is natural for us to obtain this result, as $\Delta\tau$ simply means the time elapsed as measured by a still clock while Δt is that of a moving clock.

Proper Distance

The proper distance $\Delta\sigma$ between two space-like separated events is defined as the distance between the two events, as observed in an inertial frame in which they are simultaneous. Since the interval between two events is invariant and because their temporal separation in that particular inertial frame is zero,

$$\Delta\sigma = \sqrt{-\Delta s^2} = \sqrt{\Delta x^2 + \Delta y^2 + \Delta z^2 - c^2\Delta t^2}$$

where the quantities on the right-hand side are measured with respect to an arbitrary frame S.

11.8 The Relativistic Speed Limit

So far, we have asserted that no information or massive particle can travel faster than the speed of light in vacuum c. Some justifications shall be provided here. Firstly, consider the expression for γ:

$$\gamma = \frac{1}{\sqrt{1 - \frac{v^2}{c^2}}}.$$

If $|v| > c$, γ is imaginary. Else if $|v| = c$, γ tends to infinity. This results in a loss in the physical meaning of our coordinate transformations and implies that these cases for $|v|$ should be rejected.

The second argument pertains to the violation of causality. As shown before, if two events A and B are space-like separated such that $|\frac{\Delta x}{\Delta t}| > c$, there exist inertial frames in which A precedes B and others in which B precedes A. This is perfectly fine in itself if these two events do not have a causality relationship mediated by physical entities or information. However, if there exists a particle or information that can travel at a velocity greater than the speed of light, it is possible for events A and B to be affected by one another due to the transmission of information via particles traveling at superluminal speeds. Then, these particles will be traveling back in time with respect to some inertial frames. Causality will be violated as the "effect" may precede the cause of an event in certain inertial frames. Here is an example. In an arbitrary frame S, observer A sends a superluminal signal to observer B. Immediately upon receiving the signal, observer B sends a superluminal signal back to observer A. This contravenes the causality relationship as there are inertial frames in which observer A receives the signal from B before he sends one himself (as the events are space-like separated)!

The last argument is relevant to the next chapter. As we shall see, it takes an infinite amount of energy for a particle with mass to travel at the speed of light — a feat that is physically infeasible.

In conclusion, the speed limit imposed by the speed of light is considered a corollary in special relativity. If this were to be breached, situations that are contrary to common experiences will arise. Therefore, it is widely accepted that neither information nor matter can travel at a speed greater than c.

Finally, there is a qualification to be made here, no matter or information can travel at a speed greater than c, the speed of light **in vacuum**. This is an important point to take note of as light propagates at different speeds in different media.

11.9 Other Effects

This section elaborates on the subsidiary effects due to the fundamental consequences of special relativity. We will adopt the conventional definitions for frames S and S' and v. In approaching this section, remember that v refers to the relative velocity between frames while the symbol u (and u') will be used to denote the velocity of a particle in a certain frame. Keep in mind that $\gamma = \dfrac{1}{\sqrt{1-\frac{v^2}{c^2}}}$ and is independent of u, as it is associated with the transformation between frames.

11.9.1 *Relativistic Velocity Addition*

Longitudinal Addition

If an observer in frame S' observes a particle to travel at a velocity u' in the direction of the x'-axis, what is the speed of the particle as observed by a person in frame S?

The relativistic result differs from the classical result of $u' + v$ due to the fundamental effects of special relativity. Assume that the particle undergoes a displacement dx' in a time period dt' in frame S', the expressions for the corresponding displacement and time interval, dx and dt, in frame S can be obtained via the Lorentz transformations.

$$dx = \gamma(dx' + vdt'),$$

$$dt = \gamma\left(dt' + \frac{v}{c^2}dx'\right).$$

The velocity of the particle in S' is $u' = \frac{dx'}{dt'}$ while the velocity of it in S is $u = \frac{dx}{dt}$. Thus, dividing the first equation by the second, we obtain

$$u = \frac{\frac{dx'}{dt'} + v}{1 + \frac{v}{c^2}\frac{dx'}{dt'}} = \frac{u' + v}{1 + \frac{u'v}{c^2}}$$

in the direction along the x-axis. The inverse transformations from frame S to S' can easily be obtained by replacing v with $-v$.

$$u' = \frac{u - v}{1 - \frac{uv}{c^2}}. \tag{11.11}$$

Transverse Addition

In a new set-up, an observer in S' now observes a particle to travel at (u'_x, u'_y). Keep in mind that S' travels at v relative to S in the positive x-direction. We would again like to determine the particle's velocity (u_x, u_y) in frame S.

Firstly, note that this motion in the y'-direction does not change the validity of the previous equation. The u and u' just need to be substituted by their corresponding x and x' components, u_x and u'_x which are the particle's velocities in the x and x'-directions in frames S and S' respectively. That is,

$$u_x = \frac{u'_x + v}{1 + \frac{u'_x v}{c^2}}. \tag{11.12}$$

Moving on, we are concerned with finding u_y. Using the Lorentz transformations once again, we obtain

$$dy = dy',$$

$$dt = \gamma\left(dt' + \frac{v}{c^2}dx'\right).$$

In this case we have $u'_y = \frac{dy'}{dt'}$, $u_y = \frac{dy}{dt}$ and $u'_x = \frac{dx'}{dt'}$. Thus,

$$u_y = \frac{\frac{dy'}{dt'}}{\gamma(1 + \frac{v}{c^2}\frac{dx'}{dt'})} = \frac{u'_y}{\gamma\left(1 + \frac{u'_x v}{c^2}\right)} \tag{11.13}$$

in the direction along the y-axis. Similarly, the inverse transformations from frame S to S' are

$$u'_y = \frac{u_y}{\gamma(1 - \frac{u_x v}{c^2})}. \tag{11.14}$$

It is paramount to note that the transformation of the time elapsed between two events between frames S and S' is independent of the relative y positions of the two events in the derivation above as the relativity of simultaneity only applies for events separated in the x- or x'-direction while the time dilation effect is only dependent on the relative velocity between frames.

11.9.2   *Acceleration*

Acceleration Transformations

It is also useful to determine how an acceleration a in inertial frame S will transform to the acceleration a' in inertial frame S'. It is known from the velocity transformations that

$$u'_x = \frac{u_x - v}{1 - \frac{u_x v}{c^2}}.$$

Then, taking the derivative of u'_x with respect to u_x and using the quotient rule,

$$\frac{du'_x}{du_x} = \frac{1 \cdot \left(1 - \frac{u_x v}{c^2}\right) + \frac{v}{c^2}(u_x - v)}{\left(1 - \frac{u_x v}{c^2}\right)^2}$$

$$= \frac{1 - \frac{v}{c^2}u_x + \frac{v}{c^2}u_x - \frac{v^2}{c^2}}{\left(1 - \frac{u_x v}{c^2}\right)^2}$$

$$= \frac{1}{\gamma^2\left(1 - \frac{u_x v}{c^2}\right)^2}$$

$$\implies du'_x = \frac{1}{\gamma^2\left(1 - \frac{u_x v}{c^2}\right)^2}du_x. \tag{11.15}$$

Furthermore, from the Lorentz transformations,

$$t' = \gamma \left(t - \frac{v}{c^2} x \right)$$

$$\frac{dt'}{dt} = \gamma \left(1 - \frac{u_x v}{c^2} \right)$$

$$dt' = \gamma \left(1 - \frac{u_x v}{c^2} \right) dt. \tag{11.16}$$

Dividing Eq. (11.15) by Eq. (11.16),

$$a'_x = \frac{a_x}{\gamma^3 \left(1 - \frac{u_x v}{c^2} \right)^3}. \tag{11.17}$$

Similarly, from the transverse velocity addition,

$$u'_y = \frac{u_y}{\gamma \left(1 - \frac{u_x v}{c^2} \right)}$$

$$du'_y = \frac{\partial u'_y}{\partial u_y} du_y + \frac{\partial u'_y}{\partial u_x} du_x$$

$$du'_y = \frac{1}{\gamma \left(1 - \frac{u_x v}{c^2} \right)} du_y + \frac{\frac{v}{c^2} u_y}{\gamma \left(1 - \frac{u_x v}{c^2} \right)^2} du_x. \tag{11.18}$$

Dividing Eq. (11.18) by Eq. (11.16),

$$a'_y = \frac{1}{\gamma^2 \left(1 - \frac{u_x v}{c^2} \right)^2} a_y + \frac{v u_y}{c^2 \gamma^2 \left(1 - \frac{u_x v}{c^2} \right)^3} a_x. \tag{11.19}$$

Therefore, we see that accelerations are no longer invariant when switching between inertial frames in special relativity — contrary to the situation in Galilean relativity.

Proper Acceleration

When objects are accelerating, there isn't one inertial frame associated with them. However, a momentarily co-moving reference frame (MCRF) is useful in analyzing its motion. An MCRF is an inertial frame that travels at the same instantaneous velocity of the particle with respect to another inertial frame S (defined as the lab frame). Thus, the instantaneous velocity of a particle is zero in an MCRF defined at that instant. As an object accelerates, we have to switch from one MCRF to another new MCRF at every instant as the velocity of the object in frame S changes. With this definition, the proper acceleration of an object is the acceleration of that object observed in the MCRF defined at that instant. The transformation from the acceleration of an object in frame S to that in its MCRF can be obtained from Eqs. (11.17) and (11.19) via adroit substitutions. Firstly, we choose the coordinate systems of S and the MCRF such that the velocity of the particle is solely along the x-direction in frame S (though with this definition, the axes may have to be modified constantly in a manner analogous to how polar unit vectors change with the angular coordinate θ). Then, we can substitute $u_y = 0$ and $v = u$ into Eqs. (11.17) and (11.19) to obtain the relevant transformations. For the following section, we will use γ_u instead of γ where

$$\gamma_u = \frac{1}{\sqrt{1 - \frac{u^2}{c^2}}},$$

to remind ourselves that $v = u$. Thus, substituting $v = u$ and $u_y = 0$ into Eqs. (11.17) and (11.19), the proper accelerations of a point particle in the x' and y'-directions, α_x and α_y are

$$\alpha_x = \gamma_u^3 a_x, \tag{11.20}$$

$$\alpha_y = \gamma_u^2 a_y. \tag{11.21}$$

Problem: A particle is undergoing circular motion with a velocity u and a radius of orbit r in the lab frame. Find the magnitude of its proper acceleration.

The centripetal acceleration of the particle in the lab frame is

$$a = \frac{u^2}{r}.$$

Since the instantaneous acceleration of the particle is constantly perpendicular to its instantaneous velocity, its proper acceleration is given by Eq. (11.21).

$$\alpha = \frac{\gamma_u^2 u^2}{r}.$$

One-Dimensional Motion Under Constant Proper Acceleration

In this section, we consider the motion of a point particle, undergoing a constant proper acceleration, as observed in the lab frame S. It is assumed that the direction of the initial velocity of the particle is parallel to that of its acceleration in frame S. As such, we do not have to constantly modify the orientation of the axes of frame S and the MCRF to ensure that $u_y = 0$ — reducing this to a one-dimensional problem in spatial terms (all quantities in this problem will be with respect to the x-direction that has been aligned with the direction of concern). From Eq. (11.20),

$$\alpha = \gamma_u^3 a$$

where α is the proper acceleration and a is the acceleration of the particle in frame S. The right-hand side can be written as $\frac{d}{dt}(\gamma_u u)$, where t is the time in S, as seen from the fact that

$$\frac{d}{dt}\left(\frac{u}{\sqrt{1-\frac{u^2}{c^2}}}\right) = \frac{1}{\sqrt{1-\frac{u^2}{c^2}}} \cdot \frac{du}{dt} + \frac{u}{\left(1-\frac{u^2}{c^2}\right)^{\frac{3}{2}}} \cdot -\frac{1}{2} \cdot -2 \cdot \frac{u}{c^2} \cdot \frac{du}{dt}$$

$$= \frac{1}{\sqrt{1-\frac{u^2}{c^2}}} a + \frac{\frac{u^2}{c^2}}{\left(1-\frac{u^2}{c^2}\right)^{\frac{3}{2}}} a$$

$$= \gamma_u^3 a.$$

Thus,

$$\alpha = \frac{d}{dt}(\gamma_u u).$$

Before we integrate this expression, we claim that we can always define an origin in time such that $u = 0$ at $t = 0$. This can be subsequently justified (after finding $u(t)$) by showing that for any given value of velocity u_0, there is a time t for which $u(t) = u_0$. Integrating and applying the proposed initial conditions, we obtain

$$\alpha t = \gamma_u u.$$

Substituting $\gamma_u = \frac{1}{\sqrt{1-\frac{u^2}{c^2}}}$,

$$\alpha t = \frac{u}{\sqrt{1-\frac{u^2}{c^2}}}$$

where the right-hand side is a monotonically increasing function in u. Thus, for a given value of u, we can always find a unique, corresponding value of t — implying that we can indeed set a temporal origin such that $u = 0$ at $t = 0$. Next, solving for u,

$$u = \frac{\alpha t}{\sqrt{\frac{\alpha^2 t^2}{c^2} + 1}}.$$

Defining $p = \frac{\alpha^2 t^2}{c^2} + 1$ and $dp = \frac{2\alpha^2}{c^2} t\, dt$ and expressing $u = \frac{dx}{dt}$,

$$\int_{x_0}^{x} dx = \int_{1}^{p} \frac{c^2}{2\alpha\sqrt{p}} dp.$$

Integrating and letting $x = x_0$ when $t = 0$,

$$x - x_0 = \frac{c^2}{\alpha}\sqrt{p} - \frac{c^2}{\alpha}.$$

Substituting the expression for p back into the equation above and rewriting,

$$\left(x - x_0 + \frac{c^2}{\alpha}\right)^2 - c^2 t^2 = \frac{c^4}{\alpha^2}. \tag{11.22}$$

We can choose a spatial origin such that $x_0 = \frac{c^2}{\alpha}$. Then,

$$x^2 - c^2 t^2 = \frac{c^4}{\alpha^2}.$$

Note that we only need to consider the $x > 0$ region of this graph if $\alpha > 0$, and the $x < 0$ region otherwise. It is always possible to choose the coordinate system of frame S such that the initial conditions above ($u = 0$ and $x(0) = x_0 = \frac{c^2}{\alpha}$ at $t = 0$) are satisfied. As we can see, the motion of the point particle is a hyperbola on the space-time diagram in frame S.

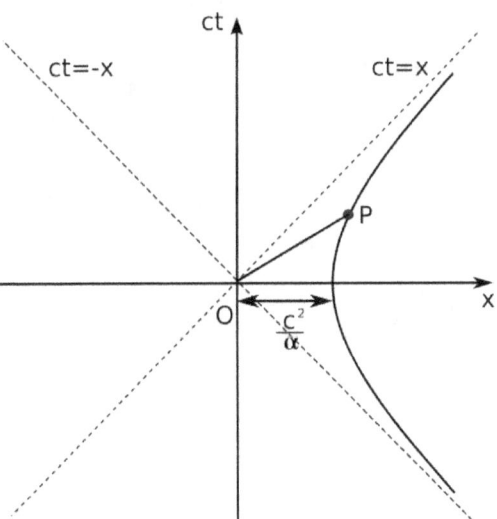

Figure 11.23: World line of particle undergoing constant proper acceleration in frame S

As $t \to \infty$, $x^2 \to c^2 t^2$ which implies that $|u| = |\frac{dx}{dt}| \to c$. This is a limit that makes sense in the context of special relativity as the particle's speed cannot surpass c. Now there are some interesting properties of this hyperbola. Consider a point P on the hyperbola shown in Fig. 11.23. The gradient of line OP can be calculated as follows. The x-coordinate of P in terms of t is

$$ x = \sqrt{\frac{c^4}{\alpha^2} + c^2 t^2} = \frac{c}{\alpha}\sqrt{c^2 + \alpha^2 t^2}. $$

The gradient is then

$$ m = \frac{ct}{x} = \frac{\alpha t}{\sqrt{c^2 + \alpha^2 t^2}} = \frac{u}{c}. $$

We see that the line OP is simply the superimposed instantaneous x'-axis of the MCRF (when the particle is at P) as it subtends an angle $\tan\theta = \frac{u}{c}$ with the positive x-axis! What this means is that a so-called pivot event, which is the origin O in this case, is always simultaneous with the instantaneous event of the particle in an MCRF. From Eq. (11.22), it can be seen that the pivot event is at $x_0 - \frac{c^2}{\alpha}$ in general for a point particle, undergoing constant proper acceleration α in the x-direction, that is located at $x = x_0$ with zero speed at $t = 0$ in frame S. If we choose the pivot event to be the origin of the particle's MCRFs too, the instantaneous event of the particle will also be at $t' = 0$ in its MCRFs. The next useful property is even a stranger one. The distance between the pivot event in an MCRF, and the instantaneous event of the particle in the MCRF is given by length contraction to be

$$ x' = \frac{x}{\gamma_u}. $$

γ_u can be computed as

$$ \gamma_u = \frac{1}{1 + \frac{u^2}{c^2}} = \frac{1}{\sqrt{1 - \frac{\alpha^2 t^2}{\alpha^2 t^2 + c^2}}} $$

$$ = \frac{\sqrt{\alpha^2 t^2 + c^2}}{c} = \frac{\alpha}{c^2} x. $$

Thus,

$$ x' = \frac{x}{\gamma_u} = \frac{c^2}{\alpha} $$

which is a constant value. A quicker way of proving this is to consider the invariant interval. Since the particle's event occurs at $t' = 0$ in the MCRF, we immediately obtain from the equation of the hyperbola

$$x'^2 = x^2 - c^2 t^2 = \frac{c^4}{\alpha^2}$$

$$\implies x' = \frac{c^2}{\alpha}$$

where we choose the positive value of x' as it corresponds to the regime of interest. We see that not only is the pivot event always simultaneous with the instantaneous event of the particle in an MCRF, the distance between them is uniform across all MCRFs[14]! The weird part is that even though the point particle accelerates away from the pivot point in frame S, the distance between them never changes as measured in its own instantaneous inertial frame. This is because the increase in the length contraction factor perfectly cancels the increase in distance between the two events as observed in frame S.

11.9.3 *Rigid Objects*

The classical definition of a rigid object is one whose particles maintain a constant separation in space. This is in fact impractical even in classical mechanics. The interactions between particles of an object are electromagnetic in nature, thus the speed of an electromagnetic wave imposes a speed limit on force propagation speed through the object. In fact, the force propagation speed in matter is the speed of sound in that medium. Thus, if one end of the object experiences a sudden change, such as an abrupt stop, the other end of the object cannot instantaneously respond to it. In the case of a sudden stop, at the next instant, the ends of the object will be closer to each other, compressing the object and thus changing the relative positions of the particles on the object. This limitation also holds in the context of special relativity, as signals cannot travel faster at a speed greater than c which is, theoretically, the maximum possible speed.

The next flaw in the classical definition of rigid objects pertains to a relativistic effect. A moving object in a certain inertial frame is length-contracted. Thus, an object does not maintain a constant separation in space across different inertial frames. In this sense, the criteria of maintaining a constant separation in space is ambiguous as there is no explicit mention of the frame of reference. Therefore, the classical definition of a rigid object makes no sense in the relativistic case.

---

[14] Actually, the invariant interval trivially and necessarily implies this.

Failure to take into account the effects above may lead to fallacious reasoning and seemingly paradoxical situations.

Problem: Consider a variant of the classic pole-and-barn paradox. A pole and barn have proper lengths L. The pole travels towards the barn at a velocity v in the barn's frame S. The end of the barn farther away from the rod, denoted as the rear end, is blocked by a massive and impenetrable door while its front end is initially open. In the barn's frame, the pole is length-contracted and is able to fit into the barn. When the back of the pole enters the barn, the door closes in the barn's frame and traps the pole. The pole then collides with the impermeable rear end of the barn and comes to a stop eventually.

In this frame, the back of the pole crosses the front of the barn. Thus, the front door of the barn can be closed. However, in the pole's frame S', the barn is length-contracted so the ladder is not able to fit inside the garage in the first place. How can the front door be closed then? In other words, does the back end of the pole really cross the front end of the barn?

The resolution to this apparent paradox is the fact that points on the pole are unable to stop instantaneously when the front end of the pole collides with the rear of the barn. Formally, we define the frame of the pole to be that of the particle at the rear tip of the pole as it is the last to stop. In frame S, the situation is depicted in Fig. 11.24.

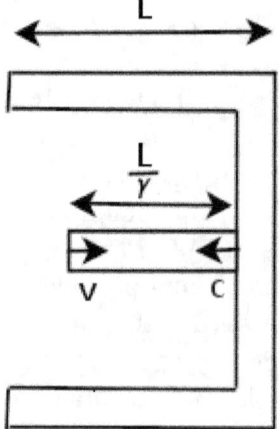

Figure 11.24:   Frame S

The pole is length-contracted to a length $\frac{L}{\gamma}$ in frame S. Thus, the back end of the pole definitely crosses the front end of the barn in frame S. In fact, the eventual distance between the rear ends of the pole and barn must be smaller than $\frac{L}{\gamma}$. Assume that a signal propagates at the speed of light c through the pole in frame S'. This is not really a physical situation but it can give us a rough notion of what happens in the boundary case by yielding an upper limit of the distance between the rear ends of the pole and barn. If the rear end of the pole crosses the front end of the barn under such an assumption, it will definitely do the same for an arbitrary signal speed which must be smaller than c. Next, it is assumed that when the signal passes by a certain part of the pole, that section immediately stops in the current frame.

Let the front end of the pole collide with the walls of the barn at $t = 0$ in frame S. The speed of the signal is still c in frame S as the speed of light is invariant across inertial frames. Then the time required for the signal to travel to the rear end of the pole in frame S is

$$t = \frac{L}{\gamma(c+v)},$$

as $c + v$ is the relative velocity between the signal and the back end of the pole in frame S while $\frac{L}{\gamma}$ is the Lorentz-contracted length of the pole in frame S. An important point to note here is that the rear of the pole does not stop traveling until it receives the signal (i.e. the rear end continues to move for a while after the front end collides with the barn). Then, the eventual distance between the rear ends of the pole and barn is simply that traveled by the signal during the time t above.

$$\Delta x = \frac{Lc}{\gamma(c+v)} = \sqrt{\frac{1-\beta}{1+\beta}} L.$$

In frame S', the situation is illustrated in Fig. 11.25.

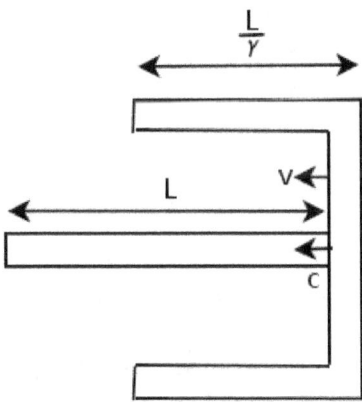

Figure 11.25: Frame S'

The length of the barn is contracted to $\frac{L}{\gamma}$ while the length of the pole is its proper length L. The barn is now approaching the stationary pole. Similarly, we define the time of the collision between the front end of the pole and the rear of the barn to be $t' = 0$ in frame S'. Then, the signal reaches the back end of the pole at

$$t' = \frac{L}{c}$$

in frame S'. During this time, the impermeable rear wall would have traveled a certain distance while compressing the pole. We assume that the velocity of the wall does not change in this process. This could approximately be attained if the mass of the rear door is large. The distance traveled by the wall during this time interval is $\frac{Lv}{c}$. Thus, the final distance between the rear ends of the barn and the pole is

$$\Delta x' = L - \frac{Lv}{c} = L(1 - \beta)$$

in frame S'. This is smaller than the length of the barn in S' as seen from the fact that

$$\frac{\Delta x'}{\frac{L}{\gamma}} = \sqrt{\frac{1 - \beta}{1 + \beta}} < 1.$$

Thus, the back end of the pole crosses the front of the barn in both frames and the front door of the barn can be closed. Incidentally, there is also another interesting result which agrees with the principle of relativity: the ratios of the eventual distance between the rear ends of the pole and barn to the observed length of the barn are identical in both frames. This can be seen from the fact that

$$\frac{\Delta x'}{\frac{L}{\gamma}} = \sqrt{\frac{1 - \beta}{1 + \beta}} = \frac{\Delta x}{L}.$$

Born Rigidity

Considering the ineptness of the classical definition of a rigid object in the context of special relativity, novel concepts of a rigid object have to be developed. Max Born proposed that rigid objects in special relativity obey the following property: the distance between all points on a rigid object is locally constant in the MCRF of any point on the object. This definition rectifies the loophole in the classical definition due to length contraction. However, this definition of a rigid object is still physically impossible as it does not circumvent the limitation of the speed of sound in a body which is the first flaw highlighted in the section above. Despite this, it is still a viable analog of the classical definition since the classical definition also idealized the propagation of "signals" within a body.

To fulfil the Born criterion, the proper accelerations of the points of the body must satisfy a certain relationship. In that sense, in order to achieve Born rigidity, the motion of an object has to be planned carefully beforehand. It is an extremely restrictive class of motions.

We shall consider the case where all points on the Born rigid body undergo a one-dimensional motion due to a constant proper acceleration in the x'-direction (where the primed frame is its MCRF). Recall that a point particle that is initially stationary at $x = x_0$ when $t = 0$ in the lab frame S and undergoing a constant proper acceleration α will follow a hyperbolic path with the pivot event at coordinate $x = x_0 - \frac{c^2}{\alpha}$. Furthermore, in any MCRF of the particle at an arbitrary time t in frame S, the difference in the x'-coordinates of the particle and the pivot event, measured simultaneously, is $\frac{c^2}{\alpha}$.

Therefore, if the pivot events of all points on an object coincide, the Born rigid condition will be satisfied! Consider two points on the object at coordinates x_1 and x_2, when $t = 0$ in frame S, undergoing proper accelerations α_1 and α_2 respectively. If their pivot events are concurrent and if we consider the MCRF of any of the two particles at any instant in time, the two particles have a difference in x'-coordinates of $\frac{c^2}{\alpha_1}$ and $\frac{c^2}{\alpha_2}$ with the pivot event, as measured simultaneously. The pivot events of the two particles are shown to be concurrent at the origin O in Fig. 11.26. The diagonal line represents a possible line of simultaneity if we were to consider the MCRF of either of the particles, defined at the corresponding point of intersection of the line with its hyperbolic path (actually, the two MCRFs are identical as the particles possess the same velocity at the points of intersection). The bold segment indirectly[15] reflects the difference in x'-coordinates between the two particles, measured simultaneously in their MCRF, as superimposed on the space-time diagram of frame S. The "proper length" between them is maintained at $\frac{c^2}{\alpha_2} - \frac{c^2}{\alpha_1}$ in the MCRF (Fig. 11.27) as the distance between each individual particle and the pivot event is constant. Furthermore, in order for

---

[15] Indirectly, in the sense that the magnitude of the difference in x'-coordinates is different from the length of the bold line in frame S (concretely, we must divide this length by the scaling factor $\sqrt{\frac{1+\beta^2}{1-\beta^2}}$, where β is the tangent of the angle of inclination of the line).

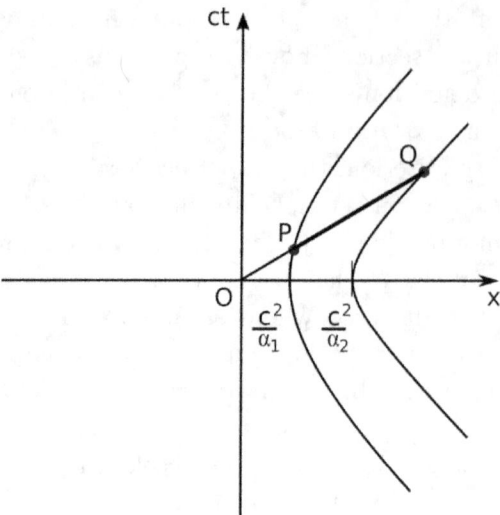

Figure 11.26:   World lines of two points on a Born rigid object in frame S

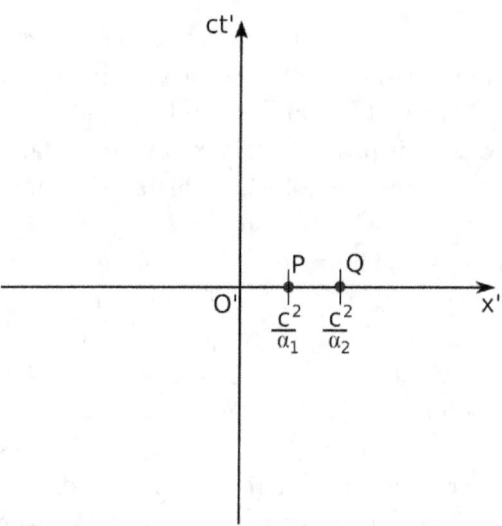

Figure 11.27:   Points P and Q in one MCRF

the pivot events to coincide in the first place,

$$x_1 - \frac{c^2}{\alpha_1} = x_2 - \frac{c^2}{\alpha_2}$$

$$\implies \frac{c^2}{\alpha_2} - \frac{c^2}{\alpha_1} = x_2 - x_1$$

which shows that the distance $x_2 - x_1$, which is also the initial distance at $t = 0$ in frame S, is maintained throughout the motion in any MCRF of either of the particles. Thus, if an object spans the entire region between coordinates $x = x_1$ and $x = x_2$ and the proper acceleration is α_0 at some x_0 ($x_1 \leq x_0 \leq x_2$), the proper acceleration, α, of a point at an arbitrary x-coordinate x ($x_1 \leq x \leq x_2$) must satisfy

$$\frac{c^2}{\alpha} = \frac{c^2}{\alpha_0} + x - x_0$$

in order for the object to be Born rigid. With this, the length of the rod in any MCRF is also maintained at $x_2 - x_1$. We see that this definition of a rigid body is extremely restrictive, as the motion of one point on the object constrains the motion of all other points if all points were to undergo constant proper accelerations. However, note that Born rigidity is merely one of the many proposed definitions of a relativistic rigid object — other less restrictive definitions have also been suggested.

11.9.4 *Relativistic Longitudinal Doppler Effect*

Consider a source, that emits waves (not necessarily electromagnetic), approaching a stationary observer at a speed v in frame S in Fig. 11.28.

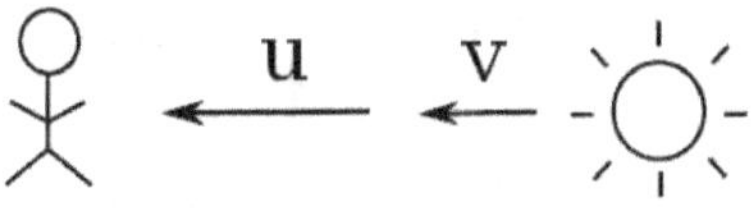

Figure 11.28: Source approaching an observer

We would like to determine the frequency of the waves received by the observer in frame S if the waves travel along the line joining the source and the observer. The frequency and the speed of the waves emitted in the frame of the source S' are f' and u' respectively.

There are two main effects which lead to a shift in the observed frequency here. The first is time dilation which causes the observed frequency of emission in the frame of the observer to differ as the source is moving. The second factor is the relative motion between the source and the observer during the time interval between consecutive emissions of wavefronts — the essence of the classical Doppler effect.

Let T and T' be the observed period of emission of the source in frames S and S' respectively. Then by time dilation, $T = \gamma T'$ as the observer sees the clock on the source running slow. In frame S, imagine a wavefront emitted at a certain instant. During the time interval between this instant and the release of the next wavefront, the source would have traveled a distance $vT = v\gamma T'$. The emitted wavefront would have traveled a displacement uT where u is the velocity of a wavefront in frame S. It can be computed via the velocity-addition formula.

$$u = \frac{u' + v}{1 + \frac{u'v}{c^2}}.$$

Thus, if we define λ and λ' to be the observed wavelengths of the waves in frames S and S' respectively, λ can be calculated as

$$\lambda = (u - v)T = \frac{u'\left(1 - \frac{v^2}{c^2}\right)}{1 + \frac{u'v}{c^2}}\gamma T' = \frac{u'}{\gamma\left(1 + \frac{u'v}{c^2}\right)}T',$$

which corresponds to the distance between consecutive wavefronts in frame S. Thus, the observed frequency of waves received by the observer is

$$f = \frac{u}{\lambda} = \frac{\gamma(u' + v)}{u'}\frac{1}{T'} = \gamma\left(1 + \frac{v}{u'}\right)f'. \tag{11.23}$$

In the case of light, $u' = c$ and we obtain

$$f = \frac{1 + \frac{v}{c}}{\sqrt{1 - \frac{v^2}{c^2}}}f' = \sqrt{\frac{1 + \beta}{1 - \beta}}f'. \tag{11.24}$$

Remember that v is defined to be positive if the source and observer are approaching each other and negative if they are retracting away from each other. When $v > 0$, the frequency of the received waves is larger in frame S than S' and the waves are said to be blue-shifted (higher frequency and thus shorter wavelength). When $v < 0$, the converse occurs and the waves are said to be red-shifted (lower frequency and thus longer wavelength). This result is truly relativistic as it only depends on the relative velocity between the source and observer as observed in the frame of one — as opposed to the non-relativistic Doppler effect which has different dependencies on the velocities of the observer and source in the lab frame.

Lastly, there is generally a distinction between an observer observing and seeing something. In the context of waves, when we refer to the frequency of the emitted waves as observed by an observer, we usually mean the frequency of the waves that are emitted at the source in the frame of the observer (i.e. the emission event is of concern). On the other hand, the frequency of waves as seen by an observer explicitly refers to the frequency of the waves that reaches him as observed in his frame (i.e. the receiving event is of concern). In the case of the longitudinal Doppler effect, the frequency of emission observed by the observer is $\frac{f'}{\gamma}$ (due to time dilation solely) while the frequency of waves seen by the observer is $\sqrt{\frac{1+\beta}{1-\beta}}f'$ (both mentioned effects have to be accounted for).

Problems

As a word of advice, it is often easier to express everything in units of c, the speed of light in a vacuum, to preclude c's from floating around everywhere. For example, instead of $v = 0.9c$, one can rewrite it as $v = 0.9$. γ then becomes $\gamma = \frac{1}{\sqrt{1-v^2}}$. Afterwards, one can add back the c's at the appropriate positions by dimensional analysis. For instance, $u' = \frac{u+v}{1+uv}$ in units of c becomes $u' = \frac{u+v}{1+\frac{uv}{c^2}}$ by observing that there is an addition between uv, which has units in $m^2 s^{-2}$, and a constant 1, which is dimensionless.

Fundamental Consequences

1. *Superluminal Travel**

Adrian shines a laser towards the Moon and forms a red spot on a crater. He claims that if he twists his wrist, the spot on the Moon will travel a great distance in a very short amount of time and thus achieve a superluminal speed — thus violating special relativity. What is wrong with his reasoning?

Now, Betty invents the following thought experiment. Suppose that you build a pair of scissors with very long blades. If you decrease the angle between the handles of the scissors during a certain time interval, the angle between the blades should also decrease by the same amount in the same time interval. Then, points arbitrarily far away from the joint should travel at superluminal speeds as the angular distance covered is fixed! Where does Betty's idea fail?

2. *Muon Decay**

Muons have a half-life of proper time t_h. They are released at a distance L above the surface of the Earth and travel at a constant velocity v towards the Earth. What is the proportion of muons that reach the surface of the Earth? Solve this problem from both the muons' frame and the Earth's frame.

3. *Rod**

Consider two frames S and S'; S' is traveling at a velocity v along the positive x-axis of frame S. A rod, of length L as measured in its rest frame S, subtends an angle θ_1 with the x-axis in frame S. Find the angle subtended by the rod and the x'-axis, θ_2, in frame S'.

4. Ladder-and-Barn Paradox*

A ladder of proper length, L, travels at a relative velocity v towards a barn with proper length L. The barn has two doors at its ends that are initially open. From the frame of the barn, frame S, the ladder is length-contracted and thus fits into the barn. However, from the frame of the ladder, S', the barn is length-contracted and thus the ladder does not fit. Resolve this apparent paradox. Consider the following: one can show that the ladder fits in the garage by closing the two doors of the barn simultaneously in the current reference frame during the brief period of time that the ladder is completely inside the barn. The doors are then opened to release the ladder when it is about to collide with the doors (so that the doors do not affect the ladder's motion).

5. Spaceships*

Consider a spaceship A of proper length L traveling towards an identical spaceship B at a relative velocity v. When the right of A reaches the right of B, a cannon is simultaneously fired from the left end of A in S', the frame of spaceship A. In frame S', spaceship B is length-contracted which causes the cannon to miss. However in ship B's frame S, spaceship A is length-contracted and the cannonball seemingly hits. Resolve this apparent paradox. Warning: misleading figure.

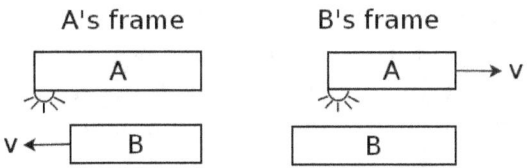

Velocity Addition and Doppler Effect

6. Stellar Aberration*

A stationary light source is situated at the origin of frame S. It emits a flash that is received by a receding observer traveling at a velocity v in the positive x-direction. Let the observer's frame be S'. If θ and θ' are the angles subtended by the path of the light and the positive x-axis in frame S and the positive x'-axis in frame S' respectively, show that

$$\cos \theta' = \frac{\cos \theta - \beta}{1 - \beta \cos \theta}$$

where $\beta = \frac{v}{c}$ is defined to be positive in the positive x or x'-direction.

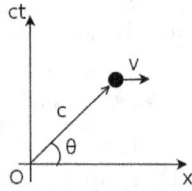

7. Relative Speed*

Consider two particles traveling at constant velocities v and u in the lab frame. The angle subtended by their velocities is θ. Find the speed of one particle in the frame of the other.

8. Another Velocity Addition Derivation*

A train of proper length L is moving longitudinally with velocity v relative to a stationary observer on the ground. A person inside the train, standing at the tail end of the train, throws a ball horizontally with constant velocity u towards the front (assume that there is no gravity) as observed in his own frame, and simultaneously sends a light signal in the same direction. The light hits the front end of the train and is reflected back, meeting the ball at some point. This meeting point of interest is a certain distance from the tail end of the train (as observed in the ground or train's frame).

(a) Find the ratio R of this distance to the proper length of the train L in the train's frame.
(b) In the ground frame, find the ratio R' of this distance to the observed length of the train. In your answer, let the length of the train and the velocity of the ball be L' and u', respectively in the ground frame.
(c) What can you say about your answers in a) and b)? Explain. Hence, derive u' in terms of u and v.

9. Two Trains**

Two identical trains are traveling at speeds $\frac{4c}{5}$ and $\frac{3c}{5}$ towards the right in frame S. The faster train is initially behind the slower train. Define events

P and Q to be the front of the faster train crossing the back of the slower train and the back of the faster train crossing the front of the slower train respectively. When event P occurs in frame S, an observer R begins walking from the back of the slower train to the front of the slower train. Coincidentally, the time during which he reaches the front of the slower train coincides with event Q. Find the velocity of the faster train in the frame of the slower train. Thus, find the speed of observer R in the frame S. ("An Introduction to Mechanics")

10. *Velocity Additions via Rapidity**

All velocities in this problem are assumed to be aligned in the x-direction. The rapidity ϕ of a particle or frame with respect to a frame S is defined as

$$\tanh \phi = \beta = \frac{v}{c},$$

where v is the velocity of the particle or frame with respect to S. $\tanh \phi = \frac{e^\phi - e^{-\phi}}{e^\phi + e^{-\phi}}$ is the hyperbolic tangent function. Show that if a particle has rapidity ϕ_1 with respect to frame 1 and frame 1 has rapidity ϕ_2 with respect to frame 2, the particle has rapidity $\phi_1 + \phi_2$ with respect to frame 2. It may be useful to know that $\tanh(\phi_1 + \phi_2) = \frac{\tanh \phi_1 + \tanh \phi_2}{1 + \tanh \phi_1 \tanh \phi_2}$.

Now, consider a particle which travels at a velocity v_1 with respect to frame S_1, which travels at velocity v_2 with respect to frame S_2, which travels at velocity v_3 with respect to frame S_3, and so on until frame S_{n-1} which travels at velocity v_n with respect to frame S_n. All of these velocities are aligned. Show that the velocity of the particle in frame S_n is

$$u = c \cdot \frac{\prod_{i=1}^N (1 + \beta_i) - \prod_{i=1}^N (1 - \beta_i)}{\prod_{i=1}^N (1 + \beta_i) + \prod_{i=1}^N (1 - \beta_i)}$$

where $\beta_i = \frac{v_i}{c}$.

11. *Collision**

In the lab frame S, a particle is traveling at a velocity v towards an identical, stationary particle. From classical mechanics, we know the resultant velocities of the two particles must be perpendicular after the imminent collision as they have equal masses. Show that it is impossible for the two particles to have non-zero velocities that are perpendicular in special relativity by considering another inertial frame where the situation is symmetric, and assuming that the dynamical laws are reversible. We do not know anything else about the dynamical laws in special relativity now. In fact, you can show that the angle of separation must be smaller than 90°. Finally, prove the classical result, that a head-on collision causes the particles to exchange velocities in frame S, holds in the context of special relativity.

12. *General Doppler Effect***

A source that emits photons at frequency f' in its own frame, S', is moving across the field of vision of a stationary observer at the origin in the frame of the observer, S. What is the observed frequency of emissions by the source in frame S? Now, what is the frequency of the photons emitted at angle θ in the left diagram below, as seen by the observer when the photons eventually reach his eyes? At the instant where the source is at the closest distance of approach to the observer, what is the frequency of the photons that enter the eyes of the observer? When the observer sees the source at the closest distance of approach, what is the frequency of the photons that enter the observer's eyes? You may find the pictures below to be useful.

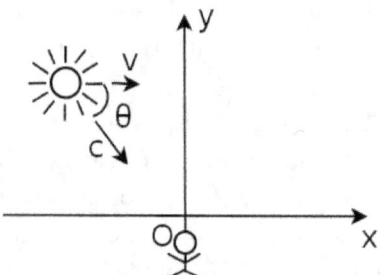

Figure 11.29:   General situation

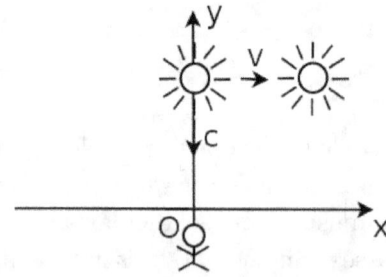

Figure 11.30:   Last situation

13. *The Twins' Paradox Revisited***

Consider the twins' paradox set-up again. Now the two twins send out a radio pulse once per second in their own frames. As before, twin A travels to the distant star, that is a distance L from the Earth in twin B's frame, and back at speed v while twin B remains on the Earth. During the entire process,

(a) How many pulses did twin A broadcast in total?
(b) How many pulses from B did A receive in total? Hence, who does twin A conclude to be younger?
(c) How many pulses did twin B broadcast in total?
(d) How many pulses from A did B receive in total? Hence, who does twin B conclude to be younger?

14. *Moving Glass***

In lab frame S, a stationary source emits light of frequency f in vacuum, in the positive x-direction. The photons then pass through a glass block of refractive index n and proper length l that is traveling at a velocity v in the positive x-direction. Determine the time taken by the light to cross the block, the frequency and wavelength of light inside the block in the frame of the block, S'. Ditto for the lab frame.

Minkowski Diagrams

15. *Simultaneous Lamps***

In the lab frame S, three lamps at coordinates x_1, x_2 and x_3 are observed to be illuminated at times t_1, t_2 and t_3. At $t = 0$ in S, a car is observed to travel from the origin at a constant velocity $v > 0$ in the positive x-direction. Under what conditions will the person P in the car observe all three lamps to be lit up simultaneously? Next, assume that P observes the events to occur at $t' = 0$ in his own frame S'. Let the time intervals between the illumination of the lamps and the receipt of the corresponding photons be Δt_1, Δt_2 and Δt_3 in frame S'. If person P observes the ratio of these intervals to be $\Delta t_1 : \Delta t_2 : \Delta t_3 = 1 : 2 : 3$, and given x_1, determine the x-coordinates of the other lamps in frame S and the times at which the lights were lit up in frame S. Solve this problem via a Minkowski diagram.

16. *Diverging Cars* **

In the lab frame, car 1 travels at speed $v_1 = \tan 15°c$ in the negative x-direction while car 2 travels at speed $v_2 = \frac{c}{\sqrt{3}}$ in the positive x-direction. The cars start from the origin O at time $t = 0$. At a certain later time, car 1 emits a light signal in the positive x-direction. If an observer in car 1 measures the time interval between the emission event and the receiving event by car 2 to be t', determine the distance that car 2 has traveled from its initial position in the lab frame when it receives the light signal with the aid of a Minkowski diagram.

Proper Time

17. *Particle's Motion* *

The velocity of a particle as a function of time t in the lab frame is given by

$$u(t) = c\sqrt{1 - \frac{1}{\left(\frac{gt}{c} + 1\right)^2}}$$

and is oriented along the positive x-axis.

(a) Show that the proper time elapsed in the particle's frame is $\tau = \frac{c}{g}\ln(\frac{gt}{c} + 1)$.
(b) Let x denote the instantaneous x-coordinate of the particle in the lab frame. If the particle starts at the origin in the lab frame originally, show that $x(\tau) = \frac{c^2}{g}[\sqrt{e^{\frac{2g\tau}{c}} - 1} - \tan^{-1}\sqrt{e^{\frac{2g\tau}{c}} - 1}]$.

18. *World-Line* *

In the standard configuration, a particle moves in the x-direction. In the lab frame, its x-coordinate is described by

$$x(\tau) = \frac{c^2}{g}\left(\cosh\left(\frac{g\tau}{c}\right) - 1\right),$$

where g is a constant with units of acceleration and τ is the proper time of the object. Define γ_u as the gamma factor ascribed to the speed of the object u in the lab frame.

(a) Express u in terms of γ_u, g, c and τ.
(b) Hence, express γ_u in terms of g, c and τ.
(c) Using the result of (b), re-express u solely in terms of g, c and τ.
(d) Express t as a function of τ and hence, $u(t)$ and $a(t)$. Show that $u(t)$ and $a(t)$ make sense for $t \to \infty$.

Solutions

1. Superluminal Travel*

Let the distance between Adrian and the Moon be L. After Adrian has shifted the direction of the laser, it takes time, on the order of $\frac{L}{c}$, for photons emitted in this new direction to form a spot on the Moon. Suppose that the direction of the laser changes by a small angle θ in time Δt. The distance that the spot moves is on the order of $L\theta$. Therefore, the "average velocity" of the spot is on the order of $\frac{L\theta}{\frac{L}{c}} = \frac{c}{\theta} > c$ in the time interval Δt, especially if θ is small. The spot on the Moon seemingly achieves superluminal travel! Well, the resolution to this paradox is that the spot is not a physical entity and is unable to carry information. The spot simply marks the location at which incident photons impinge on — its movement is no different from a fickle cartographer suddenly placing a dot on his map to define a new origin (and this requires no time). Therefore, the spot does not need to comply to physical laws such as special relativity and its speed can seemingly exceed the speed of light (but it makes no sense to define a speed for such an intangible construct anyway). However, the mediating particles, which are photons in this case, must still be unable to achieve superluminal speeds.

Betty's argument breaks down when she claims that the angle between the blades should also decrease by the same amount in the same time interval as the rigid body assumption is inherently flawed. The scissors cannot remain rigid and points on the blade do not cover the same angular distance in the same time interval (even if they rotate by the same angle so eventually). Firstly, it takes time for signals to travel from the handles to points on the blade to inform them that they should move. Therefore, points far away will begin moving at a later time and the rigid body assumption fails. Afterwards, when different points on the blade start to move, they still cannot move at a speed faster than the speed of light. Hence, they cannot "teleport" to the correct positions to maintain the rigid body property. The points on the blade do not and need not cover the same angular distance in the same time interval and hence Betty's idea fails.

2. Muon Decay*

From the frame of the muons, the distance between its initial position and the Earth is length-contracted. Thus, the total time taken for the journey in the muons' frame is

$$t' = \frac{L}{\gamma v}.$$

In the frame of the Earth, the muons require a time interval of

$$t = \frac{L}{v}$$

to reach the Earth. However, an observer on the Earth will observe that the clocks on the muons tick slower due to time dilation. Thus, the time elapsed on the clocks of the muons during this process is

$$t' = \frac{t}{\gamma} = \frac{L}{\gamma v},$$

which is consistent with the result above. The proportion of the muons that reach the Earth is then

$$2^{-\frac{L}{\gamma t_h v}},$$

as the proportion left after n half lives is 2^{-n}.

3. Rod*

The difference between the x and y-coordinates of the two ends of the rod in frame S at simultaneous times are

$$\Delta x = L \cos \theta_1,$$

$$\Delta y = L \sin \theta_1,$$

where L is the length of the rod in frame S. The difference between the x'- and y'-coordinates of the two ends of the rod in frame S' at simultaneous times are

$$\Delta x' = \frac{L \cos \theta_1}{\gamma},$$

$$\Delta y' = L \sin \theta_1,$$

due to length contraction. Remember that length contraction does not occur in the transverse direction. Then, the new angle subtended by the rod and the x'-axis is

$$\theta_2 = \tan^{-1} \frac{\Delta y'}{\Delta x'} = \tan^{-1}(\gamma \tan \theta_1).$$

4. Ladder-and-Barn Paradox*

The resolution is that it is perfectly fine for observers in the different frames to reach different conclusions as to whether the ladder will fit into the barn. In this problem, we define the fronts and backs of the ladder and barn to be the sides that are the closest to and furthest away from each other. Even though the front and back doors of the barn may trap the ladder for a period of time in frame S, they fail to do so in frame S' due to the relativity of simultaneity. Concretely, let the origins of the two frames, O and O', coincide at $t = t' = 0$. At $t = 0$, the rear end of the ladder crosses the front of the barn in frame S. Thus at this instant, both front and rear doors close simultaneously in frame S. Let events 1 to 4 be defined as that of the front door closing, rear door closing, front door opening and rear door opening respectively. The (x, ct) coordinates of these events in frame S are

$$\text{Event 1: } (0, 0),$$

$$\text{Event 2: } (L, 0),$$

$$\text{Event 3: } \left(0, \frac{Lc}{v}\left(1 - \frac{1}{\gamma}\right)\right),$$

$$\text{Event 4: } \left(L, \frac{Lc}{v}\left(1 - \frac{1}{\gamma}\right)\right).$$

The ct-coordinates of events 3 and 4 are obtained by utilizing the fact that the gap between the front end of the ladder and the rear of the barn is $L - \frac{L}{\gamma}$ at time $t = 0$ as the ladder is length-contracted in frame S. Evidently, the ladder is completely inside the barn during the time interval between $t = 0$ and $t = \frac{L}{v}(1 - \frac{1}{\gamma})$ in frame S. Applying the Lorentz transformations to these events, the corresponding coordinates in frame S' can be obtained.

$$\text{Event 1: } (0, 0),$$

$$\text{Event 2: } \left(\gamma L, -\frac{\gamma L v}{c}\right),$$

$$\text{Event 3: } \left(L(1 - \gamma), \frac{Lc}{v}(\gamma - 1)\right),$$

$$\text{Event 4: } \left(L, \frac{Lc}{v}\left(\frac{1}{\gamma} - 1\right)\right).$$

Note that the back of the rod still reaches the front of the barn at event 1 and that the front of the rod still reaches the back of the barn at event 3. This is because, they are technically the same events as they occur at the same x-coordinate and time in frame S. Rearranging these events in chronological order in S',

$$\text{Event 2: } \left(\gamma L, -\frac{\gamma L v}{c}\right),$$

$$\text{Event 4: } \left(L, \frac{Lc}{v}\left(\frac{1}{\gamma}-1\right)\right),$$

$$\text{Event 1: } (0,0),$$

$$\text{Event 3: } \left(L(1-\gamma), \frac{Lc}{v}(\gamma-1)\right).$$

This means that the rear door of the barn first closes before the front of the ladder reaches the back of the barn and then opens when the front end of the ladder reaches the rear of the barn such that the ladder is released. Afterwards, the front door of the barn closes as the rear of the ladder passes by it. Lastly, the front door opens. It can be seen that there is no moment at which the ladder is completely trapped within the closed doors in frame S' due to the relativity of simultaneity. It is perfectly fine for observers in the two frames to reach different conclusions in this set-up, as whether the ladder fits into the barn is merely a human construct and not a physical event. A brief outline of the entire process in the two frames is depicted in Figs. 11.31 and 11.32.

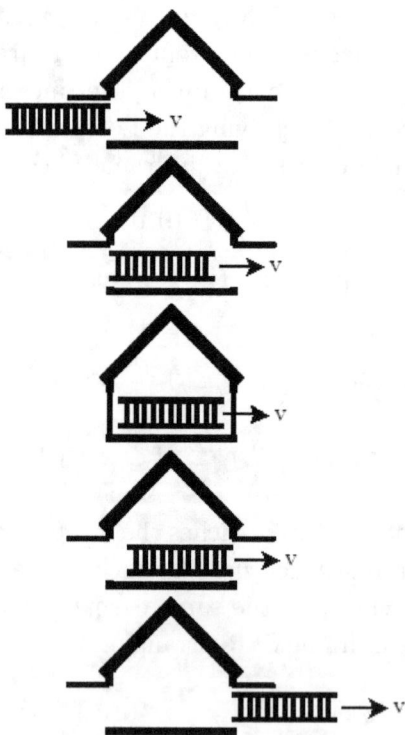

Figure 11.31:   Situation in frame S

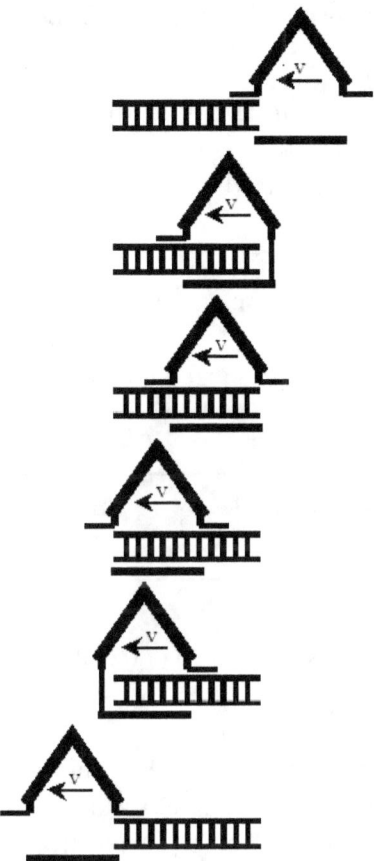

Figure 11.32:   Situation in frame S'

5. Spaceships*

The second reasoning is flawed as the event of the right of A reaching the right of B and the firing of the cannon are no longer simultaneous in frame S (the diagram is misleading). Let the origin of the two frames, O and O', coincide at time $t = t' = 0$. Let the (x', ct') coordinates of the firing event in frame S' be

$$\text{Firing Event: } (0,0),$$

and that of the right of A reaching the right of B in S' be

$$\text{Event AB: } (L,0).$$

The cannon obviously misses as the left and right ends of spaceship B are at x-coordinates $L(1 - \frac{1}{\gamma})$ and L at $t' = 0$ respectively. Then, the (x, ct)

coordinates of these two events in frame S can be obtained via the Lorentz transformations.

$$\text{Firing Event: } (0,0).$$

$$\text{Event AB: } \left(\gamma L, \frac{\gamma L v}{c}\right).$$

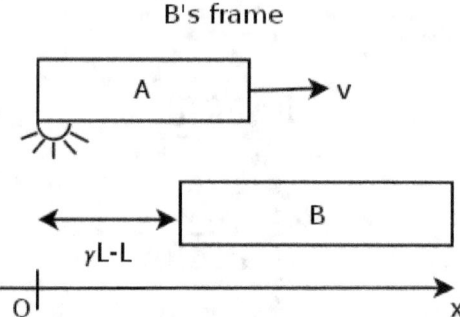

Figure 11.33:   Corrected diagram of the situation in B's frame, S

Referring to Fig. 11.33, it can be seen that the firing event occurs before the right of A reaches the right of B! This can also be concluded from the fact that the firing event is the rear clock as observed in frame S, which leads the front clock (event AB) by $\frac{\gamma L v}{c^2}$. Next, from the coordinates of event AB, we can conclude that the rear end of the stationary spaceship B is at x-coordinate

$$x = \gamma L - L > 0$$

at all times t in S, where 0 is the x-coordinate of the firing event. Hence, the cannon still misses in S.

6. Stellar Aberration*

Since the speed of light is c in both S and S', the x and x'-components of the photon's velocities in S and S' are $c\cos\theta$ and $c\cos\theta'$, respectively. By the longitudinal velocity addition formula,

$$c\cos\theta' = \frac{c\cos\theta - v}{1 - \frac{v}{c}\cos\theta}$$

$$\implies \cos\theta' = \frac{\cos\theta - \beta}{1 - \beta\cos\theta}.$$

7. Relative Speed*

Let $u_{\parallel}$ and $u_{\perp}$ be the components of $\boldsymbol{u}$ parallel and perpendicular to $\boldsymbol{v}$ respectively. Then,

$$u_{\parallel} = u \cos \theta,$$

$$u_{\perp} = u \sin \theta.$$

Thus, in the frame of the particle traveling at velocity $\boldsymbol{v}$, these two components obey the following velocity transformation rules:

$$u'_{\parallel} = \frac{u \cos \theta - v}{1 - \frac{uv \cos \theta}{c^2}},$$

$$u'_{\perp} = \frac{u \sin \theta}{\gamma_v \left(1 - \frac{uv \cos \theta}{c^2}\right)},$$

where $\gamma_v = \frac{1}{\sqrt{1 - \frac{v^2}{c^2}}}$. Thus, the speed of the other particle in this frame is

$$u' = \sqrt{u'^2_{\parallel} + u'^2_{\perp}} = \frac{\sqrt{u^2 - 2uv \cos \theta + v^2 - \frac{u^2 v^2 \sin^2 \theta}{c^2}}}{1 - \frac{uv \cos \theta}{c^2}}.$$

8. Another Velocity Addition Derivation*

(a) In the train's frame, the total distance traversed by the ball and the photon until their collision is $2L$. Therefore, the time of collision is $\frac{2L}{u+c}$ and the distance between the ball and the back end of the train at this juncture is $\frac{2Lu}{u+c}$.

$$R = \frac{2Lu}{L(u+c)} = \frac{2u}{u+c}.$$

(b) Since the distance between the front end of the train and the photon narrows at a rate $c - v$ in the ground frame, the distance traversed by the photon until it impinges the front end of the train is $\frac{L'c}{c-v}$. The total distance covered by the ball and the photon until their collision is then $\frac{2L'c}{c-v}$ such that the time of collision is $\frac{2L'c}{(c-v)(u'+c)}$. The distance between the ball and the back end of the train at this juncture is $\frac{2L'c(u'-v)}{(c-v)(u'+c)}$ as it increases at a rate of $u' - v$.

$$\Longrightarrow R' = \frac{2c(u' - v)}{(c - v)(u' + c)}.$$

(c) The two ratios must be equal, else this experiment can be used to distinguish between the two inertial frames — violating the principle of relativity.

$$R = R' \Longrightarrow \frac{2u}{u + c} = \frac{2c(u' - v)}{(c - v)(u' + c)}$$

$$u(c - v)(u' + c) = c(u' - v)(u + c)$$

$$u(c - v)u' + cu(c - v) = c(u + c)u' - c(u + c)v$$

$$(c^2 + uv)u' = c^2(u + v)$$

$$u' = \frac{u + v}{1 + \frac{uv}{c^2}}.$$

9. Two Trains**

Let us define the faster and slower trains to be A and B respectively. By the relativistic longitudinal velocity addition formula, the velocity of train A in the frame of B is

$$v_A' = \frac{\frac{4}{5}c - \frac{3}{5}c}{1 - \frac{12}{25}} = \frac{5}{13}c.$$

Let us now consider the frame of the observer. In the frame of the observer, only the two trains are moving and he or she is stationary. The two trains must travel at velocities of equal magnitudes and opposite directions in order for events P and Q to occur at the location of the observer. In search of a contradiction, suppose that the trains traveled at different speeds — the faster train would be length-contracted to a greater extent while traveling at a greater speed, causing its end to reach the observer in a shorter time. Let the magnitude of these velocities be v. Thus, train A travels at speed v towards the right while B travels at speed v towards the left. The velocity of train A in the frame of train B obtained by applying the velocity addition formula should be the same as that derived earlier. Thus,

$$\frac{2v}{1 + \frac{v^2}{c^2}} = \frac{5}{13}c \Longrightarrow v = \frac{1}{5}c$$

where we have rejected the impractical solution $v = 5c$. Next, let the velocity of the observer be u in frame S. Then again, the relativistic addition of $\frac{1}{5}c$

to u should give $\frac{4}{5}c$ which corresponds to the velocity of train A in frame S. Thus,

$$\frac{\frac{1}{5}c + u}{1 + \frac{u}{5c}} = \frac{4}{5}c.$$

Solving,

$$u = \frac{5}{7}c.$$

10. Velocity Additions via Rapidity**

Let $\beta_1 = \tanh \phi_1$, $\beta_2 = \tanh \phi_2$ and β' be the β factor of the particle in frame 2. By the relativistic velocity addition formula,

$$\beta' = \frac{\beta_1 + \beta_2}{1 + \beta_1 \beta_2} = \frac{\tanh \phi_1 + \tanh \phi_2}{1 + \tanh \phi_1 \tanh \phi_2} = \tanh(\phi_1 + \phi_2).$$

Therefore, the rapidity is additive across frames. In the second part of the question, the rapidity of the particle with respect to frame S_N is

$$\phi' = \sum_{i=1}^{N} \phi_i.$$

Then,

$$u = c \tanh \phi' = c \cdot \frac{e^{\sum_{i=1}^{N} \phi_i} - e^{-\sum_{i=1}^{N} \phi_i}}{e^{\sum_{i=1}^{N} \phi_i} + e^{-\sum_{i=1}^{N} \phi_i}}.$$

Now, simple manipulations of $\tanh \phi = \frac{e^\phi - e^{-\phi}}{e^\phi + e^{-\phi}}$ yield

$$e^{\phi_i} = \frac{1 + \tanh \phi_i}{\sqrt{1 - \tanh^2 \phi_i}} = \frac{1 + \beta_i}{\sqrt{1 - \beta_i^2}}.$$

Thus,

$$u = c \cdot \frac{\prod_{i=1}^{N} \left(\frac{1+\beta_i}{\sqrt{1-\beta_i^2}} \right) - \prod_{i=1}^{N} \left(\frac{1-\beta_i}{\sqrt{1-\beta_i^2}} \right)}{\prod_{i=1}^{N} \left(\frac{1+\beta_i}{\sqrt{1-\beta_i^2}} \right) + \prod_{i=1}^{N} \left(\frac{1-\beta_i}{\sqrt{1-\beta_i^2}} \right)}$$

$$= c \cdot \frac{\prod_{i=1}^{N}(1 + \beta_i) - \prod_{i=1}^{N}(1 - \beta_i)}{\prod_{i=1}^{N}(1 + \beta_i) + \prod_{i=1}^{N}(1 - \beta_i)}.$$

11. Collision**

In the lab frame S, define our coordinate system such that v is in the positive x-direction. Number the moving particle 1 and the other particle 2. Now, consider the inertial frame S' where particle 1 travels at velocity u while particle 2 travels at velocity $-u$ in the x'-direction. Since particle 2 was stationary in S, S' must travel at velocity u relative to S. Thus, u can be determined by applying the velocity addition formula to v.

$$\frac{v-u}{1-\frac{uv}{c^2}} = u.$$

The actual value of u can be solved from the quadratic equation above but it is not particularly important here. The pivotal point is that it exists for some value smaller than c. Now in this frame, the situation exhibits mirror symmetry. Therefore, if the final velocity of particle 2 in S' makes an angle θ anti-clockwise with the positive x'-axis, the final velocity of particle 1 in S' must subtend an angle θ anti-clockwise with the negative x'-axis. Furthermore, the magnitude of their final velocities must still be u as the set-up must be reversible. Then, the final velocities of particles 1 and 2 are $(-u\cos\theta, -u\sin\theta)$ and $(u\cos\theta, u\sin\theta)$ in frame S'. The y-direction is chosen such that the velocities lie in the xy-plane. The final velocities of the two particles in frame S can then be obtained from the velocity addition formula, as S' travels at velocity u relative to S.

$$\boldsymbol{v_1} = \begin{pmatrix} \frac{u-u\cos\theta}{1-\frac{u^2\cos\theta}{c^2}} \\ \frac{-u\sin\theta}{\gamma_u\left(1-\frac{u^2\cos\theta}{c^2}\right)} \end{pmatrix},$$

$$\boldsymbol{v_2} = \begin{pmatrix} \frac{u+u\cos\theta}{1+\frac{u^2\cos\theta}{c^2}} \\ \frac{u\sin\theta}{\gamma_u\left(1+\frac{u^2\cos\theta}{c^2}\right)} \end{pmatrix},$$

where $\gamma_u = \frac{1}{\sqrt{1-\frac{u^2}{c^2}}}$. The dot product of the two velocities is

$$\boldsymbol{v_1}\cdot\boldsymbol{v_2} = \frac{u^2(1-\cos^2\theta)}{1-\frac{u^4\cos^2\theta}{c^4}} - \frac{u^2\sin\theta}{\gamma_u^2\left(1-\frac{u^4\cos^2\theta}{c^4}\right)} = \frac{u^4\sin^2\theta}{c^2\left(1-\frac{u^4\cos^2\theta}{c^4}\right)}.$$

This expression can only be zero when $\theta = 0$ or $\theta = \pi$. However, in both cases, the dot product is zero, not because the velocities are perpendicular but because one of the velocities is zero. Therefore, it is impossible for the particles to leave with perpendicular non-zero velocities in frame S. In fact, if we let ϕ denote the angle of separation between v_1 and v_2,

$$\cos \phi = \frac{v_1 \cdot v_2}{|v_1||v_2|} > 0$$

which implies that $|\phi| < \frac{\pi}{2}$. Next, when the particles undergo a head-on collision, $\theta = 0$ which yields

$$v_1 = \begin{pmatrix} 0 \\ 0 \end{pmatrix},$$

$$v_2 = \begin{pmatrix} v \\ 0 \end{pmatrix}.$$

We do not need to take an intermediate step in explicitly evaluating u to determine $\frac{2u}{1+\frac{u^2}{c^2}}$ in the x-component of v_2, as it is simply the inverse transformation, from S' to S, of the transformation from v in S to u in S'.

12. General Doppler Effect**

In all scenarios, the observer observes the source to emit a wavefront every $T = \gamma T'$ seconds. Thus, the observer observes the source to emit at a lower frequency $f = \frac{f'}{\gamma}$. Moving onto the general problem, consider Fig. 11.34.

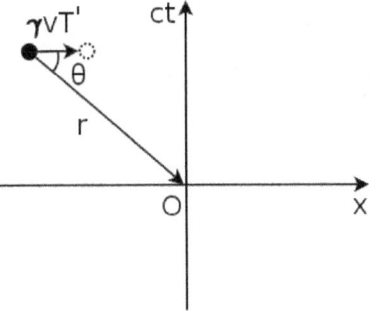

Figure 11.34: Transverse Doppler effect at an arbitrary angle in frame S

Consider two beams emitted by the moving light source in frame S. The time interval between these two beams is $T = \gamma T'$ in S. In this time interval, the source travels $v \cos \theta T = \gamma v T' \cos \theta$ along the first beam — thus narrowing the wavelength. The perceived wavelength of these light waves is then

$$\lambda = (c - v \cos \theta)T = \gamma c T'(1 - \beta \cos \theta).$$

Thus, the perceived frequency of these wavefronts by the observer is

$$f = \frac{c}{\lambda} = \frac{f'}{\gamma(1 - \beta \cos \theta)}.$$

Now, we shall analyze how to substitute θ for the last two situations in the problem. In the first scenario, the light that reaches the eyes of the observer must have been emitted before the source reaches the point of closest approach in frame S. In the second situation, the wavefront of concern is emitted when the light source reaches the closest point of approach in frame S. Thus, the wavefront is emitted vertically downwards in frame S. Evidently, $\theta = \frac{\pi}{2}$ radians in the second case — leading to a perceived frequency of

$$f = \frac{f'}{\gamma}.$$

Now, in the first situation, $\cos \theta = \frac{v}{c}$ as the ratio between the distance traveled by the beam (emitted before crossing the y-axis) and the distance traveled by the source must be $c : v$. Substituting this expression into the formula for f,

$$f = \gamma f'.$$

13. The Twins' Paradox Revisited**

(a) Again, there are two different inertial frames associated with twin A during his outbound and inbound journeys. In both of twin A's frames, the distance between the Earth and the star is length-contracted. Thus, the whole process takes

$$t_A = \frac{2L}{\gamma v},$$

and he releases

$$N_A = \frac{2L}{\gamma v}$$

pulses.

(b) During the first half of the journey, twin A receives pulses from B at a frequency of $\sqrt{\frac{1-\beta}{1+\beta}}$ due to the relativistic Doppler effect as the Earth travels away from twin A in A's frame. During the second half of the journey, the frequency of the pulses instantaneously becomes $\sqrt{\frac{1+\beta}{1-\beta}}$ when twin A switches frames. Therefore, the total number of pulses from B received by A is

$$N_B = \frac{L}{\gamma v} \cdot \left(\sqrt{\frac{1-\beta}{1+\beta}} + \sqrt{\frac{1+\beta}{1-\beta}} \right) = \frac{L}{\gamma v} \cdot \gamma (1 - \beta + 1 + \beta) = \frac{2L}{v}.$$

Hence, twin A concludes that he is younger than twin B by a factor of $\frac{1}{\gamma}$.

(c) The total time of the journey in B's frame is simply

$$t_B = \frac{2L}{v}.$$

Thus, he emits

$$N_B = \frac{2L}{v}$$

pulses.

(d) The important point to take note of is that twin B does not immediately receive pulses of a higher frequency when twin A turns around as there are already pulses that are en route to twin B. The total time during which B receives low-frequency pulses from twin A is $\frac{L}{v} + \frac{L}{c}$ where the first term arises during the first half of the journey of twin A and the second term corresponds to the subsequent time taken for the last low-frequency pulse to reach B. For the rest of the $\frac{L}{v} - \frac{L}{c}$ time, B receives high-frequency pulses. Therefore, the total number of pulses emitted by A that is received by B during the entire process is

$$N_A = \left(\frac{L}{v} + \frac{L}{c} \right) \cdot \sqrt{\frac{1-\beta}{1+\beta}} + \left(\frac{L}{v} - \frac{L}{c} \right) \cdot \sqrt{\frac{1+\beta}{1-\beta}} = \frac{2L}{\gamma v}.$$

Therefore, twin B similarly concludes that twin A is younger than him by a factor of $\frac{1}{\gamma}$.

14. Moving Glass**

In the frame of the block, S', the length of the block is just its proper length l and the velocity of light with respect to the block is just $\frac{c}{n}$ but the light is Doppler-shifted. The time taken for the light ray to exit the block in S' is just $t' = \frac{nl}{c}$. The frequency of light in the block $f_{S'}$ is related to the frequency in vacuum f by

$$f_{S'} = \sqrt{\frac{1-\beta}{1+\beta}} f$$

where $\beta = \frac{v}{c}$ — this is the longitudinal relativistic Doppler formula as the source now moves in the frame of the block. Note that the frequencies of light in vacuum and in the block must match as the block is stationary. The wavelength inside the block in frame S' is then

$$\lambda_{S'} = \frac{\frac{c}{n}}{f_{S'}} = \frac{c}{nf}\sqrt{\frac{1+\beta}{1-\beta}}.$$

Now, consider things in the lab frame — the situation is much trickier. We know that light travels at speed $\frac{c}{n}$ with respect to the glass block. Therefore, the velocity of light within the block in the lab frame is given by the velocity addition formula as

$$c' = \frac{\frac{c}{n}+v}{1+\frac{v}{nc}}.$$

Next, the length of the glass block in the lab frame is $\frac{l}{\gamma}$ where $\gamma = \frac{1}{\sqrt{1-\frac{v^2}{c^2}}}$, as it is Lorentz-contracted. Now, notice that the block also travels at v, in an attempt to chase after the light. Therefore, the relative velocity between the light and the glass block in the lab frame is $c' - v$. The total time taken for the light to escape the glass block is

$$t = \frac{l}{\gamma(c'-v)} = \frac{\gamma l\left(n+\frac{v}{c}\right)}{c}.$$

To determine the frequency of light inside the block, first observe that light from the source impinges on the closer end of the block at a frequency $\frac{c-v}{c}f$, as the block retracts at a velocity v (this is just the classical Doppler effect as there is no time dilation). However, note that this is not the frequency of light inside the block. When the edge of the block receives light, its atoms re-emit light into the block — that is, the end of the block now acts a source. Observe that the end of the block is moving at speed v. The light inside the block is then Doppler-shifted again as the "source" (the edge of the block) is now moving in the direction of the waves it emits. Thus, we need to multiply $\frac{c-v}{c}f$ by a correction factor of $\frac{c'}{c'-v}$, as the wavelength inside the block decreases by a factor of $\frac{c'-v}{c'}$ in the lab frame. The frequency in the block in the lab frame is thus

$$fs = \frac{c-v}{c} f \cdot \frac{c'}{c'-v} = (1-\beta) f \cdot \frac{\frac{c}{n}+v}{\frac{c}{n}(1-\beta^2)} = \frac{(1+n\beta) f}{1+\beta}.$$

The wavelength of light inside the block in the lab frame is

$$\lambda_S = \frac{c'}{\frac{(1+n\beta)f}{1+\beta}} = \frac{c(1+\beta)}{(n+\beta)f}.$$

15. Simultaneous Lamps*

Consider a Minkowski diagram in the lab frame S (depicted in Fig. 11.35) and plot the illumination events of the lamps as points A_1, A_2 and A_3.

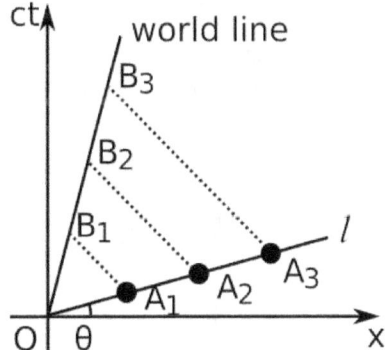

Figure 11.35: Minkowski diagram in frame S

A line of simultaneity of the car, when superimposed on the current diagram, is inclined at an angle $\theta = \tan^{-1}\beta$ anti-clockwise from the positive x-axis. Therefore, the events must be collinear through a line l that subtends an angle θ with respect to the horizontal. This requires

$$\frac{c(t_2 - t_1)}{x_2 - x_1} = \beta$$

and

$$\frac{c(t_3 - t_2)}{x_3 - x_2} = \beta.$$

Since we know that a line $ct = \frac{x}{\beta}$ represents the world line of the car and the ct'-axis when superimposed on the current Minkowski diagram, line l must cross the origin of frame S, as shown in the figure above (in order for the lights to be observed at $t' = 0$ by person P). Now draw three 45° lines, which represent the path of photons, from the corresponding illumination events and label their intersections with the world line of the car as B_1, B_2 and B_3 respectively. These events denote person P receiving the photons. Since $\overline{OB_1} : \overline{OB_2} : \overline{OB_3} = 1 : 2 : 3$ and $\triangle OA_1B_1 \sim \triangle OA_2B_2 \sim \triangle OA_3B_3$ (AA), $\overline{OA_1} : \overline{OA_2} : \overline{OA_3} = 1 : 2 : 3$ by similar triangles. Since $x_i = \overline{OA_i} \cos\theta$, $x_1 : x_2 : x_3 = 1 : 2 : 3$. Thus,

$$x_2 = 2x_1,$$

$$x_3 = 3x_1.$$

Finally, the respective times of the illumination events can be calculated via $ct_i = \tan\theta x_i = \beta x_i$.

16. Diverging Cars**

Figure 11.36 is a Minkowski diagram in the lab frame S. We first draw the world lines of cars 1 and 2, which are straight lines that subtend $15°$ and $30°$ with the ct-axis, respectively.

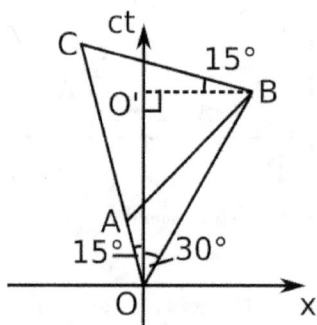

Figure 11.36:   Minkowski diagram in frame S

Let event A be the emission of the light signal from car 1 and event B be the receiving event by car 2. Then, the photons take the path AB on the Minkowski diagram, which makes a $45°$ angle with the vertical. Now, we superimpose the x' and ct'-axes of car 1 onto the current diagram. Draw a line that subtends $15°$ with the x-axis that crosses through B — this is a line of simultaneity with respect to car 1. Define the intersection of this line and the ct'-axis (which is just the world line of car 1) as C. Since the time between the emission event and the receiving event in frame S' is t' and since a time ct' in frame S' corresponds to a "length" of $\sqrt{\frac{1+\beta_1^2}{1-\beta_1^2}} ct' = \sqrt{\frac{1+\tan^2 15°}{1-\tan^2 15°}} ct' =$

$$\sqrt{\frac{1}{\cos^2 15° - \sin^2 15°}}\, ct' = \frac{ct'}{\sqrt{\cos 30°}} = \frac{\sqrt{2}}{\sqrt[4]{3}} ct' \text{ on the superimposed ct'-axis in the}$$

Minkowski diagram of frame S,

$$\overline{AC} = \frac{\sqrt{2}}{\sqrt[4]{3}} ct'.$$

Now, let us consider a few angles. Firstly, $\angle O'BO = 60°$. Since $\angle O'BC = 15°$ (recall that line BC is a line of simultaneity with respect to car 1), $\angle CBO = 75°$. Next, since lines AC and AB subtend $15°$ and $45°$ with respect to the vertical respectively, $\angle CAB = 60°$ and $\angle OAB = 120°$. Then, $\angle ABO = 180° - 45° - 120° = 15°$. This implies that $\angle CBA = \angle CBO - \angle ABO = 60°$. Since $\angle CAB = \angle ABC = 60°$, $\triangle ABC$ is equilateral. Therefore,

$$\overline{AB} = \overline{AC} = \frac{\sqrt{2}}{\sqrt[4]{3}} ct'.$$

Then, $\overline{OB}$ is given by the sine rule.

$$\overline{OB} = \frac{\overline{AB}\sin 120°}{\sin 45°} = \sqrt[4]{3}\, ct'.$$

The x-coordinate of event B is

$$x_B = \overline{OB}\sin 30° = \frac{\sqrt[4]{3}}{2} ct'.$$

17. Particle's Motion*

(a) The infinitesimal proper time interval is

$$d\tau = \frac{dt}{\gamma_u}$$

with

$$\gamma_u = \frac{1}{\sqrt{1 - \frac{u^2}{c^2}}} = \frac{gt}{c} + 1$$

$$\implies d\tau = \frac{dt}{\frac{gt}{c} + 1}.$$

$$\tau = \int_0^t \frac{c}{g} \cdot \frac{1}{t + \frac{c}{g}} dt.$$

$$= \frac{c}{g} \left[\ln \left| t + \frac{c}{g} \right| \right]_0^t$$

$$= \frac{c}{g} \ln \left(\frac{gt}{c} + 1 \right).$$

(b) Adopt the substitutions $\sec y = \frac{gt}{c} + 1$ and $\tan y \sec y \, dy = \frac{g}{c} dt$.

$$x(t) = \int_0^t u(t) dt$$

$$= \int_0^t c \sqrt{1 - \frac{1}{\left(\frac{gt}{c} + 1 \right)^2}} \, dt$$

$$= \int_0^{\sec^{-1}\left(\frac{gt}{c} + 1 \right)} \frac{c^2}{g} \sqrt{1 - \cos^2 y} \, \tan y \sec y \, dy$$

$$= \int_0^{\sec^{-1}\left(\frac{gt}{c} + 1 \right)} \frac{c^2}{g} \frac{\sin^2 y}{\cos^2 y} \, dy$$

$$= \int_0^{\sec^{-1}\left(\frac{gt}{c} + 1 \right)} \left(\frac{c^2}{g} \sec^2 y - \frac{c^2}{g} \right) dy$$

$$= \left[\frac{c^2}{g} \tan y - \frac{c^2}{g} y \right]_0^{\sec^{-1}\left(\frac{gt}{c} + 1 \right)}$$

$$= \frac{c^2}{g} \left[\sqrt{\left(\frac{gt}{c} + 1 \right)^2 - 1} - \sec^{-1} \left(\frac{gt}{c} + 1 \right) \right].$$

Observe that $\sec^{-1}(\frac{gt}{c} + 1) = \tan^{-1} \sqrt{(\frac{gt}{c} + 1)^2 - 1}$. Furthermore, from the result of a), we can write $\frac{gt}{c} + 1 = e^{\frac{g\tau}{c}}$ so

$$x(\tau) = \frac{c^2}{g} \left[\sqrt{e^{\frac{2g\tau}{c}} - 1} - \tan^{-1} \sqrt{e^{\frac{2g\tau}{c}} - 1} \right].$$

18. World Line*

(a) Since $dt = \gamma_u d\tau$,

$$u = \frac{dx}{dt} = \frac{1}{\gamma_u} \frac{dx}{d\tau} = \frac{c}{\gamma_u} \sinh \left(\frac{g\tau}{c} \right).$$

(b)

$$\gamma_u^2 \left(1 - \frac{u^2}{c^2}\right) = 1$$

$$\implies \gamma_u^2 - \sinh^2\left(\frac{g\tau}{c}\right) = 1$$

$$\gamma_u = \sqrt{1 + \sinh^2\left(\frac{g\tau}{c}\right)} = \cosh\left(\frac{g\tau}{c}\right),$$

where we have chosen the positive root since $\gamma_u \geq 1$.

(c)

$$u = \frac{c\sinh\left(\frac{g\tau}{c}\right)}{\gamma_u} = c\tanh\left(\frac{g\tau}{c}\right).$$

(d)

$$t = \int_0^\tau \gamma_u d\tau$$

$$= \int_0^\tau \cosh\left(\frac{g\tau}{c}\right) d\tau$$

$$= \frac{c}{g} \sinh\left(\frac{g\tau}{c}\right).$$

From the above,

$$\sinh\left(\frac{g\tau}{c}\right) = \frac{gt}{c}$$

$$\cosh\left(\frac{g\tau}{c}\right) = \sqrt{1 + \frac{g^2 t^2}{c^2}}$$

$$u(t) = c\tanh\left(\frac{g\tau}{c}\right) = \frac{gt}{\sqrt{1 + \frac{g^2 t^2}{c^2}}}$$

$$a(t) = \frac{du}{dt} = \frac{g}{\sqrt{1 + \frac{g^2 t^2}{c^2}}} - \frac{\frac{g^3}{c^2} t^2}{\left(1 + \frac{g^2 t^2}{c^2}\right)^{\frac{3}{2}}} = \frac{g}{\left(1 + \frac{g^2 t^2}{c^2}\right)^{\frac{3}{2}}}.$$

As $t \to \infty$, $u(t) = \frac{g}{\sqrt{\frac{1}{t^2} + \frac{g^2}{c^2}}} \to c$ while $a(t) \to 0$. This limit makes sense as the particle's speed in the lab frame cannot exceed c.

Chapter 12

Relativistic Dynamics

The previous chapter analyzed how particles "move" in space and time without considering the interactions that led to their motion. In this chapter, relativistic formulations of various physical concepts such as momentum and
energy will be introduced. The elegant 4-vector formulation, which captures
the quintessence of relativistic dynamics, in simple matrices with just four
entries, will also be explored. The prefix "relativistic" that appears in front
of many concepts in this chapter is misleading in certain aspects as the principle of relativity also exists in the classical regime — with the caveat that
Galilean relativity is assumed instead. However, this prefix shall still be used
to distinguish quantities in this chapter from their classical counterparts.

12.1   Momentum

Classical Definition

In classical mechanics, the momentum of a particle in a particular inertial
frame S is defined as

$$p = mv,$$

where m is the mass of the particle and v is the velocity of the particle
in frame S. There is no ambiguity about which frame m is measured with
respect to as the mass of a particle is assumed to be an intrinsic property that
is invariant across inertial frames. The importance of this formulation lies
in the law of conservation of momentum. It is empirically observed that the
total momentum of a system that is not under the influence of net external
forces is conserved. In an isolated system of particles, even if the particles
interacted with one another in a certain manner, the total momentum of the
particles remains constant.

The combination of the classical definition of momentum, mass invariance and the principle of relativity (that all laws hold similar forms in all inertial frames) implies the conservation of mass in closed systems. Take note of the distinction between mass invariance and the conservation of mass. Mass invariance means that if a particle is observed to have a certain mass m in an inertial frame S, its observed mass in another inertial frame S' is also m. On the other hand, the conservation of mass means that the total mass in a closed system remains the same, regardless of the inner workings of the system.

Consider a closed system of n particles in which the ith particle has an initial mass m_i and an initial velocity $\boldsymbol{u}_i$ as observed in an inertial frame S. The particles may undergo arbitrary interactions with one another. Non-conservative forces such as friction may exist such that the total mechanical energy of the system is not conserved. Furthermore, there may also be changes in the mass of each individual particle and the total number of particles as atoms may be scraped off during collisions, particles may stick together or decay. After the particles are allowed to interact for a certain amount of time, there are n' particles and the ith particle has a mass m_i' and a velocity v_i in the same frame S. By the conservation of momentum in this frame,

$$\sum_{i=1}^{n} m_i \boldsymbol{u}_i = \sum_{j=1}^{n'} m_j' \boldsymbol{v}_j. \tag{12.1}$$

Now if we were to switch to another inertial frame S' that moves at a velocity $\boldsymbol{V}$ relative to frame S, the law of the conservation of momentum should also be valid in frame S' by the principle of relativity as all inertial frames are "equivalent." Based on Galilean relativity, if a particle is observed to have a velocity $\boldsymbol{u}$ in frame S, it will be observed to possess a velocity $\boldsymbol{u} - \boldsymbol{V}$ in frame S'. Furthermore, since the mass of a particle is assumed to be invariant across inertial frames (that is, the particle with mass m_i in S still has mass m_i in S'), the conservation of momentum in frame S' becomes

$$\sum_{i=1}^{n} m_i (\boldsymbol{u}_i - \boldsymbol{V}) = \sum_{j=1}^{n'} m_j' (\boldsymbol{v}_j - \boldsymbol{V}). \tag{12.2}$$

Subtracting Eq. (12.1) from Eq. (12.2) and simplifying,

$$\sum_{i=1}^{n} m_i = \sum_{j=1}^{n'} m_j'. \tag{12.3}$$

Equation (12.3) states the conservation of mass in a closed system. If a closed system has a certain amount of total mass at a certain instance, it will also contain the same amount of mass at the next instance. This is what allows us to conclude that the perfectly inelastic collision of a particle of mass m and another particle of mass M produces a combined particle of mass $m + M$.

Relativistic Momentum

It turns out that the classical definition of momentum is not quite conserved in an isolated system. Instead, the relativistic momentum is conserved and is defined for a particle with respect to an inertial frame S as

$$p = \gamma_u m u \tag{12.4}$$

where

$$\gamma_u = \frac{1}{\sqrt{1 - \frac{u^2}{c^2}}} = \frac{1}{\sqrt{1 - \frac{u_x^2 + u_y^2 + u_z^2}{c^2}}}.$$

u is the velocity of the particle in frame S and m refers to the mass of the particle observed in a frame in which it is at rest — a quantity denoted as the rest mass of the particle. Henceforth, the term "mass" will refer to the rest mass, unless explicitly stated otherwise. Again, the rest mass of a particle is presumed to be an intrinsic property of the particle and is invariant across inertial frames.

The total relativistic momentum of particles in a system that is free from a net external force, is conserved. This assertion, similar to the classical conservation of momentum, cannot be proven and should be regarded as an axiom. However, it has been empirically verified by rigorous test and hence shall be believed to be true.

An immediate consequence of this new postulate is that the total (rest) mass of a closed system may not be conserved! The premise of the previous section (Eq. (12.1) and the Galilean velocity transformation) is inaccurate. To illustrate the mutability of the total rest mass, consider the set-up in Fig. 12.1.

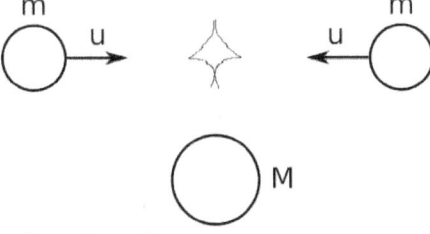

Figure 12.1:   Two particles in frame S

Two identical particles, of rest mass m, initially travel at speed u in opposite directions in an inertial frame S. They collide with one another and stick together to form a resultant particle of rest mass M, which is not necessarily $2m$. Due to the symmetry of this set-up, there is zero total momentum and the resultant particle remains stationary in this frame S by the conservation of momentum (analogous to the right-hand side of Eq. (12.1) being zero). Similar to how we proceeded from Eq. (12.1) to (12.2), consider the inertial frame S' that travels at the initial velocity of the particle on the right. In this frame, the situation is depicted by Fig. 12.2.

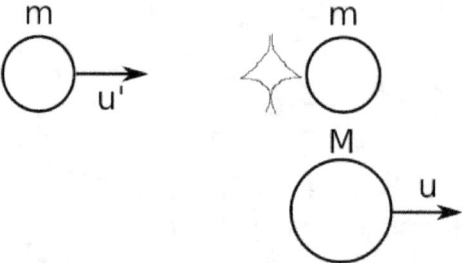

Figure 12.2:   Two particles in frame S'

The velocity u' of the left mass in frame S' can be computed via the relativistic velocity addition formula as $\frac{2u}{1+\frac{u^2}{c^2}}$. Again, the invariance of the rest mass allows us to conclude that the rest masses of these particles are the same in frame S'. By the principle of relativity, the total momenta of the system before and after the collision are identical.

$$\frac{1}{\sqrt{1-\left(\frac{\frac{2u}{1+\frac{u^2}{c^2}}}{c}\right)^2}}m\frac{2u}{1+\frac{u^2}{c^2}} = \frac{1}{\sqrt{1-\frac{u^2}{c^2}}}Mu.$$

Solving for M,

$$M = \frac{2}{\sqrt{1-\frac{u^2}{c^2}}}m.$$

It can be seen that the rest mass of the resultant particle is larger than the rest mass of its constituents! It is natural to question where this additional mass comes from. Answering this shall be the goal of the next section.

At this point, we underscore the fact that we will adopt the same conventional definitions as the previous chapter. The velocity of a particle or wave in frame S will be denoted by $\boldsymbol{u}$ by default. Usually, we will be concerned with switching to another inertial frame S'. Hence, we will reserve v to be the velocity that S' travels with respect to S in the positive x-direction in general. A similar statement holds for $\beta = \frac{v}{c}$ which is only associated with the transformations. Sometimes, we will switch to the frame of the particle and will thus substitute a quantity related to $\boldsymbol{u}$ into v.

In a general inertial frame S, we will append the prefix "coordinate" to the measurements to explicitly indicate that they are measured with respect to a general frame. For example, the coordinate time refers to time measured in S. Often, we will be concerned with quantities observed in the frame of a particle. We then append the prefix "proper" to such measurements.

12.2 Relativistic Energy

It is postulated that the total relativistic energy in an isolated system is conserved. The total energy of a particle, which includes both its kinetic energy and internal energy, in a particular inertial frame S is proposed to be

$$E = \gamma_u mc^2, \tag{12.5}$$

where

$$\gamma_u = \frac{1}{\sqrt{1 - \frac{u^2}{c^2}}}.$$

m and u are again the rest mass and the speed of the particle in frame S, respectively. Once again, this is another axiom which cannot be derived from first principles.[1] However, it has also been extensively tested by experimentalists as it establishes a fundamental basis in many branches of physics such as nuclear physics. Next, let us analyze the constituents of this energy in greater detail.

$$E = \frac{1}{\sqrt{1 - \frac{u^2}{c^2}}} mc^2$$

$$\approx mc^2 + \frac{1}{2}mu^2 + \cdots,$$

---

[1] The expression for the kinetic energy of the particle (in the section after this) can be deduced from integrating the rate of change of relativistic momentum (relativistic force) with respect to displacement, which is the relativistic analog of work. However, the "rest energy" is indeed a bold assertion.

where we have expanded the Taylor series of γ_u. We begin to see a familiar $\frac{1}{2}mu^2$ term followed by other higher order terms in $\frac{u^2}{c^2}$ that are not shown. However, there is an enormous mc^2 term in the expression as well, depending only on the rest mass of the particle. This is known as the rest energy of the particle, E_0.

$$E_0 = mc^2. \tag{12.6}$$

The rest energy of a particle is equivalent to its internal energy and is an intrinsic property of the particle. The rest energy is eponymously the energy of the particle when it is at rest and remains constant regardless of the particle's motion. Furthermore, since the rest mass is invariant across inertial frames, the rest energy of a particle is also invariant. In general, the internal energy of a particle or system consists of the (microscopic) kinetic and rest energy of its constituents as well as the potential energy associated with its constituents due to interactions between its constituents or fields produced by its constituents (this excludes fields generated by sources external to the system). A consequence of this postulate is that heating a system increases its rest mass, as its internal energy is increased.

Next, the kinetic energy of a particle is then the remaining portion of energy associated with the motion of the particle.

$$\text{KE} = (\gamma_u - 1)mc^2. \tag{12.7}$$

As seen from the previous Taylor series expansion, this expression indeed reduces to the familiar formula for kinetic energy in the classical limit.

Lastly, note that the potential energy of a particle by virtue of its position in an external field is not included in the particle's total energy. This is because this potential energy is "associated" with the particle and not possessed by it. The concept of potential energy is merely a "book-keeping" device that simplifies our calculations. When the kinetic energy of a particle increases as it is acted upon by a force due to an external field, the gain in kinetic energy should not be ascribed to its loss in potential energy. Rather, it should be understood that the field itself loses an equivalent amount of energy. Potential energy is an imaginary construct that helps us to keep track of the total energy of a system without taking into account where this energy "belongs" to. The "location" of energy matters in the context of relativity as it manifests itself in the local distortion of space and time. Hence, the potential energy due to an external field cannot be ascribed to a particle as a "real" form of energy and is forgone in special relativity.

Armed with the knowledge of relativistic energy and the conservation of relativistic energy (we will drop the "relativistic" prefix henceforth), let us revisit the previous question and verify if the total energy of the system is indeed conserved in Fig. 12.1. The total energy in the system of two particles before the collision in frame S is

$$E = \frac{2}{\sqrt{1 - \frac{u^2}{c^2}}} mc^2.$$

As we have previously computed that the rest mass of the resultant particle is $M = \frac{2}{\sqrt{1-\frac{u^2}{c^2}}}$, the energy of the system after the collision is

$$E' = Mc^2 = \frac{2}{\sqrt{1 - \frac{u^2}{c^2}}} mc^2$$

as the resultant particle is at rest. It can be seen that they are indeed equal. But wait! How can the total energy in this perfectly inelastic collision be conserved? Furthermore, we have yet to answer the question regarding the origin of the additional mass. Well, the resolution to these problems is that the kinetic energy of the particles is converted into their internal energy due to the heat released during the collision. Hence, the total energy, which includes the internal energy of the particles, of the system is still conserved. Furthermore, this additional internal energy also "shows up" as the additional mass of the resultant particle. In this particular sense, relativistic dynamics may actually be simpler than its classical counterpart, as the total energy of an isolated system is always conserved. In real life, you would expect the total energy of the resultant particle to be less than the sum of the original two. However, this deviation is due to heat transfer with the surroundings which means that the system of particles is no longer isolated and that the conservation of energy is inapplicable (but not violated). Lastly, be cautious that though the total energy of an isolated system is definitely conserved, the total kinetic energy may not necessarily be conserved — evident from the situation above.

The conservation of energy and momentum can be directly applied to solve many problems in a manner similar to the classical situation.

Problem: In Fig. 12.3, a particle of rest mass M initially travels at a velocity u in the x-direction in inertial frame S. It then decays into two identical particles of rest mass m that travel at a certain velocity v that makes a certain angle θ with the x-axis. Determine v and θ.

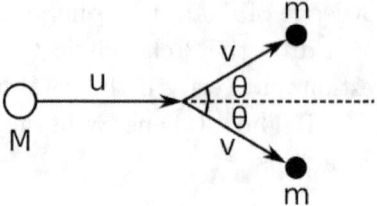

Figure 12.3:　Decay

The conservation of momentum in the x-direction implies

$$\gamma_u M u = 2\gamma_v m v_x,$$

where v_x is the component of the resultant particles' velocities in the x-direction. By the conservation of energy,

$$\gamma_u M c^2 = 2\gamma_v m c^2.$$

Dividing the two equations above and simplifying,

$$v_x = u.$$

Substituting this into the first equation,

$$\gamma_v = \frac{M}{2m}\gamma_u$$

$$\frac{1}{\sqrt{1 - \frac{v^2}{c^2}}} = \frac{M}{2m}\gamma_u$$

$$v = c\sqrt{1 - \frac{4m^2}{M^2\gamma_u^2}}$$

$$\theta = \cos^{-1}\frac{v_x}{v} = \cos^{-1}\frac{u}{c\sqrt{1 - \frac{4m^2}{M^2\gamma_u^2}}}.$$

Useful Identities

In light of how the velocity u is horrendously coupled in the γ factors in the definitions of momentum and energy, there are a few neat identities that are commonly exploited in problem-solving. Firstly, consider the expression

$$E^2 - (\boldsymbol{p} \cdot \boldsymbol{p})c^2 = E^2 - p^2 c^2$$

$$= \frac{1}{1 - \frac{u^2}{c^2}} m^2 c^4 - \frac{1}{1 - \frac{u^2}{c^2}} m^2 u^2 c^2$$

$$= \frac{1}{1 - \frac{u^2}{c^2}} m^2 c^4 \left(1 - \frac{u^2}{c^2}\right)$$

$$= m^2 c^4$$

$$E^2 - p^2 c^2 = m^2 c^4. \tag{12.8}$$

This is a convenient identity that can be used to relate the energy of a particle to its momentum. Furthermore, it can be used to isolate and eliminate the dynamical properties of a particle (energy or momentum) which is not of concern. This will be illustrated in the next example. What's more, notice that the right-hand side of the equation is frame-independent! That is, regardless of the inertial frame in which the energy E and momentum p of the particle are measured, substituting them into the equation above will always produce $m^2 c^4$ where m is the rest or invariant mass of the particle! Perhaps, the deeper reason behind this invariance can be understood once the method of four-vectors is introduced.

The next useful identity is obtained by dividing p by E.

$$\frac{p}{E} = \frac{u}{c^2}. \tag{12.9}$$

The equation above is especially helpful in determining the speed of a particle in a certain inertial frame given its momentum and energy in that frame. Note that in general, we do not wish to work in terms of u as it is usually entangled with annoying surds that are cumbersome to isolate. Hence, the momenta and energies will be the main avenues through which a dynamical problem can be solved.

Next, Eqs. (12.8) and (12.9) are particularly enlightening in the case of massless particles such as photons, which are inherently relativistic. Equations (12.5) and (12.6) are less so as γ_u tends to infinity while m tends to zero in the case of such massless particles which travel at the speed of light (we will soon see that all massless particles must travel at c) — leaving the values of those expressions indeterminate. Substituting $m = 0$ into Eq. (12.8),

$$E = pc. \tag{12.10}$$

Applying Eq. (12.9) with this relationship would show that $u = c$ in the case of massless particles. Similarly, it is not difficult to show that the speed of a massive particle must be less than c in any given inertial frame. Moving on, from quantum mechanics, the energy of a photon in an inertial frame S is

$$E = hf = \frac{hc}{\lambda}, \tag{12.11}$$

where f and λ are its frequency and wavelength in frame S, related by $c = f\lambda$. Correspondingly, the momentum of a photon in frame S is

$$p = \frac{E}{c} = \frac{h}{\lambda}, \tag{12.12}$$

which is just the de Broglie relationship.

Problem: In inertial frame S, a photon of wavelength λ, that is initially traveling the x-direction, collides with a stationary electron with rest mass m. If the photon scatters at an angle θ from the x-axis, determine the resultant wavelength of the photon. This effect is known as Compton scattering.

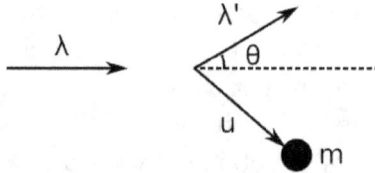

Figure 12.4:   Compton scattering

Referring to Fig. 12.4, let u be the resultant speed of the electron and let u_x and u_y be its components in the horizontal and vertical directions, positive rightwards and downwards. By the conservation of momentum and energy,

$$\frac{h}{\lambda} = \frac{h}{\lambda'} \cos\theta + \gamma_u m u_x,$$

$$\frac{h}{\lambda'} \sin\theta = \gamma_u m u_y,$$

$$\frac{hc}{\lambda} + mc^2 = \frac{hc}{\lambda'} + \gamma_u mc^2.$$

These equations appear tricky to solve because of u_x and u_y which are coupled in γ_u. However, notice that the resultant momentum and energy of the electron are not germane. Hence, we can eliminate them by using Eq. (12.8) astutely. Rewriting the equations in terms of the components of the momentum and the energy of the electron (via $p_x = \gamma_u m u_x$, $p_y = \gamma_u m u_y$ and $E = \gamma_u mc^2$),

$$p_x = \frac{h}{\lambda} - \frac{h}{\lambda'} \cos\theta,$$

$$p_y = \frac{h}{\lambda'} \sin\theta,$$

$$E = \frac{hc}{\lambda} + mc^2 - \frac{hc}{\lambda'}.$$

Applying $E^2 - p^2 c^2 = m^2 c^4$ by taking the square of the last equation and subtracting it by the first and second equations squared and multiplied by c^2,

$$\left(\frac{hc}{\lambda} + mc^2 - \frac{hc}{\lambda'}\right)^2 - c^2 \left(\frac{h}{\lambda} - \frac{h}{\lambda'}\cos\theta\right)^2 - \frac{c^2 h^2}{\lambda'^2}\sin^2\theta = E - p_x^2 - p_y^2 = m^2 c^4.$$

Simplifying,

$$\frac{2h^2 c^2 \cos\theta}{\lambda\lambda'} + \frac{2hmc^3}{\lambda} - \frac{2hmc^3}{\lambda'} - \frac{2h^2 c^2}{\lambda\lambda'} = 0$$

$$\left(hmc^3 + \frac{h^2 c^2}{\lambda} - \frac{h^2 c^2}{\lambda}\cos\theta\right)\frac{1}{\lambda'} = \frac{hmc^3}{\lambda}.$$

Multiplying both sides of the equation by $\frac{\lambda\lambda'}{hmc^3}$,

$$\lambda' = \lambda + \frac{h}{mc}(1 - \cos\theta).$$

Rest Energy and Mass of a System

The rest or invariant mass of a system m_{sys} is related to the total energy of the particles, combined with the potential energy due to internal interactions between the constituent particles (this component was excluded from the definition of the energy of a particle), denoted as $E_{tot,CM}$, in the inertial frame in which the total momentum of the system is zero — this frame is known as the center-of-momentum frame. By definition,

$$E_{tot,CM} = m_{sys} c^2. \tag{12.13}$$

Evidently, there are two factors that can affect the invariant mass of a system. Firstly, the total energy of each particle in the center-of-momentum frame may increase in a non-isolated system. In the case of an ideal gas whose particles lack potential energy, heating the gas causes the rest mass of the system to increase as the kinetic energy of the particles increases.

Another factor that affects the internal energy and thus the rest mass of the system would be the microscopic potential energy of its constituents due to their interactions (this was excluded from the total energy of each particle). This is the reason behind the large discrepancy between the mass of a proton and the sum of the individual rest masses of the component quarks.

As the rest mass of a system is dependent on both microscopic kinetic and potential energy, it is generally not equal to the sum of the rest masses of its constituents. In the case of non-interacting particles (collisions and decays are not counted here), the potential energy of the constituents is zero and the left-hand side of Eq. (12.13) is simply the sum of the energies of the constituent particles in the center-of-momentum (CoM) frame, E_{CM}.

$$E_{CM} = m_{sys}c^2. \tag{12.14}$$

Furthermore, we claim that E_{CM} can be expressed in terms of the dynamical properties observed in a general inertial frame in the following manner.

$$E_{CM}^2 = E_{tot}^2 - p_{tot}^2 c^2,$$

where E_{tot} and p_{tot} are the total energy and momentum of the system of particles in an **arbitrary** inertial frame S. This leap is not obvious now as we have yet to discuss how energy and momentum transform between inertial frames. However, the reader should just accept this for now. We will deduce this result and examine why the "invariant mass of a system" is indeed invariant later. Then,

$$E_{CM}^2 = E_{tot}^2 - p_{tot}^2 c^2 = m_{sys}^2 c^4. \tag{12.15}$$

Let us consider the example in Fig. 12.5 to convince ourselves that the rest mass of a system indeed deviates from the sum of the rest masses of its constituents. In inertial frame S, a particle of mass $2m$ travels at a speed u in the positive x-direction while another particle of mass m travels at a speed u in the negative x-direction.

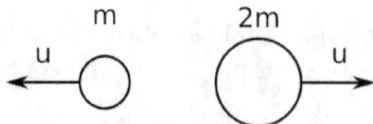

Figure 12.5:   Two particles

Then the rest mass of the system comprising the two particles is given by Eq. (12.15).

$$9\gamma_u^2 m^2 c^4 - \gamma_u^2 m^2 u^2 c^2 = m_{sys}^2 c^4$$

$$m_{sys} = \gamma_u m \sqrt{9 - \frac{u^2}{c^2}}$$

$$= \sqrt{\frac{9c^2 - u^2}{c^2 - u^2}} m$$

which differs from the sum of the individual rest masses, $3m$. In general, since the total energy and momentum is conserved in an isolated system, the rest mass of an isolated system remains unchanged too by Eq. (12.15). This is the conservation rule that replaces the classical conservation of mass. However, this is merely tautology as we have only created a new definition for $E_{tot}^2 - p_{tot}^2 c^2$. The important part lies in the fact that this quantity is in fact invariant across all inertial frames (and hence we term it the "invariant mass of a system"), as we shall prove. In practice, the rest mass of a system is not particularly useful as it is easily superseded by the formulation of four-vectors (as we shall see).

12.3 Force and Coordinate Acceleration

In the relativistic case, a net force on a system still leads to a rate of change of relativistic momentum. The forces are still of the same form as their classical counterparts (e.g. the elecromagnetic force is given by the Lorentz force law). However, the rate of change of relativistic momentum of a massive particle is no longer $m\boldsymbol{a}$ where $\boldsymbol{a}$ is its acceleration. The net external force $\boldsymbol{f}$ on a particle, as observed in an inertial frame S, engenders a rate of change of relativistic momentum.

$$\boldsymbol{f} = \frac{d\boldsymbol{p}}{dt} = \frac{d(\gamma_u m \boldsymbol{u})}{dt}, \tag{12.16}$$

where $\boldsymbol{u}$ and t are the coordinate velocity of the particle and coordinate time as observed in frame S. The lower-case letter shall be used to avoid confusion with the four-force four-vector which will later be defined with the upper-case letter. Note that there are two time-dependent terms in the expression above, γ_u and $\boldsymbol{u}$. The time derivative of γ_u shall be evaluated first.

$$\frac{d\gamma_u}{dt} = \frac{d}{dt} \frac{1}{\sqrt{1 - \frac{u_x^2 + u_y^2 + u_z^2}{c^2}}}$$

$$= -\frac{1}{2\left(1 - \frac{u^2}{c^2}\right)^{\frac{3}{2}}} \cdot -\frac{2(u_x \dot{u}_x + u_y \dot{u}_y + u_z \dot{u}_z)}{c^2}$$

$$= \frac{\gamma_u^3}{c^2}(u_x \dot{u}_x + u_y \dot{u}_y + u_z \dot{u}_z),$$

where a dot is used to denote a derivative with respect to coordinate time. The coordinate acceleration $\boldsymbol{a}$ of a particle is the derivative of its coordinate velocity $\boldsymbol{u}$ with respect to coordinate time t.

$$\boldsymbol{a} = \frac{d\boldsymbol{u}}{dt}.$$

Hence, the expression above can be rewritten as

$$\frac{d\gamma_u}{dt} = \frac{\gamma_u^3}{c^2}(\boldsymbol{a} \cdot \boldsymbol{u}).$$

If the x-axis of the Cartesian coordinate system in frame S is defined to be along the direction of the particle's instantaneous coordinate velocity $\boldsymbol{u}$,

$$\frac{d\gamma_u}{dt} = \frac{\gamma_u^3 u a_x}{c^2}.$$

Now, the chain rule can be applied to Eq. (12.16) to obtain a simpler expression for $\boldsymbol{f}$. Assuming that the rest mass of the particle remains constant,

$$\boldsymbol{f} = \frac{d\gamma_u}{dt} m\boldsymbol{u} + \gamma_u m \frac{d\boldsymbol{u}}{dt}$$

$$= \begin{pmatrix} \frac{\gamma_u^3 m u^2 a_x}{c^2} + \gamma_u m a_x \\ \gamma_u m a_y \\ \gamma_u m a_z \end{pmatrix}$$

$$= \begin{pmatrix} \gamma_u^3 m a_x \left(\frac{u^2}{c^2} + 1 - \frac{u^2}{c^2}\right) \\ \gamma_u m a_y \\ \gamma_u m a_z \end{pmatrix}$$

$$\boldsymbol{f} = \begin{pmatrix} f_x \\ f_y \\ f_z \end{pmatrix} = \begin{pmatrix} \gamma_u^3 m a_x \\ \gamma_u m a_y \\ \gamma_u m a_z \end{pmatrix}, \tag{12.17}$$

it can be seen that the force on a particle $\boldsymbol{f}$ in an inertial frame S is not proportional to the coordinate acceleration of the particle $\boldsymbol{a}$ in the same inertial frame S. It is in fact easier to accelerate a particle in the transverse direction rather than the longitudinal direction! As a result, the force vector $\boldsymbol{f}$ is no longer necessarily parallel to the coordinate acceleration $\boldsymbol{a}$.

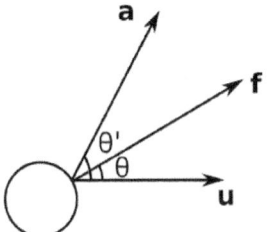

Figure 12.6:   Force and acceleration vectors

In Fig. 12.6, a force of magnitude f is exerted on a particle traveling at speed u in the x-direction, at an angle θ anti-clockwise relative to the positive x-axis in the lab frame S. Then, the angle θ' that the coordinate acceleration makes with the x-axis in S is given by

$$\tan \theta' = \frac{a_y}{a_x} = \gamma_u^2 \frac{f_y}{f_x} = \gamma_u^2 \tan \theta.$$

Furthermore, not only does the force not point along the direction of coordinate acceleration in most cases, the force vector varies across inertial frames as the coordinate velocity $\boldsymbol{u}$ in Eq. (12.16) changes in a way that affects its coordinate time derivative — contrary to the case in Galilean relativity. The transformation rules for force and coordinate acceleration will be derived in a later section.

Moving on, the confluence of the conservation of momentum and the definition of force implies a relativistic analog of Newton's third law. If a particle A exerts a force on another particle B, A also experiences an equal and opposite force such that the total momentum of the two particles is conserved.

Another fact that one needs to get used to would be that the velocities of a particle in different directions are no longer independent. If a particle is initially traveling in the positive x-direction and a constant force is exerted on it in the y-direction, the x component of the velocity of the particle must decrease — without which, γ_u will increase as u_y increases, leading to a violation of the conservation of momentum in the x-direction. However, the momentum of the particle in the x-direction in this case will still remain the same. The key takeaway from this is that one should focus on dynamical properties such as momentum and energy which often describe a system in a fashion that is more elegant than kinematic quantities such as coordinate velocities and accelerations directly.

Impulse-Momentum Theorem

Considering the definition of force, the impulse-momentum theorem can be expressed as

$$\int \boldsymbol{f} dt = \Delta \boldsymbol{p} = \Delta(\gamma_u m \boldsymbol{u}). \tag{12.18}$$

Power

Taking the derivative of $E^2 = p^2 c^2 + m^2 c^4 = (\boldsymbol{p} \cdot \boldsymbol{p})c^2 + m^2 c^4$ with respect to coordinate time and assuming that the rest mass m remains constant,

$$2E \frac{dE}{dt} = 2 \frac{d\boldsymbol{p}}{dt} \cdot \boldsymbol{p} c^2.$$

Since the force $\boldsymbol{f} = \frac{d\boldsymbol{p}}{dt}$ and $\frac{\boldsymbol{p} c^2}{E} = \boldsymbol{u}$, where $\boldsymbol{u}$ is the coordinate velocity of the particle,

$$\frac{dE}{dt} = \boldsymbol{f} \cdot \boldsymbol{u}. \tag{12.19}$$

It can be observed that the dot product of force and the particle's coordinate velocity in frame S is equal to the power delivered by the net force in S, analogous to the classical scenario.

Work-Energy Theorem

If the dt in the denominator of Eq. (12.19) is shifted to the other side and the entire equation is integrated, the work-energy theorem is obtained.

$$\int \boldsymbol{f} \cdot \boldsymbol{u} dt = \int \boldsymbol{f} \cdot d\boldsymbol{r} = \Delta E = \Delta \text{KE} \tag{12.20}$$

where $d\boldsymbol{r}$ is an infinitesimal displacement of the particle in frame S. Again, this equation is built on the assumption that the rest mass of the particle remains constant. Lastly, if the instantaneous velocity of the particle is defined to be in the x-direction, Eq. (12.19) becomes

$$\frac{dE}{dt} = f_x u = f_x \frac{dx}{dt}.$$

Hence,

$$f_x = \frac{dE}{dx}. \tag{12.21}$$

At a certain instant in frame S, the force exerted on the particle in the direction of its instantaneous velocity is the change in the energy of particle due to an infinitesimal displacement, which must be along the direction of its instantaneous velocity.

Problem: In inertial frame S, a constant force f is exerted on an initially stationary particle of rest mass m. Find the time required for the particle to travel a distance s.

Well, this looks like an innocuous and typical kinematics question about a particle undergoing a one-dimensional acceleration. However, solving this problem by analyzing the equations of motion is incredibly tedious due to the γ_u terms. Instead, the dynamical equations should be used in an elegant manner. By the work-energy theorem, the final energy E of the particle is related to its initial energy (the rest energy) by

$$f \cdot s = \Delta E = E - mc^2$$

$$E = mc^2 + fs.$$

By the impulse-momentum theorem, the final momentum of the particle when it has traveled a distance s is

$$p = f \cdot t,$$

where t is the time interval between the start of the particle's motion and the juncture at which it has traveled a distance s. Lastly, as $p = \sqrt{E^2 - m^2c^4}$,

$$ft = \sqrt{(mc^2 + fs)^2 - m^2c^4}$$

$$t = \sqrt{\frac{2mc^2s}{f} + s^2}.$$

By now, you may have realized that the additional c's popping up everywhere are extremely frustrating. Hence, we shall adopt the units $c = 1$ for the rest of the chapter to maintain our sanity and to simplify the equations. The c's can always be added back to the expressions via dimensional analysis. So far, we have endured with the c's to present a more "formal" formulation of the various dynamical properties of a particle so that one can clearly distinguish the relationship between these and the speed of light.

In the sections above, the definitions of various dynamical properties of a particle in a certain inertial frame S have been covered. However, since the chapter is on relativity after all, it is interesting to determine how these properties transform between inertial frames. The formulation of four-vectors encapsulates these transformations in a terse manner while also keeping the quintessential conservation laws. As such, the next few sections will elaborate on four-vectors and how these properties vary across inertial frames.

12.4   Four-Vectors

A four-vector is a matrix consisting of four entries that transforms between inertial frames according to the Lorentz transformations, in a manner similar to (ct, x, y, z). Four-vectors will be denoted by capital letters and units of $c = 1$ will be adopted henceforth. Consider a four-vector A, of the form:

$$A = \begin{pmatrix} A_0 \\ A_1 \\ A_2 \\ A_3 \end{pmatrix}.$$

A_0 is known as the time-like component and is similar to t in the four-vector (t, x, y, z) while A_1, A_2 and A_3 are known as the space-like components that correspond to x, y and z respectively. In an inertial frame S, the entries of A are shown above. The corresponding values as observed in an inertial frame S' that travels at a velocity v in the positive x-direction relative to S are obtained by the Lorentz transformations.

$$A'_0 = \gamma_v(A_0 - \beta A_1),$$
$$A'_1 = \gamma_v(A_1 - \beta A_0),$$
$$A'_2 = A_2,$$
$$A'_3 = A_3,$$

where $\beta = v$ in units of c. The above expressions can be represented more compactly via a matrix equation.

$$A' = \begin{pmatrix} \gamma_v & -\gamma_v\beta & 0 & 0 \\ -\gamma_v\beta & \gamma_v & 0 & 0 \\ 0 & 0 & 1 & 0 \\ 0 & 0 & 0 & 1 \end{pmatrix} A.$$

The matrix above will be referred to as the Lorentz transformation matrix $\mathcal{L}$. The inverse transformation matrix $\mathcal{L}^{-1}$ can be obtained by adding a negative sign in front of the β's while retaining the magnitude of v, which is equivalent to substituting v for $-v$.

Property 1: Multiplying a four-vector by a constant or Lorentz scalar produces another four-vector. A Lorentz scalar is a quantity that has the same value in all inertial frames (e.g. the invariant interval $(\Delta s)^2$).

Proof: Consider $X = cA$ where A is a four-vector and c is a constant or Lorentz scalar. Then, X' in another inertial frame obeys

$$X' = c'A'$$
$$= c\mathcal{L}A$$
$$= \mathcal{L}(cA)$$
$$= \mathcal{L}X,$$

where we have used the fact that $c' = c$. This shows that X is a four-vector as it transforms according to the Lorentz transformations. Now, it is tempting to exploit the seemingly distributive nature of the Lorentz transformations and claim that any linear combination of four-vectors produces another four-vector. However, we have to understand that four-vectors are generally associated with physical properties of particles. A linear combination of four-vectors may be the sum of four-vectors (multiplied by constants or scalars) evaluated at the same time in a frame S but applying a Lorentz transformation to it and using the distributive rule would result in a linear combination of four-vectors evaluated at different times! This is the result of the loss of simultaneity between events that are spatially separated (this occurs as the four-vectors usually correspond to properties of different particles). Therefore, the linear combination of four-vectors is, foremost, meaningless. Its value at a certain instance in another inertial frame (i.e. all of its component four-vectors are determined simultaneously) most definitely cannot be computed via a Lorentz transformation of its value at a certain instance in a precedent inertial frame.

Definition: The inner product of two four-vectors A and B is defined as

$$A \cdot B = A_0 \cdot B_0 - A_1 \cdot B_1 - A_2 \cdot B_2 - A_3 \cdot B_3.$$

Note that the inner product is commutative and distributive. In other words,

$$A \cdot B = B \cdot A$$

and

$$A \cdot (B + C) = A \cdot B + A \cdot C,$$

where A, B and C are four-vectors.

Property 2: The inner product of any two four-vectors is Lorentz invariant (i.e. a scalar).

Proof: Consider a Lorentz transformation in the x-direction.

$$A' \cdot B' = \gamma_v(A_0 - \beta A_1) \cdot \gamma_v(B_0 - \beta B_1) - \gamma_v(A_1 - \beta A_0)$$
$$\cdot \gamma_v(B_1 - \beta B_0) - A_2 \cdot B_2 - A_3 \cdot B_3$$
$$= \gamma_v^2(1 - \beta^2)A_0 \cdot B_0 - \gamma_v^2(1 - \beta^2)A_1 \cdot B_1 - A_2 \cdot B_2 - A_3 \cdot B_3$$
$$= A \cdot B.$$

A similar statement can be made for Lorentz transformations in the other spatial directions.

Corollary: The inner product of a four-vector with itself, $|A|^2$, is Lorentz-invariant and is defined as its squared norm. This immediately follows from above.

$$A' \cdot A' = A \cdot A = |A|^2.$$

An apt illustration of this invariance would be the invariant interval $(\Delta s)^2$ introduced in the previous chapter which is basically the norm of $(\Delta t, \Delta x, \Delta y, \Delta z)$.

Property 3: If the inner product of A and B produces the same scalar in all inertial frames while A is a four-vector, then B must also be a four-vector.

The premise is basically stating that

$$A \cdot B = A' \cdot B'$$

for all pairs of frames. Now, we know from the proof in Property 2 that

$$A \cdot B = (\mathcal{L}A) \cdot (\mathcal{L}B) = A' \cdot (\mathcal{L}B).$$

Therefore,

$$A' \cdot B' = A' \cdot (\mathcal{L}B).$$

Now, we can prove that $B' = \mathcal{L}B$ by astutely substituting appropriate values for A' (since it can be tweaked). Substituting $A' = (1, 0, 0, 0)$ would show that the first entries of B' and $\mathcal{L}B$ are equal. Repeating this for similar "unit vectors" would prove that

$$B' = \mathcal{L}B.$$

Therefore, B must be a four-vector.

Leveraging Properties 1 and 3, we can develop a repository of four-vectors that will be immensely expeditious in the problems that we will encounter.

12.4.1 Four-Coordinate

The four-coordinate vector is defined as

$$X = \begin{pmatrix} t \\ x \\ y \\ z \end{pmatrix}$$

where t, x, y and z are the temporal and spatial coordinates of an event in an arbitrary inertial frame S. Its squared norm guarantees the invariance of $s^2 = t^2 - x^2 - y^2 - z^2$.

If the coordinates describe the world line of a particle, then the invariant interval must be time-like (i.e. $(\Delta s)^2 > 0$). Now, recall that the infinitesimal proper time interval between two events separated by an infinitesimal segment along the world line of a particle is defined as the infinitesimal time between them as measured in the frame of the particle. Since the particle remains still in its own rest frame, $ds^2 = d\tau^2$. Expressing this in terms of the coordinates observed in a general inertial frame S,

$$d\tau = \sqrt{dt^2 - dx^2 - dy^2 - dz^2} = \frac{dt}{\gamma_u}.$$

Note that the infinitesimal proper time interval $d\tau$ is measured in the frame of the particle while the infinitesimal coordinate time interval dt is measured in the current inertial frame S. The last equality is obtained from extracting dt from the brackets in the second last expression (γ_u is associated with the velocity of the particle in frame S, u). The proper time elapsed between two events is then obtained from integrating the above expression.

The concept of proper time is particularly useful in two areas. Firstly, it presents another way to describe the motion of a particle by considering the proper coordinates, which are coordinates as measured in its own rest frame. Then, the corresponding coordinates in a general inertial frame can be obtained via the Lorentz transformations. Secondly and more importantly, the proper time interval is a Lorentz scalar — evident from the fact that it is directly related to the invariant interval. Invariant quantities are sacrosanct in the context of special relativity. Utilising the invariance of proper time, many other four-vectors can be formulated via the following procedure.

Property 4: If $A(t)$ is a four-vector ascribed to a particle where t is the coordinate time in the current inertial frame, $X(t) = \frac{dA}{d\tau}(t)$ — where τ is the proper time elapsed in the particle's rest frame — is a four-vector.

Proof: From the first principles of calculus,

$$X(t) = \lim_{d\tau \to 0} \frac{A(t + dt) - A(t)}{d\tau}$$

where the coordinate time $t(\tau)$ is a function of proper time τ such that $dt = \frac{dt}{d\tau} d\tau$. We can deem the above as dividing $A(t + dt) - A(t)$ by $d\tau$. Since $d\tau$ has been shown to be Lorentz invariant, we just have to show that $A(t + dt) - A(t)$ is a valid four-vector to prove that X is a four-vector (as we can subsequently apply Property 1).

Now, even though $A(t+dt) - A(t)$ is a linear combination of four-vectors, the issue highlighted in the comments of Property 1 does not crop up here because $A(t+dt)$ and $A(t)$ describe the same particle, and the time interval dt between them is infinitesimal such that the loss of simultaneity (due to the particle being at possibly different locations at t and $t + dt$) in a new inertial frame S' is infinitesimal[2] and can be absorbed into the infinitesimal time interval dt' in S'. With this clarification, it is easy to prove that $A(t + dt) - A(t)$ is a four-vector. $A'(t' + dt') - A'(t')$ in another inertial frame S' obeys

$$A'(t' + dt') - A'(t') = \mathcal{L}A(t + dt) - \mathcal{L}A(t) = \mathcal{L}[A(t + dt) - A(t)].$$

Thus, $X(t) = \frac{dA}{d\tau}$ is a valid four-vector.

12.4.2 *Four-Velocity*

By Property 4, taking the derivative of the four-coordinate of a particle in an arbitrary inertial frame S with respect to its proper time produces a new four-vector, known as the four-velocity U of the particle. Since $\frac{1}{d\tau} = \gamma_u \frac{1}{dt}$,

$$U = \frac{1}{d\tau} \begin{pmatrix} dt \\ dx \\ dy \\ dz \end{pmatrix} = \gamma_u \begin{pmatrix} \frac{dt}{dt} \\ \frac{dx}{dt} \\ \frac{dy}{dt} \\ \frac{dz}{dt} \end{pmatrix} = \begin{pmatrix} \gamma_u \\ \gamma_u u \end{pmatrix},$$

where u is the velocity of the particle in frame S. Notice that the spatial component of the four-velocity does not describe the velocity of the particle in frame S directly. It is the derivative of spatial coordinates observed in frame S with respect to the proper time interval which is observed in the frame of

---

[2]This is also partly due to the finite speed of the particle which causes the separation between its positions at $t + dt$ and t to be infinitesimal.

the particle. Hence, the four-vector itself lacks physical meaning. However, it is indirectly related to the coordinate velocity $\boldsymbol{u}$. The sole purpose of such a definition is its utility as a four-vector — namely, its transformations and inner product invariance. This will be a recurring theme for many of the other four-vectors in the following sections.

An example of the utility of the four-velocity would pertain to the derivation of the velocity-addition formulae. Let $\boldsymbol{u}$ and $\boldsymbol{u}'$ be the three-velocities in inertial frames S and S' respectively. Then, the four-velocity in S' can be obtained from that in S via a Lorentz transformation.

$$U' = \mathcal{L}U$$

$$
\begin{pmatrix}
\gamma_{u'} \\
\gamma_{u'} u'_x \\
\gamma_{u'} u'_y \\
\gamma_{u'} u'_z
\end{pmatrix}
=
\begin{pmatrix}
\gamma_v & -\gamma_v \beta & 0 & 0 \\
-\gamma_v \beta & \gamma_v & 0 & 0 \\
0 & 0 & 1 & 0 \\
0 & 0 & 0 & 1
\end{pmatrix}
\begin{pmatrix}
\gamma_u \\
\gamma_u u_x \\
\gamma_u u_y \\
\gamma_u u_z
\end{pmatrix}.
$$

Comparing first entries, we obtain the relationship between the gamma factors in both frames.

$$\gamma_{u'} = \gamma_v(\gamma_u - \beta \gamma_u u_x) = \gamma_v \gamma_u (1 - \beta u_x). \tag{12.22}$$

Comparing the second entries, the longitudinal velocity addition formula can be obtained.

$$u'_x = \frac{\gamma_v \gamma_u (u_x - \beta)}{\gamma_{u'}} = \frac{u_x - v}{1 - v u_x}$$

where we have used the previous result. Comparing the third entries, the transverse velocity addition formula can be obtained.

$$u_{y'} = \frac{\gamma_u u_y}{\gamma_{u'}} = \frac{u_y}{\gamma_v(1 - v u_x)}.$$

A similar statement can be made for $u_{z'}$.

Next, the squared norm of the four-velocity can be easily computed by considering the rest frame of the particle as the squared norm is Lorentz invariant. In the rest frame of the particle,

$$
U =
\begin{pmatrix}
1 \\
0 \\
0 \\
0
\end{pmatrix},
$$

as the velocity of the particle in its own frame is zero. Hence, the squared norm of the four-velocity is

$$|U|^2 = 1.$$

Another important property is the inner product of the four-velocities of two different particles, U_1 and U_2. This can be evaluated in the rest frame of one of the particles. Then, let $\boldsymbol{u}_{rel}$ denote the velocity of the other particle in this rest frame. The inner product of the two four-velocities in this frame is then

$$U_1 \cdot U_2 = \begin{pmatrix} 1 \\ 0 \end{pmatrix} \cdot \begin{pmatrix} \gamma_{u_{rel}} \\ \gamma_{u_{rel}} \boldsymbol{u}_{rel} \end{pmatrix} = \gamma_{u_{rel}}. \tag{12.23}$$

It can be observed that this inner product is minimized when the relative velocity $\boldsymbol{u}_{rel}$ is zero (i.e. the two particles travel at the same velocity in any arbitrary inertial frame). When considering the rest frame of a particle, it was implicitly assumed that the particle traveled at a subluminal speed. If not, there would not have been a rest frame for that particle. However, this is perfectly fine as the four-velocity is ill-defined for massless particles which travel at the speed of light.

Next, the consideration of the rest frame of the particle in computing the inner product of its four-velocity with another four-vector enables the isolation of the first entry of the other four-vector, as the first entry of the four-velocity is one while the others are all zero in the rest frame of the particle. The utility of this will be illustrated in a later section.

12.4.3 *Four-Acceleration*

Once again, the four-velocity can be differentiated with respect to the proper time τ to produce yet another four-vector which is termed as the four-acceleration. The four-acceleration in an arbitrary inertial frame S is

$$A = \frac{d}{d\tau} \begin{pmatrix} \gamma_u \\ \gamma_u \boldsymbol{u} \end{pmatrix} = \gamma_u \begin{pmatrix} \frac{d\gamma_u}{dt} \\ \frac{d\gamma_u}{dt} \boldsymbol{u} + \gamma_u \frac{d\boldsymbol{u}}{dt} \end{pmatrix}.$$

It has been shown that

$$\frac{d\gamma_u}{dt} = \gamma_u^3 (\boldsymbol{a} \cdot \boldsymbol{u})$$

where $\boldsymbol{a}$ is the coordinate acceleration, $\boldsymbol{a} = \frac{d\boldsymbol{u}}{dt}$. Hence, the four-acceleration can be expressed as

$$A = \gamma_u \begin{pmatrix} \gamma_u^3 (\boldsymbol{a} \cdot \boldsymbol{u}) \\ \gamma_u^3 (\boldsymbol{a} \cdot \boldsymbol{u}) \boldsymbol{u} + \gamma_u \boldsymbol{a} \end{pmatrix}.$$

If the x-axis of frame S is chosen such that it is aligned with the instantaneous coordinate velocity of the particle $\boldsymbol{u}$, the four-acceleration becomes

$$A = \begin{pmatrix} \gamma_u^4 u a_x \\ \gamma_u^4 a_x \\ \gamma_u^2 a_y \\ \gamma_u^2 a_z \end{pmatrix},$$

where we have used the fact that

$$A_1 = \gamma_u^4 u^2 a_x + \gamma_u^2 a_x = \gamma_u^4 a_x \left(u^2 + \frac{1}{\gamma_u^2} \right) = \gamma_u^4 a_x.$$

Once again, individual entries of the four-acceleration do not have obvious physical meanings. However, the four-acceleration provides a convenient pathway to derive how different components of acceleration transform between inertial frames.

Sometimes, a need to relate the coordinate acceleration in an inertial frame S to the proper acceleration in the rest frame of the particle may arise. Let the four-accelerations in frame S and the rest frame of the particle be A and A' respectively. Assuming that the particle travels at a speed u in the positive x-direction in frame S,

$$A = \mathcal{L}^{-1} A'$$

$$\begin{pmatrix} \gamma_u^4 u a_x \\ \gamma_u^4 a_x \\ \gamma_u^2 a_y \\ \gamma_u^2 a_z \end{pmatrix} = \begin{pmatrix} \gamma_u & \gamma_u \beta & 0 & 0 \\ \gamma_u \beta & \gamma_u & 0 & 0 \\ 0 & 0 & 1 & 0 \\ 0 & 0 & 0 & 1 \end{pmatrix} \begin{pmatrix} 0 \\ \alpha_x \\ \alpha_y \\ \alpha_z \end{pmatrix} = \begin{pmatrix} \gamma_u \beta \alpha_x \\ \gamma_u \alpha_x \\ \alpha_y \\ \alpha_z \end{pmatrix},$$

where we have substituted $v = u$ to switch between S and the rest frame of the particle. Comparing the corresponding terms, the proper accelerations, which are denoted by the symbol α, are given by

$$\alpha_x = \gamma_u^3 a_x,$$
$$\alpha_y = \gamma_u^2 a_y,$$
$$\alpha_z = \gamma_u^2 a_z.$$

12.4.4 *Four-Momentum*

Multiplying the four-velocity of a particle by its rest mass produces the four-momentum of the particle in an arbitrary inertial frame S.

$$P = mU = m \begin{pmatrix} \gamma_u \\ \gamma_u \boldsymbol{u} \end{pmatrix} = \begin{pmatrix} \gamma_u m \\ \gamma_u m \boldsymbol{u} \end{pmatrix}.$$

Observe that the first entry is simply the energy of the particle in frame S while the three-vector below corresponds to the momentum of the particle in frame S. Hence,

$$P = \begin{pmatrix} E \\ \boldsymbol{p} \end{pmatrix}.$$

Note that even though the four-velocity is ill-defined for massless particles, the four-momentum remains well-defined. Since the total energy and momentum in an isolated system is conserved, the sum of all relevant four-momenta should be equal at two different instances in the same frame — encapsulating the two conservation laws into a single four-vector equation. This combined with the squared norm of a four-momentum can greatly simplify calculations. The squared norm of a four-momentum can be computed in the rest frame of the particle (if it is massive). In this frame,

$$P = \begin{pmatrix} m \\ 0 \end{pmatrix}.$$

Hence, the squared norm of the four-momentum of a massive particle is

$$|P|^2 = E^2 - p^2 = m^2,$$

which is basically Eq. (12.8). Actually, the above equation is also valid for massless particles ($m = 0$) for which $E = p$ such that $E^2 - p^2 = 0$ — it is therefore entirely general. Next, we make a rather bold claim that the sum of four-momenta in a system undergoing purely local interactions is another four-vector. This seemingly contradicts what we have said in the comments of Property 1 of four-vectors but we are saved by the conservation of momentum and energy here. Let the total four-momentum of a system be

$$P_{tot} = \sum_{i=1}^{N} P_i.$$

When evaluating P_{tot} at a certain time $t = t_0$ in a certain frame S, we mean to sum up all P_i's evaluated at time $t = t_0$. However, note that the actual value of P_{tot} should be irrespective of time t, as it is conserved. Now, consider another inertial frame S' with the conventional definition. When calculating P'_{tot} at a certain time $t' = t'_0$, we similarly add all P'_i's evaluated at time $t' = t'_0$ but these cannot possibly correspond to events that are simultaneous in frame S, due to the spatial separations of particles. Then, the constituent P'_i's cannot simply be obtained from the Lorentz transformations, $P'_i(t' = t'_0) \neq \mathcal{L}P_i(t = t_0)$ in general. However, by asserting that P_{tot} is a four-vector, we are claiming that

$$P'_{tot} = \mathcal{L}P_{tot}$$

as a whole. Again, P'_{tot} should be independent of time, as energy and momentum are also conserved in S' by the principle of relativity.

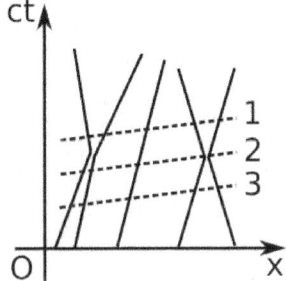

Figure 12.7: World lines of particles and lines of simultaneity

If a system of particles only undergoes short-range interactions, the energies and momenta of individual particles (and possibly new particles formed) can only change when they come into the immediate vicinity of one another (i.e. when their world lines intersect in Fig. 12.7). Then, when evaluating P_{tot}, we can deliberately choose to evaluate P_i at non-simultaneous times in S but simultaneous times in S' while capitalizing on the fact that the individual energies and momenta of particles can only vary at space-time junctions and will remain constant at other times. The existence of a valid set of times that should be chosen is best visualized by the Minkowski diagram depicting the world lines of various particles in frame S in Fig. 12.7. A line of simultaneity in frame S' superimposed on the diagram is a line that subtends an angle smaller than $45°$ from the horizontal. Since the slope of a world line must be larger than or equal to $45°$, it is impossible for a line of simultaneity to cut across the world lines of interacting particles and divide the intersected events into two groups — those before their interaction and those after their interaction. The times of the intersected events must always lie on one temporal side, with respect to the time of interaction of the particles. Therefore, it is always possible to choose a set of events that both correctly represent P_{tot} collectively and are simultaneous in S'. For the diagram in Fig. 12.7, three possible lines of simultaneity are drawn and the intersections along a single line form a possible set of events at which the individual P_i's can be evaluated. Then, applying a Lorentz transformation to each of these distinct four-momenta would yield four-momenta evaluated concurrently in S', after which they can be summed to determine P'_{tot}. Therefore, P'_{tot} is simply obtained from the Lorentz transformation of P_{tot}.

$$P'_{tot} = \mathcal{L} P_{tot}.$$

This proves that P_{tot} is a valid four-vector for collisions (and decays). Another perspective to this is that we can gradually rotate a horizontal line — anchored about a certain point — into the final line of simultaneity (akin to considering the lines of simultaneity of a continuous set of frames). The sum of the individual energies and momenta of the particles along an intermediate line of simultaneity during this rotation cannot change. This is because the individual energy and momentum of a particle recorded during this rotation will only change during an interaction. However, even during such interactions which occur at space-time intersections between particles, the total energy and momentum of the system of particles do not change by the conservation laws.

Armed with this machinery, the square of the total energy of the particles in the center-of-momentum frame of a system E^2_{CM}, can be proven to be $E^2_{tot} - p^2_{tot}$ where E_{tot} and p_{tot} are the total energy and momentum of the system of particles in an arbitrary inertial frame S; this was used to derive Eq. (12.15). Let the sum of all four-momenta of the particles in the system in frame S and the center-of-momentum frame be P_{tot} and P'_{tot} respectively. Then,

$$P_{tot} = \begin{pmatrix} E_{tot} \\ \boldsymbol{p}_{tot} \end{pmatrix} \quad P'_{tot} = \begin{pmatrix} E_{CM} \\ 0 \end{pmatrix}.$$

The squared norm of the two matrices above should be equal as they are the same four-vector observed in different inertial frames. Thus,

$$E^2_{CM} = E^2_{tot} - p^2_{tot} = m^2_{sys}.$$

The last equality stems from the fact that the total mass of a system m_{sys} of non-interacting particles is, by definition, given by $E^2_{CM} = m^2_{sys}c^4$. This relationship also proves the invariant nature of the invariant mass of a system, m_{sys}.

Problem: A particle of rest mass m_1 and initial momentum p_1 collides with another stationary particle of rest mass m_2. It is known that the final velocities of these particles are perpendicular to each other and non-zero. If the rest masses of the particles remain constant, determine the magnitudes of the resultant momenta of the two particles.

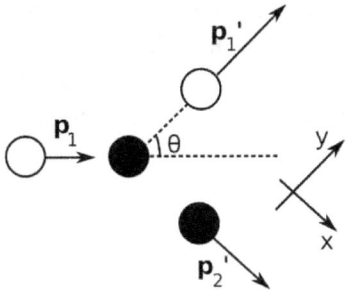

Figure 12.8: Collision with perpendicular final velocities

Define the coordinate axes to be parallel to the final velocities and define angle θ as shown in Fig. 12.8. Then, let the four-momenta of particles m_1 and m_2, before and after collision, be

$$P_1 = \begin{pmatrix} E_1 \\ p_1 \sin\theta \\ p_1 \cos\theta \\ 0 \end{pmatrix}, \quad P_2 = \begin{pmatrix} m_2 \\ 0 \\ 0 \\ 0 \end{pmatrix}, \quad P_1' = \begin{pmatrix} E_1' \\ 0 \\ p_1 \cos\theta \\ 0 \end{pmatrix}, \quad P_2' = \begin{pmatrix} E_2' \\ p_1 \sin\theta \\ 0 \\ 0 \end{pmatrix}.$$

By the conservation of energy and momentum,

$$P_1 + P_2 = P_1' + P_2'.$$

Now, we may be tempted to equate each of the rows of the four-vectors and solve for the required expressions. However, this will be extremely tedious due to the energies E_1, E_1' and E_2' being surds in terms of their corresponding momenta. A better method would entail taking the squared norm of both sides of the equations.

$$(P_1 + P_2) \cdot (P_1 + P_2) = (P_1' + P_2') \cdot (P_1' + P_2')$$

$$P_1 \cdot P_1 + 2P_1 \cdot P_2 + P_2 \cdot P_2 = P_1' \cdot P_1' + 2P_1' \cdot P_2' + P_2' \cdot P_2'.$$

Since the rest masses of the particles remain unchanged after the collision,

$$P_1 \cdot P_1 = P_1' \cdot P_1' = m_1^2,$$
$$P_2 \cdot P_2 = P_2' \cdot P_2' = m_2^2.$$

The equation above then becomes

$$P_1 \cdot P_2 = P_1' \cdot P_2'.$$

Up till now, we have not assumed anything about the exact expression of any of the four-momenta in the equation above. Hence, this equation is valid for all collisions between two particles that are rest-mass preserving — such collisions are known as elastic collisions in the context of special relativity. The reason behind this terminology will be explicated immediately after this problem. Substituting the expressions for the four-momenta into the equation above,

$$E_1 m_2 = E'_1 E'_2,$$

where we have deliberately tweaked the coordinate axes to elucidate the orthogonality of the three-momenta of P'_1 and P'_2. Using the fact that $E^2 = p^2 + m^2$ for a particle,

$$m_2 \sqrt{p_1^2 + m_1^2} = \sqrt{p_1^2 \cos^2 \theta + m_1^2} \cdot \sqrt{p_1^2 \sin^2 \theta + m_2^2}.$$

Squaring and simplifying,

$$p_1^2 \sin^2 \theta \left(p_1^2 \cos^2 \theta - m_2^2 + m_1^2 \right) = 0.$$

Since the final momentum of the second particle must be non-zero, $\sin \theta \neq 0$.

$$p_1^2 \cos^2 \theta = m_2^2 - m_1^2$$

$$|p_1 \cos \theta| = \sqrt{m_2^2 - m_1^2}$$

$$|p_1 \sin \theta| = \sqrt{p_1^2 + m_1^2 - m_2^2}.$$

Notice that the condition for the resultant configuration of velocities after the collision to be possible is $m_1^2 \leq m_2^2 \leq p_1^2 + m_1^2$.

Inner Product of Two Four-Momenta

The inner product of the four-momenta of two different particles of rest masses $m_1 > 0$ and $m_2 > 0$ can be evaluated in the rest frame of one particle as

$$P_1 \cdot P_2 = m_1 m_2 U_1 \cdot U_2 = m_1 m_2 \gamma_{u_{rel}}, \qquad (12.24)$$

where u_{rel} is the speed of one particle in the rest frame of the other particle. The second equality is obtained from applying Eq. (12.23). Hence in the case of a rest-mass preserving collision described in the previous example, the speed of one particle in the rest frame of the other particle must be the same before and after the collision (note that these are two different rest frames as the particle's velocity may have changed). Since this "relative speed" remains unchanged, such a collision is known as an elastic collision.

When $m_1 > 0$ and $m_2 \geq 0$, we can evaluate the inner product in the rest frame of the first particle. In this frame, $P_1 = (m_1, \mathbf{0})$ and $P_2 = (E_{rel}, \mathbf{p}_{rel})$, where E_{rel} and $\mathbf{p}_{rel}$ are the energy and momentum of the second particle.

$$P_1 \cdot P_2 = m_1 E_{rel}. \tag{12.25}$$

Moving on, Eq. (12.24) implies that two massive particles should have zero relative velocity to minimize the inner product of their four-momenta. On another note, it can be easily shown, by using the fact that $E = pc$ for a massless particle, that the inner product of the four-momenta of a massless particle and a massive particle or of two massless particles, is similarly minimized when both particles move in the same direction in an inertial frame.

Threshold Energy

Often, reactions are initiated by bombarding a stationary particle with another particle, producing new particles of various rest masses. The problem of finding the threshold energy entails determining the minimum amount of energy that the incoming particle must possess to spark off the reaction. Note that the required condition in such situations is not that the kinetic energy of the incoming particle must be equal to the sum of the additional rest masses of the final configuration as the product particles must still possess a certain amount of kinetic energy by the conservation of momentum. Let us derive a general formula for the threshold energy of a reaction that produces only massive products. Let the four-momenta of the incident and stationary particles be P_a and P_b respectively, and let there be k final particles with the ith particle having a four-momentum P_i and mass $m_i > 0$. The incident particle is possibly massless (but if it is, it must be absorbed as all products are massive) while the receiving particle is massive. By the conservation of momentum and energy,

$$P_a + P_b = \sum_{i=1}^{k} P_i.$$

Taking the inner product of both sides of the equation,

$$P_a \cdot P_a + 2 P_a \cdot P_b + P_b \cdot P_b = \sum_{i=1}^{k} P_i \cdot P_i + 2 \sum_{i<j} P_i \cdot P_j$$

$$m_a^2 + m_b^2 + 2 E_{rel} m_b = \sum_{i=1}^{k} m_i^2 + 2 \sum_{i<j} m_i m_j \gamma_{u_{ij}},$$

where E_{rel} is the energy of the incident particle in the rest frame of the receiving particle — it is simply the energy E_a of the incident particle in the current inertial frame, since the receiving particle is stationary in this frame! Meanwhile, u_{ij} is the relative speed between the ith and jth particle in one of their rest frames. Using $E_{rel} = E_a$ and rearranging,

$$E_a = \frac{\left(\sum_{i=1}^{k} m_i^2\right) + 2\left(\sum_{i<j} m_i m_j \gamma_{u_{ij}}\right) - m_a^2 - m_b^2}{2m_b}.$$

Note that all quantities are predetermined except $\gamma_{u_{ij}}$. To minimize E_a, $\gamma_{u_{ij}}$ must attain its minimum value of unity, which occurs when $u_{ij} = 0$, for all i and j. This implies that all resultant particles travel together in a "blob" after the process, akin to a perfectly inelastic collision in classical mechanics. Then, the threshold energy is

$$E_a = \frac{\left(\sum_{i=1}^{k} m_i\right)^2 - m_a^2 - m_b^2}{2m_b}. \tag{12.26}$$

Transformations of Energy and Momentum

The combination of the three-momentum and energy of a particle as a four-vector elucidates their transformations across inertial frames. If the components of the momentum of a particle in the x, y and z-directions are p_x, p_y and p_z respectively and if its energy is E in an inertial frame S, the corresponding quantities, denoted by appending a prime, in an inertial frame S' that travels at a velocity v in the x-direction are obtained from the Lorentz transformations.

$$E' = \gamma(E - \beta p_x),$$

$$p'_x = \gamma(p_x - \beta E),$$

$$p'_y = p_y,$$

$$p'_z = p_z.$$

Four-Frequency

As illustrated previously, the energy and momentum of a photon in an inertial frame are $E = hf$ and $p = \frac{h}{\lambda} = \frac{hf}{c}$ respectively. Hence the four-momentum of a photon in an inertial frame S is

$$P = h\begin{pmatrix} f \\ f\hat{k} \end{pmatrix},$$

where f is the frequency of the photon as observed in frame S. $\hat{k}$ is a unit vector along the velocity of the photon in frame S. For the sake of convenience, we define a new four-frequency four-vector as

$$F = \frac{P}{h} = \begin{pmatrix} f \\ f\hat{k} \end{pmatrix},$$

since multiplying a four-vector by a constant generates another four-vector. Notably, the squared norms of the four-momentum of a photon and the four-frequency are zero.

Surprisingly, the particulate nature of light allows us to deduce many relativistic effects pertaining to the wave nature[3] of light via the four-frequency vector. Consider a photon, of frequency f, that is traveling at an angle θ relative to the x-axis in an inertial frame S. We define the y-axis such that the motion of the particle is solely confined to the xy-plane.

We would like to determine the frequency f' of this photon and the angle θ' it makes with the x'-axis in an inertial frame S' that is traveling at a velocity u in the positive x-direction with respect to frame S. Let the four-frequencies of the photon in S and S' be

$$F = \begin{pmatrix} f \\ f\cos\theta \\ f\sin\theta \\ 0 \end{pmatrix}, \quad F' = \begin{pmatrix} f' \\ f'\cos\theta' \\ f'\sin\theta' \\ 0 \end{pmatrix}.$$

F' can be obtained from F via a Lorentz transformation.

$$F' = \begin{pmatrix} \gamma_u & -\gamma_u\beta & 0 & 0 \\ -\gamma_u\beta & \gamma_u & 0 & 0 \\ 0 & 0 & 1 & 0 \\ 0 & 0 & 0 & 1 \end{pmatrix} \begin{pmatrix} f \\ f\cos\theta \\ f\sin\theta \\ 0 \end{pmatrix} = \begin{pmatrix} \gamma_u f(1 - \beta\cos\theta) \\ \gamma_u f(\cos\theta - \beta) \\ f\sin\theta \\ 0 \end{pmatrix}.$$

Comparing the first entries,

$$f' = \gamma_u f(1 - \beta\cos\theta).$$

This is the formula for the relativistic Doppler effect! Consider Fig. 12.9, in which a photon emanates from a source and reaches the eye of an observer. The left diagram shows frame S, which is the rest frame of the source. In this frame, the observer is traveling at a speed u in the positive x-direction

---

[3]The proofs of these effects begin from different premises but are coherent. Hence, the four-frequency is a useful way to recall wave-like effects of light by considering its particulate nature.

and f is the frequency of the photon. On the other hand, the right diagram depicts the rest frame of the observer, S'. The source now travels at the same speed in the opposite direction while the frequency of the photon is now f'.

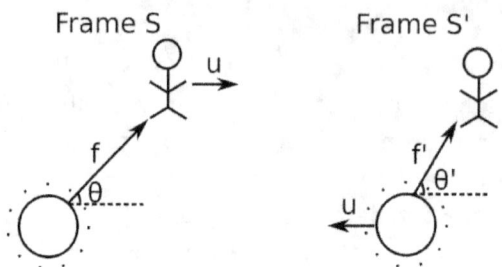

Figure 12.9:   Source emitting photons in different frames

The longitudinal case can be easily derived by setting $\theta = 0$.

$$f' = \gamma_u f(1 - u) = \sqrt{\frac{1-u}{1+u}} f.$$

Similar to Problem 12 in the previous chapter, there are two cases to consider in the transverse situation — namely, the frequencies of the received photons, observed by the observer, emitted as the source crosses the line of sight of the observer and that of photons reaching the eyes of the observer when the source crosses the vertical line of sight of the observer. In the second case, $\theta = \frac{\pi}{2}$, as the photon reaching the observer's eyes would have taken a vertical path in frame S (i.e. the photons were already emitted before the observer and the source formed a vertical line). Then,

$$f' = \gamma_u f.$$

In the first case, the photons are emitted when the observer crosses the vertical line in frame S (see Fig. 12.10) and thus reach the observer after the observer crosses the vertical line.

In this case, $\cos\theta$ will be u. Substituting this back into the equation,

$$f' = \gamma_u f(1 - u^2) = \frac{f}{\gamma_u}.$$

The four-frequency can also be used to derive the aberration formula (Problem 6 in Chapter 11). Comparing the second entries,

$$\cos'\theta = \frac{\gamma_u f(\cos\theta - \beta)}{f'} = \frac{\cos\theta - \beta}{1 - \beta\cos\theta}.$$

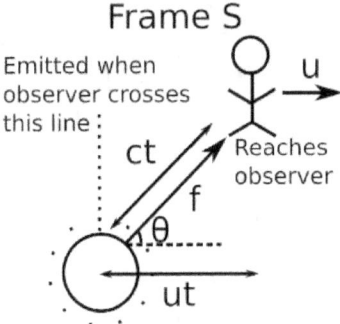

Frame S

Emitted when observer crosses this line

Reaches observer

Figure 12.10:   Emission of photon when observer crosses line of sight

Problem: A photon of frequency f is normally incident on an infinitely massive and perfectly reflective mirror that is retracting at a velocity u in inertial frame S. Determine the frequency f' of the photon after the reflection in frame S.

The initial four-frequency of the photon in frame S is

$$F_1 = \begin{pmatrix} f \\ f \end{pmatrix}.$$

Let S' be the frame that travels at a velocity u relative to S (i.e. the mirror is initially at rest in this frame). The initial four-frequency in S' is obtained from the Lorentz transformations.

$$F_1' = \begin{pmatrix} \gamma_u & -\gamma_u \beta \\ -\gamma_u \beta & \gamma_u \end{pmatrix} \begin{pmatrix} f \\ f \end{pmatrix} = \begin{pmatrix} \gamma_u f (1 - \beta) \\ \gamma_u f (1 - \beta) \end{pmatrix}$$

where $\beta = u$. After the reflection, the four-frequency in frame S' is

$$F_2' = \begin{pmatrix} \gamma_u f (1 - \beta) \\ \gamma_u f (\beta - 1) \end{pmatrix},$$

as the photon just reverses its momentum after its collision with the stationary massive mirror (for energy to be conserved, since the infinitely massive mirror gains negligible kinetic energy). Lastly, an inverse Lorentz transformation can be applied to return to the original frame S.

$$F_2 = \begin{pmatrix} f' \\ -f' \end{pmatrix} = \begin{pmatrix} \gamma_u & \gamma_u \beta \\ \gamma_u \beta & \gamma_u \end{pmatrix} \begin{pmatrix} \gamma_u f (1 - \beta) \\ \gamma_u f (\beta - 1) \end{pmatrix}$$

$$= \begin{pmatrix} \gamma_u^2 f (1 - \beta) + \gamma_u^2 \beta f (\beta - 1) \\ \gamma_u^2 \beta f (1 - \beta) + \gamma_u^2 f (\beta - 1) \end{pmatrix} = \begin{pmatrix} \frac{1-u}{1+u} f \\ \frac{u-1}{1+u} f \end{pmatrix}.$$

Comparing the terms,

$$f' = \frac{1-u}{1+u} f.$$

12.4.5   Four-Force

The four-force is obtained by differentiating the four-momentum with respect to proper time.

$$F = \frac{d}{d\tau}\begin{pmatrix} E \\ \boldsymbol{p} \end{pmatrix} = \gamma_u \begin{pmatrix} \frac{dE}{dt} \\ \boldsymbol{f} \end{pmatrix},$$

as the definition of the three-force is $\boldsymbol{f} = \frac{d\boldsymbol{p}}{dt}$. To simplify the first entry, consider the inner product of the four-force with the four-velocity in the rest frame of the particle (which exists because the four-force is usually only defined for massive particles).

$$F \cdot U = \begin{pmatrix} \frac{dE}{d\tau} \\ \boldsymbol{f'} \end{pmatrix} \cdot \begin{pmatrix} 1 \\ 0 \end{pmatrix} = \frac{dE}{d\tau} = \frac{dm}{d\tau}.$$

It can be seen that $F \cdot U$ is just the rate at which the internal energy of the particle changes in its rest frame, which is also equal to the rate of change of its rest mass in its rest frame. This expression is Lorentz invariant as it is an inner product of two four-vectors. Evaluating this in a general inertial frame S, in which the velocity of the particle is $\boldsymbol{u}$,

$$F \cdot U = \gamma_u \begin{pmatrix} \frac{dE}{dt} \\ \boldsymbol{f} \end{pmatrix} \cdot \gamma_u \begin{pmatrix} 1 \\ \boldsymbol{u} \end{pmatrix}$$

$$= \gamma_u^2 \left(\frac{dE}{dt} - \boldsymbol{f} \cdot \boldsymbol{u} \right)$$

$$= \frac{dm}{d\tau}.$$

Usually, the problems that we will encounter do not involve a change in the rest mass of the particle. In such cases, $\frac{dm}{d\tau} = 0$ and

$$\frac{dE}{dt} = \boldsymbol{f} \cdot \boldsymbol{u},$$

as expected from the previous discussion on the three-force. The four-force can then be rewritten as

$$F = \gamma_u \begin{pmatrix} \boldsymbol{f} \cdot \boldsymbol{u} \\ \boldsymbol{f} \end{pmatrix}.$$

Furthermore, observe that when the rest mass of the particle remains constant,

$$F = m \frac{d}{d\tau} \left(\frac{\gamma_u}{\gamma_u \boldsymbol{u}} \right) = mA$$

where A is the four-acceleration. This is the relativistic counterpart of the $F = ma$ equation in classical mechanics, except that the respective terms are now four-vectors instead of three-vectors.

Force Transformations

Equipped with the four-force, the transformation of forces between a lab frame S and a particle's rest frame S' can be derived. It is assumed that the particle travels at a velocity u in the positive x-direction in frame S and that the rest mass of the particle remains unchanged. The four-force in S' is

$$F' = \begin{pmatrix} 0 \\ f'_x \\ f'_y \\ f'_z \end{pmatrix}.$$

The first entry is zero as the velocity of the particle is zero in its own rest frame. The four-force in S is related to that in S' by a Lorentz transformation.

$$F = \mathcal{L}^{-1} F' = \begin{pmatrix} \gamma_u f'_x \beta \\ \gamma_u f'_x \\ f'_y \\ f'_z \end{pmatrix} = \gamma_u \begin{pmatrix} f_x u \\ f_x \\ f_y \\ f_z \end{pmatrix}.$$

Comparing the terms,

$$f'_x = f_x,$$
$$f'_y = \gamma_u f_y,$$
$$f'_z = \gamma_u f_z.$$

These are the transformation rules for forces in an arbitrary frame S and the rest frame of the particle.

12.4.6 *Four-Wave Vector*

In general, a traveling plane wave can be described by the equation

$$\psi(\boldsymbol{r}, t) = A(\boldsymbol{r}) \cos(\boldsymbol{k} \cdot \boldsymbol{r} - \omega t + \phi_0).$$

Setting the source of the wave as the origin, ψ is the displacement at a point whose position vector is $\boldsymbol{r}$, at time t in frame S. $\boldsymbol{k}$ is the wave-vector and

ω is the angular frequency. A is the amplitude of the wave which varies as some function of $\boldsymbol{r}$ while ϕ_0 is the constant phase offset of the source. The phase of the wave, which is enclosed in the brackets, is

$$\phi = \boldsymbol{k} \cdot \boldsymbol{r} - \frac{\omega}{c} \cdot ct + \phi_0$$

$$\implies \begin{pmatrix} \frac{\omega}{c} \\ k_x \\ k_y \\ k_x \end{pmatrix} \cdot \begin{pmatrix} ct \\ x \\ y \\ z \end{pmatrix} = \phi_0 - \phi.$$

The important observation here is that the phase of a wave at a particular point in time and space in an inertial frame S should be the same as that at the corresponding time and space, after a Lorentz transformation, in another inertial frame S', as the phase corresponds to a physical event. It measures the state of the displacement at a certain point in space, relative to all other points, at a given time. If an event is a peak in frame S, the corresponding event will still be a peak in frame S'. Hence, the phase must be a Lorentz scalar. Then, as the second vector is a four-vector and the right-hand side is a Lorentz scalar,

$$K = \begin{pmatrix} \omega \\ \boldsymbol{k} \end{pmatrix}$$

is a four-vector in units of c by Property 3 — referred to as the four-wave vector. Note that the relationship

$$\frac{\omega}{k} = u$$

still holds in any inertial frame S where u is the phase velocity of the wave. The fact that the angular frequency and the wave numbers of a plane wave form a four-vector is extremely convenient in determining the transformation of properties related to waves.

For example, the longitudinal Doppler effect for a wave with a frequency f and speed u in an inertial frame S can be computed relatively easily. Let f' be the frequency of the wave as observed in the frame S' that is traveling at a velocity v longitudinally relative to frame S. Then, the four-wave vectors in S and S' are

$$K = \begin{pmatrix} \omega \\ k \\ 0 \\ 0 \end{pmatrix} \quad K' = \begin{pmatrix} \omega' \\ k'_x \\ k'_y \\ k'_z \end{pmatrix}$$

respectively. From the Lorentz transformations,

$$K' = \begin{pmatrix} \gamma_v(\omega - \beta k) \\ \gamma_v(k - \beta \omega) \\ 0 \\ 0 \end{pmatrix}.$$

Hence,

$$\omega' = \gamma_v(\omega - \beta k),$$

$$f' = \gamma_v \left(f - \frac{\beta}{\lambda} \right).$$

Since $u = f\lambda$,

$$f' = \gamma_v \left(1 - \frac{\beta}{u} \right) f = \gamma_v \left(1 - \frac{v}{u} \right) f.$$

12.5 Transformation of Electric and Magnetic Fields

One of Einstein's motivations behind special relativity was how similar effects in electromagnetism were attributed to the different entities — electric and magnetic fields. Consider a charged particle and a stationary magnet in frame S. The moving charge experiences a magnetic force due to the magnetic field but no electric force due to the absence of an external electric field. Now consider the instantaneous rest frame of the particle S', in which the magnet is now moving. Though there is still a magnetic field, there is no magnetic force on the particle as it remains stationary. However, there is now a non-conservative electric field induced by the time-varying magnetic field which causes the charge to experience an electric force. Einstein firmly believed that these seemingly disparate effects were linked by a more general theory.

It turns out that electric and magnetic fields are essentially the same entity, as observed in different frames. The electric and magnetic fields at corresponding points in space and time in inertial frames S and S' (which travels at v relative to S) are related by

$$E'_{\parallel} = E_{\parallel}, \qquad\qquad B'_{\parallel} = B_{\parallel},$$

$$E'_{\perp} = \gamma_v(E_{\perp} + v \times B), \ \ B'_{\perp} = \gamma_v \left(B_{\perp} - \frac{v}{c^2} \times E \right),$$

where $\parallel$ and $\perp$ denote directions parallel and perpendicular to the velocity v, respectively. Note that the equations on the second row are vectors. It can be seen that the perpendicular components of the electric and magnetic field in frame S' need not lie along the same direction as their counterparts in frame S. If frame S' travels at a speed v in the positive x-direction relative to frame S, the transformations can be expressed in Cartesian coordinates as

$$E'_x = E_x, \qquad\qquad B'_x = B_x,$$

$$E'_y = \gamma_v(E_y - vB_z), \qquad B'_y = \gamma_v\left(B_y + \frac{v}{c^2}E_z\right),$$

$$E'_z = \gamma_v(E_z + vB_y), \qquad B'_z = \gamma_v\left(B_z - \frac{v}{c^2}E_y\right).$$

Proof: The transformations above can be deduced from the four-force vector corresponding to the Lorentz force on a charged particle in an electromagnetic region. Note that the transformations should actually be independent of the physical existence of such a particle as the external fields[4] themselves are physical entities which persist without regard of anything besides their sources. The charged particle is merely a construct — a stepping stone — in identifying these transformations.

The Lorentz force law exerts that the force on a charge q in frame S is

$$\boldsymbol{f} = q(\boldsymbol{E} + \boldsymbol{u} \times \boldsymbol{B}),$$

where $\boldsymbol{u}$ is the velocity of the particle in S. The idea here is to relate the forces (and power) in two inertial frames to deduce the electromagnetic field transformations.

Firstly, consider a charged particle, with unit charge[5] for the sake of convenience, and velocity $\boldsymbol{u} = (u, 0, 0)$ in frame S. The Lorentz force on the particle is

$$\boldsymbol{f} = \begin{pmatrix} E_x \\ E_y - uB_z \\ E_z + uB_y \end{pmatrix}.$$

The four-force vector in frame S is then

$$F = \begin{pmatrix} \gamma_u \boldsymbol{f} \cdot \boldsymbol{u} \\ \gamma_u \boldsymbol{f} \end{pmatrix} = \begin{pmatrix} \gamma_u E_x u \\ \gamma_u E_x \\ \gamma_u(E_y - uB_z) \\ \gamma_u(E_z + uB_y) \end{pmatrix}.$$

---

[4]We are only considering the external fields and are ignoring the fields generated by the hypothetical charged particle.

[5]Note that charge is presumed to be invariant across inertial frames. Therefore, the particle possesses unit charge in all inertial frames.

Now, we proceed to an inertial frame S' which travels at velocity u in the x-direction relative to S such that the charge is stationary in this frame. The force on the charge in S' is then

$$\boldsymbol{f'} = \boldsymbol{E'},$$

where $\boldsymbol{E'}$ is the electric field in frame S'. The four-force vector in S' is thus

$$F' = \begin{pmatrix} 0 \\ E'_x \\ E'_y \\ E'_z \end{pmatrix}.$$

Furthermore, we know that F' and F are related by the Lorentz transformations.

$$F' = \mathcal{L}F = \begin{pmatrix} \gamma_u & -\gamma_u \beta & 0 & 0 \\ -\gamma_u \beta & \gamma_u & 0 & 0 \\ 0 & 0 & 1 & 0 \\ 0 & 0 & 0 & 1 \end{pmatrix} \begin{pmatrix} \gamma_u E_x u \\ \gamma_u E_x \\ \gamma_u (E_y - u B_z) \\ \gamma_u (E_z + u B_y) \end{pmatrix}$$

$$= \begin{pmatrix} 0 \\ E_x \\ \gamma_u (E_y - u B_z) \\ \gamma_u (E_z + u B_y) \end{pmatrix}$$

with $\beta = u$. Comparing the two expressions for F',

$$E'_x = E_x,$$
$$E'_y = \gamma_v (E_y - v B_z),$$
$$E'_z = \gamma_v (E_z + v B_y),$$

where we have substituted v for u as we have the prerogative to choose u (i.e. because u is arbitrary). The transformations for the magnetic field can in fact be obtained from the above via some astute manipulations. Firstly, the above implies that the inverse transformations are

$$E_x = E'_x,$$
$$E_y = \gamma_v (E'_y + v B'_z),$$
$$E_z = \gamma_v (E'_z - v B'_y).$$

From the second equation,

$$B'_z = \frac{\frac{E_y}{\gamma_v} - E'_y}{v}$$

$$= \frac{\frac{E_y}{\gamma_v} - \gamma_v(E_y - vB_x)}{v}$$

$$= \gamma_v\left(B_z - \frac{v}{c^2}E_y\right).$$

Similarly, from the third equation,

$$B'_y = \gamma_v\left(B_y + \frac{v}{c^2}E_z\right).$$

Now, we need to consider a separate charged particle to determine B'_x as the force on a charged particle that is traveling solely along the x-direction will be independent of the x-component of the magnetic field. Consider a unit charge which travels at $\boldsymbol{u} = (0, 0, u)$ in S. The four-force in S is

$$F = \begin{pmatrix} \gamma_u E_z u \\ \gamma_u(E_x - uB_y) \\ \gamma_u(E_y + uB_x) \\ \gamma_u E_z \end{pmatrix}.$$

In frame S' which travels at velocity v relative to S, in the x-direction, the velocity of the particle is $\boldsymbol{u'} = (-v, 0, \frac{u}{\gamma_v})$ by the velocity addition formula. The y-component of the four-force F' in S' is thus

$$F'_2 = \gamma_{u'}\left(E'_y + \frac{uB'_x}{\gamma_v} + vB'_z\right).$$

Another expression for F'_2 can be obtained from applying a Lorentz transformation to the four-force F.

$$F'_2 = \gamma_u(E_y + uB_x).$$

Since $\gamma_{u'} = \gamma_v\gamma_u$ by Eq. (12.22) with $u_x = 0$,

$$\gamma_v E'_y + uB'_x + \gamma_v v B'_z = E_y + uB_x.$$

Since $E_y = \gamma_v(E'_y + vB'_z)$,

$$B'_x = B_x.$$

We have hence completed the derivation of the field transformations.

Finally, note that the laws of electromagnetism (Maxwell's equations) are still valid in the relativistic case. Meanwhile, the Lorentz force law now engenders a rate of change of relativistic momentum. Moreover, a crucial property when switching between inertial frames is that the quantity of electric charge is invariant, though charge densities may differ due to length contraction.

Let us verify the transformation rules using the following example. Consider a standard parallel plate capacitor that consists of two large plates with width w and length l, separated by a distance d ($d \ll w$ and $d \ll l$). Each plate carries a charge Q. In inertial frame S, the plates are stationary at the coordinates depicted in Fig. 12.11.

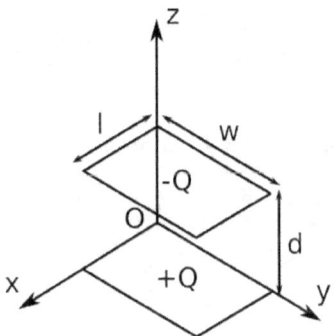

Figure 12.11: Capacitor plates in frame S

In this frame, there is an electric field (by Gauss' law) between the plates with components:

$$E_x = 0,$$

$$E_y = 0,$$

$$E_z = \frac{\sigma}{\varepsilon_0},$$

where $\sigma = \frac{Q}{wl}$ is the surface charge density of the positive plate. The magnetic field in frame S is zero everywhere due to the absence of moving charge. Now, consider a frame S' that travels at a velocity v in the x-direction relative to frame S. The configuration of the plates as observed in S' is depicted in Fig. 12.12.

Firstly, the new surface charge density σ' is larger than that in S by a factor of γ_v due to length contraction.

$$\sigma' = \gamma_v \sigma.$$

The electric field between the capacitor plates in this frame at any instant is then

$$E'_x = 0,$$

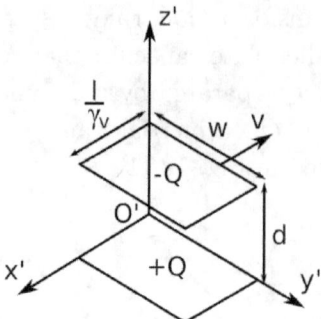

Figure 12.12:   Capacitor plates in frame S'

$$E'_y = 0,$$

$$E'_z = \frac{\gamma_v \sigma}{\varepsilon_0},$$

in correspondence with the electric field transformations stated at the start of this section. However, we are not done here. The charged plates are now traveling at a speed v in the negative x'-direction. Hence, they constitute a current similar to an infinite current sheet (the y'z'-plane is akin to the cross section). Drawing an Amperian loop, with edges parallel to the y' and z' axes, that cuts only one plate and extends to infinity, we obtain

$$B'_y \cdot w = \mu_0 I_{enc}.$$

The magnetic field in the region between the plates is taken to be uniform, as the plates are large. The total enclosed current in this case is the charge crossing the loop per unit time and is given by

$$I_{enc} = \sigma' w v = \gamma_v \sigma w v.$$

Hence,

$$B'_y = \gamma_v \mu_0 \sigma v.$$

Let us verify that this is consistent with the purported electric and magnetic field transformations.

$$B'_y = \gamma_v \left(B_y + \frac{v}{c^2} E_z \right) = \frac{\gamma \sigma v}{\varepsilon_0 c^2}.$$

Next, by using the fact that $c^2 = \frac{1}{\mu_0 \varepsilon_0}$,

$$B'_y = \gamma_v \mu_0 \sigma v,$$

which is the answer that we had obtained from applying Ampere's law in frame S'. One can also check that the transformations imply that there is no component of the magnetic field in the x' and z'-directions.

12.5.1   *Fields of a Moving Charge*

In this section, the electric and magnetic fields due to a point charge traveling at a constant velocity v in frame S will be determined by considering the corresponding quantities in the rest frame S' of the charge; this is a very important result. We define the vectors shown in Fig. 12.13 (the diagrams are drawn in 2-D and the velocity v is depicted to be along the x-axis for convenience).

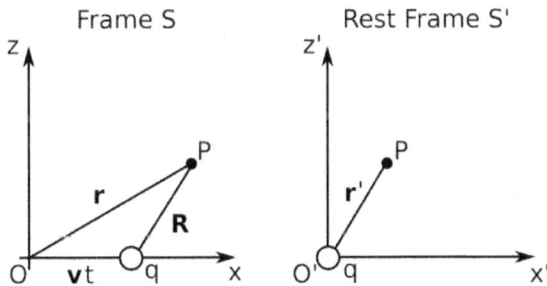

Figure 12.13:   Frame S and rest frame S'

The charge lies at the origin O in frame S at $t = 0$. Hence, the position vector of the charge q at time t in frame S is given by vt. r and r' are the position vectors of point P, the location at which the electromagnetic field is of interest, with respect to the origins of S and S' respectively. R is the vector pointing from the instantaneous position of the charge to point P in frame S. The components of the electric field in frame S' are given by Coulomb's law as

$$E'_\| = \frac{kq}{r'^3} r'_\|,$$

$$E'_\perp = \frac{kq}{r'^3} r'_\perp,$$

where $k = \frac{1}{4\pi\varepsilon_0}$. The magnetic field due to the charge is zero everywhere in frame S'. Hence, the transformations of the electric fields give

$$E_\| = \frac{kq}{r'^3} r'_\|,$$

$$E_\perp = \gamma_v E'_\perp = \frac{\gamma_v kq}{r'^3} r'_\perp.$$

Therefore,

$$E = \frac{kq}{r'^3}(r'_\parallel + \gamma_v r'_\perp).$$

Ideally, we wish to express this in terms of the quantities in frame S. From the Lorentz transformations,

$$r'_\parallel = \gamma_v(r_\parallel - vt) = \gamma_v R_\parallel,$$

$$r'_\perp = r_\perp = R_\perp,$$

as $R_\parallel = r_\parallel - vt$ and $R_\perp = r_\perp$, as illustrated in Fig. 12.13. Then,

$$r'_\parallel + \gamma_v r'_\perp = \gamma_v \left(\frac{r'_\parallel}{\gamma_v} + r'_\perp \right) = \gamma_v R$$

$$r'^2 = r'_\parallel \cdot r'_\parallel + r'_\perp \cdot r'_\perp$$
$$= \gamma_v^2 R_\parallel^2 + R_\perp^2.$$

The electric field in frame S can then be expressed as

$$E = \frac{\gamma_v kq R}{(\gamma_v^2 R_\parallel^2 + R_\perp^2)^{\frac{3}{2}}}$$

$$= \frac{\gamma_v kq R}{\gamma_v^3 (R_\parallel^2 + (1 - \beta^2) R_\perp^2)^{\frac{3}{2}}}$$

$$= \frac{kq R}{\gamma_v^2 (R^2 - \beta^2 R_\perp^2)^{\frac{3}{2}}}.$$

If we define θ to be the angle subtended by the vectors vt and R when they are placed tail to tail,

$$R_\perp = R \sin \theta,$$

$$E = \frac{kq R}{\gamma_v^2 R^3 (1 - \beta^2 \sin^2 \theta)^{\frac{3}{2}}} = \frac{q R}{4\pi\varepsilon_0 \gamma_v^2 R^3 (1 - \beta^2 \sin^2 \theta)^{\frac{3}{2}}},$$

with $k = \frac{1}{4\pi\varepsilon_0}$. In the non-relativistic limit $\beta \to 0$, this expression reduces to the familiar Coulomb's law.

$$E = \frac{q R}{4\pi\varepsilon_0 R^3}.$$

Now, the magnetic field can similarly be obtained from the transformations.

$$B_\parallel = B'_\parallel = 0,$$

$$B_\perp = \gamma_v \left(B'_\perp + \frac{v}{c^2} \times E' \right)$$

$$= \frac{v}{c^2} \times \gamma_v E'$$

$$= \frac{v}{c^2} \times \gamma_v E'_\perp,$$

as $E'_\parallel$ is defined to be in the direction of v, causing $v \times E'_\parallel = 0$. Then,

$$B = B_\parallel + B_\perp$$

$$= 0 + \frac{v}{c^2} \times \gamma_v E'_\perp$$

$$= \frac{v}{c^2} \times E_\parallel + \frac{v}{c^2} \times E_\perp$$

$$= \frac{v}{c^2} \times E.$$

Therefore, the magnetic field at point P in frame S is simply

$$B = \frac{kqv \times R}{c^2 \gamma_v^2 R^3 (1 - \beta^2 \sin^2 \theta)^{\frac{3}{2}}} = \frac{\mu_0 qv \times R}{4\pi \gamma_v^2 R^3 (1 - \beta^2 \sin^2 \theta)^{\frac{3}{2}}}$$

as $k = \frac{1}{4\pi\varepsilon_0}$ and $\frac{1}{c^2} = \mu_0\varepsilon_0$. In the non-relativistic limit, $\beta \to 0$,

$$B = \frac{\mu_0 qv \times R}{4\pi R^3}.$$

This expression is similar to substituting qv for $\int I ds$ in the Biot-Savart law. However, they arise from different premises. The Biot-Savart law is only valid for steady currents and a single moving charge most definitely does not constitute a steady current (as the net charge at various positions varies with time) and is an empirical law. The above derivation is a consequence of Coulomb's law, charge invariance and special relativity. It just happens that in the non-relativistic limit, the above expression reduces to something similar to the Biot-Savart law. The derivation of this result from purely classical tools (specifically, the Ampere–Maxwell law) was the subject of Problem 25 (Chapter 8).

Problems

Problems without Four-Vectors

1. "Infinite" Energy Generator*

Tom proposes the following mechanism to generate "infinite" energy. Orient two perfectly reflective mirrors (of arbitrary masses) such that they are mutually parallel and stationary initially. Now, place a photon between the two mirrors such that it impinges the mirrors normally. As the photon bounces back and forth between the two mirrors, it imparts momentum and thus kinetic energy to the two mirrors. Furthermore, the photon can always catch up with the mirrors, even if the mirrors begin to pick up speed so this process will continue indefinitely — producing "infinite" kinetic energy. What is wrong with Tom's reasoning? Now, consider a new set-up where the two mirrors are identical and initially stationary. Two photons, with initial velocities in opposite directions and initial frequency f each, impinge normally on the two mirrors repeatedly. What is the final kinetic energy of each mirror after a long time?

2. Exploding Particle*

A particle of rest mass m is traveling in the positive x-direction at velocity u in the lab frame. It then disintegrates into two identical particles of rest mass $\frac{m}{\sqrt{6}}$ each. Determine the velocities of the product particles in the lab frame if they are aligned with the x-axis.

3. Available Energy*

A particle of rest mass m_1 is bombarded at another stationary particle of rest mass m_2 at initial velocity u. If this collision triggers the production of a third particle (while retaining the rest masses of the other two), determine the maximum rest mass of the third particle.

4. Unknown Mass Decay*

A particle of unknown mass M decays into two particles of known masses $m_a = 0.5\text{GeV}/c^2$ and $m_b = 1.0\text{GeV}/c^2$, whose momenta are measured to be $p_a = 2.0\text{GeV}/c$ directed along the y-axis and $p_b = 1.5\text{GeV}/c$ directed along the x-axis. Find the unknown mass M and its speed (in units of c).

5. *Positron-Electron Collision* *

Consider the reaction

$$e^+ + e^- \rightarrow e^+ + e^- + \psi.$$

Determine the minimum initial energy of the electron or positron for this reaction to occur in the center-of-momentum (CoM) frame in terms of the rest masses m_ψ and m_e. Hence, find the threshold energy of the positron if the positron is bombarded at the stationary electron in the lab frame, without the aid of Eq. (12.26).

6. *Relativistic Photon Rocket* *

A rocket is initially stationary with a rest mass M_i in the lab frame S. The rocket then begins to convert mass into photons and ejects them from the back. When the rest mass of the rocket is M_f, prove that its speed u in frame S fulfils

$$\frac{M_i}{M_f} = \left(\frac{1+u}{1-u}\right)^{\frac{1}{2}}.$$

7. *Relativistic Mass Rocket* **

In the lab frame S, a relativistic rocket of initial rest mass M_i is initially stationary. The rocket then begins to eject mass continuously in minuscule amounts at one instant, towards the back, at a velocity u relative to its instantaneous rest frame. When the rest mass of the rocket is M_f, prove that the velocity v of the rocket in frame S satisfies

$$\frac{M_i}{M_f} = \left(\frac{1+v}{1-v}\right)^{\frac{1}{2u}}.$$

Hint: Consider the conservation of energy and momentum in the instantaneous rest frame and their relationship to the rest mass of the object.

8. *Bucket* **

A bucket of initial rest mass M_0 has an initial velocity u_0 in frame S. It begins to collect sand aligned in a line with a linear mass density λ in S. Assuming that the line of sand extends forever,

(a) Find the rate of rest mass increase of the bucket when the bucket has speed u. Why is this greater than λu? Find the rest mass of the bucket, $M(t)$, as a function of time.

(b) Find the energy and velocity of the bucket as functions of time, $E(t)$ and $u(t)$.

(c) Find the energy and velocity of the bucket as functions of the displacement of the bucket relative to the bucket's initial position, $E(x)$ and $u(x)$.

(Adapted from "Introduction to Mechanics")

9. *Leaking Bucket***

Referring to the previous scenario, the bucket-and-contained-sand system now loses a fraction f of its remaining rest mass per unit distance traveled. Find $E(x)$, $p(x)$ and $t(x)$. (Adapted from "Introduction to Mechanics")

Problems with Four-Vectors

A common trick in solving equations involving four-momenta involves isolating a single four-momentum that is not of interest and then taking the squared norm of both sides to eliminate the irrelevant four-momentum (as its squared norm produces the mass of the particle). This will be a common denominator in many problems.

10. *Four-Vectors**

(a) Let A be a four-vector, and suppose that one component of A is found to be zero in all inertial frames. Show that all four components of A are zero in all frames. This is known as the zero-component theorem.

(b) The four-momentum of a particle in the lab frame S is P while the four-velocity of an observer is U with respect to S. Show that the particle's energy in the rest frame of the observer is $P \cdot U$.

(c) Prove that the inner product of the four-velocity U and the four-acceleration A of a massive particle is zero.

(d) Show that the instantaneous charge density ρ and the instantaneous current density j at a particular location forms a four-vector $J = (\rho c, j)$ in an arbitrary inertial frame S. J is known as the four-current. Hint: consider the four-velocity.

11. *Chasing Particles**

A particle m_a with speed v_a is pursuing another particle m_b with v_b ($v_b < v_a$) along the x-axis of an inertial frame S. When particle a catches up with particle b, they collide and coalesce to form a single particle of mass m. Show that

$$m^2 = m_a^2 + m_b^2 + 2m_a m_b \gamma_{v_a} \gamma_{v_b} \left(1 - \frac{v_a v_b}{c^2}\right).$$

12. Disintegration*

A particle of rest mass m_1 is initially stationary in the lab frame S. It then disintegrates into a photon and another particle of rest mass $m_2 < m_1$. Find the energies of the photon and the final particle.

13. Reflected Photon*

In an inertial frame S, a photon of frequency f is normally incident on a perfect plane mirror, of mass m, retracting at a velocity u from the photon. Find the frequency f' of the reflected photon in frame S.

14. Proton Collision*

A proton of energy E collides elastically with a second proton of rest energy E_0 that is initially stationary. Subsequently, the two protons are directed at angles $\pm \frac{\phi}{2}$ relative to the initial velocity of the incident proton. Find $\cos \phi$.

15. Electron-Photon Collision*

A massive particle with energy E_0 and speed $\beta_0 c$ undergoes a head-on elastic collision with a photon with energy $E_{\gamma 0}$. Show that the final energy E_γ of the photon is

$$E_\gamma = E_0 \frac{1 + \beta_0}{2 + (1 - \beta_0)\frac{E_0}{E_{\gamma 0}}}.$$

Show that $E_\gamma < E_0$ but if $\beta_0 \to 1$, $\frac{E_\gamma}{E_0} \to 1$. That is, a high-energy particle loses most of its energy to the photon (in the ideal $\beta_0 \to 1$ limit, the particle retains energy $E_{\gamma 0}$ by the conservation of energy).

16. Emission by Excited Atom*

An excited atom A^* at rest drops to its ground state A by emitting a photon. In atomic physics, it is usually assumed that the energy E_γ of the emitted photon is equal to the difference in energies of the two atomic states, $\Delta E = (M^* - M)c^2$, where M and M^* are the rest masses of the ground and excited states of the atom. This cannot be exactly true, since the recoiling atom X must carry away part of ΔE. Show that in fact

$$E_\gamma = \Delta E \left(1 - \frac{\Delta E}{2M^* c^2} \right).$$

Given that ΔE is of order eV while the lightest atom has M of order GeV/c^2. Discuss the validity of the approximation $E_\gamma = \Delta E$.

17. Neutrino Beam*

High-energy neutrino beams are produced by allowing an excited pion π^+ to decay into an excited muon μ^+ and neutrino ν according to the equation

$$\pi^+ \to \mu^+ + \nu.$$

The masses of a pion and muon are $140\text{MeV}/c^2$ and $106\text{MeV}/c^2$. The mass of a neutrino is negligible.

(a) Find the energy of the neutrino in the rest frame of the pion.
(b) In the lab frame, the pion has an energy of 200GeV. If the momentum of the neutrino is aligned with the initial momentum of the pion, determine the energy of the neutrino.
(c) Referring to (b), let θ be the angle between the momentum of the neutrino and the initial momentum of the pion. Find the value of θ for which the neutrino's energy is half of its maximum possible energy.

18. Mad Scientist*

A mad scientist asserts to have observed the decay of a particle of mass M into two identical particles of mass $m > 0$, with $M < 2m$. He dismisses objections about the violation of the conservation of energy by this process with the claim that if M were traveling fast enough, its energy could easily exceed $2mc^2$ and could hence decay into the two particles of mass m. Prove that he is wrong and qualitatively describe the flaw in his rebuttal.

19. Photon Decay*

Show that a photon cannot spontaneously decay into a particle with a non-zero rest mass, accompanied by an arbitrary number of other particles of arbitrary masses, which may be zero.

20. *Compton Scattering* **

In frame S, a photon of frequency f is incident on an electron with momentum p_1 and energy E_1. Determine the minimum frequency of the scattered photon and the directions of the velocities of the incident photon and the scattered photon in such a situation. The rest mass of the electron is constant.

21. *Pion Photoproduction* **

Consider the reaction

$$\gamma + p \to p + \pi.$$

The rest energies of a proton and pion are 938 MeV and 135 MeV respectively.

(a) If the proton is initially at rest in the laboratory, find the laboratory threshold photon energy for this reaction to occur.
(b) If the photon's energy is 10^{-3}eV, find the minimum proton energy that can spark off this reaction.

22. *Energy Transfer* **

In the lab frame S, particle 1 of rest mass m_1 is traveling at velocity u in the positive x-direction and subsequently collides with an initially stationary particle 2 of rest mass m_2. If the final velocities of the particles are still aligned with the x-direction and their rest masses remain constant, find the final energy of particle 2. Determine the fraction of the total energy in the lab frame that is possessed by particle 2 in the limit where u tends to c. Hint: Consider the center-of-momentum frame. You will discover the ratio in fact tends to unity. Here's another subtler way of proving this. Let P_1' and P_2 be the final four-momentum of particle 1 and the initial four-momentum of particle 2 in the center-of-momentum frame. Firstly, show that the inner product of $P_1' - P_2$ with itself is larger than zero in the center-of-momentum frame (be wary that $P_1' - P_2$ is not a four-vector and its inner product is not invariant). Armed with this inequality, show that the final energy of particle 1 must not exceed $\frac{m_1^2 + m_2^2}{2m_2}$. This value is independent of u and hence shows that particle 2 absorbs most of the energy if u is large.

23. *Maximum Frequency* **

In inertial frame S, a particle of rest mass m is incident at a speed u on a nucleus of rest mass M. During the collision, a photon is emitted. The rest masses of the particle and the nucleus remain unchanged. Show that the maximum energy of the photon in frame S is

$$hf = \frac{Mm(\gamma_u - 1)}{M + \gamma_u m(1 - u)}.$$

Show that this occurs when the direction of the photon is parallel to the initial velocity of the particle u and when the nucleus and the particle "stick together" after the collision in frame S. Hint: from the four-vector equation $P_1 + P_2 = P'_1 + P'_2 + P_p$, where P_1, P_2, P'_1, P'_2 are the four-momenta of the particle and the nucleus before and after the collision and P_p is the four-momentum of the photon, obtain $P_1 + P_2 - P_p = P'_1 + P'_2$ and consider the squared norm of both sides.

Force, Impulse-Momentum and Work-Energy

24. Two Particles Connected by String*

Two particles of rest masses m and M are connected by a flexible rope of constant tension T. If the two particles are initially at rest and are separated by an initial distance L in frame S, what is the distance x between the point at which the particles meet and the initial position of the particle of rest mass m? ("Introduction to Mechanics" by David Morin)

25. Projectile Motion**

A particle initially possesses an x-component of momentum p_0 and energy E_0 in the lab frame. It is then acted upon by a constant $-F$ force in the y-direction. Determine the trajectory of the particle in the lab frame. Show that the resultant expression reduces to the familiar parabolic path in the non-relativistic limit.

26. Pulling a Leaking Bucket**

In frame S, there is an initially stationary bucket with zero initial rest mass that is pulled along by a string of constant tension T. The bucket begins to gather sand of linear mass density λ. Find the velocity u of the bucket. ("Introduction to Mechanics")

Waves and Electromagnetism

27. *Light in Moving Glass* **

In the lab frame S, a source emits light of frequency f towards a glass block of refractive index n that is retracting at velocity v. By considering the four-wave vector and the rest frame of the block S', determine the frequency and wavelength of light inside the block in the lab frame S.

28. *Crossed Fields* **

In the lab frame S, there are uniform electric and magnetic fields $\boldsymbol{E} = (0, E, 0)$ and $\boldsymbol{B} = (0, 0, B)$ where $E < 0$ and $B > 0$. A charged particle of rest mass m and charge q initially possesses velocity v_0 in the x-direction when it is at the origin. Determine the maximum x-coordinate that the particle attains in its subsequent motion. Hint: Consider another inertial frame.

29. *Magnetic Field* ***

In this problem, we shall see how the magnetic field is the manifestation of the electric field in another frame. In frame S, an infinitely long wire carries a current I in the positive x-direction. The current comprises electrons of linear charge density $-\lambda$ traveling at a velocity u in the negative x-direction in frame S ($I = \lambda u$). In frame S, there are also stationary positive ions of linear charge density λ in the wire such that the wire is neutral. A point charge q travels at a velocity v in the positive x-direction, at a distance r from the wire. Without any knowledge of the existence of a magnetic field and by considering the electric field in the rest frame of the charge S', show that the charge experiences a force of the form

$$f = qv \times A$$

in frame S where $\boldsymbol{A}$ is a certain vector. Now let $\boldsymbol{A}$ be defined as the magnetic field $\boldsymbol{B}$. Verify that the expression for the magnetic field is consistent with that obtained from Ampere's law in frame S. Do not use the transformations for the electromagnetic fields in this problem.

Solutions

1. "Infinite" Energy Generator*

Tom's reasoning is fallacious because the frequency (and thus energy) of the photon will decrease after reflection from a retracting mirror. Therefore, the photon imparts less momentum with each reflection until it ultimately disappears (or becomes insignificant) when it loses all of its initial energy. The total kinetic energy imparted to the two mirrors will be equal to the initial energy of the photon by the conservation of energy.

Because the second set-up is symmetrical, the final kinetic energies of the mirrors are identical and are each equal to the initial energy of one photon hf, by the conservation of energy.

2. Exploding Particle*

This problem is beckoning for us to consider it in the rest frame of the initial particle, S'. In this frame, the initial energy of the system is m while the initial momentum is zero. By symmetry, the final energy of each particle in this rest frame is $\frac{m}{2}$. Then, the magnitude of their momenta is

$$p = \sqrt{\frac{m^2}{4} - \frac{m^2}{6}} = \frac{m}{2\sqrt{3}}.$$

They are directed in opposite directions, parallel to the x'-axis. The magnitude of the velocities of the product particles in frame S' is

$$u_{CM} = \frac{p}{E} = \frac{1}{\sqrt{3}}$$

and the velocities point in opposite directions. The velocities of these particles in the lab frame S can then be obtained from the velocity addition formula.

$$u_1 = \frac{u - u_{CM}}{1 + u_{CM}u} = \frac{\sqrt{3}u - 1}{\sqrt{3} - u},$$

$$u_2 = \frac{u + u_{CM}}{1 + u_{CM}u} = \frac{\sqrt{3}u + 1}{\sqrt{3} + u}.$$

An equivalent method is to transform the individual four-momenta of the particles and subsequently use the relationship $u = \frac{p}{E}$.

3. Available Energy*

In the center-of-momentum frame S', energy and momentum must also be conserved. By definition, the total momentum in S' should be zero. Therefore, the most energy can be channeled into the creation of the new particle if all particles possess zero final momentum in S' (so that no energy is wasted on kinetic energy). The maximum rest mass of the new particle m' is thus the total energy in the center of momentum frame E_{CM}, minus m_1 and m_2.

$$m' = E_{CM} - m_1 - m_2.$$

Applying Eq. (12.15),

$$E_{CM} = \sqrt{E_{tot}^2 - p_{tot}^2}$$

where E_{tot} and p_{tot} are the total energy and momentum in any arbitrary inertial frame, as the above expression is a scalar. Substituting the corresponding expressions in the lab frame,

$$E_{CM} = \sqrt{(\gamma_u m_1 + m_2)^2 - \gamma_u^2 m_1^2 u^2} = \sqrt{m_1^2 + m_2^2 + 2\gamma_u m_1 m_2}.$$

Then,

$$m' = \sqrt{m_1^2 + m_2^2 + 2\gamma_u m_1 m_2} - m_1 - m_2.$$

4. Unknown Mass Decay*

Let the momentum of M be p. Since $\boldsymbol{p}_a \cdot \boldsymbol{p}_b = 0$,

$$p^2 = p_a^2 + p_b^2.$$

Meanwhile, the energies of particles a and b are

$$E_a = \sqrt{p_a^2 + m_a^2},$$
$$E_b = \sqrt{p_b^2 + m_b^2}.$$

The energy of M is thus

$$E = E_a + E_b = \sqrt{p_a^2 + m_a^2} + \sqrt{p_b^2 + m_b^2},$$
$$M = \sqrt{E^2 - p^2}$$

$$= \sqrt{m_a^2 + m_b^2 + 2\sqrt{(p_a^2 + m_a^2)(p_b^2 + m_b^2)}}$$

$$= \sqrt{m_a^2 + m_b^2 + 2\sqrt{\left(\frac{p_a^2}{c^2} + m_a^2\right)\left(\frac{p_b^2}{c^2} + m_b^2\right)}}$$

$$= 2.9 \text{GeV}/c^2 \, (\text{2sf}),$$

where the c's have been added back in the second-last expression. The velocity of M is

$$u = \frac{pc^2}{E} = \frac{\sqrt{p_a^2 + p_b^2}}{\sqrt{\frac{p_a^2}{c^2} + m_a^2} + \sqrt{\frac{p_b^2}{c^2} + m_b^2}} = 0.65c \, (\text{2sf}).$$

5. Positron-Electron Collision*

In the CoM frame, let the initial energies of the electron and positron be E. The minimum final total energy occurs when all products are stationary and is of value $2m_e + m_\psi$. By the conservation of energy,

$$2E_{min} = 2m_e + m_\psi$$

$$E_{min} = m_e + \frac{m_\psi}{2}.$$

Next, suppose that the total energy in the CoM frame is $E_{tot} = 2E$ and that the electron and positron travel towards each other at speed u. By the energy-momentum transformations, the total energy in the rest frame of the electron is

$$E'_{tot} = \gamma_u E_{tot} = 2\gamma_u E.$$

Therefore, the initial energy of the positron in this frame is

$$E_+ = E'_{tot} - m_e = 2\gamma_u E - m_e.$$

Observe that γ_u decreases as E decreases (since the electron and positron possess less kinetic energy). Therefore, the minimum E_+ (i.e. the threshold energy) occurs when $E = E_{min}$.

$$E_{thres} = 2\gamma_u E_{min} - m_e.$$

From the definition of the relativistic energy, we have

$$E_{min} = \gamma_u m_e \implies \gamma_u = \frac{E_{min}}{m_e} = 1 + \frac{m_\psi}{2m_e}.$$

Thus,

$$E_{thres} = 2\left(1 + \frac{m_\psi}{2m_e}\right)\left(m_e + \frac{m_\psi}{2}\right) - m_e$$

$$= m_e + 2m_\psi + \frac{m_\psi^2}{2m_e}.$$

6. Relativistic Photon Rocket*

The initial energy of the entire system is M_i. If we let the total energy of the radiated photons in frame S be E,

$$M_i = \gamma_u M_f + E.$$

Since the momentum of the photons differs from their energy only by a factor of c, the conservation of momentum gives

$$\gamma_u M_f u = E.$$

Hence,

$$\gamma_u M_f u + \gamma_u M_f = M_i$$

$$\frac{M_i}{M_f} = \gamma_u(1 + u) = \left(\frac{1+u}{1-u}\right)^{\frac{1}{2}}.$$

7. Relativistic Mass Rocket**

Figure 12.14 depicts an ejection event in the instantaneous rest frame of the rocket, S'. This rest frame travels at a velocity v relative to S.

Immediately after the ejection, let the velocity of the rocket be dv' in S' and let the rest mass of the rocket be $m + dm$. The fuel travels at a velocity $-u$ in the new instantaneous rest frame of the rocket which travels at dv' relative to the first instantaneous rest frame. Using the velocity addition formula, the fuel travels at speed $\frac{u-dv'}{1-udv'}$ in the original instantaneous rest frame S'.

$$\frac{u - dv'}{1 - udv'} \approx (u - dv')(1 + udv') \approx u - dv' + u^2 dv'.$$

Then, let γ_f and dm_f denote the gamma factor associated with the velocity of the fuel in S' and the rest mass of the fuel. By the conservation of energy

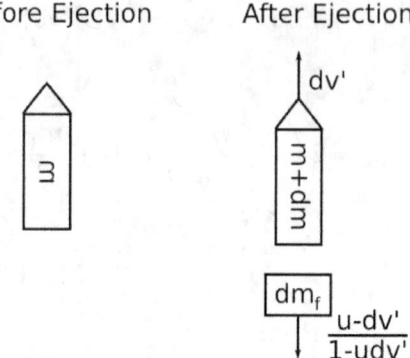

Figure 12.14:   Ejection event in instantaneous rest frame

in frame S', the total final energy of the system consisting of the ejected fuel and the rocket must be equal to m which is the initial energy.

$$m = \frac{1}{\sqrt{1-dv'^2}}(m+dm) + \gamma_f dm_f.$$

Ignoring second order terms,

$$\gamma_f dm_f = -dm.$$

Furthermore, by the conservation of momentum,

$$\frac{1}{\sqrt{1-dv'^2}}(m+dm)dv' - \gamma_f dm_f(u - dv' + u^2 dv') = 0.$$

Discarding the preponderance of second order terms,

$$mdv' = \gamma_f dm_f u = -dmu$$

$$-\frac{dm}{m} = \frac{1}{u}dv'.$$

Now, we cannot directly integrate this expression in the current instantaneous rest frame as this expression for dv' will no longer be valid as the rocket begins to acquire a non-negligible velocity in this frame. Hence, there is a need to transform back to frame S. Let the speed of the rocket in frame S after the ejection event be $v + dv$. By the velocity addition formula,

$$v + dv = \frac{dv' + v}{1 + dv'v}.$$

Performing a binomial expansion $\frac{dv'+v}{1+dv'v} \approx (dv' + v)(1 - dv'v)$,

$$dv' = \frac{dv}{1 - v^2}.$$

Substituting this into the previous relevant equation,

$$-\int_{M_i}^{M_f} \frac{dm}{m} = \frac{1}{u} \int_0^v \frac{dv}{1 - v^2} = \frac{1}{2u} \int_0^v \left(\frac{1}{1 - v} + \frac{1}{1 + v} \right) dv$$

$$\ln \left| \frac{M_i}{M_f} \right| = \frac{1}{2u} \ln \frac{1 + v}{1 - v}$$

$$\frac{M_i}{M_f} = \left(\frac{1 + v}{1 - v} \right)^{\frac{1}{2u}}.$$

8. Bucket**

(a) Consider a collision event between the bucket, with an instantaneous mass M and instantaneous velocity u, and an infinitesimal segment of sand of rest mass dm. The total energy before and after the collision are $\gamma_u M$ and $\gamma_u M + dm$ respectively. The total momenta before and after collision are still $\gamma_u M u$. Hence, the new rest mass $M + dM$ after the collision is

$$M + dM = \sqrt{(\gamma_u M + dm)^2 - (\gamma_u M u)^2}$$

$$= \sqrt{M^2 + 2\gamma_u M dm + dm^2}.$$

Ignoring the second order infinitesimal terms,

$$M + dM = M\sqrt{1 + \frac{2\gamma_u}{M} dm}$$

$$\approx M \left(1 + \frac{\gamma_u}{M} dm \right)$$

$$= M + \gamma_u dm$$

$$\implies dM = \gamma_u dm = \gamma_u \lambda dx$$

where dx is the length of an infinitesimal segment of sand swept by the bucket in time dt in the lab frame. Since $\frac{dx}{dt} = u$,

$$\frac{dM}{dt} = \gamma_u \lambda u.$$

This is greater than λu due to the increase in internal energy of the combined bucket-and-sand system during the inelastic collision, as part of the kinetic energy of the bucket is converted into heat. Next, we can use the fact that the total momentum of the bucket always remains at $p = \gamma_{u_0} M_0 u_0$ to rewrite the differential equation above as

$$\frac{dM}{dt} = \frac{p\lambda}{M}$$

$$\int_{M_0}^{M} M \, dM = \int_0^t p\lambda \, dt$$

$$M = \sqrt{2p\lambda t + M_0^2}$$

where $p = \gamma_{u_0} M_0 u_0$.

(b) The increase in the total energy of the bucket is due to the rest energy of the additional sand collected. Hence,

$$\frac{dE}{dt} = \lambda u.$$

Next, since $u = \frac{p}{E}$,

$$\int_{E_0}^{E} E \, dE = \int_0^t p\lambda \, dt$$

$$E = \sqrt{2p\lambda t + \frac{p^2}{u_0^2}}$$

where the initial energy E_0 has been expressed as $\frac{p}{u_0}$. Lastly,

$$u = \frac{p}{E}$$

$$= \frac{p}{\sqrt{2p\lambda t + \frac{p^2}{u_0^2}}}$$

$$= \frac{u_0}{\sqrt{1 + \frac{2\lambda u_0^2}{p}t}}$$

where $p = \gamma_{u_0} M_0 u_0$.

(c) The energy of the bucket as a function of x can be computed via

$$\frac{dE}{dx} = \lambda$$

$$E = \lambda x + \frac{p}{v_0}.$$

Then,

$$u = \frac{p}{E} = \frac{v_0}{1 + \frac{\lambda v_0 x}{p}}$$

where $p = \gamma_{u_0} M_0 u_0$.

9. Leaking Bucket**

This time, the change in the total energy per unit distance traveled is both due to the absorption of the sand, which has a certain amount of rest energy, and the amount of rest mass lost by the bucket-and-sand system. Since the bucket losses rest mass (which is assumed to be homogeneous) at a fraction per unit distance of f, its fractional rate of energy loss is also f. Hence,

$$\frac{dE}{dx} = \lambda - fE$$

$$\int_{E_0}^{E} \frac{1}{E - \frac{\lambda}{f}} dE = -\int_{0}^{x} f \, dx$$

$$\ln \left| \frac{E - \frac{\lambda}{f}}{E_0 - \frac{\lambda}{f}} \right| = -fx.$$

Note that $E - \frac{\lambda}{f}$ (at all times) must have the same sign as $E_0 - \frac{\lambda}{f}$. This is evident from the $\frac{dE}{dx}$ equation. E will always tend towards $\frac{\lambda}{f}$, after which it stops changing. Therefore, the term in the absolute value brackets is definitely positive.

$$E = \left(E_0 - \frac{\lambda}{f} \right) e^{-fx} + \frac{\lambda}{f}$$

where E_0 is the initial energy of the bucket, $E_0 = \gamma_{u_0} M_0$. On the other hand, the change in momentum of the bucket per unit distance is similarly,

$$\frac{dp}{dx} = -fp.$$

p is now the instantaneous momentum of the bucket-and-sand system. Solving this differential equation,

$$p = p_0 e^{-fx}$$

where $p_0 = \gamma_{u_0} M_0 u_0$. Then, using $u = \frac{p}{E}$,

$$u = \frac{p_0}{E_0 - \frac{\lambda}{f} + \frac{\lambda}{f} e^{fx}}.$$

Using the fact that $u = \frac{dx}{dt}$ and separating variables,

$$\int_0^x \left(E_0 - \frac{\lambda}{f} + \frac{\lambda}{f}e^{fx} \right) dx = \int_0^t p_0 dt$$

$$t = \frac{\left(E_0 - \frac{\lambda}{f} \right) x + \frac{\lambda}{f^2}e^{fx} - \frac{\lambda}{f^2}}{p_0}.$$

10. Four-Vectors*

(a) Suppose that the ith component of A is found to be zero in all inertial frames. Given A_i in frame S, we can compute A_i' in frame S' in the following manner, under a Lorentz transformation in an appropriate direction.

$$A_i' = \gamma(A_i - \beta A_j)$$

where A_j is the time-like component if A_i is a space-like component and vice-versa, A_j is a space-like component if A_i is the time-like component. Evidently, for $A_i' = 0$, given $A_i = 0$, we must have $A_j = 0$. Since the definition of the inertial frame S is arbitrary, the jth component of A must be zero in all inertial frames. As such, if A_i is the time-like component, we can perform Lorentz transformations along all three spatial directions to deduce that all spatial components of A are zero in all inertial frames.

Otherwise, if A_i is a space-like component, we can first apply the same argument to conclude that the time-like component A_0 is zero in all inertial frames. Then, by using the final result of the previous paragraph, we conclude that all components of A are zero in all inertial frames.

(b) Let the four-momentum of the particle and the four-velocity of the observer be P' and U' in the rest frame of the observer, S', respectively.

$$P' = \begin{pmatrix} E' \\ \boldsymbol{p}' \end{pmatrix} \quad U' = \begin{pmatrix} 1 \\ 0 \end{pmatrix}$$

where E' and $\boldsymbol{p}'$ are the energy and momentum of the particle in S'. Evidently,

$$P' \cdot U' = E'.$$

By the invariance of the inner product of two four-vectors across Lorentz transformations,

$$P \cdot U = P' \cdot U' = E'$$

which is our desired result.

(c) In the rest frame S' of the particle (which exists because it is massive and cannot travel at the speed of light in any inertial frame),

$$U' = \begin{pmatrix} 1 \\ \mathbf{0} \end{pmatrix} \quad A' = \begin{pmatrix} 0 \\ \boldsymbol{\alpha} \end{pmatrix}$$

where $\boldsymbol{\alpha}$ is the proper acceleration of the particle. Observe that

$$U' \cdot A' = 1 \cdot 0 - \mathbf{0} \cdot \boldsymbol{\alpha} = 0.$$

Since the inner product of two four-vectors is Lorentz invariant,

$$U \cdot A = U' \cdot A' = 0$$

where U and A are the particle's four-velocity and four-acceleration with respect to an arbitrary inertial frame S.

(d) Suppose for now that the charges at the location P of interest all possess the same velocity $\boldsymbol{u}$ in inertial frame S. Since the longitudinal length of the moving charges are shrunk by a factor of $\frac{1}{\gamma_u}$, as compared to that in their rest frame S', due to length contraction and because the total amount of charge in a given volume must be Lorentz invariant,

$$\rho = \gamma_u \rho_0$$

where ρ_0 is the proper charge density at the corresponding location P' in S'. Next, the current density $\boldsymbol{j}$ in S is simply

$$\boldsymbol{j} = \rho \boldsymbol{u} = \gamma_u \rho_0 \boldsymbol{u}.$$

Now, observe that the four-current in this case is merely the four-velocity U of the charges multiplied by the constant ρ_0!

$$J = \rho_0 \begin{pmatrix} \gamma_u c \\ \gamma_u \boldsymbol{u} \end{pmatrix} = \rho_0 U.$$

By Property 1 of four-vectors, J is hence a four-vector in this case. In the more general case where there are a total of k classes of charge velocities at P, with the ith class having a charge density ρ_i traveling at velocity $\boldsymbol{u}_i$ in S, we can consider the four-current $J_i = (\rho_i c, \rho_i \boldsymbol{u}_i)$ associated with each individual class. Since J_i is a four-vector for all $1 \le i \le k$ by the above proof,

$$J = \begin{pmatrix} \sum_{i=1}^{k} \rho_i c \\ \sum_{i=1}^{k} \rho_i \boldsymbol{u}_i \end{pmatrix} = \sum_{i=1}^{k} J_i$$

is a valid four-vector because

$$J' = \sum_{i=1}^{k} J'_i = \sum_{i=1}^{k} \mathcal{L} J_i = \mathcal{L} \left(\sum_{i=1}^{k} J_i \right) = \mathcal{L} J$$

in an arbitrary inertial frame S', where $\mathcal{L}$ is the Lorentz transformation matrix. Note that the addition of four-vectors to produce another four-vector is valid in this case only because the addition is performed at the same location and at the same time in every inertial frame — the loss of simultaneity across inertial frames has been precluded.

11. Chasing Particles*

Let the initial four-momenta of the particles be P_a and P_b. Denote the four-momentum of the final particle as P. The four-vector equation associated with this process is

$$P_a + P_b = P.$$

Taking the inner product of both sides,

$$m^2 = m_a^2 + m_b^2 + 2P_a \cdot P_b$$
$$= m_a^2 + m_b^2 + 2(E_a E_b - p_a p_b)$$
$$= m_a^2 + m_b^2 + 2(\gamma_{v_a} \gamma_{v_b} m_a m_b - \gamma_{v_a} \gamma_{v_b} m_a m_b v_a v_b)$$
$$= m_a^2 + m_b^2 + 2 m_a m_b \gamma_{v_a} \gamma_{v_b} \left(1 - \frac{v_a v_b}{c^2} \right),$$

where we have added back the c's in the last equation.

12. Disintegration*

The conservation of energy and momentum implies:

$$\begin{pmatrix} m_1 \\ 0 \end{pmatrix} = \begin{pmatrix} \gamma_u m_2 \\ \gamma_u m_2 \boldsymbol{u} \end{pmatrix} + \begin{pmatrix} hf \\ hf\hat{\boldsymbol{k}} \end{pmatrix}$$

$$\begin{pmatrix} m_1 \\ 0 \end{pmatrix} - \begin{pmatrix} hf \\ hf\hat{\boldsymbol{k}} \end{pmatrix} = \begin{pmatrix} \gamma_u m_2 \\ \gamma_u m_2 \boldsymbol{u} \end{pmatrix}.$$

Taking the squared norm of both sides,

$$m_1^2 - 2m_1 hf = m_2^2$$

$$hf = \frac{m_1^2 - m_2^2}{2m_1}$$

$$\gamma_u m_2 = m_1 - hf = \frac{m_1^2 + m_2^2}{2m_1}.$$

13. Reflected Photon*

Let the final four-momentum of the mirror be (E, p) where the only spatial direction of concern has been defined to be along the x-axis. Then, the conservation of energy and momentum entails

$$\begin{pmatrix} hf \\ hf \end{pmatrix} + \begin{pmatrix} \gamma_u m \\ \gamma_u mu \end{pmatrix} = \begin{pmatrix} hf' \\ -hf' \end{pmatrix} + \begin{pmatrix} E \\ p \end{pmatrix}.$$

This can be rewritten in terms of four-vectors.

$$P_1 + P_2 = P_3 + P_4.$$

Shifting P_3 to the left-hand side and finding the squared norm of both sides,

$$(P_1 + P_2 - P_3) \cdot (P_1 + P_2 - P_3) = P_4 \cdot P_4$$

$$m^2 - 2 \cdot 2h^2 ff' - 2 \cdot (\gamma_u m + \gamma_u mu)hf' + 2 \cdot (\gamma_u m - \gamma_u mu)hf = m^2.$$

Simplifying,

$$f' = \frac{(\gamma_u m - \gamma_u mu)hf}{(\gamma_u m + \gamma_u mu)h + 2h^2 f}$$

$$= \frac{\sqrt{\frac{1-u}{1+u}} mf}{\sqrt{\frac{1+u}{1-u}} m + 2hf}.$$

14. Proton Collision*

Let the initial four-momenta of the incident and stationary protons be P_1 and P_2 respectively. Denote their final four-momenta as P_1' and P_2'.

$$P_1 + P_2 = P_1' + P_2'.$$

Taking the inner product of both sides,

$$P_1 \cdot P_1 + P_2 \cdot P_2 + 2P_1 \cdot P_2 = P_1' \cdot P_1' + P_2' \cdot P_2' + 2P_1' \cdot P_2'$$

$$\implies P_1 \cdot P_2 = P_1' \cdot P_2'$$

since $P_1 \cdot P_1 = P_1' \cdot P_1'$ and $P_2 \cdot P_2 = P_2' \cdot P_2'$ (they are in fact, the squared mass of a proton E_0^2). $P_1 \cdot P_2$ can be easily computed as EE_0 since the three-momentum in P_2 is the null vector. $P_1' \cdot P_2'$ requires more work, but observe that the two protons must possess identical final momentum and thus final energy to fulfil the conservation of momentum in directions perpendicular to the incident photon's initial velocity. The final energy and squared momentum of each proton is thus $\frac{E+E_0}{2}$ and $p'^2 = (\frac{E+E_0}{2})^2 - E_0^2 = \frac{E^2}{4} + \frac{EE_0}{2} - \frac{3E_0^2}{4}$. Since $P_1' \cdot P_2' = (\frac{E+E_0}{2})^2 - p'^2 \cos \phi$,

$$EE_0 = \left(\frac{E + E_0}{2} \right)^2 - p'^2 \cos \phi$$

$$\cos \phi = \frac{\left(\frac{E+E_0}{2} \right)^2 - EE_0}{p'^2}$$

$$= \frac{\frac{E^2}{4} - \frac{EE_0}{2} + \frac{E_0^2}{4}}{\frac{E^2}{4} + \frac{EE_0}{2} - \frac{3E_0^2}{4}}$$

$$= \frac{(E - E_0)^2}{(E - E_0)(E + 3E_0)}$$

$$= \frac{E - E_0}{E + 3E_0}$$

since $E \neq E_0$.

15. Electron-Photon Collision*

Let the initial and final four-momenta of the massive particle be P and P' and those of the photon be P_γ and P_γ'. The germane four-vector equation is

$$P + P_\gamma = P' + P_\gamma'.$$

Since the particle is irrelevant after the collision, we isolate P'.

$$P + P_\gamma - P_\gamma' = P'.$$

Taking the inner product of both sides,

$$2P \cdot P_\gamma - 2P \cdot P_\gamma' - 2P_\gamma \cdot P_\gamma' = 0$$

$$P \cdot P_\gamma = P \cdot P_\gamma' + P_\gamma \cdot P_\gamma'$$

where we have canceled the squared rest mass of the particle on both sides. Since this problem is solely one-dimensional, we have $P = (E_0, E_0\beta_0, 0, 0)$,

$P_\gamma = (E_{\gamma 0}, -E_{\gamma 0}, 0, 0)$ and $P'_\gamma = (E_\gamma, E_\gamma, 0, 0)$, where we note that the velocity of the photon must be reversed after the collision because it cannot penetrate the particle.

$$E_0 E_{\gamma 0} + E_0 \beta_0 E_{\gamma 0} = E_\gamma (E_0 - E_0 \beta_0 + 2E_{\gamma 0})$$

$$E_\gamma = \frac{E_0 E_{\gamma 0}(1 + \beta_0)}{2E_{\gamma 0} + (1 - \beta_0)E_0} = E_0 \frac{1 + \beta_0}{2 + (1 - \beta_0)\frac{E_0}{E_{\gamma 0}}}.$$

Since $\beta_0 < 1$, $1 + \beta_0 < 2 < 2 + (1 - \beta_0)\frac{E_0}{E_{\gamma 0}}$ which implies that $E_\gamma < E_0$. When $\beta_0 \to 1$, $\frac{E_\gamma}{E_0} \to \frac{1+1}{2} = 1$.

16. Emission by Excited Atom*

Let the four-momenta of the excited and ground states of the atom be P^* and P. Denote the four-momentum of the emitted photon as P_γ. The relevant four-vector equation is

$$P^* = P + P_\gamma$$

$$P^* - P_\gamma = P.$$

Taking the inner product of both sides,

$$M^{*2} - 2P^* \cdot P_\gamma = M^2.$$

Since $P^* = (M^*, \mathbf{0})$ while $P_\gamma = (E_\gamma, E_\gamma \hat{\mathbf{k}})$,

$$M^{*2} - 2M^* E_\gamma = M^2$$

$$E_\gamma = \frac{M^{*2} - M^2}{2M^*}$$

$$= (M^* - M)\left(\frac{M^* + M}{2M^*}\right)$$

$$= \Delta E \left(\frac{2M^* - \Delta E}{2M^*}\right)$$

$$= \Delta E \left(1 - \frac{\Delta E}{2M^* c^2}\right),$$

where we have added back the c's in the last step. Evidently, the fractional error in assuming $E_\gamma = \Delta E$ is maximum when $M^* \to M$. Even then, the fractional error is $\frac{\frac{\Delta E}{2Mc^2}}{1 - \frac{\Delta E}{2Mc^2}} \approx \frac{10^{-9}}{1 - 10^{-9}} \approx 10^{-9}$ for ΔE of order eV and the minimum M (and thus maximum error) of order GeV$/c^2$. Therefore, $E_\gamma = \Delta E$ is typically a good approximation.

17. Neutrino Beam*

(a) Let the four-momenta of the pion, muon and neutrino be P_π, P_μ and P_ν respectively. In the rest frame of the pion,

$$P_\pi = \begin{pmatrix} m_\pi \\ \mathbf{0} \end{pmatrix} \quad P_\mu = \begin{pmatrix} m_\pi - p \\ -\mathbf{p} \end{pmatrix} \quad P_\nu = \begin{pmatrix} p \\ \mathbf{p} \end{pmatrix}$$

where p is the unknown momentum of the neutron. The four-vector equation is

$$P_\pi = P_\mu + P_\nu.$$

Since the muon is not of concern, it is customary to isolate P_μ even though it is not really necessary here.

$$P_\pi - P_\nu = P_\mu.$$

Taking the inner product of each side with itself,

$$m_\pi^2 - 2P_\pi \cdot P_\nu = m_\mu^2$$

$$m_\pi^2 - 2m_\pi p = m_\mu^2$$

$$p = \frac{m_\pi^2 - m_\mu^2}{2m_\pi} c$$

where we have added the c back. The energy of the neutrino is

$$E_\nu = pc = \frac{m_\pi^2 - m_\mu^2}{2m_\pi} c^2 = \frac{140^2 - 106^2}{2 \times 140} = 29.9 \text{MeV (3sf)}.$$

(b) There are multiple ways of solving this. The direct method is to repeat the process of (a) with

$$P_\pi = \begin{pmatrix} E_\pi \\ \mathbf{p}_\pi \end{pmatrix}, \quad P_\mu = \begin{pmatrix} E_\pi - p' \\ \mathbf{p}_\pi - \mathbf{p}' \end{pmatrix}, \quad P_\nu = \begin{pmatrix} p' \\ \mathbf{p}' \end{pmatrix}$$

$$P_\pi - P_\nu = P_\mu.$$

Since $\mathbf{p}_\pi \cdot \mathbf{p}' = p_\pi p'$, the inner products of both sides yield

$$m_\pi^2 - 2(E_\pi p' - p_\pi p') = m_\mu^2$$

$$p'c = \frac{(m_\pi^2 - m_\mu^2)c^4}{2(E_\pi - p_\pi c)}$$

$$= \frac{(m_\pi^2 - m_\mu^2)c^4}{2(E_\pi - \sqrt{E_\pi^2 - m_\pi^2 c^4})}$$

$$= \frac{140^2 - 106^2}{2(200 \times 10^3 - \sqrt{200^2 \times 10^6 - 140^2})}$$

$$= 8.53 \times 10^4 \text{MeV (3sf)}$$

$$= 85.3 \text{GeV (3sf)},$$

where we have added back the c's. Alternatively, we can apply the energy-momentum transformations to the momentum p computed in (a) to obtain

$$p'c = \gamma_\pi(1 + \beta_\pi)pc$$

$$= \gamma_\pi(1 + \beta_\pi)\frac{m_\pi^2 - m_\mu^2}{2m_\pi}c^2$$

$$= \sqrt{\frac{1 + \beta_\pi}{1 - \beta_\pi}}\frac{m_\pi^2 - m_\mu^2}{2m_\pi}c^2$$

$$= \frac{m_\pi^2 - m_\mu^2}{2\gamma_\pi(1 - \beta_\pi)m_\pi}c^2$$

$$= \frac{m_\pi^2 - m_\mu^2}{2(E_\pi - p_\pi c)}c^4$$

$$= 85.3 \text{GeV (3sf)}.$$

(c) Denoting the momentum of the neutrino as $\boldsymbol{p}_\nu$ and repeating the same process in (b) with $\boldsymbol{p}_\pi \cdot \boldsymbol{p}_\nu = p_\pi p_\nu \cos\theta$, one would obtain

$$p_\nu = \frac{m_\pi^2 - m_\mu^2}{2(E_\pi - p_\pi \cos\theta)}.$$

Evidently, the maximum neutrino energy occurs when $\theta = 0$.

$$E_{max} = \frac{m_\pi^2 - m_\mu^2}{2(E_\pi - p_\pi)}.$$

For $\frac{m_\pi^2 - m_\mu^2}{2(E_\pi - p_\pi \cos\theta)} = \frac{1}{2}E_{max}$,

$$\cos\theta = 2 - \frac{E_\pi}{p_\pi} = 2 - \frac{E_\pi}{\sqrt{E_\pi^2 - m_\pi^2}}$$

$$\implies \theta = \cos^{-1}\left(2 - \frac{E_\pi}{\sqrt{E_\pi^2 - m_\pi^2 c^4}}\right)$$

$$= \cos^{-1}\left(2 - \frac{200 \times 10^3}{\sqrt{200^2 \times 10^6 - 140^2}}\right) = 0.0401° \text{ (3sf)}.$$

18. Mad Scientist*

A simple proof for $M > 0$ (such that its rest frame exists) is to observe the situation in the rest frame of M. Presuming that such a reaction is possible, the initial total energy is only Mc^2 while the final total energy is at least $2mc^2 > Mc^2$, contradicting the conservation of energy.

For a more general rebuttal which works even when $M = 0$, let the four-momentum of M be P_1 and the four-momenta of the products be P_2 and P_3. The four-vector equation is

$$P_1 = P_2 + P_3.$$

Taking the inner product of each side with itself,

$$M^2 = 2m^2 + 2\gamma_{rel}m^2,$$

where γ_{rel} is the gamma factor associated with the velocity of one particle of mass m, as observed in the other. As $\gamma_{rel} \geq 1$, the right-hand side obeys the inequality

$$2m^2 + 2\gamma_{rel}m^2 \geq 4m^2 > M^2,$$

which establishes a contradiction. Therefore, the scientist is wrong. Qualitatively, his refutation of the violation of energy conservation is flawed because when M has a certain velocity and thus momentum in the lab frame, the products must possess some kinetic energy in addition to their rest energies, by the conservation of momentum. The total energy of the proposed products is always larger than the energy of M.

19. Photon Decay*

Let P_p be the four-momentum of the photon and let the photon disintegrate into k particles with the ith particle having a four-momentum P_i. Then,

$$P_p = \sum_{i=1}^{k} P_i.$$

Taking the squared norm of both sides,

$$P_p \cdot P_p = \left(\sum_{i=1}^{k} P_i \right) \cdot \left(\sum_{i=1}^{k} P_i \right)$$

$$0 = \sum_{i=1}^{k} m_i^2 + 2 \left(\sum_{i,j\, i\neq j} P_i \cdot P_j \right).$$

One can show that $P_i \cdot P_j \geq 0$ for all possible combinations of particles. If at least one particle is massive (suppose that it is the ith particle without the loss of generality), $P_i \cdot P_j = m_i E_{jrev} \geq 0$ (as both rest mass and energy must be non-negative) where E_{jrev} is the energy of the jth particle as observed in the rest frame of the ith particle. If both particles are massless, $P_i \cdot P_j = p_i p_j - \boldsymbol{p_i} \cdot \boldsymbol{p_j} \geq 0$ as $E = p$ for a massless particle. Since the left-hand side of the previous equation is zero while the right-hand side is a sum of non-negative numbers, all terms on the right-hand side must be zero — implying that a particle with a non-zero rest mass cannot be produced.

20. Compton Scattering**

By the conservation of energy and momentum,

$$\begin{pmatrix} E_1 \\ \boldsymbol{p_1} \end{pmatrix} + \begin{pmatrix} hf_1 \\ hf_1\boldsymbol{\hat{k}_1} \end{pmatrix} = \begin{pmatrix} hf_2 \\ hf_2\boldsymbol{\hat{k}_2} \end{pmatrix} + \begin{pmatrix} E_3 \\ \boldsymbol{p_3} \end{pmatrix}.$$

Shifting the first term on the right-hand side to the left-hand side and taking the squared norm of both sides,

$$m^2 + 2(E_1 hf_1 - hf_1 \boldsymbol{p_1} \cdot \boldsymbol{\hat{k}_1}) - 2h^2 f_1 f_2 (1 - \boldsymbol{\hat{k}_1} \cdot \boldsymbol{\hat{k}_2})$$
$$-2(E_1 hf_2 - hf_2 \boldsymbol{p_1} \cdot \boldsymbol{\hat{k}_2}) = m^2$$

$$f_2 = \frac{E_1 f_1 - f_1 \boldsymbol{p_1} \cdot \boldsymbol{\hat{k}_1}}{hf_1(1 - \boldsymbol{\hat{k}_1} \cdot \boldsymbol{\hat{k}_2}) + E_1 - \boldsymbol{p_1} \cdot \boldsymbol{\hat{k}_2}}.$$

Evidently, the minimum f_2 occurs when $\boldsymbol{p_1} \cdot \boldsymbol{\hat{k}_1} = p_1$, $\boldsymbol{\hat{k}_1} \cdot \boldsymbol{\hat{k}_2} = -1$ and $\boldsymbol{p_1} \cdot \boldsymbol{\hat{k}_2} = -p_1$. That is, $\boldsymbol{\hat{k}_1}$ is parallel to $\boldsymbol{p_1}$ while $\boldsymbol{\hat{k}_2}$ is anti-parallel to $\boldsymbol{p_1}$. Then,

$$f_2 = \frac{(E_1 - p_1)f_1}{2hf_1 + E_1 + p_1}.$$

21. Pion Photoproduction**

(a) Applying Eq. (12.26), the threshold photon energy is

$$E_\gamma = \frac{(m_p + m_\pi)^2 - m_p^2}{2m_p}c^2 = m_\pi c^2 + \frac{(m_\pi c^2)^2}{2m_p c^2}$$

$$= 135 + \frac{135^2}{2 \times 938} = 145 \text{MeV (3sf)}.$$

(b) Let the initial four-momenta of the proton be $P_p = (E, \boldsymbol{p})$ and the photon be $P_\gamma = (E_\gamma, E_\gamma \hat{\boldsymbol{k}})$ where $\hat{\boldsymbol{k}}$ is the unit vector along the photon's velocity. The four-vector equation associated with the reaction is

$$P_\gamma + P_p = P_p' + P_\pi.$$

Taking the inner product of both sides with itself,

$$m_p^2 + 2P_\gamma \cdot P_p = m_p^2 + m_\pi^2 + 2P_p' \cdot P_\pi.$$

Noting that $P_\gamma \cdot P_p = E_\gamma(E - \boldsymbol{p} \cdot \boldsymbol{k})$ and $P_p' \cdot P_\pi = 2\gamma_{p\pi} m_p m_\pi$ where $\gamma_{p\pi}$ is the gamma factor associated with the final velocity of the proton as observed in the rest frame of the pion,

$$2E_\gamma(E - \boldsymbol{p} \cdot \boldsymbol{k}) = m_\pi^2 + 2\gamma_{p\pi} m_p m_\pi$$

$$E = \frac{m_\pi^2 + 2\gamma_{p\pi} m_p m_\pi}{2E_\gamma} + \boldsymbol{p} \cdot \hat{\boldsymbol{k}}.$$

Since $\gamma_{p\pi} \geq 1$, the minimum E occurs when $\gamma_{p\pi} = 1$ and $\boldsymbol{p} \cdot \hat{\boldsymbol{k}} = -p$. Then,

$$E - b = p$$

where $b = \frac{m_\pi^2 + 2m_p m_\pi}{2E_\gamma}$. Squaring both sides and using the identity $E^2 = p^2 + m_p^2$,

$$2bE = m_p^2 + b^2.$$

$$E = \frac{m_p^2}{2b} + \frac{b}{2}$$

$$= \frac{m_p^2 E_\gamma}{m_\pi^2 + 2m_p m_\pi} + \frac{m_\pi^2 + 2m_p m_\pi}{4E_\gamma}$$

$$= \frac{938^2 \cdot 10^{-9}}{135^2 + 2 \cdot 135 \cdot 938} + \frac{135^2 + 2 \cdot 135 \cdot 938}{4 \cdot 10^{-9}}$$

$$= 6.79 \times 10^{13} \text{MeV (3sf)}.$$

22. Energy Transfer**

In the lab frame, the total energy and momentum are given by $\gamma_u m_1 + m_2$ and $\gamma_u m_1 u$. As the energy and momentum obey the Lorentz transformations, the momentum in another inertial frame S' is

$$p' = \gamma_v(\gamma_u m_1 u - (\gamma_u m_1 + m_2)v).$$

Therefore, in order for $p' = 0$, the center-of-momentum frame S' must travel at $v = \frac{\gamma_u m_1 u}{\gamma_u m_1 + m_2}$ relative to S. v must also be the initial speed of particle 2 in S'. In the center-of-momentum frame, the momenta of the two particles must simply reverse after the collision for both energy and momentum to be conserved (if one momentum increases or decreases in magnitude, the other must follow suit by the conservation of momentum, and hence lead to a violation of the conservation of energy). Therefore, the final velocity of particle 2 in S' is v in the positive x'-direction. The final velocity of particle 2 in the lab frame S is then $u' = \frac{2v}{1+v^2}$ by the velocity addition formula. The resultant energy of particle 2 in S is then

$$E = \gamma_{u'} m_2$$
$$= \frac{1+v^2}{1-v^2} m_2$$
$$= \left(1 + \frac{2\gamma_u^2 m_1^2 u^2}{m_1^2 + m_2^2 + 2\gamma_u m_1 m_2}\right) m_2.$$

The ratio of E to the total energy in S is

$$f = \frac{E}{\gamma_u m_1 + m_2} = \left(1 + \frac{2\gamma_u^2 m_1^2 u^2}{m_1^2 + m_2^2 + 2\gamma_u m_1 m_2}\right) \frac{m_2}{\gamma_u m_1 + m_2}.$$

As $u \to 1$ and $\gamma_u \to \infty$, $m_1^2 + m_2^2$ in the denominator $m_1^2 + m_2^2 + 2\gamma_u m_1 m_2$ and m_2 in the denominator $\gamma_u m_1 + m_2$ become negligible comparatively. The ratio then tends to

$$f \to \frac{2\gamma_u^2 m_1^2 u^2 m_2}{2\gamma_u m_1 m_2 \cdot \gamma_u m_1} = 1.$$

Moving on to the more indirect approach, the three-momentum components of P_1' and P_2 are identical in the center-of-momentum frame. Therefore, the inner product of $P_1' - P_2$ with itself in the center-of-momentum frame must be non-zero as it is simply the square of the difference of the two energies in the center-of-momentum frame. Then,

$$(P_1' - P_2) \cdot (P_1' - P_2) \geq 0$$

$$P_1' \cdot P_1' + P_2' \cdot P_2' \geq 2P_1' \cdot P_2.$$

Substituting the squared norm of a four-momentum and Eq. (12.24),

$$m_1^2 + m_2^2 \geq 2\gamma_{u_{rel}} m_1 m_2,$$

where $\gamma_{u_{rel}}$ is the γ factor associated with the final velocity of particle 1 in the frame that is at rest with the initial velocity of the particle 2, u_{rel}. However, notice that the initial velocity of particle 2 is v in the negative x'-direction in the center-of-momentum frame. Therefore, the final velocity of particle 1 in the frame that is at rest with the initial velocity of the particle 2 effectively performs a $-v$ velocity addition to the final velocity of particle 1 in the center-of-momentum frame — transforming it back to the final value in the lab frame u_{lab}. Since the final energy of particle 1 is $E_{lab} = \gamma_{u_{lab}} m_1$ in the lab frame,

$$E_{lab} = \gamma_{u_{lab}} m_1 \leq \frac{m_1^2 + m_2^2}{2m_2},$$

which sets a fixed upper bound on the energy retained by particle 1.

23. Maximum Frequency**

By the conservation of energy and momentum,

$$\begin{pmatrix} E_1 \\ \boldsymbol{p}_1 \end{pmatrix} + \begin{pmatrix} M \\ \boldsymbol{0} \end{pmatrix} - \begin{pmatrix} hf \\ hf\hat{\boldsymbol{k}} \end{pmatrix} = \begin{pmatrix} E_1' \\ \boldsymbol{p}_1' \end{pmatrix} + \begin{pmatrix} E_2 \\ \boldsymbol{p}_2' \end{pmatrix},$$

where $\hat{\boldsymbol{k}}$ is a unit vector in the direction of the photon's velocity. Taking the squared norm of both sides,

$$m^2 + M^2 + 2E_1 M - 2hfM - 2(E_1 hf - \hat{\boldsymbol{k}} \cdot \boldsymbol{p}_1 hf) = m^2 + M^2 + 2P_1' \cdot P_2'$$

where P_1' and P_2' are the final four-momenta of the particle and the nucleus respectively. Solving for hf,

$$hf = \frac{E_1 M - P_1' \cdot P_2'}{M + E_1 - \hat{\boldsymbol{k}} \cdot \boldsymbol{p}_1}.$$

Evidently, hf is maximized when $P_1' \cdot P_2'$ is minimized and when $\hat{\boldsymbol{k}} \cdot \boldsymbol{p}_1$ is maximized. The former expression is

$$P_1' \cdot P_2' = \gamma_{u_{rel}} mM,$$

where u_{rel} is the velocity of one particle in the rest frame of the other. This quantity is minimized when there is no relative velocity between the nucleus and the particle (i.e. $\gamma_{u_{rel}} = 1$). Next, $\hat{\boldsymbol{k}} \cdot \boldsymbol{p}_1$ is maximized when these two vectors are parallel. In other words, the ejected photon travels in the same direction as the initial velocity of the particle. Then the maximum energy of the photon in frame S is

$$hf = \frac{E_1 M - mM}{M + E_1 - p_1} = \frac{mM(\gamma_u - 1)}{M + \gamma_u m(1 - u)}.$$

24. Two Particles Connected by String*

Let E_1 and E_2 be the final energies of the particles of rest masses m and M respectively. Then, by the work-energy theorem,

$$E_1 = m + Tx,$$

$$E_2 = M + T(L - x).$$

Let the magnitudes of the final momenta of the particles, which must be equal by the conservation of momentum or by Newton's third law, be p. Then,

$$p = \sqrt{E_1^2 - m^2} = \sqrt{E_2^2 - M^2}$$

$$\sqrt{T^2 x^2 + 2mTx} = \sqrt{T^2(L - x)^2 + 2MT(L - x)}$$

$$T^2 x^2 + 2mTx = T^2(L - x)^2 + 2MT(L - x)$$

$$x = \frac{TL^2 + 2ML}{2(m + M + TL)}.$$

25. Projectile Motion**

Let x, y and t denote the spatial and temporal coordinates of the particle in the lab frame S. Define the origin at the initial position of the particle. The x and y components of the particle's momentum are

$$p_x = p_0,$$

$$p_y = -Ft.$$

Furthermore, the work-energy theorem states that the energy of the particle at (x, y) is

$$E = E_0 - Fy.$$

Then, $y(t)$ can first be solved for

$$\frac{dy}{dt} = \frac{p_y}{E} = -\frac{Ft}{E_0 - Fy}$$

$$\int_0^y \left(y - \frac{E_0}{F}\right) dy = \int_0^t t \, dt$$

$$y^2 - \frac{2E_0}{F} y - t^2 = 0.$$

Solving for y,

$$y = \frac{E_0}{F} \pm \sqrt{\frac{E_0^2}{F^2} + t^2}.$$

The y-coordinate must be negative for $t > 0$. Thus, we choose the negative expression.

$$y = \frac{E_0}{F} - \sqrt{\frac{E_0^2}{F^2} + t^2}.$$

Armed with $y(t)$, $x(t)$ can be determined.

$$\frac{dx}{dt} = \frac{p_x}{E} = \frac{p_0}{E_0 - Fy} = \frac{p_0}{\sqrt{E_0^2 + F^2 t^2}},$$

$$x = \int_0^t \frac{p_0}{F\sqrt{\frac{E_0^2}{F^2} + t^2}} dt$$

$$= \int_0^{\tan^{-1} \frac{Ft}{E_0}} \frac{p_0}{F \cdot \frac{E_0}{F} \sec \theta} \cdot \frac{E_0}{F} \sec^2 \theta \, d\theta$$

$$= \int_0^{\tan^{-1} \frac{Ft}{E_0}} \frac{p_0}{F} \sec \theta \, d\theta$$

$$= \left[\frac{p_0}{F} \ln |\sec \theta + \tan \theta|\right]_0^{\tan^{-1} \frac{Ft}{E_0}}$$

$$= \frac{p_0}{F} \ln \left(\sqrt{1 + \frac{t^2 F^2}{E_0^2}} + \frac{tF}{E_0}\right),$$

where we have adopted the trigonometric substitution $t = \frac{E_0}{F} \tan \theta$ along the way. After some algebraic manipulation, we can show that

$$\sqrt{1 + \frac{t^2 F^2}{E_0^2}} = \frac{e^{\frac{Fx}{p_0}} + e^{-\frac{Fx}{p_0}}}{2}.$$

For readers familiar with hyperbolic functions, we can instead adopt the substitutions $t = \frac{E_0}{F} \sinh\theta$, $dt = \frac{E_0}{F} \cosh\theta d\theta$ such that

$$x = \int_0^t \frac{p_0}{F\sqrt{\frac{E_0^2}{F^2} + t^2}} dt$$

$$= \int_0^{\sinh^{-1}\left(\frac{Ft}{E_0}\right)} \frac{p_0}{F \cdot \frac{E_0}{F}\cosh\theta} \cdot \frac{E_0}{F}\cosh\theta d\theta$$

$$= \int_0^{\sinh^{-1}\left(\frac{Ft}{E_0}\right)} \frac{p_0}{F} d\theta$$

$$= \frac{p_0}{F}\sinh^{-1}\left(\frac{Ft}{E_0}\right),$$

where we have used the identity $\cosh^2\theta = 1 + \sinh^2\theta$.

$$\implies \frac{Ft}{E_0} = \sinh\left(\frac{Fx}{p_0}\right)$$

$$\sqrt{1 + \frac{F^2t^2}{E_0^2}} = \sqrt{1 + \sinh^2\left(\frac{Fx}{p_0}\right)} = \cosh\left(\frac{Fx}{p_0}\right) = \frac{e^{\frac{Fx}{p_0}} + e^{-\frac{Fx}{p_0}}}{2}.$$

Then,

$$y = \frac{E_0}{F}\left(1 - \frac{e^{\frac{Fx}{p_0 c}} + e^{-\frac{Fx}{p_0 c}}}{2}\right).$$

Note that in SI units, the exponents are actually $\frac{Fx}{p_0 c}$. Therefore in the non-relativistic limit where $\frac{Fx}{p_0 c} \ll 1$, we can perform a Maclaurin expansion for the exponential functions.

$$e^x \approx 1 + x + \frac{1}{2}x^2 + \cdots$$

$$y \approx \frac{E_0}{F}\left(1 - \frac{1}{2} - \frac{Fx}{2p_0 c} - \frac{F^2x^2}{4p_0^2 c^2} - \frac{1}{2} + \frac{Fx}{2p_0 c} - \frac{F^2x^2}{4p_0^2 c^2}\right) = -\frac{E_0 F x^2}{2p_0^2 c^2}$$

which is a parabola. What's more, in the non-relativistic limit, E_0 is dominated by the rest energy ($E_0 \approx mc^2$) where m is the rest mass of the particle, while $p_0 \approx mv_0$ where v_0 is the initial velocity of the particle. Then,

$$y = -\frac{Fx^2}{2mv_0^2}.$$

When F is of the form of gravity, $F = mg$ classically.

$$y = -\frac{gx^2}{2v_0^2},$$

which is the trajectory of a particle with initial x-velocity v_0 under free fall. The latter set-up is described by the equations

$$y = -\frac{1}{2}gt^2,$$

$$x = v_0t \implies t = \frac{x}{v_0}.$$

Hence,

$$y = -\frac{gx^2}{2v_0^2}.$$

26. Pulling a Leaking Bucket**

The only external force on the bucket is the tension in the string. Hence,

$$\frac{dp}{dt} = T,$$

$$p = Tt,$$

as the initial momentum of the bucket is zero. Next, the rate of increase of the energy of the bucket is equal to the sum of the rate of the rest mass of sand gathered by the bucket and the power delivered by the tension.

$$\frac{dE}{dt} = (\lambda + T)u.$$

Next, using the relationship $u = \frac{p}{E} = \frac{Tt}{E}$,

$$\frac{dE}{dt} = (\lambda + T)\frac{Tt}{E}$$

$$\int_0^E E\,dE = \int_0^t (\lambda + T)Tt\,dt$$

$$E = \sqrt{(\lambda + T)Tt}$$

$$u = \frac{p}{E} = \frac{Tt}{\sqrt{(\lambda + T)Tt}} = \sqrt{\frac{T}{\lambda + T}},$$

which is independent of time except for the discontinuity in velocity at $t = 0$.

27. Light in Moving Glass**

In the rest frame of the block S', the block is stationary while the light source emits light of Doppler-shifted frequency $f' = \sqrt{\frac{1-\beta}{1+\beta}} f$. The speed of light inside the block is $c' = \frac{1}{n}$ in units of c. Hence, the wavelength of light inside the block in S' is $\lambda' = \frac{c'}{f'} = \frac{1}{n\sqrt{\frac{1-\beta}{1+\beta}} f}$. Thus, the four-wave vector inside the glass block in S' is

$$K' = \begin{pmatrix} 2\pi f' \\ \frac{2\pi}{\lambda'} \end{pmatrix} = \begin{pmatrix} 2\pi\sqrt{\frac{1-\beta}{1+\beta}} f \\ 2\pi n\sqrt{\frac{1-\beta}{1+\beta}} f \end{pmatrix}.$$

The four-wave vector K inside the glass block in S can be obtained from applying the inverse Lorentz transformation to K'.

$$K = \begin{pmatrix} 2\pi f_{lab} \\ \frac{2\pi}{\lambda_{lab}} \end{pmatrix} = \begin{pmatrix} \gamma_v & \gamma_v \beta \\ \gamma_v \beta & \gamma_v \end{pmatrix} \begin{pmatrix} 2\pi\sqrt{\frac{1-\beta}{1+\beta}} f \\ 2\pi n\sqrt{\frac{1-\beta}{1+\beta}} f \end{pmatrix}.$$

Equating the corresponding entries,

$$f_{lab} = \frac{1 + \beta n}{1 + \beta} f,$$

$$\lambda_{lab} = \frac{c(1 + \beta)}{(\beta + n)f},$$

where we have added back the c's.

28. Crossed Fields**

One can show that in an inertial frame S' that travels at $v = -\frac{E}{B}$ in the negative x-direction relative to S (we define v this way such that $v > 0$), the electric field is null while the magnetic field is $\boldsymbol{B'} = (0, 0, \frac{B}{\gamma_v})$. In this frame S', the particle's kinetic energy cannot change as the magnetic force does no work. Furthermore, the magnetic force is perpetually perpendicular to the particle's velocity — implying that the particle undergoes circular motion (this holds in the classical case as well). The initial speed of the particle in

S' is given by the velocity addition formula as $u = \frac{v_0 + v}{1 + v_0 v}$ (directed initially along the x'-axis), and remains constant afterwards. The magnetic force quB' provides the centripetal force. Since this force is perpendicular to the instantaneous velocity of the particle,

$$quB' = \gamma_u m a_c$$

by Eq. (12.17) where $a_c = \frac{u^2}{R}$ is the centripetal acceleration and R is the radius of rotation. The radius of rotation can then be computed as

$$R = \frac{\gamma_u m u}{qB'}$$

while the angular velocity is

$$\omega = \frac{u}{R} = \frac{qB'}{\gamma_u m}.$$

Now, if we define the initial position of the particle in S' to be at the origin O', the center of rotation must lie at $x' = 0$ and $y' = -R$. Therefore, the x-coordinate of the particle as a function of time t' in S' is

$$x' = R \sin \omega t'.$$

The x-coordinate of the particle in S is then obtained from the Lorentz transformations.

$$x = \gamma_v(x' - vt') = \gamma_v(R \sin \omega t' - vt').$$

The maximum value of x occurs when $\frac{dx}{dt'} = 0$. This yields

$$\frac{dx}{dt'} = \gamma_v(\omega R \cos \omega t' - v) = 0$$

$$\cos \omega t' = \frac{v}{\omega R} = \frac{v}{u}$$

$$t' = \frac{\cos^{-1} \frac{v}{u}}{\omega} = \frac{\gamma_u m \cos^{-1} \frac{v}{u}}{qB'}.$$

Note that $\cos^{-1} \frac{v}{u}$ exists as $\frac{v}{u} = \frac{v + v_0 v^2}{v_0 + v} < 1$ when $v_0 < 1$ and $v < 1$ (just shift the terms around). When the above condition is satisfied,

$$\sin \omega t' = \sqrt{1 - \frac{v^2}{u^2}}.$$

We choose the positive value such that the second derivative $\frac{d^2 x}{dt'^2} = -\gamma_v \omega^2 R \sin \omega t'$ is negative at this stationary point — ensuring that it is a maximum point. Observe from $x = \gamma_v(R \sin \omega t' - v t')$ that the global maxima evidently occurs for the smallest t' that satisfies $\cos \omega t' = \frac{v}{u}$. Hence, our choice of $t' = \frac{\cos^{-1} \frac{v}{u}}{\omega}$ above is justified. Then, the maximum x value is

$$x = \frac{\gamma_v \gamma_u m}{qB'} \left(\sqrt{u^2 - v^2} - v \cos^{-1} \frac{v}{u} \right)$$

$$= \frac{\gamma_v^2 \gamma_u m}{qB} \left(\sqrt{u^2 - v^2} - v \cos^{-1} \frac{v}{u} \right),$$

where $v = -\frac{E}{B}$ and $u = \frac{v_0 + v}{1 + v_0 v}$.

29. Magnetic Field***

The crux of this problem is to observe that the wire is no longer neutral in frame S' due to the different extents of length contraction of positive ions and electrons (as they were moving at different speeds in S). Let us determine the charge density of the electrons in frame S', λ'_e first. Let L be the proper distance between adjacent electrons. Then, the distance between adjacent electrons in frame S is given by the length contraction formula:

$$l = \frac{L}{\gamma_u}.$$

The distance between adjacent electrons as observed in frame S' is

$$l' = \frac{L}{\gamma_{u'}}$$

where u' is the velocity of the electrons in frame S'. Referring to Eq. (12.22),

$$\gamma_{u'} = \gamma_v \gamma_u \left(1 + \frac{\beta}{c} u \right),$$

where $\beta = \frac{v}{c}$ and we have included the c's for clarity. Then, the charge density of the electrons in frame S' is given by

$$\lambda_{e'} = -\lambda \cdot \frac{l}{l'} = -\lambda \gamma_v \left(1 + \frac{\beta}{c} u \right).$$

A similar argument can be made to conclude that the charge density of the positive ions in frame S' is

$$\lambda'_p = \lambda \cdot \gamma_v.$$

Hence, the total linear charge density of the wire in frame S' is

$$\lambda' = \lambda'_e + \lambda'_p = -\lambda \gamma_v \frac{\beta u}{c}.$$

A much more efficient alternative is to consider the four-current $J = (\rho c, \boldsymbol{j})$ introduced in part (d) of Problem 10, where ρ is the charge density and $\boldsymbol{j}$ is the current density. The four-current everywhere along the infinite wire in S is $J = (0, \lambda u, 0, 0)$, for one-dimensional charge distributions. Meanwhile, the four-current in S' is $J' = (\lambda' c, I'_x, I'_y, I'_z)$, where I'_x, I'_y and I'_z are the current components in S'. λ' can then be computed from the Lorentz transformations.

$$\lambda' c = \gamma_v (0 - \beta \lambda u)$$

$$\implies \lambda' = -\lambda \gamma_v \frac{\beta u}{c}$$

everywhere along the wire in S'. Applying Gauss' law to the wire (by drawing a cylindrical Gaussian surface whose axis coincides with the wire), the electric field in frame S' is

$$\boldsymbol{E'} = \frac{\lambda'}{2\pi r' \varepsilon_0} \hat{\boldsymbol{r}}' = \frac{-\lambda \gamma_v u v}{2\pi r' c^2 \varepsilon_0} \hat{\boldsymbol{r}}'$$

where $\boldsymbol{r}'$ is the vector pointing perpendicularly outwards from the closest point on the axis of the wire to the charge in frame S'. Hence, the force on the charge in frame S' (note that there is no velocity-dependent force, such as the magnetic force, as the charge is stationary) is

$$\boldsymbol{f'} = q\boldsymbol{E'} = -\frac{q \lambda \gamma_v u v}{2\pi r' c^2 \varepsilon_0} \hat{\boldsymbol{r}}'.$$

Notice that the net force on the charge in frame S' is purely radial and is perpendicular to the velocity of frame S relative to frame S'. Hence, from the force transformations, the net force on the charge in frame S is

$$\boldsymbol{f} = \frac{\boldsymbol{f'}}{\gamma_v} = -\frac{q \lambda u v}{2\pi r' c^2 \varepsilon_0} \hat{\boldsymbol{r}}'.$$

Furthermore, notice that since $\boldsymbol{r}'$ is also perpendicular to the velocity v, $\boldsymbol{r}' = \boldsymbol{r}$ where $\boldsymbol{r}$ is the radial vector from the wire to the charge in frame S. Hence,

$$\boldsymbol{f} = -\frac{q \lambda u v}{2\pi r c^2 \varepsilon_0} \hat{\boldsymbol{r}} = q\boldsymbol{v} \times \left(\frac{\lambda u}{2\pi r c^2 \varepsilon_0} \hat{\boldsymbol{r}}_\theta \right)$$

where $\hat{\boldsymbol{r}}_\theta$ is the azimuthal unit vector, whose positive direction is given by the right-hand-grip rule (applied to the positive current). Hence, it can be seen that the force takes a form similar to the magnetic force. We can show that the last term indeed gives the correct expression for the magnetic field. λu is simply the current I and $\mu_0 = \frac{1}{c^2 \varepsilon_0}$. Hence,

$$\boldsymbol{B} = \frac{\mu_0 I}{2\pi r} \hat{r}_\theta,$$

which is consistent with Ampere's law.

Appendix

Michelson–Morley Experiment

In 1864, James Clerk Maxwell laid the foundations of electromagnetism with his set of equations, known as Maxwell's equations, which collectively described all knowledge in that field. From his equations, it can be proven that the speed of an electromagnetic wave is a certain value c. However, there was no mention of which frame this speed was measured with respect to, which raised suspicion as the speed of light was presumed to vary across different inertial frames according to the widely-accepted Galilean transformations then. Furthermore, Maxwell's equations looked neat in a particular inertial frame but was convoluted in another inertial frame after a Galilean transformation. Then, it was proposed that the frame in which the Maxwell's equations looked nice be called the frame of ether. Ether was hypothesized to be the medium in which light propagates; it furnished an explanation for the ability of light, as a wave,[1] to seemingly propagate in a vacuum. After all, if sound waves required compressible media such as air or water to propagate in, why not electromagnetic waves too? It is then said light propagates at a speed c in ether.

Afterwards, two experimentalists, Albert A. Michelson and Edward W. Morley, set out to measure the relative speed of ether with respect to matter. It was presumed that ether was a transparent medium that filled all space then and was stationary with respect to absolute space. Michelson invented a set-up, that vaunted unprecedented accuracy, to measure the speed of ether with respect to the Earth by leveraging the different times taken by light to travel in perpendicular directions in a moving medium. Luminiferous ether, in this case, is hypothesized to move relative to the Earth as the Earth is revolving about the Sun at roughly 30km/s. Figure A.1 is a rough depiction of Michelson's set-up.

---

[1]All waves were thought to require a medium to propagate in then.

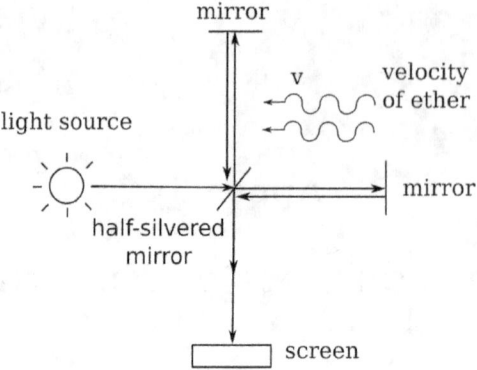

Figure A.1:   Michelson's Interferometer

Light from a single source is first divided into two rays by a partially-silvered mirror (we shall call it a beam splitter to avoid confusion with the other mirrors). The two rays then travel in perpendicular directions and are reflected by mirrors that are placed a distance L from the beam splitter, measured with respect to the frame of the earth. They are then recombined and impinge on the screen. In this case, the velocity of ether is directed purely in the horizontal direction. The velocity of "ether wind" will lead to a discrepancy in the times traveled by the two beams.

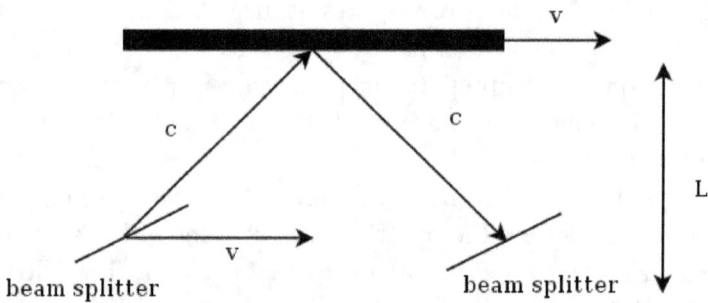

Figure A.2:   Vertical beam

In the frame of ether, the top mirror and the beam splitter both move at a speed v towards the right. The initially vertical light ray now obtains a component of velocity in the horizontal direction. Note that the speed of light (the magnitude of the diagonal vector in Fig. A.2) is c in the frame of ether by definition. Then, the total time taken for the light ray to travel back and forth is

$$t_V = \frac{2L}{\sqrt{c^2 - v^2}},$$

where we have used Pythagoras' theorem to calculate the velocity of light in the vertical direction, $\sqrt{c^2 - v^2}$.

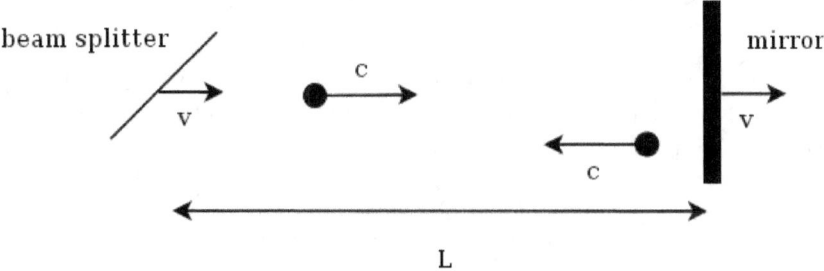

beam splitter

mirror

L

Figure A.3: Horizontal beam

Similarly, we analyze the motion of the horizontal beam in the frame of ether shown in Fig. A.3. In the case of the horizontal light beam, when it propagates forward, it is chasing the mirror which is retracting at a speed v. After its rebound, the beam then moves towards the beam splitter that is approaching at a speed v. Thus, the relative speeds between the light beam and the mirror and the light beam and the beam splitter are $c - v$ and $c + v$ in the frame of ether, respectively. Thus, the total duration of the horizontal beam's journey is

$$t_H = \frac{L}{c - v} + \frac{L}{c + v} = \frac{2Lc}{c^2 - v^2}.$$

As long as $v \neq 0$, there is a non-zero difference in duration of the journeys of the two beams, of

$$\Delta t = \frac{2Lc}{c^2 - v^2} - \frac{2L}{\sqrt{c^2 - v^2}} = \frac{2L}{c} \left(\frac{1}{1 - \frac{v^2}{c^2}} - \frac{1}{\sqrt{1 - \frac{v^2}{c^2}}} \right).$$

Though there can be multiple orientations of the set-up with respect to the direction of "ether wind", it can be proven that t_V is the smallest possible duration of the journey while t_H is the largest. Thus, the set-up can be gradually rotated until the time difference between the arrivals of the reflected beams produces the most pronounced observable differences.

However, there is still a problem. The time difference between the two beams is so minute that it severely impedes an accurate measurement by even the most precise timer. Michelson circumvented this limitation with his ingenious idea. The wave nature of light engenders an interference pattern due to the phase difference caused by the path difference between two incident light waves. Thus, if the set-up is first rotated into the above orientation and then rotated another 90 degrees (such that the vertical mirror becomes horizontal and vice versa), a fringe shift of the greatest magnitude should be observed. The sensitivity of the equipment arises from the relatively short wavelength of the light rays which causes a small change in the path difference to lead to a significant phase difference and thus, a conspicuous change in the positions of the fringes. The path difference between the horizontal and vertical light waves in the configuration above is

$$\Delta_1 = c\Delta t = 2L \left(\frac{1}{1 - \frac{v^2}{c^2}} - \frac{1}{\sqrt{1 - \frac{v^2}{c^2}}} \right).$$

If the entire set-up (light source, mirrors and screen) is then rotated by 90 degrees (clockwise or anti-clockwise), the two mirrors exchange roles — leading to a path difference that is negative of that before.

$$\Delta_2 = -c\Delta t = 2L \left(\frac{1}{\sqrt{1 - \frac{v^2}{c^2}}} - \frac{1}{1 - \frac{v^2}{c^2}} \right).$$

As the set-up is gradually rotated in this process, the fringes on the interference pattern will shift as the phase difference between the projected light beams changes. The fringe shift, which is the fraction of the distance between adjacent bright fringes that the interference pattern has moved, can be calculated as

$$n = \frac{\Delta_1 - \Delta_2}{\lambda} = \frac{4L}{\lambda} \left(\frac{1}{1 - \frac{v^2}{c^2}} - \frac{1}{\sqrt{1 - \frac{v^2}{c^2}}} \right)$$

$$\approx \frac{4L}{\lambda} \left(1 + \frac{v^2}{c^2} - 1 - \frac{v^2}{2c^2} \right) = \frac{2Lv^2}{\lambda c^2},$$

where we have used the binomial expansion for $(1 + x)^n$ and neglected second order and above terms. This is because a difference in path length by λ causes the interference pattern to move into a configuration that is identical to its original one (i.e. an initially bright fringe will become a dark fringe and return to a bright fringe). That is, the interference pattern must have "traveled" a distance equal to that between adjacent bright fringes. Michelson and Morley conducted their experiment with multiple reflections of the horizontal and vertical beams in order to extend the path of light and thus reduce the percentage uncertainty of their measurements. During their actual experiments, their physical parameters were

$$L \approx 11\text{m},$$

$$\lambda \approx 532\text{nm},$$

which would lead to an expected phase shift of 0.4 (note that v was presumed to be 30 km/s). However, the measured fringe shift was in fact less than 0.005! This suggested that if ether did exist, it had no velocity relative to the Earth. However, when the experiment was repeated half a year later, during which the Earth's velocity in revolving around the Sun was in the opposite direction, the exact same results were obtained! Some scientists, who wanted to cling onto the hitherto theory of ether, proposed that the Earth dragged ether along with its motion — causing ether to constantly have zero relative velocity near the surface of the Earth. However, this notion was then dismissed due to its inconsistency with other empirical observations.

Hendrik Antoon Lorentz, who firmly believed in the existence of ether, proposed that in the frame of ether that was moving relative to the earth, the distance between two points along ether's velocity in the Earth's frame is contracted by a factor of $\frac{1}{\sqrt{1-\frac{v^2}{c^2}}}$. This is in fact the correct conclusion (length contraction)! However, Lorentz's hypothesis was deemed to be too ad-hoc and thus was largely overlooked by the scientific committee. Lorentz, in fact, discovered the FitzGerald–Lorentz transformations, a pivotal transformation rule in special relativity, before Einstein formally formulated his theory of special relativity in 1905.

In contrast to the rather haphazard hypothesis by Lorentz, Einstein proposed a much simpler solution. Ether simply does not exist! Light does not require a medium to propagate in but instead travels at a constant speed in a vacuum with respect to observers in all inertial frames! This renowned experiment is known as the **Michelson–Morley experiment** which conferred Michelson his well-deserved Nobel Prize in 1907 and provided a solid experimental basis for the second postulate of special relativity. Then, the development of a revolutionary theory ensued.

Index

www.ingramcontent.com/pod-product-compliance
Lightning Source LLC
Chambersburg PA
CBHW082102220526
45472CB00009B/2014